ENCYCLOPEDIA OF PHYSICS

EDITED BY

S. FLÜGGE

VOLUME XXXIX

STRUCTURE OF ATOMIC NUCLEI

WITH 119 FIGURES

SPRINGER-VERLAG BERLIN HEIDELBERG GMBH
1957

HANDBUCH DER PHYSIK

HERAUSGEGEBEN VON

S. FLÜGGE

BAND XXXIX

BAU DER ATOMKERNE

MIT 119 FIGUREN

SPRINGER-VERLAG BERLIN HEIDELBERG GMBH
1957

ISBN 978-3-642-45874-3 ISBN 978-3-642-45872-9 (eBook)
DOI 10.1007/978-3-642-45872-9

Ursprünglich erschienen bei Springer-Verlag OHG Berlin Gottingen Heidelberg 1957
Softcover reprint of the hardcover 1st edition

Contents.

Page

The Two-Nucleon Problem. By LAMEK HULTHÉN, Professor of Mathematical Physics at the Royal Institute of Technology, Stockholm (Sweden) and MASAO SUGAWARA, Assistant Professor, Department of Physics, Hokkaido University, Sapporo (Japan) (With 26 Figures) 1

I. Introduction 1
II. Summary of experimental data 2
III. Meson theory and two-nucleon interaction 14
a) Meson field theory 14
b) Deduction of static two-nucleon interaction 17
c) Static interaction 23
IV. Low energy phenomena interpreted through central forces 32
a) Deuteron problem 32
b) Neutron-proton scattering 36
c) Proton-proton scattering 57
V. Low energy phenomena with tensor forces 64
a) Deuteron problem 64
b) Scattering problem with tensor forces 77
VI. Radiative processes 97
a) Interaction with electromagnetic field 97
b) Radiative neutron-proton capture 103
c) Photodisintegration of deuteron 108
d) Exchange current effect 114
VII. High energy phenomena 116
a) High energy nucleon-nucleon scattering 116
b) High energy photodisintegration 128
VIII. Mathematical appendix 131
General references 143

The Three-body Problem in Nuclear Physics. By MARIO VERDE, Professor for Theoretical Physics at the University of Torino (Italy). (With 4 Figures) 144

I. Spin eigenstates and equations of motion 145
II. The bound states of H^3 and He^3 149
III. The continuum states of three nucleons 158
IV. Effects of the electromagnetic interaction 175

Matter and Charge Distribution within Atomic Nuclei. By DAVID LAWRENCE HILL, Staff Member and Group Leader in Theoretical Nuclear Physics, Los Alamos Scientific Laboratory (USA). (With 17 Figures) 178

A. Electric measures of nuclear size and radial shape 186
I. Electron scattering 186
a) Low energy scattering 188
b) High energy scattering 191
II. Bound electrons 199
a) X-ray fine structure 199
b) Optical fine structure 202
III. μ-mesons 204
IV. COULOMB energy 211
B. Nuclear force measures of nuclear size and radial shape 212
I. Neutron scattering 213
II. Proton scattering 214

Page
III. Alpha particle scattering 216
IV. Charged particle reaction cross sections 217
V. Alpha decay 218
VI. Deuteron pick-up and stripping 218
VII. Pion scattering and bound pions 218
C. Electric measures of nuclear angular shape 219
I. Quadrupole moments 219
II. Electric quadrupole transition rates 222
III. Isotope shifts 224
IV. μ-mesonic x-rays 227
V. Electron scattering 228
D. Nonelectric measures of nuclear angular shape 229
E. Theories of nuclear density 231
I. Radial shape 231
II. Angular shape 237
F. Summary 238
General references 239

The Nuclear Shell-Model. By Dr. JAMES PHILIP ELLIOTT, Senior Scientific Officer and Dr. ANTHONY MILNER LANE, Principal Scientific Officer, Atomic Energy Research Establishment, Harwell, Berkshire (Great Britain). (With 28 Figures) 241
I. Introduction 241
II. The single particle model 245
III. Applications of the single particle model to medium and heavy nuclei . . 257
IV. Refinements of the single particle model 287
V. The individual particle model 317
VI. Application of the individual particle model to light nuclei 335
VII. Configuration mixing and other methods for improving the individual particle model 357
VIII. The theoretical basis of the shell-model 368
Acknowledgement 410

Models of Nuclear Structure. By STEVEN A. MOSZKOWSKI, Assistant Professor of Physics, University of California, Los Angeles (USA). (With 44 Figures) 411
Introduction and brief survey of the nuclear models 411
A. The liquid drop model 414
B. The Fermi gas model 421
I. Nuclear saturation and the nuclear potential 422
II. Further applications 432
III. Refinements of the model 438
C. The optical model 445
I. Multiple scattering and the complex potential 449
II. Relation between the nuclear density and the nuclear potential 453
D. The alpha particle model 460
E. The shell model 464
I. The single particle shell model 468
II. The many particle shell model 472
F. The unified model 476
I. Nuclear rotational states 481
II. Intrinsic structure of deformed nuclei 497
III. Collective vibrational excitations 516
IV. Relation between intrinsic and collective motions 525
Appendix I. Properties of the $\mathfrak{D}$ functions 543
Appendix II. Mathematical relations applicable to nuclear retational states 544
Appendix III. The unified model applied to a simple problem 547
General references 549

Sachverzeichnis (Deutsch-Englisch) 551
Subject Index (English-German) 559

The Two-Nucleon Problem.

By

LAMEK HULTHÉN and MASAO SUGAWARA.

With 26 Figures.

With a Mathematical Appendix.

By

MASAO SUGAWARA.

I. Introduction.

The history of the two-nucleon problem dates only from 1932 when the neutron was discovered by CHADWICK (the existence of the proton had already been established by RUTHERFORD in 1919) and the first serviceable nuclear force theory was advanced by HEISENBERG and MAJORANA. The reader who is interested in what might be called the "pre-meson" era of the two-nucleon problem is referred to the classical article by BETHE and BACHER [*1*] (1936), while the work of the first decade of the meson era is thoroughly reviewed in the book of ROSENFELD [*2*] (1948). We hope that the development of the later years will be reasonably accounted for by the present article, but we should also like to draw the reader's attention to other reviews and books [*3* to *11*], among which we mention those of BLATT and WEISSKOPF [*10*] and SACHS [*11*] as being most comprehensive and up to date.

In the following pages the two-nucleon problem is treated assuming a static nuclear potential and using non-relativistic quantum mechanics. We begin with a summary of the available experimental data (Chap. II) which are necessary for the study of the two-nucleon problem. This part has been compiled for the convenience of research workers and the theoretician who prefers to go right to the later chapters and look at the data whenever he needs them will find no difficulty in doing so.

We have tried to keep as close a connection as possible with the present meson field theory and, therefore, begin the theoretical exposé with a brief introduction of the meson theory and the ensuing static nuclear potential, accompanied by phenomenological considerations (Chap. III).

Having thus got a general idea about the static nuclear interaction, we enter on the actual analysis of the data, starting with the simple assumption of a central force (Chap. IV), then generalizing to a tensor force (Chap. V). In both cases only the low energy data are treated. As for the interaction of a two-nucleon system with an electromagnetic field (Chap. VI), we set out from the relativistic field theory in order to define the relativistic and meson exchange corrections to the phenomenological treatment in an unambiguous way.

The high energy data are considered in Chap. VII which tries to summarize all the available investigations, which are, of course, quite insufficient in many respects.

In the Mathematical Appendix (Chap. VIII), for which M. SUGAWARA alone is responsible, he gives a survey of some subjects which have not been treated in the foregoing chapters because they have no direct bearing on the interpretation of the experiments. They are, however, of great theoretical interest and are likely to be important for the future development of our subject, maybe also in other connections.

II. Summary of experimental data.

The spectrum of a neutron-proton system is simple and characteristic of short-range forces: there is only one bound state (ground state of the deuteron) besides the continuous spectrum. All information, therefore, either concerns the ground state (Sect. 1), or the continuous spectrum (scattering experiments, Sects. 2 and 4), or transitions between the two kinds of states (Sects. 3 and 5). In the proton-proton case only scattering data exist (no bound state). As to the interaction between two neutrons no direct evidence is available.

1. Deuteron data. *α) Binding energy.* The binding energy of the deuteron can be obtained in various ways. One may measure the threshold γ-energy which is sufficient to break up a deuteron into a proton and a neutron, using γ-rays produced by monoenergetic electrons. The most accurate measurements of this type give the following values of the deuteron binding energy ε:

$$\varepsilon = (2.227 \pm 0.003)\ \text{Mev}^{1}, \quad (2.226 \pm 0.003)\ \text{Mev}^{2}.$$

We also mention the more indirect method based on the "Q-values" of various nuclear reactions which gives

$$\varepsilon = (2.225 \pm 0.002)\ \text{Mev}^{3}.$$

In this article the following value is accepted:

$$\varepsilon = (2.226 \pm 0.002)\ \text{Mev} = 2.226\,(1 \pm 0.001)\ \text{Mev}. \tag{1.1}$$

β) Spin and magnetic moment. It is well known—e.g. from the intensities of the D_2 band spectrum—that the deuteron has a total spin 1 (unit $\hbar$). The obvious interpretation is that the spins of neutron and proton are parallel in the ground state, which would thus be described as a $^3S_1$-state. In consequence the magnetic moment μ_D of the deuteron would be expected to equal the sum of neutron and proton moments. This is very nearly so; the most recent experiments yield

$$\mu_P = 2.79276 \pm 0.00006^{4}; \quad \mu_N = -1.91304 \pm 0.00010^{5,*};$$
$$\mu_D = 0.857411 \pm 0.000019^{6,*},$$

in units of $\frac{e\hbar}{2Mc}$, where M is the proton mass. There is a small but definite difference between μ_D and $\mu_P + \mu_N$:

$$\mu_P + \mu_N - \mu_D = 0.02231 \pm 0.00012. \tag{1.2}$$

[1] J. C. NOYES, J. E. VAN HOOMISSEN, W. C. MILLER and B. WALDMAN: Phys. Rev. **95**, 396 (1954).
[2] R. C. MOBLEY and R. A. LAUBENSTEIN: Phys. Rev. **80**, 309 (1950).
[3] C. W. LI, W. WHALING, W. A. FOWLER and C. C. LAURITSEN: Phys. Rev. **83**, 512 (1951).
[4] H. SOMMER, H. A. THOMAS and J. A. HIPPLE: Phys. Rev. **82**, 697 (1951).
[5] F. BLOCH, D. NICODEMUS and H. H. STAUB: Phys. Rev. **74**, 1025 (1948).
[6] T. F. WIMETT: Phys. Rev. **91**, 499 (1953).
* These authors report the ratios of μ_N and μ_D to μ_P.

A possible explanation is that the ground state is not a pure S-state but mixed up with states of higher orbital angular momenta.

γ) Quadrupole moment. A further indication of such an admixture is given by the electric quadrupole moment of the deuteron[7]:

$$Q = (2.738 \pm 0.014) \times 10^{-27}\,\text{cm}^2, \tag{1.3}$$

where the positive sign refers to the spin direction ("cigar shape"). The simplest way to understand this feature—and at the same time the deviation of μ_D from $\mu_P + \mu_N$—is to assume a ground state of the type ${}^3S_1 + {}^3D_1$.

2. Low energy scattering data. *α) Zero energy neutron-proton scattering.* While no bound singlet state of two nucleons has been found, the singlet state is very effective in the zero energy scattering. Introducing here the triplet and singlet scattering lengths (amplitudes at zero energy), a_t and a_s, the cross section σ_0 for scattering of very slow neutrons by free protons ("free" of course also implies a random distribution of position and spin) can be written $\sigma_0 = \pi(3a_t^2 + a_s^2)$. The most precise determination[8] was carried out on protons bound in H_2, n-butane and water with neutrons of 0.8 to 15 eV. The data gave, combined with a theoretical analysis (given in Sect. 17), the following value:

$$\sigma_0 = 20.36 \pm 0.10 = 20.36\,(1 \pm 0.005)\ \text{barns}^*. \tag{2.1}$$

β) Coherent neutron-proton scattering. The coherent scattering length f is defined by $f = \frac{1}{2}(3a_t + a_s)$. The first measurements of this quantity were made by a study of interference effects in the scattering of slow neutrons by ortho- and parahydrogen. In this way it was shown that the singlet state of the deuteron is "virtual", i.e. no bound singlet state exists. However, the most accurate determination of f has been made by liquid mirror reflection, using a series of pure liquid hydrocarbons exposed to neutrons from the thermal column of a heavy water reactor at Argonne[9], which gave

$$\left.\begin{aligned} f &= -(3.78 \pm 0.02) \times 10^{-13}\,\text{cm}, \\ &= -3.78\,(1 \pm 0.005) \times 10^{-13}\,\text{cm}. \end{aligned}\right\} \tag{2.2}$$

From (2.1) and (2.2) we have, taking into account that a bound triplet state exists ($a_t > 0$), whereas the singlet state is virtual ($a_s < 0$),

$$a_t = (5.38 \pm 0.02) \times 10^{-13}\,\text{cm}, \qquad a_s = -(23.69 \pm 0.06) \times 10^{-13}\,\text{cm}.$$

For comparison we quote the latest results obtained from ortho- and para-hydrogen scattering[9a];

$$a_t = (5.37 \pm 0.04) \times 10^{-13}\,\text{cm}, \qquad a_s = -(23.73 \pm 0.07) \times 10^{-13}\,\text{cm}.$$

γ) Low energy neutron-proton scattering. From the extreme case of the zero energy, we now turn to the region of low energies in a wider sense (roughly 0 to 20 Mev). Table 1 shows some recent experimental data, which appear to be very accurate, on the total cross section for the scattering of neutrons by protons. In all data, except that for 14 Mev, the indicated errors of the cross section include the uncertainties due to those of the incident neutron energy.

* 1 barn $= 10^{-24}$ cm^2.

[7] Numerical value of the quadrupole moment after H. G. Kolsky, T. E. Phipp jr., N. F. Ramsey and H. B. Silsbee: Phys. Rev. **87**, 395 (1952).

[8] E. Melkonian: Phys. Rev. **76**, 1744 (1949).

[9] M. T. Burgy, G. R. Ringo and D. J. Hughes: Phys. Rev. **84**, 1160 (1951).

[9a] G. L. Squires and A. T. Stewart: Proc. Roy. Soc. Lond., Ser. A **230**, 19 (1955).

δ) Low energy proton-proton scattering. The proton-proton experiments consist in measuring the differential scattering cross sections accurately enough to determine the so-called nuclear *S*-wave phase shift δ_0 and *P*-wave phase shift δ_1 etc. For the sake of convenience the data are divided into two groups: the data obtained with VAN DE GRAAFF generators and those obtained with cyclotrons. The VAN DE GRAAFF data are, in general, more precise than the cyclotron data and are naturally confined to the energy region below about 4 Mev. These low energy data are especially important in the analysis. A selection of the most accurate data published so far is given in Table 2.

Table 1. *Total neutron-proton scattering cross section for some values of the incident neutron energy in laboratory system*.*

Incident neutron energy (Mev)	Total cross section (barns)	Reference number
1.005	4.228 ± 0.018	10
1.315	3.675 ± 0.016	11
2.540	2.525 ± 0.009	10
4.749	$1.690 \pm 0.006_6$	12
14.10 ± 0.05	0.689 ± 0.005	13

Table 2. VAN DE GRAAFF *data for the nuclear S- and P-phase shifts and the derived quantity K (defined in Sect. 22) for various incident proton energies in laboratory system.*

Energy (Mev)	δ_0 (degrees)	δ_1 (degrees)	*K*	Reference number
0.3828	14.583 ± 0.067	0	3.917 ± 0.016	14
0.2	6.66 ± 0.04	0	3.872 ± 0.020	15
0.3	11.16 ± 0.04	0	3.861 ± 0.011	15
0.4	15.02 ± 0.04	0	3.954 ± 0.009	15
0.45	17.16 ± 0.04	0	3.927 ± 0.007	15
0.5	18.82 ± 0.04	0	3.976 ± 0.007	15
0.6	22.32 ± 0.04	0	3.998 ± 0.006	15
0.7	25.13 ± 0.04	0	4.071 ± 0.005	15
0.8	27.97 ± 0.04	0	4.098 ± 0.005	15
0.9	30.35 ± 0.04	0	4.149 ± 0.005	15
0.860	29.28 ± 0.40	0	4.15 ± 0.03	16
1.200	35.94 ± 0.40	0	4.32 ± 0.03	16
1.390	38.76 ± 0.40	0	4.41 ± 0.03	16
1.830	44.02 ± 0.40	0	4.59 ± 0.02	16
2.105	46.18 ± 0.40	0	4.72 ± 0.03	16
2.392	48.08 ± 0.40	0	4.85 ± 0.03	16
1.855	44.212 ± 0.023	-0.049 ± 0.020	4.599 ± 0.002	17
1.858	44.218 ± 0.028	-0.057 ± 0.024	4.603 ± 0.003	17
2.425	48.318 ± 0.029	-0.075 ± 0.018	4.858 ± 0.003	17
3.037	50.971 ± 0.040	-0.082 ± 0.022	5.149 ± 0.004	17
3.527	52.475 ± 0.046	-0.094 ± 0.023	5.368 ± 0.006	17
3.899	53.257 ± 0.057	-0.109 ± 0.020	5.544 ± 0.007	17
4.203	53.808 ± 0.081	-0.074 ± 0.023	5.681 ± 0.011	17

* For the range 60 to 550 kev, see W. D. ALLEN and A. T. G. FERGUSON: Proc. Phys. Soc. Lond. A **68**, 1077 (1955).

[10] R. E. FIELDS, R. L. BECKER and R. K. ADAIR: Phys. Rev. **94**, 389 (1954).

[11] C. L. STORRS and D. H. FRISCH: Phys. Rev. **95**, 1252 (1954).

[12] E. M. HAFNER, W. F. HORNYAK, C. E. FALK, G. SNOW and T. COOR: Phys. Rev. **89**, 204 (1953).

[13] H. L. POSS, E. O. SALANT, G. A. SNOW and L. C. L. YUAN: Phys. Rev. **87**, 11 (1952).

[14] D. I. COOPER, D. H. FRISCH and R. L. ZIMMERMAN: Phys. Rev. **94**, 1209 (1954).

[15] M. C. YOVITS, R. L. SMITH jr., M. H. HULL jr., J. BENGSTON and G. BREIT: Phys. Rev. **85**, 540 (1952) (measurements of N. P. HEYDENBURG and J. L. LITTLE).

[16] R. G. HERB, D. W. KERST, D. B. PARKINSON and G. J. PLAIN: Phys. Rev. **55**, 998 (1939). Cf. also J. D. JACKSON and J. M. BLATT: Rev. Mod. Phys. **22**, 77 (1950).

[17] H. R. WORTHINGTON, J. N. MCGRUER and D. E. FINDLEY: Phys. Rev. **90**, 899 (1953). — H. H. HALL and J. L. POWELL: Phys. Rev. **90**, 912 (1953).

Cyclotron data from 4 to about 30 Mev are gathered in Table 3. Table 3 contains also the D-wave phase shift δ_2 besides δ_0 and δ_1. It is also implied in Tables 2 and 3 that errors due to inadequate knowledge of the energy are already included in the final results if they are not explicitly stated.

Table 3. *Cyclotron data for the nuclear S-, P- and D-phase shifts and the derived quantity K (defined in Sect. 22) for various incident proton energies in laboratory system.*

Energy (Mev)	δ_0 (degrees)	δ_1 (degrees)	δ_2 (degrees)	K	Reference number
4.2	52.7 ± 2.0	— *	—	5.83 ± 0.30	18
4.96 ± 0.08	54.7 ± 1.0	0	—	6.02 ± 0.18	19
5.07	54.5 ± 0.6	− 0.05 ± 0.09	—	6.13 ± 0.09	20
5.14 ± 0.14	53.9 ± 0.6	0 ± 1.5	—	6.26 ± 0.15	21
5.77	55.29 ± 0.30	− 0.08 ± 0.07	—	6.39 ± 0.05	22
3.44	52.46 ± 0.6	0.01 ± 0.1	—	5.302 ± 0.067	23
6.85	55.82 ± 0.6	0.03 ± 0.1	—	6.859 ± 0.122	23
7.51	55.78 ± 0.6	0.03 ± 0.1	—	7.177 ± 0.130	23
9.73	56.5 ± 0.5	− 0.55 ± ?	0	7.931 ± 0.125	24
12.4				9.68 ± 0.13	25, **
14.5 ± 0.7	52.2 ± 3.5	—	—	10.78 ± 0.80	26
18.2 ± 0.2	54.1 ± 0.3	+ 1.0 ± 0.8	+ 0.4 ± 0.1	11.42 ± 0.1	27
18.8				11.67 ± 0.27	28, **
21.9				12.91 ± 0.27	28, **
25.2				14.65 ± 0.33	28, **
25.45				14.84 ± 0.27	28, **
31.45				16.73 ± 0.30	28, **
31.8				16.76 ± 0.25	28, **
29.4 ± 0.1	50.22 ± 0.16	0	—	15.84 ± 0.39	29

3. Neutron-proton capture and low energy photodisintegration. α) *Thermal neutron-proton capture.* The capture of neutrons by protons is noticeable in the thermal region only, the cross section being inversely proportional to the neutron velocity. Experimental data are usually referred to a velocity of 2200 m/sec. There are some determinations of the hydrogen cross section relative to that of boron, which is known very accurately: (755 ± 4) b. The most recent ones give

$$\sigma_{\text{cap}} = (0.329 \pm 0.004)\ \text{b}^{30}; \quad (0.332 \pm 0.007)\ \text{b}^{31}.$$

* The horizontal lines mean that there are no definite reports.

** These authors measured only the differential scattering cross sections at 90° (center of mass system) and calculated K assuming pure S-wave scattering so there are no values reported for phase shifts.

[18] A. N. MAY and C. F. POWELL: Proc. Roy. Soc. Lond., Ser. A **190**, 170 (1947).

[19] R. E. MEAGHER: Phys. Rev. **78**, 667 (1950). — JACKSON and BLATT: Ref. 16, p. 91.

[20] K. B. MATHER: Phys. Rev. **82**, 133 (1951).

[21] R. O. BONDELID, C. H. BRADEN, M. E. BATTAT and P. BOHLMAN: Phys. Rev. **87**, 699 (1952).

[22] E. J. ZIMMERMAN, R. O. KERMAN, S. SINGER, P. G. KRUGER and W. J. JENTSCHKE: Phys. Rev. **96**, 1322 (1954).

[23] J. ROUVINA: Phys. Rev. **81**, 593 (1951).

[24] B. CORK and W. HARTSOUGH: Phys. Rev. **94**, 1300 (1954). — H. H. HALL: Phys. Rev. **95**, 424 (1954).

[25] F. E. FARIS and B. WRIGHT: Phys. Rev. **79**, 577 (1950).

[26] R. R. WILSON, E. J. LOFGREN, J. R. RICHARDSON, B. T. WRIGHT and R. S. SHANKLAND: Phys. Rev. **72**, 1131 (1947). — JACKSON and BLATT: Ref. 16, p. 91.

[27] J. L. YNTEMA and G. M. WHITE: Phys. Rev. **95**, 1226 (1954).

[28] B. CORK: Phys. Rev. **80**, 321 (1950).

[29] W. K. H. PANOFSKY and F. L. FILLMORE: Phys. Rev. **79**, 57 (1950). Cf. R. S. CHRISTIAN and H. P. NOYES: Phys. Rev. **79**, 85 (1950).

[30] B. HAMERMESH, G. R. RINGO and S. WEXLER: Phys. Rev. **90**, 603 (1953).

[31] S. P. HARRIS, C. O. MUEHLHAUSE, D. ROSE, H. P. SCHROEDER, G. E. THOMAS jr. and S. WEXLER: Phys. Rev. **91**, 125 (1953).

Direct measurements imply a study of the mean life of thermal neutrons in water and yield

$$\sigma_{\text{cap}} = (0.323 \pm 0.008)\ \text{b}^{32}; \quad (0.333 \pm 0.003)\ \text{b}^{33}.$$

The value which we have used later in this article, is

$$\sigma_{\text{cap}} = (0.329 \pm 0.006)\ \text{b}, \tag{3.1}$$

at a neutron velocity 2200 m/sec. (All errors quoted include the uncertainty of the neutron velocity.)

β) Total photodisintegration cross section. Table 4, which quotes an earlier summary[34] with some additions, shows how the total cross section for this reaction depends on the γ-energy.

Table 4. *Total photodisintegration cross section versus the incident γ-ray energy in laboratory system.*

γ-energy (Mev)	Total cross section (10^{-28} cm²)	Reference number	γ-energy (Mev)	Total cross section (10^{-28} cm²)	Reference number
2.504	11.9 ± 0.8	34	7.39 ± 0.15	18.4 ± 1.5	34
2.504	10.6 ± 1.1	34	8.14 ± 0.08	18.0 ± 1.3	34
2.615	13.0 ± 0.28	35	12.50 ± 0.21	10.4 ± 1.0	34
2.757	15.9 ± 0.6	34	17.6 ± 0.2	7.7 ± 0.9	34
2.757	14.3 ± 1.1	34	17.6 ± 0.2	7.1 ± 1.5	36
4.45 ± 0.04	24.3 ± 1.7	34	17.6 ± 0.2	7.1 ± 2.0	37
6.14 ± 0.01	21.9 ± 1.0	34	20	5.1 ± 0.16	*

γ) Angular distribution of photo-protons. In the very low energy region, the differential scattering cross section $d\sigma/d\Omega$ of photo-protons in the center of mass (c.m.) system of two nucleons can be described by the formula

$$d\sigma/d\Omega = a + b \sin^2\vartheta,$$

where ϑ is the angle between the directions of the photo-proton and the incident γ-ray. The photo-magnetic cross section, σ_m, and the photo-electric one, σ_e, are determined by the relations, $\sigma_m = 4\pi a$ and $\sigma_e = 8\pi b/3$. Table 5 gives a summary[34] of experimental data of the ratio σ_m/σ_e in the very low energy region.

In other energy regions, however, data are consistent with a differential scattering cross section in c.m. system given by

$$d\sigma/d\Omega = a + (b + c\cos\vartheta)\sin^2\vartheta. \tag{3.2}$$

The efforts have been concentrated on careful measurements of the isotropy coefficient a/b and forward asymmetry coefficient c/b. Although the experiments are not yet so accurate, all the available data are summarized in Table 6. Anyhow a gradual increase of a/b with the γ-energy seems to be confirmed.

* See footnote 42, p. 7.

[32] F. R. SCOTT, D. B. THOMSON and W. WRIGHT: Phys. Rev. **95**, 582 (1954).

[33] G. VON DARDEL and N. G. SJÖSTRAND: Phys. Rev. **96**, 1245 (1954).

[34] L. HULTHÉN and B. NAGEL: Phys. Rev. **90**, 62 (1953). This paper gives the original references.

[35] P. MARIN, G. R. BISHOP and H. HALBAN: Proc. Phys. Soc. Lond. A **67**, 1113 (1954).

[36] P. V. C. HOUGH: Phys. Rev. **80**, 1069 (1950).

[37] H. WÄFFLER: Report to Chicago Conference 1951.

Table 5. *Ratio of the photo-magnetic cross section σ_m to the photo-electric one σ_e versus the incident γ-ray energy in laboratory system.*

γ-energy (Mev)	σ_m/σ_e	Reference number
2.504	0.600 ± 0.02	34
2.615	0.360 ± 0.008	34
2.71	0.49 ± 0.07	34
2.757	0.247 ± 0.007	34
2.757	0.317 ± 0.012	34
2.757	0.265 ± 0.065	34
2.757	0.265 ± 0.07	34
2.89	0.27 ± 0.06	34

Table 6. *Isotropy coefficient and forward asymmetry coefficient of photo-protons for some ranges of γ-energy in laboratory system.*

γ-energy (Mev)	Isotropy coefficient a/b	Forward asymmetry coefficient c/b	Reference number
5–11	0.04 ± 0.03		38
11–13	0.24 ± 0.07		38
8–13		0.24 ± 0.09	38
14.8–17.6	0.12 ± 0.07	0.24	39
6–7	0.15 ± 0.06	$0.4 \begin{smallmatrix}+0.2 \\ -0.3\end{smallmatrix}$	40
18–22	0.132 ± 0.041	0.2	41
20	0.09 ± 0.01	0.18	42

4. High energy scattering data. α) *High energy proton-proton scattering.* As the proton-proton scattering is symmetrical in the c.m. system with respect to 90°, only the angular range between 0° and 90° need be considered. It is observed that the Coulomb force gives rise to a very strong forward scattering within about 10° from the direction of incidence, which is followed by an interference

Table 7. *The differential proton-proton scattering cross section in the isotropic region and the corresponding isotropic angular range in c.m. system against the incident proton energy in laboratory system*

Incident proton energy (Mev)	Differential scattering cross section (mb/sterad.)	Isotropic angular range in c.m. system (degrees)	Reference number
30.14 ± 0.08	15.02 ± 0.33	40–90	43
75	6.6 ± 1.3	40–90	44
105	5.4 ± 1.1	40–90	44
147	4.94 ± 0.28	25–90	45
240	4.97 ± 0.42	26.8–90	46
240	4.66 ± 0.39	13.0–90	47
120	3.95 ± 0.12	at 90	48, *
164	3.60 ± 0.17	20–90	48, 49
250	3.28 ± 0.16	20–90	48, 49
345	3.8 ± 0.3	30–90	48
225	3.56 ± 0.15	at 90	50, *
330	3.72 ± 0.15	18–90	50, 51
144 ± 5	3.21 ± 0.11	at 90	52, *
271 ± 9	3.67 ± 0.34	at 90	52, *
429 ± 14	3.3 ± 0.2	28–90	52

[38] V. E. Krohn jr. and E. F. Shrader: Phys. Rev. **86**, 391 (1952).
[39] H. Wäffler and S. Younis: Helv. phys. Acta **24**, 483 (1951).
[40] G. Goldhaber: Phys. Rev. **81**, 930 (1951).
[41] J. Halpern and E. V. Weinstock: Phys. Rev. **91**, 934 (1953).
[42] Lew Allen jr.: Phys. Rev. **98**, 705 (1955).
[43] F. L. Filmore: Phys. Rev. **83**, 1253 (1951).
[44] R. W. Birge, U. E. Kruse and N .F. Ramsey: Phys. Rev. **83**, 274 (1951).
[45] J. M. Cassels, T. G. Pickavance and G. H. Stafford: Proc. Roy. Soc. Lond., Ser. A **214**, 262 (1952).
[46] C. L. Oxley and R. D. Schamberger: Phys. Rev. **85**, 416 (1952).
[47] O. A. Towler jr.: Phys. Rev. **85**, 1024 (1952).
[48] O. Chamberlain, E. Segrè and C. Wiegand: Phys. Rev. **83**, 923 (1951).
[49] O. Chamberlain and J. D. Garrison: Phys. Rev. **95**, 1349 (1954).
[50] O. Chamberlain, G. Pettengill, E. Segrè and C. Wiegand: Phys. Rev. **93**, 1424 (1954).
[51] D. Fischer and G. Goldhaber: Phys. Rev. **95**, 1350 (1954).
[52] J. Marshall, L. Marshall, and V. A. Nedzel: Phys. Rev. **92**, 834 (1953).
* Measurements at 90° only.

minimum owing to the destructive concurrence of the (repulsive) COULOMB and the (attractive) nuclear force at about 10° to 20°, in the energy region concerned. In the remaining angular range the nuclear scattering predominates. The most remarkable characteristics are: the angular distribution is isotropic over a wide angle (20° to 90°) and this isotropic cross section is nearly independent of the incident proton energy from perhaps 100 up to 400 Mev.

Table 8. *The integrated nuclear elastic, total and inelastic proton-proton scattering cross sections (incident proton energy in laboratory system).*

Energy (Mev)	Elastic (mb)	Total (mb)	Inelastic (mb)
440	24 ± 2	27 ± 2	3 ± 3
590	25 ± 2	36 ± 3	11 ± 3.5
800	21 ± 2	47 ± 2	26 ± 3
1000	19 ± 3	48 ± 2	29 ± 3.5

Table 7 summarizes experimental data of this cross section in mb (10^{-27} cm²), together with the angular range in which this isotropy has been established. Fig. 1 presents some differential scattering cross sections graphically.

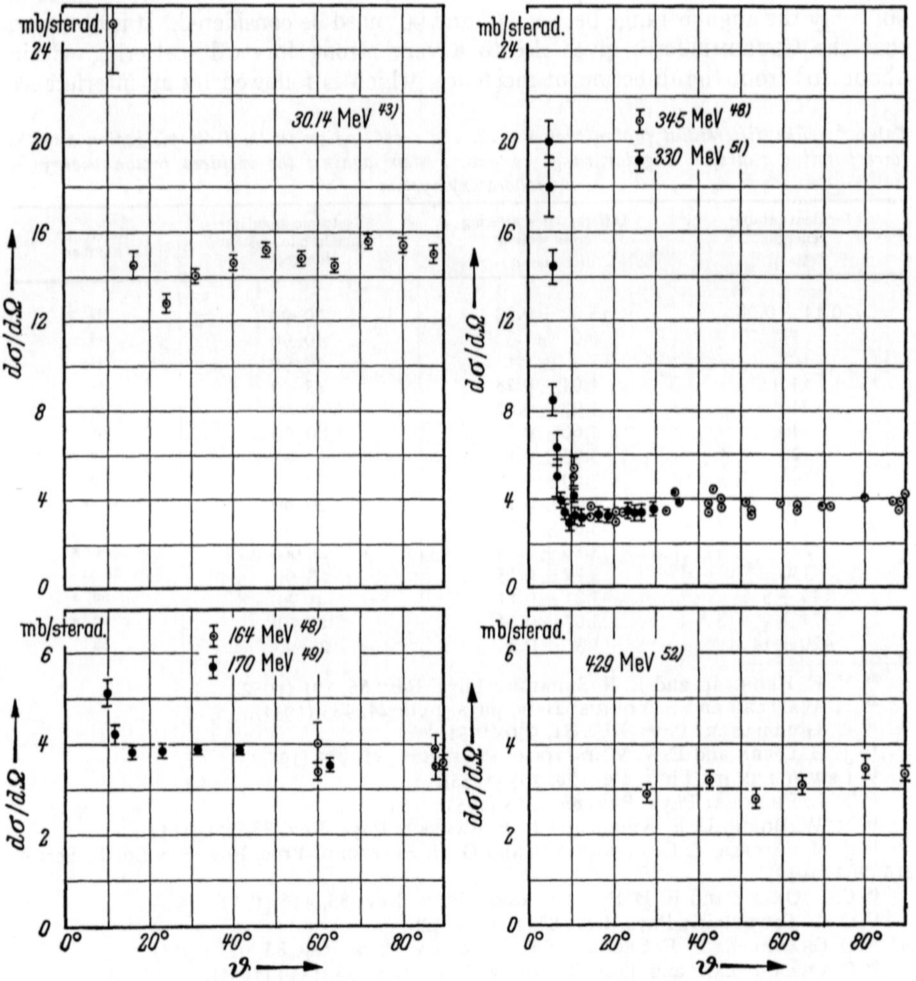

Fig. 1. Differential proton-proton scattering cross section against c.m. scattering angle ϑ. The incident proton energy in laboratory system and the reference number are given in the figures.

At still higher energies the angular isotropy is destroyed and the forward scattering becomes more and more predominant. Four examples of differential scattering cross sections have been plotted in Fig. 2, the extrapolated total elastic cross section being given in Table 8[53,54]. At these energies, inelastic scattering (accompanied by meson production) can occur, the cross section of which is obtained by subtracting the integrated elastic cross section from the total cross section[55], both of which are also contained in Table 8.

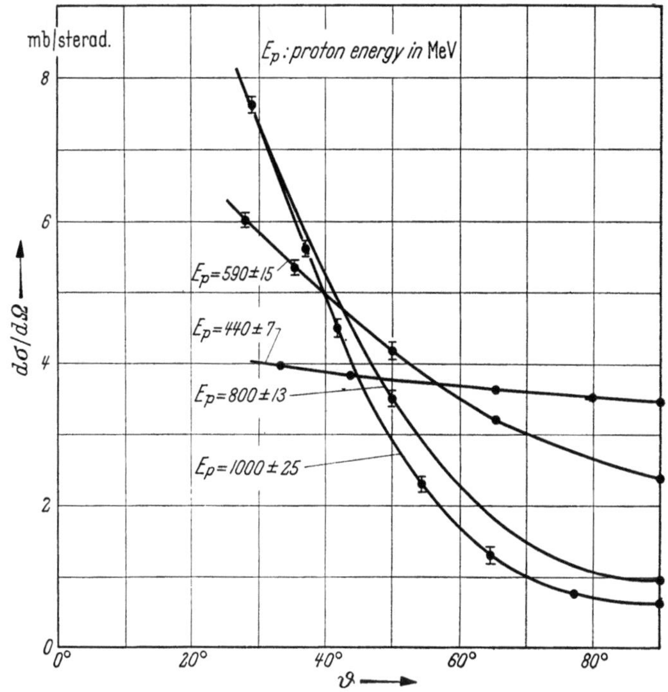

Fig. 2. The differential elastic proton-proton scattering cross section against c.m. scattering angle ϑ (incident proton energy in laboratory system).

β) High energy neutron-proton scattering. All data up to about 20 Mev (incident neutron energy, laboratory system) show isotropic scattering in the c.m. system within the experimental errors. About 30 Mev, the experiments seem to indicate some anisotropy[56] [for instance, at 27.2 Mev[57] $\sigma(180°)/\sigma(90°) = 1.28 \pm 0.10$]. At 40 Mev, the backward scattering becomes dominant, as shown in Fig. 3, which also includes data for 90 Mev[58–60]. The distributions are normalized to the total cross sections: $\sigma_{\text{tot}} = 0.170$ barns (40 Mev)[58], 0.076 barns (90 Mev)[58], respectively. Other authors report $\sigma_{\text{tot}} = (0.203 \pm 0.007)$ barn (42 Mev)[61], (0.083 ± 0.004) barns (90 Mev)[62] and (0.073 ± 0.002) barns (95 Mev)[63].

---

[53] R. B. S. Sutton, T. H. Fields, J. G. Fox, J. A. Kane, W. E. Mott and R. A. Stallwood: Phys. Rev. **97**, 783 (1955).
[54] L. W. Smith, A. W. McReynolds and G. Snow: Phys. Rev. **97**, 1186 (1955).
[55] A. H. Shapiro, C. P. Leavitt and F. F. Chen: Phys. Rev. **95**, 663 (1954).
[56] M. E. Remley, W. K. Jentschke and P. G. Kruger: Phys. Rev. **89**, 1194 (1953).
[57] J. E. Brolley jr., J. H. Coon and J. E. Fowler: Phys. Rev. **82**, 190 (1951).
[58] J. Hadley, E. Kelly, C. Leith, E. Segrè, C. Wiegand and H. York: Phys. Rev. **75**, 351 (1949).
[59] O. Chamberlain and J. W. Easley: Phys. Rev. **94**, 208 (1954).
[60] W. Selove, K. Strauch and F. Titus: Phys. Rev. **92**, 724 (1953).
[61] R. H. Hildebrand and C. E. Leith: Phys. Rev. **80**, 842 (1950).
[62] L. J. Cook, E. M. McMillan, J. M. Peterson and D. C. Sewell: Phys. Rev. **75**, 7 (1949).
[63] J. DeJuren and N. Knabe: Phys. Rev. **77**, 606 (1950).

Other experiments[64] confirmed the 90 Mev data in Fig. 3. An important characteristic is the similarity with respect to the forward and backward directions, the minimum, however, being attained at an angle slightly smaller than 90°. At higher energies the approximate symmetry around 90° is gradually being lost. Some of these data are summarized in Fig. 4, which includes 4 curves at (156 ± 3) Mev[65], 220[66] and 260 Mev[67], 300 Mev[68] and 400 Mev[69], respectively. The most important general feature is that they show strong backward peaks

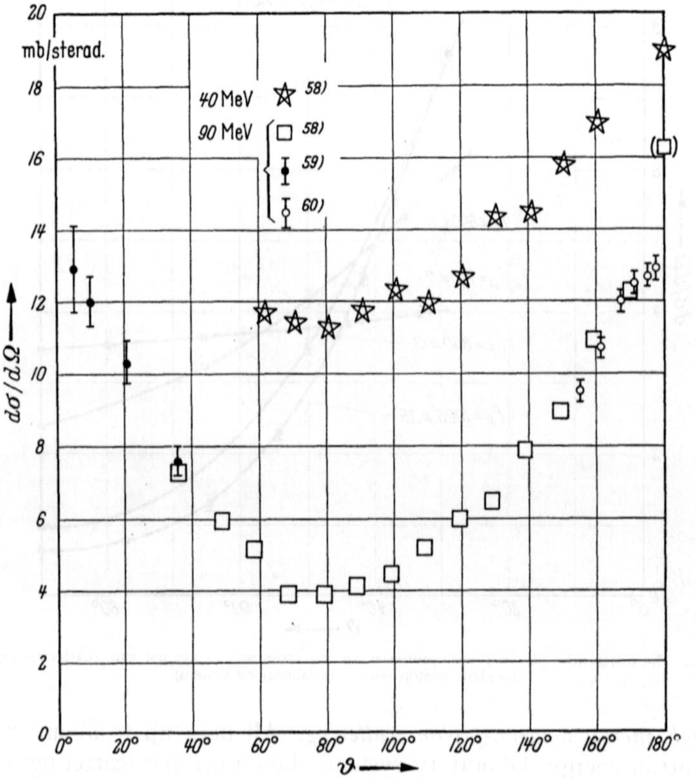

Fig. 3. Differential neutron-proton scattering cross sections at 40 and 90 Mev against c.m. scattering angle ϑ.

and smooth rises in the forward direction. The minimum points move gradually towards the backward direction as the energy increases (minimum at 80° for 156 Mev, at 100° for 400 Mev).

γ) Polarization measurements. In all the scattering experiments thus far mentioned, both the beam and the target are unpolarized. It is found, however, that the scattered beam is, in general, polarized in a direction perpendicular to the plane of scattering, even though the initial beam and the target are both unpolarized. This polarization can be detected experimentally by observing

[64] R. H. STAHL and N. F. RAMSEY: Phys. Rev. **96**, 1310 (1954). For somewhat higher energies see also J. J. THRESHER, R. G. P. VOSS and R. WILSON: Proc. Roy. Soc. Lond., Ser. A **229**, 492 (1955).

[65] T. C. RANDLE, A. E. TAYLOR and E. WOOD: Proc. Roy. Soc. Lond., Ser. A **213**, 392 (1952).

[66] G. GUERNSEY, G. MOTT and B. K. NELSON: Phys. Rev. **88**, 15 (1952).

[67] E. KELLY, C. L. LEITH, E. SEGRÈ and C. WIEGAND: Phys. Rev. **79**, 96 (1950).

[68] J. DE PANGHER: Phys. Rev. **95**, 578 (1954).

[69] J. A. HARTZLER, R. T. SIEGEL and W. OPITZ: Phys. Rev. **95**, 591 (1954).

the azimuthal asymmetry in the subsequent scattering of the beam by a second unpolarized target. Let us denote the scattering angles Θ_1, Φ_1 and Θ_2, Φ_2, respectively, where Θ_2 and Φ_2 are measured with respect to the first scattering direction, Φ_2 being zero in the first scattering plane (Fig. 5). The experimentally measured quantity is the azimuthal asymmetry e defined by

$$e = \frac{\sigma(\Theta_2, \Phi_2 = 0) - \sigma(\Theta_2, \Phi_2 = 180^\circ)}{\sigma(\Theta_2, \Phi_2 = 0) + \sigma(\Theta_2, \Phi_2 = 180^\circ)}, \tag{4.1}$$

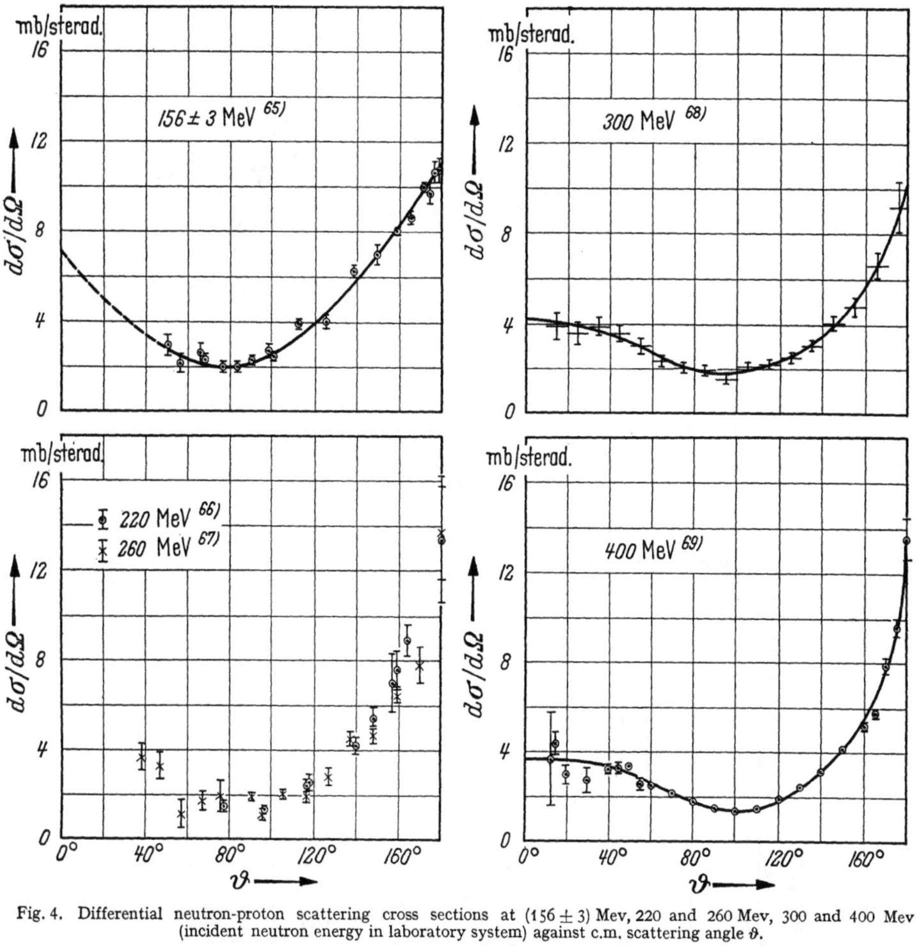

Fig. 4. Differential neutron-proton scattering cross sections at (156 ± 3) Mev, 220 and 260 Mev, 300 and 400 Mev (incident neutron energy in laboratory system) against c.m. scattering angle ϑ.

where $\sigma(\Theta_2, \Phi_2)$ is the differential cross section for the second scattering. Wouters[70] observed the asymmetry in the double neutron-proton scattering near 150 Mev for $\Theta_1 \approx \Theta_2 \approx 30^\circ$ and found something which might be interpreted as $e \approx (9 \pm 6)\%$. Data for the double proton-proton scattering have been obtained by Oxley, Cartwright and Rouvina[71]: $e = (4.8 \pm 1.8)\%$ for $\Theta_1 = 19^\circ$ and $\Theta_2 = 27^\circ$ in the energy range near 200 Mev.

It has later been recognized that a nucleon beam elastically scattered by certain nuclei is in general more strongly polarized than one scattered by nucleons.

[70] L. F. Wouters: Phys. Rev. **84**, 1069 (1951).

[71] C. L. Oxley, W. F. Cartwright and J. Rouvina: Phys. Rev. **93**, 806 (1954).

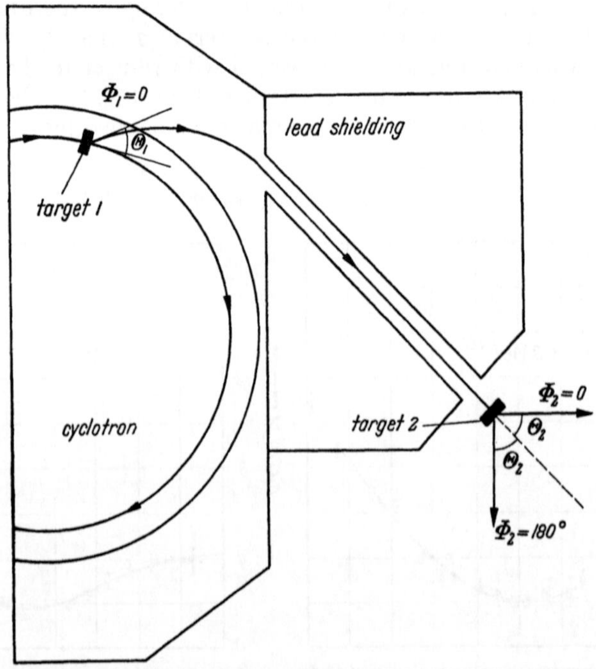

Fig. 5. Plane view of the polarization measurement, showing the scattering angles Θ_1, Φ_1 and Θ_2, Φ_2 in the first and second scattering, respectively.

The degree of polarization can be measured by observing the azimuthal asymmetry in the subsequent scattering with the same target and scattering angle

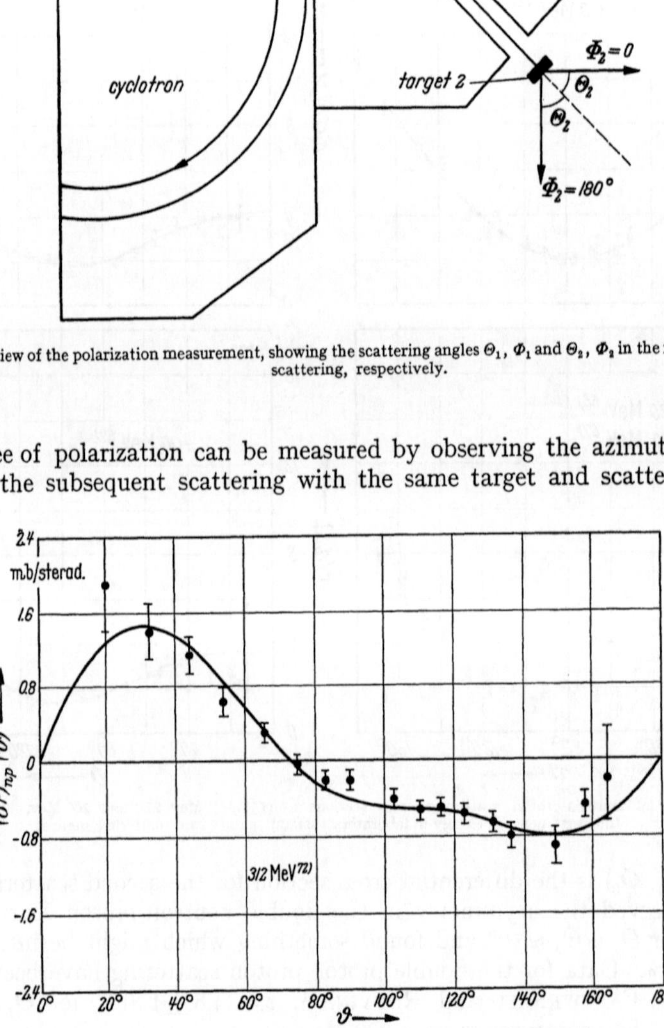

Fig. 6. $\sigma_{np}(\vartheta)\, P_{np}(\vartheta)$ as a function of c.m. scattering angle ϑ.

as in the first one. Using such nucleon beams with known degrees of polarization, one can measure the "polarization" $P(\Theta)$ (cf. Sect. 45) in the nucleon-nucleon scattering. Several experiments have been made[72-74] the results of which can be approximately described by the following analytic expressions, where $\sigma(\vartheta)$ is the unpolarized differential scattering cross section in mb/sterad. and ϑ is the c.m. scattering angle:

n-p, 312 Mev[72] (cf. also Fig. 6):

$$\sigma(\vartheta)\,P(\vartheta) = -0.016\sin\vartheta + 0.958\sin 2\vartheta + 0.324\sin 3\vartheta + 0.366\sin 4\vartheta, \quad (4.2)$$

p-p, 312 Mev[73]:

$$P(\vartheta) = 0.3595\sin 2\vartheta + 0.0645\sin 4\vartheta + 0.0309\sin 6\vartheta, \quad (4.3)$$

pp, 415 Mev[74]:

$$\sigma(\vartheta)\,P(\vartheta) = K\sin\vartheta\cos\vartheta \times (1 + b\cos^2\vartheta + c\cos^4\vartheta),$$

$$\left.\begin{array}{l} K = 0.62 \pm 0.14, \quad b = 1.0 \pm 0.7, \\ c = 0.63 \pm 0.77. \end{array}\right\} \quad (4.4)$$

[$\sigma(\vartheta)$ is normalized to 1 at 90°. b and c are connected by: $0.98\,b + c = 1.6 \pm 0.3$.] One important characteristic of these data is that the angular dependence is very roughly of the form $\sin\vartheta\cos\vartheta$ at lower energies*; it deviates from this simple form rather markedly at higher energies. It should be added that only the elastic scattering has been measured in these experiments.

5. High energy photodisintegration data. α) *Angular distribution.* All the existing data seem to be consistent with the formula

$$\left.\begin{array}{l} \dfrac{d\sigma(\vartheta)}{d\Omega} = (A + B\sin^2\vartheta) \times \\ \qquad \times (1 + 2\beta\cos\vartheta), \end{array}\right\} \quad (5.1)$$

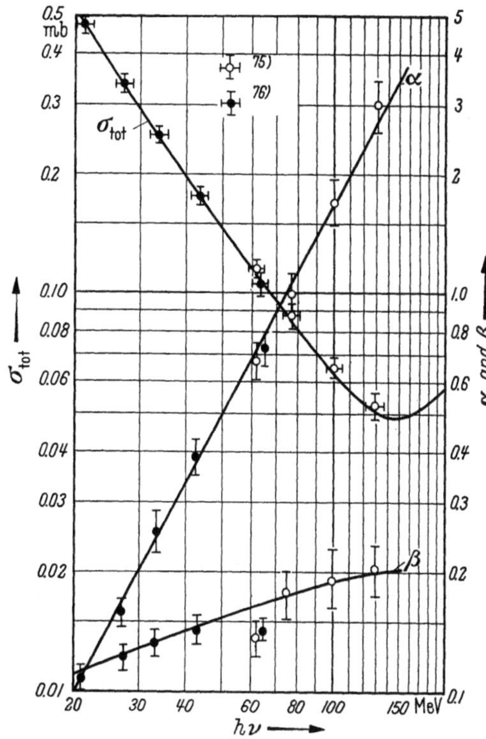

Fig. 7. The isotropy factor α and the retardation factor β, together with σ_{tot}, as a function of c.m. incident photon energy $h\nu$.

where $d\sigma/d\Omega$ is the differential photo-proton cross section as a function of c.m. emission angle ϑ. Eq. (5.1) is just a slight modification of the formula (3.2), which is used to fit the low energy data. It is to be noted that the pure $\sin^2\vartheta$ component which has been the major term in the low energy region and is still appreciable at 62.9 Mev[75] is observed to disappear, practically, at 130.6 Mev[76].

* According to D. FISCHER and J. BALDWIN: Phys. Rev. **100**, 1445 (1955), this is not the case even at 170 Mev.

[72] O. CHAMBERLAIN, R. DONALDSON, E. SEGRÈ, R. TIPP, C. WIEGAND and T. YPSILANTIS: Phys. Rev. **95**, 850 (1954).

[73] O. CHAMBERLAIN, G. PETTENGILL, E. SEGRÈ and C. WIEGAND: Phys. Rev. **95**, 1348 (1954).

[74] J. A. KANE, R. A. STALLWOOD, R. B. SUTTON, T. H. FIELDS and J. G. FOX: Phys. Rev. **95**, 1694 (1954).

[75] E. A. WHALIN jr.: Phys. Rev. **95**, 1362 (1954).

[76] LEW ALLEN jr.: Phys. Rev. **98**, 705 (1955).

At higher energies the forward emission becomes predominant due to the term $\beta \cos\vartheta$. The isotropic term A, which is quite negligible at low energies but tends to increase at somewhat higher energies, is now quite dominant. Fig. 7 gives the experimental data on the isotropy factor $\alpha \equiv A/B$ and the retardation factor β together with σ_{tot} as a function of the incident photon energy in c.m. system. It is seen that α increases linearly in the logarithmic plot of the data. Reference numbers are given in the figure.

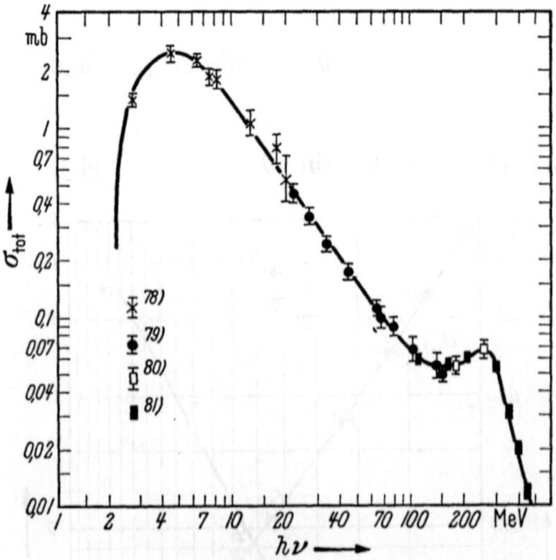

Fig. 8. Total photodisintegration cross section versus incident photon energy $h\nu$ in laboratory system.

β) Total cross section versus energy. All available data have been plotted in Fig. 8[77] against the incident photon energy in laboratory system, together with the low energy data cited before. The most characteristic feature is the appearance of the minimum at about 150 Mev and the maximum at about 250 Mev, which is then reduced by a factor of about 5 at 450 Mev. Reference numbers are given in the figure.

III. Meson theory and two-nucleon interaction.

a) Meson field theory*.

6. Pseudoscalar meson theory. Although the meson field theory was proposed by YUKAWA as early as 1935, a definite quantitative success is still wanting. It is, however, generally accepted that the basic idea is correct. Among the different versions of meson theory which fulfill the requirements for relativistic invariance, positive definite field energy etc., the pseudoscalar theory stands out as the only one which is not excluded by fundamental empirical evidence. The formal difference between a pseudoscalar and a scalar field is that a space reflection** changes the sign of a pseudoscalar variable***, whereas it does not affect a scalar. Other transformation properties are identical (invariance under rotation). Therefore, a one-component field variable $\varphi(x_\lambda)$ at each space-time point x_λ† is sufficient to describe the pseudoscalar meson field. $\varphi(x_\lambda)$ is then assumed to satisfy the LORENTZ invariant equation

$$(\Box - \mu^2)\,\varphi(x_\lambda) = 0, \tag{6.1}$$

[77] J. S. LEVINGER: Phys. Rev. **97**, 970 (1955).

[78] Low energy data cited before.

[79] L. ALLAN: Phys. Rev. **98**, 705 (1955). — E. A. WHALIN: Phys. Rev. **95**, 1362 (1954).

[80] J. KECK, R. M. LITTAUER, G. K. O'NEILL, A. M. PERRY and W. M. WOODWARD: Phys. Rev. **93**, 827 (1954).

[81] J. C. KECK, A. V. TOLLESTRUP and W. R. SMYTHE: Phys. Rev. **96**, 850 (1954).

* Cf. also the articles on meson physics in Vol. XLIII of this Encyclopedia, especially the contribution by M. LÉVY on the meson theory of nuclear forces.

** The space reflection is a coordinate transformation in which an odd number of space axes are reversed at the same time.

*** Example; volume in 3 dimensions.

† We use the notations $x_1 = x$, $x_2 = y$, $x_3 = z$ and $x_4 = i\,x_0 = ict$. Here we consider the local field theory only.

without interaction with nucleons, where

$$\square = \sum_{\nu=1}^{4} \frac{\partial^2}{\partial x_\nu^2} = \frac{\partial^2}{\partial \boldsymbol{x}^2} - \frac{1}{c^2}\frac{\partial^2}{\partial t^2}, \tag{6.2}$$

and the constant μ is related to the meson mass m by the relation

$$\mu = m\,c/\hbar. \tag{6.3}$$

It is known that a real $\varphi_0(x_\lambda)$ corresponds to a neutral meson field, while a complex $\varphi(x_\lambda)$ can describe a charged meson field. The field Eq. (6.1) can be derived by the usual variational principle from the LORENTZ invariant Lagrangian

$$L_0 = -\frac{1}{2}\left(\sum_\lambda \left(\frac{\partial \varphi_0}{\partial x_\lambda}\right)^2 + \mu^2 \varphi_0^2\right) \tag{6.4}$$

for a real field and

$$L_0 = -\left(\sum_\lambda \frac{\partial \varphi^*}{\partial x_\lambda}\frac{\partial \varphi}{\partial x_\lambda} + \mu^2 \varphi^* \varphi\right) \tag{6.5}$$

for a complex field, where φ and its complex or hermite conjugate φ^* are to be varied independently. The factor $\frac{1}{2}$ in (6.4) is introduced for later convenience.

To represent the interaction with nucleons, we have only to form a LORENTZ invariant additional term L' from $\varphi(x_\lambda)$ and the spinor field variable $\psi(x_\lambda)$ representing the nucleons. Recalling the pseudoscalar nature of $\varphi(x_\lambda)$, the two simplest forms for L' are given by*

$$L' = -i f \overline{\psi} \gamma_5 \psi \varphi \tag{6.6}$$

and

$$L' = -i \frac{g}{\mu} \sum_\lambda \overline{\psi} \gamma_5 \gamma_\lambda \psi \frac{\partial \varphi}{\partial x_\lambda}, \tag{6.7}$$

where $\gamma_5 = \gamma_1 \gamma_2 \gamma_3 \gamma_4$, $\overline{\psi} = \psi^* \gamma_4$ and f and g are real quantities with the dimension of an electric charge, which are called the pseudoscalar-pseudoscalar (ps-ps) and pseudoscalar-pseudovector (ps-pv) coupling constants, respectively. The factors i were introduced to make (6.6) and (6.7) hermitian for a real φ. In order to ensure the conservation of the total electric charge, we must be a little more careful in constructing L'. According to the usual quantum mechanical interpretation of the field, ψ contains operators for the annihilation of a nucleon and creation of an anti-nucleon, φ operators for the annihilation of a positive meson and creation of a negative meson (and φ_0 annihilation and creation operators of a neutral meson). In order not to violate the charge conservation law, therefore, L' in (6.6) must be written, using ψ_p and ψ_n to represent the proton and neutron fields, respectively,

$$L' = -i f \overline{\psi}_p \gamma_5 \psi_p \varphi_0 - i f \overline{\psi}_n \gamma_5 \psi_n \varphi_0 \tag{6.8}$$

for a real field and

$$L' = -i f \sqrt{2}\, \overline{\psi}_p \gamma_5 \psi_n \varphi - i f \sqrt{2}\, \overline{\psi}_n \gamma_5 \psi_p \varphi^* \tag{6.9}$$

for a complex field, and analogous expressions instead of (6.7). The factor $\sqrt{2}$ in (6.9) corresponds to the factor $\frac{1}{2}$ in (6.4). The Lagrangian for the nucleon field

* Cf. S. S. SCHWEBER, H. A. BETHE and F. DE HOFFMANN: Mesons and Fields, Vol. I, Fields. Evanston, Ill.: Row, Peterson & Co. 1955.

is given by

$$L = -\hbar c \left[\overline{\psi}_p \left(\sum_\mu \gamma_\mu \frac{\partial}{\partial x_\mu} + \varkappa\right) \psi_p + \overline{\psi}_n \left(\sum_\mu \gamma_\mu \frac{\partial}{\partial x_\mu} + \varkappa\right) \psi_n\right], \tag{6.10}$$

where $\varkappa$ is related to the nucleon mass M by the relation

$$\varkappa = M c/\hbar. \tag{6.11}$$

7. Charge coordinate. Instead of considering proton and neutron fields separately, we may introduce a single nucleon field with "charge spin" $\frac{1}{2}$. The charge spin is assumed to have the same mathematical properties as the ordinary spin, although pertaining to "charge space" instead of ordinary space. Thus, introducing the charge spin operator $\boldsymbol{\tau}$ having three components

$$\tau_1 = \begin{bmatrix} 0 & 1 \\ 1 & 0 \end{bmatrix}, \quad \tau_2 = \begin{bmatrix} 0 & -i \\ +i & 0 \end{bmatrix}, \quad \tau_3 = \begin{bmatrix} 1 & 0 \\ 0 & -1 \end{bmatrix}, \tag{7.1}$$

a proton-neutron field can be described by a single nucleon field ψ with two components corresponding to the eigenvalues of τ_3:

$$\psi = \begin{bmatrix} \psi_p \\ \psi_n \end{bmatrix}, \tag{7.2}$$

where we have identified a nucleon belonging to an eigenvalue $+1$ of τ_3 as a proton and one to -1 as a neutron. Furthermore, introducing two real fields φ_1 and φ_2, corresponding to two kinds of charge*, instead of φ and φ^*,

$$\varphi = \frac{1}{\sqrt{2}} (\varphi_1 - i\,\varphi_2) \quad \text{and} \quad \varphi^* = \frac{1}{\sqrt{2}} (\varphi_1 + i\,\varphi_2), \tag{7.3}$$

and remarking

$$\tfrac{1}{2}(\tau_1 + i\,\tau_2) = \begin{bmatrix} 0 & 1 \\ 0 & 0 \end{bmatrix} \quad \text{and} \quad \tfrac{1}{2}(\tau_1 - i\,\tau_2) = \begin{bmatrix} 0 & 0 \\ 1 & 0 \end{bmatrix}, \tag{7.4}$$

the Lagrangians (6.5), (6.8) to (6.10) can be rewritten:

$$L_0 = -\frac{1}{2} \sum_{i=1}^{2} \left[\sum_\lambda \left(\frac{\partial \varphi_i}{\partial x_\lambda}\right)^2 + \mu^2 \varphi_i^2\right], \tag{7.5}$$

$$L' = -i f \overline{\psi} \gamma_5 \psi \varphi_0 \quad \text{or} \quad -i \frac{g}{\mu} \sum_\lambda \overline{\psi} \gamma_5 \gamma_\lambda \psi \frac{\partial \varphi_0}{\partial x_\lambda}, \tag{7.6}$$

$$L' = -i f \sum_{i=1}^{2} \overline{\psi} \gamma_5 \tau_i \psi \varphi_i \quad \text{or} \quad -i \frac{g}{\mu} \sum_{i=1}^{2} \sum_\lambda \overline{\psi} \gamma_5 \gamma_\lambda \tau_i \psi \frac{\partial \varphi_i}{\partial x_\lambda}, \tag{7.7}$$

$$L = -\hbar c \left[\overline{\psi} \left(\sum_\mu \gamma_\mu \frac{\partial}{\partial x_\mu} + \varkappa\right) \psi\right]. \tag{7.8}$$

Now we introduce another neutral field φ_3 interacting with the nucleon field by

$$L' = -i f \overline{\psi} \gamma_5 \tau_3 \psi \varphi_3 \tag{7.9}$$

or more explicitly

$$L' = -i f \overline{\psi}_p \gamma_5 \psi_p \varphi_3 + i f \overline{\psi}_n \gamma_5 \psi_n \varphi_3, \tag{7.10}$$

which is a little different from (6.8). φ_3 represents a purely neutral meson field just as φ_0, but it differs in its interaction with the nucleon field. Mixing φ_3

* Neither φ_1 nor φ_2 represents a pure charged field in the usual sense, but together they describe a charged field containing positive and negative mesons.

with the charged meson field φ_1 and φ_2, we obtain the symmetrical meson field, the Lagrangian of which is given by

$$L = -\frac{1}{2}\sum_{i=1}^{3}\left[\sum_{\lambda}\left(\frac{\partial \varphi_i}{\partial x_\lambda}\right)^2 + \mu^2 \varphi_i^2\right], \tag{7.11}$$

$$L' = -i f \sum_{i=1}^{3} \bar{\psi}\gamma_5 \tau_i \psi \varphi_i \quad \text{or} \quad -i\frac{g}{\mu}\sum_{i=1}^{3}\sum_{\lambda}\bar{\psi}\gamma_5\gamma_\lambda\tau_i\psi\frac{\partial \varphi_i}{\partial x_\lambda}. \tag{7.12}$$

From the above Lagrangian we can readily go over to the charged theory by restricting the summation to $i=1$ and 2. The neutral theory corresponds to a single term $i=0$, with

$$\tau_0 = 1 = \begin{bmatrix} 1 & 0 \\ 0 & 1 \end{bmatrix}. \tag{7.13}$$

We have thus far tacitly assumed that the masses of a charged and a neutral meson are the same and the coupling constants of these mesons with the nucleon field are also identical. Obviously there is a great deal of freedom in constructing a meson field theory which can describe neutral and charged mesons. The symmetrical meson theory given by (7.11) and (7.12), and the simple addition of the charged and neutral meson theories which can be constructed by extending the sum from $i=0$ to 2, are nothing but the two simplest possibilities. It should be added that the existence of a neutral meson is now experimentally well established.

Existing empirical evidence seems to favour the symmetrical meson theory. As is seen from the expressions (7.11) and (7.12), the theory is invariant under rotation in charge space. This implies charge-independence of the interaction between mesons and nucleons, or more specifically that the interaction between two nucleons depends only on the total charge spin of the two-nucleon system and not on the more detailed specification of charge to the individual nucleons. The empirical evidence on this point is discussed in Sects. 32 and 47. It may, however, be argued that the empirical masses of charged and neutral mesons, or of a proton and a neutron, are definitely different, although we have assumed above that they are identical. It is to be noted here that the interaction of mesons and nucleons with an electromagnetic field is never charge-independent, since only the charged particles can interact with it. Therefore, the present field theory cannot yield rigorously charge-independent results, even though we employ the symmetrical meson theory. For example, the observed masses of charged and neutral mesons, or of a proton and a neutron, are predicted different. However, the present field theory is far from satisfactory in explaining the above mentioned mass differences; we shall assume the symmetrical meson theory throughout this article.

b) Deduction of static two-nucleon interaction.

8. Classical treatment. In this section the static nuclear potential is derived classically, starting from the Lagrangians (7.11) and (7.12). Following the usual prescription, the Hamiltonian is given by $H+H'$, where

$$H = \hbar c\left[\bar{\psi}\left(\sum_{i=1}^{3}\gamma_i\frac{\partial}{\partial x_i} + \varkappa\right)\psi\right] + \frac{1}{2}\sum_{\alpha}\left[c^2\pi_\alpha^2 + (\operatorname{grad}\varphi_\alpha)^2 + \mu^2\varphi_\alpha^2\right], \tag{8.1}$$

and

$$H' = i f \sum_{\alpha} \bar{\psi}\gamma_5\tau_\alpha\psi\varphi_\alpha$$

or

$$i\frac{g}{\mu}\sum_{\alpha}\sum_{i=1}^{3}\overline{\psi}\gamma_5\gamma_i\tau_\alpha\psi\frac{\partial\varphi_\alpha}{\partial x_i}+c\frac{g}{\mu}\sum_{\alpha}\overline{\psi}\gamma_5\gamma_4\tau_\alpha\psi\pi_\alpha+\frac{1}{2}\sum_{\alpha}\left(\frac{g}{\mu}\overline{\psi}\gamma_5\gamma_4\tau_\alpha\psi\right)^2, \quad (8.2)$$

π_α being the canonically conjugate momentum of φ_α.

If we make the static approximation for nucleons by dropping the odd DIRAC matrices* and adopting the configuration space representation, e.g. for an operator Θ:

$$\psi^*\Theta\psi=\sum_i\Theta^{(i)}\delta(\boldsymbol{x}-\boldsymbol{x}^{(i)}), \quad (8.3)$$

where $\boldsymbol{x}^{(i)}$ is the coordinate vector of the i-th nucleon, the integrated interaction Hamiltonian H' becomes

$$\overline{H}'=\int H'\,d^3x=\frac{g}{\mu}\sum_i\sum_\alpha\sigma^{(i)}\tau_\alpha^{(i)}\operatorname{grad}\varphi_\alpha(\boldsymbol{x}^{(i)}), \quad (8.4)$$

where $\sigma^{(i)}$ and $\tau_\alpha^{(i)}$ are the ordinary and charge spin operators of the i-th nucleon. For ps-ps coupling, the static approximation must be treated more carefully, which will be done in the following section. In this section we consider the simpler ps-pv case only. The equation of motion for φ_α can be directly obtained from (7.11) and (7.12) and reads

$$(\Box-\mu^2)\varphi_\alpha=-i\frac{g}{\mu}\sum_\lambda\frac{\partial}{\partial x_\lambda}(\overline{\psi}\gamma_5\gamma_\lambda\tau_\alpha\psi), \quad (8.5)$$

or in the static approximation for nucleons

$$(\Delta-\mu^2)\varphi_\alpha=-\frac{g}{\mu}\sum_i\sigma^{(i)}\tau_\alpha^{(i)}\operatorname{grad}\delta(\boldsymbol{x}-\boldsymbol{x}^{(i)}). \quad (8.6)$$

The static nuclear potential between two nucleons can now be obtained by calculating the interaction energy of the first nucleon with the meson field produced by the second nucleon. This field is nothing but the solution of (8.6), retaining only one term $i=2$ on the right-hand side. Remarking that the function $e^{-\mu r}/r$ satisfies the equation

$$(\Delta-\mu^2)\frac{e^{-\mu r}}{r}=-4\pi\delta(\boldsymbol{x}), \quad (8.7)$$

the meson field induced by the second nucleon can be obtained as

$$\varphi_\alpha(\boldsymbol{x})=+\frac{g}{4\pi\mu}\sigma^{(2)}\tau_\alpha^{(2)}\operatorname{grad}\frac{e^{-\mu|\boldsymbol{x}-\boldsymbol{x}^{(2)}|}}{|\boldsymbol{x}-\boldsymbol{x}^{(2)}|}, \quad (8.8)$$

which in turn gives the static potential energy between two nucleons, simply by substituting (8.8) on the right-hand side of (8.4) with one term ($i=1$) only:

$$V(\boldsymbol{r})=\frac{g^2}{4\pi\mu^2}(\boldsymbol{\tau}^{(1)}\cdot\boldsymbol{\tau}^{(2)})(\sigma^{(1)}\operatorname{grad})(\sigma^{(2)}\operatorname{grad})\frac{e^{-\mu r}}{r}, \quad (8.9)$$

with $\boldsymbol{r}=\boldsymbol{x}^{(1)}-\boldsymbol{x}^{(2)}$, for the symmetrical meson theory. For the charged and neutral theories, only the charge spin factor is modified, viz. in the following

---

* Odd DIRAC matrices are such matrices that have non-vanishing matrix elements between states with different signs of energy.

way:

$$\left.\begin{array}{ll}\text{Symmetrical:} & (\boldsymbol{\tau}^{(1)}\cdot\boldsymbol{\tau}^{(2)}),\\ \text{Charged:} & \tau_1^{(1)}\tau_1^{(2)}+\tau_2^{(1)}\tau_2^{(2)},\\ \text{Neutral:} & \tau_0^{(1)}\tau_0^{(2)}=1.\end{array}\right\} \tag{8.10}$$

Carrying out the differentiations in (8.9), the static nuclear potential can be written as

$$V(\boldsymbol{r})=m c^2\left(\frac{g^2}{4\pi\hbar c}\right)(\boldsymbol{\tau}^{(1)}\cdot\boldsymbol{\tau}^{(2)})\left[\frac{1}{3}(\boldsymbol{\sigma}^{(1)}\cdot\boldsymbol{\sigma}^{(2)})\frac{e^{-\mu r}}{\mu r}+\frac{1}{3}S_{12}\left(1+\frac{3}{\mu r}+\frac{3}{(\mu r)^2}\right)\frac{e^{-\mu r}}{\mu r}\right], \tag{8.11}^*$$

where

$$S_{12}=\frac{3\,(\boldsymbol{\sigma}^{(1)}\cdot\boldsymbol{r})\,(\boldsymbol{\sigma}^{(2)}\cdot\boldsymbol{r})}{r^2}-(\boldsymbol{\sigma}^{(1)}\cdot\boldsymbol{\sigma}^{(2)}). \tag{8.12}$$

In the next section it is shown that the same nuclear potential can be derived by a quantum mechanical calculation to the second order in the sense of perturbation theory.

9. Quantum procedure. *α) Second order.* In order to derive the static nuclear potential quantum-mechanically we must calculate the interaction energy between two nucleons due to the virtual exchange of meson field quanta (mesons), the presence of which is a consequence of the quantization of the field with the commutation relations

$$[\varphi_\alpha(\boldsymbol{x}),\pi_\beta(\boldsymbol{x}')\,]=i\hbar\,\delta_{\alpha\beta}\,\delta(\boldsymbol{x}-\boldsymbol{x}'); \tag{9.1}$$

other commutators equal zero.

Before we enter into further calculations, we must discuss the static approximation in the case of ps-ps coupling. Several approaches have been proposed[82,83], among which the so-called Dyson transformation[83] is the most convenient for our purpose. This method consists in applying to our Hamiltonian (8.1) and (8.2) for ps-ps coupling a canonical transformation (Dyson transformation) defined by

$$e^{iS},\quad S=\frac{f}{2\hbar\varkappa c}\sum_\alpha\int\bar{\psi}\gamma_4\gamma_5\tau_\alpha\psi\,\varphi_\alpha\,d^3x, \tag{9.2}$$

which transforms the ps-ps interaction term into such a form as to allow the static approximation. Using the formula

$$\exp(i\,S)\,\overline{H}\exp(-i\,S)=\overline{H}+i\,[S,\overline{H}]+\frac{i^2}{2!}\left[S,[S,\overline{H}]\right]+\cdots, \tag{9.3}$$

with $[A,B]=AB-BA$, and recalling the commutation relations (9.1) as well as those of the nucleon field;

$$[\psi_\lambda(\boldsymbol{x}),\psi_\mu^*(\boldsymbol{x}')]_+=\delta_{\lambda\mu}\,\delta(\boldsymbol{x}-\boldsymbol{x}') \tag{9.4}$$

---

* This is valid for $r>0$. If we include $r=0$, we should add an extra term,

$$-m c^2\left(\frac{g^2}{4\pi\hbar c}\right)(\boldsymbol{\tau}^{(1)}\cdot\boldsymbol{\tau}^{(2)})\,(\boldsymbol{\sigma}^{(1)}\cdot\boldsymbol{\sigma}^{(2)})\,\frac{4\pi}{3\mu^3}\,\delta(\boldsymbol{r}).$$

[82] L. L. Foldy: Phys. Rev. **84**, 168 (1951). — J. M. Berger, L. L. Foldy and R. K. Osborn: Phys. Rev. **87**, 1061 (1952). — G. Wentzel: Phys. Rev. **86**, 802 (1952).

[83] J. V. Lepore: Phys. Rev. **88**, 750 (1952). — S. D. Drell and E. M. Henley: Phys. Rev. **88**, 1053 (1952). — F. J. Dyson: Phys. Rev. **73**, 929 (1948).

(other anticommutators = 0), where $[A, B]_+ = AB + BA$ and μ, λ specify different charge and spinor components of ψ, we get the transformed Hamiltonian

$$\left.\begin{aligned} \overline{H}_t &= \int H_t\, d^3x = e^{iS}\, \overline{H}\, e^{-iS}, \\ H_t &= \hbar c \left[\overline{\psi}\left(\sum_{i=1}^{3} \gamma_i \frac{\partial}{\partial x_i} + \varkappa\right)\psi\right] + \frac{1}{2}\sum_\alpha \left[c^2 \pi_\alpha^2 + (\operatorname{grad}\varphi_\alpha)^2 + \mu^2 \varphi_\alpha^2\right] + \\ &+ \frac{i f}{2\varkappa} \sum_\alpha \overline{\psi}\gamma_5 \tau_\alpha \left(\sum_{i=1}^{3} \gamma_i \frac{\partial}{\partial x_i}\varphi_\alpha - i c \gamma_4 \pi_\alpha\right)\psi + \\ &+ \frac{1}{2}\left(\frac{f}{2\varkappa}\right)^2 \sum_\alpha [\overline{\psi}\gamma_5\gamma_4\tau_\alpha\psi]^2 + \frac{f^2}{2\varkappa\hbar c}\sum_\alpha \overline{\psi}\psi\varphi_\alpha^2 + \\ &+ \frac{1}{2\hbar c}\left(\frac{f}{2\varkappa}\right)^2 \sum_{\alpha,\beta} \overline{\psi}(\tau_\alpha\tau_\beta - \tau_\beta\tau_\alpha)\varphi_\alpha\left[\sum_{i=1}^{3}\gamma_i \frac{\partial\varphi_\beta}{\partial x_i} - i c\gamma_4\pi_\beta\right]\psi + \cdots, \end{aligned}\right\} \tag{9.5}$$

where only the terms up to the second order in the coupling constant f are given, because the higher order terms are unnecessary for the later calculations.

Comparing (9.5) with (8.1) and (8.2) for ps-pv coupling, we see that the ps-ps and ps-pv couplings are equivalent, as far as the lowest order perturbation is concerned, because the interaction terms which are linear in f and g, respectively, are the same in (9.5) and (8.2), if the relation

$$\frac{g}{\mu} = \frac{f}{2\varkappa} \tag{9.6}$$

is valid. Thus, for example, the second order static nuclear potential is the same for the two kinds of coupling. The transformed form (9.5) now allows the static approximation. Repeating the procedure which gave (8.4), we obtain as the integrated static interaction Hamiltonian for ps-ps coupling*

$$\overline{H}_t' = \frac{f}{2\varkappa}\sum_i \sum_\alpha \sigma^{(i)} \tau_\alpha^{(i)} \operatorname{grad}\varphi_\alpha(\boldsymbol{x}^{(i)}) + \frac{f^2}{2\varkappa\hbar c}\sum_i\sum_\alpha \varphi_\alpha^2(\boldsymbol{x}^{(i)}). \tag{9.7}$$

The second term in (9.7) is the characteristic meson pair term in the symmetrical ps-ps theory and contributes to the fourth order nuclear potential only.

The static nuclear potential can now be calculated as the expectation value of (8.4) or (9.7) in the state of two nucleons which are at rest with a distance $\boldsymbol{r}$. According to the perturbation theory, we have, up to the fourth order (first order vanishing),

$$\left.\begin{aligned} V(\boldsymbol{r}) &= \sum_l \frac{\langle i|\overline{H}'|l\rangle\langle l|\overline{H}'|i\rangle}{(E_i - E_l)} + \sum_{l,m} \frac{\langle i|\overline{H}'|m\rangle\langle m|\overline{H}'|l\rangle\langle l|\overline{H}'|i\rangle}{(E_i - E_m)(E_i - E_l)} + \\ &+ \sum_{l,m,n} \frac{\langle i|\overline{H}'|n\rangle\langle n|\overline{H}'|m\rangle\langle m|\overline{H}'|l\rangle\langle l|\overline{H}'|i\rangle}{(E_i - E_n)(E_i - E_m)(E_i - E_l)}, \end{aligned}\right\} \tag{9.8}$$

where i and l, m, n mean initial and intermediate states, respectively. In order to calculate the above matrix elements, we introduce the annihilation and creation operators $a_{\boldsymbol{k}}$ and $a_{\boldsymbol{k}}^*$ of a meson with wave vector $\boldsymbol{k}$, expanding the field variables

* The static approximation means here, not only dropping the odd DIRAC matrices, but also neglecting π_α, the time derivative of φ_α.

φ_α and π_α as follows:

$$\left.\begin{aligned}\varphi_\alpha(\boldsymbol{x}) &= \frac{1}{\sqrt{V}}\sum_{\boldsymbol{k}}\sqrt{\frac{\hbar c}{2\omega_{\boldsymbol{k}}}}\,(a_{\boldsymbol{k}}^{(\alpha)} + a_{-\boldsymbol{k}}^{(\alpha)*})\,e^{i\boldsymbol{k}\boldsymbol{x}},\\ \pi_\alpha(\boldsymbol{x}) &= \frac{1}{\sqrt{V}}\sum_{\boldsymbol{k}}\sqrt{\frac{\hbar\omega_{\boldsymbol{k}}}{2c}}\,i\,(a_{\boldsymbol{k}}^{(\alpha)*} - a_{-\boldsymbol{k}}^{(\alpha)})\,e^{-i\boldsymbol{k}\boldsymbol{x}},\end{aligned}\right\} \tag{9.9}$$

where V is the normalization volume, $\omega_{\boldsymbol{k}} = \sqrt{\boldsymbol{k}^2 + \mu^2}$ and the commutation relations $[a_{\boldsymbol{k}}^{(\alpha)}, a_{\boldsymbol{k}'}^{(\beta)*}] = \delta_{\alpha\beta}\,\delta_{\boldsymbol{k}\boldsymbol{k}'}$ etc. follow from (9.1).

The second order static nuclear potential is obtained from the first term of (9.8). In the intermediate state l there is, besides the original two nucleons, one meson of wave vector $\boldsymbol{k}$, which may be emitted by either of the two nucleons and subsequently absorbed by either of the two. Since such processes in which a meson is emitted and absorbed by the same nucleon may be neglected because they cannot contribute to the interaction energy, there are two virtual processes left; one in which a meson is emitted by nucleon 1 and absorbed by nucleon 2, and the reverse. In the former case, the two matrix elements can easily be evaluated as

$$\langle l\,|\overline{H}'|\,i\rangle = -i\,\frac{f}{2\varkappa}\,(\boldsymbol{k}\,\sigma^{(1)})\,\tau_\alpha^{(1)}\,\frac{1}{\sqrt{V}}\sqrt{\frac{\hbar c}{2\omega_{\boldsymbol{k}}}}\,e^{-i\boldsymbol{k}\boldsymbol{x}^{(1)}}, \tag{9.10}$$

$$\langle i\,|\overline{H}|'\,l\rangle = i\,\frac{f}{2\varkappa}\,(\boldsymbol{k}\,\sigma^{(2)})\,\tau_\alpha^{(2)}\,\frac{1}{\sqrt{V}}\sqrt{\frac{\hbar c}{2\omega_{\boldsymbol{k}}}}\,e^{+i\boldsymbol{k}\boldsymbol{x}^{(2)}}, \tag{9.11}$$

for the emission of the φ_α-meson. In the latter case, we have only to interchange the indices 1 and 2. It can easily be shown that after the summation over l, or more explicitly α and $\boldsymbol{k}$, the two processes give the same contribution. The second order potential is then given by

$$V_2(\boldsymbol{r}) = -\left(\frac{f}{2\varkappa}\right)^2\frac{\boldsymbol{\tau}^{(1)}\boldsymbol{\tau}^{(2)}}{(2\pi)^3}\int(\sigma^{(1)}\boldsymbol{k})\,(\sigma^{(2)}\boldsymbol{k})\,\frac{e^{-i\boldsymbol{k}\boldsymbol{r}}}{\omega_{\boldsymbol{k}}^2}\,d^3k, \tag{9.12}$$

where we have put $E_i - E_l = -\hbar c\,\omega_{\boldsymbol{k}}$, $\boldsymbol{r} = \boldsymbol{x}^{(1)} - \boldsymbol{x}^{(2)}$, and the sum $\frac{1}{V}\sum_{\boldsymbol{k}}$ has been replaced by the integral $\frac{1}{(2\pi)^3}\int d^3k$. Furthermore, substituting i grad for $\boldsymbol{k}$ in the integrand (9.12) and observing

$$\int\frac{e^{-i\boldsymbol{k}\boldsymbol{r}}}{\omega_{\boldsymbol{k}}^2}\,d^3k = 2\pi^2\,\frac{e^{-\mu r}}{r}, \tag{9.13}$$

we get the same result (8.9) as was derived classically, only with $f/2\varkappa$ instead of g/μ.

β) Fourth order. As is well known, the coupling of mesons with nucleons is not particularly small, which makes it necessary to calculate the higher order terms. Here we are faced with the convergency problem of such a perturbation expansion. As far as the ps-pv coupling term or its equivalent term in ps-ps theory, i.e. the first term in (9.7), is concerned, the perturbation expansion might be a good approximation, because the expansion parameter $g^2/4\pi\hbar c \approx (f^2/4\pi\hbar c)\times(\mu/2\varkappa)^2$ is empirically known to be at most 0.1. On the other hand, the expansion parameter for the characteristic meson pair term in (9.7) is $(f^2/4\pi\hbar c)\,(\mu/2\varkappa)$ and this is certainly larger than 1, which would invalidate the perturbation expansion entirely. There are some theoretical arguments[84], however, which suggest that

[84] S. DRELL and E. M. HENLEY: Phys. Rev. **88**, 1053 (1952). — G. WENTZEL: Phys. Rev. **86**, 802 (1953). — K. A. BRUECKNER, M. GELL-MANN and M. GOLDBERGER: Phys. Rev. **90**, 476 (1953). — A. KLEIN: Phys. Rev. **95**, 1061 (1954).

this pair term is strongly damped through the overall higher order effects, perhaps by a factor of ≈ 10. If that should be so, then the expansion parameter of this term would be of the same order as that of the ps-pv term. For detailed discussions about this pair damping, however, we refer to the existing literature[84]. Here we assume that the perturbation expansion is valid and introduce a damping parameter λ in the meson pair term of (9.7) to account for the possible damping effect.

The fourth order static potential can be calculated from the second and third terms in (9.8), using the interaction Hamiltonian (9.7), in the same way as $V_2(\boldsymbol{r})$. Here we only give the result[85], putting $x = \mu r$,

$$\left.\begin{aligned}
V_4(\boldsymbol{r}) = & - m c^2 \lambda^2 \left(\frac{\mu}{2\varkappa}\right)^2 \left(\frac{f^2}{4\pi\hbar c}\right)^2 \frac{6}{\pi} \frac{1}{x^2} K_1(2x) + \\
& + m c^2 \lambda \left(\frac{\mu}{2\varkappa}\right)^3 \left(\frac{f^2}{4\pi\hbar c}\right)^2 6 \left(\frac{1+x}{x^2}\right)^2 e^{-2x} - \\
& - m c^2 \left(\frac{\mu}{2\varkappa}\right)^4 \left(\frac{f^2}{4\pi\hbar c}\right)^2 \left[R_1(x) + (\sigma^{(1)}\cdot\sigma^{(2)}) R_2(x) + S_{12} R_3(x)\right], \\
\text{where} & \\
R_1(x) = & \frac{2}{\pi} \Big[(\boldsymbol{\tau}^{(1)}\cdot\boldsymbol{\tau}^{(2)}) \Big\{ \left(\frac{12}{x^2} + \frac{23}{x^4}\right) K_1(2x) + \left(\frac{4}{x} + \frac{23}{x^3}\right) K_0(2x) \Big\} + \\
& + (3 - 2\boldsymbol{\tau}^{(1)}\cdot\boldsymbol{\tau}^{(2)}) \Big\{ \left(\frac{1}{x^2} + \frac{4}{x^3} + \frac{4}{x^4}\right) K_1(x) + \left(\frac{1}{x} + \frac{2}{x^2} + \frac{2}{x^3}\right) K_0(x) \Big\} e^{-x} \Big], \\
R_2(x) = & \frac{2}{\pi} \Big[- \Big\{ \left(\frac{8}{x^2} + \frac{12}{x^4}\right) K_1(2x) + \frac{12}{x^3} K_0(2x) \Big\} + \\
& + \frac{2}{3} (3 - 2\boldsymbol{\tau}^{(1)}\cdot\boldsymbol{\tau}^{(2)}) \Big\{ \left(\frac{1}{x^2} + \frac{2}{x^3} + \frac{2}{x^4}\right) K_1(x) + \left(\frac{1}{x^2} + \frac{1}{x^3}\right) K_0(x) \Big\} e^{-x} \Big], \\
R_3(x) = & \frac{2}{\pi} \Big[\Big\{ \left(\frac{4}{x^2} + \frac{15}{x^4}\right) K_1(2x) + \frac{12}{x^3} K_0(2x) \Big\} - \\
& - \frac{1}{3} (3 - 2\boldsymbol{\tau}^{(1)}\cdot\boldsymbol{\tau}^{(2)}) \Big\{ \left(\frac{1}{x^2} + \frac{5}{x^3} + \frac{5}{x^4}\right) K_1(x) + \left(\frac{1}{x^2} + \frac{1}{x^3}\right) K_0(x) \Big\} e^{-x} \Big],
\end{aligned}\right\} \quad (9.14)$$

and $K_n(x)$ is a modified HANKEL function of the first kind defined by[86]

$$K_n(x) = \frac{\pi}{2} i^{n+1} H_n^{(1)}(i x),$$

which behaves at the origin as

$$K_0(x) \approx -\log\left(\frac{x}{2}\right) \quad \text{and} \quad K_1(x) \approx \frac{1}{x} \tag{9.15}$$

and asymptotically as

$$K_n(x) \approx \left(\frac{\pi}{2x}\right)^{\frac{1}{2}} e^{-x}. \tag{9.16}$$

If we put $\lambda = 0$ in (9.14), it gives the result for ps-pv coupling, taking (9.6) into account.

It must be added that there are different opinions[87] about the third term (which does not include λ) of $V_4(\boldsymbol{r})$ given by (9.14). The problem is whether a non-static correction to the second order static potential (8.11) should be neglected[85] or considered as a contribution to the fourth order static potential[87].

---

[85] K. A. BRUECKNER and K. M. WATSON: Phys. Rev. **92**, 1023 (1953).

[86] G. N. WATSON: Theory of Bessel Functions, pp. 78, 80 and 202. Cambridge 1952.

[87] M. M. LÉVY: Phys. Rev. **88**, 72, 725 (1952). — A. KLEIN: Phys. Rev. **90**, 1101 (1953). — M. TAKETANI, S. MACHIDA and S. OHNUMA: Progr. Theor. Phys. **6**, 638 (1951); **7**, 45 (1952). — K. NISHIJIMA: Progr. Theor. Phys. **6**, 815, 911 (1951). — D. FELDMAN: Phys. Rev. **98**, 1456 (1955).

There is also another opinion about $V_4(\boldsymbol{r})$[88]. Although this diversity of opinions has not yet been settled, we mainly consider the fourth order potential given by (9.14), because it is interesting from the phenomenological point of view; other versions of $V_4(\boldsymbol{r})$ can not even predict the existence of a bound state for a neutron-proton system, as is shown in detail in Sects. 11 and 33.

As regards the higher order effects, which have so far been neglected, there are some investigations, for example, on a presumably important contribution of this kind[85] or about the effect of the renormalization procedure upon the fourth order static potential[89]. The results are that such higher order effects can reasonably be neglected. There is also a preliminary calculation on the effect of the possible nucleon isobar state upon the static nuclear potential[90]. Some quantitative considerations about these effects will be deferred to Sects. 11 and/or 33, where the discussion about the nuclear potentials (8.11) and (9.14) and some numerical results will also be given.

c) Static interaction.

Thus far we have derived the nuclear potential in the static approximation for nucleons. Therefore, the nuclear potential depends only on the relative coordinates of two nucleons but not upon the nucleon velocities. Such interactions are called static. As regards the non-static interaction between two nucleons, we have so far no direct empirical evidence for its existence. Perhaps it is more correct to say that the present experimental and theoretical ambiguities are still too large to allow a discussion of the non-static part of the nuclear interaction. In this article, therefore, we do not intend to discuss this topic.

10. Central force and tensor force. *α) The most general nuclear potential.* From (8.11) and (9.14), we see that the meson theoretical nuclear potential has the general form

$$V(\boldsymbol{r}) = V_0(r) + V_\sigma(r)\,(\boldsymbol{\sigma}^{(1)}\cdot\boldsymbol{\sigma}^{(2)}) + V_T(r)\,S_{12} \tag{10.1}$$

with

$$S_{12} = \frac{3\,(\boldsymbol{\sigma}^{(1)}\cdot\boldsymbol{r})\,(\boldsymbol{\sigma}^{(2)}\cdot\boldsymbol{r})}{r^2} - (\boldsymbol{\sigma}^{(1)}\cdot\boldsymbol{\sigma}^{(2)}), \tag{8.12}$$

where special attention is paid only to its dependence on $\boldsymbol{\sigma}^{(1)}$ and $\boldsymbol{\sigma}^{(2)}$. The first two terms in (10.1) are called "central force" since they do not depend on the orientation of $\boldsymbol{r}$, while the last term containing S_{12} is named "tensor force". The operator is well known from various kinds of dipole-dipole interaction and is sometimes called the "tensor operator". The tensor force is a kind of non-central (spin-orbit) force, depending on the direction of $\boldsymbol{r}$ with respect to the two spin vectors $\boldsymbol{\sigma}^{(1)}$ and $\boldsymbol{\sigma}^{(2)}$.

So far as we confine ourselves to conservative forces, independent of the nucleon velocities, it can be shown that the most general potential can be written in the form (10.1). The reason for this limited choice comes from the requirement that the potential must be invariant against rotations and space reflections of the coordinate system which we use to describe the relative motion of a two-nucleon system. Such invariant quantities which can be made from $\boldsymbol{r}$ alone or $\boldsymbol{\sigma}^{(1)}$ and $\boldsymbol{\sigma}^{(2)}$ alone are functions of r only or $(\boldsymbol{\sigma}^{(1)}\cdot\boldsymbol{\sigma}^{(2)})$, because all other invariant

[88] N. Fukuda, K. Sawada and M. Taketani: Progr. Theor. Phys. **12**, 156 (1954).
[89] M. M. Lévy: Phys. Rev. **88**, 72, 725 (1952).
[90] T. Matsumoto, T. Hamada and M. Sugawara: Progr. Theor. Phys. **10**, 199 (1953). — T. Matsumoto and M. Sugawara: Progr. Theor. Phys. **12**, 553 (1954).

higher products of $\boldsymbol{\sigma}^{(1)}$ and $\boldsymbol{\sigma}^{(2)}$ can be reduced to some pure number plus some multiple of $(\boldsymbol{\sigma}^{(1)}\cdot\boldsymbol{\sigma}^{(2)})$. Therefore, the most general central force is the combination of a spin-independent potential $V_0(r)$ and a spin-dependent potential $V_\sigma(r)\ (\boldsymbol{\sigma}^{(1)}\cdot\boldsymbol{\sigma}^{(2)})$.

To form the general non-central force, we have only to make products of $\boldsymbol{r}$ and $\boldsymbol{\sigma}$'s. It must be remarked here that $(\boldsymbol{r}\cdot\boldsymbol{\sigma})$ is invariant against rotations, but not against space reflections, in which $\boldsymbol{r}\to-\boldsymbol{r}$ and $\boldsymbol{\sigma}\to\boldsymbol{\sigma}$ since $\boldsymbol{\sigma}$ behaves like an angular momentum. Thus invariant non-central forces must contain even powers of $\boldsymbol{\sigma}$. As regards the higher products in $\boldsymbol{\sigma}$, they can again be reduced to an expression linear in $\boldsymbol{\sigma}$ or without $\boldsymbol{\sigma}$. The only possible forms, therefore, are $(\boldsymbol{\sigma}^{(1)}\boldsymbol{r})\ (\boldsymbol{\sigma}^{(2)}\boldsymbol{r})$ and $(\boldsymbol{\sigma}^{(1)}\times\boldsymbol{r})\cdot(\boldsymbol{\sigma}^{(2)}\times\boldsymbol{r})$, the latter of which is a linear combination of the former and $(\boldsymbol{\sigma}^{(1)}\cdot\boldsymbol{\sigma}^{(2)})$. Hence $(\boldsymbol{\sigma}^{(1)}\boldsymbol{r})\ (\boldsymbol{\sigma}^{(2)}\boldsymbol{r})$ is essentially the only relevant combination, which implies that the most general non-central force is given by the last term $V_T(r)\ S_{12}$ in (10.1). The special form (8.12) of the tensor operator is so devised that its average over the directions of $\boldsymbol{r}$ may vanish. It has thus been proven that the most general velocity-independent static nuclear interaction has the form (10.1).

With respect to the charge spin factor, we cannot restrict it within such a limited scope, since we have no corresponding a priori requirement for the invariance of the static nuclear potential against rotations and space reflections of the charge spin space. All polynomials containing various components of $\boldsymbol{\tau}^{(1)}$ and $\boldsymbol{\tau}^{(2)}$ in various ways must be allowed. If we demand, however, the invariance of the potential against rotations and space reflections of the charge spin space, the most general static nuclear potential must have the form (10.1), with

$$V_i(r) = V_{i0}(r) + V_{i\tau}(r)\ (\boldsymbol{\tau}^{(1)}\cdot\boldsymbol{\tau}^{(2)})\,; \qquad (i = 0, \sigma, T)\,, \tag{10.2}$$

where $V_{i0}(r)$ and $V_{i\tau}(r)$ are quite arbitrary functions of r. It should be stressed that in (10.2) there is no term corresponding to the tensor operator S_{12} in (10.1), because the charge spin space has nothing to do with ordinary space and therefore they must be transformed quite independently of each other. The static nuclear potential (8.11) and (9.14) derived from the symmetrical meson theory, which is originally devised so as to satisfy the above invariance requirement (see Sect. 7), in fact has the general form of (10.1) and (10.2). The charged meson theory predicts the form (10.1), but not (10.2), as can readily be seen from the substitution (8.10). We shall later see that the form (10.1) and (10.2) seems to be supported by the empirical evidence (Sects. 32 and 47).

β) Classification of states of two nucleons. In this subsection we consider the possible states of a two-nucleon system in the case of the nuclear potential (10.1) and (10.2). First we confine ourselves to a central force case. Then the static nuclear potential is invariant against rotations of the space and spin coordinates separately, which means that all the components of the orbital angular momentum operator $\boldsymbol{L}$ and the spin angular momentum operator $\boldsymbol{S}$ (both in units of $\hbar$) are separately commutable with the total Hamiltonian of a two-nucleon system. Thus we can specify the states using the usual quantum numbers L, S, L_z and S_z of the above operators and their z-components. From the same argument, the square of the total charge spin "angular momentum" operator

$$\boldsymbol{T} = \tfrac{1}{2}(\boldsymbol{\tau}_1 + \boldsymbol{\tau}_2)$$

and one of its components, say T_3, are constants of motion, and we can use the corresponding quantum numbers T and T_3 to specify the states.

Here we recall the exclusion principle: The states of a two-nucleon system must be antisymmetric with respect to the interchange of two nucleons. Thus, remarking that the orbital wave functions of even L and odd L are symmetric and antisymmetric, triplet ($S=1$) and singlet ($S=0$) spin functions symmetric and antisymmetric, respectively, when interchanging two nucleons, the assignment of L and S automatically determines T, which can only be 0 or 1, and we need not specify T. Also we need not fix L_z, S_z and T_3, because the state does not depend upon the special choice of the coordinate system. We have, therefore, only to assign L and S and can classify the states, using the spectroscopic notations, as follows:

$$^1S,\ ^3S,\ ^1P,\ ^3P,\ ^1D,\ ^3D,\ ^1F,\ ^3F,\ \ldots, \tag{10.3}$$

where 1 and 3 mean the spin singlet and triplet states and S, P, D, F etc. mean $L=0, 1, 2, 3$ etc., respectively, of which ^1S, ^3P, ^1D, ^3F etc. are charge triplet states ($T=1$) and ^3S, ^1P, ^3D, ^1F etc. are charge singlet states ($T=0$). The values of the operators $\boldsymbol{\sigma}^{(1)}\cdot\boldsymbol{\sigma}^{(2)}$ and $\boldsymbol{\tau}^{(1)}\cdot\boldsymbol{\tau}^{(2)}$ for these states are easily evaluated as*

$$\left.\begin{aligned} \boldsymbol{\sigma}^{(1)}\cdot\boldsymbol{\sigma}^{(2)} &= \boldsymbol{\tau}^{(1)}\cdot\boldsymbol{\tau}^{(2)} = 1, && \text{for triplet states;} \\ \boldsymbol{\sigma}^{(1)}\cdot\boldsymbol{\sigma}^{(2)} &= \boldsymbol{\tau}^{(1)}\cdot\boldsymbol{\tau}^{(2)} = -3, && \text{for singlet states.} \end{aligned}\right\} \tag{10.4}$$

In the case of the charge dependent potential predicted, for example, by the charged meson theory, T and T_3 would not in general be good quantum numbers. Even in these more general cases, the Hamiltonian is symmetric with respect to the interchange of the charge coordinates only, since the Hamiltonian must be symmetric with respect to the interchange of two nucleons and also symmetric with respect to the interchange of the space and spin coordinates only, as is seen from (10.1). From this, it follows that the states must be either symmetric or antisymmetric with respect to the interchange of the charge coordinates. Thus the states must be either charge singlet or charge triplet, and T is still a good quantum number, even though T_3 is not. To specify the states, we need, besides L and S, T_3 for charge spin triplet states, since T_3 takes the value 1, 0 or -1, corresponding to two-proton, proton-neutron or two-neutron systems, respectively.

In the case of the most general potential (10.1) and (10.2) including a tensor force, the Hamiltonian is invariant only against the coupled rotations of space and spin coordinates, which means the physical rotation of the coordinate system used to determine the general form of the potential. Thus only the square of the total angular momentum operator $\boldsymbol{J}=\boldsymbol{L}+\boldsymbol{S}$ (in units of $\hbar$) and its z-component J_z are constants of motion and the corresponding quantum numbers J and J_z can be used to specify the states, while L and S cannot in general be used. We add that the Hamiltonian is still symmetric with respect to the interchange of the spin coordinates only, as is seen from (10.1), which again means that S is still a good quantum number, even though S_z is not.

Having shown that S and also T are still good quantum numbers, we remark that there is another important constant of motion, i.e. the even or odd *parity* of the state. Even or odd parity means that the space wave functions of the system are even or odd with respect to space inversion, i.e. the replacement of $\boldsymbol{r}$ by $-\boldsymbol{r}$. The above property readily follows from the invariance of the Hamiltonian with respect to the space inversion. It is well known that the states with even L have an even parity and those with odd L have an odd parity. Thus states of even and odd L never mix and we have four quantum numbers J, J_z, S and parity (automatically including T) to specify the states.

* These values are most easily calculated using the formula $\boldsymbol{S}^2=\frac{1}{4}(\boldsymbol{\sigma}^{(1)}+\boldsymbol{\sigma}^{(2)})^2=S(S+1)$, which gives $\boldsymbol{\sigma}^{(1)}\cdot\boldsymbol{\sigma}^{(2)}=2S(S+1)-3=1$ for $S=1$ and -3 for $S=0$.

Now S is either 0 or 1. For $S=0$, $L=J$ and L is in this case a good quantum number. On the other hand, for $S=1$, the law of addition of angular momenta permits $L=J-1, J, J+1$, excluding the case $J=0$. However, the state $L=J$ has a parity opposite to those of $L=J-1$ and $J+1$, so that the state $S=1$, $L=J$, defines a pure state by itself, and the other states of opposite parity with $S=1$ are mixtures of states with $L=J+1$ and $J-1$. The special case of $S=1$ and $J=0$ is a pure state with $L=1$. To summarize, the possible states are classified as follows, using the same notations as in (10.3):

$$\left.\begin{aligned} &J=0;\ {}^1S_0,\ {}^3P_0,\\ &J=1;\ {}^1P_1,\ {}^3S_1+{}^3D_1,\ {}^3P_1,\\ &J=2;\ {}^1D_2,\ {}^3P_2+{}^3F_2,\ {}^3D_2,\\ &\text{etc.,}\end{aligned}\right\}\tag{10.5}$$

where the subscript denotes J. We need not specify J_z or T_3, again for the charge-independent potential, but otherwise we must indicate T_3 besides those given above. The values of the operators $\boldsymbol{\sigma}^{(1)}\cdot\boldsymbol{\sigma}^{(2)}$ and $\boldsymbol{\tau}^{(1)}\cdot\boldsymbol{\tau}^{(2)}$ are again given by (10.4), since S and T are good quantum numbers. The value of the tensor operator S_{12} for these states can be calculated, using the ordinary spin functions of a two-particle system, and is shown to vanish identically in a spin singlet state; thus the addition of a tensor force has no effect on a spin singlet two-nucleon system. For spin triplet states S_{12} does not take any simple value; the matrix elements of S_{12} with respect to the three triplet spin states do not form a diagonal matrix like $\boldsymbol{\sigma}^{(1)}\cdot\boldsymbol{\sigma}^{(2)}$ but are represented by a non-diagonal matrix

$$\frac{2\sqrt{2\pi}}{\sqrt{15}}\begin{bmatrix}\sqrt{6}\,Y_{20} & -3\sqrt{2}\,Y_{2-1} & 6\,Y_{2-2}\\ 3\sqrt{2}\,Y_{21} & -2\sqrt{6}\,Y_{20} & 3\sqrt{2}\,Y_{2-1}\\ 6\,Y_{22} & -3\sqrt{2}\,Y_{21} & \sqrt{6}\,Y_{20}\end{bmatrix},\tag{10.6}$$

where Y_{lm} are the normalized spherical harmonics*, which have the property $Y_{lm}^*=(-1)^m Y_{l-m}$.

Since the deuteron has spin 1 and its magnetic moment approximates the sum of the moments of a proton and a neutron (Sect. 1), it is safe to assume that the proton and neutron spins are parallel in the deuteron. Accordingly the deuteron ground state is a spin triplet state and is denoted by ${}^3S$ if the nuclear force is central.

However, the deuteron is known to have a positive electric quadrupole moment (Sect. 1), which means that its charge distribution is not spherically symmetric, but is elongated along the direction of the total spin, so that the deuteron ground state is not a pure ${}^3S$ state. Thus the quadrupole moment supplies an experimental evidence for "tensor force", which would make the deuteron ground state a mixture of ${}^3S$ and ${}^3D$ states, according to (10.5). An admixture of ${}^3D$ state can certainly produce a quadrupole moment.

As is seen from (8.12), S_{12} will assume a positive value for an elongated state along the direction of the total spin**. Therefore, a nuclear potential (10.1)

* The definition is

$$Y_{lm}(\vartheta,\varphi)=\sqrt{\frac{2l+1}{4\pi}}\,\frac{(-1)^l}{2^l\,l!}\,\sqrt{\frac{(l-m)!}{(l+m)!}}\,\sin^m\vartheta\,\frac{d^{l+m}}{(d\cos\vartheta)^{l+m}}\,\sin^{2l}\vartheta\,e^{im\varphi}$$

for any (positive or negative) m.

** This can most easily be seen by considering $\boldsymbol{\sigma}^{(1)}$ and $\boldsymbol{\sigma}^{(2)}$ as ordinary unit vectors which are parallel with each other.

with negative $V_T(r)$ predicts a positive quadrupole moment, whereas a positive $V_T(r)$ leads to a negative moment. Thus the tensor force must have a negative coefficient of S_{12}. Such tensor forces are said to have the "right sign". On the other hand, negative and positive central forces are usually called attractive and repulsive, respectively. However, we cannot classify a tensor force by calling it attractive or repulsive, since positive as well as negative tensor forces can act as binding agents in the deuteron ground state.

11. General behavior of meson theoretical interaction and introduction of cut-off procedure. The nuclear potential (8.11) and (9.14) derived from the symmetrical pseudoscalar meson theory contains the most general types of spin and charge dependence. As to the distance dependence, the asymptotic behavior of the second order term $V_2(\boldsymbol{r})$ is $e^{-\mu r}/r$, while the fourth order potential $V_4(\boldsymbol{r})$ behaves asymptotically as $e^{-2\mu r}/r$, the range of which is one half of that of $V_2(\boldsymbol{r})$. Near the origin $V_2(\boldsymbol{r})$ goes to infinity as r^{-3} and $V_4(\boldsymbol{r})$ as r^{-5}. These singularities do not allow any stationary states of finite binding energies.

The spin and charge dependence was considered in some detail in the previous section. According to (10.5), we need only know the values of the potential in four states; singlet and triplet even states, singlet and triplet odd states. The values of the potential in these four states are given in Fig. 9* as functions of $x = \mu r$, where the central force and the tensor force, which is now defined as the coefficient of S_{12} in the nuclear potential, are plotted separately. $V_2(\boldsymbol{r})$ is shown by the bold-faced line, the sum of $V_2(\boldsymbol{r})$ and $V_4(\boldsymbol{r})$ by the thin line. The solid lines represent forces in triplet states, the dotted ones those in singlet states. For even states, the second order central force is by chance the same for both singlet and triplet states, thus there is no bold-faced dotted line in the left half of the figure. The dimensionless ps-ps coupling constant $(f^2/4\pi\hbar c)$ is chosen to be 15 or the equivalent ps-pv one $(\mu/2\varkappa)^2(f^2/4\pi\hbar c) = 0.083$ and the unit of the potential energy is the meson rest energy $mc^2 = 139.5$ Mev, m being taken 273 times the electron mass. The corresponding unit of length $1/\mu$ is 1.414×10^{-13} cm. The damping parameter λ, which can affect only the central force, according to (9.14), is chosen as 0, 0.1 and 0.2. $\lambda = 0$ gives the forces with ps-pv coupling, and in the ps-ps case there are some arguments that a value between 0.1 and 0.2 may be reasonable from the theoretical as well as the experimental aspect[91].

Some general features of the curves in Fig. 9 should be emphasized. Roughly speaking, forces in even states are attractive and of right sign, i.e. consistent with the positive quadrupole moment of the deuteron, while those in odd states are repulsive and of "wrong sign". As regards the triplet even state (the deuteron ground state), we notice that the present meson theory seems to predict a combination of a rather strong tensor force with weaker central force, which may even happen to be repulsive in the ps-ps case. As is shown in Sect. 28, such a strong tensor force is actually necessary, although the apparent degree of ^3D-state admixture in the deuteron ground state is small. As regards the singlet even state, we see that the second order force is very weak, while the fourth order addition makes it very strong. Finally, forces in odd states are seen to be smaller in magnitude than those in even states, with the exception of the singlet second plus fourth order forces (dotted thin lines).

---

* The curves in the original paper by BRUECKNER and WATSON[85] are somewhat inaccurate. The curves in Fig. 9 are corrected.

[91] G. WENTZEL: Phys. Rev. **86**, 802 (1953). — K. A. BRUECKNER, M. GELL-MANN and M. GOLDBERGER: Phys. Rev. **90**, 476 (1953). — A. KANAZAWA and M. SUGAWARA: Progr. Theor. Phys. **10**, 399 (1953). — A. KLEIN: Phys. Rev. **95**, 1061 (1954).

As regards the asymptotic behaviour, the fact that the force range of $V_4(\boldsymbol{r})$ is one half of the second order range $1/\mu$ is a general feature of the perturbation calculation. For example, the sixth order potential $V_6(\boldsymbol{r})$ is shown to have a force range of only one third of $1/\mu$. This might to some extent justify our assumption that the higher order effects are negligible, at least as far as we are

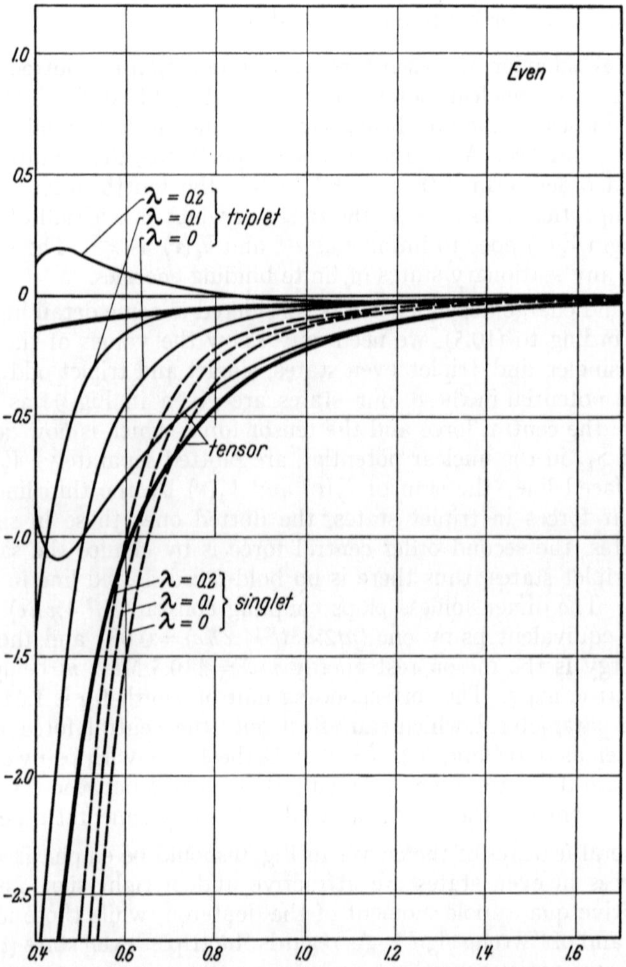

Fig. 9 a.

Fig. 9a and b. The static nuclear potential in the symmetrical pseudoscalar meson theory (8.11) and (9.14), plotted against $x = \mu r$ in units of the meson rest energy 139.5 Mev. Bold-faced lines represent the second order potential, thin lines the second plus fourth order potential; solid lines give forces in triplet states, dotted lines those in singlet states. The dimensionless ps-ps coupling constant $(f^2/4\pi\hbar c)$ is assumed to be 15. λ is the damping parameter.

not interested in the region very near the origin. As is seen from Fig. 9, however, $V_4(\boldsymbol{r})$ is sometimes far from small compared with $V_2(\boldsymbol{r})$ even at larger distances than $1/\mu$, which exemplifies the bad convergence of the perturbation expansion in the meson theory. At smaller distances than $1/\mu$, $V_4(\boldsymbol{r})$ is generally a very large correction to $V_2(\boldsymbol{r})$, thus making us suspicious whether the potential in question can be relied upon. There is, however, not yet any other satisfactory way of deriving a static nuclear potential, and we must be satisfied with some theoretical estimates which suggest that the potential considered here may not be quite wrong down to distances of, very roughly, $\frac{1}{2}\mu$ or $x = 0.5$[85,89]. At smaller distances, the current form of field theory is quite powerless and there is at present no reasonable way of getting theoretical estimates.

A clear indication that our potential (8.11) and (9.14) fails near the origin are the strong singularities at $r=0$ mentioned before. We have just seen that there is no reliable estimate of the nuclear potential at distances smaller than, say, $x=0.5$. Then we must resort to phenomenological arguments. The usual

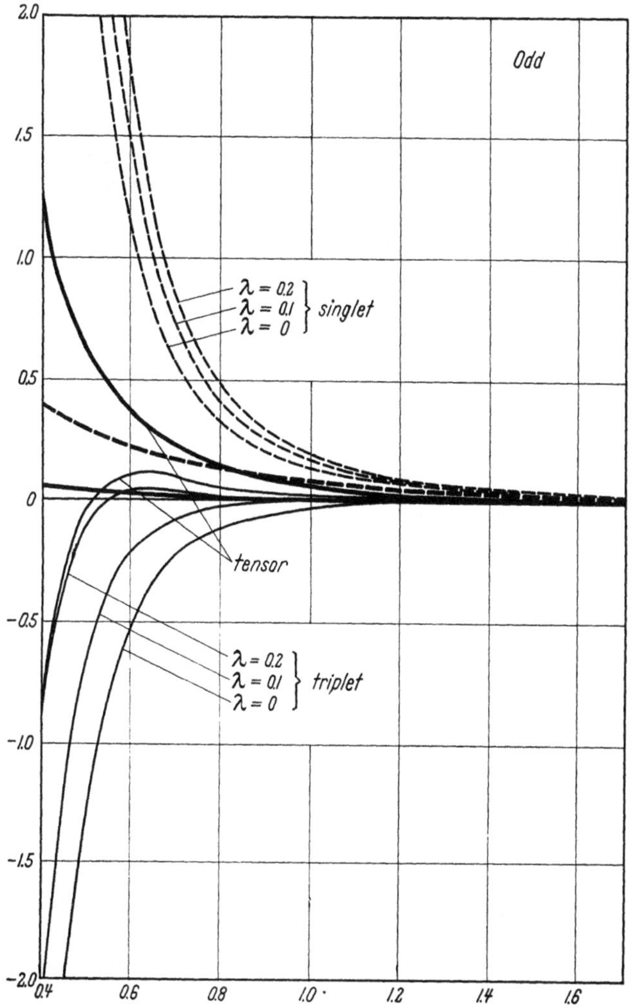

Fig. 9b.

approach is to modify the above meson theoretical potential at small distances quite arbitrarily so as to allow bound states with finite binding energies and to adapt the modification to the empirical facts. The simplest way would be to cut off the potential at some small distance and to replace it by an arbitrary constant value inside. If we choose this value as the original potential at the cut-off radius, the procedure is called a "straight cut-off". If, on the other hand, we put it equal to zero, we have a "zero cut-off". In all cases where we want to use the meson theoretical potential (8.11) and (9.14), we must employ some kind of cut-off method.

As is shown later, the low energy data alone cannot, as a principle, give us information about what kinds of cut-off we must use. From the analysis of the high energy data (Sect. 43 and 44), the only cut-off which seems to be passable is that proposed by JASTROW[92]. His model is called a "hard core cut-off", since it consists in replacing the nuclear potential by an infinite repulsive potential inside a small distance which must be determined so as to fit the empirical facts. At present it seems that the introduction of a hard core is inevitably necessary only in the singlet even state, the radius being about $0.4\frac{1}{\mu}$ while in other states there is no definite evidence on this point. From the theoretical side, it might be reasonable to expect that a hard core also exists in these other states. Again we recall that the meson theoretical potential up to the fourth order may be valid at distances larger than $x \approx 0.5$. It is, therefore, usual in current work to assume hard cores in all states, with radii, say, $x = 0.3$ to 0.4 and employ the meson theoretical potential (8.11) and (9.14) in the outer region, although other types of cut-off are never definitely excluded.

Thus far we have considered only the nuclear potential due to BRUECKNER and WATSON[85]. As regards the versions of other authors[87], one of the most important differences from the potential plotted in Fig. 9 is that the central force in the triplet even state is very strongly repulsive, while the tensor force is almost unchanged. This repulsive central force is, however, too strong to allow any bound state (Sect. 33), even though the tensor force is strong. According to a preliminary calculation of the effect of the possible nucleon isobar states[90], a very strongly attractive central force is expected in the triplet even state, while the tensor force is again unaffected. This central force is almost as strong as the difference of the central forces due to the two theoretical approaches mentioned above. We can, therefore, sum up the status of the meson theoretical nuclear potential up to the fourth order, in the triplet even state, by saying that a large tensor force of the right sign seems to be well established, while the central force is still quite uncertain; even its sign is indefinite, but it seems that the meson theory predicts a central force weaker than the tensor force (or possibly repulsive).

12. Phenomenological nuclear potential. As explained in the previous section, the meson theoretical potential (8.11) and (9.14) is neither definitive nor satisfactory—it cannot even be uniquely derived from the current meson field theory. Moreover it is mathematically complicated and involves some arbitrary cut-off procedure. Thus it is not always appropriate to use this potential everywhere in the analysis of the two-nucleon problem. In many cases it is more reasonable to use some phenomenological nuclear potential with simpler mathematical properties, still containing some of their essential features, since the particular analytic form appearing in (8.11) and (9.14) has no deeper physical meaning, as far as nuclear problems are concerned.

In such phenomenological nuclear potentials, there are always two kinds of ambiguity, one in the radial dependence and the other in the relative magnitude of central to tensor force.

With respect to the radial dependence, the following forms are the typical specimens which have thus far been investigated most extensively:

Square well: $V(r) = -V_0,\ r < r_0;\ V(r) = 0,\ r > r_0,$ (12.1)

GAUSS: $V(r) = -V_0 \exp(-\mu^2 r^2),$ (12.2)

Exponential: $V(r) = -V_0 \exp(-\mu r),$ (12.3)

[92] R. JASTROW: Phys. Rev. 81, 165 (1951).

$$\text{YUKAWA:} \quad V(r) = -V_0 \exp(-\mu r)/(\mu r), \tag{12.4}$$

$$\text{HULTHÉN:} \quad V(r) = -V_0 \exp(-\mu r)/\{1 - \exp(-\mu r)\}, \tag{12.5}$$

Common features of these are that they have a short range, contain two adjustable parameters, the depth V_0 and the range r_0 or $1/\mu$, and behave regularly near the origin, which makes it needless to introduce any cut-off procedures. The first point (short range) relates to a well established empirical fact and must be inherent in every radial function used for our purpose. As regards the second point, it is quite obvious that such a radial dependence implies at least two adjustable parameters. Concerning the final point, a regular form has always been considered as a matter of course; the cut-off procedure has never been anything but an emergency exit. In particular the YUKAWA potential (12.4) has long been regarded as the most reasonable one since it is the characteristic radial function of the meson field theory. The SCHRÖDINGER wave equation, however, cannot be solved analytically for the YUKAWA potential even in the central force case. On the other hand, this can be done with the HULTHÉN potential which resembles the YUKAWA potential especially near the origin, as is seen from (12.4) and (12.5). Therefore, the HULTHÉN potential has been used rather frequently as a "stand in" for the YUKAWA function.

It should be recalled that the high energy data seem to support a hard core hypothesis. We also add that according to the analysis of the polarization measurements (Sect. 46), a more singular radial dependence than in the YUKAWA potential, perhaps of the kind appearing in the meson theoretical potential (8.11) and (9.14), seems to be required by the observed high polarization. For these reasons it is important to consider, for example, the following phenomenological nuclear potentials:

$$\text{Singular type:} \quad \left.\begin{aligned} V(r) &= -V_0 \exp(-\mu r)/(\mu r)^n; \; r > r_c, \\ &= +\infty; \; r < r_c, \end{aligned}\right\} \tag{12.6}$$

where $n = 0$ and 1 give the combination of a hard core with the exponential and YUKAWA well, respectively, and $n \geq 2$ gives a singular well; the cut-off value $+\infty$ may of course be replaced by something else. Some combinations of the singular type potentials with various n together with those given before are also quite interesting. Although the singular type is very important, there are few results reported thus far which can be cited in this article.

As regards the relative magnitude of central to tensor force, to which the present meson theory does not give any definite answer, it seems possible to summarize existing investigations[93] by saying that, in the triplet even state, the central force must be attractive and roughly of the same magnitude as the tensor force, especially in order to fit the quadrupole moment. We therefore assume that the central force is attractive and the tensor force of right sign in the triplet even state. In the singlet even state, the meson theory predicts an attractive force, which is also consistent with the empirical facts. We therefore assume an attractive central force in the singlet even state. As regards the forces in odd states, the empirical data are not yet accurate enough to allow an estimate of the potentials in these states. In this article the theoretical predictions will be reported in a few cases which deserve some attention.

Although the tensor force is relatively large compared with the central force in the deuteron ground state, its apparent effect seems to be small as shown,

[93] See especially M. SUGAWARA: Progr. Theor. Phys. **5**, 920 (1950). — T. MATSUMOTO and M. SUGAWARA: Progr. Theor. Phys. **12**, 553 (1954).

for instance, by the small value of the quadrupole moment. This is related to the fact that the deuteron problem and other two-nucleon problems can rather well be explained in terms of a central force only. Also for the sake of mathematical simplicity, let us begin our analysis assuming an attractive central force, and then enter on the necessary modifications due to a tensor force.

IV. Low energy phenomena interpreted through central forces.

a) Deuteron problem.

13. Ground state. Empirically it is known that the deuteron spin is 1 and the deuteron magnetic moment nearly equal to the sum of proton and neutron moments (Sect. 1), which suggests that the deuteron ground state is a ^3S-state. It is indeed a natural supposition that the ground state is an S-state, since otherwise the nuclear force must be rather peculiar; because of the centrifugal force due to the angular momentum, the attraction in a P-state, for example, would have to be much stronger than in an S-state in order to make the corresponding energy lower than any possible S-level. For such a force there is, however, no evidence, neither experimental nor theoretical. We can, therefore, conclude that the deuteron ground state is a ^3S-state, using the terminology given in Sect. 10. As pointed out there, the assumption of a pure ^3S-state does not suffice to explain the deuteron quadrupole moment (Sect. 1), which makes it necessary to introduce a tensor force. This will be discussed in the next chapter.

Let us assume a central potential energy $V(r)$. The SCHRÖDINGER wave equation for the relative motion of a proton-neutron system is given by

$$\left[-\frac{\hbar^2}{M}\Delta + V(r)\right]\psi(\boldsymbol{r}) = E\,\psi(\boldsymbol{r}), \tag{13.1}$$

where M is the nucleon mass (or more exactly twice the reduced mass of a proton and neutron), $\boldsymbol{r}$ is the relative coordinate and E is the energy of the relative motion. For a bound S-state, we may rewrite it in the form

$$\left[\frac{d^2}{dr^2} + (-\alpha^2 + v(r))\right]u(r) = 0 \tag{13.2}$$

where we have put

$$\psi(\boldsymbol{r}) = \psi(r) = u(r)/r, \tag{13.3}$$

$$\alpha^2 = -E M/\hbar^2, \quad v(r) = -V(r)\,M/\hbar^2, \tag{13.4}$$

and $u(r)$ must satisfy the boundary conditions

$$u(0) = u(\infty) = 0. \tag{13.5}$$

In order to get a general picture of the deuteron ground state, let us assume a square well potential which has a constant value $-V_0$ for $r<r_0$ and vanishes for $r>r_0$. Then we easily get a relation between V_0 and r_0 valid for bound states:

$$\cot\left(\sqrt{v_0-\alpha^2}\cdot r_0\right) = \alpha/\sqrt{v_0-\alpha^2}. \tag{13.6}$$

For the ground state, $-E$, the binding energy of the deuteron, is known to be 2.226 Mev from (1.1), which gives

$$1/\alpha = 4.315_7\times 10^{-13}\text{ cm}. \tag{13.7}$$

Now we assume $V_0 \gg -E$ or equivalently $v_0 \gg \alpha^2$, which will later be shown to be the case for the ground state. Then, neglecting α^2 in (13.6), we see that $\sqrt{v_0}\,r_0$

is very near by $\pi/2$, because the value of $\sqrt{v_0}\,r_0$ which is, for instance, close to $3\pi/2$, is not the correct solution, for the ground state wave function cannot have any radial node. From $\sqrt{v_0}\,r_0 \approx \pi/2$, we get

$$V_0 r_0^2 \approx \pi^2 \hbar^2/4M, \tag{13.8}$$

which gives, for example, with $r_0 = 1.7 \times 10^{-13}$ cm,

$$V_0 \approx 35.2 \text{ Mev}, \tag{13.9}$$

which justifies the previous assumption $V_0 \gg -E$.

The solution of (13.2) outside the nuclear force range which satisfies the boundary condition (13.5) is proportional to $e^{-\alpha r}$. This exponential decrease of the deuteron wave function shows that the quantity $1/\alpha$ can be regarded as a measure of the size of the deuteron. It should be noticed that $1/\alpha$ given by (13.7) is considerably larger than the nuclear force range $\approx 2 \times 10^{-13}$ cm, which means that the deuteron is a loosely bound system. It is also to be noted that the exponential form outside the nuclear force range is determined only by the empirically known deuteron binding energy—a feature characteristic of short-range forces. This fact enables us to construct an approximate deuteron wave function, quite independently of the detailed shape of the nuclear potential. Thus the simple function

$$u(r) = 2\alpha\, e^{-\alpha r} \tag{13.10}$$

is sometimes useful. It does not satisfy the boundary condition at $r=0$ (13.5) but this can easily be remedied by modifying it to

$$u(r) = \sqrt{\frac{2\alpha(\alpha+\mu)(2\alpha+\mu)}{\mu^2}}\, e^{-\alpha r}(1 - e^{-\mu r}), \tag{13.11}$$

which tends to (13.10) as $\mu \to \infty$ and is called the HULTHÉN wave function, because it is the exact wave function for the HULTHÉN potential (12.5). The wave function for the YUKAWA potential (12.4) has been investigated in detail by HULTHÉN and LAURIKAINEN[94], who have expanded the corresponding solution in the form

$$u(r) = N e^{-\alpha r}(1 - e^{-\mu r})(h_0 + h_1 e^{-\mu r} + h_2 e^{-2\mu r} + h_3 e^{-3\mu r} + \cdots) \tag{13.12}$$

and shown that the convergence of the above series* is satisfactory; $h_1 \approx 0.6\, h_0$, $h_2 = h_3 = \cdots = 0$ is already a good approximation. A more detailed discussion of the deuteron wave function will be given in Sects. 28 and 33.

14. Excited states. Although there is no experimental evidence for excited states of the deuteron, we shall now extend the theoretical discussion of the previous section to excited states.

We anticipate that, for a short range potential, the number of discrete states is always finite. For a COULOMB potential it is well known that there are infinitely many bound states. By short range we mean that the force practically vanishes outside some finite distance (force range) from the force center. Such a situation can be represented mathematically by an asymptotic form $e^{-\mu r}$ of the potential. Let us first assume that the force vanishes exactly outside of r_0. The ground state wave function ψ_0 and the first excited one ψ_1 belonging to the energies E_0

[94] L. HULTHÉN and K. V. LAURIKAINEN: Rev. Mod. Phys. **23**, 1 (1951).

* Or rather $u(r) = N e^{-\alpha r} \sum_{\nu=1}^{\infty} \omega_\nu (1 - e^{-\mu r})^\nu$.

and E_1, respectively, behave as shown in Fig. 10. In general the n-th excited state wave function has n zeros between $r=0$ and $r=r_0$, which readily shows that there cannot be infinitely many bound states, since a wave function cannot oscillate infinitely many times within a finite distance unless the potential is infinitely attractive for at least part of the range between $r=0$ and $r=r_0$.

In the case of a potential with an exponentially decreasing tail, we can safely confine ourselves to the HULTHÉN potential (12.5), since it has the longest tail $e^{-\mu r}$ and the strongest permissible singularity $1/r$ at the origin, among the potential types considered above. Then the wave equation

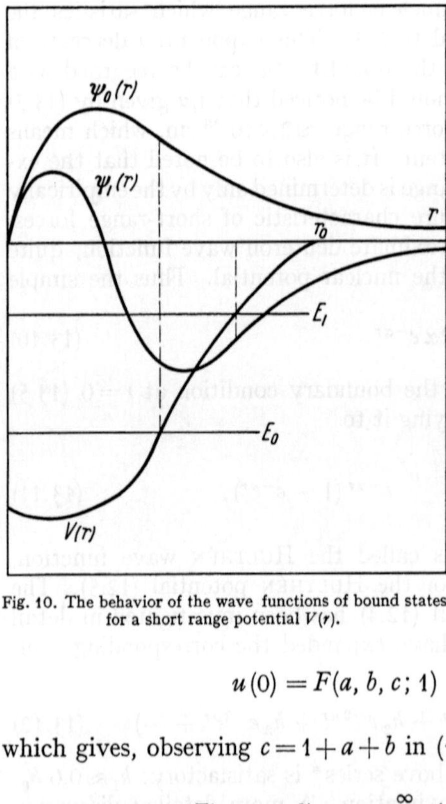

Fig. 10. The behavior of the wave functions of bound states for a short range potential $V(r)$.

$$\left.\begin{aligned}\left[\frac{d^2}{dr^2}+k^2\right]u(r)\\ =-v_0\frac{e^{-\mu r}}{1-e^{-\mu r}}u(r),\quad \mu>0,\end{aligned}\right\}\tag{14.1}$$

with

$$k^2=\frac{EM}{\hbar^2},\quad v_0=\frac{V_0 M}{\hbar^2}>0,\tag{14.2}$$

can be solved exactly. The solution of (14.1) which has the asymptotic form $\exp(-ikr)$ was shown by JOST[95] to be

$$u(r)=\exp(-ikr)F(a,b,c;z),\tag{14.3}$$

where F is the hypergeometric function with

$$a=\frac{ik}{\mu}\left[1-\sqrt{1-(v_0/k^2)}\right],$$

$$b=\frac{ik}{\mu}\left[1+\sqrt{1-(v_0/k^2)}\right],$$

$$c=1+(2ik/\mu),\quad z=\exp(-\mu r).\tag{14.4}$$

The value of $u(r)$ at $r=0$ is[96]

$$u(0)=F(a,b,c;1)=\frac{\Gamma(c)\,\Gamma(c-a-b)}{\Gamma(c-a)\,\Gamma(c-b)},\tag{14.5}$$

which gives, observing $c=1+a+b$ in (14.4),

$$u(0)=\frac{\Gamma(1+a+b)}{\Gamma(1+a)\,\Gamma(1+b)}=\prod_{n=1}^{\infty}\frac{(a+n)\,(b+n)}{n\,(a+b+n)}=\prod_{n=1}^{\infty}\frac{(k+i\alpha_n)}{\left(k-\dfrac{in\mu}{2}\right)},\tag{14.6}$$

where

$$\alpha_n=\frac{1}{2}\left(\frac{v_0}{n\mu}-n\mu\right).\tag{14.7}$$

For the bound states, k must be purely imaginary and $u(r)$ must satisfy two boundary conditions (13.5), which can be done at the same time only by making k equal to one of the quantities $-i\alpha_n$, for every positive α_n. As seen from (14.7), there is a finite number N of positive α_n for a non-vanishing μ, N being determined by $N^2 \leq v_0/\mu^2$. Each positive α_n corresponds to a bound state with

[95] R. JOST: Helv. phys. Acta 20, 256 (1947). — S. T. MA: Austral. J. Phys. 7, 365 (1954).

[96] E. T. WHITTAKER and G. N. WATSON: Modern Analysis, p. 282. Cambridge 1950. For Eq. (14.6) cf. also p. 239 of this reference.

$-E_n = \alpha_n^2 \hbar^2/M$, $n = 1, 2, 3, \ldots, N$. In the case of a force extending to infinity ($\mu \to 0$, v_0/μ finite), like the COULOMB potential, it is seen from (14.7) that all α_n are positive, whence there are infinitely many bound states.

Next we consider the problem whether excited states of the deuteron are possible or not. Again we first assume the HULTHÉN potential. Then putting $-E_1 = 2.226$ Mev and using (13.4) and (13.7), we get

$$\frac{1}{2}\left(\frac{v_0}{\mu} - \mu\right) = 0.2317 \times 10^{13}\ \text{cm}^{-1}, \tag{14.8}$$

which is too small to make $\alpha_2 = \frac{1}{2}\left(\frac{v_0}{2\mu} - 2\mu\right)$ positive, because μ is at least about 0.5×10^{13} cm^{-1}, showing that no other bound ^3S-state can exist besides the ground state of the deuteron. The same conclusion can easily be reached, assuming the square well potential. In this case we need only remember Eq. (13.8), which forbids the existence of other bound states, since two such states, for instance, would require the approximate equality $\sqrt{v_0}\, r_0 \gtrsim 3\pi/2$, derivable from (13.6). These considerations, mainly based on the fact that the "deuteron radius" $1/\alpha$ is considerably larger than the nuclear force range, indicate that there are no excited levels in the ^3S-state of the deuteron.

It would not be appropriate to discuss the ^1S-state of the deuteron in this section, since we have not yet learnt anything about the nuclear force in this state. Anyway it will be shown in Sect. 20 that there is no bound ^1S-state.

Finally let us consider the odd states of the neutron-proton system. There is some evidence, experimental and theoretical, that the nuclear force is in general smaller in magnitude in odd states than in even states. Let us, however, assume that the nuclear force is the same in odd states as in the ^3S-state. The wave Eq. (13.2) is generalized to

$$\frac{d^2 u(r)}{dr^2} = \left[\alpha^2 - v(r) + \frac{l(l+1)}{r^2}\right] u(r), \tag{14.9}$$

for a state with orbital angular momentum quantum number l. Assuming, for the sake of simplicity, the square well potential with range r_0 and depth V_0, we get the solution satisfying the boundary conditions (13.5) for $l = 1$, save for normalization factors,

$$u = \frac{\sin\left(\sqrt{v_0 - \alpha^2}\, r\right)}{\sqrt{v_0 - \alpha^2}\, r} - \cos\left(\sqrt{v_0 - \alpha^2}\, r\right) \quad \text{for } r < r_0, \tag{14.10}$$

$$u = e^{-\alpha r}\left[1 + \frac{1}{\alpha r}\right] \quad \text{for } r > r_0, \tag{14.11}$$

which must be joined at $r = r_0$. Here we can safely neglect α^2 compared with v_0, since it was already allowed for the ground state. From the equation of joining $\frac{du}{dr}\Big/ u$ at $r = r_0$, omitting α^2 and considering αr_0 as small, we get $\sin\left(\sqrt{v_0}\, r_0\right) \approx 0$, which demands $\sqrt{v_0}\, r_0 \approx \pi$. This condition, however, cannot be satisfied because of the relation (13.8), which shows that the nuclear potential is too weak to allow bound states with $l \neq 0$.

All the considerations given in this section are solely based upon a central force assumption. Even though a tensor force exists, we can reasonably introduce the idea of an effective potential, which can be visualized as a central force acting in the deuteron state (Sect. 28). Therefore, we may expect that such semi-quantitative arguments as have been developed so far are valid also in the more general case of a tensor force. Thus we can finally conclude that the deuteron ground state is the only possible bound state of a proton-neutron system.

b) Neutron-proton scattering.

15. Phase shift analysis. In analysing the scattering data, the phase shift is of fundamental importance. Therefore, we shall briefly mention the definition and the properties of the phase shift.

In a scattering problem, we want a solution of the SCHRÖDINGER equation of the relative motion (13.1) which has the asymptotic form

$$\psi(\boldsymbol{r}) \xrightarrow[r\to\infty]{} \left[e^{ikz} + f(\vartheta,\varphi)\,\frac{e^{ikr}}{r}\right], \tag{15.1}$$

$$k^2 = EM/\hbar^2. \tag{15.2}$$

The differential scattering cross section $\sigma(\vartheta,\varphi)$ is given by

$$\sigma(\vartheta,\varphi) = |f(\vartheta,\varphi)|^2. \tag{15.3}$$

Assuming a central force, we need not consider the φ dependence of $f(\vartheta,\varphi)$ or $\psi(\boldsymbol{r})$, and the general solution of (13.1) can be expanded, using LEGENDRE polynomials, as

$$\psi(r,\vartheta) = \sum_{l=0}^{\infty} \frac{u_l(r)}{r}\, P_l(\cos\vartheta), \tag{15.4}$$

where $u_l(r)$ satisfies [cf. (13.3) and (13.4)]

$$\left[\frac{d^2}{dr^2} + \left(k^2 + v(r) - \frac{l(l+1)}{r^2}\right)\right] u_l(r) = 0, \tag{15.5}$$

with the boundary condition

$$u_l(0) = 0. \tag{15.6}$$

In the absence of the nuclear potential $v(r)$, $\psi(\boldsymbol{r})$ reduces to the incident plane wave $\exp(ikz)$, which can also be expanded, in terms of the LEGENDRE polynomials, as

$$e^{ikz} = e^{ikr\cos\vartheta} = \sum_{l=0}^{\infty} (2l+1)\, i^l j_l(kr)\, P_l(\cos\vartheta), \tag{15.7}$$

where $j_l(x)$ is the spherical BESSEL function, defined as[98]

$$j_l(x) = \left(\frac{\pi}{2x}\right)^{\frac{1}{2}} J_{l+\frac{1}{2}}(x).$$

The asymptotic form is

$$e^{ikz} \xrightarrow[r\to\infty]{} \sum_{l=0}^{\infty} (2l+1)\, i^l\, \frac{\sin(kr - \frac{1}{2}l\pi)}{kr}\, P_l(\cos\vartheta). \tag{15.8}$$

Now let us consider the effect of the nuclear potential $v(r)$. As the nuclear force has a short range, the exact solution of (15.5) satisfies asymptotically the same differential equation as $j_l(kr)$. Thus, introducing two constants, A_l and δ_l, for each partial wave, the asymptotic solution of (15.5) can in general be written

$$u_l(r) \xrightarrow[r\to\infty]{} (2l+1)\, i^l\, \frac{A_l}{k} \sin\left(kr - \frac{1}{2}l\pi + \delta_l\right), \tag{15.9}$$

[98] L. I. SCHIFF: Quantum Mechanics, pp. 77 and 78. New York 1949, or J. MEIXNER, Vol. I of this Encyclopedia.

and therefore the asymptotic solution of (13.1) is given by

$$\psi(r,\vartheta) \xrightarrow[r\to\infty]{} \sum_{l=0}^{\infty} (2l+1)\, i^l A_l \frac{\sin(k r - \frac{1}{2} l \pi + \delta_l)}{k r} P_l(\cos\vartheta), \tag{15.10}$$

where the factors in (15.9) and (15.10) are chosen for later convenience. The constant δ_l, which we can assume as real without loss of generality, is called the nuclear phase shift of the l-th partial wave, since it is just the shift of the phase of the force-free solution (15.8) due to the nuclear force $v(r)$.

We now require that the general solution (15.10) has the form of (15.1) (no incoming spherical wave), in order to represent a scattering situation. Substituting (15.8) and (15.10) into (15.1) and comparing the two sides, we get

$$A_l = e^{i\delta_l} \tag{15.11}$$

and

$$f(\vartheta) = \frac{1}{2ik} \sum_{l=0}^{\infty} (2l+1)(e^{2i\delta_l} - 1) P_l(\cos\vartheta), \tag{15.12}$$

which gives the differential scattering cross section

$$\sigma(\vartheta) = \frac{1}{k^2} \left| \sum_{l=0}^{\infty} (2l+1)\, e^{i\delta_l} \sin\delta_l P_l(\cos\vartheta) \right|^2 \tag{15.13}$$

and the total scattering cross section σ:

$$\sigma = 2\pi \int_0^{\infty} \sigma(\vartheta) \sin\vartheta\, d\vartheta = \frac{4\pi}{k^2} \sum_{l=0}^{\infty} (2l+1) \sin^2\delta_l. \tag{15.14}$$

Thus $\sigma(\vartheta)$ and σ have been expressed in terms of the phase shifts δ_l only. The theoretical problem has, therefore, been reduced to the calculation of δ_l, starting from the differential equation (15.5) with the boundary conditions (15.6) and (15.9) for a given nuclear potential $v(r)$.

Finally some simple but important characteristics of these phase shifts are summarized. Owing to the increasing predominance of the centrifugal force in (15.5) as l increases, the effect of $v(r)$ becomes smaller and smaller as l increases. Thus the phase shifts of higher waves can usually be neglected, so we need only consider a small number of terms in (15.13) and (15.14). In the low energy region the S-wave phase shift δ_0 is quite predominant. This fact can be experimentally verified by observing the spherical symmetry of the scattering in the c.m. system at low energies; (15.13) is independent of ϑ, if other phase shifts than δ_0 vanish. Another important point is that positive and negative signs of the phase shift indicate attractive and repulsive potentials, respectively. This is easily understood by considering qualitatively how the potential $v(r)$ affects the wave function.

16. Effective range theory. *α) Scattering length and effective range.* It was pointed out in Sect. 15 that the low energy scattering cross section is expressed in terms of the S-wave phase shift δ_0 only. Experimentally we can measure the cross section at different energies of the incident neutron, which gives the behaviour of δ_0 as depending on k^2. The usual way to analyse this relation is to employ the elegant approximation that $k\cot\delta_0$ is a linear function of k^2;

$$k\cot\delta_0 = -\frac{1}{a} + \frac{1}{2} r_0 k^2, \tag{16.1}$$

where the two constants a and r_0 with the dimension of length are called the scattering length and the effective range, respectively. The above property of δ_0 is explained in the effective range theory, which was first developed by SCHWINGER[99] using the variational method (Sect. 51). It was further exploited by BLATT and JACKSON[100] in their comprehensive analysis of the low energy data. A simpler deduction was proposed by BETHE[101], whom we follow in this section. Necessary modifications in the proton-proton case and the tensor force case are given in Sects. 22 and 30, respectively.

We start from a differential equation (15.5) with $l=0$ and consider two such equations belonging to energies E_1 and E_2, or k_1^2 and k_2^2, respectively. Let us drop the index 0 (for $l=0$) and use subscripts 1 and 2 in order to denote quantities pertaining to k_1^2 and k_2^2, respectively. These equations are

$$\left[\frac{d^2}{dr^2} + \left(k_1^2 + v(r)\right)\right] u_1(r) = 0, \tag{16.2}$$

$$\left[\frac{d^2}{dr^2} + \left(k_2^2 + v(r)\right)\right] u_2(r) = 0. \tag{16.3}$$

For a triplet state, there is a bound level, for which (13.2) is valid:

$$\left[\frac{d^2}{dr^2} + \left(-\alpha^2 + v(r)\right)\right] u_g(r) = 0, \tag{16.4}$$

using a subscript g to denote ground state quantities. All these functions satisfy boundary conditions

$$u_1(0) = u_2(0) = u_g(0) = 0. \tag{16.5}$$

We then consider their asymptotic form $\bar{u}(r)$, which is given by (15.9) or (13.10). Let us normalize by requiring

$$\bar{u}_1(0) = \bar{u}_2(0) = \bar{u}_g(0) = 1, \tag{16.6}$$

which determines them as follows:

$$\bar{u}_i(r) = \sin(k_i r + \delta_i)/\sin\delta_i, \tag{16.7}$$

$$\bar{u}_g(r) = e^{-\alpha r}. \tag{16.8}$$

These barred functions satisfy differential equations (16.2), (16.3) and (16.4), without the nuclear potential term $v(r)$. It is important to note that $\bar{u}(r)$ as given by (16.7) and (16.8) fixes the normalization of $u(r)$ uniquely.

If we multiply (16.2) by $u_2(r)$ and (16.3) by $u_1(r)$, subtract one equation from the other and integrate over r from 0 to infinity, we get

$$\left[u_2(r)\frac{du_1(r)}{dr} - u_1(r)\frac{du_2(r)}{dr}\right]\Bigg|_0^\infty = (k_2^2 - k_1^2)\int_0^\infty u_1(r)\, u_2(r)\, dr. \tag{16.9}$$

The same procedure gives for the barred functions

$$\left[\bar{u}_2(r)\frac{d\bar{u}_1(r)}{dr} - \bar{u}_1(r)\frac{d\bar{u}_2(r)}{dr}\right]\Bigg|_0^\infty = (k_2^2 - k_1^2)\int_0^\infty \bar{u}_1(r)\, \bar{u}_2(r)\, dr. \tag{16.10}$$

---

[99] J. SCHWINGER: Harward lecture notes (hectographed only). Quoted by J. BLATT: Phys. Rev. 74, 92 (1948).

[100] J. M. BLATT and J. D. JACKSON: Phys. Rev. 76, 18 (1949).

[101] H. A. BETHE: Phys. Rev. 76, 38 (1949).

Taking the difference of these two equations, we have no contribution from infinity on the left-hand side, since barred and unbarred functions must coincide at infinity. At the lower limit, only the barred ones can contribute. Using (16.7), we finally get

$$k_2 \cot \delta_2 - k_1 \cot \delta_1 = (k_2^2 - k_1^2) \int\limits_0^\infty \left(\bar{u}_1(r)\,\bar{u}_2(r) - u_1(r)\,u_2(r)\right) dr. \tag{16.11}$$

If we start from (16.2) and (16.4) instead of (16.2) and (16.3), the same procedure gives

$$k_1 \cot \delta_1 + \alpha = (k_1^2 + \alpha^2) \int\limits_0^\infty \left(\bar{u}_g(r)\,\bar{u}_1(r) - u_g(r)\,u_1(r)\right) dr. \tag{16.12}$$

These two identities are exact and form the fundamental equations of the effective range theory. In a singlet state we have only one equation, (16.11).

We define the scattering length a, according to (16.1), by

$$\lim_{k^2 \to 0} \left[k \cot \delta_0\right] = -\frac{1}{a}, \tag{16.13}$$

where we have replaced the index 0 ($l=0$). It is to be noted that δ_0 belongs to an energy k^2. From (15.14) we get for the total scattering cross section σ_0 at zero energy:

$$\sigma_0 = \lim_{k^2 \to 0} \sigma = \lim_{k^2 \to 0} \frac{4\pi \sin^2 \delta_0}{k^2} = \lim_{k^2 \to 0} \frac{4\pi}{k^2 + k^2 \cot^2 \delta_0} = 4\pi a^2, \tag{16.14}$$

because all other phase shifts vanish faster than k. Dropping the index 1 in (16.11) and (16.12) and making k_2^2 approach zero, we get

$$k \cot \delta_0 = -\frac{1}{a} + \frac{1}{2} k^2 \varrho(0, E) \tag{16.15}$$

and

$$k \cot \delta_0 = -\alpha + \frac{1}{2}(k^2 + \alpha^2)\,\varrho(-\varepsilon, E) \tag{16.16}$$

where we have defined

$$\varrho(0, E) = 2 \int\limits_0^\infty \left(\bar{u}_0(r)\,\bar{u}(r) - u_0(r)\,u(r)\right) dr, \tag{16.17}$$

$$\varrho(-\varepsilon, E) = 2 \int\limits_0^\infty \left(\bar{u}_g(r)\,\bar{u}(r) - u_g(r)\,u(r)\right) dr; \tag{16.18}$$

an index 0 means the zero energy wave function. Furthermore, making E approach zero in (16.16), we get

$$\frac{1}{a} = \alpha - \frac{1}{2}\alpha^2 \varrho(-\varepsilon, 0), \tag{16.19}$$

where we have introduced the "mixed effective range"

$$\varrho(0, -\varepsilon) = \varrho(-\varepsilon, 0) = 2 \int\limits_0^\infty \left(\bar{u}_g(r)\,\bar{u}_0(r) - u_g(r)\,u_0(r)\right) dr. \tag{16.20}$$

In order to verify (16.1) we have to prove that $\varrho(0, E)$ in (16.15) is a constant. We can indeed show that all ϱ's defined above can be regarded as constants in a good approximation, as far as we are interested only in the low energy region. For this purpose, we first remark that $\bar{u}(r)$ and $u(r)$ differ from each other only inside the nuclear force range. Thus, the main contribution to such integrals as (16.17),

(16.18) and (16.20) comes from the inside region, where the potential energy is numerically much larger than the total energy. Thus the above three ϱ's depend only slightly upon E and we can approximate them by

$$r_0 \equiv \varrho(0,0) = 2\int_0^\infty \left(\bar{u}_0^2(r) - u_0^2(r)\right) dr. \tag{16.21}$$

In this approximation, we obtain (16.1) or

$$k \cot \delta_0 = -\alpha + \frac{1}{2}(k^2 + \alpha^2)\, r_0, \tag{16.22}$$

and

$$\frac{1}{a} = \alpha - \frac{1}{2}\alpha^2 r_0, \tag{16.23}$$

for the triplet state.

We emphasize two consequences of the effective range theory. First of all it yields (16.1), which means that we can only determine two parameters a and r_0 for each spin state from the low energy scattering data. Now $v(r)$ contains at least two adjustable parameters. We can, therefore, conclude that all forms of $v(r)$ considered in Sect. 12 can be adapted to the low energy scattering data.

Another important point is that the effective range theory predicts a simple relation between the scattering parameters a and r_0 and the deuteron radius $1/\alpha$. According to (16.23) the shape of the nuclear potential cannot be decided even by the combined knowledge of the deuteron binding energy and the low energy scattering data. The approximation in which (16.1), (16.22) and (16.23) are valid is, therefore, characterized as "shape independent". The fact that this is such a good approximation is one of the main reasons why the problem of the potential shape is still unsolved. It is also of interest to consider the following quantity:

$$\varrho(-\varepsilon, -\varepsilon) = 2\int_0^\infty \left(\bar{u}_g^2(r) - u_g^2(r)\right) dr = 2\left[\frac{1}{2\alpha} - \int_0^\infty u_g^2(r)\, dr\right], \tag{16.24}$$

which is sometimes called the "deuteron effective range", since the above argument shows that it is also equal to r_0 in the present approximation. Then, if a deuteron wave function is adjusted to the correct binding energy and its effective range is equal to r_0, it is also consistent with a and therefore with all low energy scattering data. This fact is exploited in Sects. 28 and 33 to construct reasonable deuteron wave functions.

β) *Shape dependent parameter.* Let us now turn to the higher terms in the effective range expansion (16.1). For this purpose, we expand $u(r)$ and $\bar{u}(r)$ in terms of k^2, which is always possible because k^2 appears linearly in the differential equations. It can be done, for example, by iteration:

$$u(r) = u_0(r) + k^2 u_0'(r) + \cdots, \qquad \bar{u}(r) = \bar{u}_0(r) + k^2 \bar{u}_0'(r) + \cdots. \tag{16.25}$$

As regards the deuteron wave function, the situation is somewhat different since a bound state is allowed only for a specified energy. We recall that in the integrals (16.18), (16.20) and (16.24) the principal contribution comes from inside the nuclear force range. Let us, therefore, neglect for the time being the correct boundary condition at infinity and consider the transition of α^2 approaching zero. Then, remembering (16.5) and (16.6), we can expand $u_g(r)$ and $\bar{u}_g(r)$ in the same way as (16.25) with k^2 replaced by $-\alpha^2$:

$$u_g(r) = u_0(r) - \alpha^2 u_0'(r) + \cdots, \qquad \bar{u}_g(r) = \bar{u}_0(r) - \alpha^2 \bar{u}_0'(r) + \cdots, \tag{16.26}$$

with the understanding that the above expansions are valid only inside the nuclear force range and that the individual terms in the expansion need not behave correctly at infinity. The boundary conditions (16.5) require that

$$u_0(0) = u_0'(0) = 0, \ldots, \quad \bar{u}_0(0) = 1, \quad \bar{u}_0'(0) = 0, \ldots. \tag{16.27}$$

From (16.7) and (16.8) we get

$$\left.\begin{aligned} \bar{u}(r) &= 1 - \tfrac{1}{2}k^2 r^2 + \cot\delta_0 (k r - \tfrac{1}{6}k^3 r^3) + O(k^4 r^4), \\ \bar{u}_g(r) &= 1 - \alpha r + \tfrac{1}{2}\alpha^2 r^2 - \tfrac{1}{6}\alpha^3 r^3 + O(\alpha^4 r^4). \end{aligned}\right\} \tag{16.28}$$

Here we can substitute (16.1) or (16.23) for $k\cot\delta_0$ or α, in order to eliminate the odd powers of k or α, since those equations are accurate up to k^3 or α^3, respectively. Then we get the expansions (16.25) and (16.26) with

$$\bar{u}_0(r) = 1 - \frac{r}{a}, \tag{16.29}$$

$$\bar{u}_0'(r) = \frac{1}{2} r (r_0 - r) + \frac{1}{6a} r^3, \tag{16.30}$$

which satisfy (16.27). It is noted that $\bar{u}_g(r)$ given by (16.26) never behaves correctly at infinity. The simple form (16.29) of $\bar{u}_0(r)$ is also notable.

It is then easy to compute the next order term in (16.1): We get from (16.15) and (16.17)

$$k\cot\delta_0 = -\frac{1}{a} + \frac{1}{2} k^2 r_0 - P k^4 r_0^3 + \cdots, \tag{16.31}$$

where we have defined a dimensionless quantity P by

$$P = -\frac{1}{r_0^3}\int_0^\infty [\bar{u}_0(r)\,\bar{u}_0'(r) - u_0(r)\,u_0'(r)]\,dr, \tag{16.32}$$

which is called the shape dependent parameter, since it is connected with the shape of the nuclear potential. We obtain from (16.20) and (16.24)

$$\varrho(-\varepsilon, -\varepsilon) = \varrho(0, -\varepsilon) + 2P\alpha^2 r_0^3 + \cdots, \tag{16.33}$$

with the same P as (16.32). Further we get from (16.19)

$$\frac{1}{a} = \alpha - \frac{1}{2}\alpha^2 r_0 - P\alpha^4 r_0^3 + \cdots, \tag{16.34}$$

which is sometimes used to calculate P from a and r_0.

The relations of the empirical scattering quantities a, r_0 with P and the potential shape parameters have been extensively investigated by Blatt and Jackson[100] for most of the regular types of potential given in Sect. 12. We quote here their main results. Consider a nuclear potential $V(r)$ and modify it by a factor s: $V(r) = s V'(r)$, so that $V'(r)$ gives a ground state with zero binding energy. We call s the well depth parameter of $V(r)$. Then $s > 1$ implies that $V(r)$ allows bound states, while $s < 1$ makes $V(r)$ too weak to bring about binding. Then the intrinsic range b of $V(r)$ is defined as the effective range of $V'(r)$. It can be shown that, for the square well potential, the intrinsic range b is just the same as its ordinary force range. Some regular potentials are then expressed in terms of s and b as follows:

Square well (S):
$$\left.\begin{aligned} V(r) &= -s\,(102.276)\,b^{-2}; & r &< b, \\ &= 0; & r &> b, \end{aligned}\right\} \tag{16.35}$$

GAUSS (G): $V(r) = -s\,(229.208)\,b^{-2} \exp[-2.0604\,(r/b)^2]$, (16.36)

Exponential (E): $V(r) = -s\,(751.541)\,b^{-2} \exp[-3.5412\,(r/b)]$, (16.37)

YUKAWA (Y): $V(r) = -s\,(147.585)\,b^{-2}\,(b/r) \exp[-2.1196\,(r/b)]$, (16.38)

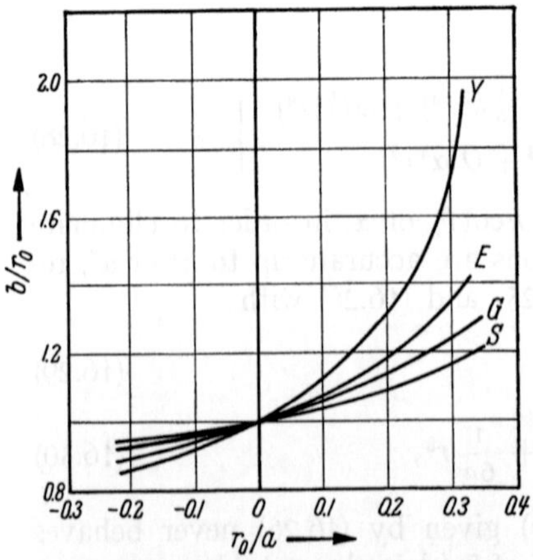

Fig. 11. The ratio b/r_0 (intrinsic range/effective) range (plotted against r_0/a (effective range/scattering length) for the well shapes (16.35) to (16.38).

where $V(r)$ is expressed in Mev, b and r in 10^{-13} cm. For the YUKAWA well, the ordinary range $1/\mu$ and its corresponding meson mass m are given by

$$\left.\begin{aligned} 1/\mu &= (b/2.1196)\times 10^{-13}\ \text{cm},\\ m &= (818.42/b)\,m_e, \end{aligned}\right\} \quad (16.39)$$

where m_e is the electron mass.

Let us suppose that a and r_0 are given empirically. Then b and s are first determined, after which P can be calculated for individual well shapes, the results being plotted in Figs. 11, 12 and 13. Actual calculations are suitably made with the aid of the interpolation formulas given in the original paper[100]. We add that the really interesting region of r_0/a is $-0.2 < r_0/a < +0.3$. It is therefore seen that P lies between -0.05 and $+0.15$. If $k r_0$ and αr_0 are already small numbers, as in the low energy case, the terms with P in (16.31), (16.33) and (16.34) are in fact negligible.

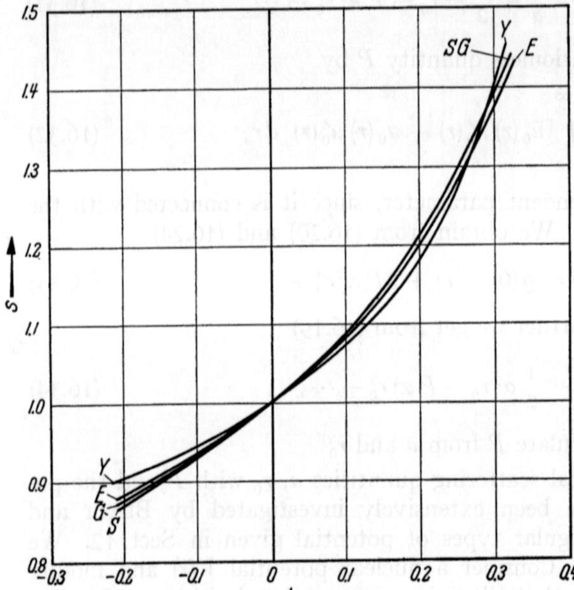

Fig. 12. The well depth parameter s plotted against r_0/a (effective range/scattering length) for the well shapes (16.35) to (16.38).

The main contribution to the integral (16.32) again comes from inside the nuclear force range, where its integrand would, however, be quite small, except in the potential tail region, since the derivative of the wave function with respect to k^2 would in general be appreciably different from zero only near and outside the tail region. The expression (16.30) shows this feature explicitly (notice that r_0 is nearly equal to and a much larger than the force range). This explains why P is in general a small quantity and at the same time suggests that P mainly depends on the potential tail. Indeed Fig. 13 tells us that P is small and negative for the square well, having the shortest tail, while it is positive and takes its largest value for the YUKAWA potential, which has the longest tail. Thus it seems possible to say that P is a measure of the extension of the potential outside its intrinsic range: P is larger for a longer tail.

As regards other potentials, especially such as the singular (12.6), there are no systematic investigations. It is, however, relatively easy to estimate the effect of a hard core on the shape of the outside potential. For the sake of simplicity, take the potential of square well type (Fig. 14). The zero energy wave function $u_0(r)$ and its asymptotic form $\bar{u}_0(r)$ given by (16.29) are also plotted in the figure. These two touch each other at the force range and coincide in the outer region. Now introduce a hard core. We naturally require that its range and depth are so modified as to give the same r_0 and a as before. Then $\bar{u}_0(r)$ is unchanged since it depends only on a. In order to give the same r_0, it is seen from (16.21), the

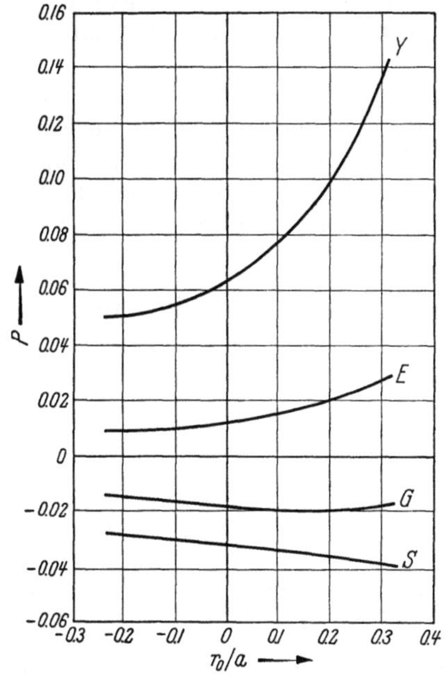

Fig. 13. The shape dependent parameter P plotted against r_0/a (effective range/scattering length) for the well shapes (16.35) to (16.38).

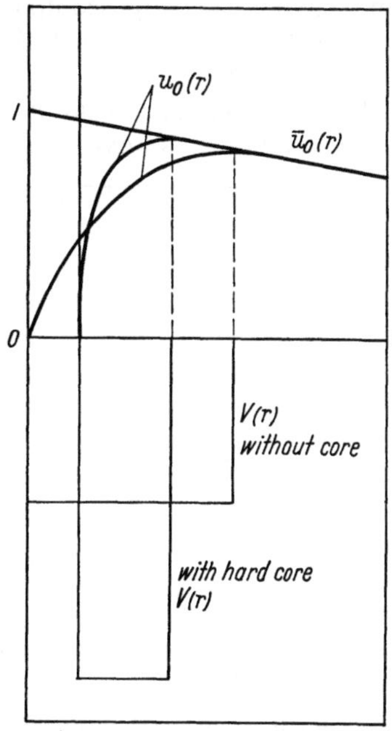

Fig. 14. Effect of a hard core on the zero energy wave function $u_0(r)$ and the outside potential, scattering data unchanged.

zero energy wave function $u_0(r)$ must be modified as in Fig. 14. This means that the introduction of a hard core always makes the range shorter and the well deeper.

Let us recall here that P seems to depend mainly on the potential tail. We may, therefore, conclude that the introduction of a hard core would in general reduce P. Accrding to Fig. 16 (Sect. 19), this is actually the case.

17. Zero energy cross section and effect of chemical binding*. The total scattering cross section at zero energy, σ_0, defined in (16.14), relates to the scattering of a neutron by a free proton. In the actual experiments the total cross section σ varies appreciably near the zero energy limit, as is seen from Fig. 15. It is observed that σ is nearly constant from 2500 ev downward but begins to rise just below 10 ev, reaching rapidly up to about 4 times this plateau value at very low energies. This is due to the chemical binding of the protons which apparently increases the effective proton mass. As the chemical binding energy is about 0.1 to 1 ev, this effect appears only at very low energies. In order to determine σ_0 we must first investigate this chemical binding effect.

* For subject-matter treated in this section, cf. E. AMALDI's article in Vol. XXXVIII of this Encyclopedia.

The task posed essentially involves a many-body problem, since we must pay regard to the motion of the residual part of the molecule, besides that of the proton and neutron. We can, however, make an essential simplification which is due to the fact that the nuclear force range is extremely small compared with the relevant molecular dimension and the incident neutron wave length. Let us, therefore, assume a zero range approximation for the neutron-proton interaction, which means that it is infinite when the particles touch each other and vanishes otherwise. Mathematically this approximation implies the replacement of the neutron-proton interaction term in the Hamiltonian by some boundary condition imposed upon the SCHRÖDINGER wave function ψ of the whole system in the neighbourhood of $\boldsymbol{r}=0$ where $\boldsymbol{r}$ is the relative coordinate of the proton and neutron.

The behaviour of ψ near $\boldsymbol{r}=0$ must be the same as $\bar{u}_0(r)$ given by (16.29), since we are now in an extremely low energy region and outside the nuclear force range, because of the above assumptions. Thus, recalling the definition (15.4), we infer that ψ has such a singularity at $r=0$ that

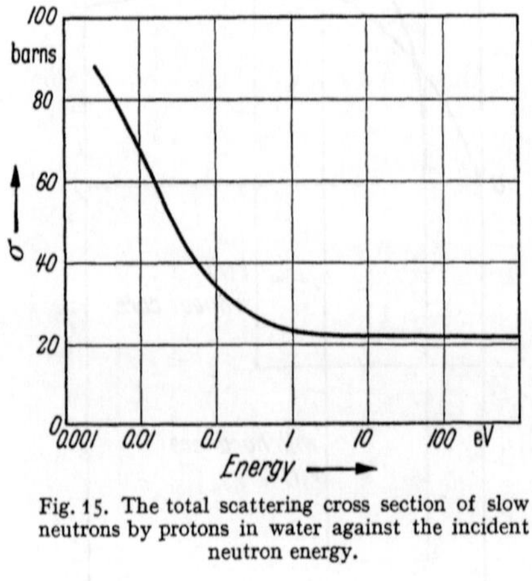

Fig. 15. The total scattering cross section of slow neutrons by protons in water against the incident neutron energy.

$$\psi \xrightarrow[r\to 0]{} \frac{1}{r}\left(1-\frac{r}{a}\right) \times \text{finite term}, \tag{17.1}$$

which can just as well be represented by

$$\frac{\partial \psi}{\partial r} \xrightarrow[r\to 0]{} +\frac{a}{r^2}\frac{\partial (r\psi)}{\partial r}. \tag{17.2}$$

If we integrate both sides over a sphere of radius r, we get from the left hand side

$$\int r^2 \frac{\partial \psi}{\partial r}\, d\Omega = \int \frac{\partial \psi}{\partial r}\, dS = \int \Delta\psi\, dV, \tag{17.3}$$

where the last integral is extended over the volume of that sphere. The right hand side simply becomes $4\pi a\,\partial(r\psi)/\partial r$, which can be written as

$$4\pi a \int \frac{\partial (r\psi)}{\partial r}\,\delta(\boldsymbol{r})\, dV, \tag{17.4}$$

where we have used DIRAC's δ-function. Thus we get another form of the boundary condition of ψ at $r=0$

$$\Delta\psi = 4\pi a\,\delta(\boldsymbol{r})\,\frac{\partial (r\psi)}{\partial r}. \tag{17.5}$$

Now let H stand for the Hamiltonian of the whole system without the neutron-proton interaction term. If we introduce their relative coordinates, the part of H which represents the kinetic energy of this relative motion has the well-known form $-\hbar^2\Delta/M$, where M is twice the reduced mass of a proton and neutron. Thus it has been shown that ψ must satisfy the equation[102]

$$(H-E)\,\psi = -V'(\boldsymbol{r}_n-\boldsymbol{r}_p)\,\frac{\partial (r\psi)}{\partial r}, \tag{17.6}$$

with

$$V'(\boldsymbol{r}_n-\boldsymbol{r}_p) = 4\pi a\,\frac{\hbar^2}{M}\,\delta(\boldsymbol{r}_n-\boldsymbol{r}_p), \tag{17.7}$$

where $\boldsymbol{r}_n$ and $\boldsymbol{r}_p$ are the coordinates of neutron and proton, respectively.

---

[102] G. BREIT: Phys. Rev. **72**, 215 (1947). — J. M. BLATT and V. F. WEISSKOPF: Theoretical Nuclear Physics, p. 74. New York 1952.

In order to solve (17.6) let us expand ψ in a power series with respect to V':

$$\psi = \psi_0 + \psi_1 + \psi_2 + \cdots, \tag{17.8}$$

where ψ_0 is the wave function for a vanishing neutron-proton interaction. Equating terms of the same order in V' in (17.6), we get

$$(H - E)\,\psi_n = -\,V'(\boldsymbol{r}_n - \boldsymbol{r}_p)\,\frac{\partial(r\,\psi_{n-1})}{\partial r}, \tag{17.9}$$

the first of which has a specially simple form[103]:

$$(H - E)\,\psi_1 = -\,V'(\boldsymbol{r}_n - \boldsymbol{r}_p)\,\psi_0, \tag{17.10}$$

because ψ_0 has no singularity at $r=0$. Here it is to be noted that (17.10) is just what would be obtained if we let the pseudopotential $V'(\boldsymbol{r}_n - \boldsymbol{r}_p)$ represent the potential energy between two nucleons and could furthermore use the first BORN approximation to solve the scattering problem due to $V'(\boldsymbol{r}_n - \boldsymbol{r}_p)$. It must be remarked, however, that an exact solution of the SCHRÖDINGER equation with the pseudopotential $V'(\boldsymbol{r}_n - \boldsymbol{r}_p)$ would not yield the true scattering at all. Rather, any improvement over the first BORN approximation must be based directly on (17.9), since $\lim_{r\to 0} \partial(r\psi_{n-1})/\partial r$ is not equal to ψ_{n-1} for $n \geqq 2$.

With respect to the validity of the expansions (17.8), we refer to existing calculations of the higher order terms. They show that the term ψ_2 gives a contribution of about 0.3% only to the cross section for scattering by the H_2O molecule[104] and the parahydrogen molecule[105]. We can, therefore, trust the approximation in which we calculate the cross section for the scattering of a very slow neutron by a bound proton, employing the pseudopotential $V'(\boldsymbol{r}_n - \boldsymbol{r}_p)$, (17.7), and confining ourselves to the first BORN approximation. This procedure is called the FERMI approximation[103].

If we regard the three-body problem relating to an incident neutron, a bound proton and the residual part of its molecule, whose coordinates are denoted by $\boldsymbol{r}_n$, $\boldsymbol{r}_p$ and $\boldsymbol{r}$, respectively, the SCHRÖDINGER equation reads

$$\left[-\frac{\hbar^2}{2M}\Delta_n - \frac{\hbar^2}{2M}\Delta_p - \frac{\hbar^2}{2(M_m - M)}\Delta + V'(\boldsymbol{r}_n - \boldsymbol{r}_p) + V(\boldsymbol{r}_p - \boldsymbol{r})\right]\psi = E\,\psi, \tag{17.11}$$

where $V'(\boldsymbol{r}_n - \boldsymbol{r}_p)$ is the pseudopotential, $V(\boldsymbol{r}_p - \boldsymbol{r})$ the potential by which the proton is bound and M_m the molecular mass including the proton. We introduce a relative coordinate $\bar{\boldsymbol{r}}'_p = \boldsymbol{r}_p - \boldsymbol{r}$, the c.m. coordinate $\boldsymbol{R}'$ of the whole molecule, further $\bar{\boldsymbol{r}}_n = \boldsymbol{r}_n - \boldsymbol{R}'$ and the c.m. coordinate $\boldsymbol{R}$ of the whole system; finally a modified relative coordinate of the proton $\bar{\boldsymbol{r}}_p = \bar{\boldsymbol{r}}'_p (M_m - M)/M_m$. (17.11) then transforms into

$$\left[-\frac{\hbar^2}{2(M + M_m)}\Delta_{\boldsymbol{R}} - \frac{\hbar^2}{2M'}\Delta_{\bar{n}} - \frac{\hbar^2}{2M''}\Delta_{\bar{p}} + V'(\bar{\boldsymbol{r}}_n - \bar{\boldsymbol{r}}_p) + V(\bar{\boldsymbol{r}}'_p)\right]\psi = E\,\psi, \tag{17.12}$$

where $M' = (M \cdot M_m)/(M + M_m)$ and $M'' = (M \cdot M_m)/(M_m - M)$.

Separating the c.m. motion of the whole system and introducing the eigenfunctions $u_k(\bar{\boldsymbol{r}}_p)$ of the internal motion of the proton inside the molecule, the first BORN approximation gives the differential scattering cross section $d\sigma_{if}$ of the neutron with the initial wave vector $\boldsymbol{k}_i$ into the solid angle $d\Omega$ around the final wave vector $\boldsymbol{k}_f$ (in the center-of-total-mass system) with the simultaneous

---

[103] E. FERMI: Ric. sci. 7, 13 (1936).
[104] G. BREIT and P. R. ZILSEL: Phys. Rev. 71, 232 (1947).
[105] B. A. LIPPMANN: Phys. Rev. 79, 481 (1950).

transition of the internal motion from $u_i(\overline{\boldsymbol{r}}_p)$ to $u_f(\overline{\boldsymbol{r}}_p)$:

$$d\sigma_{if} = \frac{v_f}{v_i}\left[\frac{2M_m}{M+M_m}\right]^2 a^2 |I_{if}|^2 d\Omega, \tag{17.13}$$

with

$$I_{if} = \int e^{i(\boldsymbol{k}_i-\boldsymbol{k}_f)\overline{\boldsymbol{r}}_p} u_f(\overline{\boldsymbol{r}}_p)\, u_i(\overline{\boldsymbol{r}}_p)\, d\overline{\boldsymbol{r}}_p, \tag{17.14}$$

where we have used (17.7) for $V'(\overline{\boldsymbol{r}}_n - \overline{\boldsymbol{r}}_p)$; v_i and v_f are the initial and final velocities of the neutron relative to the center-of-mass of the molecule.

Take the simplest case where the neutron energy approaches zero and the proton is initially in the ground state. Excitation is impossible and the neutron wavelength is large compared with the molecular dimensions, whence $v_i = v_f$ and $I_{if} = 1$. Thus we get the total cross section

$$\sigma_{ii} = 4\pi a^2 \left[\frac{2M_m}{M+M_m}\right]^2; \tag{17.15}$$

this tells us that the chemical binding makes the zero energy cross section about 4 times as large as that of a free proton, which is confirmed by experiment (Fig. 15). Another characteristic of this effect is that the angular distribution is modified, since the scattering is now isotropic in the c.m. system of neutron and molecule, rather than neutron and proton.

To proceed further, we must assume some special form for $V(\overline{\boldsymbol{r}}_p')$ and calculate the internal states of the molecule. Furthermore the distribution on these states as well as the thermal agitation of the molecules must be taken into account. For details the reader is referred to the original papers[102, 104, 105, 106] and a forthcoming treatise by S. BRIMBERG*.

Let us now come back to our original problem how to get at the free proton cross section σ_0 at zero energy. One might, for example, carry out the experiments in the energy region where the effect of the chemical binding is negligible. The most accurate measurement cited in Sect. 2, however, makes use of this chemical binding effect in the energy range where it just becomes appreciable. It is shown by PLACZEK[107] that if the free neutron-proton cross section σ_0 is constant and the incident neutron energy E is much larger than the energy associated with the hydrogen bond, then the neutron cross section σ for bound protons is given by

$$\sigma = \sigma_0\left(1 + \frac{\varepsilon}{4E}\right), \tag{17.16}$$

where ε is the mean vibrational energy of the bound proton. In the scattering experiment cited[8], this prediction was confirmed with hydrogen gas, gaseous n-butane and H_2O as scatterer in the energy range between 1 and 15 ev; σ was observed to rise much more steeply below 1 ev. The cited value $\sigma_0 = (20.36 \pm 0.10)$ barns in (2.1) was measured in this way. It is interesting to note that the main contribution to the gradual rise of σ in (17.16) is due to the quantum mechanical zero point energy of the hydrogen bond oscillator.

We have also to make sure that the cross section σ_0 can really be regarded as constant between 1 and 15 ev. To get an estimate, we use the theoretical predictions (15.14) and (16.1), which give

$$\sigma = \frac{4\pi}{k^2 + k^2\cot^2\delta_0} = 4\pi a^2\left[1 - \left(1 - \frac{r_0}{a}\right)k^2 a^2\right], \tag{17.17}$$

* On the Scattering of Slow Neutrons by Hydrogen Molecules, Thesis, Royal Institute of Technology, Stockholm, 1956.

106 M. HAMERMESH and J. SCHWINGER: Phys. Rev. 69, 145 (1946).

107 G. PLACZEK: Phys. Rev. 86, 377 (1952).

retaining the terms of the lowest order in k^2. The error comes out as 10^{-4} times $4\pi a^2$ at the highest energy 15 ev, if we use the singlet scattering data summarized at the end of Sect. 18; for the triplet states it is shown to be less than 10^{-5} times $4\pi a^2$. Comparing with the experimental error which amounts to $5 \cdot 10^{-3}$ times the cross section, according to (2.1), we can safely regard the empirical value (2.1) as the zero energy cross section.

If the scattering is independent of the spin, we can assess σ_0 through (16.23). As a crude estimate, we neglect the second term in (16.23), as the deuteron radius $1/\alpha$ is known to be larger than the effective range r_0. Then, putting $a \approx 1/\alpha = 4.3 \times 10^{-13}$ cm, we get $\sigma_0 \approx 2.3$ barns, which is much smaller than the observed value 20.36 barns. This great difference is explained as due to the spin dependence of the nuclear scattering. As (16.23) holds only in the triplet state, this rough value is nothing but the triplet zero energy cross section. In the actual measurement, we do not specify any spin orientations, which implies that the observed σ_0 must be a statistical average of singlet and triplet cross sections:

$$\sigma_0 = \tfrac{3}{4}\sigma_{0t} + \tfrac{1}{4}\sigma_{0s} = 3\pi a_t^2 + \pi a_s^2, \tag{17.18}$$

where we have used subscripts s and t to denote singlet and triplet quantities. From the observed σ_0 we can estimate that

$$|a_s| \approx 24.4 \times 10^{-13}\,\text{cm} \gg a_t \approx 4.3 \times 10^{-13}\,\text{cm},$$

where the sign of a_s cannot be determined by σ_0 alone.

In the previous discussion, we have neglected the above spin dependence. To amend this, it is convenient to use the projection operators P_t and P_s of the triplet and singlet states, respectively. It is readily seen from (10.4) that they are given by

$$P_t = \tfrac{1}{4}[3 + (\boldsymbol{\sigma}_n \cdot \boldsymbol{\sigma}_p)], \qquad P_s = \tfrac{1}{4}[1 - (\boldsymbol{\sigma}_n \cdot \boldsymbol{\sigma}_p)], \tag{17.19}$$

which have naturally the properties $P_t^2 = P_t$, $P_s^2 = P_s$ and $P_t P_s = 0$, where $\boldsymbol{\sigma}_n$ and $\boldsymbol{\sigma}_p$ are PAULI's spin operators of neutron and proton, respectively. The only modification we must make is to replace a in (17.7) by

$$a = a_s P_s + a_t P_t \tag{17.20}$$

and distinguish between different initial and final spin states in the matrix element I_{if} in (17.14).

18. Scattering length and coherent scattering. In the previous section we have tacitly assumed that there is only one proton in the molecule and that different molecules scatter incoherently. In practice, there may be more than one proton in the molecule and, like in the case of scattering by a solid substance, different protons can scatter coherently. As the distance between two protons in the hydrogen molecule is about 0.75×10^{-8} cm, we must use neutron wavelengths of this size or larger to detect the interference effect. This criterion demands a neutron energy below about 0.1 ev. Thus the interference effect is negligible in the energy range between 1 and 15 ev, where the actual measurement of σ_0 was performed. This argument at the same time shows that the theory developed in the previous section must be supplemented to account for the interference in the energy range below 0.1 ev.

As an example take the scattering of slow neutrons by hydrogen molecules. The SCHRÖDINGER equation is obtained from (17.11) simply by adding another pseudopotential $V'(\boldsymbol{r}_n - \boldsymbol{r})$, because there is another proton situated at $\boldsymbol{r}$, and putting $M_m = 2M$. The coordinate transformations used to derive (17.12) now give, observing $\boldsymbol{r}_n - \boldsymbol{r} = \overline{\boldsymbol{r}}_n + \overline{\boldsymbol{r}}_p$,

$$\left[-\frac{\hbar^2}{6M}\Delta_{\boldsymbol{R}} - \frac{3\hbar^2}{4M}\Delta_{\overline{n}} - \frac{\hbar^2}{4M}\Delta_{\overline{p}} + V'(\overline{\boldsymbol{r}}_n - \overline{\boldsymbol{r}}_p) + V'(\overline{\boldsymbol{r}}_n + \overline{\boldsymbol{r}}_p) + V(\overline{\boldsymbol{r}}'_p)\right]\psi = E\,\psi, \quad (18.1)$$

where $\boldsymbol{R}$ is the c.m. coordinate of the whole system, $\overline{\boldsymbol{r}}_n$ the coordinate of the neutron relative to the c.m. of the hydrogen molecule, $\overline{\boldsymbol{r}}'_p$ the distance between the two protons and $\overline{\boldsymbol{r}}_p = \overline{\boldsymbol{r}}'_p/2$. What we must do next is again to separate the c.m. motion of the total system, find the internal motion of the molecule under the potential $V(\overline{\boldsymbol{r}}'_p)$ and then calculate the scattering cross section using the standard BORN approximation. For details we again refer to the papers cited in the previous section.

We can, however, easily observe some characteristic points in this problem. The pseudopotential $V'(\boldsymbol{r})$, according to (17.7), (17.19), and (17.20), can be written

$$V'(\boldsymbol{r}) = \frac{4\pi\hbar^2}{M}\,\delta(\boldsymbol{r})\left[\frac{1}{4}(a_s + 3a_t) + \frac{1}{4}(a_t - a_s)(\boldsymbol{\sigma}_n\boldsymbol{\sigma}_p)\right], \quad (18.2)$$

which in turn gives the following expression for the scattering potential of the hydrogen molecule in (18.1)

$$\left.\begin{aligned} &V'(\overline{\boldsymbol{r}}_n - \overline{\boldsymbol{r}}_p) + V'(\overline{\boldsymbol{r}}_n + \overline{\boldsymbol{r}}_p) = \\ &= \frac{4\pi\hbar^2}{M}\left[\left\{\frac{1}{4}(a_s + 3a_t) + \frac{1}{4}(a_t - a_s)(\boldsymbol{\sigma}_n\boldsymbol{S}_H)\right\}\{\delta(\overline{\boldsymbol{r}}_n - \overline{\boldsymbol{r}}_p) + \delta(\overline{\boldsymbol{r}}_n + \overline{\boldsymbol{r}}_p)\} + \right. \\ &\quad \left. + \frac{1}{4}(a_t - a_s)(\boldsymbol{\sigma}_n\boldsymbol{S}'_H)\{\delta(\overline{\boldsymbol{r}}_n - \overline{\boldsymbol{r}}_p) - \delta(\overline{\boldsymbol{r}}_n + \overline{\boldsymbol{r}}_p)\}\right], \end{aligned}\right\} \quad (18.3)$$

where $\boldsymbol{S}_H = \frac{1}{2}(\boldsymbol{\sigma}_{p_1} + \boldsymbol{\sigma}_{p_2})$ is the total spin of the hydrogen molecule and $\boldsymbol{S}'_H = \frac{1}{2}(\boldsymbol{\sigma}_{p_1} - \boldsymbol{\sigma}_{p_2})$, $\boldsymbol{\sigma}_{p_1}$ and $\boldsymbol{\sigma}_{p_2}$ being the PAULI matrices of the two protons, respectively. In the case of the scattering of extremely slow neutrons whose wavelengths are much larger than the distance of the two protons in the molecule, the phase difference of the incident neutron wave between two protons becomes negligible, which means that the two δ-functions, $\delta(\overline{\boldsymbol{r}}_n - \overline{\boldsymbol{r}}_p)$ and $\delta(\overline{\boldsymbol{r}}_n + \overline{\boldsymbol{r}}_p)$, can be replaced by $\delta(\overline{\boldsymbol{r}}_n)$ if the origin is chosen somewhere near the hydrogen molecule. Thus the scattering potential (18.3) becomes

$$V'(\overline{\boldsymbol{r}}_n - \overline{\boldsymbol{r}}_p) + V'(\overline{\boldsymbol{r}}_n + \overline{\boldsymbol{r}}_p) \approx \frac{4\pi\hbar^2}{M}\left[2\left(\frac{1}{4}a_s + \frac{3}{4}a_t\right) + \frac{1}{2}(a_t - a_s)(\boldsymbol{\sigma}_n\boldsymbol{S}_H)\right]\delta(\overline{\boldsymbol{r}}_n). \quad (18.4)$$

Here we define the coherent (neutron-proton) scattering length f by

$$f = 2\left(\tfrac{1}{4}a_s + \tfrac{3}{4}a_t\right). \quad (18.5)$$

The differential scattering cross section becomes [cf. Eqs. (17.13) and (17.14)]

$$d\sigma = \tfrac{16}{9}\left[f + \tfrac{1}{2}(a_t - a_s)(\boldsymbol{\sigma}_n\boldsymbol{S}_H)\right]^2 d\Omega, \quad (18.6)$$

since for extremely slow neutrons $I_{if} = 1$ and $v_i = v_f$, if the molecule is initially in the ground state. Averaging (18.6) over the initial neutron spin directions, we finally get

$$d\sigma = \tfrac{16}{9}\left[f^2 + \tfrac{1}{4}(a_t - a_s)^2 S_H(S_H + 1)\right] d\Omega, \quad (18.7)$$

where S_H is the total spin quantum number of the hydrogen molecule.

It is well-known that there are two kinds of hydrogen; para- and ortho-hydrogen characterized by $S_H = 0$ and 1, respectively. Experimentally it is possible to prepare these separately at very low temperature, which allows the separate measurements of extremely slow neutron scattering by para- and orthohydrogen, respectively. From (18.7), these two cross sections are given by

$$\sigma_{\text{para}} = 4\pi \tfrac{16}{9} f^2 \quad \text{and} \quad \sigma_{\text{ortho}} = 4\pi \tfrac{16}{9} \left[f^2 + \tfrac{1}{2}(a_t - a_s)^2\right], \tag{18.8}$$

respectively. Using the values $|a_s| \approx 24.4 \times 10^{-13}$ cm $\gg a_t \approx 4.3 \times 10^{-13}$ cm obtained in the previous section, we have the following estimate: a positive a_s gives $\sigma_{\text{para}} \approx 78$ barns and $\sigma_{\text{ortho}}/\sigma_{\text{para}} \approx 1.6$, while a negative a_s predicts $\sigma_{\text{para}} \approx 7.6$ barns and the ratio ≈ 13.1. These values show a difference large enough to allow experimental determination of the correct sign. In this way the negative sign of a_s has long been established.

There is one particular thing which we must refine in the simple theoretical prediction (18.7), since the assumption that the molecule is initially in its ground state is true for the parahydrogen but not for the orthohydrogen. In the latter case we must take into account that the neutron can gain energy by a transition from ortho- to parahydrogen. This process gives a value of v_f/v_i in the cross section (17.13), which is the larger, the more the neutron energy decreases. Besides we must pay attention to the radiative neutron-proton capture process in which a deuteron is formed, the effect of the small phase difference of a neutron wave travelling between two protons, as the neutron energy is not exactly zero, and the velocity distribution of the gas molecules which becomes more important as the neutron energy approaches zero. The details of these effects cannot be discussed here. Anyway, owing to the experimental difficulty in determining ortho-contamination of parahydrogen, measurements of para- and orthohydrogen cross sections have not been able to determine the scattering lengths a_s and a_t with the same precision as the liquid mirror reflection mentioned below.

We then consider the coherent scattering by a solid substance containing hydrogen. In this case all the protons are virtually fixed and their effective masses are infinite. Thus the effective scattering length is twice that of a free proton as is seen from (17.13) or (17.15). As all the nuclei have random spin orientation, the effective coherent scattering length is eventually twice the statistical average of a_t and a_s, which is just f defined by (18.5). Thus, the measurement of the coherent scattering in crystals containing hydrogen can determine the coherent scattering length f. However, such a measurement [108] can at present determine f somewhat less accurately than experiments on para- and orthohydrogen cross sections[109]. The main difficulty is the uncertainty of the correction for the thermal vibration of the crystal, which reduces the observed coherent scattering.

It was suggested by HAMERMESH[110] that f can be measured by observing the critical angle of the total reflection of very slow neutrons upon a hydrogen mirror. In order to discuss this problem, we must first consider the behaviour of the neutron wave function inside a crystal. It is a combination of the incident neutron wave $e^{i\boldsymbol{k}\boldsymbol{r}}$, where $\boldsymbol{k}$ is the wave vector of the incident neutron, and the scattered wave which has the form indicated in (15.1), $\sum_i f_i(\vartheta, \varphi)\, e^{ik|\boldsymbol{r}-\boldsymbol{r}_i|}/|\boldsymbol{r}-\boldsymbol{r}_i|$, where $\boldsymbol{r}_i$ is the coordinate of the i-the scatterer. For very slow neutrons $f(\vartheta, \varphi)$ is, according to (15.12), given by $e^{i\delta_0} \sin\delta_0/k$. Following (16.1), at extremely low energies δ_0 approaches 180° if $a > 0$ and 0° if $a < 0$. In both cases $f(\vartheta, \varphi)$ is

[108] C. G. SHULL, E. O. WOLLAN, G. A. MORTON and W. L. DAVIDSON: Phys. Rev. **73**, 842 (1948).

[109] R. B. SUTTON, T. HALL, E. E. ANDERSON, H. S. BRIDGE, J. W. DE WIRE, L. S. LAVATELLI, E. A. LONG, T. SNYDER and R. W. WILLIAMS: Phys. Rev. **72**, 1147 (1947). — G. L. SQUIRES and A. T. STEWART: Proc. Roy. Soc. Lond., Ser. A **230**, 19 (1955).

[110] M. HAMERMESH: Phys. Rev. **77**, 140 (1950).

shown to approach $-a$. In crystal scattering, however, we must use f instead of a, as was just remarked. For the neutron wave function $\psi(\boldsymbol{r})$ inside the crystal we assume[111]

$$\psi(\boldsymbol{r}) = e^{i\boldsymbol{k}'\cdot\boldsymbol{r}} \cdot \chi(\boldsymbol{r}), \tag{18.9}$$

where $\chi(\boldsymbol{r})$ is periodic with respect to the lattice points, and behaves as $1 - \frac{f}{|\boldsymbol{r}-\boldsymbol{r}_i|}$ near each lattice point. Our aim is to determine $\boldsymbol{k}'$.

The differential equation satisfied by $\psi(\boldsymbol{r})$ is [cf. (17.6) and (17.7)]

$$(\varDelta + k^2)\,\psi(\boldsymbol{r}) = 4\pi f \cdot \sum_i \delta(\boldsymbol{r}-\boldsymbol{r}_i) \cdot e^{i\boldsymbol{k}'\cdot\boldsymbol{r}_i}. \tag{18.10}$$

We can expand the periodic functions $\chi(\boldsymbol{r})$ and $\sum_i \delta(\boldsymbol{r}-\boldsymbol{r}_i)$ as

$$\left.\begin{aligned} \chi(\boldsymbol{r}) &= \sum_{\boldsymbol{l}} \chi_{\boldsymbol{l}} \cdot e^{i\boldsymbol{l}\cdot\boldsymbol{r}}, \\ \sum_i \delta(\boldsymbol{r}-\boldsymbol{r}_i) &= \sum_{\boldsymbol{l}} \delta_{\boldsymbol{l}} \cdot e^{i\boldsymbol{l}\cdot\boldsymbol{r}}, \end{aligned}\right\} \tag{18.11}$$

where the summation is to be carried out over all vectors $\boldsymbol{l}$, the components of which are zero or integer multiples of $2\pi/d$, d being the lattice constant. It is easily shown that $\delta_{\boldsymbol{l}} = N$, where N is the number of lattice points or scattering centers per cm³. Inserting (18.9) and (18.11) in (18.10), we readily get

$$\chi_{\boldsymbol{l}} = \frac{4\pi N f}{k^2 - (\boldsymbol{k}'+\boldsymbol{l})^2}, \quad \chi(\boldsymbol{r}) = 4\pi N f \sum_{\boldsymbol{l}} \frac{e^{i\boldsymbol{l}\cdot\boldsymbol{r}}}{k^2 - (\boldsymbol{k}'+\boldsymbol{l})^2}. \tag{18.12}$$

The boundary condition for $\chi(\boldsymbol{r})$ near every lattice point is

$$\chi(\boldsymbol{r}) + \frac{f}{|\boldsymbol{r}-\boldsymbol{r}_i|} \xrightarrow[\boldsymbol{r}\to\boldsymbol{r}_i]{} 1, \quad \text{or} \quad \chi(\boldsymbol{r}) + 4\pi N f \sum_{\boldsymbol{l}}{}' \frac{e^{i\boldsymbol{l}\cdot\boldsymbol{r}}}{l^2} \xrightarrow[\boldsymbol{r}\to\boldsymbol{r}_i]{} 1, \tag{18.13}$$

where ′ denotes omission of $\boldsymbol{l}=0$ in the summation. Then (18.12) and (18.13) give

$$4\pi N f \left\{ \frac{1}{k^2 - k'^2} + \sum_{\boldsymbol{l}}{}' \left[\frac{1}{k^2 - (\boldsymbol{k}'+\boldsymbol{l})^2} + \frac{1}{l^2} \right] \right\} = 1, \tag{18.14}$$

where we have put $\boldsymbol{r} = \boldsymbol{r}_i$, as $\sum'$ is convergent. Here we are interested only in such cases where $\boldsymbol{k}^2 \approx \boldsymbol{k}'^2 \approx 0$ and all nonvanishing vectors $\boldsymbol{l}$ are much larger than $\boldsymbol{k}$ and $\boldsymbol{k}'$. Further remarking that the index of refraction n of the crystalline medium is defined by $k' = n k$, we get[111] from (18.14), neglecting $\sum'$ compared to $\frac{1}{k^2 - k'^2}$,

$$n^2 = 1 - \frac{4\pi N f}{k^2}, \quad \text{or} \quad n = 1 - \frac{2\pi N f}{k^2}, \tag{18.15}$$

if $n \approx 1$. Thus it has been shown that, if f is positive, n is smaller than unity, which allows total reflection, the critical angle ϑ being given from (18.15) by

$$\vartheta = \lambda \sqrt{\frac{N f}{\pi}}, \tag{18.15a}$$

where we have assumed $n \approx 1$ and introduced the incident neutron wavelength $\lambda = 2\pi/k$. In most cases $n-1$ is of the order of 10^{-6} and the critical angle ϑ is in the region of 10′[112]. Anyway the presence or absence of total reflection can readily determine the sign of the coherent scattering length f.

---

111 M. L. GOLDBERGER and F. SEITZ: Phys. Rev. **71**, 294 (1947).
112 E. FERMI and L. MARSHALL: Phys. Rev. **71**, 666 (1947).

It is already known that a_s is negative and $|a_s| \gg a_t$, which means that the hydrogen coherent scattering length f_H is negative, so a crystal containing only hydrogen atoms cannot produce total reflection. We can, however, use hydrocarbon substances, for example tri-ethyl benzene ($C_{12}H_{18}$), because the coherent scattering length of a carbon nucleus f_C is positive and accurately known and $12 f_C + 18 f_H$ is positive, though small. The critical angle is given by (18.15a), in which f is now replaced by $12 f_C + 18 f_H$ and N is the number of the above molecules per cm³. The incident neutron wavelength λ can also be determined by total reflection using a crystal containing nuclei with known f values. Thus the experiment has been performed with neutron wavelengths from 8 to 15 Å, employing liquid hydrocarbons as reflectors, which led to the accurate value $f_H = -(3.78 \pm 0.02) \times 10^{-13}$ cm, cited in (2.2).

Now we can determine the two scattering lengths a_s and a_t from (2.1) and (2.2): $\sigma_0 = \pi a_s^2 + 3\pi a_t^2 = 20.36\,(1 \pm 0.005)$ barns and $f = 2(\frac{1}{4} a_s + \frac{3}{4} a_t) = -3.78\,(1 \pm 0.005) \times 10^{-13}$ cm, which give

$$a_t = (5.377 - 2.05\,\varepsilon_f + 3.72\,\varepsilon_\sigma) \times 10^{-13}\text{ cm}$$

and

$$a_s = (-23.69 - 1.40\,\varepsilon_f - 11.15\,\varepsilon_\sigma) \times 10^{-13}\text{ cm},$$

where ε_f and ε_σ are the fractional errors of f and σ_0, respectively. Thus we get*

$$a_t = (5.377 \pm 0.021) \times 10^{-13}\text{ cm} = 5.377\,(1 \pm 0.0040) \times 10^{-13}\text{ cm}, \tag{18.16}$$

$$a_s = (-23.69 \pm 0.055) \times 10^{-13}\text{ cm} = -23.69\,(1 \pm 0.0023) \times 10^{-13}\text{ cm}, \tag{18.17}$$

or

$$\left.\begin{aligned} 1/a_t &= 0.1860\,(1 \pm 0.0040) \times 10^{13}\text{ cm}^{-1}, \\ 1/a_s &= -0.04221\,(1 \pm 0.0023) \times 10^{13}\text{ cm}^{-1}. \end{aligned}\right\} \tag{18.18}$$

The above values give the zero energy cross sections

$$\sigma_{0t} = 4\pi a_t^2 = 3.63\,(1 \pm 0.0080)\text{ barns}$$

and

$$\sigma_{0s} = 4\pi a_s^2 = 70.52\,(1 \pm 0.0046)\text{ barns}.$$

19. Determination of effective ranges. *α) Triplet effective range.* The orthodox way of determining the effective range r_0 is to analyse the low energy scattering data summarized in Table 1 in terms of the shape independent expansion (16.1). This is, however, not so suitable for the neutron-proton scattering data because there are two spin states involved; moreover the data are at present neither abundant nor accurate enough. We rather start from the theoretical prediction (16.19) which is fit to determine the mixed effective range $\varrho(-\varepsilon, 0)$ since α is known much more accurately than the scattering data; from its definition (13.4) $\alpha^2 = \varepsilon M/\hbar^2$** and (1.1), we have

$$1/\alpha = 4.3157\,(1 \pm 0.0005) \times 10^{-13}\text{ cm}, \quad \alpha = 0.23171\,(1 \pm 0.0005) \times 10^{13}\text{ cm}^{-1}. \tag{19.1}$$

* *Note added in proof.* Brimberg, starting from his theoretical analysis of the scattering of slow neutrons by hydrogen molecules (see footnote * p. 46), has reexamined the results of Melkonian[8], using the f-value of Burgy, Ringo and Hughes[9], and found

$$a_t = (5.415 \pm 0.012) \times 10^{-13}\text{ cm},$$
$$a_s = -(23.806 \pm 0.028) \times 10^{-13}\text{ cm},$$

corresponding to $\sigma_0 = (20.57 \pm 0.04)$ barns. This would also change the value of $\varrho(-\varepsilon, 0)$ to $(1.752 \pm 0.020) \times 10^{-13}$ cm.

** M is twice the neutron-proton reduced mass.

From (16.19) and (18.16), we then get*

$$\left.\begin{aligned}\varrho(-\varepsilon, 0) &= (1.704 + 2.61\,\delta_\varepsilon - 2.64\,\delta_f + 4.79\,\delta_\sigma)\times 10^{-13}\,\text{cm}\\ &= (1.704 \pm 0.028)\times 10^{-13}\,\text{cm} = 1.704\,(1 \pm 0.016)\times 10^{-13}\,\text{cm},\end{aligned}\right\}\tag{19.2}$$

where δ_ε, δ_f and δ_σ are the fractional errors of ε, f and σ_0, respectively. Since ε is known much more accurately than f and σ_0, the latter two are the main sources of error in $\varrho(-\varepsilon, 0)$, being comparably effective.

In order to determine the triplet effective range r_{0t}, let us use (16.34), from which

$$\varrho(-\varepsilon, 0) = r_{0t} + 2P_t\,\alpha^2 r_{0t}^3,\tag{19.3}$$

where P_t is the triplet shape dependent parameter. We first put $P_t=0$, which makes $r_{0t}=\varrho(-\varepsilon, 0)$ and gives a corrected value of P_t from the curves in Fig. 13 for individual potentials since a_t is known. This P_t then determines an improved value of r_{0t} by (19.3) and so forth. The iteration procedure converges very rapidly, for in most cases P_t is practically unaffected by a small change in r_{0t}. Having determined r_{0t} and P_t, we can obtain the well depth parameter s_t and the intrinsic range b_t of the triplet potential by means of the curves in Figs. 11 and 12 and/or some interpolation formulas obtained by BLATT and JACKSON[100]. The results are summarized in Table 9 for three types of potential, and the fractional error of r_{0t} ($\approx 1.6\%$) is the same as that of $\varrho(-\varepsilon, 0)$. The table also contains the corresponding meson mass value for the YUKAWA well, which agrees with the empirical π-meson mass $(273 \pm 0.8)\,m_e$[113]. This fact, however, cannot be taken too seriously, since we have neglected the tensor force in the above analysis. It is to be noted that P_t is exceptionally large for the YUKAWA well (cf. Sect. 16). This makes the effective range expansion (19.3) relatively inaccurate for the YUKAWA well.

Table 9. *The effective range, shape dependent parameter and potential shape parameters in the triplet neutron-proton system, determined from zero energy scattering data for three types of potential.*

Well shape	r_{0t} (10^{-13} cm)	b_t (10^{-13} cm)	s_t	P_t	Meson mass
Square well	1.726 ± 0.028	2.043 ± 0.038	1.441 ± 0.009	-0.040	
Shape independent	1.704 ± 0.028			0	
Exponential	1.689 ± 0.027	2.351 ± 0.053	1.417 ± 0.010	$+0.029$	
YUKAWA	1.639 ± 0.026	2.919 ± 0.121	1.420 ± 0.017	$+0.137$	$(280.4 \pm 11.6)\,m_e$

As regards the accuracy of the expansion (19.3) for the YUKAWA well, we quote a more accurate calculation[114] according to which the contribution from all the neglected higher terms on the right hand side of (19.3) never exceeds about 20% of the second term $2P_t\alpha^2 r_{0t}^3$ which amounts to 0.065×10^{-13} cm in the above analysis. This implies that the error of (19.3) is only 0.013×10^{-13} cm, which is half the corresponding experimental error in (19.2). It can, therefore, be said that the expansion (19.3) is not so very accurate for the YUKAWA potential, but its error is still smaller than the experimental error of the effective range.

* See footnote *, p. 51.

113 F. M. SMITH, W. BIRNBAUM and W. H. BARKAS: Phys. Rev. **91**, 765 (1953).

114 L. HULTHÉN and K. V. LAURIKAINEN: Rev. Mod. Phys. **23**, 1 (1951). — L. HULTHÉN and S. SKAVLEM: Phys. Rev. **87**, 297 (1952). — L. HULTHÉN and B. NAGEL: Phys. Rev. **90**, 62 (1953).

With respect to the effect of a hard core upon P_t, there are calculations[12,*] based on "displaced" potentials the radial behavior of which is given by

$$\left.\begin{array}{ll} \text{Square well} \propto \text{const.}, & \text{Gauss} \propto \exp\{-\mu^2(r-r_c)^2\}, \\ \text{Exponential} \propto \exp\{-\mu(r-r_c)\}, & \text{Yukawa} \propto \exp\{-\mu(r-r_c)\}/\mu(r-r_c), \end{array}\right\} \quad (19.4)$$

for $r > r_c$ = radius of the hard core. For a given r_c, we first determine their well depth and force range parameters so as to give a scattering length and effective range fitted to the experimental data. It is then possible to calculate the shape dependent parameter as a function of r_c. The calculations have been

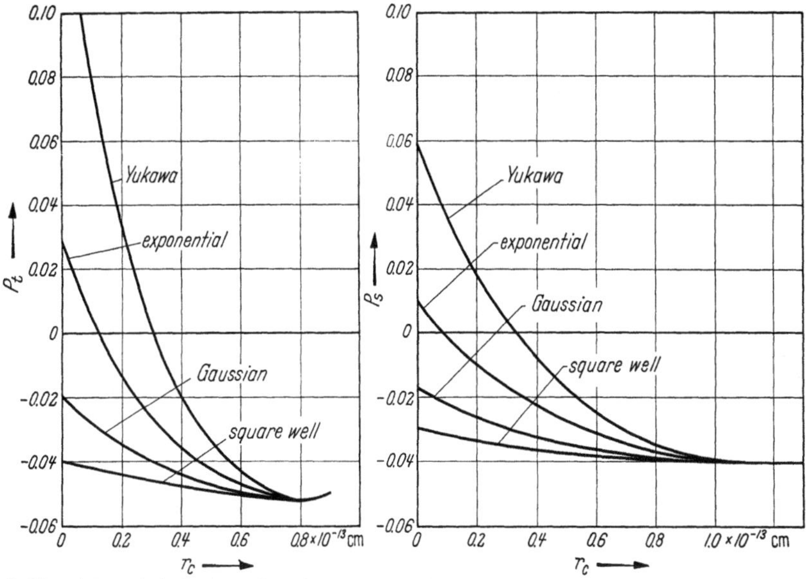

Fig. 16. The triplet and singlet shape dependent parameters P_t and P_s as functions of a hard core radius r_c for various types of potentials (19.4), fitted to $a_t = 5.28 \times 10^{-13}$ cm, $r_{0t} = 1.56 \times 10^{-13}$ cm, $a_s = -23.68 \times 10^{-13}$ cm and $r_{0s} = 2.6 \times 10^{-13}$ cm.

made for triplet and singlet systems (the latter will be treated later in this article), for which the values $a_t = 5.28 \times 10^{-13}$ cm, $r_{0t} = 1.56 \times 10^{-13}$ cm and $a_s = -23.68 \times 10^{-13}$ cm, $r_{0s} = 2.6 \times 10^{-13}$ cm are assumed. The results are plotted in Fig. 16. It is seen that the effect of a hard core on P_t and P_s is quite large; even a small core radius ($\approx 0.3 \times 10^{-13}$ cm) can reduce the large shape dependent parameter of the Yukawa potential to zero.

For other types of potential, in particular such as the singular (12.6), there are no calculations of the corresponding P_t value. Of course, if a certain potential should predict the same P_t for relevant values of r_{0t} and a_t as one of the potentials in Table 9, then the same r_{0t} (but not necessarily b_t and s_t) would be obtained for both: In determining r_{0t}, the essential point is not the particular well shape but the connection between P_t and the two parameters r_{0t} and a_t. Although the particular potentials treated in Table 9 have no direct physical meaning, as stressed in Sect. 12 (tensor force neglected, for instance), Table 9 can still have some sense because of this fact, in particular as it is shown in Sects. 30 and 31 that the essential effect of a tensor force upon the low energy data is to modify the relation between P_t and the two parameters r_{0t} and a_t.

* M. A. Preston and R. L. Preston: Private communication. We are indebted to Dr. M. A. Preston for permission to use these results. — *References added in proof.* J. R. Bird and M. A. Preston: Canad. J. Phys. **33**, 399 (1955). — J. Shapiro and M. A. Preston: Canad. J. Phys. **34**, 451 (1956).

All the potentials determined in Table 9 can equally well account for the deuteron binding energy and the zero energy scattering data (f and σ_0) within the accuracy of the expansion (19.3). Since different potentials predict different r_{0t}, they can in principle be distinguished by low energy scattering. It is, however, seen in Table 9 that owing to rather large experimental errors all r_{0t} are partially overlapping each other. Only the two extremes, square well and YUKAWA potential, are separated and could in principle be distinguished by sufficiently accurate low energy data. This is discussed in the next subsection.

β) Singlet effective range. We now proceed to the analysis of the low energy data summarized in Table 1 of Sect. 2. The total cross section σ can be obtained from (15.14) and (16.15), assuming pure S-wave scattering,

$$\sigma = \frac{3\pi}{k^2 + \left[\frac{1}{a_t} - \frac{1}{2}\varrho_t(0, E)\,k^2\right]^2} + \frac{\pi}{k^2 + \left[\frac{1}{a_s} - \frac{1}{2}\varrho_s(0, E)\,k^2\right]^2}, \tag{19.5}$$

where k^2 is related to the incident neutron energy E in laboratory system by $k^2 = (2M_n E/\hbar^2)\,(M_p/(M_n+M_p))^2$, where M_n and M_p are neutron and proton mass, respectively. If we further employ the expansion (16.31), we may put

$$\varrho_t(0, E) = r_{0t}(1 - 2P_t r_{0t}^2 k^2), \qquad \varrho_s(0, E) = r_{0s}(1 - 2P_s r_{0s}^2 k^2). \tag{19.6}$$

As we have determined a_t, r_{0t}, P_t and a_s, the data in Table 1 can now be used to fix the two remaining parameters r_{0s} and P_s, again using an iteration procedure together with the curves in Fig. 13 for individual potentials. The results are given in Table 10. For an energy as large as 14.1 Mev, our method is too simple, since higher waves (P, D ...) must be considered and the expansion (19.6) tends to fail, at least for the YUKAWA well. SNOW[115] has analysed this case, taking these points into account and assuming $\varepsilon = (2.225 \pm 0.002)$ Mev, with the following result

$$\left.\begin{array}{llll} & \text{Square well} & \text{Exponential} & \text{YUKAWA} \\ r_{0s} = & 2.22 \pm 0.24\times 10^{-13}\,\text{cm}, & 2.24 \pm 0.33\times 10^{-13}\,\text{cm}, & 2.11 \pm 0.40\times 10^{-13}\,\text{cm}. \end{array}\right\} \tag{19.7}$$

Table 10. *The effective range and shape dependent parameter in the singlet neutron-proton system determined from low energy data for three types of potential.* r_{0t} *is in* 10^{-13} *cm.*

Incident neutron energy E		1.005 Mev	1.315 Mev	2.540 Mev	4.749 Mev	Weighted mean
σ (barns)		4.228 ± 0.018	3.675 ± 0.016	2.525 ± 0.009	1.690 ± 0.007	
Square well	r_{0s}	2.540 ± 0.245	2.324 ± 0.253	2.509 ± 0.215	2.467 ± 0.195	2.47 ± 0.11
	P_s	-0.03	-0.03	-0.03	-0.03	-0.03
Shape independent	r_{0s}	2.494 ± 0.245	2.276 ± 0.253	2.435 ± 0.215	2.368 ± 0.200	2.40 ± 0.11
	P_s	0	0	0	0	0
Exponential	r_{0s}	2.498 ± 0.245	2.280 ± 0.253	2.409 ± 0.215	2.294 ± 0.200	2.37 ± 0.11
	P_s	$+0.01$	$+0.01$	$+0.01$	$+0.01$	$+0.01$
YUKAWA	r_{0s}	2.456 ± 0.245	2.196 ± 0.253	2.235 ± 0.217	2.011 ± 0.204	2.21 ± 0.11
	P_s	$+0.057$	$+0.058$	$+0.058$	$+0.058$	$+0.058$
$\Delta'\sigma$ (barns)		0.014	0.012	0.008	0.005	
σ_t (barns)		2.190	2.064	1.675	1.238	
σ_s (barns)		2.038	1.611	0.850	0.453	

[115] G. SNOW: Phys. Rev. **87**, 21 (1952).

In Table 10, σ_t and σ_s are the triplet and singlet contributions to σ, in the shape independent approximation, which are modified by at most 3% for other well shapes; $\Delta'\sigma$ is the uncertainty of σ due to the errors of a_t, a_s and r_{0t} only. It is seen that $\Delta'\sigma$ is slightly smaller than the corresponding experimental error of σ, which shows that all the scattering data (zero and low energy ones) are now about equally accurate as far as their relation to the scattering parameters is concerned.

According to the effective range theory, data at different energies must be explained by the same r_{0s}; this requirement could in principle be used to find out such potentials as are compatible with low energy scattering. However, the values of r_{0s} always overlap within the estimated errors at the five energies in Table 10 and (19.7), for all potentials investigated. Thus we can say that the present low energy data are not accurate enough to fix any potential shape*.

We note that the errors of r_{0s} in Table 10 and (19.7) are much larger than those of r_{0t} in Table 9, which clearly shows that the low energy scattering data alone do not suffice to determine the effective range more accurately than the method employed in the previous subsection. The latter cannot, however, be applied to r_{0s}. It is added that in Table 10 the particular potential shapes are not important; it is the relation between P_s and the two parameters r_{0s} and a_s that is essential in utilizing Tables 10 and 11. As regards the approximation inherent in (19.6), it is certainly unimportant since P_s is smaller than P_t and the maximum energy in Table 10 corresponds to about 2.4 Mev in the c.m. system; furthermore the error of r_{0s} is much larger than that of $\varrho(-\varepsilon, 0)$, which was discussed in the previous subsection.

The standard error of the weighted average of r_{0s} in Table 10 is considered too small compared with the overlapping experimental errors. For safety we take instead doubled standard errors which are 0.23×10^{-13} cm for all cases in Table 10. The corresponding potential shape parameters b_s and s_s can then be determined by means of the curves in Figs. 11 and 12 and/or some interpolation formulas obtained by Blatt and Jackson[100]. The results are summarized in Table 11, together with r_{0s} and P_s from Table 10. The corresponding meson mass value is also included for the Yukawa well. If we compare Tables 9 and 11, the following points are observed; P_s is generally smaller than P_t, while r_{0s} is larger than r_{0t}, which does not imply that b_s is larger than b_t. As regards s_t and s_s, we can say that they are insensitive to the potential shape and definitely fulfill $s_t > 1 > s_s$, the implication of which is discussed in the next section. It is also noted that the error of r_{0s} is nearly ten times as large as that of r_{0t}.

Table 11. *The effective and intrinsic ranges, well depth and shape dependent parameters in the singlet neutron-proton system determined from low energy scattering data for three potential types.*

Well shape	r_{0s} (10^{-13} cm)	b_s (10^{-13} cm)	s_s	P_s	Meson mass
Square well . . .	2.47 ± 0.23	2.37 ± 0.22	0.925 ± 0.006	-0.03	
Shape independent	2.40 ± 0.23			0	
Exponential . .	2.37 ± 0.23	2.23 ± 0.22	0.935 ± 0.006	$+0.01$	
Yukawa	2.21 ± 0.23	2.06 ± 0.21	0.949 ± 0.005	$+0.058$	$(398.2\pm41.5)\,m_e$

* This statement must not be confused with the previous assertion, relating to the shape independent approximation, that the combined knowledge of the deuteron binding energy and the low energy scattering data cannot in principle distinguish between the potential shapes (Sect. 16). We are now analysing the data, not in the shape independent approximation (16.1), (16.22) and (16.23), but retaining the shape dependent parameters P_t and P_s in (19.3) and (19.6), respectively.

The "meson mass" in Table 11 is certainly much larger than the empirical value $(273 \pm 0.8)\, m_e$[113]. This situation is not so surprising since recent high energy scattering data suggest a hard core to be introduced in singlet even states of a neutron-proton system (Sect. 44), which makes it natural to accept such singular potentials as are predicted by the present meson field theory (Sect. 11). In fact the potential (12.6) with $n \geq 2$ has a smaller actual range than the YUKAWA potential (12.4) even though the same μ is used. The above discrepancy can, therefore, be regarded as an evidence for the meson theoretical nuclear potential characterized by a singular radial dependence. Here it is interesting to quote a paper[116] which employs a singular potential (12.6) with $n=2$ to calculate singlet scattering parameters, adjusted to $a_s = -23.69 \times 10^{-13}$ cm:

r_c (10^{-13} cm)	V_0 (Mev)	r_{0s} (10^{-13} cm)	$1/\mu$ (10^{-13} cm)	Meson mass (m_e)
0.014	3.50	2.33	2.60	148.5
0.007	3.07	2.35	2.70	143.0

(19.8)

We note that the "meson mass" values in (19.8) are only one half of the π-meson mass. These calculations suggest that the meson theoretical nuclear potential which is a linear combination of YUKAWA and singular potentials (12.6) with $n \geq 2$ can give correct singlet scattering parameters, even though its meson mass value is fixed as $273\, m_e$, if we adjust the coupling constant and the hard core radius properly. This is quantitatively shown in Sect. 33 [see the results denoted by (33.10)].

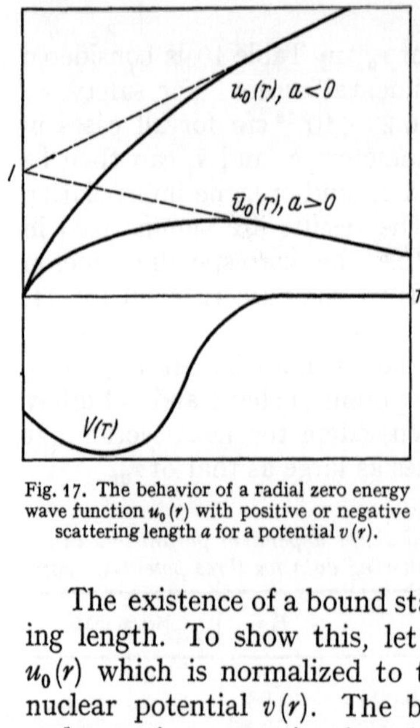

Fig. 17. The behavior of a radial zero energy wave function $u_0(r)$ with positive or negative scattering length a for a potential $v(r)$.

20. Virtual ^1S-state of deuteron. In Sect. 14 it was shown that there is only one bound state in the triplet neutron-proton system.

The absence of any bound ^1S-state can now be readily concluded from the finding in the previous section that s_s is smaller than 1; it follows immediately from the definition of s in Sect. 16. With respect to states of higher orbital angular momentum, we recall the general evidence that forces in odd states seem to be weaker than in even states and may have opposite sign (Sect. 11), not to mention the effect of the centrifugal force. Thus we can infer that there are no bound states in the singlet neutron-proton system.

The existence of a bound state can also be decided from the sign of the scattering length. To show this, let us consider a radial zero energy wave function $u_0(r)$ which is normalized to the asymptotic form $\bar{u}_0(r)$ given by (16.29) for a nuclear potential $v(r)$. The behavior of $u_0(r)$ is shown in Fig. 17 for positive and negative scattering length a. We need only observe that the curvature of the radial wave function corresponding to a negative energy is smaller than that of $u_0(r)$ inside the nuclear force range, which means that bound states are possible for a positive but not for a negative a*.

[116] M. A. PRESTON and J. SHAPIRO: Phys. Rev. 96, 813 (1954).

* The argument, of course, breaks down if the wave function has one or more nodes, which implies two or more bound states.

In Sect. 17, we mentioned that the singlet scattering length a_s must be much larger in magnitude than the triplet one a_t, in order to explain the zero energy scattering cross section σ_0, although the sign of a_s is not decided by σ_0 alone. A large value of $|a_s|$ means that $u_0(r)$ approaches a nearly horizontal line outside the nuclear force range. If a_s should be positive, then a bound level with almost zero binding energy would be possible in the singlet even state. If this should be the case, we could explain a large singlet scattering cross section as due to a kind of resonance with the loosely bound state. Now a_s is negative, and there is no bound state, but the large singlet scattering is sometimes ascribed to the so-called virtual ^1S-state of the deuteron, with an energy just above zero (0.0382 Mev). For a detailed discussion we refer to an article by MA[116a].

c) Proton-proton scattering.

Before entering on the analysis of the slow proton-proton scattering, we briefly mention some characteristic differences between slow proton-proton and neutron-proton scattering. First of all, the experimental accuracy is much higher than in the neutron-proton case, because charged particles are far easier to detect than neutral particles. Thus the proton-proton scattering parameters are much more accurately known. Secondly, a two-proton system must be spin-singlet in even states, spin-triplet in odd states, to make the wave function as a whole antisymmetric. In analysing slow proton-proton data, we have, therefore, singlet scattering only*. This greatly simplifies the analysis. Furthermore, since a tensor force vanishes identically in the spin singlet state, the results of the following sections are exact and need not be modified by the introduction of a tensor force in the next chapter. Naturally the analysis of the proton-proton scattering is complicated by the presence of the COULOMB force, but its action is very well known and causes some complication but no ambiguity. The necessary modifications of the scattering theory developed in Sects. 15 and 16 are given below.

21. Phase shift analysis of combined COULOMB and nuclear scattering. The basic differential equation (13.1) is generalized for a two-proton system to

$$\left[-\frac{\hbar^2}{M}\Delta + V(r) + \frac{e^2}{r}\right]\psi(\mathbf{r}) = E\,\psi(\mathbf{r}), \tag{21.1}$$

where M is exactly the proton mass, e the elementary charge, E the relative kinetic energy which is just half the incident proton energy in the laboratory system, and $V(r)$ is the nuclear potential which is now assumed to be central. As the resultant force is central as before, we can expand $\psi(\mathbf{r}) = \psi(r, \vartheta)$ in the same way as (15.4):

$$\psi(r, \vartheta) = \sum_{l=0}^{\infty} \frac{u_l(r)}{r} P_l(\cos\vartheta), \tag{21.2}$$

where $u_l(r)$ satisfies the differential equation

$$\left[\frac{d^2}{dr^2} + \left(k^2 + v(r) - \frac{1}{\varrho r} - \frac{l(l+1)}{r^2}\right)\right] u_l(r) = 0, \tag{21.3}$$

with $k^2 = ME/\hbar^2$, $v(r) = -MV(r)/\hbar^2$ and $\varrho = \hbar^2/(Me^2) = 2.88 \times 10^{-12}$ cm, and the boundary condition $u_l(0) = 0$. It can easily be shown that the asymptotic solution

* More correctly we must say that there is no triplet nuclear scattering in the low energy region, since singlet as well as triplet COULOMB scattering is effective down to zero energy because of the long range of the force.

[116a] S. T. MA: Rev. Mod. Phys. **25**, 853 (1953).

of (21.3) is a linear combination of $\sin(kr-\eta\log 2kr)$ and $\cos(kr-\eta\log 2kr)$, where $\eta=1/(2k\varrho)$. The appearance of this logarithmic term in the argument is characteristic of the COULOMB scattering. In the pure COULOMB case, it is known[117] that the solution of (21.3) which vanishes at $r=0$ behaves asymptotically as

$$u_l(r) \xrightarrow[r\to\infty]{} (2l+1)\, i^l \frac{e^{i\sigma_l}}{k} \sin\left(kr - \frac{1}{2} l\pi - \eta\log 2kr + \sigma_l\right), \tag{21.4}$$

where σ_l is the COULOMB phase shift, a real number given by $\exp(2i\sigma_l) = \Gamma(l+1+i\eta)/\Gamma(l+1-i\eta)$. For convenience we have put the same factors in (21.4) as in (15.9) and (15.11).

Since the nuclear force has a finite range, the addition of $v(r)$ cannot modify the radial dependence of the asymptotic form (21.4) but only introduces a real phase shift δ_l into the argument. This nuclear phase shift should not be confused with the COULOMB phase shift σ_l. The requirement that the total wave function (21.2) must consist of incident and scattered waves is easily shown to be satisfied by the choice of (21.4) in which σ_l is replaced by $\sigma_l+\delta_l$, using the expansion (15.8). The result is

$$\psi(\boldsymbol{r}) \xrightarrow[r\to\infty]{} [I(\boldsymbol{r}) + f(\vartheta)\, S(\boldsymbol{r})], \tag{21.5}$$

with

$$\left.\begin{aligned} I(\boldsymbol{r}) &= [1-(\eta/k\,r)\log 2k\,r]\, e^{i(kr-\eta\log 2kr)\cos\vartheta}, \\ S(\boldsymbol{r}) &= e^{i(kr-\eta\log 2kr)}, \end{aligned}\right\} \tag{21.6}$$

and

$$f(\vartheta) = \frac{1}{2ik}\sum_{l=0}^{\infty}(2l+1)\,(e^{2i(\sigma_l+\delta_l)}-1)\,P_l(\cos\vartheta), \tag{21.7}$$

where $I(\boldsymbol{r})$ and $S(\boldsymbol{r})$ are the incident plane and scattered spherical waves, modified by the COULOMB force. Both approach their respective counterparts in (15.1), in the limit of a vanishing electric charge. Because of the normalization of $I(\boldsymbol{r})$ and $S(\boldsymbol{r})$, the differential scattering cross section $\sigma(\vartheta)$ is again given by (15.3), $\sigma(\vartheta)=|f(\vartheta)|^2$.

Here is one important point to note: scattered and recoiled particles are both protons and can contribute to the scattering cross section. The real scattering amplitude is, therefore, not $f(\vartheta)$ given by (21.7) but the superposition of $f(\vartheta)$ and $f(\pi-\vartheta)$, for if a particle is scattered through an angle ϑ, then the recoil particle is ejected in the opposite direction $\pi-\vartheta$, in the c.m. system. Because of the exclusion principle, the superposition must be $f(\vartheta)+f(\pi-\vartheta)$ or $f(\vartheta)-f(\pi-\vartheta)$, depending upon whether the spin state is singlet or triplet. Since in the actual scattering experiment no spin direction is specified, the observed differential scattering cross section $\sigma(\vartheta)$ is given by

$$\left.\begin{aligned} \sigma(\vartheta) &= \tfrac{1}{4}|f(\vartheta)+f(\pi-\vartheta)|^2 + \tfrac{3}{4}|f(\vartheta)-f(\pi-\vartheta)|^2 \\ &= |f(\vartheta)|^2 + |f(\pi-\vartheta)|^2 - \tfrac{1}{2}[f^*(\vartheta)\, f(\pi-\vartheta) + f(\vartheta)\, f^*(\pi-\vartheta)], \end{aligned}\right\} \tag{21.8}$$

where $f(\vartheta)$ is given by (21.7).

With respect to the nuclear phase shift δ_l, the statements made at the end of Sect. 15 are still valid; (i) because of the short range of the nuclear force all phase shifts except δ_0 and possibly δ_1 are negligible in the lower energy region and (ii) the positive or negative sign of δ_l immediately implies attractive or repulsive nuclear force, respectively, in the state of azimuthal quantum number l.

[117] J. JACKSON and J. BLATT: Rev. Mod. Phys. 22, 77 (1950).

The first statement, of course, does not apply to the COULOMB phase shift σ_l because of the long range, which makes it necessary to keep all terms in the expansion (21.7) even when we are interested in the low energy data. It is, however, known that the summation in (21.7) can be carried out analytically for the pure COULOMB scattering and gives

$$f_c(\vartheta) = -\frac{e^2}{M v^2} \operatorname{cosec}^2\left(\frac{\vartheta}{2}\right) \exp\left[-2i\eta \log \sin\left(\frac{\vartheta}{2}\right) + 2i\sigma_0\right], \tag{21.9}$$

where v is the velocity of the incident proton and $\exp[2i\sigma_0] = \Gamma(1+i\eta)/\Gamma(1-i\eta)$, σ_0 being the S-wave COULOMB phase shift. Then we can rewrite (21.7) as

$$f(\vartheta) = f_c(\vartheta) + \frac{1}{2ik} \sum_{l=0}^{\infty} (2l+1)\, e^{2i\sigma_l}(e^{2i\delta_l} - 1)\, P_l(\cos\vartheta), \tag{21.10}$$

where we need in general retain only the first few terms of the sum. If we can neglect all the nuclear phase shifts beside δ_0, the differential scattering cross section in the c.m. system is given by

$$\left.\begin{aligned} \sigma(\vartheta) = \left(\frac{e^2}{M v^2}\right)^2 &\left[\left\{\operatorname{cosec}^4\frac{\vartheta}{2} + \sec^4\frac{\vartheta}{2} - \operatorname{cosec}^2\frac{\vartheta}{2}\sec^2\frac{\vartheta}{2}\cos\left(2\eta\log\tan\frac{\vartheta}{2}\right)\right\} - \right.\\ &- \frac{2}{\eta}\sin\delta_0\left\{\operatorname{cosec}^2\frac{\vartheta}{2}\cos\left(\delta_0 + 2\eta\log\sin\frac{\vartheta}{2}\right) + \right.\\ &\left.\left. + \sec^2\frac{\vartheta}{2}\cos\left(\delta_0 + 2\eta\log\cos\frac{\vartheta}{2}\right)\right\} + \frac{4}{\eta^2}\sin^2\delta_0\right]. \end{aligned}\right\} \tag{21.11}$$

The first term in (21.11) accounts for the COULOMB scattering and is called the MOTT formula, the second term is due to the interference between the COULOMB and nuclear S-wave scattering and the final term is the nuclear S-wave scattering.

The proton-proton differential scattering cross section is naturally symmetric with respect to 90° in the c.m. system. Therefore, we need only speak of the angular range from 0° to 90°. It has been observed that the COULOMB scattering is predominant in the forward direction only, especially as the energy increases, and that the nuclear S-phase shift δ_0 is positive (Tables 2 and 3), which means that the nuclear force between two protons in a singlet even state is attractive. Then the interference term in (21.11) becomes negative, except at small angles, because of cancellation between the repulsive COULOMB and attractive singlet nuclear forces. This causes a characteristic interference minimum in the angular range where the nuclear scattering begins to show up. Beyond this interference minimum, the nuclear scattering predominates.

22. Effective range theory of proton-proton scattering. As in the analysis of slow neutron-proton data, it is again most convenient to employ the predictions of the effective range theory. The necessary modification of the theory developed in Sect. 16 is given here according to BETHE[101].

We start from Eq. (21.3) for $l=0$:

$$\left[\frac{d^2}{dr^2} + \left(k^2 + v(r) - \frac{1}{\varrho r}\right)\right] u(r) = 0, \tag{22.1}$$

while its asymptotic solution $\bar{u}(r)$ satisfies

$$\left[\frac{d^2}{dr^2} + \left(k^2 - \frac{1}{\varrho r}\right)\right] \bar{u}(r) = 0. \tag{22.2}$$

The normalization is fixed by requiring that

$$u(r) \xrightarrow[r\to\infty]{} \bar{u}(r) \xrightarrow[r\to\infty]{} C \sin(kr - \eta\log 2kr + \sigma_0 + \delta_0)/\sin\delta_0, \tag{22.3}$$

where $C=[2\pi\eta/(e^{2\pi\eta}-1)]^{\frac{1}{2}}$. It is known[117] that the regular and irregular solutions of (22.2), $F(r)$ and $G(r)$, behave asymptotically as $\sin(kr-\eta\log 2kr+\sigma_0)$ and $\cos(kr-\eta\log 2kr+\sigma_0)$, respectively. Then $\bar{u}(r)$ can be expressed, for all r, as

$$\bar{u}(r) = C\,G(r) + C\cot\delta_0\,F(r)\,, \tag{22.4}$$

which satisfies $\bar{u}(0)=1$ just as (16.6), since $F(0)=0$ and $CG(0)=1$[117].

Starting from (22.1) and (22.2), the same manipulation as in Sect. 16 gives

$$\frac{d\bar{u}_2(r)}{dr} - \frac{d\bar{u}_1(r)}{dr} = (k_2^2 - k_1^2)\int\limits_r^\infty [\bar{u}_1(r)\,\bar{u}_2(r) - u_1(r)\,u_2(r)]\,dr\,, \tag{22.5}$$

where the only difference from Sect. 16 is that we take a small but finite lower limit r, which is due to the fact that $d\bar{u}(r)/dr$ is logarithmically infinite at $r=0$. From the known expressions of $F(r)$ and $G(r)$[117] we can expand $d\bar{u}(r)/dr$ near $r=0$ as follows:

$$\frac{d\bar{u}(r)}{dr} = \frac{1}{\varrho} + \frac{1}{\varrho}\left[\log\frac{r}{\varrho} + 2\gamma - 1 + h(\eta) + \frac{\pi\cot\delta_0}{e^{2\pi\eta}-1}\right] + \cdots, \tag{22.6}$$

where γ is EULER's constant, 0.57721.. and

$$h(\eta) = -\log\eta - \gamma + \eta^2\sum_{n=1}^\infty \frac{1}{n(n^2+\eta^2)}\,.$$

It is seen that the troublesome logarithmic term in $d\bar{u}(r)/dr$ is cancelled on the left-hand side of (22.5), which allows putting the lower limit r to zero. Thus we finally get

$$K = \varrho\left[-\frac{1}{a} + k^2\int\limits_0^\infty (\bar{u}_0(r)\,\bar{u}(r) - u_0(r)\,u(r))\,dr\right], \tag{22.7}$$

where

$$K \equiv \frac{\pi\cot\delta_0}{e^{2\pi\eta}-1} + h(\eta), \qquad -\frac{\varrho}{a} \equiv \lim_{k^2\to 0} K \tag{22.8}$$

and $u_0(r)$, $\bar{u}_0(r)$ are the zero energy radial wave functions of (22.1) and (22.2), respectively. δ_0 is the nuclear S-wave phase shift.

The same argument as in Sect. 16 can also be applied to the integral in (22.7), since the main contribution comes from inside the nuclear force range, where the repulsive COULOMB force as well as the relevant incident proton energy are known to be much smaller than the nuclear force. We can, therefore, expand the integral in (22.7) in terms of k^2, which gives

$$K = \varrho\left[-\frac{1}{a} + \frac{1}{2}r_0 k^2 - P r_0^3 k^4 + Q r_0^5 k^6 + \cdots\right], \tag{22.9}$$

where we have used the same expansions of $u(r)$ and $\bar{u}(r)$ as (16.25), which in turn give the same expressions for r_0 and P as (16.21) and (16.32), and a corresponding one for Q.

Eq. (22.9) implies that if the function K defined by (22.8) is plotted against the energy or k^2, then a straight line is obtained, which determines two parameters, the scattering length a and the effective range r_0. Again the low energy data alone cannot determine the potential shape. The shape dependent parameter P causes a deviation from the straight line downward or upward, according as P is positive or negative, which can give us some information on the potential shape.

JACKSON and BLATT[117] have given interpolation formulas connecting a and r_0 with the potential shape parameters b and s, defined in Sect. 16, for some potential shapes. They have also calculated the values of P and Q for these potentials. The results are summarized in Table 12, which concerns only a narrow range of values for a and r_0. This is, however, quite sufficient in actual calculations because a and r_0 are already known very accurately.

Table 12. *The interpolation formulas between s, b and a, r_0 together with the values of P and Q for three potentials.*

Well shape	Relations between s, b and a, r_0 (all lengths are in 10^{-13} cm)	P	Q
Square well	$s=0.890-0.02712(a+7.793)-0.05946(r_0-2.639)$ $b=2.626-0.04369(a+7.793)+0.95716(r_0-2.639)$	-0.03313	$+0.00179$
Exponential	$s=0.900-0.02543(a+7.424)-0.05056(r_0-2.678)$ $b=2.500-0.06333(a+7.424)+0.83354(r_0-2.678)$	$+0.00907$	$+0.00089$
YUKAWA	$s=0.924-0.02162(a+7.651)-0.04070(r_0-2.676)$ $b=2.400-0.07051(a+7.651)+0.76353(r_0-2.676)$	$+0.05540$	$+0.019$

We have thus shown that the same effective range expansion is valid for the proton-proton scattering as in the neutron-proton case, if we employ the quantity K defined by (22.8) instead of $k\cot\delta_0$. Most statements given in Sect. 16 apply to this case without any alterations. It must, however, be stressed that the same nuclear potential never predicts the same a, r_0 and P in the proton-proton as in the neutron-proton case, although the definitions of r_0 and P are apparently identical. This is due to the fact that the radial wave functions are now generated by the combined action of COULOMB and nuclear forces and accordingly are distinct from those of the neutron-proton case even if the nuclear potential is the same. In the expressions defining r_0 and P, we notice that the main contribution comes from the interior of the nuclear force range, where the COULOMB force is small compared with the nuclear force. This implies that r_0 and P are nearly the same in both cases. For a we must expect a great difference since it depends explicitly on the S-wave nuclear phase shift for zero energy, which is strongly affected by the COULOMB force. This makes it necessary to carry out the explicit calculation of the scattering parameters for a given potential if we want to compare the proton-proton and proton-neutron forces in the ^1S-state very carefully.

23. Analysis of slow proton-proton scattering data. The most accurate data are collected in Tables 2 and 3, where the nuclear S-, P- and D-phase shifts are given against the incident proton energy in the laboratory system. It is noticed that up to about 30 Mev the S-wave phase shift δ_0 is predominant and that the P-shift δ_1 seems to be negative, although the experimental error is still large. This implies that the force in the triplet P-state is repulsive, in accordance with the general theoretical ideas (Sect. 11)*.

In order to analyse the energy dependence of δ_0, it is most convenient to examine the quantity K defined by (22.8), which is also contained in Tables 2 and 3. To determine the scattering length a and the effective range r_0, we can use the accurate low energy data in Table 2. The best values of a and r_0 have been calculated by the least square method, the results of which are given in Table 13. Here we have ascribed equal weight to all data in Table 2 since there

* This statement is concerned with the "effective central force", not the central part of the triplet odd nuclear force.

is no definite criterion for their relative accuracy. We have also assumed the values of P given in Table 12 for the respective potentials and simply put $Q=0$ throughout, since the effect of this term would certainly be within the present experimental error. The values of b and s are then calculated using the interpolation formulas in Table 12. For the YUKAWA well, Table 13 also gives the corresponding meson mass value.

Table 13. *The scattering parameters and potential shape parameters in the singlet proton-proton system determined from the low energy scattering data for three types of potential.*

Well shape	r_0 (10^{-13} cm)	a (10^{-13} cm)	b (10^{-13} cm)	s	P	Meson mass
Square well	2.559 ± 0.017	-7.680 ± 0.012	2.583 ± 0.016	0.8893 ± 0.0011	-0.003313	
Shape independent	2.656 ± 0.017	-7.694 ± 0.012	—	—	0	
Exponential	2.673 ± 0.017	-7.698 ± 0.012	2.513 ± 0.014	0.9073 ± 0.0009	$+0.00907$	
YUKAWA	2.774 ± 0.017	-7.723 ± 0.012	2.480 ± 0.013	0.9216 ± 0.0008	$+0.05540$	$(330.0 \pm 1.7)\, m_e$

In Table 13 all sets of a, r_0 and P are equally fitted to all the data, which implies that the "present best" values of a and r_0 depend on the choice of P; this is, of course, against the spirit of the expansion (22.9). The point is that no very accurate data are available in the very low energy region, as is seen from Table 2 or Fig. 18; in fact data in Table 2 are in general less accurate for lower than for higher energies which is naturally due to the experimental difficulty of the extreme low energy scattering. This makes it difficult to determine the two parameters a and r_0 from the scattering data alone, without any ambiguity arising from the potential shape.

It is seen that the figures in Table 13 are much more accurate than those given for the neutron-proton system. They are also free from any theoretical ambiguities—apart from the one just mentioned—since the force is rigorously central in this case. Of course we have implied that the pure COULOMB law is valid down to $r=0$, since the calculations of JACKSON and BLATT [117] (Table 12) are based on this assumption. We can reasonably expect that a deviation from the COULOMB law may occur at very short distances because of the meson fields surrounding the nucleons. As the COULOMB law is well established at larger distances, the expected deviation would be maximal at the origin and vanish outside some distance which is not yet definitely known. This finite range makes it possible to treat the deviation as part of the nuclear potential. Thus the phase shift δ_0 appearing in K (22.8) is not a pure nuclear phase shift but the S-phase shift due to the combined action of the nuclear force and the deviation from the pure COULOMB law, which is the measurable quantity. If the deviation had a shorter range than the nuclear force, the values of P in Table 12 would not be modified practically, since P mainly depends on the potential tail*. If this should be so, the same a and r_0 would be obtained as in Table 13 for the respective potentials but a small change could be expected for s and a still smaller for b, since b is more related to the range and s more to the depth of the potential. In the case of a hard core, for which there is strong evidence in the singlet even neutron-proton system and therefore also in this case, the effect of the above deviation would be much reduced. Summing up, we may expect that it is not likely to be a major effect.

* A deviation with a range much larger than that of the nuclear force has been discussed by L. L. FOLDY and E. ERIKSEN, Phys. Rev. **98**, 775 (1955), concluding that the behavior of K against k^2 can be quite different from that given in this article.

It is noted in Table 13 that s is certainly smaller than 1, which implies, by definition, that no bound state would be possible in a two-proton system even without the repulsive COULOMB force.

Comparing Tables 11 and 13, we note that the force parameters b and s are almost equal, which implies that the nuclear force is nearly the same in both cases, although the scattering lengths are quite different. In Sect. 32 the equality of the forces is discussed in detail. The larger meson mass value in Table 13 (330 instead of 273 m_e) can again be regarded as an indication of the singular radial dependence, predicted by the meson field theory, such as (12.6) with $n \geqq 2$ (cf. end of Sect. 19). Let us, therefore, quote the results of a calculation[116] of the proton-proton singlet effective range r_0 for the same singular potentials as in (19.8):

r_c (10^{-13} cm)	V_0 (Mev)	$1/\mu$ (10^{-13} cm)	r_0 (10^{-13} cm)
0.014	3.50	2.60	2.60
0.007	3.07	2.70	2.65

(23.1)

where all cases are adjusted to the scattering length $a = -7.68 \times 10^{-13}$ cm and a pure COULOMB law is assumed. If we compare the figures in (23.1) with those in (19.8), it is noticed that the same nuclear potential (12.6) with $n=2$ gives a proton-proton effective range which is larger than that of a singlet neutron-proton system by about 0.3×10^{-13} cm.

Having determined the two scattering parameters, we can now compare the higher energy data of Table 3 with the theoretical prediction (22.9). Using the values of P and Q in Table 12, the curves of K against energy can readily be drawn. This is, however, not a good approximation for the YUKAWA potential since the convergence of the effective range expansion (22.9) is somewhat poor at higher energies due to the long tail of the potential. This case was investigated in detail by BREIT and others[118]. The curve in Fig. 18 for the YUKAWA potential is due to these authors, the rest are calculated from (22.9), including the Q-term and using Tables 12 and 13. The data summarized in Tables 2 and 3 are also plotted in Fig. 18. The beautiful behaviour of the lower energy data (Table 2) is remarkable.

Although the present higher energy data have still relatively large errors, it seems certain that they deviate downward from the straight line determined by the lower energy data. This favours a positive shape dependent parameter P and seems to exclude potentials with short or no tail, such as the square well. It is also likely that potentials with larger P than the YUKAWA well are excluded. This indication that the shape dependent parameter P in a singlet proton-proton system is definitely positive and lies roughly between $+0.01$ and $+0.05$ is very important because it can also be applied to a singlet neutron-proton system, if we assume charge independence of the nuclear forces, P being almost unaffected by the COULOMB force (cf. Tables 11 and 12) except for the square well. In fact this is the only reliable information we have about the potential shape in the ^1S-state.

Finally we remark that according to Fig. 16 (Sect. 19) a hard core can reduce P very much; if we introduce a hard core of radius 0.6×10^{-13} cm, as suggested by high energy data (Sect. 44), the singlet neutron-proton shape dependent parameter P_s is always negative (around -0.03), even for the YUKAWA potential.

[118] M. C. YOVITS, R. L. SMITH jr., H. HULL jr., J. BENGSTON and G. BREIT: Phys. Rev. **85**, 540 (1952).

However, comparing the P-values in Tables 11 and 13 for the square well, it seems possible that a potential with short tail might give a proton-proton P smaller in magnitude than that of the singlet neutron-proton system. This may improve the situation somewhat, but still it appears that the positive P-value suggested by the proton-proton scattering data would be rather difficult to explain if we assume a hard core of radius, say, 0.6×10^{-13} cm in the ^1S-state.

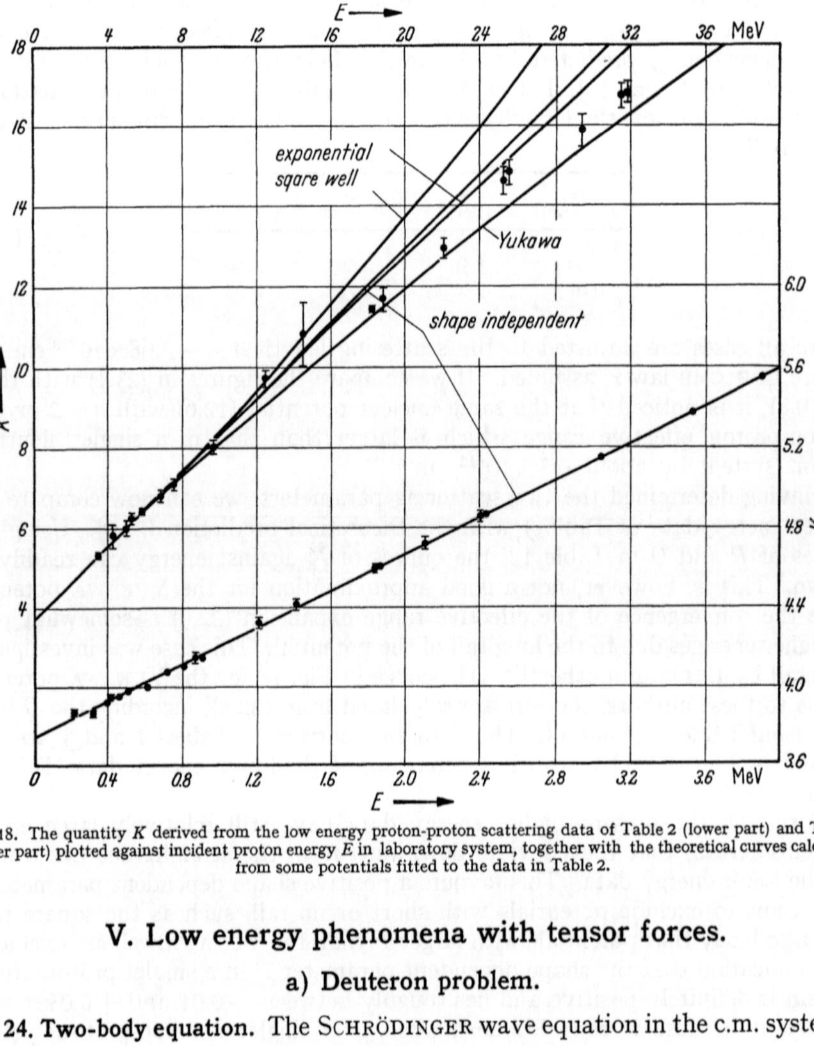

Fig. 18. The quantity K derived from the low energy proton-proton scattering data of Table 2 (lower part) and Table 3 (upper part) plotted against incident proton energy E in laboratory system, together with the theoretical curves calculated from some potentials fitted to the data in Table 2.

V. Low energy phenomena with tensor forces.

a) Deuteron problem.

24. Two-body equation. The SCHRÖDINGER wave equation in the c.m. system is

$$\left[-\frac{\hbar^2}{M}\Delta+V(\boldsymbol{r})\right]\psi(\boldsymbol{r})=E\,\psi(\boldsymbol{r}), \tag{24.1}$$

where $V(\boldsymbol{r})$ consists of a central force $V_C(r)$ and a tensor force $V_T(r)$:

$$V(\boldsymbol{r})=V_C(r)+S_{12}V_T(r), \tag{24.2}$$

with the definition (8.12) of S_{12}. According to Sect. 10, we have four quantum numbers J, J_z, parity and S, the last of which we can fix at 1 (spin triplet states) throughout this chapter since S_{12} vanishes identically in a spin singlet state. Because of the fact that S_{12} is a spin-orbit operator, spin coordinates are not separable in $\psi(\boldsymbol{r})$. We study in this section how spin coordinates are involved in $\psi(\boldsymbol{r})$ and how we can eliminate them in order to get the equation for the radial part of $\psi(\boldsymbol{r})$.

Let us denote the three triplet spin eigenfunctions by χ_i:

$$\left.\begin{aligned} \chi_1 &= \alpha(1)\,\alpha(2), \quad \chi_{-1} = \beta(1)\,\beta(2), \\ \chi_0 &= \frac{1}{\sqrt{2}}\{\alpha(1)\,\beta(2) + \beta(1)\,\alpha(2)\}, \end{aligned}\right\} \tag{24.3}$$

where $\alpha(i)$ and $\beta(i)$ are the eigenfunctions of σ_z of the i-th nucleon belonging to the eigenvalue $+1$ and -1, respectively. The orbital angular momentum eigenfunctions are the normalized spherical harmonics, which we denote by $Y_{LL_z}(\vartheta,\varphi)$, the definition of which is the same as in (10.6). Then the eigenfunction $\Phi_{J,J_z,L}$ pertaining to eigenvalues J and J_z and the quantum number L is given by

$$\Phi_{J,J_z,L} = \sum_{L_z=J_z-1}^{J_z+1} C_{JJ_zLL_z}\, Y_{LL_z}(\vartheta,\varphi)\,\chi_{J_z-L_z}, \tag{24.4}$$

where $L=J+1$, J or $J-1$ and the numerical coefficients are determined by the general addition law of angular momenta[119]. We normalize Φ, which implies that the C's satisfy the relation

$$\sum_{L_z=J_z-1}^{J_z+1} |C_{JJ_zLL_z}|^2 = 1. \tag{24.5}$$

Then the eigenfunction ψ belonging to the eigenvalues J, J_z and parity $(-1)^L$ can be written in terms of Φ, defined by (24.4), as

Table 14. *The values of $S_{JLL'}$ defined by* (24.9)

	$L'=J+1$	$L'=J$	$L'=J-1$
$L=J+1$	$\frac{-2(J+2)}{2J+1}$	0	$\frac{+6\sqrt{J(J+1)}}{2J+1}$
$L=J$	0	$+2$	0
$L=J-1$	$\frac{+6\sqrt{J(J+1)}}{2J+1}$	0	$\frac{-2(J-1)}{2J+1}$

$$\psi = \frac{1}{r}\sum_L u_L(r)\,\Phi_{JJ_zL} \times (\vartheta,\varphi,\text{ spin variables}), \tag{24.6}$$

where the expansion coefficients u_L are functions of r only. The summation in (24.6) is over two terms $L=J\pm1$ or a single term $L=J$ (cf. Sect. 10). Because of (24.5), the normalization of ψ is reduced to

$$\sum_L \int_0^\infty u_L^2(r)\,dr = 1. \tag{24.7}$$

Substituting the expansion (24.6) into (24.1), we get

$$\frac{d^2u_L}{dr^2} - \left[\alpha^2 + \frac{L(L+1)}{r^2} - v_C(r)\right]u_L(r) + v_T(r)\sum_{L'} S_{JLL'}\,u_{L'}(r) = 0, \tag{24.8}$$

where we have put $\alpha^2 = -EM/\hbar^2$, $v_i(r) = -V_i(r)M/\hbar^2$ and defined $S_{JLL'}$ by

$$S_{JLL'} = \int (\Phi_{JJ_zL}, S_{12}\Phi_{JJ_zL'})\,d\Omega, \tag{24.9}$$

which does not depend on J_z because of the symmetry; $\int d\Omega$ in (24.9) implies integration over ϑ and φ as well as summation with respect to spin coordinates. The values of $S_{JLL'}$ are given in Table 14[120]. Eq. (24.8) is the required equation system for the radial wave functions.

[119] E. U. Condon and G. H. Shortley: Theory of Atomic Spectra, Cambridge, 1953, p. 76, where the definition of $Y_{LL_z}(\vartheta,\varphi)$ is different from ours in (10.6) by the factor $(-1)^{L_z}$.

[120] H. A. Bethe: Phys. Rev. **57**, 390 (1940), p. 393.

Let us examine the deuteron ground state in more detail. As it is known to be a $({}^3S_1+{}^3D_1)$-state, we need only put $J=1$ and sum over the two terms $L=0$ and $L=2$ in (24.8). Using $u(r)$ and $w(r)$ instead of $u_0(r)$ and $u_2(r)$, the deuteron wave function ψ can be written as

$$\psi=\frac{1}{r}\,u(r)\,\Phi_{1J_z0}+\frac{1}{r}\,w(r)\,\Phi_{1J_z2}, \tag{24.10}$$

where

$$\left.\begin{aligned}\Phi_{110}&=Y_{00}(\vartheta,\varphi)\,\chi_1=\frac{1}{\sqrt{4\pi}}\,\chi_1,\qquad \Phi_{100}=\frac{1}{\sqrt{4\pi}}\,\chi_0,\\ \Phi_{1-10}&=\frac{1}{\sqrt{4\pi}}\,\chi_{-1},\end{aligned}\right\} \tag{24.11}$$

and

$$\left.\begin{aligned}\Phi_{112}&=\sqrt{\frac{1}{10}}\,Y_{20}(\vartheta,\varphi)\,\chi_1+\sqrt{\frac{3}{10}}\,Y_{21}(\vartheta,\varphi)\,\chi_0+\sqrt{\frac{6}{10}}\,Y_{22}(\vartheta,\varphi)\,\chi_{-1},\\ \Phi_{102}&=-\sqrt{\frac{3}{10}}\,Y_{2-1}(\vartheta,\varphi)\,\chi_1-\sqrt{\frac{4}{10}}\,Y_{20}(\vartheta,\varphi)\,\chi_0-\sqrt{\frac{3}{10}}\,Y_{21}(\vartheta,\varphi)\,\chi_{-1},\\ \Phi_{1-12}&=\sqrt{\frac{6}{10}}\,Y_{2-2}(\vartheta,\varphi)\,\chi_1+\sqrt{\frac{3}{10}}\,Y_{2-1}(\vartheta,\varphi)\,\chi_0+\sqrt{\frac{1}{10}}\,Y_{20}(\vartheta,\varphi)\,\chi_{-1}.\end{aligned}\right\} \tag{24.12}$$

The normalization condition (24.7) becomes

$$\int\limits_0^\infty [u^2(r)+w^2(r)]\,dr=1 \tag{24.13}$$

and the coupled equations for the S- and D-components, $u(r)$ and $w(r)$, of the radial wave function are obtained from (24.8) and Table 14:

$$\left.\begin{aligned}&\frac{d^2u(r)}{dr^2}-[\alpha^2-v_C(r)]\,u(r)+2\sqrt{2}\,v_T(r)\,w(r)=0,\\ &\frac{d^2w(r)}{dr^2}-\left[\alpha^2+\frac{6}{r^2}-v_C(r)+2v_T(r)\right]w(r)+2\sqrt{2}\,v_T(r)\,u(r)=0.\end{aligned}\right\} \tag{24.14}$$

We note that the action of the tensor operator S_{12} on the triplet spin eigenfunctions χ_i given by (24.3) can be expressed as

$$S_{12}\chi_i=\sum_j C_{ji}\,\chi_j, \tag{24.15}$$

where the coefficients C_{ji} are given by (10.6) in a matrix form. It is readily seen by comparing (10.6) with the Eqs. (24.12) that the following relation is valid:

$$S_{12}\chi_i=4\sqrt{2\pi}\,\Phi_{1i2}. \tag{24.16}$$

Thus we can rewrite (24.10) as

$$\psi=\frac{1}{\sqrt{4\pi}}\left\{\frac{u(r)}{r}+\frac{1}{\sqrt{8}}\,S_{12}\,\frac{w(r)}{r}\right\}\chi_{J_z} \tag{24.17}$$

for the respective quantum numbers J_z. This simple form of ψ is valid only for the $({}^3S_1+{}^3D_1)$-state. It was derived by RARITA and SCHWINGER[121] by a different argument and is very convenient in carrying out practical calculations with the deuteron wave function.

---

[121] W. RARITA and J. SCHWINGER: Phys. Rev. 59, 436 (1941).

25. Electric quadrupole moment of deuteron. *α) Multipole moments of nuclei.* Consider a nucleus under the action of an external electromagnetic field, which we assume too weak to disturb the charge distribution of the nucleus. The field in question is usually caused by the electrons surrounding the nucleus or by other electrons and nuclei in the molecule. The charge distribution of the nucleus is characterized by a current density $\boldsymbol{i}(\boldsymbol{x})$ and a charge density $\varrho(\boldsymbol{x})$. If $\boldsymbol{A}(\boldsymbol{x})$ and $\varphi(\boldsymbol{x})$ denote the vector and scalar potentials of the external field, the interaction energy E with the nucleus is expressed by

$$E = -\frac{1}{c}\int \boldsymbol{A}(\boldsymbol{x})\, \boldsymbol{i}(\boldsymbol{x})\, dv + \int \varphi(\boldsymbol{x})\, \varrho(\boldsymbol{x})\, dv. \tag{25.1}$$

The integration is performed over the whole volume of the nucleus which is in general small compared with the wave lengths involved, whence the external field can be considered as slowly varying. In such a case, it is well-known that the expression (25.1) can be expanded in terms of the electromagnetic field quantities and their derivatives at some point, conveniently chosen as the center-of-mass of the nucleus, and the successive multipole moments of the nucleus with respect to that point.

Taking the center-of-mass of the nucleus for origin and expanding $\boldsymbol{A}(\boldsymbol{x}$ and $\varphi(\boldsymbol{x})$ around this point, we have

$$\left.\begin{aligned} \boldsymbol{A}(\boldsymbol{x}) &= \boldsymbol{A}_0 + (\boldsymbol{x}\,\mathrm{grad})\,\boldsymbol{A}_0 + \cdots,\\ \varphi(\boldsymbol{x}) &= \varphi_0 + (\boldsymbol{x}\,\mathrm{grad})\,\varphi_0 + \tfrac{1}{2}\left(\boldsymbol{x}\,(\boldsymbol{x}\,\mathrm{grad})\right)\varphi_0 + \cdots, \end{aligned}\right\} \tag{25.2}$$

neglecting higher terms; the subscript zero means that the values at the origin should be inserted after the differentiation. Substituting the above expansions in (25.1) and combining suitable terms, we get

$$\left.\begin{aligned} E &= q\,\varphi_0 - \boldsymbol{P}\boldsymbol{E}_0 - \mu\boldsymbol{H}_0 - \frac{1}{6}\sum_{ik} Q'_{ik}\,(\mathrm{grad}_i\,\boldsymbol{E}_k)_0 - \\ &\quad - \frac{1}{c}\frac{\partial}{\partial t}\left[\boldsymbol{P}\boldsymbol{A}_0 + \frac{1}{6}\sum_{ik} Q'_{ik}\,(\mathrm{grad}_i\,\boldsymbol{A}_k)_0\right] + \cdots, \end{aligned}\right\} \tag{25.3}$$

where we have defined

$$\left.\begin{aligned} &\boldsymbol{E}_0 = \left[-\,\mathrm{grad}\,\varphi - \frac{1}{c}\frac{\partial \boldsymbol{A}}{\partial t}\right]_{\boldsymbol{x}=0}, \qquad \boldsymbol{H}_0 = [\mathrm{rot}\,\boldsymbol{A}]_{\boldsymbol{x}=0},\\ &q = \int \varrho(\boldsymbol{x})\, dv, \quad \boldsymbol{P} = \int \boldsymbol{x}\,\varrho(\boldsymbol{x})\, dv, \quad \mu = \frac{1}{2c}\int \boldsymbol{x}\times\boldsymbol{i}(\boldsymbol{x})\, dv,\\ &Q'_{ik} = \int 3\,x_i\,x_k\,\varrho(\boldsymbol{x})\, dv \end{aligned}\right\} \tag{25.4}$$

and used the formula $\int \boldsymbol{i}(\boldsymbol{x})\, dv = \frac{\partial}{\partial t}\int \boldsymbol{x}\,\varrho(\boldsymbol{x})\, dv$, which is valid owing to the continuity equation $\mathrm{div}\,\boldsymbol{i} + \frac{\partial}{\partial t}\varrho(\boldsymbol{x}) = 0$. The term containing the time derivative in (25.3) can be omitted, since it contributes neither to the expectation value of the interaction energy in stationary states nor to the matrix elements of transitions between states of the same total energy. Then we finally get

$$E = q\,\varphi_0 - \boldsymbol{P}\boldsymbol{E}_0 - \mu\boldsymbol{H}_0 - \tfrac{1}{6}\sum_{ik} Q_{ik}\,(\mathrm{grad}_i\,\boldsymbol{E}_k)_0 + \cdots, \tag{25.5}$$

where we have replaced Q'_{ik} by

$$Q_{ik} = \int (3\,x_i\,x_k - \delta_{ik}\,r^2)\,\varrho(\boldsymbol{x})\, dv, \tag{25.6}$$

which is allowed since the external field satisfies $\mathrm{div}\,\boldsymbol{E} = 0$ inside the nucleus.

The quantity q appearing in the first term of (25.5) is the total electric charge, the two vectors $\boldsymbol{P}$ and $\boldsymbol{\mu}$ are the electric and magnetic dipole moments and the tensor Q_{ik} the electric quadrupole moment of the nucleus. The higher multipole moments can be obtained in an analogous way.

β) Electric quadrupole moment. The first term in (25.5) simply represents the electrostatic interaction of the whole nuclear charge q, considered as a point charge, with the surrounding electrons and other charged particles and explains the ordinary atomic and molecular energy levels. The remaining terms with higher nuclear moments give rise to the hyperfine structure of the levels, which in turn enables us to measure these moments.

It is well known, however, that the second term containing the electric dipole moment or more generally odd moments vanish identically in all stationary states; the wave functions have either even or odd parity and therefore the expectation values of odd operators vanish identically. Then the first non-vanishing electric moment is the quadrupole moment given by (25.6).

Now the quantity (25.6) is a symmetric tensor, with the trace (spur) zero and components depending on the state considered. In the case of the deuteron, there is only one charged particle, the wave function of which we denote as ψ. Then we can write $\varrho(\boldsymbol{x}) = e\psi^*\psi$. From a group theoretical argument it follows that the matrix element defined by

$$(J J_z | Q_{ik} | J J_z') = e \int \psi_{J J_z}^* (3 x_i x_k - \delta_{ik} r^2) \psi_{J J_z'} \, dv, \tag{25.7}$$

$\psi_{J J_z'}$ being an eigenfunction belonging to quantum numbers J and J_z, can be written as[122]

$$\frac{1}{e} (J J_z | Q_{ik} | J J_z') = C \left(J J_z \left| 3 \frac{\boldsymbol{J}_i \boldsymbol{J}_k + \boldsymbol{J}_k \boldsymbol{J}_i}{2} - \delta_{ik} \boldsymbol{J}^2 \right| J J_z' \right), \tag{25.8}$$

where $\boldsymbol{J}\hbar$ is the total angular momentum operator and C is a constant, which is readily determined by making a special matrix element between two top states; for example, by

$$\frac{1}{e} (J J | Q_{33} | J J) = C (J J | 3 \boldsymbol{J}_z^2 - \boldsymbol{J}^2 | J J) = C J (2J - 1). \tag{25.9}$$

Thus we get

$$(J J_z | Q_{ik} | J J_z') = \frac{e Q}{J(2J-1)} \left(J J_z \left| 3 \frac{\boldsymbol{J}_i \boldsymbol{J}_k + \boldsymbol{J}_k \boldsymbol{J}_i}{2} - \delta_{ik} \boldsymbol{J}^2 \right| J J_z' \right), \tag{25.10}$$

where we have defined

$$Q = \frac{1}{e} (J J | Q_{33} | J J) = \int \psi_{JJ}^* (3 z^2 - r^2) \psi_{JJ} \, dv, \tag{25.11}$$

which is conventionally called the nuclear quadrupole moment. It has, therefore, been shown that Q alone is sufficient to determine the electric quadrupole interaction energy with the external field.

Eq. (25.9) tells us that the nuclear quadrupole moment vanishes if the nuclear spin is 0 or $\frac{1}{2}$ ($J = 0$ or $\frac{1}{2}$). The deuteron ground state is known to have spin 1 and thus can possess an electric quadrupole moment. In fact the deuteron has a positive quadrupole moment (Sect. 1). Its experimental value (1.3) has been determined by observing the interaction energy given by the last term in (25.5) and using the theoretically calculated value of the gradient of the external

122 E. SEGRÈ: Experimental Nuclear Physics, Vol. 1, p. 378. New York 1953.

electric field strength at the nucleus. The positive value of Q means, according to the definition (25.11), that the charge distribution is elongated along the direction of the nuclear spin.

Before we can substitute the deuteron wave function (24.10) into (25.11), we must remark that the coordinate used in (25.11) is referred to the center-of-mass of the deuteron. Thus the deuteron quadrupole moment is calculated as the expectation value of $(3z^2-r^2)/4=(3\cos^2\vartheta-1)r^2/4$ in terms of the deuteron wave function (24.10) belonging to $J_z=1$. The result is

$$Q=\frac{\sqrt{2}}{10}\int_0^\infty r^2\left(u\,w-\frac{1}{2\sqrt{2}}\,w^2\right)dr. \tag{25.12}$$

This expression shows that only equal sign of $w(r)$ and $u(r)$ can give a positive quadrupole moment, which in turn is attained by a negative coefficient of the tensor operator S_{12} in the nuclear potential (24.2), as seen from the coupled Eqs. (24.14). This was already pointed out in Sect. 10 using an intuitive argument.

We assumed above that one of the two particles in the deuteron is definitely a proton, which is all right in the case of an ordinary (non-exchange) force between the proton and neutron but is not correct when employing a charge spin formalism. In the latter case, we cannot say which of the two is a proton, and both nucleons can contribute to Q. Introducing a proton projection operator $\frac{1}{2}(\tau_3+1)$, we have

$$Q=\sum_{i=1}^{2}\int\psi_{JJ}^*(\boldsymbol{r}_1,\boldsymbol{r}_2)\,\frac{\tau_3^{(i)}+1}{2}\,(3z_i^2-r_i^2)\,\psi_{JJ}(\boldsymbol{r}_1,\boldsymbol{r}_2)\,dv_1\,dv_2 \tag{25.13}$$

or, introducing the relative coordinate $\boldsymbol{r}$,

$$Q=\int\psi_{JJ}^*(\boldsymbol{r})\left[1+\frac{1}{2}\,(\tau_3^{(1)}+\tau_3^{(2)})\right]\frac{3z^2-r^2}{4}\,\psi_{JJ}(\boldsymbol{r})\,dv, \tag{25.14}$$

where $\psi_{JJ}(\boldsymbol{r})$ is now supposed to contain the charge spin function. But the effect of the operator $\frac{1}{2}(\tau_3^{(1)}+\tau_3^{(2)})$ on a neutron-proton wave function is zero and we get the same equation as (25.12).

26. Magnetic moment of deuteron. We next consider the magnetic dipole interaction term in (25.5). As is seen from its definition (25.4), the magnetic dipole moment is an expectation value of an even operator and in general need not vanish. Here we must also pay attention to the magnetic dipole moment due to the nucleon spin, which we denote as $\mu_N\sigma$ or $\mu_P\sigma$ for a neutron or proton, in units of a nuclear magneton $e\hbar/2M_P$, M_P being the proton mass. Thus the magnetic dipole moment of the deuteron is, in the same unit, the expectation value of the operator

$$\mu=\sum_{i=1}^{2}\left[\frac{1+\tau_3^{(i)}}{2}\,\mu_P\,\sigma^{(i)}+\frac{1-\tau_3^{(i)}}{2}\,\mu_N\,\sigma^{(i)}+\frac{1+\tau_3^{(i)}}{2}\,\frac{1}{\hbar}\,\boldsymbol{r}_i\times\boldsymbol{p}_i\right], \tag{26.1}$$

using the charge spin formalism (Sect. 7). The above expression can be written as

$$\left.\begin{aligned}\mu=(\mu_P+\mu_N)\,&\frac{\sigma^{(1)}+\sigma^{(2)}}{2}+(\mu_P-\mu_N)\,\frac{\tau_3^{(1)}+\tau_3^{(2)}}{2}\,\frac{\sigma^{(1)}+\sigma^{(2)}}{2}+\\&+(\mu_P-\mu_N)\,\frac{\tau_3^{(1)}-\tau_3^{(2)}}{2}\,\frac{\sigma^{(1)}-\sigma^{(2)}}{2}+\\&+\left(\frac{M_N}{M_N+M_P}+\frac{1}{4}\,(\tau_3^{(1)}+\tau_3^{(2)})+\frac{1}{4}\,(1+\tau_3^{(1)}\,\tau_3^{(2)})\cdot\frac{M_P-M_N}{M_P+M_N}\right)\frac{\boldsymbol{r}\times\boldsymbol{p}}{\hbar},\end{aligned}\right\} \tag{26.2}$$

where we have introduced relative coordinates

$$\boldsymbol{r}_1 = \frac{m_2 \boldsymbol{r}}{m_1 + m_2}, \qquad \boldsymbol{r}_2 = -\frac{m_1 \boldsymbol{r}}{m_1 + m_2}, \qquad \boldsymbol{p}_1 = \boldsymbol{p} = -\boldsymbol{p}_2,$$

with

$$m_i = \tfrac{1}{2}(M_N + M_P) + \tfrac{1}{2}\tau_3^{(i)}(M_P - M_N).$$

The terms containing the factor $\frac{1}{2}(\tau_3^{(1)} + \tau_3^{(2)})$ vanish identically as the deuteron ground state is a charge singlet. The term with $\frac{1}{2}(\sigma^{(1)} - \sigma^{(2)})$ cannot contribute either because its matrix elements vanish identically for spin triplets. Introducing the orbital angular momentum operator $\boldsymbol{L} = \frac{1}{\hbar}(\boldsymbol{r} \times \boldsymbol{p})$ and the total angular momentum operator $\boldsymbol{J} = \boldsymbol{L} + \frac{1}{2}(\sigma^{(1)} + \sigma^{(2)})$, we get

$$\left.\begin{aligned} \mu &= (\mu_P + \mu_N)\boldsymbol{J} - \left(\mu_P + \mu_N - \frac{M_N}{M_N + M_P}\right)\boldsymbol{L} \\ &= (\mu_P + \mu_N)\boldsymbol{J} - (\mu_P + \mu_N - \tfrac{1}{2})\boldsymbol{L}, \end{aligned}\right\} \tag{26.3}$$

since the difference between neutron and proton mass can be neglected in this connection.

We then take its expectation value using the deuteron wave function (24.10) to (24.12). It is readily seen that in a state with a definite value of J_z the only component of μ which does not vanish is μ_z. Then the expectation value of μ_z in the substate belonging to the quantum number J_z is

$$\left[\mu_P + \mu_N - \tfrac{3}{2}(\mu_P + \mu_N - \tfrac{1}{2})\int_0^\infty w^2(r)\,dr\right]J_z, \tag{26.4}$$

where the coefficient of J_z or the expectation value of μ_z in the top state $J_z = J = 1$ is usually called the deuteron magnetic moment μ_D. The integral appearing in μ_D is just the probability with which the D-state will occur in the deuteron ground state, usually denoted as P_D. Then

$$P_D = \int_0^\infty w^2(r)\,dr \tag{26.5}$$

and

$$\mu_D = \mu_P + \mu_N - \tfrac{3}{2}(\mu_P + \mu_N - \tfrac{1}{2})P_D. \tag{26.6}$$

Thus the deviation of the deuteron magnetic moment from the simple sum of proton and neutron moments gives a direct measure of the D-state probability. The experimental values of the μ's given in Sect. 1 show a definite negative deviation (1.2), which gives

$$P_D = 0.039. \tag{26.7}$$

This quantity and the electric quadrupole moment defined by (25.12) [experimental value (1.3)] are very important by indicating a tensor force in the deuteron. Here we must note that the D-state probability is determined by the difference $\mu_D - (\mu_P + \mu_N)$ which is a very small quantity compared with the moments themselves. Hence a change of μ_D by only 1%, for example, can cause a change of P_D of ± 0.015 or 38% of its value (26.7). Although the experimental values cited are quite accurate, there is strong theoretical evidence that the deuteron moment must be subject to small corrections before we can use (26.5) and (26.6); this will be discussed in the next section.

27. Further considerations on electromagnetic properties of the deuteron. We have already stressed that the D-state probability is very sensitive to the deuteron magnetic moment. In the foregoing two sections, we have based our considerations upon the non-relativistic quantum mechanics, in order to describe the interactions of nucleons with an electromagnetic field. More correctly we must start from the relativistic field theory for the coupled nucleon, meson and electromagnetic fields, a brief sketch of which was given in Sects. 6 and 7. Thus, we can expect several other sources of deuteron moments besides those considered above.

Quite conventionally we can classify the expected corrections into the following three groups. First of all, charged mesons which are interchanged between nucleons by emission and subsequent absorption can, as well as recoiling nucleons, contribute to the deuteron moments; this is commonly called the *exchange current correction*. Secondly, we have so far tacitly assumed that a proton and a neutron have the same magnetic moments in the deuteron as in their free states. According to the meson field theory, the nucleon magnetic moment is mainly due to its surrounding meson field. It is reasonable to expect that such fields will interfere with each other when the nucleons are close together, thus causing a modification of the magnetic moments of bound nucleons relative to those in a free state, which we may call the *non-additivity correction*. Finally, we can foresee a purely *relativistic correction* to the deuteron magnetic moment which is determined by the motion of nucleons in the ground state.

Let us denote the four-vector potential of an external electromagnetic field as A_μ, the interaction of which with the meson-nucleon system can be described by replacing $\partial/\partial x_\mu$ by $\partial/\partial x_\mu - (ie/\hbar c) A_\mu$ in the original Lagrangians (6.5) and (6.10). Thus, introducing the charge coordinate together with the decomposition of the meson field by (7.3) and passing to the Hamiltonian formalism, we get the following expression for the electromagnetic interaction energy of the meson-nucleon system:

$$\left.\begin{aligned} H'' = &-i e \sum_\mu \overline{\psi} \frac{1+\tau_3}{2} \gamma_\mu \psi A_\mu + \frac{e}{\hbar c}\Big[\sum_i A_i \Big(\varphi_1 \frac{\partial \varphi_2}{\partial x_i} - \varphi_2 \frac{\partial \varphi_1}{\partial x_i}\Big) - \\ &- i c A_4 (\varphi_1 \pi_2 - \varphi_2 \pi_1)\Big] - \frac{1}{2}\Big(\frac{e}{\hbar c}\Big)^2 A_4^2 (\varphi_1^2 + \varphi_2^2), \end{aligned}\right\} \quad (27.1)$$

where the last term which is quadratic in A_4^2 can be neglected provided the electromagnetic field is weak enough. If we want to introduce the so called PAULI-term in order to account for the anomalous magnetic moment of the nucleon, we have only to add an extra term

$$\sum_{\mu\nu} \frac{i}{2} \overline{\psi} \frac{e}{2\varkappa} \Big[(\mu_P - 1)\frac{1+\tau_3}{2} + \mu_N \frac{1-\tau_3}{2}\Big] \gamma_\mu \gamma_\nu \psi \Big(\frac{\partial A_\nu}{\partial x_\mu} - \frac{\partial A_\mu}{\partial x_\nu}\Big) \quad (27.2)$$

to the above interaction Hamiltonian.

Confining ourselves to a constant magnetic field $\boldsymbol{H}$ by putting $\boldsymbol{A} = -\frac{1}{2}\boldsymbol{x} \times \boldsymbol{H}$ and $A_4 = 0$, we get from (27.1) the magnetic moment operator $\boldsymbol{\mu}$ in units of $e/2\varkappa$:

$$\boldsymbol{\mu} = \varkappa \int \psi^* \frac{1+\tau_3}{2} \boldsymbol{x} \times \boldsymbol{\alpha} \psi \, d^3x - \frac{\varkappa}{\hbar c} \int \Big(\varphi_1 \boldsymbol{x} \times \frac{\partial}{\partial \boldsymbol{x}} \varphi_2 - \varphi_2 \boldsymbol{x} \times \frac{\partial}{\partial \boldsymbol{x}} \varphi_1\Big) d^3x, \quad (27.3)$$

$\boldsymbol{\alpha}$ being the ordinary DIRAC matrix. If, on the other hand, we introduce a pure electric field $\boldsymbol{E}$, varying slowly inside the meson-nucleon system, by putting $\boldsymbol{A} = 0$, $A_4 = i\varphi$ and $\varphi(\boldsymbol{x}) = \varphi(0) - (\boldsymbol{x}\boldsymbol{E}) - \frac{1}{2}\sum_{ij} x_i x_j \frac{\partial E_j}{\partial x_i}$, we get the following electric quadrupole moment operator Q:

$$Q = \int (3z^2 - r^2) \Big[\psi^* \frac{1+\tau_3}{2} \psi + \frac{1}{\hbar}(\varphi_1 \pi_2 - \varphi_2 \pi_1)\Big] d^3x. \quad (27.4)$$

These two expressions, (27.3) and (27.4), are now to be compared with (25.13) and (26.1). The PAULI-term (27.2) gives corresponding additional terms to μ and Q, which are not given here.

As we can regard the interaction term H'' (27.1) as a small perturbation, we first treat the meson-nucleon problem starting from the Hamiltonian $H+H'$ given by (8.1) and (8.2). In order to solve the SCHRÖDINGER equation of the meson-nucleon system

$$\int (H+H')\, d^3x\, |\Psi\rangle = E\, |\Psi\rangle, \tag{27.5}$$

let us use the TAMM-DANCOFF formalism, i.e. we expand $|\Psi\rangle$ as follows:

$$\left.\begin{aligned} |\Psi\rangle = \iint c(\boldsymbol{p}_1, \boldsymbol{p}_2)\, |\boldsymbol{p}_1 \boldsymbol{p}_2\rangle\, d^3p_1\, d^3p_2 + {}\\ + \iiint c(\boldsymbol{p}_1, \boldsymbol{p}_2, \boldsymbol{k})\, |\boldsymbol{p}_1 \boldsymbol{p}_2 \boldsymbol{k}\rangle\, d^3p_1\, d^3p_2\, d^3k + \cdots, \end{aligned}\right\} \tag{27.6}$$

where $|\boldsymbol{p}_1 \boldsymbol{p}_2\rangle$ represents a state of two nucleons of momenta $\boldsymbol{p}_1$ and $\boldsymbol{p}_2$, $|\boldsymbol{p}_1 \boldsymbol{p}_2 \boldsymbol{k}\rangle$ two nucleons and one meson with wave vector $\boldsymbol{k}$. The c's are then solved, neglecting all the higher coefficients which contain many mesons besides the nucleon pair. The deuteron moments can then be calculated as the expectation values of μ (27.3) and Q (27.4) in terms of the solution $|\Psi\rangle$ of (27.5). In these expressions several terms appear, among which those corresponding to the emission and reabsorption of mesons by the same nucleon are called self-meson terms. These are the terms which should explain the anomalous part of the nucleon magnetic moment and therefore give rise to the non-additivity correction to the deuteron moment. As there is, however, no satisfactory way to treat these terms, we first neglect them entirely, instead of introducing the PAULI-term (27.2) into H''.

If we disregard all amplitudes except the first one (no meson) in (27.6) and make the non-relativistic approximation in evaluating μ and Q, we get the same expressions as (26.6) and (25.12), respectively, identifying the "no meson" amplitude with the FOURIER transform of the deuteron wave function. Thus it is natural to define as the relativistic correction those contributions which arise if we push the calculation one step further with respect to the nucleon velocities. Such corrections as are due to the amplitudes with mesons are defined as the exchange current correction. This is by itself a small quantity, so we keep the non-relativistic approximation in evaluating it. For the details we refer to the existing literature[123] and only quote the results here. For the deuteron magnetic moment, we get

$$\frac{\langle\Psi|\, \mu_z\, |\Psi\rangle}{\langle\Psi|\Psi\rangle} = \mu_D + \Delta_{\text{rel}}\, \mu_D + \Delta_{\text{exch}}\, \mu_D, \tag{27.7}$$

where μ_D is the phenomenological expression (26.6) and the corrections are given by

$$\left.\begin{aligned} \Delta_{\text{rel}}\, \mu_D = \frac{\mu_P + \mu_N + 3}{6\varkappa^2} \int_0^\infty u_g(r)\, u_g''(r)\, dr + {}\\ + \sqrt{2}\, \frac{(\mu_P + \mu_N - \frac{3}{2})}{6\varkappa^2} \int_0^\infty w_g(r) \left[u_g''(r) - \frac{3}{r}\, u_g'(r) + \frac{3}{r^2}\, u_g(r)\right] dr \end{aligned}\right\} \tag{27.8}$$

---

[123] M. SUGAWARA: Phys. Rev. **99**, 1601 (1955). — Ark. Fysik **10**, 113, (1955). — Progr. Theor. Phys. **14**, 535 (1955).

and

$$\left.\begin{aligned}\Delta_{\text{exch}}\,\mu_D = \frac{2}{\pi}\,\alpha_g^2(\mu_P+\mu_N)\Bigg\{&\int_0^\infty u_g^2(r)\left[K_0(x)-\frac{K_1(x)}{x}\right]dr + \\ &+\sqrt{2}\int_0^\infty u_g(r)\,w_g(r)\left[K_0(x)+\frac{2K_1(x)}{x}\right]dr+\frac{1}{2}\int_0^\infty w_g^2(r)\left[K_0(x)+\frac{5K_1(x)}{x}\right]dr\Bigg\}+ \\ &+\frac{2}{\pi}\,\alpha_g^2\Bigg\{\frac{3\sqrt{2}}{2}\int_0^\infty u_g(r)\,w_g(r)\left[K_0(x)+\frac{2K_1(x)}{x}\right]dr - \\ &-\frac{3}{4}\int_0^\infty w_g^2(r)\left[K_0(x)+\frac{5K_1(x)}{x}\right]dr\Bigg\}-\mu_D\,\Delta N_g,\end{aligned}\right\}\qquad(27.9)$$

where

$$\left.\begin{aligned}\Delta N_g = \frac{2}{\pi}\,\alpha_g^2\Bigg\{&\int_0^\infty u_g^2(r)\left[K_0(x)-\frac{K_1(x)}{x}\right]dr + \\ &+4\sqrt{2}\int_0^\infty u_g(r)\,w_g(r)\left[K_0(x)+\frac{2K_1(x)}{x}\right]dr-\int_0^\infty w_g^2(r)\left[K_0(x)+\frac{5K_1(x)}{x}\right]dr\Bigg\},\end{aligned}\right\}\qquad(27.10)$$

$K_0(x)$ and $K_1(x)$ are the Hankel functions of imaginary arguments, $x=\mu r$, α_g^2 is $(f^2/4\pi\hbar c)\,(\mu/2\varkappa)^2$ and dashes mean derivatives with respect to r*. In (27.8) we have neglected the D-D cross term, since the S-D cross term turns out to be quite small. The corresponding expression for Q [123] is not quoted here, since it is much less important.

The numerical values of these corrections are then evaluated using the phenomenologically adjusted deuteron wave functions which are given in Sect. 33. It is shown[123] that the relativistic correction is almost insensitive to the D-state probability and increases very much if a hard core is introduced, while the exchange current correction is very sensitive to the D-admixture (it vanishes in a pure S-state) and is reduced very much by a hard core. Both corrections are definitely negative. Their sum turns out to be rather insensitive to a hard core and may be estimated to be -2% of the empirical deuteron moment, if we assume $f^2/4\pi\hbar c=10$ and the D-state probability is chosen as 3%. In the above figure, more than one half is due to the relativistic correction.

As regards the non-additivity correction, there are two investigations. In one of them[123], the effect of the deuteron binding energy appearing in the energy denominator of the Tamm-Dancoff solution has been estimated in the same approximation as above and is shown to be $+0.26\%$ of the empirical deuteron moment, the result being quite insensitive to the so-called cut-off momentum. In another paper[124], the effect of the two-meson exchange processes on the non-additivity correction has been estimated, and it is shown that a correction of

* This explicit appearance of the second derivative with respect to r is characteristic of the present method of estimating the relativistic correction. The conventional method has been to replace this second derivative (or the kinetic energy term) by the potential energy acting between two nucleons, using the fact that the total (relative) energy is the sum of the (relative) kinetic and potential energies. This method, however, suffers from the difficulty caused by the ambiguity of the nuclear force. As we have determined reasonable deuteron wave functions in Sect. 33, we need not eliminate the second derivative and have thus evaded this conventional difficulty.

[124] H. Miyazawa: Progr. Theor. Phys. 7, 207 (1952).

roughly $+(1 \pm 1)\%$ of the empirical moment would be expected from the current field theory. Although the above calculations need improving, we may reasonably expect that there may be a small positive non-additivity correction in addition to those mentioned previously.

Thus, we may conclude that the resulting theoretical corrections to the deuteron magnetic moment amount to about $-(1 \pm 1)\%$ of the empirical moment. As the total moment given by (27.7) must be compared with the empirical deuteron moment, we must subtract the above correction from the empirical value in order to get the phenomenological value μ_D given by (26.6). The negative correction, therefore, implies the reduction of the D-state probability P_D from its phenomenological value 3.9%. It is essential here to note that the above theoretical estimates give a definitely negative correction, which is rather much reduced if the deuteron wave function is fitted to a smaller P_D than 3.9%. These two features make it possible to get a final estimate of

$$P_D = (3 \pm 1)\%, \tag{27.11}$$

which seems rather definite, notwithstanding the ambiguities.

For the quadrupole moment, we can naturally expect a few percent correction of the same origin[123]. Such a correction is, however, quite unimportant compared with the uncertainty of P_D given by (27.11); the quadrupole moment is a much more well defined quantity than the D-state probability. We neglect, therefore, the correction entirely in the analysis of the two-nucleon problem since there are so many other ambiguities which are more serious.

28. Deuteron wave functions. It is shown in Sect. 24 that the deuteron problem with a tensor force is to solve the coupled differential equations (24.14):

$$\frac{d^2 u(r)}{dr^2} - [\alpha^2 - v_C(r)]\, u(r) + 2\sqrt{2}\, v_T(r)\, w(r) = 0, \tag{28.1}$$

$$\frac{d^2 w(r)}{dr^2} - \left[\alpha^2 + \frac{6}{r^2} - v_C(r) + 2 v_T(r)\right] w(r) + 2\sqrt{2}\, v_T(r)\, u(r) = 0, \tag{28.2}$$

with the normalization condition $\int\limits_0^\infty [u^2(r) + w^2(r)]\, dr = 1$ and the ordinary boundary conditions at $r=0$ and ∞. From the discussions about the deuteron moments, it is clear that $u(r)$ and $w(r)$ must have the same sign in order to give a positive Q [see Eq. (25.12)]; furthermore $w(r)$, which determines P_D by (26.5), is very roughly 0.2 times $u(r)$, judging from (27.11).

To see how these requirements are satisfied, it is very convenient to employ the idea of the effective (central) potential for $u(r)$ and $w(r)$, respectively, which are given, from (28.1) and (28.2), by

$$v_C(r) + 2\sqrt{2}\, v_T(r) \frac{w(r)}{u(r)} \tag{28.3}$$

and

$$v_C(r) - 2\, v_T(r) + 2\sqrt{2}\, v_T(r) \frac{u(r)}{w(r)}, \tag{28.4}$$

respectively. It is obvious that positive and negative $v_T(r)$ can both act as attractive or binding forces but only a positive $v_T(r)$, i.e. negative $V_T(r)$ [see definition after Eq. (24.8)], can give rise to the same sign of $u(r)$ and $w(r)$.

In order to allow a bound state, the above two effective potentials must be sufficiently strong. In particular the latter (28.4) must be much stronger than the other because of the repulsive term $6/r^2$ in (28.2). In fact the argument in Sect. 14 shows that Eq. (14.9) with $l=1$ can allow a bound state of a small binding energy, assuming a square well of depth v_0 and range r_0, only if $\sqrt{v_0} r_0 \approx \pi$, which means that the depth must be about four times larger than the depth of a force which allows an S-bound state of the same binding energy. This indicates that the force (28.4) must be at least four times stronger than (28.3), which demands that $v_T(r)$ is at least as large as $v_C(r)$ since we know approximately $w(r)/u(r) \approx 0.2$. It is important to note that a small ratio $w(r)/u(r)$ does not imply that $v_T(r)$ is small compared with $v_C(r)$. We also recall that the meson theoretical nuclear potential discussed in Sect. 11 is characterized by a large tensor force.

The above argument excludes a small $v_T(r)$ but not a $v_C(r)$ much smaller than $v_T(r)$ or even a repulsive $v_C(r)$, since (28.3) and (28.4) can be sufficiently strong even for such a $v_C(r)$ if only $v_T(r)$ is large enough. In fact such cases have been investigated[120] and the results show that bound state solutions of (28.1) and (28.2) are possible even for a repulsive $v_C(r)$. It seems, however, that an attractive $v_C(r)$ is necessary to explain the deuteron moments quantitatively[93], if we assume reasonable potential shapes for $v_C(r)$ and $v_T(r)$, as mentioned in Sects. 12 and 33.

The statement in Sect. 13 that the deuteron is a loosely bound physical system is still valid since the deuteron radius $1/\alpha$ is known to be larger than the ranges of $v_C(r)$ and $v_T(r)$, or the two effective potentials (28.3) and (28.4). Thus the asymptotic solutions of (28.1) and (28.2) which are given by

$$u(r) \propto e^{-\alpha r}, \qquad w(r) \propto e^{-\alpha r}\left(1 + \frac{3}{\alpha r} + \frac{3}{(\alpha r)^2}\right), \tag{28.5}$$

could be regarded as reasonable deuteron wave functions since they are independent of any detailed assumptions about the nuclear forces and depend only upon α, the binding energy. We cannot, however, adopt the above expressions without any modifications as was done in (13.10), since $w(r)$ has too strong a singularity at the origin even to be normalized. The following expressions, obtained by neglecting all the singular terms in $w(r)$, presumably make the simplest possible example of approximate wave functions:

$$u(r) = \sqrt{2\alpha}\, e^{-\alpha r}, \qquad w(r) = \sqrt{2\alpha}\, n\, e^{-\alpha r}, \tag{28.6}$$

where $u(r)$ alone is normalized to unity since $w(r)$ does not contribute much to the normalization integral because of the small P_D. The factor n is then determined so as to give the empirical value (1.3) of Q:

$$n = 0.22. \tag{28.7}$$

This gives a P_D of 4.8%, which is not so bad an approximation, considering the crudeness of the above functions.

From (28.1) we infer that $u(r)$ is proportional to r near the origin, while (28.2) shows that $w(r)$ vanishes as r^2 or r^3, depending on whether $v_T(r)$ behaves as r^{-1} or r^0 near the origin, respectively, because of the last term in (28.2)*. Generally speaking, therefore, $w(r)$ increases much more slowly near the origin, while decreasing much faster at larger distances than $u(r)$ [see Eq. (28.5)], thus making a relative peak somewhere near the range of the nuclear force.

In order to make the functions (28.5) behave correctly near the origin, we can use the same procedure which gave (13.11). With an eye to the possibility

* The appearance of logarithmic terms in the expansion of $u(r)$ and $w(r)$ does not affect the present argument.

of introducing a hard core of radius r_c, let us consider, among many alternatives, the following set of functions:

$$u(r) = N e^{-\alpha r}\left[1 - e^{-\beta(r-r_c)}\right], \tag{28.8}$$

$$w(r) = N\xi e^{-\alpha r}\left[1 - e^{-\gamma(r-r_c)}\right]^n \left[1 + \frac{3(1-e^{-\gamma r})}{\alpha r} + \frac{3(1-e^{-\gamma r})^2}{\alpha^2 r^2}\right], \tag{28.9}$$

where n is fixed at 2 or 3 when $r_c = 0$, while all positive n are allowed in the case $r_c \neq 0$ because of the change of the boundary conditions. The value of α is fixed if they are fitted to the correct binding energy, while the remaining parameters ξ, β and γ could be determined so as to fit other empirical properties of the deuteron. If they have been so adjusted, we may expect that the deuteron wave functions (28.8) and (28.9) are reasonable approximations. This is indeed the most rational and unambiguous way to construct deuteron wave functions, since we do not yet know much about the shape of the nuclear forces. The actual method to determine the parameters is postponed to Sect. 33, together with the results and relating discussions, since we must also fit them to the scattering data and the scattering theory with tensor forces is treated later on.

Contrary to the central force case, the coupled Eqs. (28.1) and (28.2) cannot be solved analytically even for the square well potential. It is practically always necessary to apply numerical integration. To get rid of the time-consuming and tedious numerical work, special automatic electronic computing machines have recently been employed to solve the coupled Eqs. (28.1) and (28.2). The first to be used was "Illiac" (University of Illinois Electronic Digital Computer), which has already contributed greatly in this field. Some of the numerical results are quoted in Sects. 31 and 33. Details about Illiac are found in some mimeographed notes circulated by the University of Illinois.

There have been some interesting attempts to obtain approximate analytic solutions of equation systems like (28.1) and (28.2) by variational methods[124a]. One slightly deviating procedure will be indicated here. First we note that in actual cases we do not want to calculate the eigenvalue α^2, but to determine the coupling constant or to adjust a numerical factor of the potential so as to give the correct deuteron binding energy. Suppose, for example, that we want to fix $v_C(r)$ and adjust the strength of $v_T(r)$ so as to give the correct binding energy. First we replace $v_T(r)$ by $\lambda v_T(r)$ in (28.1) and (28.2), which then read

$$\left.\begin{aligned}\frac{d^2u(r)}{dr^2} - [\alpha^2 - v_C(r)]\,u(r) &= -2\sqrt{2}\lambda\, v_T(r)\,w(r),\\ \frac{d^2w(r)}{dr^2} - \left[\alpha^2 + \frac{6}{r^2} - v_C(r)\right]u(r) &= -\lambda\left[2\sqrt{2}v_T(r)\,u(r) - 2v_T(r)\,w(r)\right].\end{aligned}\right\} \tag{28.10}$$

Defining I_1 and I_2 by

$$\left.\begin{aligned}I_1 &= \int\limits_0^\infty\left[\left(\frac{du(r)}{dr}\right)^2 + \left(\frac{dw(r)}{dr}\right)^2 + (\alpha^2 - v_C(r))\,(u^2(r) + w^2(r)) + \frac{6}{r^2}w^2(r)\right]dr,\\ I_2 &= \int\limits_0^\infty w(r)\left[4\sqrt{2}\,v_T(r)\,u(r) - 2v_T(r)\,w(r)\right]dr,\end{aligned}\right\} \tag{28.11}$$

we see that the coupled differential equations (28.10) are equivalent to the variational problem

$$\delta(I_1/I_2) = 0, \tag{28.12}$$

[124a] K. V. LAURIKAINEN and E. K. EURANTO: Ann. Univ. Turk. A **18**, 2 (1955). — K. V. LAURIKAINEN: Ann. Acad. Sci. fenn. A **1**, Nr. 208 (1955).

with independent variations of $u(r)$ and $w(r)$. It is also seen that the required λ is just equal to the stationary value of I_1/I_2:

$$\lambda = I_1/I_2. \tag{28.13}$$

If, on the other hand, we want to fix $v_T(r)$ instead of $v_C(r)$, then we have only to rewrite $v_C(r)$ as $\lambda v_C(r)$ in (28.1) and (28.2) and define the integrals

$$\left.\begin{aligned} I_1 &= \int_0^\infty \Big[\Big(\frac{du(r)}{dr}\Big)^2 + \Big(\frac{dw(r)}{dr}\Big)^2 + \alpha^2 u^2(r) + \Big(\alpha^2 + \frac{6}{r^2} + 2v_T(r)\Big) w^2(r) - \\ &\qquad - 4\sqrt{2}\, v_T(r)\, u(r)\, w(r)\Big] dr, \\ I_2 &= \int_0^\infty [2v_C(r)\, u(r)\, w(r)]\, dr. \end{aligned}\right\} \tag{28.14}$$

Then we can proceed in the same way as before. As trial functions we can use such expressions as (28.8) and (28.9), which allow us to calculate the integrals I_1 and I_2. If we minimize the quantity I_1/I_2 with respect to the parameters contained in $u(r)$ and $w(r)$, we can determine the required λ approximately and at the same time obtain the deuteron wave functions, which can be used to evaluate other physical quantities.

b) Scattering problem with tensor forces.

29. Phase shift analysis. The theory developed in Sect. 15 will now be extended to the tensor force case, following a method of BLATT and BIEDENHARN[125]. For more detailed discussion we refer to their original work and other papers quoted there.

Our purpose is eventually to construct a solution of (24.1) and (24.2) that has an asymptotic form (15.1), the differential cross section being again given by (15.3). The presence of a tensor force, however, makes the scattering depend also on φ, which makes it impossible to expand $\psi(\mathbf{r})$ in terms of LEGENDRE functions alone as is done in (15.4). We must instead use eigenfunctions Φ_{JJ_zL} defined by (24.4) to expand $\psi(\mathbf{r})$:

$$\psi(\mathbf{r}) = \sum_{J=0}^{\infty} \sum_{L=J-1}^{J+1} \frac{1}{r}\, u_{JL}(r)\, \Phi_{JJ_zL}(\vartheta, \varphi, \text{spin variables}), \tag{29.1}$$

since only the total (not the orbital) angular momentum is a constant of motion. In (29.1) J_z is given from the beginning, because J_z can be fixed if the incident wave is assumed to proceed along the z-axis. The radial wave functions $u_{JL}(r)$ satisfy the coupled Eqs. (24.8). Let us use in the following the notations $u_J(r)$, $v_J(r)$ and $w_J(r)$ instead of the three $u_{JL}(r)$, for $L=J-1$, J and $J+1$, respectively, as we did in (24.10) to represent the deuteron wave function. The incident plane wave e^{ikz} is expanded as in (15.7):

$$e^{ikz}\chi_{S_z} = \sum_{L=0}^{\infty} \sqrt{4\pi(2L+1)}\, i^L j_L(kr)\, Y_{L0}(\vartheta, \varphi)\, \chi_{S_z}, \tag{29.2}$$

where a spin function χ_{S_z} is inserted. The eigenvalue J_z in (29.1) must be equal to S_z in (29.2) since it is a constant of motion. It is now convenient to go over from (29.2) to an expansion analogous to (29.1), by expanding $Y_{L0}(\vartheta, \varphi)\, \chi_{S_z}$ in

[125] J. M. BLATT and L. C. BIEDENHARN: Phys. Rev. **86**, 399 (1952).

terms of the three Φ_{JS_zL}'s with $J=L-1$, L and $L+1$, which is just the inverse transformation of (24.4):

$$Y_{L0}(\vartheta,\varphi)\,\chi_{S_z}=\sum_{J=L-1}^{L+1} C_{JS_zL0}\,\Phi_{JS_zL}, \tag{29.3}$$

where the values of C_{JS_zL0} are explicitly given in Table 15. We then rewrite (29.2) as

$$e^{ikz}\chi_{S_z}=\sum_{J=0}^{\infty}\sum_{L=J-1}^{J+1}\sqrt{4\pi(2L+1)}\,i^L j_L(kr)\,C_{JS_zL0}\,\Phi_{JS_zL}. \tag{29.4}$$

In (29.1) and (29.4), the summation over L is restricted to $L=0$ and 1 in the exceptional case $J=0$. Asymptotically (29.4) behaves as

$$e^{ikz}\chi_{S_z}\xrightarrow[r\to\infty]{}\sum_{J=0}^{\infty}\sum_{L=J-1}^{J+1}\sqrt{4\pi(2L+1)}\,i^L\,\frac{\sin\left(kr-\frac{1}{2}L\pi\right)}{kr}\,C_{JS_zL0}\,\Phi_{JS_zL}. \tag{29.5}$$

Table 15. *The values of* C_{JS_zL0} *in* (29.3).

	$J_z=S_z=1$	$J_z=S_z=0$	$J_z=S_z=-1$
$J=L+1$	$\sqrt{\dfrac{L+2}{2(2L+1)}}$	$\sqrt{\dfrac{L+1}{2L+1}}$	$\sqrt{\dfrac{L+2}{2(2L+1)}}$
$J=L$	$-\dfrac{1}{\sqrt{2}}$	0	$+\dfrac{1}{\sqrt{2}}$
$J=L-1$	$\sqrt{\dfrac{L-1}{2(2L+1)}}$	$-\sqrt{\dfrac{L}{2L+1}}$	$\sqrt{\dfrac{L-1}{2(2L+1)}}$

Let us begin with the case $L=J$. It is shown in Sect. 10 that this state by itself is an eigenstate and no other states can mix in; this is also seen from the values of $S_{JLL'}$ in Table 14 which imply that $v_J(r)$ satisfies an ordinary uncoupled differential equation. Asymptotically $v_J(r)$ behaves as a force free solution and its most general asymptotic form is given by a linear combination of $\sin\left(kr-\frac{1}{2}J\pi\right)$ and $\cos\left(kr-\frac{1}{2}J\pi\right)$, or of an incoming and an outgoing wave:

$$v_J(r)\xrightarrow[r\to\infty]{}A\,e^{-i(kr-\frac{1}{2}J\pi)}-B\,e^{+i(kr-\frac{1}{2}J\pi)}. \tag{29.6}$$

The relative value of B (outgoing amplitude) to A (incoming amplitude) is determined by the boundary condition $v_J(r)=0$ at $r=0$. It is useful to define S by

$$B=SA, \tag{29.7}$$

where S is called the scattering matrix, which is just an ordinary number (in general complex) in this case. Since in a pure elastic scattering the flux of the outgoing wave is equal to that of the incoming wave, S must satisfy $|S|^2=1$ whence S can be written in the form

$$S=e^{2i\delta_{J,J}}, \tag{29.8}$$

where the real quantity $\delta_{J,J}$ is just the phase shift which is defined in Sect. 15, for the partial wave $J=L$. Indeed, substituting (29.8) into (29.6), one gets

$$v_J(r)\xrightarrow[r\to\infty]{}-2i\,A\,e^{i\delta_{J,J}}\sin\left(kr-\tfrac{1}{2}J\pi+\delta_{J,J}\right), \tag{29.9}$$

which is the same as (15.9).

The remaining two cases $L=J-1$ and $J+1$ are mixed up because of the tensor force and the two radial functions $u_J(r)$ and $w_J(r)$ satisfy the coupled differential equations (24.8), with k^2 for $-\alpha^2$. For large values of r, however, $u_J(r)$ and $w_J(r)$ satisfy the force free equations one by one and their asymptotic forms are expressed as

$$\left.\begin{aligned} u_J(r) &\xrightarrow[r\to\infty]{} A_1 e^{-i(kr-\frac{1}{2}(J-1)\pi)} - B_1 e^{+i(kr-\frac{1}{2}(J-1)\pi)}, \\ w_J(r) &\xrightarrow[r\to\infty]{} A_2 e^{-i(kr-\frac{1}{2}(J+1)\pi)} - B_2 e^{+i(kr-\frac{1}{2}(J+1)\pi)}. \end{aligned}\right\} \tag{29.10}$$

The outgoing amplitudes (B_1 and B_2) are again determined by the boundary conditions $u_J(r)=w_J(r)=0$ at $r=0$ for specified incoming amplitudes (A_1 and A_2). The scattering matrix is now a 2×2 matrix, defined by

$$B = SA, \tag{29.11}$$

where

$$B=\begin{bmatrix} B_1 \\ B_2 \end{bmatrix}, \quad S=\begin{bmatrix} S_{11} & S_{12} \\ S_{21} & S_{22} \end{bmatrix} \quad \text{and} \quad A=\begin{bmatrix} A_1 \\ A_2 \end{bmatrix}. \tag{29.12}$$

Now let us consider some general properties of the scattering matrix S. If we use the notation $\dagger$ to represent the hermite conjugate matrix, then the conservation of incident and outgoing fluxes demands the equality $A^\dagger A = B^\dagger B$ or $S^\dagger S = 1$, which implies that S is an unitary matrix. Since $u_J(r)$ and $w_J(r)$ satisfy the coupled Eqs. (24.8) with real coefficients, the pair of $u_J^*(r)$ and $w_J^*(r)$ must describe the same scattering as the original pair. This requires $A^* = SB^*$, since A^* and B^* represent the outgoing and incoming amplitudes, respectively, in the solutions $u_J^*(r)$ and $w_J^*(r)$. From the requirement $A^*=SB^*$, together with $S^\dagger S=1$, one gets $\tilde{S}=S$, where $\tilde{S}$ is the transpose of S, which implies that S is symmetric. A simple algebraic consideration shows that the most general unitary and symmetric 2×2 matrix contains 3 real parameters. Thus the scattering matrix can be written in a form

$$S = U^{-1} e^{2i\Delta} U, \tag{29.13}$$

where U is an unitary matrix containing only one real parameter and Δ is a diagonal matrix with real elements:

$$U=\begin{bmatrix} \cos\varepsilon_J & \sin\varepsilon_J \\ -\sin\varepsilon_J & \cos\varepsilon_J \end{bmatrix} \quad \text{and} \quad \Delta=\begin{bmatrix} \delta_{J,\alpha} & 0 \\ 0 & \delta_{J,\gamma} \end{bmatrix}. \tag{29.14}$$

From (29.13) and (29.14) the two eigenstates $A^{(\alpha)}$ and $A^{(\gamma)}$ of S, corresponding to eigenvalues $e^{2i\delta_{J,\alpha}}$ and $e^{2i\delta_{J,\gamma}}$, are obtained as

$$(A_2^{(\alpha)}/A_1^{(\alpha)}) = \tan\varepsilon_J, \quad (A_2^{(\gamma)}/A_1^{(\gamma)}) = -\cot\varepsilon_J, \tag{29.15}$$

for which the outgoing amplitudes are given, respectively, by

$$B^{(\alpha)} = e^{2i\delta_{J,\alpha}} A^{(\alpha)}, \quad B^{(\gamma)} = e^{2i\delta_{J,\gamma}} A^{(\gamma)}. \tag{29.16}$$

It has thus been shown that these particular mixtures of two states $L=J-1$ and $L=J+1$ are eigenstates of the scattering problem, in the sense that the scattering force produces only a shift of the phase without admixing any other states. We call the real constants $\delta_{J,\alpha}$ and $\delta_{J,\gamma}$ the phase shifts of the α- and γ-eigenstates belonging to the quantum number J. The real quantity ε_J is named mixing parameter in the state J. We may correspondingly denote the pure state $L=J$ as β and use the notation $\delta_{J,\beta}$ in (29.8) and (29.9) instead of $\delta_{J,J}$.

Substituting (29.15) and (29.16) into (29.10), we get, writing A_α and A_γ instead of $A_1^{(\alpha)}$ and $A_2^{(\gamma)}$, respectively,

$$\left.\begin{aligned} u_{J,\alpha}(r) &\xrightarrow[r\to\infty]{} -2i\,A_\alpha e^{i\delta_{J,\alpha}} \sin\left(kr - \tfrac{1}{2}(J-1)\pi + \delta_{J,\alpha}\right), \\ w_{J,\alpha}(r) &\xrightarrow[r\to\infty]{} -2i\,A_\alpha \tan\varepsilon_J\, e^{i\delta_{J,\alpha}} \sin\left(kr - \tfrac{1}{2}(J+1)\pi + \delta_{J,\alpha}\right) \end{aligned}\right\} \tag{29.17}$$

and

$$\left.\begin{aligned} u_{J,\gamma}(r) &\xrightarrow[r\to\infty]{} +2i\,A_\gamma \tan\varepsilon_J\, e^{i\delta_{J,\gamma}} \sin\left(kr - \tfrac{1}{2}(J-1)\pi + \delta_{J,\gamma}\right), \\ w_{J,\gamma}(r) &\xrightarrow[r\to\infty]{} -2i\,A_\gamma e^{i\delta_{J,\gamma}} \sin\left(kr - \tfrac{1}{2}(J+1)\pi + \delta_{J,\gamma}\right). \end{aligned}\right\} \tag{29.18}$$

We construct the following expressions with explicit subscripts S_z,

$$\left.\begin{aligned} \psi_{J,S_z,\alpha} &= \frac{1}{r}\, u_{J,S_z,\alpha}(r)\, \Phi_{J,S_z,J-1} + \frac{1}{r}\, w_{J,S_z,\alpha}(r)\, \Phi_{J,S_z,J+1}, \\ \psi_{J,S_z,\gamma} &= \frac{1}{r}\, u_{J,S_z,\gamma}(r)\, \Phi_{J,S_z,J-1} + \frac{1}{r}\, w_{J,S_z,\gamma}(r)\, \Phi_{J,S_z,J+1}, \end{aligned}\right\} \tag{29.19}$$

where only the factors A_α and A_γ in u and w depend on S_z, and get

$$\left.\begin{aligned} \psi_{J,S_z,\alpha} &\xrightarrow[r\to\infty]{} \frac{-2i}{r} A_{J,S_z,\alpha}\, e^{i\delta_{J,\alpha}} \sin\left(kr - \frac{1}{2}(J-1)\pi + \delta_{J,\alpha}\right) \times \\ &\qquad \times \left(\Phi_{J,S_z,J-1} - \tan\varepsilon_J\, \Phi_{J,S_z,J+1}\right), \\ \psi_{J,S_z,\gamma} &\xrightarrow[r\to\infty]{} \frac{-2i}{r} A_{J,S_z,\gamma}\, e^{i\delta_{J,\gamma}} \sin\left(kr - \frac{1}{2}(J+1)\pi + \delta_{J,\gamma}\right) \times \\ &\qquad \times \left(\tan\varepsilon_J\, \Phi_{J,S_z,J-1} + \Phi_{J,S_z,J+1}\right). \end{aligned}\right\} \tag{29.20}$$

The phase shifts $\delta_{J,\alpha}$ and $\delta_{J,\gamma}$ and the mixing parameter ε_J are uniquely determined by the requirements $u_J(0) = w_J(0) = 0$. In both cases u and w have the same real phase shift, as is seen from (29.17) and (29.18).

Corresponding to (29.19), let us define

$$\psi_{J,S_z,\beta} = \frac{1}{r}\, v_{J,S_z}(r)\, \Phi_{J,S_z,J}, \tag{29.21}$$

which behaves asymptotically as

$$\psi_{J,S_z,\beta} \xrightarrow[r\to\infty]{} \frac{-2i}{r} A_{J,S_z,\beta}\, e^{i\delta_{J,\beta}} \sin\left(kr - \frac{1}{2} J\pi + \delta_{J,\beta}\right) \Phi_{J,S_z,J}. \tag{29.22}$$

We then write (29.1) alternatively as

$$\left.\begin{aligned} \psi(\mathbf{r}) &= \sum_{J=0}^{\infty} \sum_{\alpha,\beta,\gamma} \psi_{J,S_z,\alpha} \\ &\xrightarrow[r\to\infty]{} \frac{-2i}{r} \sum_J A_{J,S_z,\alpha}\, e^{i\delta_{J,\alpha}} \sin\left(kr - \frac{1}{2}(J-1)\pi + \delta_{J,\alpha}\right) \left(\Phi_{J,S_z,J-1} - \right. \\ &\left. -\tan\varepsilon_J\, \Phi_{J,S_z,J+1}\right) + \frac{-2i}{r} \sum_J A_{J,S_z,\beta}\, e^{i\delta_{J,\beta}} \sin\left(kr - \frac{1}{2} J\pi + \delta_{J,\beta}\right) \Phi_{J,S_z,J} + \\ &+ \frac{-2i}{r} \sum_J A_{J,S_z,\gamma}\, e^{i\delta_{J,\gamma}} \sin\left(kr - \frac{1}{2}(J+1)\pi + \delta_{J,\gamma}\right) \times \\ &\qquad \times \left(\tan\varepsilon_J\, \Phi_{J,S_z,J-1} + \Phi_{J,S_z,J+1}\right), \end{aligned}\right\} \tag{29.23}$$

where the three angular and spin functions are mutually orthogonal. It must be emphasized here that $\delta_{J,\alpha}$ and $\delta_{J,\gamma}$ are not to be considered as the phase shifts in the states $L = J-1$ and $J+1$, respectively. There are no such phase shifts, since neither state is an eigenstate, although we may sometimes use the notations $\delta_{J,J-1}$ and $\delta_{J,J+1}$ instead of $\delta_{J,\alpha}$ and $\delta_{J,\gamma}$, for the sake of simplicity.

Thus far no prescription has been given for calling one of two eigenstates α and the other γ. By definition, we require that ε_J approaches zero in the limit of a vanishing tensor force; in general ε_J tends either to zero or to $\pi/2$ in this limit, since $\Phi_{J,S_z,J-1}$ and $\Phi_{J,S_z,J+1}$ themselves become eigenstates. Thus, as far as the effect of the tensor force is small, the α-state is mainly of the character $L=J-1$ and the γ-state $L=J+1$. The same situation is attained when we approach zero energy of the incident particle, since the different partial waves are decoupled in this limit because of the long range centrifugal force. Another ambiguity which is always inherent, viz. the possibility of adding any integral multiple of π to all phase shifts independently, can be removed by requiring all phase shifts to approach zero in the limit of a vanishing nuclear interaction.

Thus we have shown that for every J there are three phase shifts $\delta_{J,\alpha}$, $\delta_{J,\beta}$ and $\delta_{J,\gamma}$ and one mixing parameter ε_J. $\delta_{J,\beta}$ alone is the phase shift of a pure state $L=J$. In a certain sense we can also say that $\delta_{J,\alpha}$ and $\delta_{J,\gamma}$ are phase shifts of the states $L=J-1$ and $L=J+1$, respectively. The exceptional case $J=0$ is easily included by formally requiring that $\varepsilon_0=\delta_{0,\alpha}=\delta_{0,\beta}=0$, since for $J=0$ there is only one triplet state: $L=1$.

To obtain an explicit expression for the differential cross section in terms of the parameters defined above, we have only to break up the asymptotic form of $\psi(\boldsymbol{r})$, Eq. (29.23), into an incident plane wave and an outgoing spherical wave, using the relation

$$\left.\begin{aligned} &e^{i\delta_{J,\alpha}}\sin\left(kr-\tfrac{1}{2}(J-1)\pi+\delta_{J,\alpha}\right)=\\ &\quad=\sin\left(kr-\tfrac{1}{2}(J-1)\pi\right)+(-i)^{J-1}e^{i\delta_{J,\alpha}}\sin\delta_{J,\alpha}\,e^{ikr}.\end{aligned}\right\}\qquad(29.24)$$

The former part of $\psi(\boldsymbol{r})$ is then put equal to the corresponding asymptotic form of the incident wave (29.5), which determines the amplitudes $A_{J,S_z,\alpha}$ uniquely. Then the scattering amplitude (15.1) is given by

$$\left.\begin{aligned} f_{S_z}(\vartheta,\varphi)=\frac{\sqrt{4\pi}}{k}\sum_{J=0}^{\infty}&\left[A'_{J,S_z,\alpha}\,e^{i\delta_{J,\alpha}}\sin\delta_{J,\alpha}\left(\cos\varepsilon_J\,\Phi_{J,S_z,J-1}-\right.\right.\\ &-\sin\varepsilon_J\,\Phi_{J,S_z,J+1})+A'_{J,S_z,\beta}\,e^{i\delta_{J,\beta}}\sin\delta_{J,\beta}\,\Phi_{J,S_z,J}+\\ &\left.+A'_{J,S_z,\gamma}\,e^{i\delta_{J,\gamma}}\sin\delta_{J,\gamma}\left(\sin\varepsilon_J\,\Phi_{J,S_z,J-1}+\cos\varepsilon_J\,\Phi_{J,S_z,J+1}\right)\right],\end{aligned}\right\}\qquad(29.25)$$

The coefficients $A'_{J,S_z,\alpha}$ are introduced for convenience:

$$\left.\begin{array}{llll} & S_z=1 & S_z=0 & S_z=-1\\ {}_{,S_z,\alpha}= & \cos\varepsilon_J\sqrt{\frac{J+1}{2}}-\sin\varepsilon_J\sqrt{\frac{J}{2}}, & \cos\varepsilon_J\sqrt{J}+\sin\varepsilon_J\sqrt{J+1}, & \cos\varepsilon_J\sqrt{\frac{J+1}{2}}-\sin\varepsilon_J\sqrt{\frac{J}{2}},\\ {}_{,S_z,\beta}= & -\sqrt{\frac{2J+1}{2}}, & 0 & +\sqrt{\frac{2J+1}{2}},\\ {}_{,S_z,\gamma}= & \sin\varepsilon_J\sqrt{\frac{J+1}{2}}+\cos\varepsilon_J\sqrt{\frac{J}{2}}, & \sin\varepsilon_J\sqrt{J}-\cos\varepsilon_J\sqrt{J+1}, & \sin\varepsilon_J\sqrt{\frac{J+1}{2}}+\cos\varepsilon_J\sqrt{\frac{J}{2}};\end{array}\right\}\qquad(29.26)$$

where the values of $C_{J,S_z,L,0}$ are already inserted from Table 15. It is easily checked that the coefficients $A'_{J,S_z,\alpha}$ satisfy the relations

$$\sum_{S_z=-1}^{1}A'_{J,S_z,\alpha}A'_{J,S_z,\alpha'}=\delta_{\alpha\alpha'}(2J+1),\quad\sum_{\alpha'=\alpha}^{\beta,\gamma}A'_{J,S_z,\alpha'}A'_{J,S'_z,\alpha'}=\delta_{S_z,S'_z}(2J+1).\qquad(29.27)$$

The triplet differential scattering cross section for a polarized beam (fixed S_z but unspecified final spin state) is $\sum_{\substack{\text{spin}\\ \text{coordinates}}} |f_{S_z}(\vartheta, \varphi)|^2$. For an unpolarized beam, we get by averaging over S_z

$$\sigma_t(\vartheta, \varphi) = \tfrac{1}{3} \sum_{S_z=-1}^{+1} \sum_{\substack{\text{spin}\\ \text{coordinates}}} |f_{S_z}(\vartheta, \varphi)|^2, \tag{29.28}$$

which is finally given in terms of the phase shifts, the mixing parameters and the Legendre functions only and does not depend on φ, which is as it should be. The resulting expression is, however, rather complicated and the reader is referred to original papers[126].

The total triplet cross section σ_t takes, on the other hand, a very simple form. If we integrate (29.28) over ϑ and φ, then all cross terms vanish because of the orthogonality of $\Phi_{J,S_z,L}$, which makes the mixing parameters disappear in the total cross section. We get, using (29.27),

$$\sigma_t = \frac{4\pi}{k^2} \sum_{J=0}^{\infty} \frac{2J+1}{3} \left(\sin^2 \delta_{J,\alpha} + \sin^2 \delta_{J,\beta} + \sin^2 \delta_{J,\gamma}\right). \tag{29.29}$$

In a pure central force case, $\delta_{J,\alpha} = \delta_{J,J-1}$ and $\delta_{J,\gamma} = \delta_{J,J+1}$, which makes (29.29) change into

$$\sigma_t = \frac{4\pi}{k^2} \sum_{J=0}^{\infty} \sum_{L=J-1}^{J+1} \frac{2J+1}{3} \sin^2 \delta_{J,L} = \frac{4\pi}{k^2} \sum_{L=0}^{\infty} \sum_{J=L-1}^{L+1} \frac{2J+1}{3} \sin^2 \delta_{J,L}. \tag{29.30}$$

This proves identical with (15.14), if we remark that the phase shifts $\delta_{J,L}$ are independent of J and equal to the phase shift δ_L.

It should be stressed that in the presence of a tensor force the differential scattering cross section can no longer be expressed in terms of phase shifts only; the mixing parameters appear explicitly besides. This is due to the fact that the eigenfunctions of the scattering problem in question cannot be uniquely determined from general symmetry arguments, e.g. the rotational invariance of the scattering force. However, the total cross section can always be expressed in terms of phase shifts alone, because they are directly related to the eigenvalues S_n of the scattering matrix S by

$$S_n = \exp(2i\,\delta_n), \tag{29.31}$$

and the corresponding eigenstates are mutually orthogonal*. For this reason the unknown parameters in the eigenfunctions never appear explicitly in the total cross section.

If, however, we are interested only in the low energy region, we can neglect the mixing parameters and all phase shifts except $\delta_{1,\alpha}$, since this is the only one which is related to an S-state. Then the total triplet cross section is given by

$$\sigma_t = \frac{4\pi}{k^2} \sin^2 \delta_{1,\alpha} \tag{29.32}$$

[126] F. Rohrlich and J. Eisenstein: Phys. Rev. **75**, 705 (1949). — M. Matsumoto: Progr. Theor. Phys. **13**, 329 (1955).

* The relevant eigenfunctions $\Phi_{J,S_z,L}$ are those of the total angular momentum and therefore mutually orthogonal if we combine the integral over angular variables with the sum over spin, as in the total cross section. In the differential cross section, on the other hand we take only the sum over spin, so the orthogonality does not apply.

and the differential scattering cross section is spherically symmetric. Indeed, retaining only $\delta_{1,\alpha}$ and ε_1, (29.28) can be written

$$\sigma_t(\vartheta) = \frac{1}{k^2}\sin^2\delta_{1,\alpha}\left[1+\sin^2\varepsilon_1\left(2\cos\varepsilon_1+\frac{1}{\sqrt{2}}\sin\varepsilon_1\right)^2 P_2(\cos\vartheta)\right], \qquad (29.33)$$

which shows that ε_1 is negligible to the extent that the scattering is isotropic. We can show[125,127] that

$$\varepsilon_1 \approx Q\,k^2 \qquad (29.34)$$

near zero energy, Q being the electric quadrupole moment of the deuteron. Then, from (1.3), the coefficient of $P_2(\cos\vartheta)$ or $4\sin^2\varepsilon_1$ in (29.33) is estimated at 1×10^{-3} for an incident particle energy of 5 Mev, which is to be compared with the corresponding experimental error ($\approx 4\times10^{-3}$) in Table 1*. It turns out, therefore, that the low energy scattering data determine the energy dependence of the single phase shift $\delta_{1,\alpha}$; this is just the same conclusion obtained in Sect. 15 for a central force. The question how to analyse the energy dependence of $\delta_{1,\alpha}$ is discussed in the next section.

All the above arguments hold for more general nuclear interactions, since they are based on a universal property of the scattering matrix. We also add that the sign of the phase shift is no longer a direct measure of the sign of the nuclear force, since the central and non-central parts of the force can act in different ways.

30. Effective range theory. The discussion in the previous section has shown that a single quantity $\delta_{1,\alpha}$ is sufficient to determine the low energy neutron-proton scattering data even when there is a tensor force. The energy dependence of $\delta_{1,\alpha}$ is again most conveniently analysed in terms of the effective range expansion. The necessary modification of the theory in Sect. 16 is given here according to BIEDENHARN and BLATT[127].

Let us denote the α-wave functions at energies E_1 and E_2 by $(u_{\alpha 1}, w_{\alpha 1})$ and $(u_{\alpha 2}, w_{\alpha 2})$, respectively, each set of which satisfies the coupled differential equations (24.8) or (24.14), replacing α^2 by $-k_1^2 = -E_1 M/\hbar^2$ and $-k_2^2 = -E_2 M/\hbar^2$, respectively. In the same way as before, we multiply these two equations for E_1 by $u_{\alpha 2}$ and $w_{\alpha 2}$, those for E_2 by $u_{\alpha 1}$ and $w_{\alpha 1}$, respectively, and subtract one from the other, which gives

$$\left.\begin{aligned}\frac{d}{dr}\left[u_{\alpha 2}\frac{du_{\alpha 1}}{dr}+w_{\alpha 2}\frac{dw_{\alpha 1}}{dr}-u_{\alpha 1}\frac{du_{\alpha 2}}{dr}-w_{\alpha 1}\frac{dw_{\alpha 2}}{dr}\right]&=\\ =(k_2^2-k_1^2)\,(u_{\alpha 1}u_{\alpha 2}+w_{\alpha 1}w_{\alpha 2})&.\end{aligned}\right\} \qquad (30.1)$$

Let us then introduce the force free solutions u_α^0 and w_α^0, which are normalized as

$$\left.\begin{aligned}u_\alpha \xrightarrow[r\to\infty]{} u_\alpha^0 &\xrightarrow[r\to\infty]{} \cos\varepsilon\sin(k\,r+\delta_\alpha)/\sin\delta_\alpha,\\ w_\alpha \xrightarrow[r\to\infty]{} w_\alpha^0 &\xrightarrow[r\to\infty]{} \sin\varepsilon\sin(k\,r-\pi+\delta_\alpha)/\sin\delta_\alpha,\end{aligned}\right\} \qquad (30.2)$$

corresponding to (16.7), where δ_α and ε are the α-wave phase shift $\delta_{1,\alpha}$ and the mixing parameter ε_1 belonging to $J=1$. If we define $F_l(kr)$ and $G_l(kr)$

[127] L. C. BIEDENHARN and J. M. BLATT: Phys. Rev. **93**, 1387 (1954). The same results were obtained independently by B. C. H. NAGEL and P. O. OLSSON (unpublished manuscript).

* As already pointed out by BLATT and BIEDENHARN[125] the approximation employed here is not quite consistent: one ought to take P-states and coupled D-states into account. Their effect is, however, hard to estimate since they depend more on the unknown aspects of the nuclear forces.

as the regular and irregular solutions of the radial wave equation for a free particle of orbital angular momentum l and wave vector k, which we normalize as

$$F_l(kr) \xrightarrow[kr\to\infty]{} \sin\left(kr - \tfrac{1}{2} l\pi\right), \qquad G_l(kr) \xrightarrow[kr\to\infty]{} \cos\left(kr - \tfrac{1}{2} l\pi\right), \tag{30.3}$$

the complete expressions of u_α^0 and w_α^0 are given by

$$u_\alpha^0 = \cos\varepsilon\,[\cot\delta_\alpha F_0(kr) + G_0(kr)], \qquad w_\alpha^0 = \sin\varepsilon\,[\cot\delta_\alpha F_2(kr) + G_2(kr)], \tag{30.4}$$

where

$$\left.\begin{aligned} F_0(x) &= \sin x, \quad F_1(x) = \frac{\sin x}{x} - \cos x, \quad F_2(x) = \left(\frac{3}{x^2} - 1\right)\sin x - \frac{3}{x}\cos x,\\ G_0(x) &= \cos x, \quad G_1(x) = \frac{\cos x}{x} + \sin x, \quad G_2(x) = \left(\frac{3}{x^2} - 1\right)\cos x + \frac{3}{x}\sin x. \end{aligned}\right\} \tag{30.5}$$

These free solutions are, however, inconvenient for our present purpose, since w_α^0 given by (30.4) diverges at $r=0$. Let us, therefore, define modified asymptotic functions $\bar{u}_\alpha$ and $\bar{w}_\alpha$, which are finite at $r=0$ and approach u_α^0 and w_α^0 or u_α and w_α asymptotically:

$$\left.\begin{aligned} \bar{u}_\alpha &= u_\alpha^0 = \cos\varepsilon\,[\cot\delta_\alpha F_0(kr) + G_0(kr)],\\ \bar{w}_\alpha &= w_\alpha^0 - \frac{3\sin\varepsilon}{(kr)^2} = \sin\varepsilon\left[\cot\delta_\alpha F_2(kr) + G_2(kr) - \frac{3}{(kr)^2}\right], \end{aligned}\right\} \tag{30.6}$$

which are shown to satisfy

$$\left.\begin{aligned} &\left(\frac{d^2}{dr^2} + k^2\right)\bar{u}_\alpha = 0,\\ &\left(\frac{d^2}{dr^2} - \frac{6}{r^2} + k^2\right)\bar{w}_\alpha = -3\sin\varepsilon\,\frac{1}{r^2}. \end{aligned}\right\} \tag{30.7}$$

Then repeating the procedure which gave (30.1), we have

$$\left.\begin{aligned} &\frac{d}{dr}\left[\bar{u}_{\alpha 2}\frac{d\bar{u}_{\alpha 1}}{dr} + \bar{w}_{\alpha 2}\frac{d\bar{w}_{\alpha 1}}{dr} - \bar{u}_{\alpha 1}\frac{d\bar{u}_{\alpha 2}}{dr} - \bar{w}_{\alpha 1}\frac{d\bar{w}_{\alpha 2}}{dr}\right] =\\ &\quad = (k_2^2 - k_1^2)(\bar{u}_{\alpha 1}\bar{u}_{\alpha 2} + \bar{w}_{\alpha 1}\bar{w}_{\alpha 2}) - \frac{3}{r^2}(\sin\varepsilon_1\bar{w}_{\alpha 2} - \sin\varepsilon_2\bar{w}_{\alpha 1}), \end{aligned}\right\} \tag{30.8}$$

where ε_1 and ε_2 are the mixing parameters at energies E_1 and E_2, respectively. Using the power series of $F_2(kr)$ and $G_2(kr)$, we can show that the last term of (30.8) is finite at $r=0$.

The next step is, just as in Sect. 16, to integrate both Eqs. (30.1) and (30.8) from zero to infinity and subtract one from the other, which gives

$$\left.\begin{aligned} &\left[\bar{u}_{\alpha 2}\frac{d\bar{u}_{\alpha 1}}{dr} + \bar{w}_{\alpha 2}\frac{d\bar{w}_{\alpha 1}}{dr} - \bar{u}_{\alpha 1}\frac{d\bar{u}_{\alpha 2}}{dr} - \bar{w}_{\alpha 1}\frac{d\bar{w}_{\alpha 2}}{dr}\right]_{r=0} =\\ &\quad = (k_1^2 - k_2^2)\int_0^\infty [\bar{u}_{\alpha 1}\bar{u}_{\alpha 2} + \bar{w}_{\alpha 1}\bar{w}_{\alpha 2} - u_{\alpha 1}u_{\alpha 2} - w_{\alpha 1}w_{\alpha 2}]\,dr +\\ &\qquad + 3\int_0^\infty \frac{1}{r^2}(\sin\varepsilon_1\bar{w}_{\alpha 2} - \sin\varepsilon_2\bar{w}_{\alpha 1})\,dr, \end{aligned}\right\} \tag{30.9}$$

since the contribution from $r=\infty$ on the left-hand side of (30.9) vanishes and only the barred functions contribute at $r=0$. The left-hand side, by (30.6) and (30.5), becomes simply

$$\cos\varepsilon_1\cos\varepsilon_2(k_1\cot\delta_{\alpha 1} - k_2\cot\delta_{\alpha 2}), \tag{30.10}$$

since the derivative of $\overline{w}_\alpha$ vanishes at $r=0$. In order to evaluate the second term of the right-hand side of (30.9), we use the identities

$$\frac{d}{dx}\left[\frac{F_l(x)}{x^{l+1}}\right] = -\frac{F_{l+1}(x)}{x^{l+1}} \quad \text{and} \quad \frac{d}{dx}\left[\frac{G_l(x)}{x^{l+1}}\right] = -\frac{G_{l+1}(x)}{x^{l+1}}. \tag{30.11}$$

After a plain calculation, we get

$$\left.\begin{aligned} &3\int_0^\infty \frac{1}{r^2}(\sin\varepsilon_1 \overline{w}_{\alpha 2} - \sin\varepsilon_2 \overline{w}_{\alpha 1})\,dr = \\ &\qquad = -\sin\varepsilon_1 \sin\varepsilon_2 (k_1\cot\delta_{\alpha 1} - k_2\cot\delta_{\alpha 2}).\end{aligned}\right\} \tag{30.12}$$

Combining (30.12), (30.10), and (30.9), we finally get

$$\left.\begin{aligned} &\cos(\varepsilon_1-\varepsilon_2)(k_1\cot\delta_{\alpha 1} - k_2\cot\delta_{\alpha 2}) \\ &\quad = (k_1^2-k_2^2)\int_0^\infty [\overline{u}_{\alpha 1}\overline{u}_{\alpha 2} + \overline{w}_{\alpha 1}\overline{w}_{\alpha 2} - u_{\alpha 1}u_{\alpha 2} - w_{\alpha 1}w_{\alpha 2}]\,dr,\end{aligned}\right\} \tag{30.13}$$

which corresponds to (16.11).

Let us now recall that the mixing parameter approaches zero as the energy tends to zero. Defining the scattering length a_t in the same way as (16.13) by

$$\lim_{k^2\to 0}[k\cot\delta_\alpha] = -\frac{1}{a_t}, \tag{30.14}$$

we get from (30.13), making E_2 approach zero,

$$k\cot\delta_\alpha = -\frac{1}{a_t} + \frac{k^2}{\cos\varepsilon}\int_0^\infty [\overline{u}_\alpha\overline{u}_{\alpha 0} + \overline{w}_\alpha\overline{w}_{\alpha 0} - u_\alpha u_{\alpha 0} - w_\alpha w_{\alpha 0}]\,dr, \tag{30.15}$$

where we have dropped the subscript 1 and put zero to denote the zero energy wave functions. The integral appearing in (30.15) is shown to be insensitive to the energy, using the same argument as in Sect. 16, so far as we are interested in the low energy region. Thus, expanding all quantities in terms of the energy or k^2, we finally get the same effective range expansion as (16.31):

$$k\cot\delta_\alpha = -\frac{1}{a_t} + \frac{1}{2}k^2 r_{0t} - P_t k^4 r_{0t}^3 + \cdots, \tag{30.16}$$

where

$$r_{0t} \equiv \varrho_t(0,0) = 2\int_0^\infty [\overline{u}_{\alpha 0}^2 - u_{\alpha 0}^2 - w_{\alpha 0}^2]\,dr, \qquad \overline{u}_{\alpha 0} = 1 - \frac{r}{a_t}, \tag{30.17}$$

$$P_t = -\frac{1}{r_{0t}^3}\int_0^\infty [\overline{u}_{\alpha 0}\overline{u}'_{\alpha 0} - u_{\alpha 0}u'_{\alpha 0} - w_{\alpha 0}w'_{\alpha 0}]\,dr, \tag{30.18}$$

remarking that $\overline{w}_{\alpha 0}=0$, as is seen from (30.6), and ε is proportional to k^2 and, therefore, neglected in the above expansion up to k^4. The constants a_t, r_{0t} and P_t are the triplet scattering length, effective range and shape dependent parameter, respectively.

In order, then, to derive a relation between the scattering parameters defined above and the deuteron constants, let us start from the coupled Eqs. (24.14) satisfied by the deuteron wave functions u_g and w_g, which are now normalized as follows:

$$\left.\begin{aligned} u_g &\xrightarrow[r\to\infty]{} \cos\varepsilon_g e^{-\alpha r}, \\ w_g &\xrightarrow[r\to\infty]{} \sin\varepsilon_g e^{-\alpha r}\left[1 + \frac{3}{\alpha r} + \frac{3}{(\alpha r)^2}\right],\end{aligned}\right\} \tag{30.19}$$

where ε_g is the mixing parameter in the deuteron ground state. The asymptotic functions $\bar{u}_g$ and $\bar{w}_g$ are defined, corresponding to (30.6), as

$$\bar{u}_g = \cos\varepsilon_g e^{-\alpha r}, \quad \bar{w}_g = \sin\varepsilon_g \left[e^{-\alpha r}\left(1 + \frac{3}{\alpha r} + \frac{3}{(\alpha r)^2}\right) - \frac{3}{(\alpha r)^2}\right], \tag{30.20}$$

both of which are finite at $r=0$ and satisfy the same equations as (30.7), replacing k^2 and ε by $-\alpha^2$ and $-\varepsilon_g$, respectively. Even when we combine the α-wave functions at energy E, defined above, and the deuteron wave functions just introduced*, we can proceed in the same way as before, since the derivative of $\bar{w}_g$ also vanishes at $r=0$ and there is no difficulty in performing the integral $\int\limits_0^\infty \bar{w}_g dr/r^2$. Thus we finally get, instead of (30.13),

$$\cos(\varepsilon + \varepsilon_g)\cdot(k\cot\delta_\alpha + \alpha) = (k^2+\alpha^2)\int\limits_0^\infty [\bar{u}_\alpha\bar{u}_g + \bar{w}_\alpha\bar{w}_g - u_\alpha u_g - w_\alpha w_g]\,dr, \tag{30.21}$$

which corresponds to (16.12) in the central force case.

Making k approach zero in (30.21), we get the exact formula

$$\frac{1}{a_t} = \alpha - \frac{1}{2}\alpha^2\varrho(0, -\varepsilon)\frac{1}{\cos\varepsilon_g}, \tag{30.22}$$

where

$$\varrho(0, -\varepsilon) = 2\int\limits_0^\infty [\bar{u}_{\alpha 0}\bar{u}_g - u_{\alpha 0}u_g - w_{\alpha 0}w_g]\,dr. \tag{30.23}$$

Obviously it is $\varrho(0, -\varepsilon)/\cos\varepsilon_g$ and not $\varrho(0, -\varepsilon)$ that can be determined directly from the empirical values of a_t and α. From the argument given in Sect. 16, we may expand the deuteron wave functions in terms of α^2, neglecting for the time being the correct boundary condition at infinity. We then have, since ε_g can be considered as proportional to α^2,

$$\varrho(0, -\varepsilon)/\cos\varepsilon_g = \varrho(0, -\varepsilon) = \varrho(0, 0) + 2\alpha^2 r_{0t}^3 P_t + \cdots \tag{30.24}$$

with the same P_t as (30.18); this agrees with (16.34) or (19.3) on which the actual analysis was based in the previous chapters. The deuteron effective range (16.24) is now redefined as follows:

$$\varrho(-\varepsilon, -\varepsilon) = 2\int\limits_0^\infty [\bar{u}_g^2 + \bar{w}_g^2 - u_g^2 - w_g^2]\,dr = 2\left[\frac{1}{2\alpha} - \int\limits_0^\infty (u_g^2 + w_g^2)\,dr\right], \tag{30.25}$$

where the latter version can be obtained by an explicit integration using the expressions (30.20). From the expansion of the deuteron wave functions we can deduce that

$$\left.\begin{aligned}\varrho(-\varepsilon, -\varepsilon) &= \varrho(0, -\varepsilon) + 2\alpha^2 r_{0t}^3 P_t + \cdots\\ &= \varrho(0, -\varepsilon)/\cos\varepsilon_g + 2\alpha^2 r_{0t}^3 P_t + \cdots\end{aligned}\right\} \tag{30.26}$$

again with the same P_t as (30.18), which agrees with (16.33).

It has, therefore, been shown that the energy dependence of $\delta_{1,\alpha}$ (which is the counterpart of δ_0 in Sect. 16) and its connection with the deuteron constants α and $\varrho(-\varepsilon, -\varepsilon)$ are governed by the same formulas for a tensor force as in the case of a pure central force. All the results of the analysis of the low energy scattering data obtained in the previous chapters can, therefore, be applied without

* We apologize for using "α" in two different senses: an index α always refers to an α-wave function at energy E or zero, whereas α by itself means the inverse deuteron radius.

change. The only modification due to a tensor force is that the relation between the two triplet scattering parameters a_t and r_{0t} and the triplet shape dependent parameter P_t is not so simple as before. This problem is discussed in the next section.

31. Effect of tensor force on low energy scattering. In order to get a general idea about the influence of a tensor force on the triplet shape dependent parameter P_t, let us use the concept of the effective potential given by (28.3) and consider how it affects the S-wave function $u_\alpha(r)$, while the D-wave $w_\alpha(r)$ may be neglected for the time being.

To be definite, suppose that the central force $v_C(r)$ is attractive and of YUKAWA type. Since $w_\alpha(r)$ approaches zero faster than $u_\alpha(r)$ near the origin, the second term in (28.3), $2\sqrt{2}v_T(r)\,(w_\alpha(r)/u_\alpha(r))$, in general has the form of a lump. Here let us recall the general statement of Sect. 16 that P_t may be regarded as a measure of the potential tail. If the intrinsic range b_T of $v_T(r)$ is chosen large compared with that of $v_C(r)$, b_C, the addition of a tensor force is equivalent to introducing a lump on $v_C(r)$ at larger distances, which certainly extends its tail and is expected to make P_t larger. If, on the other hand, b_T is nearly equal to b_C, the effective potential (28.3) becomes more similar in shape to an exponential well, for example, which may be expected to reduce P_t relative to a central YUKAWA force.

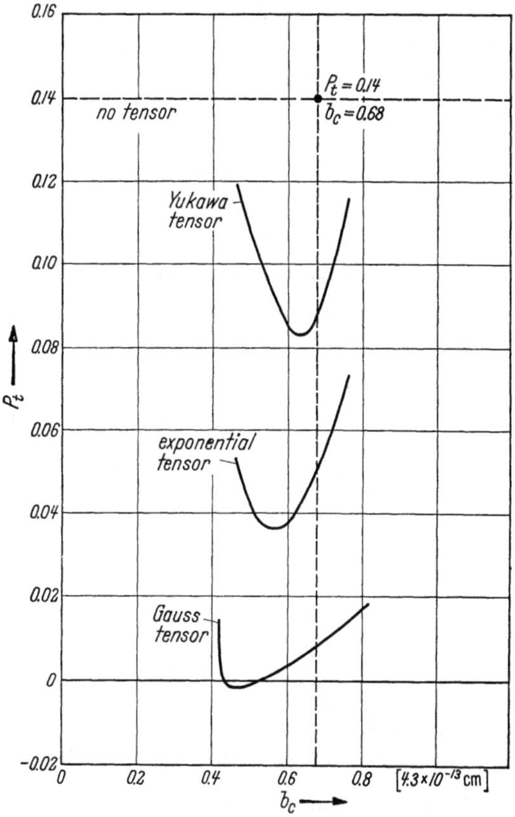

Fig. 19. The triplet shape dependent parameter P_t as a function of the intrinsic range b_C of a central YUKAWA potential, combined with various types of tensor forces, fitted to all triplet data except the D-state probability. A point at the top of the figure indicates a pure central YUKAWA well.

In order to get a quantitative result, let us imagine the four potential parameters of $v_C(r)$ and $v_T(r)$ — two well depths and two intrinsic ranges — adapted to the correct values of the deuteron binding energy, quadrupole moment and triplet effective range, which automatically include the scattering length in the shape independent approximation. Choose the undetermined parameter as b_C. One can then calculate P_t as a function of b_C, the result[128] being given in Fig. 19, in which a point for a pure YUKAWA central potential is also included. The value of P_D, empirically known as $(3\pm1)\%$, Eq. (27.11), and the connection between b_C and b_T have not been reported. It is, however, seen that for a reasonable value of b_C, which is presumably around 0.68, P_t is much reduced, viz. from the original $+0.14$ down to nearly zero. The authors mention that the minimum values occur when $b_T \approx b_C$. For values of b_C larger or smaller than 0.68 we see that P_t increases very rapidly.

Let us then consider central potentials of other types than that of YUKAWA. In these cases, $v_C(r)$ and the additional term in (28.3) due to the tensor force have more similar shapes, which implies that no great reductions of P_t would occur. Although there are no systematic calculations, we may quote some examples[128], all of which are but approximately fitted to the triplet data and relate only to a few reasonable values of b_C and b_T. They are summarized in Table 16 which also includes the data for pure central forces. The opposite extreme in which only tensor forces are assumed is also contained in Table 16 although these cases cannot be fitted to all triplet data.

As a general conclusion we may say that the introduction of reasonable types of tensor forces would reduce the big value of P_t attached to a pure central

Table 16. *The triplet shape dependent parameter P_t calculated for some combinations of central and tensor forces which are only approximately fitted to the triplet data, the intrinsic ranges being measured in units of 4.3×10^{-13} cm.*

Central well shape	Tensor well shape	b_C	b_T	P_t
GAUSS	no	0.5	—	−0.02
GAUSS	GAUSS	0.5	0.5	−0.02
GAUSS	YUKAWA	0.5	0.8	−0.004
GAUSS	YUKAWA	0.7	0.5	+0.02
Exponential	no	0.55	—	+0.03
Exponential	Exponential	0.5	0.7	+0.01
Exponential	YUKAWA	0.5	0.8	+0.027
Exponential	YUKAWA	0.7	0.6	+0.06
no	GAUSS	—	0.5	−0.03
no	Exponential	—	0.6	−0.02
no	YUKAWA	—	0.6	+0.02

YUKAWA well, while for other types of central forces the situation is too complicated to allow a simple statement. It seems possible to vary P_t within rather wide limits by introducing suitable tensor forces. As regards the effect of a hard core, the general consideration given in Sect. 16 is valid in this case, too. We may, therefore, add that P_t would always be further reduced, if we assume a hard core.

32. Charge independence hypothesis. In this section we enter upon a more detailed discussion of the figures in Table 11 and (19.8) compared with those in Table 13 and (23.1), which give the potential parameters adjusted to low energy data in singlet even neutron-proton and proton-proton systems and suggest that the nuclear force is the same for both, i.e. "charge independent". In Sect. 7 we have taken this charge independence for granted and introduced the symmetrical meson theory. The evidence carried by the low energy scattering is the most important support for this hypothesis, one reason being that the analysis of low energy data is relatively unambiguous.

A closer study of the tables referred to suggests that the neutron-proton and proton-proton range parameters could well be equal within existing errors, as long as we consider potentials with short tails [square well, exponential and the singular force (12.6) with $n=2$]. This is, however, not so for the long-tailed YUKAWA well. For all potentials the well depth parameter is a little smaller in the proton-proton case than in the neutron-proton case.

[128] M. H. KALOS and J. M. BLATT (unpublished calculations made on Illiac). Attention is drawn to a recent paper by these authors and L. C. BIEDENHARN: Nuclear Phys. 1, 233 (1956).

With respect to the first point, let us recall that the results in Table 11 are based on the triplet data in Table 9, which were obtained under a pure central force assumption. But a hard core (Sect. 19) as well as a tensor force (Sect. 31), if introduced in a triplet neutron-proton system, can reduce P_t very much for the YUKAWA potential. Almost any positive value from $+0.14$ downward could be attained if we assume a suitable tensor force and/or a reasonable hard core. As a quantitative example, take $P_t=0$ or $+0.03$ in Table 9 for the YUKAWA well, which gives $r_{0s}=(2.44\pm 0.23)$ or $(2.40\pm 0.23)\times 10^{-13}$ cm, respectively, instead of (2.21 ± 0.23) in Table 11. Then the potential parameters for a singlet n-p system are $b_s=(2.26\pm 0.22)$ or $(2.22\pm 0.22)\times 10^{-13}$ cm and $s_s=0.944\pm 0.006$ or 0.945 ± 0.006, respectively, for the YUKAWA potential, which should be compared with the corresponding values in Table 13. It can, therefore, be said that even in the YUKAWA case there is no definite evidence against the charge independence hypothesis, as far as the range parameters are concerned.

As to the second point, the difference is only about 3%, but we note that the well depth parameter is much more sharply defined than the effective range and much less sensitive to the potential shape (as exemplified by the calculation just quoted), which suggests that the observed difference may be real. It could, of course, be argued that the well depth parameter may be sensitive to a small deviation from the COULOMB law (see Sect. 23). Such a deviation would arise from the self meson field around the nucleons, which also induces the anomalous magnetic moments. The electromagnetic interaction between two nucleons due to these magnetic moments, however, has not been taken into account in the foregoing discussions. It is expected that this magnetic moment interaction will be no less important in this connection than a small deviation from the pure COULOMB law.

In order to estimate the effect of the magnetic moment interaction, let us define the magnetic moment density operator $\boldsymbol{\mu}_i(\boldsymbol{r})$ by

$$\boldsymbol{\mu}_i(\boldsymbol{r}) = \boldsymbol{\mu}_i\,\delta(\boldsymbol{r}-\boldsymbol{r}_i), \tag{32.1}$$

$\boldsymbol{\mu}_i$ and $\boldsymbol{r}_i$ being the magnetic moment and the coordinate of the i-th nucleon. The magnetic moment interaction energy $V^{(\mathrm{mag})}$ is then given by

$$V^{(\mathrm{mag})} = -\int \frac{\mathrm{rot}_1\,\boldsymbol{\mu}_1(\boldsymbol{r}_1)\cdot\mathrm{rot}_2\,\boldsymbol{\mu}_2(\boldsymbol{r}_2)}{|\boldsymbol{r}_1-\boldsymbol{r}_2|}\,d^3r_1\,d^3r_2, \tag{32.2}$$

which can be rewritten, after partial integrations, putting $\boldsymbol{r}\equiv\boldsymbol{r}_1-\boldsymbol{r}_2$, as

$$V^{(\mathrm{mag})} = -\frac{8\pi}{3}\,\boldsymbol{\mu}_1\cdot\boldsymbol{\mu}_2\,\delta(\boldsymbol{r}) - \frac{1}{r^3}\left[\frac{3(\boldsymbol{\mu}_1\boldsymbol{r})(\boldsymbol{\mu}_2\boldsymbol{r})}{r^2} - \boldsymbol{\mu}_1\cdot\boldsymbol{\mu}_2\right]. \tag{32.3}$$

With $\boldsymbol{\mu}_p=\mu_p\,\boldsymbol{\sigma}_p$ and $\boldsymbol{\mu}_n=\mu_n\,\boldsymbol{\sigma}_n$, we get

$$V_{np}^{(\mathrm{mag})} = 8\pi\,\mu_n\,\mu_p\,\delta(\boldsymbol{r}) \quad\text{and}\quad V_{pp}^{(\mathrm{mag})} = 8\pi\,\mu_p^2\,\delta(\boldsymbol{r}), \tag{32.4}$$

assuming a spin singlet S-state. It is readily seen that $V_{np}^{(\mathrm{mag})}$ is negative, while $V_{pp}^{(\mathrm{mag})}$ is positive, thus making the force between two protons a little weaker than that between a proton and a neutron. SCHWINGER[129] has investigated the problem thoroughly and shown that the effect on the force ranges is negligible, while there is an observable effect on the well depth parameters, especially when we assume the YUKAWA well, where it is just sufficient to make the above mentioned difference vanish; if we assume the square well, the effect is still too small to make the difference disappear.

---

[129] J. SCHWINGER: Phys. Rev. **78**, 135 (1950).

This is certainly a reasonable explanation. It is, however, to be noted that the above magnetic interaction energy (32.4) contains $\delta(\boldsymbol{r})$ and, therefore, vanishes if we introduce a hard core. This point has been investigated in detail by SALPETER[130]: He assumed a soft repulsive core, viz. a finite rectangular repulsive potential of radius r_c and height about $\frac{1}{2}Mc^2$ and at the same time replaced the δ-function in (32.4) by one extended over a distance of the order of r_c, for example by a function which is $3/(4\pi r_c^3)$ for $r<r_c$ and zero for $r>r_c$. Adjusting the outside potential so as to fit the low energy data, he got the following results: The introduction of a hard core of radius $r_c \gtrsim 0.3\times 10^{-13}$ cm makes the magnetic interaction effect vanish entirely, irrespective of the potential shapes assumed outside the hard core. According to Sect. 44, a hard core radius of the order of 0.6×10^{-13} cm seems to be necessary in the singlet even state, from the analysis of high energy scattering data. Thus we cannot apply SCHWINGER'S results as far as we keep to the hard core model.

There is another important effect which may cause a small well depth difference between the singlet forces. It is due to the observed charged and neutral meson mass difference[131]:

$$m_{\pi^\pm} = (273 \pm 0.5)\, m_e, \qquad m_{\pi^0} = (262 \pm 2)\, m_e. \tag{32.5}$$

This implies that, even assuming the symmetrical meson theory, the nuclear force is charge dependent, if we use the observed mass values (32.5) for charged and neutral mesons. More correctly, as the electromagnetic interaction is essentially charge dependent, even the symmetrical theory with equal masses for charged and neutral mesons does not predict exactly charge independent results. On the contrary, it is generally believed that the correct symmetrical meson theory will ultimately explain, for example, the magnetic moments of the nucleons, the proton-neutron mass difference and the charged-neutral meson mass difference etc., all of which are certainly charge dependent quantities. Although the present field theory is quite insufficient to explore these problems, let us dwell a little on this apparent lack of charge independence.

If we assume the charge symmetrical interaction (8.2), the second order static nuclear potential contains the factor $\tau_\alpha^{(1)}\tau_\alpha^{(2)}$ and the fourth order $\tau_\alpha^{(1)}\tau_\beta^{(1)}\tau_\alpha^{(2)}\tau_\beta^{(2)}$ or $\tau_\alpha^{(1)}\tau_\beta^{(1)}\tau_\beta^{(2)}\tau_\alpha^{(2)}$. Let us introduce phenomenologically the observed mass values (32.5) and estimate the effect of their inequality on the static nuclear potentials. In the second order potential, we remark that the last term in $\tau_\alpha^{(1)}\tau_\alpha^{(2)} = \tau_1^{(1)}\tau_1^{(2)} + \tau_2^{(1)}\tau_2^{(2)} + \tau_3^{(1)}\tau_3^{(2)}$ is due to a neutral meson exchange and, therefore, its force range is a little (about 4%) longer than that of the previous two terms. If we denote the COMPTON wavelength of a charged meson as $1/\mu$, the second order static nuclear potential (8.11) can be written, for a spin singlet state;

$$V(r) = -m_\pi c^2 \frac{g^2}{4\pi\hbar c}(\boldsymbol{\tau}^{(1)}\cdot\boldsymbol{\tau}^{(2)})\frac{e^{-\mu r}}{\mu r} + m_\pi c^2 \frac{g^2}{4\pi\hbar c}(\tau_3^{(1)}\tau_3^{(2)})\left(\frac{2}{\mu r}-1\right)e^{-\mu r}\frac{\Delta\mu}{\mu} \tag{32.6}$$

where $\Delta\mu/\mu = \frac{11}{273} = 0.04$.

In the derivation of (32.6) it must be remembered that a factor $1/\mu^2$ in (8.11) comes from the definition of the ps-pv coupling constant as g/μ [see Eq. (6.7)] and consequently should not be influenced by the change in m_π. This is perhaps more clearly seen from the transformed ps-ps interaction in (9.5), where $f/2\varkappa$ takes the place of g/μ. The second term in (32.6) is charge dependent and is positive for a proton-proton system, negative for a proton-neutron system if

[130] E. E. SALPETER: Phys. Rev. **91**, 994 (1953).
[131] K. M. CROWE and R. H. PHILLIPS: Phys. Rev. **96**, 470 (1954).

$\mu r < 2$. This is a rather large correction to the original second order force, and works in the right direction.

It should be remarked that the singlet second order force is too weak to be considered as satisfactory (see Sect. 33). Thus we must proceed to the fourth order. This has been done by SUGIE[132], who showed that the symmetrical ps-pv theory gives an effect of correct sign, which, combined with the correction to the second order potential, gives a result of the required sign and magnitude. Thus it may be possible that the small disparity of the singlet forces could be, at least partly, explained by the charged-neutral meson mass difference, although the indications are still preliminary. It should also be pointed out that we have here assumed the same coupling constants for charged and neutral mesons, whereas one should expect the electromagnetic interaction to make their effective values differ a little.

As a summary, we cannot say that there is a definite difference between the neutron-proton and proton-proton forces in singlet even states, nor can we show the rigorous equality of the forces, owing to ambiguities on the experimental as well as the theoretical side. Remembering, however, that an exact symmetrical meson theory can predict rather large apparent deviations from charge independence, the difference deduced from the analysis of the low energy scattering data seems to be small enough to admit the charge independence hypothesis. Thus we conclude that the charge independence of the nuclear forces is reasonably well established by the low energy scattering data, which favours the symmetrical meson theory.

33. Numerical results. α) *Phenomenological deuteron wave functions.* As pointed out in Sect. 28, the deuteron is a loosely bound physical system and therefore reasonable deuteron wave functions may be constructed by assuming suitable functional forms containing several parameters and adjusting these so as to fit the existing empirical information on the neutron-proton system.

The empirical facts which can be used for this purpose are the deuteron binding energy ε, the quadrupole moment Q, the D-state probability P_D and, in addition, the triplet scattering length a_t and the triplet effective range r_{0t}. The deuteron binding energy ε determines the correct asymptotic behaviour which is given by (28.5). All deuteron wave functions which behave correctly at larger distances must be regarded as fitted to the correct binding energy. Thus it is convenient to introduce $x \equiv \alpha r$ as a unit of length, where $1/\alpha = 4.316 \times 10^{-13}$ cm for $\varepsilon = 2.226$ Mev.

As remarked at the end of subsection 16α, the only requirement from the two scattering parameters a_t and r_{0t} is that the wave function should give a reasonable value of the deuteron effective range $\varrho(-\varepsilon, -\varepsilon)$, defined by (30.25), as far as it is already fitted to the correct binding energy. The reasonable value of $\varrho(-\varepsilon, -\varepsilon)$ can be estimated from (30.26). According to Sect. 31, P_t would be rather small and such a high value (+0.14) as is attained by a pure YUKAWA central force would not be actual. We therefore consider the following two values of $\varrho(-\varepsilon, -\varepsilon)$:

$$\varrho(-\varepsilon, -\varepsilon) = 1.704 \quad \text{and} \quad 1.734 \times 10^{-13}\,\text{cm}, \tag{33.1}$$

corresponding to $P_t = 0$ and $+0.0565$. As regards the other two deuteron constants, we assume the following values

$$Q = 2.738 \times 10^{-27}\,\text{cm}^2, \tag{33.2}$$

$$P_D = 3\%, 4\%, 5\%. \tag{33.3}$$

[132] A. SUGIE: Progr. Theor. Phys. **11**, 333 (1954).

This value of Q is just the empirical value (1.3) and we neglect its small theoretical ambiguity (Sect. 27) which is only a few percent. As regards P_D, we consider the above three values, although our theoretical estimate (27.11) favours a smaller figure, since 4% is the phenomenological value (26.7) and a larger P_D can not be definitely excluded.

This is all the empirical information which can be extracted from the low energy data. With regard to the high energy data (Sect. 44), we allow the possibility for a hard core of radius r_c to be introduced. Then, as a simple, but reasonable example, let us consider such deuteron wave functions as are given by (28.8) and (28.9). They behave correctly at $r=0$ as well as at larger distances and can in principle be fitted to the above three requirements for a given r_c, which we fix as

$$\left.\begin{aligned} x_c \equiv \alpha r_c &= 0, \quad 0.10, \quad 0.13, \\ r_c &= 0, \quad 0.4316, \quad 0.5610\times 10^{-13}\,\text{cm} \end{aligned}\right\} \tag{33.4}$$

from the considerations in Sect. 44.

In order to use the definition (30.25), it is important to note the normalization conditions (30.19). We therefore drop the common factor N and employ ε_g instead of ξ in (28.8) and (28.9), writing

$$\left.\begin{aligned} u(x) &= \cos\varepsilon_g\left[1-e^{-\beta(x-x_c)}\right]e^{-x}, \\ w(x) &= \sin\varepsilon_g\left[1-e^{-\gamma(x-x_c)}\right]^2 e^{-x}\left[1+\frac{3\,(1-e^{-\gamma x})}{x}+\frac{3\,(1-e^{-\gamma x})^2}{x^2}\right], \quad x=\alpha r \geq x_c, \end{aligned}\right\} \tag{33.5}$$

$u=w=0$, $x<x_c$, where we have chosen $n=2$, although $n=3$ is also allowed, including the case $x_c=0$. The normalization factor N is obtained from (30.25):

$$N^2 = \frac{2}{\left[1-\alpha\varrho(-\varepsilon,-\varepsilon)\right]} = 3.3047 \quad\text{and}\quad 3.3433, \tag{33.6}$$

Table 17. *The parameters in the deuteron wave function (33.5) fitted to all the triplet low energy data and a hard core radius x_c*.*

x_c (4.316×10^{-13} cm)	$\varrho(-\varepsilon,-\varepsilon)$ (10^{-13} cm)	P_D (%)	β	γ	$\sin\varepsilon_g$
0	1.704	3	4.860	2.494	0.03232
		4	4.751	2.922	0.02928
		5	4.647	3.275	0.02754
	1.734	3	4.741	2.505	0.03192
		4	4.637	2.936	0.02891
		5	4.536	3.289	0.02720
0.10	1.704	3	8.237	3.155	0.02942
		4	7.961	3.798	0.02666
		5	7.699	4.346	0.02514
	1.734	3	7.933	3.175	0.02901
		4	7.675	3.814	0.02634
		5	7.431	4.364	0.02487
0.13	1.704	3	10.223	3.413	0.02873
		4	9.814	4.144	0.02611
		5	9.433	4.771	0.02471
	1.734	3	9.774	3.436	0.02832
		4	9.397	4.170	0.02577
		5	9.045	4.799	0.02438
errors of calculation			$\Delta\beta=0.0005$	$\Delta\gamma=0.002$	$\Delta\sin\varepsilon_g=0.00002$

* Numerical calculations by L. T. HEDIN and P. H. L. CONDÉ.

for the two cases in (33.1), respectively. The calculated values of β, γ and $\sin\varepsilon$ are summarized in Table 17. Some of the deuteron wave functions determined in this way are plotted in Figs. 20 and 21.

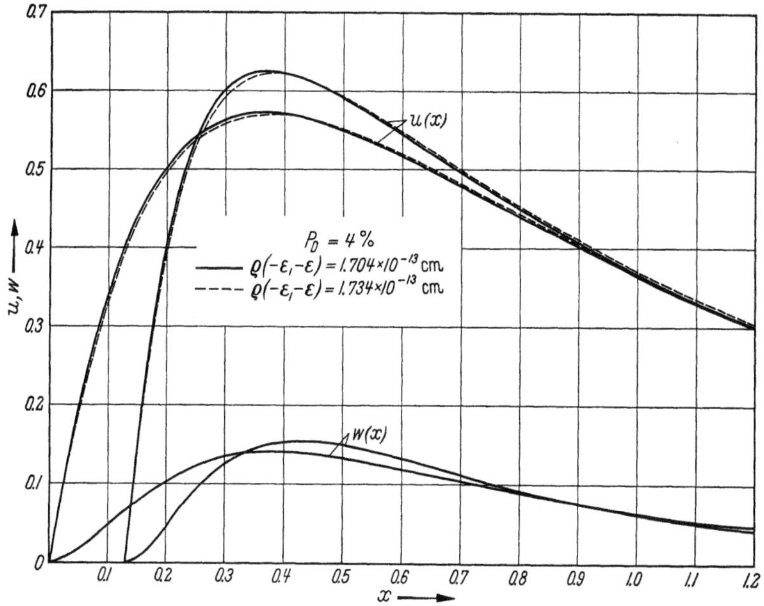

Fig. 20. The phenomenologically adjusted deuteron wave functions (33.5) fitted to the values $P_D = 4\%$, $\varrho(-\varepsilon, -\varepsilon) = 1.704$ and 1.734×10^{-13} cm, $x_c = 0$ and 0.13, besides the correct values of the binding energy and the quadrupole moment. [The difference between the two uppermost curves (for $x_c = 0.13$) is somewhat exaggerated in the outer region; it should be only 0.6 %.]

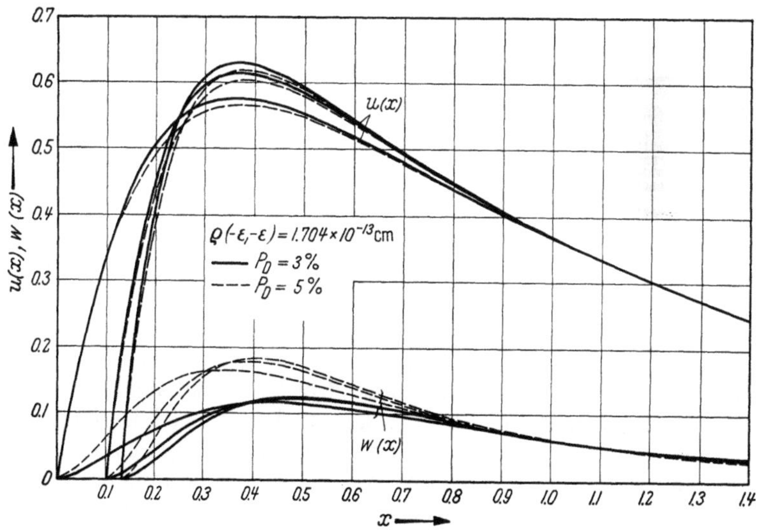

Fig. 21. The phenomenologically adjusted deuteron wave functions (33.5) fitted to the values $\varrho(-\varepsilon, -\varepsilon) = 1.704 \times 10^{-13}$ cm, $P_D = 3\%$ and 5%, $x_c = 0$, 0.10 and 0.13, besides the correct values of the binding energy and the quadrupole moment.

It is seen from the figures that among the three quantities $\varrho(-\varepsilon, -\varepsilon)$, P_D and x_c, the ambiguity arising from $\varrho(-\varepsilon, -\varepsilon)$ is a minor one, while the latter two give rise to big uncertainties in the deuteron wave functions, especially at smaller distances. Besides, there is a certain amount of arbitrariness in the particular functional form (33.5) chosen above. According to a preliminary calculation[123] the effect of changing the power $n=2$ to $n=3$ in (33.5) [cf. Eq. (28.9)] seems much less important than the effect due to the change of P_D by 1% *. It appears, therefore, that the ambiguities of P_D and x_c given by (33.3) and (33.4), respectively, are the main reasons for the uncertainty of the phenomenologically adjusted deuteron wave functions. The bearing of these ambiguities on actual applications is discussed in the paper by SUGAWARA[123].

β) Numerical results for particular potentials. In this subsection we quote some results of calculations of the deuteron constants (binding energy ε, quadrupole moment Q, D-state probability P_D) and low energy scattering parameters (a_t and r_{0t}) based on various nuclear potentials. Because of the fact that P_D is rather ambiguous and a_t is nearly correct if a potential is adjusted to ε and r_{0t}, the potential cannot be determined uniquely even if we specify functional forms of central and tensor forces, $V_C(r)$ and $V_T(r)$.

We get a general picture from a calculation[133] in which the YUKAWA well is assumed for both $V_C(r)$ and $V_T(r)$:

$$V_C(r) = -V_C \exp[-2.1196(r/b_C)]/(r/b_C) \tag{33.7}$$

and corresponding for $V_T(r)$ in terms of V_T and b_T. The results of the calculation are found in Table 18, where the potentials are fitted to $\varepsilon=2.23$ Mev and $Q=2.766\times10^{-27}$ cm². We see from the table that a combination of nearly equal central and tensor forces ($b_C\approx b_T$ and $V_T\approx V_C$) seems to give reasonable values

Table 18. *The triplet constants calculated for potentials, both central and tensor parts of which are assumed of the* YUKAWA *type and which are adjusted to the binding energy and quadrupole moment.*

b_C (10^{-13} cm)	b_T (10^{-13} cm)	V_T/V_C	P_D (%)	r_{0t} (10^{-13} cm)
2.863	2.927	1.7860	4.2	1.71
2.516	3.247	0.8359	3.8	1.71
2.342	3.247	0.7344	3.9	1.68
2.049	4.500	0.1696	2.8	1.68
2.049	5.854	0.0788	2.1	1.68

of P_D and r_{0t}. It is also interesting that these b_C and b_T are roughly the same as b_t in Table 9. Even for a very small V_T, the results are still quite satisfactory if b_T is chosen suitably large.

We then turn to a calculation[134] based on the meson theoretical potentials discussed in Sect. 11. As regards the second order potential $V_2(\boldsymbol{r})$, (8.11), it is known that if the coupling constant is adjusted to the triplet data, then the singlet force is too weak to fit the known singlet data: $V_2(\boldsymbol{r})$ consists of a large tensor force and a smaller central force, which is the same for singlet and triplet even states. In order to examine this point in detail, let us add the following

* In Table 2 of the paper by M. SUGAWARA, Ark. Fysik **10**, 113 (1955), the effect of changing n is large as regards the calculated values of $\Delta Q/Q$, which is, however, due to the fact that this is an "outside" quantity and the deuteron wave functions (55) in the above paper do not behave correctly at larger distances. The importance of changing n should be inferred from its effect upon $\Delta\mu_D/\mu_D$ which is an "inside" quantity.

[133] H. FESHBACH and J. SCHWINGER: Phys. Rev. **84**, 194 (1951).

[134] J. M. BLATT and M. H. KALOS: Phys. Rev. **92**, 1563 (1953). Cf. a recent paper by M. H. KALOS, L. C. BIEDENHARN and J. M. BLATT: Nuclear Phys. **1**, 233 (1956).

central force to $V_2(\boldsymbol{r})$:

$$V_{\text{add}}(r - r_c),$$

where

$$V_{\text{add}}(r) = -m c^2 \left(\frac{G^2}{4\pi\hbar c}\right)^2 \left(\frac{2\varkappa}{\mu}\right)^2 \frac{6}{\pi} \frac{K_1(2\mu r)}{(\mu r)^2}, \tag{33.8}$$

and r_c is a hard core radius. The above additional force is nothing but the first term of the fourth order potential $V_4(\boldsymbol{r})$ (9.14) written with $G^2/4\pi\hbar c = \lambda(\mu/2\varkappa)^2(f^2/4\pi\hbar c)$. Such an additional term tells us how the enlargement of the central force would improve the situation, while it has also a theoretical interest, because (33.8) would be the largest term in $V_4(\boldsymbol{r})$ if there is no pair damping[84]. The force range parameter μ is fixed at $1/\mu = 1.414 \times 10^{-13}$ cm by the known meson mass value[113], so we have only two adjustable parameters, the coupling constant $g^2/4\pi\hbar c$ and the hard core radius r_c. First we choose these two so as to give correct values of ε and Q without the additional term (33.8). Then we can calculate other quantities, including the singlet parameters, assuming the same hard core in both states; they are, however, seen (Table 19) to be quite unreasonable. In order to improve this, let us increase r_c and adjust the two coupling constants $g^2/4\pi\hbar c$ and $G^2/4\pi\hbar c$ so as to fit ε and Q. The results are summarized in Table 19.

Table 19. *Calculated values[134] of triplet and singlet parameters for the meson theoretical second order potential $V_2(\boldsymbol{r})$ (8.11) with an added central force (33.8), cut off by a hard core of radius r_c. All cases are fitted to correct binding energyand quadrupole moment. The same hard core is assumed in both states.*

r_c (10^{-13} cm)	$g^2/4\pi\hbar c$	$G^2/4\pi\hbar c$	P_D (%)	$\varrho(-\varepsilon, -\varepsilon)$ (10^{-13} cm)	r_{0s} (10^{-13} cm)	a_s (10^{-13} cm)
0.456	0.0752	0	7.18	1.66	—	− 0.018
0.531	0.0728	0.0563	6.55	1.749	35.324	− 0.478
0.604	0.0701	0.0563	5.91	1.835	9.659	− 1.327
0.798	0.0645	0.1017	4.41	2.068	3.520	− 16.024
1.036	0.0605	0.1628	3.08	2.334	2.974	+ 11.251
Experimental values			3 ± 1	1.704	2.40	− 23.69

It is seen from the first line of Table 19 that $V_2(\boldsymbol{r})$ can well be fitted to all triplet data except P_D if we choose a rather big r_c (0.456×10^{-13} cm) and a rather small $g^2/4\pi\hbar c$ (0.0752), while the singlet data are completely unreasonable. Too small an a_s means too weak a singlet force, which is due to the tensor force being predominant in $V_2(\boldsymbol{r})$; this is also reflected in the large P_D (7.18%). If we introduce an additional central force (33.8), singlet data and P_D are much improved, while $\varrho(-\varepsilon, -\varepsilon)$ becomes too large to be accepted. Therefore, we see that both singlet and triplet data can never be accounted for at the same time even if we allow an additional term (33.8), which has also been ascertained for other shapes of $V_{\text{add}}(r)$ (GAUSS, YUKAWA and MORSE wells have been considered). Even the triplet data seem hard to fit at the same time, because of too large a tensor force in $V_2(\boldsymbol{r})$. It should be noted that $g^2/4\pi\hbar c$ is very insensitive to r_c; $g^2/4\pi\hbar c$ is almost entirely determined by Q, which then makes either P_D or $\varrho(-\varepsilon, -\varepsilon)$ too large to be accepted, since we cannot adjust the force range parameter $1/\mu$. Characteristic of the potential investigated is that the central force varies but is always the same in singlet and triplet even states, while the large tensor force in $V_2(\boldsymbol{r})$ is exploited throughout.

As appears from Fig. 9 in Sect. 11, the effect of adding the total fourth order potential $V_4(\boldsymbol{r})$ (9.14) is to make the tensor force in a triplet even state somewhat weaker and the singlet even central force much stronger than in $V_2(\boldsymbol{r})$ alone. These features both tend to remedy the shortcomings of $V_2(\boldsymbol{r})$. We quote here the results of calculation[85] for a potential $V_2(\boldsymbol{r})+V_4(\boldsymbol{r})$ with $\lambda=0$, cut off by a hard core of radius r_c and adjusted to the correct binding energy.

$$\left.\begin{aligned} g^2/4\pi\hbar c &= 0.0851\,, \quad r_c = 0.424\times 10^{-13}\,\text{cm}, \quad Q = 2.83\times 10^{-27}\,\text{cm}^2, \\ P_D &= 6.12\%, \quad \varrho(-\varepsilon, -\varepsilon) = 1.73\times 10^{-13}\,\text{cm}. \end{aligned}\right\} \tag{33.9}$$

These may be regarded as reasonable although P_D is still somewhat large, which is due to the fact that even for $\lambda=0$ the tensor force dominates in a triplet even state (Fig. 9). As regards the singlet data, we first remark that there is no a priori reason why the same hard core should be assumed both in singlet and triplet even states. From the empirical point of view, it seems necessary to introduce a larger hard core in a singlet even state than in a triplet state (Sect. 44). Admitting this we get the following singlet parameters for the potential $V_2(\boldsymbol{r})+V_4(\boldsymbol{r})$ with $\lambda=0$[85]:

$$\left.\begin{aligned} g^2/4\pi\hbar c &= 0.0735\,, \quad r_c = 0.464\times 10^{-13}\,\text{cm}, \quad r_{0s} = 2.10\times 10^{-13}\,\text{cm}, \\ &= 0.0884, \quad\quad\; = 0.543\times 10^{-13}\,\text{cm}, \quad\quad = 2.585\times 10^{-13}\,\text{cm}, \end{aligned}\right\} \tag{33.10}$$

where both cases are adjusted to the correct singlet scattering length. In Sect. 44, it is shown that $r_c\approx 0.6\times 10^{-13}$ cm would be reasonable in a singlet even state. Recalling that the empirical value of r_{0s} is 2.40×10^{-13} cm in the shape independent approximation with an error of 10%, the values (33.10) can be regarded as satisfactory. Thus we may conclude that the situation is much improved if we include $V_4(\boldsymbol{r})$; in particular there is no difficulty with the singlet even potential if we allow a larger hard core in a singlet than in a triplet state. On the other hand the force in triplet even states still suffers from the same flaws as $V_2(\boldsymbol{r})$ itself: P_D is somewhat too large ($6\sim 7\%$) if other triplet data are correct, because of the large tensor force. This is, however, a feature common to all meson theoretical potentials.

Another version of the fourth order potential mentioned in Sect. 9 (cf. also Sect. 11) is characterized by a strongly repulsive central force combined with a large tensor force which is almost the same as that of $V_2(\boldsymbol{r})+V_4(\boldsymbol{r})$ considered above. Let us first choose a suitable value of $g^2/4\pi\hbar c$, introduce an adjustable parameter λ and write the potential as follows[90]:

$$V(\boldsymbol{r}) = \lambda\,[V_C(r) + V_T(r)\,S_{12}]\,, \tag{33.11}$$

in order to see whether the deuteron can exist under such a strong repulsive central force. Calculations by Illiac indicate that for reasonable hard core radii the deuteron can be bound only for negative λ (correct binding energy for $\lambda\approx -1$). This means that the potential can explain the deuteron binding for reasonable values of coupling constant and hard core radius only when its sign is reversed, but then the connection with the meson theory is lost and the sign of the quadrupole moment, for instance, goes wrong. From a purely phenomenological point of view, such a potential cannot be accepted.

We have seen above that the central force in triplet even states comes out fairly weak in the meson theoretical potential. It is then of interest to quote the conclusion of some authors[90] that an attractive central force of a magnitude comparable with the tensor force seems necessary to explain Q and P_D at the same time; a central force which is nearly zero or weakly repulsive would be excluded.

Thus far we have discussed only forces in even states. As regards the force in a triplet odd state, there is an interesting calculation[135] which shall be quoted here. The low energy proton-proton scattering data summarized in Table 2 suggest a negative P-wave phase shift at low energies. It is first to be noted that because of the tensor force there are three triplet P-wave phase shifts which are denoted, according to Sect. 29, by $\delta_{J,L}$ with $L=1$ and $J=0, 1, 2$. Among these, $J=0$ and 1 are pure states, while $J=2$ is a mixture of two states with $L=1$ and 3, in which, however, the share of $L=3$ can certainly be neglected in this low energy region. Thus we have three phase shifts $\delta_{0,1}$, $\delta_{1,1}$ and $\delta_{2,1}$ instead of one phase shift δ_1 given in Table 2. Now all these phase shifts are quite small, which implies that the combination $\delta_1 \equiv \frac{1}{9}(\delta_{0,1}+3\delta_{1,1}+5\delta_{2,1})$ alone appears in the expression of the differential cross section at very low energies, as follows from the theory in Sect. 29, and that this combination is just the P-wave phase shift empirically determined in Table 2. It is also important to note that at these low energies only the nuclear force at larger distances affects $\delta_{J,1}$ because of the centrifugal force. Furthermore, at these larger distances only the second order potential $V_2(\boldsymbol{r})$, Eq. (8.11), is important; the fourth order gives only slight modifications because of the shorter range (Sect. 11). This means that the ambiguities of the fourth order force do not matter here. For an incident proton laboratory energy of 3.9 Mev the results are:

$$\left.\begin{aligned} g^2/4\pi\hbar c &= \quad 0.06, & & 0.08, & & 0.10, \\ \delta_1 &= -0.11 \sim -0.09, & & -0.14 \sim -0.10, & & -0.17 \sim -0.11, \end{aligned}\right\} \quad (33.12)$$

employing the potentials $V_2(\boldsymbol{r})$ (8.11) plus $V_4(\boldsymbol{r})$ (9.14), cut off at $r_c=1.40\times \times 10^{-13}$ cm (left figure) or $r_c=0.42\times 10^{-13}$ cm (right figure) with zero values inside r_c. The fact that the results (33.12) are fairly insensitive to r_c confirms that the interior behaviour of the nuclear force does not affect δ_1. If we compare these values with those of Table 2, we may say that the agreement is reasonable, which gives a certain support to the current meson field theory.

VI. Radiative processes.

In order to define various corrections to the phenomenological theory of radiative processes unambiguously, we start in the following, not from the usual classical treatment which was given in Sect. 25 in defining multipole moments of the nucleus, but straight from the relativistic quantum field theory*. Such considerations were already made in Sect. 27 in order to discuss the corrections to the phenomenological deuteron moments.

a) Interaction with electromagnetic field.

34. General form of the interaction. The Hamiltonian for the coupled nucleon, meson and electromagnetic fields has been given in Sect. 27. We introduce the Pauli-term (27.2) from the beginning, which is equivalent to assuming the strict additivity of nucleon moments, only because there is no other simple and satisfactory way of treating the anomalous magnetic moments of nucleons.

---

[135] S. Otsuki and R. Tamagaki: Progr. Theor. Phys. **12**, 806 (1954).

* The principles of quantized field theory are treated in detail in Vol. V of this Encyclopedia in two articles by J. Schwinger and G. Källén.

We furthermore employ the simple choice of gauge $A_4 = 0$ to represent the electromagnetic field, which is allowed since in the radiative processes we may regard it as a transverse photon field. Although we are originally interested in a ps-ps meson theory, we may just as well start from a ps-pv meson theory, since both are known to be equivalent, as far as the lowest order calculation is concerned. This equivalence can be shown using the procedure given in Sect. 9α even if we include the interaction with the electromagnetic field. In order to simplify the calculation we start from a ps-pv theory, in which it is necessary to add the following term to (27.1):

$$\frac{i f e}{2\varkappa \hbar c} \sum_{\mu} \overline{\psi} \gamma_5 \gamma_\mu A_\mu (\tau_2 \varphi_1 - \tau_1 \varphi_2) \psi. \tag{34.1}$$

Then, under the special gauge chosen, we get as H'' the expression

$$\left.\begin{aligned} H'' = &- e \psi^* \frac{1+\tau_3}{2} \boldsymbol{\alpha} \psi \boldsymbol{A} - \frac{e}{2\varkappa} \psi^* \left[(\mu_P - 1) \frac{1+\tau_3}{2} + \mu_N \frac{1-\tau_3}{2} \right] \beta \boldsymbol{\sigma} \psi \operatorname{rot} \boldsymbol{A} + \\ &+ \frac{e}{\hbar c} \left(\varphi_1 \frac{\partial}{\partial \boldsymbol{x}} \varphi_2 - \varphi_2 \frac{\partial}{\partial \boldsymbol{x}} \varphi_1 \right) \boldsymbol{A} + \frac{e f}{2 \hbar c \varkappa} \psi^* \boldsymbol{\sigma} (\tau_2 \varphi_1 - \tau_1 \varphi_2) \psi \boldsymbol{A}. \end{aligned}\right\} \tag{34.2}$$

Since the above Hamiltonian can be regarded as a small perturbation, the transition probability of the radiative process is proportional to the matrix element

$$\langle f | \int H'' d^3x | i \rangle / \sqrt{\langle f | f \rangle \langle i | i \rangle}\,, \tag{34.3}$$

where $|i\rangle$ and $|f\rangle$ are the initial and final states of the meson-nucleon system, respectively. These states are now expanded as (27.6) in terms of the two-nucleon amplitude and amplitudes containing one or more mesons in addition. We identify the zero meson amplitude with the FOURIER transform of the ordinary two-nucleon wave function in order to eliminate the ambiguity in the meson theory of nuclear forces.

We are going to show that, neglecting all amplitudes containing mesons and making the non-relativistic approximation in evaluating the matrix element (34.3), we get the same expression as would be obtained by non-relativistic particle quantum mechanics. This expression will be used as the phenomenological starting point when studying the radiative processes of the deuteron in the following sections. By the relativistic correction we mean those contributions which are obtained when the calculation is carried up to the next step with respect to the nucleon velocities, while still neglecting all the meson amplitudes. We then define the exchange current corrections as those due to the amplitudes containing mesons in the matrix element (34.3). In the latter case we make the non-relativistic approximation, because the relativistic effects are small, and neglect the so-called self-meson processes, since their main contribution is supposed to be accounted for by the PAULI-term. These two corrections are considered later in Sects. 41 and 42.

To obtain the phenomenological expression, it is readily seen, only the first two terms in (34.2) need be considered. In evaluating the matrix element, we expand the nucleon field quantities $\psi(\boldsymbol{x})$ as follows:

$$\psi = \frac{1}{\sqrt{(2\pi\hbar)^3}} \int \left[a_{\boldsymbol{p}} u_{\boldsymbol{p}} \exp\left(\frac{i \boldsymbol{p} \boldsymbol{x}}{\hbar}\right) + b^*_{-\boldsymbol{p}} v_{\boldsymbol{p}} \exp\left(\frac{i \boldsymbol{p} \boldsymbol{x}}{\hbar}\right) \right] d^3 p\,, \tag{34.4}$$

where $u_{\boldsymbol{p}}$ and $v_{\boldsymbol{p}}$ are the free DIRAC spinors belonging to the momentum $\boldsymbol{p}$ and to positive and negative energies, respectively. $a_{\boldsymbol{p}}$ and $a^*_{\boldsymbol{p}}$ are the annihilation and creation operators of a nucleon of momentum $\boldsymbol{p}$, $b_{\boldsymbol{p}}$ and $b_{\boldsymbol{p}}^*$ those of an anti-nucleon of the same momentum, with which the first state vector in the expansion in (27.6) is constructed as

$$|\boldsymbol{p}_1 \boldsymbol{p}_2\rangle = \frac{1}{\sqrt{2}} a_{\boldsymbol{p}_1}^* a_{\boldsymbol{p}_2}^* |0\rangle,$$

$|0\rangle$ being the vacuum state. The evaluation of the matrix element can then be carried out using the property of the vacuum state and the commutation relations

$$[a_{\boldsymbol{p}}, a_{\boldsymbol{p}'}^*]_+ = \delta(\boldsymbol{p} - \boldsymbol{p}'), \text{etc.}$$

and remarking that*

$$\left.\begin{aligned} (u_{\boldsymbol{p}}^* \boldsymbol{\alpha}\, u_{\boldsymbol{p}'}) &= \sqrt{\frac{M c^2 + E_{\boldsymbol{p}}}{2E_{\boldsymbol{p}}}} \sqrt{\frac{M c^2 + E_{\boldsymbol{p}'}}{2E_{\boldsymbol{p}'}}} \left(\frac{(c\,\boldsymbol{\sigma}\,\boldsymbol{p})\,\boldsymbol{\sigma}}{E_{\boldsymbol{p}} + M c^2} + \frac{\boldsymbol{\sigma}\,(c\,\boldsymbol{\sigma}\,\boldsymbol{p}')}{E_{\boldsymbol{p}'} + M c^2}\right),\\ (u_{\boldsymbol{p}}^* \beta\, \boldsymbol{\sigma}\, u_{\boldsymbol{p}'}) &= \sqrt{\frac{M c^2 + E_{\boldsymbol{p}}}{2E_{\boldsymbol{p}}}} \sqrt{\frac{M c^2 + E_{\boldsymbol{p}'}}{2E_{\boldsymbol{p}'}}} \left(\boldsymbol{\sigma} - \frac{(c\,\boldsymbol{\sigma}\,\boldsymbol{p})\,\boldsymbol{\sigma}\,(c\,\boldsymbol{\sigma}\,\boldsymbol{p}')}{(E_{\boldsymbol{p}} + M c^2)\,(E_{\boldsymbol{p}'} + M c^2)}\right), \end{aligned}\right\} \tag{34.5}$$

where $E_{\boldsymbol{p}}$ is the total energy of a nucleon with momentum $\boldsymbol{p}$ and we are using the convention that $\boldsymbol{\sigma}$ still remains as an operator. Neglecting the relativistic aspects, we get the expressions

$$(u_{\boldsymbol{p}}^* \boldsymbol{\alpha}\, u_{\boldsymbol{p}'}) = \frac{1}{2Mc}\,[(\boldsymbol{p} + \boldsymbol{p}') + i\,\boldsymbol{\sigma} \times (\boldsymbol{p} - \boldsymbol{p}')], \quad (u_{\boldsymbol{p}}^* \beta\, \boldsymbol{\sigma}\, u_{\boldsymbol{p}'}) = \boldsymbol{\sigma}. \tag{34.6}$$

The electromagnetic field $\boldsymbol{A}$ in (34.2) can be expanded as

$$\boldsymbol{A}(\boldsymbol{x}) = \frac{1}{\sqrt{(2\pi)^3}} \sum_{\lambda} \int \sqrt{\frac{2\pi\hbar c}{k}}\, (a_{\boldsymbol{k}}^{(\lambda)} + a_{-\boldsymbol{k}}^{(\lambda)*})\, \boldsymbol{e}_\lambda\, e^{i\boldsymbol{k}\boldsymbol{x}}\, d^3 k, \tag{34.7}$$

where λ denotes the direction of polarization and $\boldsymbol{e}_\lambda$ its unit vector. Thus for an emission or absorption of a light quantum with wave vector $\boldsymbol{k}$ and polarization λ, $\boldsymbol{A}(\boldsymbol{x})$ can be replaced by

$$\sqrt{\frac{\hbar c}{4\pi^2 k}}\, \boldsymbol{e}_\lambda\, e^{-i\boldsymbol{k}\boldsymbol{x}} \text{ (emission)}, \qquad \sqrt{\frac{\hbar c}{4\pi^2 k}}\, \boldsymbol{e}_\lambda\, e^{+i\boldsymbol{k}\boldsymbol{x}} \text{ (absorption)}, \tag{34.8}$$

respectively.

Considering for the time being an absorption process, we get as the final expression for the required matrix element (34.3),

$$\left.\begin{aligned} \langle f|\textstyle\int H''\, d^3x\,|i\rangle = \sqrt{\frac{\hbar c}{4\pi^2 k}} \iint \psi_f^*(\boldsymbol{x}_1 \boldsymbol{x}_2)\, \boldsymbol{e}_\lambda \Big[e^{i\boldsymbol{k}\boldsymbol{x}_1}\Big(-\frac{e\,\boldsymbol{p}_1}{Mc}\,\frac{1+\tau_3^{(1)}}{2} - i\,\boldsymbol{\mu}^{(1)} \times \boldsymbol{k}\Big) + \\ + e^{+i\boldsymbol{k}\boldsymbol{x}_2}\Big(-\frac{e\,\boldsymbol{p}_2}{Mc}\,\frac{1+\tau_3^{(2)}}{2} - i\,\boldsymbol{\mu}^{(2)} \times \boldsymbol{k}\Big)\Big]\, \psi_i(\boldsymbol{x}_1, \boldsymbol{x}_2)\, d^3x_1\, d^3x_2, \end{aligned}\right\} \tag{34.9}$$

with

$$\boldsymbol{\mu}^{(i)} = \frac{e}{2\varkappa}\left[\left(\frac{1+\tau_3^{(i)}}{2}\right)\mu_P + \left(\frac{1-\tau_3^{(i)}}{2}\right)\mu_N\right] \boldsymbol{\sigma}^{(i)}, \tag{34.10}$$

where the transformation from the amplitude $c(\boldsymbol{p}_1, \boldsymbol{p}_2)$ to the two-nucleon wave function $\psi(\boldsymbol{x}_1, \boldsymbol{x}_2)$ has been made by

$$c(\boldsymbol{p}_1, \boldsymbol{p}_2) = \frac{1}{(2\pi\hbar)^3} \iint \psi(\boldsymbol{x}_1, \boldsymbol{x}_2)\, \exp\Big(-\frac{i\,\boldsymbol{p}_1 \boldsymbol{x}_1}{\hbar}\Big) \exp\Big(-\frac{i\,\boldsymbol{p}_2 \boldsymbol{x}_2}{\hbar}\Big)\, d^3x_1\, d^3x_2. \tag{34.11}$$

---

* These relations can most easily be derived by expressing $u_{\boldsymbol{p}}$ and $u_{\boldsymbol{p}}^*$ in terms of the eigenstate of ϱ_3 (DIRAC's notation) with eigenvalue $+1$.

We normalize by

$$\iint |\psi(\boldsymbol{x}_1, \boldsymbol{x}_2)|^2 d^3x_1 d^3x_2 = \iint |c(\boldsymbol{p}_1, \boldsymbol{p}_2)|^2 d^3p_1 d^3p_2 = 1, \tag{34.12}$$

which makes the denominator of (34.3) unity.

It is readily noticed that (34.9) and (34.10) agree with the predictions of the non-relativistic particle quantum mechanics, assuming that nucleons are charged mass points with intrinsic magnetic moments μ_P and μ_N, respectively. This is our starting point in the phenomenological treatment of the radiative processes.

35. Multipole transitions and selection rules. In order to bring (34.9) and (34.10) into a more convenient form, let us introduce the center-of-mass coordinate $\boldsymbol{X}$ and the relative coordinate $\boldsymbol{x}$ of two nucleons by $\boldsymbol{x}_1 = \boldsymbol{X} + \frac{1}{2}\boldsymbol{x}$ and $\boldsymbol{x}_2 = \boldsymbol{X} - \frac{1}{2}\boldsymbol{x}$. Confining our attention to an absorption process, we may assume that in the initial state the center-of-mass is at rest, while it is moving with a wave vector $\boldsymbol{K}$ in the final state. Then the two wave functions are given by

$$\psi_i(\boldsymbol{x}_1, \boldsymbol{x}_2) = \frac{1}{\sqrt{(2\pi)^3}} \psi_i(\boldsymbol{x}), \quad \psi_f(\boldsymbol{x}_1, \boldsymbol{x}_2) = \frac{1}{\sqrt{(2\pi)^3}} e^{i\boldsymbol{K}\boldsymbol{X}} \psi_f(\boldsymbol{x}), \tag{35.1}$$

where the internal wave functions $\psi_i(\boldsymbol{x})$ and $\psi_f(\boldsymbol{x})$ are normalized in the relative coordinate system. It is then easily seen that the center-of-mass motion cannot contribute to the matrix element at all owing to the transverse nature of a light quantum. Denoting the relative momentum as $\boldsymbol{p}$, the expression (34.9) can be written

$$\left.\begin{aligned} &\langle f| \int H'' d^3x |i\rangle \\ &= \delta(\boldsymbol{K}-\boldsymbol{k}) \sqrt{\frac{\hbar c}{4\pi^2 k}} \Big[\int \psi_f^*(\boldsymbol{x})\, \boldsymbol{e}_\lambda \Big(\frac{-e\boldsymbol{p}}{Mc}\Big) \Big(e^{\frac{i\boldsymbol{k}\boldsymbol{x}}{2}} \frac{1+\tau_3^{(1)}}{2} - e^{\frac{-i\boldsymbol{k}\boldsymbol{x}}{2}} \frac{1+\tau_3^{(2)}}{2} \Big) \psi_i(\boldsymbol{x})\, d^3x - \\ &\quad - i k \int \psi_f^*(\boldsymbol{x}) \Big(\frac{\boldsymbol{k}}{k} \times \boldsymbol{e}_\lambda\Big) \Big(e^{\frac{i\boldsymbol{k}\boldsymbol{x}}{2}} \mu^{(1)} + e^{\frac{-i\boldsymbol{k}\boldsymbol{x}}{2}} \mu^{(2)} \Big) \psi_i(\boldsymbol{x})\, d^3x \Big]. \end{aligned}\right\} \tag{35.2}$$

The first factor states the conservation law of the total momentum and can be omitted in the following. Finally let us make the following rearrangements in (35.2):

$$\left.\begin{aligned} &e^{\frac{i\boldsymbol{k}\boldsymbol{x}}{2}} \frac{1+\tau_3^{(1)}}{2} - e^{-\frac{i\boldsymbol{k}\boldsymbol{x}}{2}} \frac{1+\tau_3^{(2)}}{2} = \\ &\quad = \frac{1}{2} \Big\{ \Big(e^{\frac{i\boldsymbol{k}\boldsymbol{x}}{2}} - e^{-\frac{i\boldsymbol{k}\boldsymbol{x}}{2}} \Big) + \Big(e^{\frac{i\boldsymbol{k}\boldsymbol{x}}{2}} + e^{-\frac{i\boldsymbol{k}\boldsymbol{x}}{2}} \Big) \Big(\frac{\tau_3^{(1)} - \tau_3^{(2)}}{2} \Big) \Big\}, \end{aligned}\right\} \tag{35.3}$$

$$\left.\begin{aligned} &e^{\frac{i\boldsymbol{k}\boldsymbol{x}}{2}} \mu^{(1)} + e^{-\frac{i\boldsymbol{k}\boldsymbol{x}}{2}} \mu^{(2)} = \\ &= \frac{1}{2}\Big(\frac{e}{2\varkappa}\Big) \Big[\Big(e^{\frac{i\boldsymbol{k}\boldsymbol{x}}{2}} + e^{-\frac{i\boldsymbol{k}\boldsymbol{x}}{2}} \Big) \Big\{ (\mu_P + \mu_N) \frac{\boldsymbol{\sigma}^{(1)} + \boldsymbol{\sigma}^{(2)}}{2} + (\mu_P - \mu_N) \frac{\tau_3^{(1)} - \tau_3^{(2)}}{2} \frac{\boldsymbol{\sigma}^{(1)} - \boldsymbol{\sigma}^{(2)}}{2} \Big\} + \\ &\quad + \Big(e^{\frac{i\boldsymbol{k}\boldsymbol{x}}{2}} - e^{-\frac{i\boldsymbol{k}\boldsymbol{x}}{2}} \Big) \Big\{ (\mu_P + \mu_N) \frac{\boldsymbol{\sigma}^{(1)} - \boldsymbol{\sigma}^{(2)}}{2} + (\mu_P - \mu_N) \frac{\tau_3^{(1)} - \tau_3^{(2)}}{2} \frac{\boldsymbol{\sigma}^{(1)} + \boldsymbol{\sigma}^{(2)}}{2} \Big\} \Big], \end{aligned}\right\} \tag{35.4}$$

where we have dropped the terms containing $(\tau_3^{(1)} + \tau_3^{(2)})$ since this factor vanishes identically when applied to the initial state $\psi_i(\boldsymbol{x})$, which can now be assumed to be the deuteron ground state (a charge singlet state).

As the wavelength λ of the incident photon is related to the photon energy $\hbar c k$ by

$$\lambda = 1.2396 \times 10^{-10}\,\text{cm} \Big/ \Big(\frac{\hbar c k}{\text{Mev}} \Big), \tag{35.5}$$

λ is always very large compared with the deuteron radius $1/\alpha = 4.3157 \times 10^{-13}$ cm in the low energy region; even at 100 Mev λ is still 3 times $1/\alpha$. Thus it is legitimate to expand the exponential factors in (35.3) and (35.4) with respect to $\boldsymbol{k}\boldsymbol{x}$ and retain the first few terms. This procedure leads to the ordinary decomposition of the matrix element (35.2) into the multipole transition matrix elements. An attempt to treat the original form (35.2) without expanding the retardation factor $\exp(\pm i\boldsymbol{k}\boldsymbol{x}/2)$ has been made by FOLDY and BERGER[136]. We employ, however, the conventional method and refer otherwise to the cited authors, since the difference between the two methods would certainly be smaller than many other uncertainties which we encounter in the analysis of the radiative process.

In the first approximation let us put all retardation factors equal to one. Then the integrand of the first term of (35.2) is $(-e\boldsymbol{p}/Mc)(\tau_3^{(1)} - \tau_3^{(2)})/2$, which causes a transition from the ground state $({}^3S_1 + {}^3D_1)$ to an odd parity state, since it is an odd operator; it does not change the spin-configuration of $\psi_i(\boldsymbol{x})$ because no spin operators are involved. Thus $\psi_f(\boldsymbol{x})$ is a $({}^3P + {}^3F)$-state which is a charge triplet state, while $\psi_i(\boldsymbol{x})$ is charge singlet. The relevant matrix element of $\frac{1}{2}(\tau_3^{(1)} - \tau_3^{(2)})$ is just one. The result is

$$\langle f | \int H'' d^3x | i \rangle = \sqrt{\frac{\hbar c}{4\pi^2 k}} \int \psi_f^*(\boldsymbol{x})\, \boldsymbol{e}_\lambda (-e\boldsymbol{p}/Mc)\, \psi_i(\boldsymbol{x})\, d^3x, \tag{35.6}$$

with the selection rule

$$ {}^3S_1 + {}^3D_1 \to {}^3P + {}^3F \quad \text{(electric dipole transition)}. \tag{35.7}$$

If and only if we assume that the same nuclear potential acts in initial and final states, we can show that

$$\int \psi_f^* (-e\boldsymbol{p}/Mc)\, \psi_i\, d^3x = \frac{i}{\hbar c} (E_i - E_f) \int \psi_f^* (e\boldsymbol{x}/2)\, \psi_i\, d^3x, \tag{35.8}$$

with the aid of the SCHRÖDINGER wave equations satisfied by $\psi_i(\boldsymbol{x})$ and $\psi_f(\boldsymbol{x})$, E_i and E_f being the corresponding relative energies which satisfy

$$E_f - E_i = \hbar c k - \frac{\hbar^2 k^2}{4M}. \tag{35.9}$$

If (35.8) holds, we may rewrite (35.6) as

$$\langle f | \int H'' d^3x | i \rangle = \frac{i}{\hbar c} (E_i - E_f) \sqrt{\frac{\hbar c}{4\pi^2 k}} \int \psi_f^*(\boldsymbol{x})\, \boldsymbol{e}_\lambda (e\boldsymbol{x}/2)\, \psi_i(\boldsymbol{x})\, d^3x, \tag{35.10}$$

which is conventionally called the electric dipole transition matrix element. Practically all existing calculations on the radiative processes of the deuteron are based on (35.10), not (35.6). In the expression (35.6), we have carried out the matrix element with respect to the charge spin variables, which implies that $\psi_i(\boldsymbol{x})$ and $\psi_f(\boldsymbol{x})$ do not contain any charge spin functions. Then the assumption which led to (35.8) is certainly not correct, since the nuclear potential must apparently be charge spin dependent.

However, it is also possible to formulate the above treatment so that (35.10) may directly follow as the phenomenological electric dipole matrix element. This is accomplished if, before giving the definition of various contributions, we rewrite $\int H'' d^3x$ in (34.3) in a different form using the continuity equation for the total (nucleon and meson) charge current density and an expansion, or at

[136] L. L. FOLDY: Phys. Rev. **92**, 178 (1953). — J. M. BERGER: Phys. Rev. **94**, 1698 (1954).

least some decomposition, of the electromagnetic potential $\boldsymbol{A}$ in H'' (34.2). This is just the straightforward generalization of the ordinary treatment of radiative processes of nuclei to the field theoretical approach. From the point of view adopted here there is no immediate reason to prefer one of the forms (35.6) and (35.10) to the other. It may be remarked[136a], however, that if we assume a phenomenological, point mechanical approach, where the nucleons interacting with an electromagnetic field are described by a gauge invariant SCHRÖDINGER equation not involving any meson quantities, then the electric multipole moment operators are uniquely determined and in the dipole case we get (35.10). Besides, although there is some ambiguity because of the lack of knowledge of the nuclear force in the triplet odd state, calculations indicate that (35.10) gives a better fit than (35.6) to the experimental data in the low energy region. Therefore we shall give the results of the calculations based on (35.10) only. It is added that there is no difference between these two approaches with respect to the electric quadrupole or the magnetic transition matrix elements, which will be given in the following.

Let us then consider the second term of (35.2), the integrand of which becomes in the relevant approximation

$$\frac{e}{2\varkappa}\left(\frac{\boldsymbol{k}}{k}\times\boldsymbol{e}_\lambda\right)\left\{(\mu_P+\mu_N)\,\frac{\boldsymbol{\sigma}^{(1)}+\boldsymbol{\sigma}^{(2)}}{2}+(\mu_P-\mu_N)\,\frac{\tau_3^{(1)}-\tau_3^{(2)}}{2}\,\frac{\boldsymbol{\sigma}^{(1)}-\boldsymbol{\sigma}^{(2)}}{2}\right\}. \quad (35.11)$$

Since this term does not contain any orbital operators, it cannot change the orbital configuration. If the spin configuration is not changed either, the relevant matrix element must vanish identically owing to the orthogonality since it is calculated between two states belonging to the same spin and orbital configuration and different energies. Thus the spin state must be changed, which demands, in turn, the change of the charge spin state and allows us to drop the first term of (35.11) and replace the charge spin factor in the second term by one. The corresponding matrix element is finally given by

$$\langle f|\int H''\,d^3x\,|i\rangle=-i\,k\,\frac{e}{2\varkappa}\sqrt{\frac{\hbar c}{4\pi^2 k}}\int\psi_f^*(\boldsymbol{x})\left(\frac{\boldsymbol{k}}{k}\times\boldsymbol{e}_\lambda\right)(\mu_P-\mu_N)\,\frac{\boldsymbol{\sigma}^{(1)}-\boldsymbol{\sigma}^{(2)}}{2}\,\psi_i(\boldsymbol{x})\,d^3x, \quad (35.12)$$

which just represents the ordinary magnetic dipole transition due to the magnetic moments of the nucleons; the selection rule is

$$^3S_1+{}^3D_1\rightarrow{}^1S+{}^1D \quad \text{(spin magnetic dipole transition)}. \quad (35.13)$$

We pass to the linear term of the expansion of the retardation factor. The integrand of the first part of (35.2) becomes

$$-\frac{i\,e}{2Mc}(\boldsymbol{e}_\lambda\boldsymbol{p})(\boldsymbol{k}\boldsymbol{x})=-\frac{i\,e}{4Mc}\left[(\boldsymbol{k}\times\boldsymbol{e}_\lambda)(\boldsymbol{x}\times\boldsymbol{p})+(\boldsymbol{e}_\lambda\boldsymbol{p})(\boldsymbol{k}\boldsymbol{x})+(\boldsymbol{e}_\lambda\boldsymbol{x})(\boldsymbol{k}\boldsymbol{p})\right], \quad (35.14)$$

where we have used a vector identity. The first term of the right-hand side of (35.14) represents the orbital magnetic dipole transition:

$$\langle f|\int H''\,d^3x\,|i\rangle=-\frac{i\,e\,k}{2Mc}\sqrt{\frac{\hbar c}{4\pi^2 k}}\int\psi_f^*(\boldsymbol{x})\left(\frac{\boldsymbol{k}}{k}\times\boldsymbol{e}_\lambda\right)\left(\frac{\boldsymbol{x}}{2}\times\boldsymbol{p}\right)\psi_i(\boldsymbol{x})\,d^3x, \quad (35.15)$$

the selection rule of which is given by

$$^3S_1+{}^3D_1\rightarrow{}^3S+{}^3D+{}^3G \quad \text{(orbital magnetic dipole transition)}, \quad (35.16)$$

[136a] R. G. SACHS and N. AUSTERN: Phys. Rev. **81**, 705, 710 (1951).

where the spin singlet state is not allowed contrary to (35.13). As regards the remaining terms in (35.14), we can show in the same way as for (35.8) that

$$\left.\begin{aligned}\frac{i}{2}\int\psi_f^*\left[\left(\boldsymbol{e}_\lambda\frac{\boldsymbol{x}}{2}\right)\left(\boldsymbol{k}\frac{-e\boldsymbol{p}}{Mc}\right)+\left(\boldsymbol{k}\frac{\boldsymbol{x}}{2}\right)\left(\boldsymbol{e}_\lambda\frac{-e\boldsymbol{p}}{Mc}\right)\right]\psi_i\,d^3x=\\=\frac{-(E_i-E_f)}{2\hbar c}\int\psi_f^*\left(\boldsymbol{e}_\lambda\frac{\boldsymbol{x}}{2}\right)\left(\boldsymbol{k}\frac{e\boldsymbol{x}}{2}\right)\psi_i\,d^3x,\end{aligned}\right\}\tag{35.17}$$

which is exact, since both states have the same spin and parity, and gives the electric quadrupole transition:

$$\langle f|\int H''\,d^3x\,|i\rangle=-\frac{ek}{2\hbar c}(E_i-E_f)\sqrt{\frac{\hbar c}{4\pi^2k}}\int\psi_f^*(\boldsymbol{x})\left(\boldsymbol{e}_\lambda\frac{\boldsymbol{x}}{2}\right)\left(\frac{\boldsymbol{k}}{k}\frac{\boldsymbol{x}}{2}\right)\psi_i(\boldsymbol{x})\,d^3x,\tag{35.18}$$

the selection rule of which is

$$^3S_1+{}^3D_1\rightarrow{}^3S+{}^3D+{}^3G\quad\text{(electric quadrupole transition)}.\tag{35.19}$$

In the same order of expansion the second term of (35.2) leads to the spin magnetic quadrupole matrix element:

$$\left.\begin{aligned}\langle f|\int H''\,d^3x|i\rangle=k^2\frac{e}{2\varkappa}\sqrt{\frac{\hbar c}{4\pi^2k}}\int\psi_f^*(\boldsymbol{x})\left(\frac{\boldsymbol{k}}{k}\times\boldsymbol{e}_\lambda\right)\left(\frac{\boldsymbol{k}}{k}\frac{\boldsymbol{x}}{2}\right)\cdot\\\cdot\left\{(\mu_P+\mu_N)\frac{\boldsymbol{\sigma}^{(1)}-\boldsymbol{\sigma}^{(2)}}{2}+(\mu_P-\mu_N)\frac{\tau_3^{(1)}-\tau_3^{(2)}}{2}\,\frac{\boldsymbol{\sigma}^{(1)}+\boldsymbol{\sigma}^{(2)}}{2}\right\}\psi_i(\boldsymbol{x})\,d^3x,\end{aligned}\right\}\tag{35.20}$$

which gives rise to a transition

$$^3S_1+{}^3D_1\rightarrow{}^3P+{}^3F+{}^1P+{}^1F\quad\text{(spin magnetic quadrupole transition)}.\tag{35.21}$$

Quite analogously we get by steps all the multipole transition matrix elements from the consecutive terms in the expansion of the retardation factor. However, as far as we are not interested in the very high energy region, we need only consider the first few moments (dipole and at most quadrupole) and the first few orbital angular momentum states (S-, P- and D-states), which simplifies the analysis very much.

b) Radiative neutron-proton capture.

36. Capture cross section. As the first example of a radiative process, we take the thermal neutron capture process, where a neutron impinges on a proton at rest, with the result that a deuteron is formed and a photon is emitted. The relevant matrix elements are readily obtained by replacing $\boldsymbol{k}$ by $-\boldsymbol{k}$ in the expressions of the previous section. As the incident neutron is very slow (thermal energy), we may confine ourselves to the transition from an S-state to the deuteron ground state. It is seen from the selection rules given in the previous section that a major contribution comes from the spin magnetic dipole transition ($^1S\rightarrow{}^3S$) given by (35.12) and (35.13), while the contribution due to the electric quadrupole transition (35.18) and (35.19) is a minor one since it goes from the ^3S-continuum to the D-state of the deuteron. The orbital magnetic dipole transition (35.15) and (35.16) can be shown to vanish if we neglect the ^3D-continuum.

Since the spins of both nucleons are directed at random, the above two transitions are incoherent. The capture cross section is given by

$$d\sigma=\frac{2\pi}{\hbar}\sum\left|\langle f|\int H''\,d^3x|i\rangle\right|^2\frac{\varrho_F}{n}\,g\,d\Omega\tag{36.1}$$

for a photon emitted into a solid angle $d\Omega$, where $\varrho_F = k^2/\hbar c$ is the number of final states per unit energy, n is the number of incident neutrons per unit time and unit area (or rather a normalization factor of ψ_i), g is the statistical weight factor ($\frac{1}{4}$ for singlet and $\frac{3}{4}$ for triplet transitions) and Σ means the average over the initial spin directions and the sum over the final directions of spin and polarization. It is easily shown that

$$\sum\left|\int \psi_f^*(\boldsymbol{x})\left(\frac{\boldsymbol{k}}{k}\times\boldsymbol{e}_\lambda\right)\left(\frac{\boldsymbol{\sigma}^{(1)}-\boldsymbol{\sigma}^{(2)}}{2}\right)\psi_i(\boldsymbol{x})\,d^3x\right|^2 = 2\left|\int \psi_f^*(\boldsymbol{x})\,\psi_i(\boldsymbol{x})\,d^3x\right|^2, \quad (36.2)$$

where the symbols ψ_i and ψ_f in the latter expression indicate orbital wave functions. Thus the magnetic dipole capture is isotropic and the total cross section is

$$\sigma = \frac{1}{4}\left(\frac{e^2}{\hbar c}\right)\left(\frac{\varepsilon}{M c^2}\right)^2 \frac{\varepsilon}{n\hbar}(\mu_P-\mu_N)^2\left|\int \psi_f^*(\boldsymbol{x})\,\psi_i(\boldsymbol{x})\,d^3x\right|^2, \quad (36.3)$$

where we have introduced the deuteron binding energy ε instead of the photon energy $\hbar c k$ (thermal neutrons!). As the S-D term vanishes in (36.3), we may put

$$\psi_f(\boldsymbol{x}) = N_g \frac{u_g(r)}{\sqrt{4\pi}\,r}, \quad u_g(r) \xrightarrow[r\to\infty]{} \cos\varepsilon_g\, e^{-\alpha r}, \quad (36.4)$$

where N_g is given by

$$N_g^{-2} = \int_0^\infty [u_g^2(r) + w_g^2(r)]\,dr \quad (36.5)$$

and $w_g(r)$ is the deuteron D-wave function normalized to $\sin\varepsilon_g\, e^{-\alpha r}$ asymptotically. Let us choose $\psi_i(\boldsymbol{x})$ to represent, at large distances, an incident plane wave along the z-direction with a wave vector $\boldsymbol{k}$ or more specifically the form (15.1). As the S-wave alone need be considered, we may use the partial wave expansion (15.9) and (15.10):

$$\psi_i(\boldsymbol{x}) = \frac{u_s(r)}{k r}, \quad u_s(r) \xrightarrow[r\to\infty]{} \sin(k r + \delta_s), \quad (36.6)$$

where we have neglected an unimportant phase factor and δ_s is the singlet S-wave phase shift. In the zero energy limit, we may rewrite the above formulas as

$$\psi_i(\boldsymbol{x}) = -\frac{a_s u_s(r)}{r}, \quad u_s(r) \xrightarrow[r\to\infty]{} \frac{\sin(k r + \delta_s)}{\sin\delta_s} \approx 1 - \frac{r}{a_s}, \quad (36.7)$$

introducing the singlet scattering length a_s and remarking that δ_s approaches zero with k. If we adopt this normalization of $\psi_i(\boldsymbol{x})$, n is simply given by v, where v is the velocity of the incident neutron towards the proton. Then we get

$$\sigma = \pi\left(\frac{e^2}{\hbar c}\right)\left(\frac{\varepsilon}{M c^2}\right)^2\sqrt{\frac{\varepsilon}{2E}}\,(\mu_P-\mu_N)^2\,\alpha\, N_g^2\, a_s^2\left[\int_0^\infty u_g(r)\,u_s(r)\,dr\right]^2, \quad (36.8)$$

where we have introduced the incident neutron energy $E = M v^2/2$.

For the electric quadrupole matrix element (35.18), we may put

$$\psi_f(\boldsymbol{x}) = \frac{1}{\sqrt{32\pi}}\, S_{12}\frac{N_g w_g(r)}{r}\chi_m, \quad (36.9)$$

since the S-wave part vanishes under the integration. Here we have used (24.17), S_{12} is the tensor operator and χ_m is the triplet spin function belonging to the quantum number m. For $\psi_i(\boldsymbol{x})$ we may assume an expression similar to (36.7):

$$\psi_i(\boldsymbol{x}) = +\frac{a_t u_t(r)}{r}\chi_{m'}, \quad u_t(r) \xrightarrow[r\to\infty]{} 1 - \frac{r}{a_t}, \quad (36.10)$$

introducing the initial spin triplet function $\chi_{m'}$, the triplet scattering length a_t and remarking that the triplet S-wave phase shift δ_t approaches π as $k\to 0$. Then the integral in (35.18) can be carried out to give

$$\left.\begin{aligned}&\int \psi_f^*(\boldsymbol{x})\left(\boldsymbol{e}_\lambda \frac{\boldsymbol{x}}{2}\right)\left(\frac{\boldsymbol{k}}{k}\frac{\boldsymbol{x}}{2}\right)\psi_i(\boldsymbol{x})\,d^3x = \\ &= +\frac{\sqrt{2\pi}}{40} N_g a_t \int_0^\infty r^2 w_g(r) u_t(r)\,dr \left(\chi_{m}^*, \left[(\boldsymbol{e}_\lambda \boldsymbol{\sigma}^{(1)})\left(\frac{\boldsymbol{k}}{k}\boldsymbol{\sigma}^{(2)}\right) + (\boldsymbol{e}_\lambda \boldsymbol{\sigma}^{(2)})\left(\frac{\boldsymbol{k}}{k}\boldsymbol{\sigma}^{(1)}\right)\right]\chi_{m'}\right).\end{aligned}\right\} \quad (36.11)$$

The remaining spin factor can most easily be evaluated noting that the singlet-triplet matrix element vanishes and $(\boldsymbol{e}_\lambda \boldsymbol{k})=0$, further exploiting completeness of the spin functions:

$$\left.\begin{aligned}&\frac{1}{3}\sum_{mm'}\left|\left(\chi_m^*, \left[(\boldsymbol{e}_\lambda \boldsymbol{\sigma}^{(1)})\left(\frac{\boldsymbol{k}}{k}\boldsymbol{\sigma}^{(2)}\right) + (\boldsymbol{e}_\lambda \boldsymbol{\sigma}^{(2)})\left(\frac{\boldsymbol{k}}{k}\boldsymbol{\sigma}^{(1)}\right)\right]\chi_{m'}\right)\right|^2 = \\ &\quad = \frac{1}{3}\sum_{m}\left(\chi_m^*, \left[(\boldsymbol{e}_\lambda \boldsymbol{\sigma}^{(1)})\left(\frac{\boldsymbol{k}}{k}\boldsymbol{\sigma}^{(2)}\right) + (\boldsymbol{e}_\lambda \boldsymbol{\sigma}^{(2)})\left(\frac{\boldsymbol{k}}{k}\boldsymbol{\sigma}^{(1)}\right)\right]^2\chi_m\right) = \frac{8}{3}.\end{aligned}\right\} \quad (36.12)$$

Thus the electric quadrupole capture is also isotropic, with a total cross section

$$\sigma = \frac{\pi}{400}\left(\frac{e^2}{\hbar c}\right)\left(\frac{\varepsilon}{Mc^2}\right)^2\sqrt{\frac{\varepsilon}{2E}}\,\alpha^5 N_g^2 a_t^2\left[\int_0^\infty r^2 w_g(r)\,u_t(r)\,dr\right]^2. \quad (36.13)$$

In the following section it is shown that (36.13) is really negligible compared with (36.8).

37. Numerical values and comparison with experiments. The theoretical neutron-proton capture cross sections (36.8) and (36.13) should be compared with the experimental value (3.1):

$$\sigma_{\text{exp}} = (0.329 \pm 0.006)\ \text{barns}, \quad (37.1)$$

for a neutron velocity 2200 m/sec. This corresponds to a neutron energy $E = 0.02526$ ev, which guarantees the zero energy approximation employed in the previous section. First we notice that N_g^{-2} given by (36.5) can be expressed, using (30.25) and (30.26),

$$N_g^{-2} = \frac{1}{2\alpha} - \frac{1}{2}\varrho(0, -\varepsilon) - \alpha^2 r_{0t}^3 P_t + \cdots, \quad (37.2)$$

where $\varrho(0, -\varepsilon)$ is experimentally known as

$$\varrho(0, -\varepsilon) = 1.704\,(1 \pm 0.016) \times 10^{-13}\ \text{cm}, \quad (37.3)$$

according to (30.22) and (19.2), since $\cos\varepsilon_g$ may be put equal to 1, as shown later in this section. In the third term of (37.2) we may replace r_{0t} by $\varrho(0, -\varepsilon)$. Being short of empirical information about P_t, we assume the value

$$P_t = 0.048 \pm 0.089, \quad (37.4)$$

in order to cover the whole range of P_t between square well and Yukawa potential. The above choice gives, using (13.7) for $1/\alpha$

$$N_g^2 = 0.7733\,(1 \pm 0.0212) \times 10^{13}\ \text{cm}^{-1}, \quad (37.5)$$

where most of the error, almost 2%, is due to P_t or the lack of knowledge about the shape of the nuclear force.

In order to evaluate the matrix element (36.8), we assume the following functional forms:

$$u_g(r) = e^{-\alpha r} - e^{-\eta r} \quad \text{and} \quad u_s(r) = 1 - \frac{r}{a_s} - e^{-\xi r}, \tag{37.6}$$

where we have neglected the factor $\cos\varepsilon_g$. The former is the HULTHÉN wave function, discussed in Sects. 13 and 33, the latter is a similar modification of the asymptotic function (36.7), where the added term goes rapidly to zero at larger distances and makes $u_s(r)$ vanish at $r=0$, as the boundary condition demands. Since these approximate wave functions are already fitted to the deuteron binding energy and the singlet scattering length, we have only to require that they give correct effective ranges. According to (16.21), the singlet effective range r_{0s} is given by

$$r_{0s} = 2\int_0^\infty \left[\left(1-\frac{r}{a_s}\right)^2 - u_s^2\right] dr, \tag{37.7}$$

which can be used to determine ξ in (37.6). From Table 11, we fix r_{0s} as

$$r_{0s} = (2.40 \pm 0.40)\times 10^{-13}\ \text{cm}, \tag{37.8}$$

again extending the limits to cover the range between square well and YUKAWA potential. Taking a_s from (18.17) and (18.18), we have

$$\xi = (1.3040 \pm 0.2087)\times 10^{13}\ \text{cm}^{-1}. \tag{37.9}$$

From (36.5) and (37.2), it is seen that $u_g(r)$ is fitted to the triplet effective range r_{0t} if it satisfies the following relation:

$$\int_0^\infty u_g^2\, dr = N_g^{-2}(1-P_D), \tag{37.10}$$

introducing the D-state probability P_D (26.5). From (27.11) we take $P_D = 3\pm 1\%$. Then (37.5) and (37.10) give

$$\eta = (1.3321 \pm 0.0547)\times 10^{+13}\ \text{cm}^{-1}. \tag{37.11}$$

The uncertainty of ξ is mainly due to that of r_{0s}, while the error of η derives from P_t, with lesser contributions from P_D and $\varrho(0,-\varepsilon)$. The matrix element is then computed as

$$\int_0^\infty u_g u_s\, dr = 4.0555\,(1\pm 0.0157)\times 10^{-13}\ \text{cm}, \tag{37.12}$$

where the error is chiefly due to the uncertainty of r_{0s}.

As regards the effect of changing the functional forms (37.6), the most interesting point is to see the influence of a hard core. For this purpose, consider the following expressions

$$u_g(r) = \left[1 - e^{-\eta(r-R_t)}\right] e^{-\alpha r} \quad \text{and} \quad u_s(r) = \left[1 - \frac{r}{a_s}\right]\left[1 - e^{-\xi(r-R_s)}\right] \tag{37.13}$$

outside hard cores with radii R_t and R_s, for triplet and singlet states, respectively. Again the two requirements (37.7) and (37.10) determine η and ξ. As a reasonable example, take the following values of R_t and R_s:

$$\left.\begin{aligned} R_t &= (0.5656 \pm 0.1414)\times 10^{-13}\ \text{cm} \\ \text{and}\qquad R_s &= (0.6787 \pm 0.1414)\times 10^{-13}\ \text{cm}. \end{aligned}\right\} \tag{37.14}$$

After detailed calculations for various combinations of R_t and R_s inside these ranges, it is found that the matrix element itself is almost unchanged, but its error is increased from 1.6% up to 2.3%, and the main source is still the large indeterminacy of r_{0s}. Other possible forms of u_g and u_s have been discussed by AUSTERN[137] and shown to be uninfluential. Thus we may take as a final estimate

$$\int_0^\infty u_g u_s \, dr = 4.0555\,(1 \pm 0.023) \times 10^{-13}\,\text{cm}. \qquad (37.15)$$

The capture cross section (36.8) is then evaluated, using the values in Sect. 1, (13.7) and (18.17) besides those given above, as

$$\sigma_{\text{theor}} = (0.3137 \pm 0.016)\,\text{barn} = 0.3137\,(1 \pm 0.051)\,\text{barn}, \qquad (37.16)$$

where the quoted error (about 5%) is mainly due to the square of the integral (37.15) (about 4.6%) and N_g^2 (about 2%). The former is largely due to the uncertainty of r_{0s}, the latter to that of P_t. This theoretical value coincides with the experimental value (37.1) within the limits of error*. We may, however, say that a small difference might exist between them for the following reasons. We have assumed a rather large error of r_{0s} in (37.8). If we rely upon the charge independence hypothesis, we may take a smaller error; for example,

$$r_{0s} = (2.40 \pm 0.23) \times 10^{-13}\,\text{cm}, \qquad (37.17)$$

which in turn gives

$$\sigma_{\text{theor}} = (0.3137 \pm 0.010)\,\text{barn} = 0.3137\,(1 \pm 0.03)\,\text{barn}; \qquad (37.18$$

this is definitely smaller than (37.1). Thus a smaller error of r_{0s} can reduce the theoretical ambiguity rather much. It should also be added that the increase of r_{0s} reduces the theoretical cross section. Thus we may conclude that the experimental value of the capture cross section is perhaps slightly larger than the theoretical value calculated on phenomenological basis, although the evidence is still ambiguous. It is possible that the exchange current contribution, for instance, may explain the eventual difference; this is discussed in Sects. 41 and 42.

Finally we estimate the electric quadrupole contribution (36.13). It is readily shown that

$$\frac{\sigma_{\text{el. quad.}}}{\sigma_{\text{mag. dipole}}} = \frac{1}{400} \frac{\alpha^4}{(\mu_P - \mu_N)^2} \frac{a_t^2}{a_s^2} \left[\frac{\int_0^\infty w_g u_t r^2 \, dr}{\int_0^\infty u_g u_s \, dr} \right]^2. \qquad (37.19)$$

As the integral in the numerator is an "outside" quantity, we can use the zero range approximation for w_g and u_t:

$$w_g(r) = \sin\varepsilon_g \, e^{-\alpha r} \left[1 + \frac{3}{\alpha r} + \frac{3}{(\alpha r)^2} \right], \qquad u_t = 1 - \frac{r}{a_t}, \qquad (37.20)$$

which gives

$$\int_0^\infty r^2 w_g u_t \, dr = -\,4.04 \frac{\sin\varepsilon_g}{\alpha^3}, \qquad (37.21)$$

where we have used (18.16) for a_t. As regards ε_g we can look at Table 17 (Sect. 33), from which we may take

$$\sin\varepsilon_g = 0.03, \qquad \cos\varepsilon_g = 1 - 0.0005. \qquad (37.22)$$

[137] N. AUSTERN: Phys. Rev. 92, 670 (1953).

* This conclusion differs from that of AUSTERN[137], which is mainly due to his accepting a smaller error of r_{0s} and neglecting the third term on the right-hand side of (37.2).

Thus we see that $\cos\varepsilon_g$ may actually be replaced by 1 in the above calculations. Further using (37.12), we finally obtain

$$\frac{\sigma_{\text{el. quad.}}}{\sigma_{\text{mag. dipole}}} \approx 2\times 10^{-8}, \tag{37.23}$$

which shows that the higher multipole moments are quite negligible in this case.

c) Photodisintegration of deuteron.

38. Disintegration cross section. In the photodisintegration of the deuteron, the initial state is the deuteron ground state, the wave function of which is obtained from (24.17):

$$\psi_i(\boldsymbol{x}) = \frac{N_g}{\sqrt{4\pi}}\left\{\frac{u_g(r)}{r} + \frac{1}{\sqrt{8}} S_{12}\frac{w_g(r)}{r}\right\}\chi_m, \tag{38.1}$$

where we normalize $u_g(r)$ and $w_g(r)$ in the same way as (36.4) and (36.9) and m is the initial spin quantum number. The final spin state may be either singlet or triplet. The corresponding cross sections are evaluated separately and added to get the total disintegration cross section.

From the selection rules given in Sect. 35, it is seen that the electric dipole transition leads to the triplet state and the spin magnetic dipole to the singlet state. In the former we neglect the final F-state which would give only a small contribution. We furthermore disregard the tensor force in the triplet odd state. According to some theoretical investigations[138], this effect on the total cross section is always small, while the angular distribution may be affected rather much, although it is difficult to discern that effect because of many sources of uncertainty. We can then expand the final wave function $\psi_f(\boldsymbol{x})$ in terms of LEGENDRE polynomials as done in (15.10). Retaining S-, P- and D-waves, we may write

$$\psi_f(\boldsymbol{x}) = \left[e^{i\delta_0}\frac{u_0(r)}{kr} + 3i\, e^{i\delta_1}\frac{u_1(r)}{kr}\left(\frac{\boldsymbol{k}\boldsymbol{x}}{kr}\right) - 5\, e^{i\delta_2}\frac{u_2(r)}{kr}\,\frac{1}{2}\left\{3\left(\frac{\boldsymbol{k}\boldsymbol{x}}{kr}\right)^2 - 1\right\}\right]\chi_{m'} \tag{38.2}$$

for the final triplet state and

$$\psi_f(\boldsymbol{x}) = \left[e^{i\Delta_0}\frac{v_0(r)}{kr} + 3i\, e^{i\Delta_1}\frac{v_1(r)}{kr}\left(\frac{\boldsymbol{k}\boldsymbol{x}}{kr}\right) - 5 e^{i\Delta_2}\frac{v_2(r)}{kr}\,\frac{1}{2}\left\{3\left(\frac{\boldsymbol{k}\boldsymbol{x}}{kr}\right)^2 - 1\right\}\right]\chi_0, \tag{38.3}$$

for the singlet state, where $u_l(r)$ and $v_l(r)$ are normalized so as to approach asymptotically $\sin(kr - \frac{1}{2}l\pi + \delta_l)$ and $\sin(kr - \frac{1}{2}l\pi + \Delta_l)$, respectively. $\chi_{m'}$ and χ_0 are the triplet and singlet spin functions, $\hbar\boldsymbol{k}$ is the relative momentum in the c.m. system and δ_l and Δ_l are the triplet and singlet phase shifts of the l-th wave. The differential cross section for a solid angle $d\Omega$ is given by

$$d\sigma = \frac{2\pi}{\hbar}\,\frac{1}{6}\sum_{\lambda, m, m'} |\langle f\,|\int H''\, d^3x\,|\,i\rangle|^2\,\frac{\varrho_F}{n}\, d\Omega, \tag{38.4}$$

where, with the choice of the above wave function,

$$\varrho_F = \frac{M k}{2\,(2\pi)^3\,\hbar^2}, \qquad n = \frac{c}{(2\pi)^3}, \tag{38.5}$$

---

[138] W. RARITA and J. SCHWINGER: Phys. Rev. **59**, 436, 556 (1941). — T. M. HU and H. S. W. MASSEY: Proc. Roy. Soc. Lond., Ser. A **196**, 135 (1949). — N. AUSTERN: Phys. Rev. **85**, 283 (1952).

for the triplet transition. In the singlet case, we have only to drop the summation over the final spin state m'.

In Sect. 34, we have used $\boldsymbol{k}$ to represent the wave vector of the incident photon. To avoid confusion, let us introduce $\boldsymbol{\varkappa}$ as a unit vector of the direction of the incident photon and $h\nu$ as its energy. As regards the higher multipole transitions we consider only the major terms: for the orbital magnetic dipole the S-D and D-S transitions (S-S is forbidden), in the electric quadrupole case S-D and D-S (S-S is again forbidden) and for the spin magnetic quadrupole only the S-P transition. It is then easy to show that the orbital magnetic dipole does not give any contribution; according to (35.15), the relevant matrix element can be written as

$$\frac{N_g}{k\sqrt{4\pi}}\, e^{i\delta_0}\Big(\chi_{m'},\int u_0(r)\,(\boldsymbol{\varkappa}\times\boldsymbol{e}_\lambda)\left(\frac{\boldsymbol{x}}{2}\times\boldsymbol{p}\right)\frac{1}{\sqrt{8}}\,S_{12}\,w_g(r)\,dr\,d\Omega\,\chi_m\Big), \tag{38.6}$$

which, however, vanishes on carrying out the integration $d\Omega$. Thus we get

$$\left.\begin{aligned} d\sigma = \frac{e^2}{\hbar c}\frac{(E_f-E_i)^2}{h\nu}\frac{M k}{\hbar^2}\frac{d\Omega}{4\pi}\frac{1}{6}\sum_{\lambda,m,m'}\Big|\int\psi_f^*\Big[\left(\boldsymbol{e}_\lambda\frac{\boldsymbol{x}}{2}\right)+\frac{i h\nu}{2\hbar c}\left(\boldsymbol{e}_\lambda\frac{\boldsymbol{x}}{2}\right)\left(\boldsymbol{\varkappa}\frac{\boldsymbol{x}}{2}\right)+ \\ +\frac{i(h\nu)^2}{E_f-E_i}\frac{(\mu_P-\mu_N)}{2Mc^2}(\boldsymbol{\varkappa}\times\boldsymbol{e}_\lambda)\left(\frac{\boldsymbol{\sigma}^{(1)}+\boldsymbol{\sigma}^{(2)}}{2}\right)\left(\boldsymbol{\varkappa}\frac{\boldsymbol{x}}{2}\right)\Big]\psi_i\,d^3x\Big|^2 \end{aligned}\right\} \tag{38.7}$$

for the triplet transition and

$$\left.\begin{aligned} d\sigma = \frac{e^2}{\hbar c}(\mu_P-\mu_N)^2\frac{k h\nu}{4Mc^2}\frac{d\Omega}{4\pi}\frac{1}{6}\sum_{\lambda,m}\Big|\int\psi_f^*\Big[(\boldsymbol{\varkappa}\times\boldsymbol{e}_\lambda)\left(\frac{\boldsymbol{\sigma}^{(1)}-\boldsymbol{\sigma}^{(2)}}{2}\right)+ \\ +\frac{i h\nu}{\hbar c}\left(\frac{\mu_P+\mu_N}{\mu_P-\mu_N}\right)\left(\boldsymbol{\varkappa}\frac{\boldsymbol{x}}{2}\right)(\boldsymbol{\varkappa}\times\boldsymbol{e}_\lambda)\left(\frac{\boldsymbol{\sigma}^{(1)}-\boldsymbol{\sigma}^{(2)}}{2}\right)\Big]\psi_i\,d^3x\Big|^2 \end{aligned}\right\} \tag{38.8}$$

for the singlet transition. In (38.7) we have used the electric dipole expression (35.10), rather than (35.6), for the reasons given after Eq. (35.10).

The methods of evaluation of integrals and spin sums in (38.7) and (38.8) are similar to those used in (36.11) and (36.12). Here we give some useful formulas:

$$\left.\begin{aligned} &\int(\boldsymbol{A}\boldsymbol{x})(\boldsymbol{B}\boldsymbol{x})(\boldsymbol{C}\boldsymbol{x})(\boldsymbol{D}\boldsymbol{x})\,d^3x = \\ &\quad=\frac{4\pi}{15}\int r^6[(\boldsymbol{A}\boldsymbol{B})(\boldsymbol{C}\boldsymbol{D})+(\boldsymbol{A}\boldsymbol{C})(\boldsymbol{B}\boldsymbol{D})+(\boldsymbol{A}\boldsymbol{D})(\boldsymbol{B}\boldsymbol{C})]\,dr, \\ &\qquad\sum_m\chi_m^*(\boldsymbol{A}\boldsymbol{\sigma}^{(1)})(\boldsymbol{B}\boldsymbol{\sigma}^{(2)})\chi_m=(\boldsymbol{A}\boldsymbol{B}), \\ &\qquad\chi_0^*(\boldsymbol{A}\boldsymbol{\sigma}^{(1)})(\boldsymbol{B}\boldsymbol{\sigma}^{(2)})\chi_0=-(\boldsymbol{A}\boldsymbol{B}). \end{aligned}\right\} \tag{38.9}$$

We introduce the following notations for various types of radial integrals:

$$\left.\begin{aligned} I_{sp}&=\int_0^\infty r\,u_g\,u_1\,dr, & I_{sd}&=\int_0^\infty r^2\,u_g\,u_2\,dr, \\ I_{dp}&=\int_0^\infty r\,w_g\,u_1\,dr, & I_{ds}&=\int_0^\infty r^2\,w_g\,u_0\,dr, \end{aligned}\right\} \tag{38.10}$$

which appear in the triplet transition and

$$K_{ss}=\int_0^\infty u_g\,v_0\,dr,\qquad K_{dd}=\int_0^\infty w_g\,v_2\,dr,\qquad K_{sp}=\int_0^\infty r\,u_g\,v_1\,dr, \tag{38.11}$$

which are necessary in the singlet transition. The total differential cross section is then given by

$$\left.\begin{aligned}
d\sigma = \pi \frac{e^2}{\hbar c} \frac{(E_f - E_i)^2}{h\nu\varepsilon} \alpha^2 \frac{N_g^2}{k} \frac{d\Omega}{4\pi} \Big[\frac{1}{2} \sin^2\vartheta\, I_{sp}^2 + \frac{1}{50} (6 + \sin^2\vartheta)\, I_{dp}^2 + \\
+ \left(\frac{h\nu}{\hbar c}\right)^2 \left\{ \frac{1}{32} \cos^2\vartheta \sin^2\vartheta\, I_{sd}^2 + \frac{1}{1200} I_{ds}^2 \right\} + \frac{1}{6} \cos^2\vartheta \frac{(h\nu)^4}{(E_f - E_i)^2} \frac{(\mu_P - \mu_N)^2}{(M c^2)^2} I_{sp}^2 + \\
+ \left(\frac{h\nu}{\hbar c}\right) \left\{ \frac{1}{4} \cos(\delta_1 - \delta_2) \sin^2\vartheta \cos\vartheta\, I_{sp} I_{sd} - \frac{1}{50} \cos(\delta_1 - \delta_2) \cos\vartheta\, I_{dp} I_{ds} \right\} \Big] + \\
+ \frac{\pi}{3} \frac{e^2}{\hbar c} (\mu_P - \mu_N)^2 \frac{h\nu}{M c^2} \frac{N_g^2}{k} \frac{d\Omega}{4\pi} \Big[K_{ss}^2 + \frac{1}{4} (2 + 3\sin^2\vartheta)\, K_{dd}^2 - \\
- \frac{1}{\sqrt{2}} \cos(\Delta_0 - \Delta_2) (2 - 3\sin^2\vartheta)\, K_{ss} K_{dd} + \frac{1}{4} \cos^2\vartheta \left(\frac{h\nu}{\hbar c}\right)^2 \left(\frac{\mu_P + \mu_N}{\mu_P - \mu_N}\right)^2 K_{sp}^2 + \\
+ \left\{ \cos(\Delta_0 - \Delta_1) \cos\vartheta\, K_{ss} - \frac{1}{2\sqrt{2}} \cos(\Delta_2 - \Delta_1) \cos\vartheta\, (2 + 3\sin^2\vartheta)\, K_{dd} \right\} \times \\
\times \left(\frac{h\nu}{\hbar c}\right) \left(\frac{\mu_P + \mu_N}{\mu_P - \mu_N}\right) K_{sp} \Big],
\end{aligned}\right\} \tag{38.12}$$

where ϑ is the angle of the emitted neutron or proton with respect to the incident photon direction in the two-nucleon c.m. system and $h\nu$ the γ-energy in the laboratory system [cf. Eq. (35.9)]. In the above expression the first two terms are due to the electric dipole transition, the third to the electric quadrupole, the fourth to the magnetic quadrupole and the fifth to the interference between electric dipole and quadrupole transitions. The latter half of (38.12) is due to the singlet transitions: the first three terms are due to the spin magnetic dipole transition and the remaining terms to the magnetic quadrupole transition.

The total cross section is easily evaluated as

$$\left.\begin{aligned}
\sigma = \pi \frac{e^2}{\hbar c} \frac{(E_f - E_i)^2}{h\nu\varepsilon} \alpha^2 \frac{N_g^2}{k} \Big[\frac{1}{3} \left(I_{sp}^2 + \frac{2}{5} I_{dp}^2 \right) + \frac{1}{240} \left(\frac{h\nu}{\hbar c}\right)^2 \left(I_{sd}^2 + \frac{1}{5} I_{ds}^2 \right) + \\
+ \frac{1}{18} \left(\frac{(\mu_P - \mu_N)(h\nu)^2}{(E_f - E_i)\, M c^2} \right)^2 I_{sp}^2 \Big] + \frac{\pi}{3} \frac{e^2}{\hbar c} (\mu_P - \mu_N)^2 \frac{h\nu\, N_g^2}{M c^2 k} \times \\
\times \Big[K_{ss}^2 + K_{dd}^2 + \frac{1}{12} \left(\frac{h\nu}{\hbar c}\right)^2 \left(\frac{\mu_P + \mu_N}{\mu_P - \mu_N}\right)^2 K_{sp}^2 \Big].
\end{aligned}\right\} \tag{38.13}$$

It is readily noticed that higher multipole contributions are becoming more and more important as the incident photon energy $h\nu$ increases. In the low energy region which interests us in this chapter, however, the dipole terms are quite sufficient and especially the electric dipole term [the first one in Eq. (38.13)] dominates up to energies of one or two hundred Mev, i.e. as far as the conventional multipole decomposition is valid.

39. Total cross section versus energy and comparison with experiments. Now we compare our theoretical prediction with the experimental data summarized in Sect. 3. In the low energy region we can approximate (35.9) by

$$E_f - E_i = h\nu \tag{39.1}$$

and neglect all terms in (38.13) besides the leading terms in the triplet and singlet transitions, respectively, which gives

$$\sigma = \frac{\pi}{3} \frac{e^2}{\hbar c} \left(\frac{h\nu}{\varepsilon}\right) \alpha^2 \frac{N_g^2}{k} [I_{sp}^2] + \frac{\pi}{3} \frac{e^2}{\hbar c} (\mu_P - \mu_N)^2 \frac{h\nu}{M c^2} \frac{N_g^2}{k} [K_{ss}^2]. \tag{39.2}$$

The neglected terms will be estimated later. The same approximation applied to (38.12) gives

$$d\sigma = \frac{\pi}{2}\frac{e^2}{\hbar c}\left(\frac{h\nu}{\varepsilon}\right)\alpha^2\frac{N_g^2}{k}\frac{d\Omega}{4\pi}\left[\sin^2\vartheta I_{sp}^2\right] + \frac{\pi}{3}\frac{e^2}{\hbar c}(\mu_P-\mu_N)^2\frac{h\nu}{Mc^2}\frac{N_g^2}{k}\frac{d\Omega}{4\pi}\left[K_{ss}^2\right]. \quad (39.3)$$

This shows that the low energy photodisintegration cross section consists of a $\sin^2\vartheta$-term and an isotropic term, which has been verified experimentally in the very low energy region (below 3 Mev). Indeed the determination of the ratio of the photomagnetic cross section σ_m [second term of Eq. (39.2)] to the photoelectric σ_e [first term of Eq. (39.2)] is entirely based on this fact, as explained in Sect. 3γ. Thus we calculate σ_e and σ_m separately and compare their sum and ratio with the data in Tables 4 and 5, which amounts to a complete comparison of theory and experiment. It is noted, however, that the above approximation does not suffice to explain the angular distribution observed at higher energies [cf. Eq. (3.2) and Table 6], which is discussed in more detail in the next section.

The normalization factor N_g^2 was evaluated in Sect. 37. Let us now introduce a dimensionless normalization factor n_g^2 by

$$N_g{}^2 = 2\alpha\, n_g^2, \quad n_g^2 = 1.6687\,(1 \pm 0.0212), \quad (39.4)$$

from (37.5), where the error is due to the ambiguity of P_t given by (37.4). The deuteron S-wave function $u_g(r)$ was also determined in Sect. 37 as

$$u_g(r) = e^{-\alpha r} - e^{-\eta r}, \quad \eta = (1.3321 \pm 0.0547)\times 10^{13}\ \mathrm{cm}^{-1}, \quad (39.5)$$

where the error of η is again largely due to the uncertainty of P_t. As regards the singlet S-wave function $v_0(r)$, the form (37.6) can be used near the zero energy region. The corresponding modification of the asymptotic form $\sin(kr+\Delta_0)$ of $v_0(r)$ is

$$v_0(r) = \sin kr\cos\Delta_0 + \cos kr\sin\Delta_0\,(1-e^{-\xi r}), \quad (39.6)$$

where the parameter ξ is again determined by fitting (39.6) to the empirical singlet effective range r_{0s} given by (37.8). Since the effective range is determined by the zero energy wave function of (39.6), which agrees with (37.6) except for the normalization factor, ξ takes the same value as in Sect. 37 if we neglect its energy dependence:

$$\xi = (1.304 \pm 0.209)\times 10^{13}\ \mathrm{cm}^{-1}. \quad (39.7)$$

We must now construct a reasonable triplet P-wave function $u_1(r)$, which tends asymptotically to $\sin\left(kr - \frac{\pi}{2} + \delta_1\right)$. As there is some evidence that the force in a triplet odd state is weak and the integral I_{sp} in which $u_1(r)$ appears is an "outside" quantity in containing a factor r, we neglect the inside deviation of u_1 from its asymptotic form and assume the zero range approximation:

$$u_1(r) = \frac{\sin(kr+\delta_1)}{kr} - \cos(kr+\delta_1) = R\left\{e^{i(kr+\delta_1)}\frac{(1-ikr)}{ikr}\right\}, \quad (39.8)$$

where R means the real part of the quantity which follows. The above form is exact when the force vanishes in the triplet odd state ($\delta_1=0$). We estimate the influence of the force in the triplet P-state assuming (39.8), which results in overrating the effect since (39.8) does not satisfy the correct boundary condition at $r=0$. Still it will prove small enough, so it is legitimate to assume the zero range form (39.8) for $u_1(r)$.

The integrals I_{sp} and K_{ss} are easily evaluated as

$$I_{sp}=2k^2\left[\frac{1}{(\alpha^2+k^2)^2}-\frac{1}{(\eta^2+k^2)^2}\right]\left[1+\delta_1\left(\frac{\alpha(\alpha^2+3k^2)(\eta^2+k^2)^2-\eta(\eta^2+3k^2)(\alpha^2+k^2)^2}{2k^3(\eta^2-\alpha^2)(\eta^2+\alpha^2+2k^2)}\right)\right], \quad (39.9)$$

$$\left.\begin{aligned}K_{ss}=\sin\Delta_0\Big[k\cot\Delta_0\Big(\frac{1}{\alpha^2+k^2}-\frac{1}{\eta^2+k^2}\Big)+\\ +\Big(\frac{\alpha}{\alpha^2+k^2}-\frac{\eta}{\eta^2+k^2}\Big)-\Big(\frac{\alpha+\xi}{(\alpha+\xi)^2+k^2}-\frac{\eta+\xi}{(\eta+\xi)^2+k^2}\Big)\Big],\end{aligned}\right\} \quad (39.10)$$

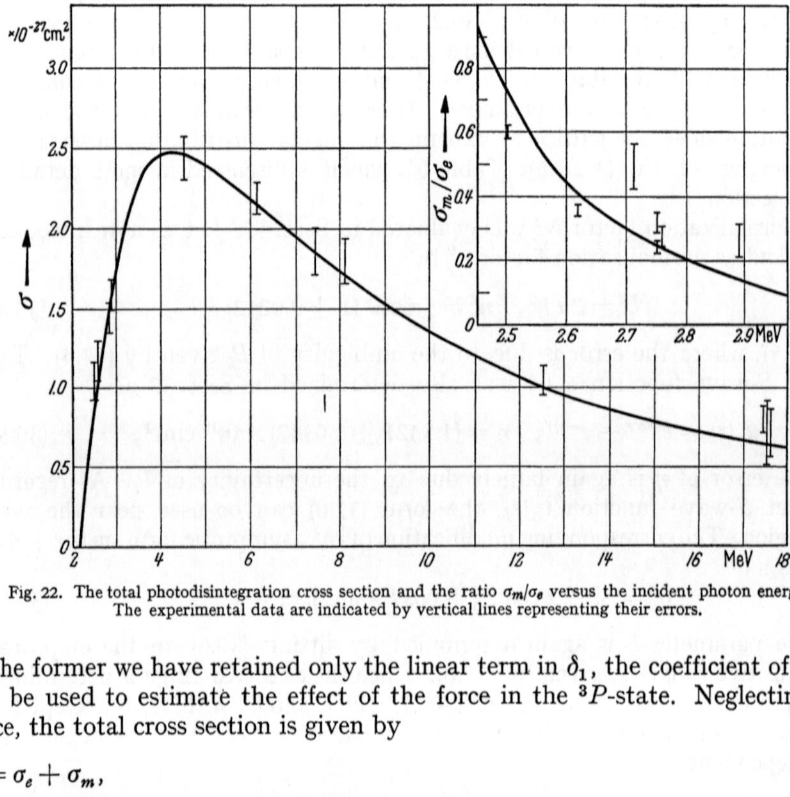

Fig. 22. The total photodisintegration cross section and the ratio σ_m/σ_e versus the incident photon energy. The experimental data are indicated by vertical lines representing their errors.

in the former we have retained only the linear term in δ_1, the coefficient of which can be used to estimate the effect of the force in the ^3P-state. Neglecting this force, the total cross section is given by

$$\sigma=\sigma_e+\sigma_m,$$

$$\sigma_e=\frac{8\pi}{3}\left(\frac{e^2}{\hbar c}\right)n_g^2\left(\frac{\alpha^2+k^2}{\alpha^2}\right)\frac{\alpha}{k}\left[\frac{\alpha^2k^2}{(\alpha^2+k^2)^2}-\frac{\alpha^2k^2}{(\eta^2+k^2)^2}\right]^2\frac{1}{\alpha^2}, \quad (39.11)$$

$$\left.\begin{aligned}\sigma_m=\frac{2\pi}{3}\left(\frac{e^2}{\hbar c}\right)(\mu_P-\mu_N)^2\frac{h\nu}{Mc^2}n_g^2\frac{\alpha}{k}\sin^2\Delta_0\Big[\frac{k\cot\Delta_0}{\alpha}\Big(\frac{\alpha^2}{\alpha^2+k^2}-\frac{\alpha^2}{\eta^2+k^2}\Big)+\\ +\Big(\frac{\alpha^2}{\alpha^2+k^2}-\frac{\alpha\eta}{\eta^2+k^2}\Big)-\Big(\frac{\alpha(\alpha+\xi)}{(\alpha+\xi)^2+k^2}-\frac{\alpha(\eta+\xi)}{(\eta+\xi)^2+k^2}\Big)\Big]^2\frac{1}{\alpha^2}.\end{aligned}\right\} \quad (39.12)$$

Here we can use the effective range expansion (16.31):

$$k\cot\Delta_0=-\frac{1}{a_s}+\frac{1}{2}k^2r_{0s}-P_sk^4r_{0s}^3, \quad (39.13)$$

where a_s is given by (18.17) and (18.18), r_{0s} by (37.8) and we can take $P_s=0.01\pm0.04$ from Table 11.

The theoretical values of $\sigma=\sigma_e+\sigma_m$ and σ_m/σ_e are plotted in Fig. 22, together with the experimental data summarized in Sect. 3. We see that the agreement between theory and experiment is excellent although the data in the very low energy region do not allow a definite conclusion.

Let us now consider the effects which have been neglected in the final theoretical cross sections (39.11) and (39.12). First of all, the influence of the force in the ^3P-state can be estimated from the term containing δ_1 in (39.9). In order to get its maximum value, take a γ-energy of 17.8 Mev, for which the relevant term in (39.9) becomes 0.3700 δ_1. δ_1 can be estimated using the BORN expression for the phase shift and assuming the square well potential, the intrinsic range of which is chosen as 2.043×10^{-13} cm, the same value as in the triplet even state for the same type of potential (cf. Table 9). Then the correction factor in (39.9) is given by $1+0.032s$, where s is the well-depth parameter, known to be 1.441 in the ^3S-state. As the well depth in the ^3P-state is likely to be several times smaller than in the ^3S-state, it may be estimated that the change in the theoretical curve in Fig. 22 would not exceed 2% even at the highest energy ($\approx$ 20 Mev)*. Above 4 or 5 Mev, σ_m is of the order of at most a few percent of σ_e, so we need not worry about the uncertainty of σ_m. From (39.11) it is readily shown that the ambiguity of η in (39.5) produces an error of at most 1.13% in σ_e, that of n_g^2 given by (39.4) an error of 2.12%. Assuming these errors to work at random, it is concluded that the theoretical total cross section would be uncertain within 3 or 4%, which is smaller than the experimental error of 5 to 10% in this energy region.

40. Angular distribution. So far we have neglected the influence of the terms due to the D-component of the deuteron ground state and the higher multipole moments on the angular distribution.

According to (3.2), all data are consistent with an angular distribution of the form $d\sigma/d\Omega=a+(b+c\cos\vartheta)\sin^2\vartheta$. At is seen from (38.12), the interference between the electric dipole and quadrupole transitions is the main reason for the asymmetry in the angular distribution. The theory thus predicts an asymmetry coefficient c/b essentially given by

$$\frac{c}{b}=\frac{1}{2}\left(\frac{h\nu}{\hbar c}\right)\cos(\delta_1-\delta_2)\frac{I_{sd}}{I_{sp}}. \tag{40.1}$$

To estimate its numerical value, we neglect the small P- and D-phase shifts δ_1 and δ_2. Then the functions $u_l(r)$ defined in (38.2) are simply given by $kr\,j_l(kr)$ from the expansion (15.7):

$$\left.\begin{aligned} u_1(r)&=\frac{\sin kr}{kr}-\cos kr,\\ u_2(r)&=\left[\frac{3}{(kr)^2}-1\right]\sin kr-\frac{3}{kr}\cos kr.\end{aligned}\right\} \tag{40.2}$$

As for the deuteron wave function, we may use the expression given in Sect. 33. The above integrals are "outside" quantities, so it is sufficient to consider one case where, for example, $x_c=0$ and $P_D=3\%$. We obtain

$$c/b=0.267 \quad \text{for} \quad h\nu=17.8\ \text{Mev}^{**}, \tag{40.3}$$

which agrees fairly well with the experimental values in Table 6.

In the previous section it was shown that the isotropic part of the cross section is large in the very low energy region and is well explained as due to the

* Cf. the results of I. F. E. HANSSON and L HULTHÉN: Phys. Rev. **76**, 1163 (1949). — HANSSON: Phys. Rev. **79**, 909 (1950).

** In this energy region, the asymmetry coefficient approximately agrees with the value of $2\frac{v}{c}$, v being the velocity of the photoproton (or -neutron) in the c.m. system, c the velocity of light.

spin magnetic dipole transition. As suggested by Fig. 22, however, this component becomes negligibly small as the energy increases. The theory predicts, for instance,

$$\frac{a}{b} = \frac{2\sigma_m}{3\sigma_e} = 0.96\ \% \quad \text{for} \quad h\nu = 17.8\ \text{Mev} \tag{40.4}$$

as the contribution due to the spin magnetic dipole transition. This figure is certainly too small compared with the experimental values in Table 6.

From (38.12), it is seen that, among the remaining terms, the $^3D \to {}^3P$ transition [the second term in Eq. (38.12)] gives the most important contribution to the isotropic ratio, viz.

$$a/b = \tfrac{6}{25}\,[I_{dp}/I_{sp}]^2, \tag{40.5}$$

the size of which is easily estimated, employing the zero force approximation in the ^3P-state and the deuteron wave function of Sect. 33. Evaluating the numerator of (40.5) by numerical quadrature, one gets

$$a/b = 1.14\,\%\,, \quad \text{for} \quad h\nu = 17.8\ \text{Mev}. \tag{40.6}$$

It is also possible to rate the contribution to the isotropy coefficient from the electric and magnetic quadrupole transitions, etc. All these, however, turn out much smaller than (40.4) or (40.6). Thus the phenomenological theory predicts an isotropy coefficient of at most 3% at the γ-energy 17.8 Mev. The above calculations also show that the terms neglected in the previous section are indeed of minor importance. We remark that the relatively large asymmetric term [cf. Eq. (40.3)] does not contribute to the total cross section.

As regards the force in the ^3P-state, which has been neglected throughout, we estimated in Sect. 39 that it would probably give a correction of at most 2% to the total cross section through the electric dipole term. According to a more detailed investigation[138], a tensor force in the ^3P-state may possibly influence the isotropic term, although its effect on the $\sin^2\vartheta$-term is insignificant. For a reasonable tensor force in the ^3P-state, however, the contribution to the isotropy coefficient would be at most 1% for 20 Mev γ-energy.

The exchange and relativistic corrections are discussed in the next section, but there are not yet any detailed numerical estimates of their influence on the photodisintegration. According to a preliminary calculation, however, they are not likely to contribute substantially to the isotropy coefficient. Thus it may be concluded that the big experimental isotropy coefficient of about 10% at 10 to 20 Mev γ-energy (Table 6) seems to be incompatible with current theory, although the experimental value is by no means definite. With this exception, the low energy photodisintegration data are very well explained by the theory.

d) Exchange current effect.

41. Relativistic and exchange current corrections. In Sect. 34, we have given a definition of the relativistic and exchange current corrections. The former consists of correction terms to the non-relativistic matrix element (34.6). From (34.5), we readily get

$$\left.\begin{aligned}
(u_{\boldsymbol p}^{*}\,\boldsymbol\alpha\,u_{\boldsymbol p'}) &= \frac{1}{2Mc}\Big[\{(\boldsymbol p+\boldsymbol p') + i\,\boldsymbol\sigma\times(\boldsymbol p-\boldsymbol p')\} - \\
&\quad - \frac{1}{8M^2c^2}\{(p^2+p'^2)(\boldsymbol p+\boldsymbol p') + 2(p^2\boldsymbol p + p'^2\boldsymbol p') + \\
&\quad + i\,\boldsymbol\sigma\times[(p^2+p'^2)(\boldsymbol p-\boldsymbol p') + 2(p^2\boldsymbol p - p'^2\boldsymbol p')]\}\Big],\\
(u_{\boldsymbol p}^{*}\,\beta\,\boldsymbol\sigma\,u_{\boldsymbol p'}) &= \boldsymbol\sigma - \frac{1}{8M^2c^2}\,[(p^2+p'^2)\,\boldsymbol\sigma + 2(\boldsymbol\sigma\boldsymbol p)\,\boldsymbol\sigma(\boldsymbol\sigma\boldsymbol p')],
\end{aligned}\right\} \tag{41.1}$$

where we have retained only terms of the next higher order with respect to the nucleon velocity, M being the nucleon mass. The contribution of the above correction terms to the matrix element (34.3) and especially to the multipole transition matrix elements given in Sect. 35 can be evaluated using the same method as in Sects. 34 and 35. It is just these terms that have given the relativistic correction to the deuteron magnetic moment in Sect. 27.

The exchange current correction arises from the amplitudes containing mesons in the matrix element (34.3). As these effects are small, we evaluate them in the non-relativistic approximation. The details of calculation are similar to those in Sect. 27, the decomposition into corrections to the individual multipole matrix elements being analogous to the procedure in Sect. 35.

The relativistic and exchange current corrections to the deuteron magnetic moment have been derived in Sect. 27 using a slightly different technique, but they can also be obtained by the formalism outlined in Sects. 34 and 35. They are important in estimating the D-state probability P_D as (27.11). This figure, however, lacks any direct experimental verification. For an empirical test of the theory developed in Sects. 27, 34, and 35, the thermal capture cross section would be the only available quantity, since the experimental and theoretical ambiguities in the photodisintegration of the deuteron appear too large beside the relevant small corrections.

As regards the thermal neutron-proton capture, we need only the corrections to the matrix element M in (37.12) (spin magnetic dipole transition), which are, to the lowest order in the meson-nucleon coupling constant and the nucleon velocity[139],

$$\left.\begin{aligned} \Delta_{\mathrm{rel}} M = {} & \frac{\mu_P - \mu_N + 3}{6\varkappa^2(\mu_P - \mu_N)} \int_0^\infty u_g(r)\, u_s''(r)\, dr - \\ & - \sqrt{2}\, \frac{\mu_P - \mu_N - \frac{3}{2}}{6\varkappa^2(\mu_P - \mu_N)} \int_0^\infty w_g(r) \left[u_s''(r) - \frac{3}{r} u_s'(r) + \frac{3}{r^2} u_s(r) \right] dr, \end{aligned}\right\} \quad (41.2)$$

$$\left.\begin{aligned} \Delta_{\mathrm{exch}} M = {} & \frac{2}{\pi} \alpha_g^2 \int_0^\infty u_s(r) \left[\left(K_0(x) - \frac{K_1(x)}{x} \right) u_g(r) + \right. \\ & \left. + \sqrt{2} \left(K_0(x) + \frac{2K_1(x)}{x} \right) w_g(r) \right] dr + \frac{(4\varkappa/\mu)}{3(\mu_P - \mu_N)} \alpha_g^2 \times \\ & \times \int_0^\infty u_s(r) \left[\left(2e^{-x} - \frac{e^{-x}}{x} \right) u_g(r) + \sqrt{2} \left(e^{-x} + \frac{e^{-x}}{x} \right) w_g(r) \right] dr - \frac{1}{2} M \Delta N_g, \end{aligned}\right\} \quad (41.3)$$

where ΔN_g is the same as (27.10), $\alpha_g^2 = (f^2/4\pi\hbar c)(\mu/2\varkappa)^2$, $x = \mu r$, $K_0(x)$ and $K_1(x)$ are the HANKEL functions of imaginary arguments, and dashes mean derivatives* with respect to r. We note that the two expressions (41.2) and (41.3) are very similar to those given by (27.8) and (27.9) for the corrections to the deuteron magnetic moment, which is due to the fact that they arise from the same physical sources and have been calculated on the same basis. The reason why no D-D terms appear in the expressions (41.2) and (41.3) is that the initial state is a pure S-state in the capture process.

42. Possible contribution to neutron-proton capture and photo-decay. The above corrections can be evaluated using the phenomenologically adjusted

[139] M. SUGAWARA: Progr. Theor. Phys. **14**, 535 (1955).

* See footnote *, p. 73.

deuteron wave functions of Sect. 33 and the analogous ^1S wave function given in Sect. 37. It is found[139] that the relativistic correction is small, which is due to the fact that the energy of the initial state in the capture process is nearly zero; it is negative and the S-S term dominates, which is the same situation as in Sect. 27 for the deuteron magnetic moment. In the exchange current correction (41.3), the situation is different from that in Sect. 27; (41.3) is definitely positive, the S-S term is predominant and the correction is increased very much if a hard core is introduced. This difference is due to the fact that, in the case of the deuteron magnetic moment, the so-called meson contributions vanish identically owing to the antisymmetric charge spin factor, while such terms do not in general vanish in radiative transitions and now happen to be dominant.

The sum of (41.2) and (41.3) is again shown[139] to be rather insensitive to a hard core, although the individual corrections are affected separately. If we assume 10 for the ps-ps coupling constant $f^2/4\pi\hbar c$ and the D-state probability is 3%, we get a final correction to the capture cross section of about $+1\%$ of the phenomenological value. This is smaller than the corresponding experimental and theoretical errors in (37.1), (37.16) and (37.18), so it is futile to compare the above theory with experiment. It seems, however, that the correction does not suffice to explain the possible discrepancy discussed in Sect. 37, although the above figure does not include the non-additivity correction, which was a positive effect in the case of the deuteron magnetic moment (Sect. 27).

As regards the photo-decay, detailed calculations of the relevant corrections are not yet available. According to Sects. 38, 39 and 40, theory and experiment agree well within their limits of error, except for the isotropic part of the cross section. We therefore give a preliminary discussion of the possible corrections to this part.

The corrections to the magnetic dipole matrix element (41.2) and (41.3) are naturally valid also for the photo-decay. Their values are easily estimated using the deuteron wave functions given in Sect. 33 and the ^1S wave function (39.6). It is found that the resulting correction increases very rapidly with the photon energy: from 1% in the very low energy region it rises to about 15% of the phenomenological value at 20 Mev. This correction is, however, far too small to remove the discrepancy in the angular distribution. The relativistic and exchange current effects naturally give rise to isotropic terms also in other multipole transitions. It seems, however, that all these corrections together will not be large enough to explain the big discrepancy found in Sect. 40.

VII. High energy phenomena.

a) High energy nucleon-nucleon scattering.

43. General survey. Available experimental data are summarized in Sect. 4, but we recapitulate here the most remarkable characteristics. In the high energy p-p scattering, the differential cross section is practically constant over a very wide angular range and its value is almost independent of the incident proton energy over a wide energy range. In the high energy n-p scattering, the scattering observed at lower energies is nearly symmetrical in the c.m. system with respect to forward and backward directions, the minimum occurring near 90°. This close symmetry is lost as the energy increases, the backward scattering becomes predominant and the minimum point moves gradually towards the backward direction. These general features have, however, been established only recently and have not yet been fully investigated.

Before entering on the details of the analysis, we briefly mention some investigations on high energy scattering, which were made during the time when the features mentioned above were being clarified. Otherwise we are only interested in finding a static potential between two nucleons which can explain the empirical facts.

At first it seemed as if the high energy *n-p* scattering were perfectly symmetrical in the c.m. system. It was, therefore, suggested that the nuclear force acts only in even parity states and vanishes identically in odd states (SERBER force), which would directly explain the symmetrical scattering. A very extensive investigation along this line has been made by CHRISTIAN and HART[140], who could very well explain the data available at that time. A difficulty was, however, soon encountered in trying to interpret the high energy *p-p* scattering data, as it was found [141] that no such forces could explain the flat angular distribution. It was, therefore, suggested that the charge-independence of the nuclear force might not be valid at high energies although it seemed a well established fact in the low energy region.

We now know that the high energy *n-p* scattering is not symmetric and, therefore, we cannot accept the SERBER force. Gradually it became clear that the peculiar behaviour of the high energy *p-p* data would be more difficult to interpret, assuming a static nuclear potential, than the *n-p* data, also because we have less freedom in the *p-p* case since the two particles are identical. Most efforts have, therefore, been concentrated on finding out how to explain the *p-p* data.

For this purpose three suggestions have been made. First of all, CHRISTIAN and NOYES[141] proposed a model in which the tensor force in a triplet odd state behaves singularly at the origin. Secondly, CASE and PAIS[142] argued that the high energy nucleon-nucleon scattering could be explained by introducing a special kind of non-static spin-orbit coupling, viz. the $\boldsymbol{L}\cdot\boldsymbol{S}$ force, into the nuclear interaction. Finally, JASTROW[143] has proposed the so-called hard core model which is characterized by a short range infinite repulsion inside an attractive nuclear interaction. CHRISTIAN and NOYES abandoned charge independence, so they did not try to explain *p-p* and *n-p* data on the same basis. CASE and PAIS, and JASTROW, on the other hand, reported that their potentials could explain *p-p* and *n-p* data in a charge independent way.

These authors all used approximate methods in their calculations (mostly the BORN approximation), since the exact numerical work is so tedious; moreover those approximate methods were believed to be accurate enough at higher energies. It was soon pointed out, however, that we have to do with very singular potentials in all three cases and that the approximate methods might not be justified. Therefore some authors[144] have taken the trouble to solve these scattering problems by more exact methods and found that the BORN approximation even with its improvements is inadequate for such singular potentials, even at rather high energies. The results show that only the hard core model[143] can give the isotropic angular distribution found in the *p-p* scattering, while the singular tensor force model of CHRISTIAN and NOYES[141] and the $\boldsymbol{L}\cdot\boldsymbol{S}$ force model of CASE and PAIS[142] give rise to anisotropic scattering.

140 R. S. CHRISTIAN and E. W. HART: Phys. Rev. **77**, 441 (1950).
141 R. S. CHRISTIAN and H. P. NOYES: Phys. Rev. **79**, 85 (1950).
142 K. M. CASE and A. PAIS: Phys. Rev. **80**, 203 (1950).
143 R. JASTROW: Phys. Rev. **81**, 165 (1951).
144 L. J. B. GOLDFARB and D. C. FELDMAN: Phys. Rev. **88**, 1099 (1952). — H. P. NOYES and H. G. CAMNITZ: Phys. Rev. **88**, 1206 (1952). — D. R. SWANSON: Phys. Rev. **89**, 740 (1953).

It seems, therefore, that the only model which can explain high energy *p-p* and *n-p* data on a charge-independent basis is the hard core model. In the next section we summarize the available investigations of the high energy scattering which are based on this model, stressing that much remains to be done.

44. Hard core hypothesis. According to the hypothesis of JASTROW[143], the nuclear forces are infinitely repulsive within a short inter-nucleonic distance. As mentioned in the previous section, this hard core hypothesis seems[144] to be compatible with the high energy *p-p* scattering data, unlike other proposals made. Theoretically we are at a loss to find arguments for or against such a hypothesis from the current meson field theory*. We have its justification only from the phenomenological side, which would, however, suffice to make us believe that it contains something true, since the high energy *p-p* data have been so difficult to interpret otherwise. Outside the hard core an ordinary static nuclear interaction is assumed between the two nucleons. The radius of the hard core and the outside static force must be chosen so as to fit the low energy properties of the two-nucleon system and at the same time to reproduce the higher energy nucleon-nucleon scattering data.

In solving the high energy scattering problem, we have no such simplified method as the effective range theory which was used extensively in the low energy case. What we must do is to calculate the phase shifts and the mixing parameters for a given nuclear potential and compute the differential scattering cross section, which can then be compared with the empirical data. In calculating these parameters we have even no reliable approximate methods for such singular potentials as are contemplated now, so the differential equations must be solved numerically. This is tedious indeed and not very interesting theoretically. We therefore quote only the results of the existing papers in the following.

Let us for the time being consider only the *p-p* scattering. This consists of singlet and triplet scattering which are not coherent with each other. Owing to the exclusion principle, the former takes place only in even parity states, the latter in odd states. It is easy to see that the singlet scattering is always predominantly forward if the potential in the singlet even state is of the same sign everywhere, as has usually been assumed thus far, and therefore the phase shifts are all of the same sign. On the other hand, the triplet scattering can be quite different since a spin-orbit interaction, such as the tensor force or the $\boldsymbol{L}\cdot\boldsymbol{S}$ force, can give rise to side-wise scattering (scattering angle $\pi/2$). CHRISTIAN and NOYES[141] and CASE and PAIS[142] have investigated these effects of the spin-orbit interaction, which were, however, later found insufficient to make the whole scattering so isotropic as is observed experimentally[144].

The hard core model, on the other hand, aims at modifying the strong forward singlet scattering by assuming a hard core in the singlet even state. JASTROW[143] has proposed the following potential:

$$\left.\begin{aligned} \text{Singlet even potential: } V_s(r) &= +\infty; & r &< r_c, \\ &= -V_s e^{-(r-r_c)/r_s}; & r &> r_c, \\ \text{Triplet odd potential: } V_t(r) &= V_t S_{12} e^{-r/r_t}; & r &> 0, \end{aligned}\right\} \tag{44.1}$$

* There is an interesting theoretical argument for the hard core hypothesis [S. GARTENHAUS, Phys. Rev. **100**, 900 (1955)], which is based on a cut-off meson theory. An important qualification is, however, that the neglected higher order effects do not change the situation, which hardly seems to be justified.

with the parameters

$$\left.\begin{array}{c} r_c = 0.6\times 10^{-13}\,\mathrm{cm}, \quad r_s = 0.4\times 10^{-13}\,\mathrm{cm}, \quad r_t = 0.75\times 10^{-13}\,\mathrm{cm}, \\ V_s = 375\ \mathrm{Mev} \quad \text{and} \quad V_t = -50.8\ \mathrm{Mev}. \end{array}\right\} \tag{44.2}$$

The singlet potential is naturally adjusted to the singlet low energy scattering data (the singlet effective range and scattering length). The effect of introducing a hard core into the triplet state has been neglected for the sake of simplicity.

Take the case of 30 Mev in Table 7 or Fig. 1. It was first pointed out that the experimental curve in Fig. 1 is very well reproduced by assuming a suitable value for the ^1S-wave phase shift and making all others vanish. It was then shown that the singlet D-wave phase shift cannot be neglected at 30 Mev if we assume an ordinary type of singlet nuclear interaction which is fitted to the low energy data. The introduction of a hard core in the singlet state reduces the intrinsic range of the force if it is kept to fit the low energy data. This, in turn, makes the D-wave phase shift decrease very much. Thus the potential (44.1) can explain the 30 Mev data very well.

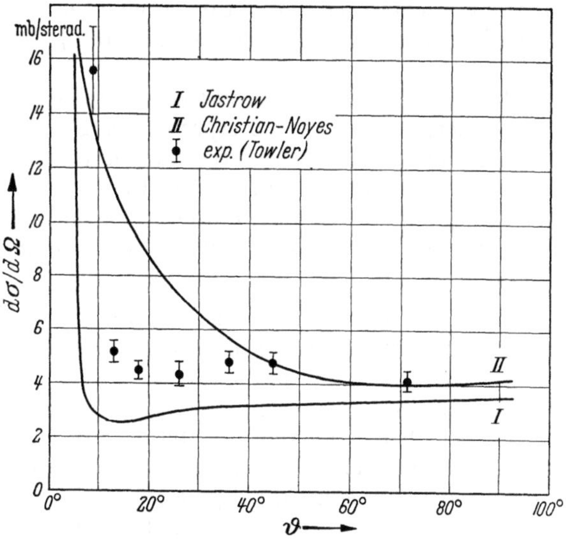

Fig. 23. The differential p-p scattering cross section at 240 Mev versus c.m. scattering angle calculated for the hard core potential (44.1) (curve I) and for the singular tensor model (curve II), together with the experimental data.

The next problem is to see whether the above potential can reproduce the higher energy data. Jastrow's original calculation (rigorous for the singlet potential and Born approximation for the triplet tensor scattering) was refined by Goldfarb and Feldman[144], using a variational procedure, and further by Noyes and Camnitz[144], taking the interference between Coulomb and nuclear scattering into account. They showed that the Born approximation is reasonable for such a well-behaved potential as $V_t(r)$ in (44.1) and the interference between Coulomb and nuclear scattering is a minor effect at higher energies, whence the potential (44.1) can in fact produce a very isotropic scattering. A detailed calculation has been made at 240 Mev, the result of which is plotted in Fig. 23 together with the experimental data[47]. The calculated curve for the singular tensor force model of Christian and Noyes[141] is also included in the figure. As for the $\boldsymbol{L}\cdot\boldsymbol{S}$ force model, it leads to a still worse disagreement with experimental data[144]. The difference between the hard core curve and the experimental data is insignificant and could presumably be removed by minor readjustments of parameters. The surprising flatness of the distribution is due to the change of sign of the ^1S-wave phase shift in the neighbourhood of 150 Mev as a consequence of the interference between repulsive and attractive regions of the potential, which then results in the appearance of a minimum in the singlet cross section between 100 and 200 Mev. This is of course partially compensated by a monotonic decrease of the triplet cross section, which explains the empirical fact that the cross section is nearly independent of the energy over a wide range. As regards the core radius, even a change as small as $\pm 0.1 \times 10^{-13}$ cm would make the agreement much worse, according to JASTROW[143]. It would, however, be necessary to make more detailed calculations before we can be sure about this point.

The question whether the high energy n-p data are consistent with such a potential as (44.1) has not yet been treated in detail. In the n-p case, the singlet odd and triplet even states are active; in particular the latter play an essential rôle, not only because of their statistical weight. As there are many degrees of freedom in fixing the forces in these states, we could reasonably expect that the potential (44.1) may be supplemented so as to reproduce the empirical high energy n-p scattering, without destroying the adjustment to the low energy data. According to a preliminary calculation by JASTROW[143], this seems indeed to be the case.

It should also be remarked that the relativistic effect will become important as the energy increases and especially so if we accept the hard core hypothesis, since it makes the depth of the nuclear potential increase very much [see, for example, the figures in Eq. (44.2)]. Besides, the velocity dependent force might not be negligible. We may, however, conclude that the high energy n-p and p-p scattering data could be explained by a charge-independent static nuclear potential in spite of their different appearance.

45. Theory of polarized nucleon beam. In the above treatment of the nucleon-nucleon scattering, nothing has been mentioned about the spin orientation of the particles in the beam or in the target. We have tacitly assumed that the different spin directions are equally probable, i.e. the beam or target is "unpolarized". If, on the other hand, the probabilities are unequal, such a beam or target has a non-vanishing expectation value of the spin in one direction and is called "polarized". To define the degree of polarization quantitatively, let us introduce the "polarization" P of a beam as the expectation value of twice the spin $\boldsymbol{S}$ of the particle, in units of $\hbar$, or the PAULI spin operator σ in the direction y, for instance:

$$P = 2\langle S_y \rangle = \langle \sigma_y \rangle = \frac{I_+ - I_-}{I_+ + I_-}, \tag{45.1}$$

where I_+ and I_- are the intensities of particles having their spins in the respective directions. According to the definition, $P=0$ for an unpolarized beam and $P=1$ for a perfectly polarized beam.

Several authors have made theoretical investigations[145] on the polarization of nucleon beams and the scattering of polarized beams. We quote in the following their main results, referring to the original papers for details. Owing to the spin-orbit interaction between two nucleons, such as a tensor force or an $\boldsymbol{L} \cdot \boldsymbol{S}$ force mentioned in Sect. 43, the scattered beam is shown to be partially polarized in a direction perpendicular to the plane of scattering, even though the incident beam and the target are both unpolarized. The polarization measurement is, therefore, a very important source of information on the nuclear spin-orbit interaction.

Suppose the incident unpolarized beam is moving along the z-direction and the scattering takes place in the xz-plane. Then the scattered beam is polarized

[145] L. WOLFENSTEIN: Phys. Rev. **75**, 1664 (1949); **76**, 541 (1949); **82**, 308 (1951). — L. WOLFENSTEIN and J. ASHKIN: Phys. Rev. **85**, 947 (1952). — R. H. DALITZ: Proc. Phys. Soc. Lond. A **65**, 175 (1952). — DON S. SWANSON: Phys. Rev. **84**, 1068 (1951); **89**, 749 (1953). — L. J. B. GOLDFARB and D. FELDMAN: Phys. Rev. **88**, 1099 (1952). — A. SJÖLANDER and S. KÖHLER: Ark. Fysik **8**, Nr. 52 (1954).

in the y-direction. Its polarization is a function of the scattering angle ϑ and the wave number k of the incident particle, both of which are now referred to the c.m. system. We mention here that the polarization effect vanishes if calculated in the BORN approximation for a tensor force as well as an $\boldsymbol{L}\cdot\boldsymbol{S}$ force. We must, therefore, use the detailed numerical method to evaluate the polarization for some assumed nuclear potential, which is the same situation as in the previous two sections. The problem is then how to detect this polarization experimentally. It can be shown that, if the polarized beam is again scattered by an unpolarized nucleon target, azimuthal asymmetry can in general appear, under the assumption of some spin-orbit interaction; this azimuthal asymmetry can be measured directly.

In order to describe the second scattering let us choose the first scattering direction as the new z-axis and take the same y-axis as before. The subscripts 1 and 2 will denote the quantities connected with the first and the second scattering, respectively. In each case the quantities are referred to the c.m. system of the interacting particles. Then the scattered intensity $I(\vartheta_2, \varphi_2)$ is shown to be

$$I(\vartheta_2, \varphi_2) \propto 1 + P(k_1, \vartheta_1)\, P(k_2, \vartheta_2) \cos\varphi_2, \tag{45.2}$$

as a function of ϑ_2 and φ_2. This equation is the most important formula for the polarization measurement (cf. Sect. 4 and Fig. 5). First of all, it tells us that the asymmetry does not appear in the direction of polarization of the beam but in the plane perpendicular to it (plane of the first scattering). Secondly, the asymmetry is proportional to the polarization of the beam. Finally, the proportional factors in the last term of (45.2) are the polarizations $P(k_1, \vartheta_1)$ and $P(k_2, \vartheta_2)$ defined by (45.1). The formula (45.2) can be derived under the general assumption that the interaction is invariant against rotation, reflection and time reversal. The measured quantity is usually the asymmetry $e(\vartheta_2)$ defined as

$$e(\vartheta_2) \equiv \frac{I(\vartheta_2, 0) - I(\vartheta_2, \pi)}{I(\vartheta_2, 0) + I(\vartheta_2, \pi)} = P(k_1, \vartheta_1)\, P(k_2, \vartheta_2), \tag{45.3}$$

where the second equality is readily obtained from (45.2). Therefore the asymmetry $e(\vartheta_2)$ gives the polarization $P(k_2, \vartheta_2)$ of the second scattered beam if the polarization of the second incident (first scattered) beam $P(k_1, \vartheta_1)$ is known.

It is thus seen that the polarization measurement is essentially a double scattering experiment, to which fact are due the main difficulties of the measurement. Only few experiments have been successful in getting the asymmetry $e(\vartheta_2)$ for double p-p and n-p scattering. The data have been given in Sect. 4, while the theoretical analysis is deferred to the next section.

We emphasize here that nucleons elastically scattered by nuclei can, under suitable conditions, be much more strongly polarized than those scattered by nucleons. Indeed, recent experiments have been successful in getting nucleon beams polarized to more than 70% ($P=0.7$). In order to treat the elastic nucleon-nucleus scattering, the most primitive way would be to choose some static interaction between nucleon and nucleus, regarded as a point, which can reasonably be assumed to contain a strong $\boldsymbol{L}\cdot\boldsymbol{S}$ force since such an interaction is required by the nuclear shell model. Then we may expect a strong polarization in the elastic nucleon-nucleus scattering. The situation is, however, never so simple as that. In most energy regions concerned, the wave length of the incident nucleon is already so small that the nucleus can no longer be regarded as a point source of force. Thus one must devise special approximations in order to treat this kind of many-body problem and also pay regard to possible resonance scattering due to the excited states of the nucleus. Many authors[146] have investigated this problem and explained why such large polarizations as are observed can occur in the elastic nucleon-nucleus scattering. It has also been shown that the formula (45.2) can be regarded as valid for reasonable interactions between nucleons and nuclei. Then it is readily seen that the polarization could be obtained by measuring the asymmetry $e(\vartheta_2)$ in a double elastic nucleon-nucleus scattering in which the same kind of target is used both times and the two scattering angles ϑ_1 and ϑ_2 are adjusted to be equal. If the small energy decrease in the successive scattering processes can be neglected, we have $P(k, \vartheta) = \sqrt{e(\vartheta)}$.

Naturally there are still many unresolved problems in the polarization by elastic nucleon-nucleus scattering. However, the above method is the most accurate one now available to measure the polarization and at the same time the most effective method to produce a highly polarized beam of nucleons. Once having got such a beam of known polarization, it is possible to measure the polarization in p-p and n-p scattering, using the Eqs. (45.2) and (45.3). The polarization data quoted in Sect. 4 have been obtained in this way.

46. Analysis of polarization measurements. Although most polarization measurements reported so far are not very definite, it seems possible to draw some important conclusions which will be outlined here.

It was mentioned in Sect. 44 that, among several proposals, JASTROW's hard core model [143] is the only one that can explain the flatness of the high energy p-p scattering cross section over a very wide angular range. We therefore first examine whether JASTROW's potential given by (44.1) can explain the observed azimuthal asymmetry in the double p-p scattering (cf. Sect. 4). Such a calculation has been made by GOLDFARB and FELDMAN[144]. The result was that the potential (44.1) predicts, at 240 Mev, a maximum asymmetry of about 0.8% at $\vartheta_1 = \vartheta_2 \approx 40°$. This figure is certainly too small to be consistent with the experimental value of (4.8 ± 1.8)% for $\Theta_1 \approx 19°$ and $\Theta_2 \approx 27°$, i.e. $\vartheta_1 \approx 40°$ and $\vartheta_2 \approx 50°$, at about 200 Mev. They also made the calculation for the singular tensor force model of CHRISTIAN and NOYES[141] and the $\boldsymbol{L} \cdot \boldsymbol{S}$ force model of CASE and PAIS[142] (cf. Sect. 44) and found that the latter predicts an asymmetry from 15 to 25% under the same conditions, which is certainly too large to be accepted. We may, therefore, conclude that the $\boldsymbol{L} \cdot \boldsymbol{S}$ force model cannot explain the high energy p-p scattering data, neither the angular distribution nor the polarization measurements.

GOLDFARB and FELDMAN[144] got, however, a very reasonable value for the singular tensor force model, which will now be expounded in some detail. CHRISTIAN and NOYES[141] proposed the following potential between two protons:

$$\left.\begin{aligned} &\text{Singlet even potential: } V_s(r) = 0; && r > 2.615 \times 10^{-13}\,\text{cm},\\ &\qquad\qquad = -13.273\ \text{Mev}; && r < 2.615 \times 10^{-13}\,\text{cm},\\ &\text{Triplet odd potential: } V_t(r) = 0; && r < 0.17664 \times 10^{-13}\,\text{cm},\\ &\qquad\qquad = 18\, S_{12} \frac{e^{-r/r_t}}{(r/r_t)^2}\,\text{Mev}; && r > 0.17664 \times 10^{-13}\,\text{cm}, \end{aligned}\right\} \quad (46.1)$$

---

[146] J. V. LEPORE: Phys. Rev. **79**, 137 (1950). — E. FERMI: Nuovo Cim. **11**, 407 (1954). — W. HECKROTTE and J. V. LEPORE: Phys. Rev. **94**, 500 (1954). — G. A. SNOW, R. M. STERNHEIMER and C. N. YANG: Phys. Rev. **94**, 1073 (1954). — S. TAMOR: Phys. Rev. **94**, 1087 (1954). — G. TAKEDA and K. M. WATSON: Phys. Rev. **94**, 1087 (1954). — W. HECKROTTE: Phys. Rev. **94**, 1797 (1954). — B. J. MALENKA: Phys. Rev. **95**, 522 (1954). — R. M. STERNHEIMER: Phys. Rev. **95**, 587 (1954). — S. KÖHLER: Nuovo Cim. **10**, 2, 911 (1955).

with $r_t = 1.6 \times 10^{-13}$ cm, in order to explain the flat angular distribution of the high energy p-p data. The characteristic of this potential is the singular radial dependence of the triplet tensor force. As mentioned in Sect. 44, the tensor force will in general cause side-wise scattering. CHRISTIAN and NOYES[141] asserted that a singular tensor force like (46.1) is required by the isotropic high energy p-p data. The potential (46.1) is adjusted to the low energy p-p data up to 30 Mev. Although they got good agreement at higher energies using BORN approximation, later and more exact calculations[144] have disproved their results as is seen from Fig. 23, where curve II is calculated for the potential (46.1); the pronounced forward scattering is due to the singlet potential. It is furthermore reported that the cross section is rather sensitive to the detail of the cut-off precept adopted and that both signs of $V_t(r)$ in (46.1) can explain the lower energy data; the positive sign in (46.1) has been shown to give a much more isotropic angular distribution at higher energies than the negative sign.

For the potential (46.1), GOLDFARB and FELDMAN[144] calculated the asymmetry $e(\vartheta)$ in the double p-p scattering as a function of the c.m. scattering angle ϑ, putting $\vartheta_1 = \vartheta_2 = \vartheta$, at 240 and 450 Mev; the results are given in Fig. 24 together with two dotted curves for the JASTROW potential (44.1) and the $\boldsymbol{L} \cdot \boldsymbol{S}$ force model at 240 Mev. It appears from the figure that the agreement with the experimental value is reasonable for the singular tensor force model, while it is too bad for the other two models.

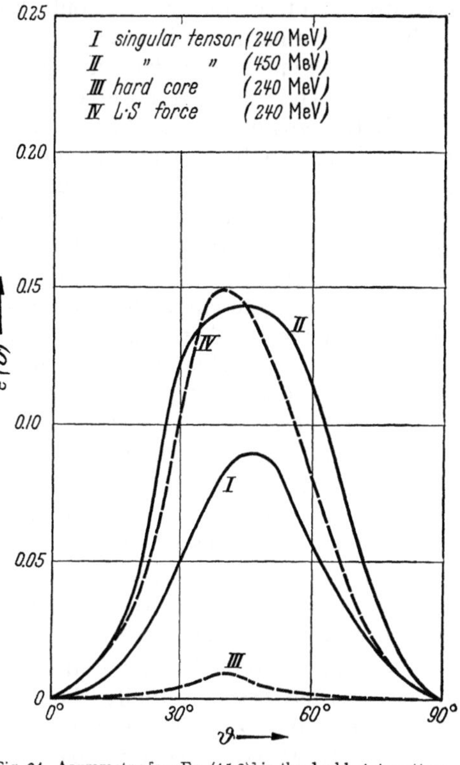

Fig. 24. Asymmetry [see Eq. (45.3)] in the double p-p scattering for the singular tensor potential (46.1) at 240 and 450 Mev (solid curves), the hard core potential (44.1) and $\boldsymbol{L} \cdot \boldsymbol{S}$ force at 240 Mev (dotted curves) as a function of the c.m. scattering angle $\vartheta = \vartheta_1 = \vartheta_2$.

The theoretical value at 450 Mev (maximum $\approx 15\%$) is also consistent with the polarization data since the empirical value would be about $(20 \pm 14)\%$ at 415 Mev, according to the data in (4.4) and Fig. 2, together with the theoretical formula (45.3).

SWANSON[147] has made a similar calculation and shown that the asymmetry is 1% at 129 Mev, 2.3% at 350 Mev with $\vartheta_1 = \vartheta_2 \approx 50°$ for a potential, which agrees with (46.1) for $V_s(r)$, while the triplet odd force is replaced by

$$\left.\begin{aligned} V_t(r) &= 15.2\, S_{12} \frac{e^{-r/r_t}}{(r/r_t)^2}\ \text{Mev}, \quad r > r_1 = 0.48 \times 10^{-13}\ \text{cm};\\ &= 15.2\, S_{12} \frac{e^{-r_1/r_t}}{(r_1/r_t)^2}\ \text{Mev} = \text{const}, \quad r_1 > r > r_0;\\ &= \infty, \qquad\qquad\qquad\qquad r < r_0 = 0.24 \times 10^{-13}\ \text{cm}; \end{aligned}\right\} \tag{46.2}$$

[147] DON R. SWANSON: Phys. Rev. **84**, 1068 (1951); **89**, 749 (1953).

with $r_t = 1.6 \times 10^{-13}$ cm as in (46.1). The above potential is also adjusted to the lower energy data and gives a little better angular distribution of the high energy p-p scattering than curve II in Fig. 23 due to the potential (46.1). The big difference between the values of the asymmetry due to (46.1) and (46.2) shows how sensitive it is to the tensor force and especially to the details of the cut-off.

To sum up, we first recall that the rather poor fit of the potentials (46.1) and (46.2) to the flat high energy p-p distribution is essentially due to the strong forward singlet scattering. The only known way to explain this distribution is to introduce a hard core in the singlet even state. It can, therefore, be concluded that the high energy p-p data give us important information on the singlet even interaction while they do not give us any clue to the triplet odd interaction. Indeed both signs of $V_t(r)$ in (44.1) can explain the data, since the BORN approximation is shown by GOLDFARB and FELDMAN[144] to be valid for such a non-singular force. Furthermore we can reasonably expect that such modifications of $V_t(r)$ as in (46.1) and (46.2) are also compatible with the high energy p-p data as far as a hard core is assumed in $V_s(r)$. On the other hand, the asymmetry in the double p-p scattering is essentially due to the triplet odd tensor interaction and its rather large empirical value seems to rule out such non-singular tensor forces as in (44.1). Apparently it demands a more singular radial dependence of the tensor force, which must not, however, be cut off too far out, since a large cut-off radius ($\approx 0.5 \times 10^{-13}$ cm) weakens the singularity so much that it cannot explain the observed asymmetry. It seems that the core radius must be equal to or smaller than 0.2×10^{-13} cm in the triplet odd state, while the singlet even hard core must be about 0.6×10^{-13} cm, according to the analysis in Sect. 44. Now, the singular tensor force is a well-known characteristic of the meson theoretical potential discussed in Sect. 11; even the second order force $V_2(\boldsymbol{r})$ given by (8.11) contains a r^{-3} singularity. It is also noted that the positive sign of $V_t(r)$ in (46.1) and (46.2) is consistent with the meson theoretical prediction, since it relates to triplet odd states.

As regards the double n-p scattering data, the analysis is still too tentative to be reported here. Actually there is no satisfactory treatment even of the single n-p scattering at high energies. It might be mentioned that the preliminary results of SWANSON[147] seem compatible with the data. There are also some reports on phase shift analysis of the polarization data, given in Sect. 4. They are quoted in the following section together with the similar analysis of the high energy nucleon-nucleon scattering data. It may, however, be said that because of the increased number of adjustable parameters in the n-p case, no very definite information can be expected on the forces in the triplet even and singlet odd states.

47. Charge independence in high energy processes. It is one of the most important and beautiful consequences of the analysis of the low energy two-nucleon data that the charge independence seems to hold well. For some time it was thought, however, that this would not be the case in the high energy region, because of the confusing data on the high energy p-p scattering. This difficulty seems to be avoided, as suggested in the previous sections, by introducing a hard core in the singlet even state and a singular radial dependence, cut off near the origin, in the triplet odd tensor interaction. Although more detailed investigations are necessary before we can make a more definite statement, it is reasonable to say that there is no direct evidence against the charge independence hypothesis in the high energy experiments. It is the purpose of this section to present two further arguments indicating that the present high energy data are not incompatible with charge independence of the nuclear forces.

One of them consists in comparing the data with the following theoretical prediction:

$$\sigma_{pp}(90°) \leq 4\sigma_{np}(90°), \tag{47.1}$$

where $\sigma_{pp}(\vartheta)$ and $\sigma_{np}(\vartheta)$ are the differential p-p and n-p scattering cross sections in terms of the c.m. scattering angle ϑ. The above inequality can be shown[148] to be a direct consequence of the charge independence hypothesis. With FELDMAN[148] we consider the following three reactions

$$\left.\begin{aligned} p_1 + p_2 &\to p_3 + p_4, \\ n_1 + p_2 &\to n_3 + p_4, \\ n_1 + p_2 &\to p_3 + n_4, \end{aligned}\right\} \tag{47.2}$$

where the notations n and p indicate neutron and proton, respectively, and the subscripts refer to the preassigned momenta and spins; we assume that the initial and final momenta and spins are being specified.

We denote the charge triplet and singlet scattering amplitudes by f and g, respectively; such a statement is allowed under the charge independence hypothesis. The first process of (47.2) relates to a pure charge triplet state; in the other two we have mixtures of triplet and singlet states with equal weights. With proper normalization of f and g, it can be shown that the differential scattering cross sections for the three processes in (47.2) are, if we neglect the COULOMB force,

$$\left.\begin{aligned} \sigma_{pp}(\vartheta) &= |f|^2, \\ \sigma_{np}(\vartheta) &= \tfrac{1}{4}|f+g|^2, \\ \sigma_{pn}(\vartheta) &= \tfrac{1}{4}|f-g|^2, \end{aligned}\right\} \tag{47.3}$$

respectively, where f and g are functions of ϑ as well as momentum and spin values. Since there are three unknowns (magnitudes of f and g and their relative phase) on the right hand sides of (47.3), we cannot derive equalities between the cross sections, but we can deduce inequalities. For example, we can prove that

$$\left.\begin{aligned} (\sigma_{np}(\vartheta))^{\frac{1}{2}} + (\sigma_{pn}(\vartheta))^{\frac{1}{2}} &\geq (\sigma_{pp}(\vartheta))^{\frac{1}{2}}, \\ (\sigma_{pn}(\vartheta))^{\frac{1}{2}} + (\sigma_{pp}(\vartheta))^{\frac{1}{2}} &\geq (\sigma_{np}(\vartheta))^{\frac{1}{2}}, \\ (\sigma_{pp}(\vartheta))^{\frac{1}{2}} + (\sigma_{np}(\vartheta))^{\frac{1}{2}} &\geq (\sigma_{pn}(\vartheta))^{\frac{1}{2}}. \end{aligned}\right\} \tag{47.4}$$

Normally the cross sections imply an average over the initial and a summation over the final spins, but the inequalities (47.4) are valid all the same. Remarking that $\sigma_{pn}(\vartheta) = \sigma_{np}(\pi - \vartheta)$ in the c.m. system, we finally get

$$\left.\begin{aligned} (\sigma_{np}(\vartheta))^{\frac{1}{2}} + (\sigma_{np}(\pi-\vartheta))^{\frac{1}{2}} &\geq (\sigma_{pp}(\vartheta))^{\frac{1}{2}}, \\ (\sigma_{np}(\pi-\vartheta))^{\frac{1}{2}} + (\sigma_{pp}(\vartheta))^{\frac{1}{2}} &\geq (\sigma_{np}(\vartheta))^{\frac{1}{2}}, \\ (\sigma_{pp}(\vartheta))^{\frac{1}{2}} + (\sigma_{np}(\vartheta))^{\frac{1}{2}} &\geq (\sigma_{np}(\pi-\vartheta))^{\frac{1}{2}}. \end{aligned}\right\} \tag{47.5}$$

At high energies, it is known that the COULOMB force plays a negligible role in the scattering except for very small angles. Now it appears from the data in Sect. 4 that $\sigma_{np}(\vartheta)$ is roughly equal to $\sigma_{np}(\pi-\vartheta)$ (approximate symmetry in a wide region around 90°), so the latter two conditions in (47.5) would not be so stringent. As regards the first equation of (47.5), it is most restrictive for $\vartheta = 90°$, since according to the empirical data $\sigma_{pp}(\vartheta)$ has a flat angular distribution, while $\sigma_{np}(\vartheta)$ is minimum near 90°. Thus we arrive at (47.1).

---

[148] B. A. JACOBSOHN: Phys. Rev. **89**, 881 (1953). — D. FELDMAN: Phys. Rev. **89**, 1159 (1953).

The available data have been taken from Sect. 4 and plotted in Fig. 25. Although the experimental errors are still rather large, we may conclude that the inequality (47.1) is satisfied by the high energy data. It must, however, be stressed that this is only a necessary condition for the validity of the charge independence hypothesis.

The second argument referred to is concerned with the analysis of phase shifts in the high energy scattering. The procedure outlined in Sect. 44 was to assume some suitable nuclear potential, calculate phase shifts and mixing parameters and compare the results with experiments. The conclusion was that it seems possible to explain the high energy data by a charge independent static nuclear potential. The present analysis does not, however, assume any special form of nuclear potential, but aims at examining whether it is possible to describe the data, including the polarization, by taking reasonable values for the phase shifts and mixing parameters, varying with energy, in a charge independent way. The last expression means to assume the same p-p and n-p values for such parameters as belong to the same states.

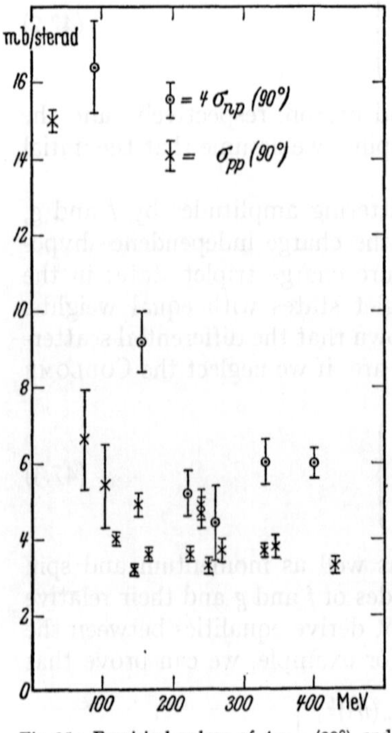

Fig. 25. Empirical values of $4\sigma_{np}$ (90°) and σ_{pp} (90°) against incident energy.

Among the results published[149], those which refer to the 300 Mev data[150] seem to give a rather definite information. According to the theory developed in Sect. 29, the triplet phase shifts δ associated with a spin-orbit interaction are classified as follows: δ is specified by two numbers J and L, the total and orbital angular momentum quantum numbers, respectively. For a given J, L can take the three values J and $J\pm1$; if $J=0$ only $L=1$ is allowed. $J=L$ is a pure state, while the states $L=J\pm1$ are coupled and, therefore, mixing parameters ε_J are necessary besides these phase shifts, in order to express the asymptotic behaviour of the scattered wave. In the singlet state, $L=J$ and one phase shift δ_L suffices. The result is that the differential scattering cross section as well as the polarization $P(\vartheta)$ defined in Sect. 45 can be expressed in terms of the phase shifts δ_L and $\delta_{J,L}$ and the mixing parameters ε_J.

THALER and BENGSTON[150] have shown that the p-p and n-p scattering data at 300 Mev (see Sect. 4) can be reasonably reproduced with seven phase shifts, viz. δ_0, δ_1, $\delta_{0,1}$, $\delta_{1,1}$, $\delta_{2,1}$, $\delta_{1,0}$ and $\delta_{1,2}=\delta_{2,2}=\delta_{3,2}$, putting $\varepsilon_1=0$. FRIED, BREIT *et al.*[150] have then pointed out that the polarization data at 300 Mev cannot be explained by such a simple assumption. They found, however, that

---

[149] A. GARREN: Phys. Rev. **92**, 213 (1953). — R. M. THALER and J. BENGSTON: Phys. Rev. **94**, 679 (1954). — R. M. THALER, J. BENGSTON and G. BREIT: Phys. Rev. **94**, 683 (1954). — B. D. FRIED: Phys. Rev. **95**, 851 (1954). — G. BREIT and J. B. EHRMAN: Phys. Rev. **96**, 805 (1954). — M. H. HULL jr., and A. M. SAPERSTEIN: Phys. Rev. **96**, 806 (1954). — G. BREIT, J. B. EHRMAN, A. M. SAPERSTEIN and M. H. HULL jr.: Phys. Rev. **96**, 807 (1954).

[150] R. M. THALER and J. BENGSTON: Phys. Rev. **94**, 679 (1954). — B. D. FRIED: Phys. Rev. **95**, 851 (1954). — G. BREIT, J. B. EHRMAN, A. M. SAPERSTEIN and M. H. HULL jr.: Phys. Rev. **96**, 807 (1954).

it would be quite sufficient to allow $\delta_{1,2} \neq \delta_{2,2} \neq \delta_{3,2}$ and a non-vanishing mixing parameter ε_1, in order to make the theory agree with experiment.

In conclusion one might say that the charge independence hypothesis has not been contradicted by the high energy data obtained and analysed thus far.

48. Summary and additional remark on relativistic effect. Up to about 400 Mev rather accurate data of the cross section and polarization of nucleon-nucleon scattering are now available. The characteristics of these can be summarized as follows: In the *p-p* case, the angular distribution is fairly isotropic and the cross section is rather insensitive to the energy over a wide range. The *n-p* data show, generally speaking, a minimum near 90° in the c.m. system and peaks in the forward and backward directions. The forward peak becomes, however, less dominant as the energy increases and the total cross section gradually decreases. A striking feature is further the fairly strong polarization in the nucleon-nucleon scattering.

The ultimate purpose of our analysis is to see whether these data can be interpreted by some charge independent static potential between nucleons. According to the analysis in Sect. 44, the isotropy of the *p-p* data forces us to introduce a hard core in the singlet even state with a radius of, say, 0.6×10^{-13} cm; otherwise a strong forward scattering resulting from the ordinary type of singlet nuclear interaction would take the place of the flat angular distribution, even though a special triplet interaction be devised.

As regards the polarization data, it is shown in Sect. 46 that the marked asymmetry observed in the double *p-p* scattering could be explained only by a singular tensor force in the triplet odd states, cut off at a small distance ($\approx 0.2 \times 10^{-13}$ cm or less). A larger cut-off radius would weaken the singularity too much to produce a high asymmetry.

The information about a hard core in the singlet even state and a singular tensor force in the triplet odd state is all that seems to be in some degree definite. Besides this no detailed evidence of the nuclear potential is obtained from the existing high energy data. As mentioned in the preceding section, there are, however, arguments which support the charge independence hypothesis in relation to the high energy data.

Finally we note that all calculations thus far quoted are based on the non-relativistic quantum mechanics. In discussing the high energy data, we cannot neglect the relativistic effect a priori. This problem is difficult indeed and has not yet been solved. What we can do here is only to recall the importance of this point and briefly relate some preliminary work on the subject.

BEARD and BETHE[151] have calculated the meson field correction to the nucleon-nucleon scattering cross section, assuming a special type of meson field theory, to avoid divergencies. It was done in the following way: first the scattering matrix element was calculated by field theoretical methods up to the fourth order of the meson-nucleon coupling constant. We then subtract those contributions which can be ascribed to the scattering by a static nuclear potential and finally identify the remainder as the relativistic correction or the meson field correction. It turns out to be of the order of μ/M [not $(\mu/M)^2$] of the cross section, where μ and M are the masses of a meson and a proton, respectively.

A phenomenological approach has been tried by SIEGEL[152]. The idea is to generalize a given static potential so that the resulting interaction becomes relativistically invariant in a certain sense and then to interpret the difference

[151] D. B. BEARD and H. A. BETHE: Phys. Rev. **83**, 1106 (1951).

[152] A. SIEGEL: Phys. Rev. **82**, 194 (1951).

between results obtained with the new and the original interactions as the relativistic effect. In the above generalization, there are, however, many ambiguities and SIEGEL posed the condition, inter alia, that the calculated relativistic correction must be small at low energies. He considered a broad class of charge independent static interactions containing various spin and space operators. In this way he got the result that the relativistic correction would not exceed about 5 to 10% of the cross section due to a given static nuclear potential at 90 Mev.

Thus it appears that the relativistic corrections would neither be large enough to invalidate the foregoing arguments nor so small as to be entirely neglected.

b) High energy photodisintegration.

49. Theoretical cross section and comparison with experiments. The differential and total cross sections of the photodisintegration of the deuteron are given by (38.12) and (38.13), respectively, including the electric and magnetic quadrupole contributions. The effect of a tensor force in the final (continuum) state has, however, been neglected. In order to evaluate the integrals appearing in these expressions, we need to know the deuteron ground state wave functions and the scattering wave functions belonging to *S*-, *P*- and *D*-states. For the former, we can use the functions given in Sect. 33. To compute the latter, it should be necessary to know the forces in those states, which are, however, rather imperfectly known. We therefore assume in the following that there are no forces in the continuum states, as was already done in evaluating the low energy cross section in Sect. 39. It may be hoped that this is a reasonable approximation, since the electric dipole term [first term in Eq. (38.12) or (38.13)] is always the dominant one, as will be shown later, and in this transition the final state is a ^3P-state, where the force is expected to be small compared with the force in an *S*-state. The effect of the ^3P-potential was estimated in Sect. 39 to be below 2% at the incident photon energy 20 Mev. However, the error of the zero-force-approximation is likely to be larger at higher energies.

With this approximation in all final states, we may put all phase shifts equal to zero in (38.12) and (38.13) and make the functions v_l equal to u_l, for which we have the expressions

$$\left.\begin{aligned} u_0(r) &= \sin k r, \qquad u_1(r) = \frac{\sin k r}{k r} - \cos k r, \\ u_2(r) &= \left[\frac{3}{(k r)^2} - 1\right] \sin k r - \frac{3}{k r} \cos k r. \end{aligned}\right\} \tag{49.1}$$

In the total cross section (38.13), we neglect the most awkward term K_{dd}^2, since it can be shown that those terms in the electric dipole and quadrupole transitions which are due to the *D*-state admixture in the deuteron [the terms containing I_{dp}^2 and I_{ds}^2 in (38.13)] are always (up to 200 Mev incident photon energy) smaller than 0.8% of the corresponding terms due to the *S*-state [the terms with I_{sp}^2 and I_{sd}^2 in (38.13)].

As regards the differential cross section (38.12), we neglect the electric quadrupole transition [the third term in (38.12)] and the second term of the interference between the electric dipole and quadrupole transitions (which is proportional to $\cos\vartheta$). In the singlet transition we keep only the dominant term due to the spin magnetic dipole transition [the term with K_{ss}^2 in (38.12)]. These omissions are made because the terms are in fact small, as is seen from Table 20, and tend to complicate the angular distribution so as to obscure the relation to the empirical data [cf. the expression (5.1)]. The isotropic terms in (38.12) are, however,

Table 20. *The total photodisintegration cross section σ_{tot}, the isotropy coefficient $\alpha \equiv A/B$ and the retardation factor β calculated for three values of $h\nu$, the incident photon energy in the laboratory system, together with their empirical values. The contributions from different multipole transitions are shown separately.*

σ (10^{-28} cm^2)		$h\nu = 50$ Mev	$h\nu = 100$ Mev	$h\nu = 200$ Mev
Triplet	elect. dipole	1.1475	0.2662	0.04780
	elect. quad.	0.0148	0.0084	0.00367
	mag. quad.	0.0123	0.0117	0.00893
Singlet	mag. dipole	0.1664	0.0613	0.01878
	mag. quad.	0.0002	0.0002	0.00014
σ_{tot} (theory)		1.3412	0.3478	0.07932
σ_{tot} (experiment)		1.5	0.64	0.58
α (theory)		0.1257	0.2688	1.0178
α (experiment)		0.48	1.14	5.6
β (theory)		0.2606	0.4454	0.9886
β (experiment)		0.15	0.17	0.21

mostly retained although they are not always larger than the neglected terms; the reason is an extreme disagreement between theory and experiment with respect to these isotropic terms (see Fig. 26). It should also be added that our theoretical prediction (38.12) implies an angular distribution.

$$d\sigma/d\Omega = A + B \times \sin^2\vartheta\,(1 + 2\beta\cos\vartheta)\,, \quad (49.2)$$

rather than the empirical formula (5.1). In the lower energy region they are almost the same since A and β are both small coefficients. The difference, however, increases rather rapidly with the energy. The calculated values* of the total cross section σ_{tot} in units of mb, the isotropy coefficient $\alpha \equiv A/B$ and the retardation factor β as functions of the incident photon energy in the laboratory system are summarized in Table 20 and Fig. 26, together with the experimental data taken from Figs. 7 and 8.

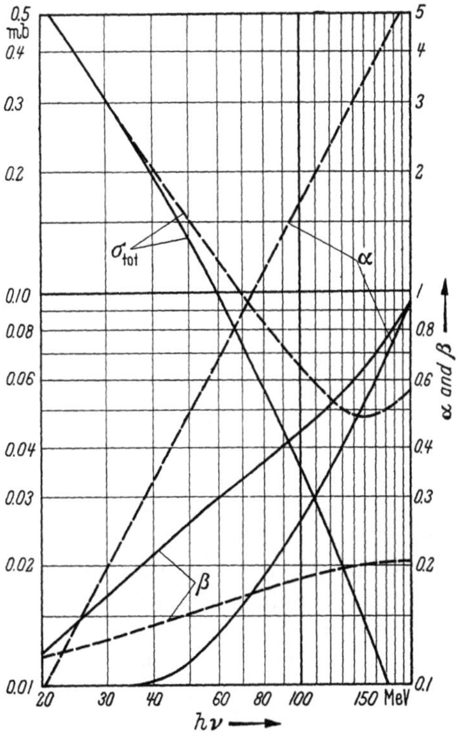

Fig. 26. The total photodisintegration cross section σ_{tot}, the isotropy coefficient $\alpha \equiv A/B$ and the retardation factor β against the incident photon energy $h\nu$ in laboratory system. The solid lines are theoretical (taken from Table 20 and low energy values given in Sects. 39 and 40), while the dotted lines represent the empirical data (taken from Figs. 7 and 8).

It is seen from Table 20 that the electric dipole term is always the major one, while the magnetic dipole term is smaller but appreciable. Among other contributions the magnetic quadrupole one is the most

* The deuteron wave function (33.5) with $\varkappa_c = 0$, $P_D = 4\%$ and $\varrho(-\varepsilon, -\varepsilon) = 1.734 \times 10^{-13}$ cm has been used.

important, especially in the higher energy region. Although our zero-force-assumption would not be so serious for the electric dipole term, we must recognize that it is not so good an approximation in the magnetic dipole transition, which implies that our theoretical values in Table 20 and Fig. 26 are not very accurate.

We note that the agreement with experiment is very poor especially in the higher energy region; this is also true of the angular distribution. The theoretical expression (49.2) is already different from the empirical formula (5.1), and even in the low energy region the agreement is very poor for the isotropy coefficient α (Sect. 40). It seems that these discrepancies are too large to be explained by any refined calculation of the phenomenological cross sections (38.12) and (38.13).

50. Further improvement of theoretical cross section. Let us begin with the refinements of our preceding calculation of the cross sections (38.12) and (38.13). As regards the special form of the deuteron wave function chosen above, we may quote some calculations[153] in which central forces with both exponential and YUKAWA wells are assumed. They indicate that the difference between these two well shapes is certainly negligible compared with other theoretical ambiguities or the difference between theory and experiment. According to another calculation introducing a hard core in the deuteron ground state[154], this has, however, an important effect and increases the total cross section a good deal between 30 Mev and 150 Mev, beyond which it makes the cross section smaller than the value without a hard core. If we choose a core radius of 0.4 to 0.6 $\times 10^{-13}$ cm, it seems possible to reach agreement with experiment in the energy range just mentioned. It is also shown that a core radius of 0.8×10^{-13} cm must be excluded since it gives too large a cross section in this energy range. We note, however, that even the former values of the hard core radius are rather too large (cf. Sect. 46); moreover, the hard core can never improve the angular distribution since its essential effect is to increase the electric dipole term.

As regards the refinement beyond the phenomenological scope, the theory outlined in Sect. 34 and further discussed in Sects. 41 and 42 shows how to estimate the relativistic and exchange current corrections. In the case of the neutron-proton capture it was shown that the relativistic effect is quite negligible and the exchange current correction increases the cross section by about 1%. This figure is small compared with the other theoretical and experimental imperfections, but this is due to the extremely low energy concerned in the capture process. In the high energy photodisintegration we may expect that these corrections would change the situation. According to a preliminary calculation the change seems, however, still too small compared with the relevant discrepancy between theory and experiment.

We must recall here that we have made another important approximation at the very outset of the theory developed in Sects. 34, 41 and 42: we have neglected all the self-meson processes, instead of which the PAULI-term (27.2) is introduced to account for the anomalous parts of nucleon magnetic moments. This may be a good approximation in the lower energy region but is expected to get worse rapidly as the energy increases, since the virtual emission and re-absorption of mesons belonging to the self-meson fields around nucleons may be more strongly modified by the incident photon than is described by the PAULI-term (27.2), although the modifications of the virtual meson processes between

[153] L. I. SCHIFF: Phys. Rev. **78**, 733 (1950). — J. F. MARSHALL and E. GUTH: Phys. Rev. **78**, 738 (1950).

[154] N. AUSTERN: Phys. Rev. **88**, 1207 (1952).

two nucleons are correctly treated by our exchange current correction. One may of course argue that the wavelength of the incident photon ($\approx 1.2\times 10^{-12}$ cm at 100 Mev) is still large compared with the extension of the self-meson field around nucleons (1 or 2×10^{-13} cm), which may allow us to use a static expression or the PAULI-term (27.2) for the interaction of the incident photon with the self meson field. Because of the divergencies, it is extremely difficult to treat this problem properly and we have practically no material to cite about this point.

Finally we quote some preliminary investigations on the meson exchange effects just mentioned. One of these[155] starts from a Hamiltonian in which nucleons are regarded as particles interacting with electromagnetic as well as meson (ps-pv) fields and calculates the photodisintegration cross section, treating the interactions of the nucleons with both fields as weak perturbations (lowest order only). Self-meson effects and the anomalous magnetic moments of the nucleons are entirely neglected. According to this work, such virtual processes in which a photon is absorbed by an exchanged meson or a meson is exchanged after a photon is absorbed by a proton can give rise to large contributions which are almost of the same size as the difference between the theoretical and experimental curves in Fig. 26, if the ps-pv coupling constant $g^2/4\pi\hbar c$ is chosen as 0.1. We must, however, notice that the calculation of BRUNO and DEPKEN[155] is based on a simple perturbation method and is inconsistent in the sense that they regard all interactions as weak perturbations, while the initial state is naturally assumed to be bound owing to a part of the interaction.

Another approach[156] is, first to construct an additional magnetic dipole expression which is due to the mesons exchanged between two nucleons, and then to calculate its effect on the photodisintegration cross section. According to the theory developed in Sects. 34, 41 and 42, the exchanged meson contribution to the spin magnetic dipole transition is given by (41.3). The present calculation, however, starts from quite a different expression and leads to the result that the discrepancy between theory and experiment seems to be well covered by the meson exchange effect. The basis of this work is, however, questionable and the favourable result is intimately connected with a strong radial singularity of the expression employed. Moreover, the contributions from the meson exchange to other multipole transitions are entirely neglected. Indeed we must be very careful in treating such a strong singularity so as not to involve unphysical elements due to unpermissible infinities. We must also expect that an effect which is essentially due to a singularity is reduced very much when a hard core of a reasonable size is introduced.

VIII. Mathematical appendix.

51. Variational principles of HULTHÉN and SCHWINGER with generalization. In Sect. 28, a variational method was developed in treating the deuteron problem. A similar procedure has been devised by HULTHÉN[157] for the scattering problem; the essence of this method is explained in the following.

For simplicity let us consider the scattering due to a central force. Then the problem is to solve the differential equation

$$\left[\frac{d^2}{dr^2}+\left\{k^2+v(r)-\frac{l(l+1)}{r^2}\right\}\right]u_l(r)=0, \tag{51.1}$$

[155] B. BRUNO and S. DEPKEN: Ark. Fysik 6, 177 (1953).

[156] Y. NAGAHARA and J. FUJIMURA: Progr. Theor. Phys. 8, 49 (1952).

[157] L. HULTHÉN: Kgl. Fysiograf. Sällskap. Lund, Förh. 14, No. 21, 257 (1944). — Den 10. Skandinaviske Matematiker Kongress 1946, Copenhagen, p. 201, 1947.

with the conditions

$$\left.\begin{aligned} &u_l(0)=0\,,\\ &u_l(r)\xrightarrow[r\to\infty]{}\sin\left(k\,r-\frac{l\pi}{2}+\eta_l\right),\end{aligned}\right\}\tag{51.2}$$

where we have normalized the amplitude to 1 and η_l is the l-th wave phase shift; other notations are explained in Sect. 15.

We define an integral

$$L=\int\limits_0^\infty u_l(r)\left[\frac{d^2}{dr^2}+\left\{k^2+v(r)-\frac{l(l+1)}{r^2}\right\}\right]u_l(r)\,dr\,,\tag{51.3}$$

which evidently vanishes for the exact solution of (51.1). We then consider small variations $\delta u_l(r)$ of $u_l(r)$ such that the function $u_l(r)+\delta u_l(r)$ still satisfies both requirements (51.2) with a corresponding small variation $\delta\eta_l$ of η_l:

$$\left.\begin{aligned} &\delta u_l(r)\xrightarrow[r\to0]{}0\,,\\ &\delta u_l(r)\xrightarrow[r\to\infty]{}\cos\left(k\,r-\frac{l\pi}{2}+\eta_l\right)\delta\eta_l\,,\end{aligned}\right\}\tag{51.4}$$

where k is fixed at a given value. The corresponding variation of L can be evaluated as

$$\delta L=2\int\limits_0^\infty \delta u_l(r)\left[\frac{d^2}{dr^2}+\left\{k^2+v(r)-\frac{l(l+1)}{r^2}\right\}\right]u_l(r)\,dr-k\,\delta\eta_l\,,\tag{51.5}$$

where we have used

$$\left.\begin{aligned}\int\limits_0^\infty u_l\frac{d^2}{dr^2}\,\delta u_l\,dr&=\int\limits_0^\infty \delta u_l\frac{d^2}{dr^2}\,u_l\,dr+\left[u_l\frac{d\,\delta u_l}{dr}-\delta u_l\frac{d u_l}{dr}\right]_0^\infty\\ &=\int\limits_0^\infty \delta u_l\frac{d^2}{dr^2}\,u_l\,dr-k\,\delta\,\eta_l\,.\end{aligned}\right\}\tag{51.6}$$

Eq. (51.5) states that, if $u_l(r)$ satisfies (51.1), the quantity $L+k\,\eta_l$ is stationary. If, furthermore, $u_l(r)$ is varied in such a way that always $L=0$, then $\delta L=0$, whence

$$\delta\eta_l=0\,,\tag{51.7}$$

which expresses that the phase shift η_l is stationary under the relevant variations of $u_l(r)$. Reversely, the above stationary requirement of η_l results in $u_l(r)$ satisfying the differential equation (51.1). Thus the variational principle is equivalent to the original SCHRÖDINGER equation.

The practical application of this variational principle is obvious. First we make a trial function which satisfies (51.2) and contains, besides the unknown η_l, some parameters $c_1, c_2, \ldots, c_n$. The first requirement then is that L should vanish:

$$L(\eta_l, c_1, c_2, \ldots, c_n)=0\,.\tag{51.8}$$

Secondly the stationary property of η_l requires that

$$\frac{\partial L}{\partial c_\nu}=0\,,\qquad \nu=1,2,\ldots,n\,.\tag{51.9}$$

These $n+1$ equations suffice to determine $c_1\ldots c_n$ and η_l.

A slightly different variational principle was proposed by KOHN[158]. Instead of the above L, let us define

$$Q = \int_0^a \left[-\left(\frac{d u_l(r)}{dr}\right)^2 + \left\{ k^2 + v(r) - \frac{l(l+1)}{r^2} \right\} u_l^2(r) \right] dr + K u_l^2(a), \quad (51.10)$$

where K is the logarithmic derivative of $u_l(r)$ at $r=a$. It can readily be shown that $Q=0$ if $u_l(r)$ satisfies (51.1) and the boundary condition $u_l(0)=0$. If a is chosen large compared with the nuclear force range,

$$K = k \cot\left(k a - \frac{l\pi}{2} + \eta_l\right).$$

We now consider such small variations of $u_l(r)$ that do not violate the condition $u_l(0)=0$ and always make $Q=0$. It is easy to show that K is stationary under such variations and that the stationary property of K under the relevant variations of $u_l(r)$ results in the original SCHRÖDINGER equation (51.1). This variational principle has two features: Firstly the trial function of $u_l(r)$ need not be correctly normalized asymptotically and secondly the unknown K, which is to be varied, appears linearly in Q, which can make it easy to solve the variational problem.

Practical applications and generalizations of the above variational principles have been made by several authors[159]; for details the reader is referred to their original papers.

Another form of the variational principle has been derived by SCHWINGER[160]. The following explanation is due to BLATT and JACKSON[100]. Let us for simplicity consider the S-wave scattering [$l=0$ in Eq. (51.1)]. The asymptotic form of $u(r)$ (dropping the subscript 0) is now defined as

$$u(r) \xrightarrow[r\to\infty]{} \sin kr + \tan\eta \cos kr, \quad (51.11)$$

the boundary condition $u(0)=0$ being the same as (51.2). Let us first replace the differential equation (51.1) by an integral equation. We rewrite (51.1) as

$$\left[\frac{d^2}{dr^2} + k^2\right] u(r) = -v(r)\, u(r), \quad (51.12)$$

and introduce a GREEN's function $G(r, r')$, which satisfies

$$\left[\frac{\partial^2}{\partial r^2} + k^2\right] G(r, r') = -\delta(r - r'), \quad (51.13)$$

$$G(0, r') = 0, \quad G(r, r') \sim \cos kr \quad \text{for} \quad r > r',$$

where $\delta(x)$ is DIRAC's δ-function. This GREEN's function is given in this case by

$$G(r, r') = \left(\frac{1}{k}\right) \sin(k r_<) \cos(k r_>), \quad (51.14)$$

where $r_<$ stands for the smaller of r and r', $r_>$ for the larger of them. This form of (51.14) explicitly shows that $G(r, r') = G(r', r)$; $G(r, r')$ is symmetric. Thus

[158] W. KOHN: Phys. Rev. **74**, 1763 (1948).

[159] L. HULTHÉN: Ark. Mat., Astronom., Fys., Ser. A **35**, No. 25 (1948). — L. HULTHÉN and S. SKAVLEM: Phys. Rev. **87**, 297 (1952). — H. FESHBACH and S. I. RUBINO: Phys. Rev. **88**, 484 (1952). — S. I. RUBINO: Phys. Rev. **96**, 218 (1954); **98**, 183 (1955). — H. E. MOSES: Phys. Rev. **92**, 817 (1953); **96**, 519 (1954). — T. REGGE and M. VERDE: Nuovo Cim. **10**, 997 (1953). — P. O. OLSSON: Unpublished calculations. — S. SKAVLEM and I. ESPE: Ark. Fysik **10**, Nr. 8 (1955).

[160] J. SCHWINGER: Phys. Rev. **72**, 742 (1947); **78**, 135 (1950).

the differential equation (51.1), together with the condition (51.11) and $u(0)=0$, is equivalent to the integral equation:

$$u(r) = \sin k r + \int_0^\infty G(r, r')\, v(r')\, u(r')\, dr'. \tag{51.15}$$

It is easy to see that this $u(r)$ satisfies (51.1). The inhomogeneous term $\sin kr$ in (51.15) is fixed by the two requirements: $u(0)=0$ and Eq. (51.11). The former is obviously satisfied. In order to check the latter requirement, we let r be much larger than the range of the nuclear force. The integral over r' in the second term of (51.15) practically extends only to r' of the order of the nuclear force range. We may, therefore, substitute $r_< = r'$, $r_> = r$ in $G(r, r')$ in (51.15). It gives

$$u(r) \xrightarrow[r\to\infty]{} \sin k r + \left[\frac{1}{k} \int_0^\infty \sin k r'\, v(r')\, u(r')\, dr'\right] \cos k r, \tag{51.16}$$

which shows, by (51.11), that

$$k \tan \eta = \int_0^\infty \sin k r'\, v(r')\, u(r')\, dr'. \tag{51.17}$$

In order then to construct a variational principle, we multiply both sides of (51.15) by $v(r)\,u(r)$ and integrate over r. A slight manipulation, together with the identity (51.17), gives the fundamental equation of SCHWINGER's variational principle:

$$k \cot \eta = \frac{\int_0^\infty v(r)\, u^2(r)\, dr - \int_0^\infty\int_0^\infty v(r)\, u(r)\, G(r, r')\, v(r')\, u(r')\, dr'\, dr}{\left[\frac{1}{k}\int_0^\infty v(r)\, u(r) \sin k r\, dr\right]^2}. \tag{51.18}$$

Now let us consider a small variation $\delta u(r)$ of $u(r)$, which is now assumed to satisfy (51.15), with η defined by (51.11). Then it is verified that $k\cot\eta$ is stationary, no matter what form is chosen for $\delta u(r)$. Conversely it can be shown that the stationary property of $k\cot\eta$ under arbitrary variations of $u(r)$ makes $u(r)$ satisfy the integral equation (51.15). Thus it has been shown that SCHWINGER's variational principle is also equivalent to the original SCHRÖDINGER equation.

This variational principle has been generalized and applied to practical problems[161] especially by BLATT and others[100,117,127], who have used it in making an extensive analysis of the low energy nucleon-nucleon scattering in terms of the effective range theory. We have already quoted many of their results in the former parts of this article.

All the variational principles mentioned above state that the phase shift is stationary under some kind of infinitesimal variation of the wave function. One special advantage of SCHWINGER's method is that the trial function need not be normalized to the correct asymptotic form, while this is definitely necessary in HULTHÉN's method. In some special examples, it has been shown that even simple trial functions can give fairly accurate numerical results. We remark that the variational principles developed here give neither an upper nor a lower limit for the varied quantities (η or $k\cot\eta$ etc.), contrary to the variational

[161] S. ALTSHULER: Phys. Rev. **89**, 1278 (1953).

principle for a discrete eigenvalue problem*. We have, therefore, no general criterion which may help us to measure how near we are to the correct value of the phase shift. We further add that, as is stressed in Sect. 43, the above mentioned variational procedures can not always give reasonable values for the phase shifts, because it is quite difficult to find suitable trial functions especially in a case when we treat singular types of the nuclear potential, which are, however, the most interesting in the present status of the two-nucleon problem. For these reasons we must recognize that they are of minor practical importance at least under the present circumstances.

Finally, it may be added that KATO[162] has given a simple inter-relation between HULTHÉN'S and SCHWINGER'S variational principles.

52. Scattering matrix and LIPPMANN-SCHWINGER equation. In order to treat the scattering problem in general, we must solve a SCHRÖDINGER equation, the Hamiltonian of which consists of H_0, the unperturbed Hamiltonian, and H_i expressing the interaction energy. Since the problem is to describe the effect due to H_i, it is convenient to remove the time dependence associated with H_0 from the SCHRÖDINGER equation

$$i\hbar\frac{\partial\Psi'}{\partial t}=(H_0+H_i)\,\Psi'(t)\,. \tag{52.1}$$

This is accomplished by the unitary transformation

$$\Psi'(t)=\exp(-i\,H_0\,t/\hbar)\,\Psi(t)\,, \tag{52.2}$$

which yields

$$i\hbar\frac{\partial\Psi}{\partial t}=H_i(t)\,\Psi(t)\,, \tag{52.3}$$

$$H_i(t)=\exp(i\,H_0\,t/\hbar)\,H_i\exp(-i\,H_0\,t/\hbar)\,. \tag{52.4}$$

This representation is called the interaction representation, for in the transformed SCHRÖDINGER equation (52.3) only the interaction Hamiltonian appears explicitly.

Now let us suppose that at $t=-\infty$ two particles (or the non-interacting parts of the system) are situated at infinite distance from each other and described by the state vector $\Psi(-\infty)$. On following the course of the interaction and the eventual separation of the two parts, we are led to the state vector $\Psi(+\infty)$ at $t=+\infty$, representing the final state of the system. These two state vectors are naturally connected by a unitary operator S:

$$\left.\begin{aligned}\Psi(+\infty)&=S\,\Psi(-\infty)\,,\\ S\,S^{\dagger}&=1\,,\end{aligned}\right\} \tag{52.5}$$

where † means the HERMITE conjugate operator. This operator includes all information about the scattering process. It is called the scattering (or collision) operator. The importance of this quantity was first stressed by HEISENBERG[163] and it is also called the HEISENBERG S-operator. To solve the scattering problem is equivalent to finding the matrix elements of this operator (the scattering matrix elements) for a given problem. A variety of methods have been devised in practical calculations of these matrix elements. For example, FEYNMAN and DYSON have developed an elegant method to evaluate the scattering matrix

* KATO[162] has got some information on this topic.

[162] T. KATO: Phys. Rev. **80**, 475 (1950).

[163] W. HEISENBERG: Z. Physik **120**, 513, 673 (1943).

elements in the complicated field theoretical problems[164]. In the following we derive the LIPPMANN-SCHWINGER[165] equation, which has a fundamental importance in several investigations of the scattering problem.

It should be noted that $\Psi(-\infty)$ and $\Psi(+\infty)$ cannot be exact eigenfunctions of H_0, since a superposition of momentum eigenstates (wave packets) is required to produce the spatial localizability involved in the definite separation of the two parts of the system. In order to get rid of the complicated discussions about the wave packet[166], we use here a mathematical trick, viz. an assumption of the adiabatic switching off the interaction at time $t=\pm\infty$. This is accomplished by introducing a factor $\exp(-\varepsilon|t|/\hbar)$ into H_i, where ε is an infinitesimal positive constant, which will be put equal to zero in the final results. Then $\Psi(-\infty)$ can be identified as an eigenfunction Φ_a of H_0 belonging to an energy eigenvalue E_a, which satisfies

$$H_0\,\Phi_a = E_a\,\Phi_a. \tag{52.6}$$

Since $S\,\Phi_a$ is the final state that emerges from the initial state Φ_a, the probability that the system will eventually be found in another eigenstate Φ_b of H_0 belonging to an energy eigenvalue E_b is

$$W_{ba} = |(\Phi_b,\ S\,\Phi_a)|^2 = |S_{ba}|^2. \tag{52.7}$$

In order to evaluate the S-matrix element S_{ba}, we define the unitary operator $U(t)$ by

$$\Psi(t) = U(t)\,\Psi(-\infty), \qquad U^\dagger(t)\,U(t) = 1, \tag{52.8}$$

from which it follows that

$$S = U(\infty). \tag{52.9}$$

The differential equation for $U(t)$,

$$i\hbar\frac{\partial U(t)}{\partial t} = H_i(t)\,U(t), \tag{52.10}$$

together with the boundary condition,

$$U(-\infty) = 1, \tag{52.11}$$

can be replaced by the integral equation

$$\left.\begin{aligned} U(t) &= 1 - \frac{i}{\hbar}\int_{-\infty}^{t} H_i(t')\,U(t')\,dt' \\ &= 1 - \frac{i}{\hbar}\int_{-\infty}^{+\infty} \eta(t-t')\,H_i(t')\,U(t')\,dt', \end{aligned}\right\} \tag{52.12}$$

where we have defined

$$\left.\begin{aligned} \eta(t-t') &= 1, \quad t > t'; \\ &= 0, \quad t < t'. \end{aligned}\right\} \tag{52.13}$$

Considering the limiting case $t\to+\infty$, we get

$$S = 1 - \frac{i}{\hbar}\int_{-\infty}^{+\infty} H_i(t)\,U(t)\,dt. \tag{52.14}$$

---

[164] R. P. FEYNMAN: Phys. Rev. **76**, 749, 769 (1949). — F. J. DYSON: Phys. Rev. **75**, 486, 1736 (1949).

[165] B. A. LIPPMANN and J. SCHWINGER: Phys. Rev. **79**, 469 (1950).

[166] F. LOW and G. F. CHEW: Lecture notes on the scattering theory at the University of Illinois, 1953.

It is slightly more convenient to deal with the operator

$$T = S - 1, \tag{52.15}$$

which represents only the change of the state vector due to the interaction. As far as we are interested in the transition into a particular final state differing from the initial one, it is obvious that

$$T_{ba} = S_{ba}, \quad W_{ba} = |T_{ba}|^2, \quad \text{for} \quad b \neq a. \tag{52.16}$$

The matrix element T_{ba} can be evaluated, from (52.14) and (52.4), as

$$\left.\begin{aligned} T_{ba} &= -\frac{i}{\hbar}\int_{-\infty}^{+\infty}(\Phi_b, H_i(t)\,U(t)\,\Phi_a)\,dt \\ &= -\frac{i}{\hbar}\int_{-\infty}^{+\infty}(\Phi_b, \exp(i H_0 t/\hbar)\,H_i \exp(-H_0 t/\hbar)\,U(t)\,\Phi_a)\,dt. \end{aligned}\right\} \tag{52.17}$$

Since Φ_b is an eigenstate of H_0, the exponential factor on the left-hand side of H_i can be replaced by a pure number $\exp(i E_b t/\hbar)$, while the other exponential cannot. We separate ϑ in the following the adiabatic switch-off-factor $\exp(-\varepsilon|t|/\hbar)$ from H_i. Then the above T_{ba} can be rewritten as

$$T_{ba} = -\frac{i}{\hbar}(\Phi_b, H_i \Psi_a(E_b)), \tag{52.18}$$

where

$$\Psi_a(E) = \int_{-\infty}^{+\infty}\exp\left(i(E - H_0)\,t/\hbar\right)\exp(-\varepsilon|t|/\hbar)\,U(t)\,\Phi_a\,dt. \tag{52.19}$$

To find the equation satisfied by $\Psi_a(E)$, let us substitute (52.12) into (52.19) and introduce a variable $\tau = t - t'$ instead of t. Then we get

$$\left.\begin{aligned} \Psi_a(E) &= \int_{-\infty}^{+\infty}\exp\left(i(E - E_a)\,t/\hbar\right)\exp(-\varepsilon|t|/\hbar)\,\Phi_a\,dt \\ &\quad -\frac{i}{\hbar}\int_0^{\infty}\exp\left(i(E - H_0)\,\tau/\hbar\right)\exp(-\varepsilon\tau/\hbar)\,H_i\Psi_a(E)\,d\tau, \end{aligned}\right\} \tag{52.20}$$

where we have replaced H_0 by E_a in the first term and used the definition (52.19) in the second term. Now the integral in the first term is just DIRAC's δ-function, $2\pi\hbar\,\delta(E - E_a)$, if we make ε vanish afterwards since there is no singularity. As regards the second integral in (52.20), we should note that the adiabatic switch-off-factor serves to eliminate the singular behavior which would appear without this factor. The direct integration gives

$$\left.\begin{aligned} &-\frac{i}{\hbar}\int_0^{\infty}\exp\left(i(E - H_0)\,\tau/\hbar\right)\exp(-\varepsilon\tau/\hbar)\,d\tau = \\ &= \frac{1}{E - H_0 + i\varepsilon} = \frac{(E - H_0)}{(E - H_0)^2 + \varepsilon^2} - i\frac{\varepsilon}{(E - H_0)^2 + \varepsilon^2} = \mathrm{P}\frac{1}{E - H_0} - i\pi\,\delta(E - H_0). \end{aligned}\right\} \tag{52.21}$$

The last expression is a symbolic statement of the following integral properties possessed by the real and imaginary parts of (52.21) in the limit of $\varepsilon \to 0$:

$$\left.\begin{aligned} &\lim_{\varepsilon\to 0}\int_{-\infty}^{+\infty}\frac{x}{x^2+\varepsilon^2}f(x)\,dx = \mathrm{P}\int_{-\infty}^{+\infty}\frac{f(x)}{x}\,dx,\\ &\lim_{\varepsilon\to 0}\int_{-\infty}^{+\infty}\frac{\varepsilon}{x^2+\varepsilon^2}f(x)\,dx = \pi f(0),\end{aligned}\right\} \tag{52.22}$$

where P denotes the principal part of the integral and $f(x)$ is an arbitrary function. Therefore we get from (52.20)

$$\Psi_a(E) = 2\pi\hbar\,\delta(E-E_a)\,\Phi_a + \frac{1}{E-H_0+i\varepsilon}H_i\Psi_a(E) \tag{52.23}$$

and, on writing

$$\Psi_a(E) = 2\pi\hbar\,\delta(E-E_a)\,\Psi_a, \tag{52.24}$$

we finally get

$$\Psi_a = \Phi_a + \frac{1}{E-H_0+i\varepsilon}H_i\Psi_a, \tag{52.25}$$

which is called the LIPPMANN-SCHWINGER equation.

Corresponding to the re-definition (52.24), we define the transition matrix element $\boldsymbol{T}_{ba}$ by

$$\boldsymbol{T}_{ba} = (\Phi_b, H_i\Psi_a), \tag{52.26}$$

which gives

$$T_{ba} = -2\pi i\,\delta(E_b-E_a)\,\boldsymbol{T}_{ba}, \tag{52.27}$$

which expresses that the transition operator $\boldsymbol{T}$ is defined only for states of equal energy or on the energy shell $E_a = E_b = E$. The probability W_{ba} is then given by

$$W_{ba} = [2\pi\,\delta(E_b-E_a)]^2\,|\boldsymbol{T}_{ba}|^2. \tag{52.28}$$

It should be remembered that the δ-function $\delta(E_b-E_a)$ originates from the time integral in the first term of (52.20):

$$\delta(E_b-E_a) = \frac{1}{2\pi\hbar}\int_{-\infty}^{+\infty}\exp\left(i\,(E_b-E_a)\,t/\hbar\right)dt, \tag{52.29}$$

where ε has already been put equal to zero. As we may confine ourselves to an energy shell, we can interpret Eq. (52.29) as saying that $2\pi\hbar\,\delta(E_b-E_a)$ represents the over all time interval in which the scattering process takes place. On the other hand, W_{ba} is the total transition probability. The transition probability per unit time w_{ba} is then given by

$$w_{ba} = \frac{2\pi}{\hbar}\,\delta(E_b-E_a)\,|\boldsymbol{T}_{ba}|^2. \tag{52.30}$$

This final result contains a well-known consequence of the perturbation theory with respect to the transition probability per unit time. It is to be noted, however, that (52.30), (52.25), and (52.26) are exact.

It is important to explain the nature of Ψ_a defined above in a little more detail. By multiplying the operator $E_a - H_0 + i\varepsilon$ into both sides of (52.25) and afterwards putting ε equal to zero, we get

$$(H_0 + H_i)\,\Psi_a = E_a\Psi_a, \tag{52.31}$$

since Φ_a is an eigenfunction of H_0. The above equation is just a time-independent SCHRÖDINGER equation belonging to the total Hamiltonian $H_0 + H_i$. It has, therefore, been shown that Ψ'_a, which satisfies (52.25), is an eigensolution of the scattering problem containing, besides the incident wave, an outgoing scattered wave. It should also be added that the infinitesimal positive constant ε in the denominator of (52.25) serves to eliminate the singularity in defining the inverse operator of $E - H_0$, and at the same time automatically to select only the outgoing scattered wave.

If we put $\Psi'_a = \Phi_a$ in (52.26), the result is just what is predicted by the lowest order perturbation theory. If the integral equation (52.25) is solved by an iterational procedure, it gives higher order corrections successively. Variational principles can also be constructed in this formulation[167]. HEITLER's damping theory[168] can be derived using the above formalism[169]. A very interesting application of the LIPPMANN-SCHWINGER equation has been made by BRUECKNER and WATSON[170]. For details we refer to the original papers, together with later work on the generalization and refinement[171] of the theory developed above.

53. Momentum space representation. Let us start from the SCHRÖDINGER equation (13.1), with notations (13.4), which is written as follows:

$$(-\Delta + \varkappa^2)\,\psi(\boldsymbol{r}) = v(\boldsymbol{r})\,\psi(\boldsymbol{r}) \tag{53.1}$$

for the bound state. The FOURIER transformations

$$\left.\begin{aligned} \psi(\boldsymbol{r}) &= \frac{1}{\sqrt{(2\pi)^3}}\int \psi(\boldsymbol{k})\, e^{i\boldsymbol{k}\boldsymbol{r}}\, d^3k, \\ v(\boldsymbol{r}) &= \int v(\boldsymbol{k})\, e^{i\boldsymbol{k}\boldsymbol{r}}\, d^3k, \end{aligned}\right\} \tag{53.2}$$

transform (53.1) into an integral equation

$$(k^2 + \varkappa^2)\,\psi(\boldsymbol{k}) = \int v(\boldsymbol{k} - \boldsymbol{k}')\,\psi(\boldsymbol{k}')\,d^3k'. \tag{53.3}$$

The wave function $\psi(\boldsymbol{k})$, which satisfies

$$\int \psi^*(\boldsymbol{k})\,\psi(\boldsymbol{k})\,d^3k = \int \psi^*(\boldsymbol{r})\,\psi(\boldsymbol{r})\,d^3r = 1, \tag{53.4}$$

can now be interpreted as the momentum space wave function which expresses the momentum distribution of the particle in the relevant bound state.

One of the characteristics in the momentum space representation is that the wave function satisfies an integral equation which is in general written as

$$(k^2 + \varkappa^2)\,\psi(\boldsymbol{k}) = \int v(\boldsymbol{k}, \boldsymbol{k}')\,\psi(\boldsymbol{k}')\,d^3k'. \tag{53.5}$$

Another important characteristic is that, only when the interaction term $v(\boldsymbol{k}, \boldsymbol{k}')$ is a function of $\boldsymbol{k} - \boldsymbol{k}'$ alone:

$$v(\boldsymbol{k}, \boldsymbol{k}') = v(\boldsymbol{k} - \boldsymbol{k}'), \tag{53.6}$$

---

[167] B. A. LIPPMANN and J. SCHWINGER: Phys. Rev. **79**, 469 (1950). — B. A. LIPPMANN: Phys. Rev. **79**, 481 (1950). — M. L. GOLDBERGER: Phys. Rev. **82**, 757 (1951); **84**, 929 (1951). — H. E. MOSES: Phys. Rev. **92**, 817 (1953); **96**, 519 (1954).

[168] W. HEITLER: Proc. Cambridge Phil. Soc. **37**, 301 (1941).

[169] M. L. GOLDBERGER: Phys. Rev. **82**, 757 (1951); **84**, 929 (1951).

[170] K. A. BRUECKNER and K. M. WATSON: Phys. Rev. **90**, 699 (1953); **92**, 1023 (1953).

[171] G. F. CHEW and M. L. GOLDBERGER: Phys. Rev. **87**, 778 (1952). — M. GELL-MANN and M. L. GOLDBERGER: Phys. Rev. **91**, 398 (1953). — M. N. HACK: Phys. Rev. **96**, 196 (1954).

this interaction can be identified with an ordinary potential $v(\boldsymbol{r})$ in the coordinate space, otherwise the interaction may be velocity-dependent or non-local in the coordinate space. The requirement (53.6) is indeed a guiding principle in deriving a static nuclear potential from a field theoretical calculation[87]. It may be possible that the momentum space representation is more convenient in treating more general types of interaction than that employed in this article. An example in which $v(\boldsymbol{k}, \boldsymbol{k}')$ can be written as a product of factors $g(\boldsymbol{k})$ and $g(\boldsymbol{k}')$, i.e. the interaction is non-local but separable, has been investigated in detail by YAMAGUCHI[172]. The momentum space representation is known to be very convenient when discussing a relativistic two-body equation such as proposed by BETHE and SALPETER[173]. For these reasons, an introductory explanation of the momentum space representation is given in the following.

From (53.2), one obtains

$$v(\boldsymbol{k}) = \frac{1}{(2\pi)^3} \int v(\boldsymbol{r})\, e^{-i\boldsymbol{k}\boldsymbol{r}}\, d^3 r, \tag{53.7}$$

which can be written, if $v(\boldsymbol{r})$ is a central force potential, $v(\boldsymbol{r}) = v(r)$, as

$$v(\boldsymbol{k}) = \frac{1}{2\pi^2 k} \int_0^\infty r\, v(r) \sin kr\, dr, \tag{53.8}$$

which implies that, if the potential is central in the coordinate space, then the potential is also central in the momentum space, and vice versa. For some typical forms of $v(r)$, $v(\boldsymbol{k})$ or $v(k)$ takes the following analytic forms:

$$\left.\begin{aligned}
v(r) &= v_0 \frac{e^{-\mu r}}{\mu r}; & v(k) &= \frac{v_0}{2\pi^2 \mu (\mu^2 + k^2)} & &\text{(YUKAWA)},\\
v(r) &= v_0\, e^{-\mu r}; & v(k) &= \frac{v_0}{\pi^2 (\mu^2 + k^2)^2} & &\text{(Exponential)},\\
v(r) &= v_0\, e^{-\mu^2 r^2}; & v(k) &= \frac{v_0}{(2\sqrt{\pi}\,\mu)^3}\, e^{-\frac{k^2}{4\mu^2}} & &\text{(GAUSS)}.
\end{aligned}\right\} \tag{53.9}$$

For the square well type, it does not take a simple form; it is not included in (53.9) for this reason. The general behaviour of $v(k)$ is seen from (53.9) as follows: It decreases fairly slowly as k increases as far as $k \lesssim \mu$, the inverse of the force range, and it decreases rapidly for $k \gtrsim \mu$; the longer the tail in the coordinate space, the shorter it is in momentum space.

As regards the method to solve the integral equation (53.3), we readily note that the iterational method may be suitable; it is in fact the only one which can generally be used. SALPETER and GOLDSTEIN[174] have investigated this point in detail. The variational principle can also be constructed in this representation[175], which may help us in solving problems. We refer, however, to the original papers for details. Only one point may be mentioned here, namely that the HULTHÉN wave function which, save for the normalization factor, reads

$$\psi(r) = e^{-\alpha r}(1 - e^{-\mu r})/r, \tag{53.10}$$

[172] Y. YAMAGUCHI: Phys. Rev. **95**, 1628, 1635 (1954).

[173] E. E. SALPETER and H. A. BETHE: Phys. Rev. **81**, 1232 (1951).

[174] E. E. SALPETER: Phys. Rev. **84**, 1226 (1951). — E. E. SALPETER and J. S. GOLDSTEIN: Phys. Rev. **90**, 983 (1953).

[175] W. KOHN: Phys. Rev. **84**, 495 (1951).

takes the following form in momentum space:

$$\psi(\boldsymbol{k}) = \sqrt{\frac{2}{\pi}}\left[\frac{1}{\alpha^2+k^2} - \frac{1}{(\alpha+\mu)^2+k^2}\right], \tag{53.11}$$

which of course shows spherical symmetry since $\psi(\boldsymbol{r})$ is spherically symmetric. It should be mentioned that, if $\psi(r)$ is a regular function, $\psi(k)$ must vanish as k approaches infinity and vice versa, since $\psi(k)$ and $\psi(r)$ are connected by such an equation as (53.8); the factor $\sin kr$ oscillates infinitely rapidly as $k\to\infty$ or $r\to\infty$.

This final remark readily implies that, if the wave function $\psi(r)$ extends to infinity, as is the case for positive energy states, the momentum space wave function $\psi(k)$ should be a singular function of k. For example, if $\psi(\boldsymbol{r})$ is a plane wave solution, $\exp(i\boldsymbol{k}_0\boldsymbol{r})$, the FOURIER transformation (53.2) gives

$$\psi(\boldsymbol{k}) = \sqrt{(2\pi)^3}\,\delta(\boldsymbol{k}-\boldsymbol{k}_0),$$

a DIRAC's δ-function. In order to discuss this singularity in more detail, let us treat the S-wave scattering problem in momentum space representation.

The radial equation for $u(r) = r\,\psi(r)$ is given by (15.5), where we put $l=0$ and replace k by k_0:

$$\left(-\frac{d^2}{dr^2} - k_0^2\right)u(r) = v(r)\,u(r), \tag{53.12}$$

with the boundary condition

$$u(0) = 0. \tag{53.13}$$

We want to find a solution that behaves asymptotically as

$$u(r) \xrightarrow[r\to\infty]{} \sin(k_0 r + \eta), \tag{53.14}$$

where η is the phase shift.

To go over to the momentum space representation, we first note that the boundary condition (53.13) excludes the mixing of any cosine functions into the expansion. Hence we expand $u(r)$ as

$$u(r) = \int_0^\infty u(k)\sin kr\,dk, \tag{53.15}$$

with the understanding that, if $u(k)$ is singular, the principal value of the integral is to be taken. We then get, upon substituting it into (53.12),

$$(k^2-k_0^2)\,u(k) = \int_0^\infty v(k,k')\,u(k')\,dk', \tag{53.16}$$

where

$$v(k,k') = v(k',k) = \frac{2}{\pi}\int_0^\infty \sin kr\,v(r)\sin k'r\,dr, \tag{53.17}$$

and we have used

$$\int_0^\infty \sin kr\sin k'r\,dr = \frac{\pi}{2}\,\delta(k-k'). \tag{53.18}$$

We can replace (53.16) by an integral equation

$$u(k) = A\,\delta(k-k_0) + \frac{1}{(k^2-k_0^2)}\int_0^\infty v(k,k')\,u(k')\,dk', \tag{53.19}$$

where the first term is due to the property $x\delta(x)=0$ and the constant A is as yet undetermined. Since $v(k, k')$ is a regular function of k and k', the integral in (53.19) is a regular function of k, even if $u(k)$ has singularities. Thus we get, denoting the integral in (53.19) as $B(k)$, the expression which represents the relevant singularity explicitly:

$$u(k) = A\,\delta(k-k_0) + \frac{B(k)}{k^2-k_0^2}, \tag{53.20}$$

where it is to be noted that $B(k)$ is free from singularities.

Let us now transform (53.20) back into the coordinate space. From (53.15), we get

$$u(r) = A\sin k_0 r + \int\limits_0^\infty \frac{B(k)}{k^2-k_0^2}\sin kr\,dk. \tag{53.21}$$

We separate the singular part of $B(k)/(k^2-k_0^2)$ by writing

$$\frac{B(k)}{k^2-k_0^2} = \frac{B(k)}{2k}\left[\frac{1}{k-k_0}+\frac{1}{k+k_0}\right]. \tag{53.22}$$

The second term in the bracket is regular and cannot give any contribution to the integral in (53.21) as $r\to\infty$. The first term can be calculated as follows:

$$\lim_{r\to\infty} \mathrm{P}\int\limits_0^\infty \frac{B(k)}{2k}\,\frac{\sin kr}{k-k_0}\,dk = \frac{B(k_0)}{2k_0}\,\mathrm{P}\int\limits_{-\infty}^{+\infty}\frac{\sin kr}{k-k_0}\,dk, \tag{53.23}$$

since the contribution comes only from $k=k_0$ and the integration domain from $-\infty$ to 0 gives no contribution. The symbol P implies that the principal value is to be taken. The right hand side can be simplified by introducing a variable $x=k-k_0$ and remarking that

$$\mathrm{P}\int\limits_{-\infty}^{+\infty}\frac{\cos ax}{x}\,dx = 0, \tag{53.24}$$

hence:

$$\lim_{r\to\infty} \mathrm{P}\int\limits_0^\infty \frac{B(k)}{2k}\,\frac{\sin kr}{k-k_0}\,dk = \frac{B(k_0)}{2k_0}(\cos k_0 r)\,\mathrm{P}\int\limits_{-\infty}^{+\infty}\frac{\sin x}{x}\,dx, \tag{53.25}$$

where the remaining integral is known to be just π. We finally get from (53.21)

$$u(r) \xrightarrow[r\to\infty]{} A\sin k_0 r + \frac{\pi B(k_0)}{2k_0}\cos k_0 r, \tag{53.26}$$

which should be identified with (53.14). The comparison gives

$$A = \cos\eta, \qquad \sin\eta = \frac{\pi B(k_0)}{2k_0}, \tag{53.27}$$

where $B(k)$ is, by definition, given as

$$B(k) = \int\limits_0^\infty v(k,k')\,u(k')\,dk'. \tag{53.28}$$

The second equation of (53.27) plays a role, in the momentum space representation, as the determining equation of the phase shift which is given by (53.14) in the coordinate space.

Acknowledgements.

We express our sincere gratitude to the Swedish Atomic Energy Commission (*Atomkommittén*) for the financial support which made our collaboration possible.

The staff of the Division of Mathematical Physics, Royal Institute of Technology, has been very helpful in many ways. In particular we thank Tekn. Lic. B. C. H. NAGEL for reading and commenting the manuscript. Valuable suggestions have also been made by Civilingenjör S. G. M. GUSTAVI. Last, but not least, we are indebted to Miss M. K. LUNDQVIST and Miss G. MOBACH for typing and retyping the troublesome manuscript.

General references.

[1] BETHE, H. A., and R. F. BACHER: Nuclear Physics. A. Stationary States of Nuclei Rev. Mod. Phys. **8**, 82 (1936).
[2] ROSENFELD, L.: Nuclear Forces. Amsterdam: North-Holland Publishing Company 1948.
[3] BETHE, H. A.: Elementary Nuclear Theory. New York: John Wiley & Sons 1947.
[4] FERMI, E.: Nuclear Physics, Revised Edition. Chicago: The University of Chicago Press 1950.
[5] FLÜGGE, S.: Das Zwei-Nucleonen-Problem. Ergebn. exakt. Naturw. **26**, 165 (1952).
[6] SQUIRES, G. L.: The Neutron-Proton Interaction. Progr. in Nucl. Phys. **2**, 89 (1952).
[7] CHRISTIAN, R. S.: The Analysis of High Energy Neutron-Proton and Proton-Proton Scattering Data, Reports on Progr. Phys. **15**, 68 (1952).
[8] RAMSEY, N. F.: Nuclear Two-Body Problems and Elements of Nuclear Structure, Part IV of Experimental Nuclear Physics, edited by E. SEGRÈ, Vol. I. New York: John Wiley & Sons 1953.
[9] BREIT, G., and R. L. GLUCKSTERN: Advances in Nucleon-Nucleon Scattering Experiments and their Theoretical Consequences. Ann. Rev. Nucl. Sci. **2**, 365 (1953).
[10] BLATT, J. M., and V. F. WEISSKOPF: Theoretical Nuclear Physics. New York: John Wiley & Sons 1952.
[11] SACHS, R. G.: Nuclear Theory. Cambridge, Mass.: Addison-Wesley Publishing Company 1953.

Note added in proof. Attention is drawn to a recent report by J. L. GAMMEL, R. S. CHRISTIAN and R. M. THALER: Calculation of Phenomenological Nucleon-Nucleon Potentials, Phys. Rev. **105**, 311 (1957).

The Three-body Problem in Nuclear Physics.

By

MARIO VERDE.

With 4 Figures.

Introduction.

The classical three-body problem in celestial mechanics has the advantage of working with a well founded notion of potential energy between two bodies. In atomic physics this advantage is not lost, since the COULOMB law constitutes a firm ground upon which the quantum mechanics has built a prodigious edifice that allows the interpretation of a very large number of phenomena including complex atomic and molecular structures.

The three-body problem in nuclear physics rests on a weak basis since the two-body nuclear interaction is still far from being adequately understood.

In most cases one accepts the phenomenological potential which seems to be successful in explaining two-body phenomena and one tries to corroborate or disprove its validity with a theoretical prediction of a particular three-body configuration.

The main purpose of studying many-body systems in nuclear physics is the search for a more complete understanding of the mutual interaction between nucleons. From this study, one may learn about the neutron-neutron force and hope to gain informations on the isotopic spin dependence of the nuclear potential. This latter important question still remains open, but charge symmetry is now a well established fact at low as well as at high energies.

The theoretical methods adopted in the nuclear three-body problems are natural extensions of those already familiar in two-body problems or in the atomic case.

The ordinary and isotopic spin dependence of the nuclear forces adds some minor difficulties of algebraic character which can be easily overcome, but the short range nature of the nuclear interaction implies the more difficult task of approximating complicated wave functions by trial functions of simple analytical character.

A systematic use of the isotopic spin formalism has been made here, because the different symmetry classes of the eigenfunctions which enter into the problem emerge from the very beginning. The theoretical treatment without isotopic spin formalism must, of course, become the same if one considers a posteriori the different components belonging to a given symmetry class. Often one of these components is by far the most important.

In the first two sections we give the classification of the spin eigenstates and the few algebraic manipulations needed to deduce the equations of motion for the spatial eigenfunctions.

Being primarily interested in the general aspects of the problem, we have avoided a detailed discussion of numerical results, for which we refer to the original literature. Unfortunately, a good deal of this numerical work ignores the presence of the tensor force which often plays a significant role.

I. Spin eigenstates and equations of motion.

1. Symmetry operators and spin eigenstates. In the three-body problem an important part is played by the symmetry operators T^s, T', T'' which are defined as follows

$$\left.\begin{aligned} T^s &= (23) + (13) + (12)\,, \\ T' &= \frac{\sqrt{3}}{2}\left[(13) - (12)\right], \\ T'' &= -(23) + \frac{1}{2}\left[(13) + (12)\right]. \end{aligned}\right\} \tag{1.1}$$

(ij) is the permutation of the coordinates of the body i with those of j. We may consider permutations of the spatial coordinates only, or of the ordinary and isotopic spin variables. This will be specified by a subscript, σ for the ordinary spin and τ for the isotopic spin. No subscript will be used if one operates on the spatial coordinates or on functions of only one kind of variables where no ambiguities arise.

Applied to a function symmetric in the coordinates of 2 and 3, the operators (1.1) generate three functions, the first of which is completely symmetrical in the coordinates of 1, 2, 3 while the other two functions transform by permutation of the coordinates according to the following orthogonal representation:

$$(23) = \begin{pmatrix} -1 & 0 \\ 0 & +1 \end{pmatrix}; \quad (13) = \begin{pmatrix} \frac{1}{2} & -\frac{\sqrt{3}}{2} \\ -\frac{\sqrt{3}}{2} & -\frac{1}{2} \end{pmatrix}; \quad (12) = \begin{pmatrix} \frac{1}{2} & \frac{\sqrt{3}}{2} \\ \frac{\sqrt{3}}{2} & -\frac{1}{2} \end{pmatrix}. \tag{1.2}$$

With the help of Eq. (1.1) we can generate the spin eigenstates for a three-nucleon configuration. The total spin S can assume the values $\frac{3}{2}$ and $\frac{1}{2}$.

To $S = \frac{3}{2}$ belong four eigenstates χ^s_μ in correspondence to the four values of the total z-component μ. They are invariant under permutations of the bodies 1, 2, 3, and can be obtained by means of T^s. Thus, for instance,

$$\chi^s_{\frac{1}{2}} = \frac{1}{\sqrt{3}}\, T^s(\alpha_2\,\alpha_3\,\beta_1) \tag{1.3}$$

where α, β are the two spin eigenstates for the single nucleon: $\sigma_z\,\alpha = +\alpha$, $\sigma_z\,\beta = -\beta$. The factor $1/\sqrt{3}$ is chosen so as to have χ^s normalized to one. Operating with the same $\frac{1}{\sqrt{3}}T^s$ on $\alpha_2\,\alpha_3\,\alpha_1$, $\beta_2\,\beta_3\,\alpha_1$, $\beta_2\,\beta_3\,\beta_1$, one obtains clearly the other three components of χ^s.

To $S = \frac{1}{2}$ belong also four eigenstates. For each value of the total z-component there is a pair which can be generated by means of T' and T''.

$$\left.\begin{aligned} \chi'_{\frac{1}{2}} &= \sqrt{\frac{2}{3}}\, T'(\alpha_2\,\alpha_3\,\beta_1)\,, & \chi'_{-\frac{1}{2}} &= \sqrt{\frac{2}{3}}\, T'(\beta_2\,\beta_3\,\alpha_1)\,, \\ \chi''_{\frac{1}{2}} &= \sqrt{\frac{2}{3}}\, T''(\alpha_2\,\alpha_3\,\beta_1)\,, & \chi''_{-\frac{1}{2}} &= \sqrt{\frac{2}{3}}\, T''(\beta_2\,\beta_3\,\alpha_1)\,. \end{aligned}\right\} \tag{1.4}$$

The factor $\sqrt{\frac{2}{3}}$ assures the normalization to 1 of each component.

Since the isotopic spin for a single nucleon is $\frac{1}{2}$, the same considerations hold for the three-body isotopic spin eigenstates. For the total isotopic spin $T = \frac{3}{2}$

we have again four symmetrical eigenstates ζ^s_μ, and for the isotopic spin $T=\frac{1}{2}$, two pairs $(\zeta'_{+\frac{1}{2}}, \zeta''_{+\frac{1}{2}})$ and $(\zeta'_{-\frac{1}{2}}, \zeta''_{-\frac{1}{2}})$ which are obtained in exactly the same way as the corresponding ordinary spin eigenfunctions.

Thus, for example, the pair $(\zeta'_{+\frac{1}{2}}, \zeta''_{+\frac{1}{2}})$ is generated as follows

$$\left.\begin{aligned} \zeta'_{\frac{1}{2}} &= \sqrt{\frac{2}{3}}\, T'(a_2\, a_3\, b_1)\,, \\ \zeta''_{\frac{1}{2}} &= \sqrt{\frac{2}{3}}\, T''(a_2\, a_3\, b_1)\,. \end{aligned}\right\} \tag{1.5}$$

where a and b are the two charge states for the single nucleon: $\tau_3\, a = +a$, $\tau_3\, b = -b$. The operator of the total charge is

$$Q = e\left(\tfrac{3}{2} + \tfrac{1}{2}\sum_k \tau_3^{(k)}\right).$$

$(\zeta'_{\frac{1}{2}}, \zeta''_{\frac{1}{2}})$ therefore describe the $T=\frac{1}{2}$ state of two protons and one neutron, while $(\zeta'_{-\frac{1}{2}}, \zeta''_{-\frac{1}{2}})$ refer to the mirror configuration of two neutrons and one proton. Three identical nucleons belong always to the ζ^s states.

The next step is now to find the eigenstates in the product space of ordinary spin and isotopic spin. We need to consider the only non-trivial case of $S=\frac{1}{2}$ and $T=\frac{1}{2}$. With the two pairs (χ', χ'') and (ζ', ζ''), we may form the combination

$$\xi^s = \frac{1}{\sqrt{2}}\,(\chi'\,\zeta' + \chi''\,\zeta'')\,. \tag{1.6s}$$

This is clearly totally symmetric with respect to permutation of both ordinary and isotopic spin coordinates, since each pair transforms according to the orthogonal representation (1.2), which leaves invariant the scalar product. We note furthermore that the pair $(\zeta'', -\zeta')$ transforms according to the same (1.2) but with opposite sign. Thus the combination:

$$\xi^a = \frac{1}{\sqrt{2}}\,(\chi'\,\zeta'' - \chi''\,\zeta') \tag{1.6a}$$

is totally antisymmetrical. The other two eigenstates are:

$$\left.\begin{aligned} \xi' &= \frac{1}{\sqrt{2}}\,(\chi'\,\zeta'' + \chi''\,\zeta')\,, \\ \xi'' &= \frac{1}{\sqrt{2}}\,(\chi'\,\zeta' - \chi''\,\zeta'')\,. \end{aligned}\right\} \tag{1.6d}$$

They belong to the same representation (1.2). All ξ's are orthogonal eigenfunctions, normalized to one.

Expectation values in the spin space are most easily evaluated by considering the symmetry classes under permutations of the same operators. We wish to illustrate this for a particular case that will be important in the discussion of the magnetic moments (Sect. 4) and for the electromagnetic transitions (Sect. 18). The operator $\sum_{k=1}^{3} \tau_3^{(k)}\, \sigma_z^{(k)}$ can be written:

$$\sum_{k=1}^{3} \tau_3^{(k)}\, \sigma_z^{(k)} = \tfrac{1}{3}\,\tau_3^s\,\sigma_z^s + \tfrac{2}{3}\,(\tau_3'\,\sigma_z' + \tau_3''\,\sigma_z'') \tag{1.7}$$

where

$$\left.\begin{aligned} \tau_3^s &= T^s\,\tau_3^{(1)} = \tau_3^{(1)} + \tau_3^{(2)} + \tau_3^{(3)}\,, \\ \tau_3' &= T'\,\tau_3^{(1)} = \frac{\sqrt{3}}{2}\left(\tau_3^{(3)} - \tau_3^{(2)}\right), \\ \tau_3'' &= T''\,\tau_3^{(2)} = -\,\tau_3^{(1)} + \tfrac{1}{2}\left(\tau_3^{(3)} + \tau_3^{(2)}\right). \end{aligned}\right\} \tag{1.8}$$

with analogous definitions for the σ's. We wish to find the expectation values of

$$\Omega = \tau_3'\,\sigma_z' + \tau_3''\,\sigma_z'' \tag{1.9}$$

in the ξ space.

Since Ω is invariant for permutations of the spin variables of two bodies, we can immediately infer that the expectation values:

$$\langle \xi^a | \Omega | \xi^a \rangle, \quad \langle \xi^s | \Omega | \xi^s \rangle, \quad \langle \xi' | \Omega | \xi' \rangle = \langle \xi'' | \Omega | \xi'' \rangle$$

are the only ones different from zero. Moreover, we can affirm that ξ^a is an eigenstate of Ω. Actually, one has

$$\tfrac{1}{2}\,\Omega \begin{pmatrix} \xi^a \\ \xi^s \\ \xi' \\ \xi'' \end{pmatrix} = \begin{pmatrix} -\,\xi^a \\ \xi^s + \sqrt{2}\,\chi^s \zeta^s \\ \chi' \zeta^s + \zeta' \chi^s \\ \chi'' \zeta^s + \chi^s \zeta'' \end{pmatrix}. \tag{1.10}$$

This is for a value $+\frac{1}{2}$ of both the total z-components of the ordinary and isotopic spin. All other components differ in sign only, since Ω is an odd operator both in σ and τ. From (1.5) and a substate $\sigma_z^s = +1$ we deduce:

$$\left.\begin{aligned} &\langle \xi^a | \Omega | \xi^a \rangle = -\,2 \langle \tau_3^s \rangle, \quad \langle \xi^s | \Omega | \xi^s \rangle = +\,2 \langle \tau_3^s \rangle, \\ &\langle \xi' | \Omega | \xi' \rangle = \langle \xi'' | \Omega | \xi'' \rangle = 0. \end{aligned}\right\} \tag{1.11}$$

2. Equations of motion for central forces. The wave function $\Psi(1,2,3)$ describing a system of three nucleons in a stationary state of total energy E is an integral of the equation of motion:

$$(\boldsymbol{T} + V)\,\Psi = E\,\Psi. \tag{2.1}$$

We restrict ourselves to non-relativistic kinematics for which the operator $\boldsymbol{T}$ of the total kinetic energy is in a SCHRÖDINGER representation

$$\boldsymbol{T} = \boldsymbol{T}_1 + \boldsymbol{T}_2 + \boldsymbol{T}_3 = -\,\frac{\hbar^2}{2M}\,(\Delta_1 + \Delta_2 + \Delta_3)\,.$$

We assume an interaction between the bodies 1 and 2 of the form:

$$U(\boldsymbol{r}_1 - \boldsymbol{r}_2)\,\boldsymbol{O}(12)$$

where $\boldsymbol{O}$ contains the dependence upon ordinary and isotopic spin and U is a function of the separation of the particles.

Admitting the invariance for rotation in the ordinary and isotopic spin space, $\boldsymbol{O}(12)$ has the form

$$\boldsymbol{O}(12) = w + b\,(12)_\sigma + h\,(12)_\tau + m\,(12)_{\sigma\tau} \tag{2.2}$$

w, b, h, m are constants and (12) is the permutation of those coordinates of 1 and 2 as indicated in the subscript. We will also introduce instead of the constants w, b, h, m, the four potential strengths corresponding to the possible triplet and singlet two-body states:

$$\langle s, t | \boldsymbol{O}(12) | s, t \rangle = \lambda_{2s+1, 2t+1} \cdot U_0 . \tag{2.3}$$

Thus, for instance, $\lambda_{3,1} U_0$ is the potential strength for a triplet ordinary spin ($s=1$) and a singlet isotopic spin ($t=0$). Choosing $\lambda_{3,1}=1$, U_0 is the central interaction in the deuteron ground state.

Expressed in terms of the constants λ, Eq. (2.2) becomes:

$$\left.\begin{aligned} \boldsymbol{O}(12) = \lambda_{3,3} \boldsymbol{P}_\sigma^{(+)}(12) \boldsymbol{P}_\tau^{(+)}(12) + \lambda_{3,1} \boldsymbol{P}_\sigma^{(+)}(12) \boldsymbol{P}_\tau^{(-)}(12) + \\ + \lambda_{1,3} \boldsymbol{P}_\sigma^{(-)}(12) \boldsymbol{P}_\tau^{(+)}(12) + \lambda_{1,1} \boldsymbol{P}_\sigma^{(-)}(12) \boldsymbol{P}_\tau^{(-)}(12) \end{aligned}\right\} \tag{2.4}$$

where $\boldsymbol{P}$ are projection operators as defined by:

$$\boldsymbol{P}_\sigma^{(\pm)} = \frac{1 \pm (12)_\sigma}{2}, \qquad \boldsymbol{P}_\tau^{(\pm)} = \frac{1 \pm (12)_\tau}{2}.$$

The total interaction V in Eq. (2.1), assuming the additivity of the two-body potentials, may now be written:

$$V = V(23) + V(12) + V(13) = U^s \boldsymbol{O}^s + U' \boldsymbol{O}' + U'' \boldsymbol{O}''$$

where

$$\left.\begin{aligned} U^s &= T^s\, U(23) = U(23) + U(12) + U(13), \\ U' &= T'\, U(23) = \frac{\sqrt{3}}{2} [U(12) - U(13)], \\ U'' &= T''\, U(23) = -U(23) + \tfrac{1}{2} [U(12) + U(13)]. \end{aligned}\right\} \tag{2.5}$$

The $\boldsymbol{O}$'s are linear combinations of the same symmetry operators T [Eq. (1.1)] operating in the spin space.

Elimination of the spin coordinates in Eq. (2.1) can be easily performed. We consider, for example, configurations with a total isotopic spin $T=\frac{1}{2}$. The form of the total eigenfunction Ψ will then be

$$\left.\begin{aligned} &\text{for } S=\tfrac{3}{2} \quad \Psi = (\psi' \zeta'' - \psi'' \zeta') \cdot \chi^s, \\ &\text{for } S=\tfrac{1}{2} \quad \Psi = \psi^a \xi^s - \psi^s \xi^a + \psi' \xi'' - \psi'' \xi'. \end{aligned}\right\} \tag{2.6}$$

ψ', ψ'', ψ^s and ψ^a depend on the spatial coordinates only. To satisfy the PAULI principle ψ^s must be totally symmetrical, ψ^a totally antisymmetrical, and finally (ψ', ψ'') must transform according to Eq. (1.2). We notice, in particular, that ψ' is antisymmetrical in 2 and 3, while ψ'' is symmetrical in these coordinates.

We have the following systems of differential equations for the spatial components of the wave function[1]:

for $S=\frac{3}{2}$, $T=\frac{1}{2}$:

$$\left.\begin{aligned} (\boldsymbol{T} - E)\psi' &= -\tfrac{1}{2}(\lambda_{3,3} + \lambda_{3,1})\, U^s \psi' + \tfrac{1}{2}(\lambda_{3,3} - \lambda_{3,1})\, (U' \psi'' + U'' \psi'), \\ (\boldsymbol{T} - E)\psi'' &= -\tfrac{1}{2}(\lambda_{3,3} + \lambda_{3,1})\, U^s \psi'' + \tfrac{1}{2}(\lambda_{3,3} - \lambda_{3,1})\, (U' \psi' - U'' \psi''), \end{aligned}\right\} \tag{2.7}$$

for $S=\frac{1}{2}$, $T=\frac{1}{2}$:

$$\left.\begin{aligned} (\boldsymbol{T}-E)\psi^a &= -\tfrac{1}{2}(\lambda_{3,3}+\lambda_{1,1})\, U^s \psi^a + \tfrac{1}{2}(\lambda_{3,3}-\lambda_{1,1})\, (U'\psi'' - U''\psi'), \\ (\boldsymbol{T}-E)\psi^s &= -\tfrac{1}{2}(\lambda_{3,1}+\lambda_{1,3})\, U^s \psi^s - \tfrac{1}{2}(\lambda_{3,1}-\lambda_{1,3})\, (U'\psi' + U''\psi''), \\ (\boldsymbol{T}-E)\psi' &= -\tfrac{1}{2}(\lambda_{3,3}-\lambda_{1,1}) U''\psi^a - \tfrac{1}{2}(\lambda_{3,1}-\lambda_{1,3}) U'\psi^s - w\, U^s \psi' + m(U'\psi'' + U''\psi'), \\ (\boldsymbol{T}-E)\psi'' &= +\tfrac{1}{2}(\lambda_{3,3}-\lambda_{1,1}) U'\psi^a - \tfrac{1}{2}(\lambda_{3,1}-\lambda_{1,3}) U''\psi^s - w\, U^s \psi'' + m(U'\psi' - U''\psi''). \end{aligned}\right\} \tag{2.8}$$

[1] M. VERDE: Helv. Phys. Acta 22, 339 (1949).

The potentials U are defined in Eq. (2.5), and:

$$w = \tfrac{1}{4}(\lambda_{3,3} + \lambda_{3,1} + \lambda_{1,3} + \lambda_{1,1}),$$
$$m = \tfrac{1}{4}(\lambda_{3,3} - \lambda_{3,1} - \lambda_{1,3} + \lambda_{1,1}).$$

For the choice of the spatial coordinates, it will be convenient to take instead of the vectors $\boldsymbol{r}_1$, $\boldsymbol{r}_2$ and $\boldsymbol{r}_3$ the linear combinations:

$$\left.\begin{aligned} \boldsymbol{q}^s &= T^s(\tfrac{1}{3}\boldsymbol{r}_1) = \tfrac{1}{3}(\boldsymbol{r}_1 + \boldsymbol{r}_2 + \boldsymbol{r}_3), \\ \boldsymbol{r} &= T'(\boldsymbol{r}_1) = \frac{\sqrt{3}}{2}(\boldsymbol{r}_3 - \boldsymbol{r}_2), \\ \boldsymbol{q} &= T''(\boldsymbol{r}_1) = -\boldsymbol{r}_1 + \tfrac{1}{2}(\boldsymbol{r}_2 + \boldsymbol{r}_3). \end{aligned}\right\} \tag{2.9}$$

$\boldsymbol{q}^s$ is the center of mass coordinate. Any pair of components of the two vectors $\boldsymbol{r}$ and $\boldsymbol{q}$ transforms according to (1.2). In this coordinate system the kinetic energy has the form:

$$\boldsymbol{T} = -\frac{\hbar^2}{6M}\Delta_{q^s} - \frac{3}{4}\frac{\hbar^2}{M}(\Delta_r + \Delta_q).$$

In the center of mass system the total energy is $\frac{2}{3}E$ and Eq. (2.1) can be written

$$\left(\Delta_r + \Delta_q + \frac{8M}{9\hbar^2}E\right)\Psi = \frac{4M}{3\hbar^2}V \cdot \Psi.$$

Thus, for example, for the quartet continuum configuration of a nucleon and a deuteron in its ground state, Eq. (2.7) is rewritten as

$$\left.\begin{aligned} (\Delta + k^2 - k_d^2)\,\psi' &= \tfrac{1}{2}(\lambda_{3,3} + \lambda_{3,1})\,U^s\psi' - \tfrac{1}{2}(\lambda_{3,3} - \lambda_{3,1})\,(U'\psi'' + U''\psi'), \\ (\Delta + k^2 - k_d^2)\,\psi'' &= \tfrac{1}{2}(\lambda_{3,3} + \lambda_{3,1})\,U^s\psi'' - \tfrac{1}{2}(\lambda_{3,3} - \lambda_{3,1})\,(U'\psi' - U''\psi'') \end{aligned}\right\} \tag{2.10}$$

where it is understood that all potentials are multiplied by $\frac{4}{3}\frac{M}{\hbar^2}$.

$\Delta = \Delta_r + \Delta_q$ is the LAPLACE operator in the six dimensional space of $\boldsymbol{r}$ and $\boldsymbol{q}$. It is completely symmetric under permutation of the coordinates of the three bodies, and is related to Δ_q as follows:

$$\Delta = \tfrac{1}{3}\left[(23) + (12) + (13)\right]\Delta_q. \tag{2.11}$$

In Eq. (2.10) we have put

$$k^2 = \frac{8}{9}\frac{M}{\hbar^2}E \quad \text{and} \quad k_d^2 = \frac{4}{3}\frac{M}{\hbar^2}|E_d|,$$

E_d being the deuteron binding energy.

The total angular momentum

$$\boldsymbol{L} = -i\hbar\sum_{k=1}^{3}[\boldsymbol{r}_k \times \nabla_{r_k}] = -i\hbar\,[\boldsymbol{r} \times \nabla_r + \boldsymbol{q} \times \nabla_q]$$

commutes with Δ. Further, since the potentials U belong to a total angular momentum zero, any one of the Eqs. (2.7) and (2.8) admits the rotation group in the space of $\boldsymbol{r}$ and $\boldsymbol{q}$. Every component ψ therefore belongs to a given total angular momentum.

II. The bound states of H^3 and He^3.

3. The ground states of H^3 and He^3. The only bound and stable configuration formed by three nucleons is the He^3 nucleus. The mirror nucleus H^3 has a slightly higher binding energy and decays into He^3 with a half-life of 12.4 years.

The experimental data referring to the ground states of these two nuclei are summarized in the following table 1. No experimental evidence exists on possible excited states.

From the table below we see that the difference in binding energy of the two mirror nuclei amounts to $\Delta E \approx 0.76$ Mev. Its origin may be ascribed entirely to the COULOMB repulsion between the two protons in He^3. We have here an argument in favor of the equality of *n-n* and *p-p* interaction and we may gain furthermore a rough indication about the size of He^3 and H^3. Assuming a uniform charge distribution in a sphere of radius R, from the equation

$$\frac{6}{5}\frac{e^2}{R} = 0.76 \text{ Mev} \quad \text{we deduce} \quad R \approx 1.6 \frac{\hbar}{\mu c}$$

where $\hbar/\mu c = 1.4 \times 10^{-13}$ cm is the COMPTON wavelength of the π meson. Since the recent measurements[1] of the charge and magnetic moment spread of a free proton indicates for this radius the value $0.5\,\hbar/\mu c$, we conclude that the nuclei He^3 and H^3 are rather tight structures in which the three particles are moving very closely together. This is in contrast with the deuteron which is a more loosely bound system. Since the magnetic moments of the proton and neutron are $\mu_P = 2.793$ n.m. and $\mu_N = -1.913$ n.m., we may tentatively interpret the H^3 and He^3 ground state as mostly $^2S_{\frac{1}{2}}$.

Table 1.

Nucleus	Nuclear spin[2]	Binding energy[3]	Magnetic moment[2]
He^3	$\frac{1}{2}$	-7.718 Mev	-2.1274 n.m.
H^3	$\frac{1}{2}$	-8.482 Mev	$+2.9786$ n.m.

We devote the next section to a more quantitative study of the magnetic moments and Sect. 5 to the binding energies. No relativistic corrections have been contemplated although they are certainly more important than in the deuteron's case[4], owing to the small nuclear radii of H^3 and He^3.

4. The magnetic moments of H^3 and He^3. The experimental values of the magnetic moments (see Table 1) exclude for the ground configurations of H^3 and He^3 the possibility of a pure $^2S_{\frac{1}{2}}$ state. The dipole magnetic moment would have in this case the form:

$$\mu = \tfrac{1}{2}(\mu_P + \mu_N) \sum_{k=1}^{3} \sigma^{(k)} + \tfrac{1}{2}(\mu_P - \mu_N) \sum_{k=1}^{3} \tau_3^{(k)} \sigma^{(k)}. \tag{4.1}$$

From Eq. (4.1) would follow:

$$\mu_{\mathrm{H}} + \mu_{\mathrm{He}} = (\mu_P + \mu_N) \sum_{k=1}^{3} \sigma^{(k)}. \tag{4.2}$$

This is because the expectation values of $\tau_3^{(k)}$, which is an odd operator with respect to an exchange of a neutron with a proton, are equal and opposite in sign for the two mirror nuclei H^3 and He^3. The sum of the experimental magnetic moments is 0.8512 n.m. (see Table 1), the value predicted by Eq. (4.2) is $\mu_P + \mu_N = 0.8797$ n.m. We may explain the small discrepancy by allowing a small admixture of states with angular momenta different from zero[5].

---

[1] R. W. McALLISTER and R. HOFSTADTER: Phys. Rev. **102**, 851 (1956).
[2] N. F. RAMSEY: Nuclear Moments, p. 78. New York: John Wiley & Sons, Inc. 1953.
[3] C. W. LI *et al.*: Phys. Rev. **83**, 512 (1951).
[4] H. PRIMAKOFF: Phys. Rev. **72**, 118 (1947).
[5] R. G. SACHS and J. SCHWINGER: Phys. Rev. **70**, 41 (1946).

The contribution to the magnetic dipole moment coming from the orbital angular momentum is:

$$\mu_{\text{orb}} = \sum_{k=1}^{3} \frac{1+\tau_3^{(k)}}{2} \boldsymbol{L}^{(k)}. \tag{4.3}$$

Thus Eq. (4.2) must be completed, and becomes

$$\mu_{\text{H}} + \mu_{\text{He}} = (\mu_P + \mu_N) \sum_{k=1}^{3} \boldsymbol{\sigma}^{(k)} + \sum_{k=1}^{3} \boldsymbol{L}^{(k)} = 2(\mu_P + \mu_N)\, \boldsymbol{S} + \boldsymbol{L}.$$

The expectation value of the z-component of this operator in the substate $I_z = I = \frac{1}{2}$ is:

$$\begin{aligned}\mu_{\text{H}} + \mu_{\text{He}} &= \frac{1}{I+1} \langle I\, L\, S |\, (\mu_{\text{H}} + \mu_{\text{He}}) \cdot \boldsymbol{I}\, |\, I\, L\, S \rangle \\ &= \frac{1}{2}\left(\mu_P + \mu_N + \frac{1}{2}\right) + \frac{2}{3}\left(\mu_P + \mu_N - \frac{1}{2}\right)[S(S+1) - L(L+1)]\end{aligned}$$

where $\boldsymbol{I} = \boldsymbol{L} + \boldsymbol{S}$.

For an admixture characterized by the weights $|C_{LS}|^2$, one obtains

$$\left.\begin{aligned}\mu_{\text{H}} + \mu_{\text{He}} = \tfrac{1}{2}(\mu_P + \mu_N + \tfrac{1}{2}) + {}\\ + \tfrac{2}{3}(\mu_P + \mu_N - \tfrac{1}{2}) \sum |C_{LS}|^2 [S(S+1) - L(L+1)].\end{aligned}\right\} \tag{4.4}$$

Since the only states compatible with the total spin $I = \frac{1}{2}$, are two doublets ${}^2S_{\frac{1}{2}}$, ${}^2P_{\frac{1}{2}}$ and two quartets ${}^4P_{\frac{1}{2}}$, ${}^4D_{\frac{1}{2}}$, Eq. (4.4) provides a single relation for three unknown parameters. We have reasons for excluding any appreciable presence of P states (see Sect. 18). Thus we are left with the admixture ${}^2S_{\frac{1}{2}} + {}^4D_{\frac{1}{2}}$ which is uniquely determined by Eq. (4.4). The ${}^4D_{\frac{1}{2}}$ state enters with a weight of 3.8%.

The evaluation of the magnetic moment for the single nucleus H^3 or He^3 requires a more detailed knowledge of their spatial eigenfunctions.

The magnetic moment of a ${}^2S_{\frac{1}{2}}$ state is [see Eqs. (4.1) and (1.7)]

$$\mu({}^2S_{\frac{1}{2}}) = \frac{1}{2}(\mu_P + \mu_N) + \frac{1}{3}(\mu_P - \mu_N)\left[\left\langle \frac{\tau_3^s}{2} \right\rangle + \langle \tau_3' \sigma_z' + \tau_3'' \sigma_z'' \rangle\right].$$

The expectation values of the operator $\Omega = \tau_3' \sigma_z' + \tau_3'' \sigma_z''$ in the substate $S_z = +\frac{1}{2}$ depend on the spatial symmetry of the configuration. We have seen in Eq. (1.11) that $\langle \Omega \rangle = -2\langle \tau_3^s \rangle$ for a totally symmetric spatial eigenfunction, $\langle \Omega \rangle = +2\langle \tau_3^s \rangle$ for the totally antisymmetric case, and finally that $\langle \Omega \rangle = 0$ for a spatial symmetry of the type (1.2). Moreover, $\langle \Omega \rangle$ vanishes also for a state belonging to a total isotopic spin $T = \frac{3}{2}$.

We obtain

$$\left.\begin{aligned}\mu({}^2S_{\frac{1}{2}}) &= \tfrac{1}{2}(1 - \langle \tau_3^s \rangle)\, \mu_P + \tfrac{1}{2}(1 + \langle \tau_3^s \rangle)\, \mu_N \\ &\qquad (T = \tfrac{1}{2} \text{ and full symmetry in space}), \\ \mu({}^2S_{\frac{1}{2}}) &= \tfrac{1}{2}(1 + \tfrac{5}{3}\langle \tau_3^s \rangle)\, \mu_P + \tfrac{1}{2}(1 - \tfrac{5}{3}\langle \tau_3^s \rangle)\, \mu_N \\ &\qquad (T = \tfrac{1}{2} \text{ and full antisymmetry in space}), \\ \mu({}^2S_{\frac{1}{2}}) &= \tfrac{1}{2}(1 + \tfrac{1}{3}\langle \tau_3^s \rangle)\, \mu_P + \tfrac{1}{2}(1 - \tfrac{1}{3}\langle \tau_3^s \rangle)\, \mu_N \\ &\qquad [T = \tfrac{1}{2} \text{ or } \tfrac{3}{2} \text{ and symmetry class (1.2)}].\end{aligned}\right\} \tag{4.5}$$

where $\langle \tau_3^s \rangle = +1$ for He^3 and $\langle \tau_3^s \rangle = -1$ for H^3. Since the ${}^2S_{\frac{1}{2}}$ state is dominant with a weight of about 96%, the experimental magnetic moments strongly favor a configuration totally symmetric in the space coordinates of the three bodies. The study of the binding energy leads to the same conclusions.

The contribution to the magnetic moment of the ${}^4D_{\frac{1}{2}}$ state is given by the expectation value of the z-component of the operator [see Eqs. (4.1) and (4.3)]

$$\boldsymbol{\mu} = (\mu_P + \mu_N)\,\boldsymbol{S} + \tfrac{1}{2}(\mu_P - \mu_N)\sum_{k=1}^{3} \tau_3^{(k)}\,\boldsymbol{\sigma}^{(k)} + \tfrac{1}{2}\sum_{k=1}^{3}(1+\tau_3^{(k)})\,\boldsymbol{L}^{(k)}$$

in the substate $I_z = \frac{1}{2}$. This is:

$$\left.\begin{aligned}\mu = \frac{1}{I+1}\langle \boldsymbol{\mu}\cdot\boldsymbol{I}\rangle = \frac{2}{3}(\mu_P+\mu_N)\langle \boldsymbol{S}\cdot\boldsymbol{I}\rangle + \frac{1}{3}(\mu_P-\mu_N)\sum_{k=1}^{3}\langle \tau_3^{(k)}(\boldsymbol{\sigma}^{(k)}\cdot\boldsymbol{I})\rangle + \\ + \frac{1}{3}\langle \boldsymbol{L}\cdot\boldsymbol{I}\rangle + \frac{1}{3}\sum_{k=1}^{3}\langle \tau_3^{(k)}\cdot(\boldsymbol{L}^{(k)}\cdot\boldsymbol{I})\rangle.\end{aligned}\right\}\tag{4.6}$$

For a ${}^4D_{\frac{1}{2}}$ state $\langle \boldsymbol{S}\cdot\boldsymbol{I}\rangle = -\frac{3}{4}$. From Eq. (1.7)

$$\sum_{k=1}^{3}\langle \tau_3^{(k)}(\boldsymbol{\sigma}^{(k)}\cdot\boldsymbol{I})\rangle = \tfrac{2}{3}\langle \tau_3^s\,\boldsymbol{S}\cdot\boldsymbol{I}\rangle = -\tfrac{1}{2}\langle \tau_3^s\rangle,$$

since $\langle(\tau_3'\sigma' + \tau_3''\sigma'')\cdot\boldsymbol{I}\rangle$ vanishes in a spin state $S=\frac{3}{2}$ because of its complete symmetry.

Thus in Eq. (4.6) the part of the magnetic moment arising from $\boldsymbol{S}$ is:

$$\mu_{\text{spin}}({}^4D_{\frac{1}{2}}) = -\tfrac{1}{2}\left[(\mu_P+\mu_N) + \tfrac{1}{3}(\mu_P-\mu_N)\langle\tau_3^s\rangle\right].\tag{4.7}$$

For the evaluation of $\mu_{\text{orb}}({}^4D_{\frac{1}{2}})$ we need the matrix elements

$$\langle \boldsymbol{L}\cdot\boldsymbol{I}\rangle = \tfrac{3}{2},\quad \sum_{k=1}^{3}\langle \tau_3^{(k)}(\boldsymbol{L}^{(k)}\cdot\boldsymbol{I})\rangle = \tfrac{1}{3}\langle \tau_3^s(\boldsymbol{L}\cdot\boldsymbol{I}) + 2(\tau_3'\boldsymbol{L}' + \tau_3''\boldsymbol{L}'')\cdot\boldsymbol{I}\rangle\tag{4.8}$$

where $\boldsymbol{L}' = T'(\boldsymbol{L}_1)$, $\boldsymbol{L}'' = T''(\boldsymbol{L}_1)$ [see Eqs. (1.1) and (1.8)] and $\boldsymbol{L}_1$ is the orbital angular momentum of nucleon 1. For a total isotopic spin $T=\frac{3}{2}$, $\langle(\tau_3'\boldsymbol{L}' + \tau_3''\boldsymbol{L}'')\cdot\boldsymbol{I}\rangle = 0$, but for $T=\frac{1}{2}$ the calculation of $\langle(\tau_3'\boldsymbol{L}' + \tau_3''\boldsymbol{L}'')\cdot\boldsymbol{I}\rangle$ needs a more detailed knowledge of the spatial configuration[1]. Assuming invariance with respect to the interchange of the coordinates $\boldsymbol{r}$ and $\boldsymbol{q}$ [see Eq. (2.9)], its value is again zero. In this case the contribution to the magnetic moment due to the orbit is [see Eqs. (4.6) and (4.8)]:

$$\mu_{\text{orb}}({}^4D_{\frac{1}{2}}) = \tfrac{1}{2} + \tfrac{1}{6}\langle\tau_3^s\rangle.\tag{4.9}$$

The total moment with the symmetry restrictions made above and an admixture $(1-w_D)\,{}^2S_{\frac{1}{2}} + w_D\,{}^4D_{\frac{1}{2}}$ is finally

$$\left.\begin{aligned}\mu = \frac{1-\langle\tau_3^s\rangle}{2}\left[\mu_P - \frac{2}{3}w_D\left(2\mu_P + \mu_N - \frac{1}{2}\right)\right] + \\ + \frac{1+\langle\tau_3^s\rangle}{2}\left[\mu_N - \frac{2}{3}w_D(2\mu_N + \mu_P - 1)\right]\end{aligned}\right\}\tag{4.10}$$

[cf. Eqs. (4.5), (4.7) and (4.9)]. Accepting the admixture $w_D = 0.038$ already established with the help of Eq. (4.4), the numerical values of

$$\mu_{\mathrm{H}^3} = \mu_P - \tfrac{2}{3}w_D(2\mu_P + \mu_N - \tfrac{1}{2}),$$
$$\mu_{\mathrm{He}^3} = \mu_N - \tfrac{2}{3}w_D(2\mu_N + \mu_P - 1)$$

deviate from the experimental values reported in Table 1 by about $\frac{1}{4}$ nuclear magnetons:

$$\left.\begin{aligned}\mu_{\text{theor}}(\mathrm{H}^3) &= \mu_{\exp}(\mathrm{H}^3) - 0.27\,\text{n.m.},\\ \mu_{\text{theor}}(\mathrm{He}^3) &= \mu_{\exp}(\mathrm{He}^3) + 0.27\,\text{n.m.}\end{aligned}\right\}\tag{4.11}$$

---

[1] R. G. SACHS: Nuclear Theory, p. 251. Cambridge, Mass.: Addison-Wesley Publishing Company, Inc. 1953.

An explanation of this discrepancy has been given in terms of the virtual meson currents which yield the interaction between two nucleons[1]. The total magnetic moment is no longer the sum of the moments of each single body. There are corrections to it, the so-called exchange-moments, which depend upon the space and spin coordinates of each pair of nucleons.

In analogy with the phenomenological theory of nuclear forces, some requirements of invariance and symmetry are sufficient to limit to a few types all possible ordinary and isotopic spin dependence of the exchange currents[2]. However, no restriction is obtained for the radial dependence of the spin-exchange currents, which alone can explain the above discrepancy. An accurate quantitative estimate of the exchange magnetic moment based on a meson theory is unfortunately not possible because no fully satisfactory meson theory of nuclear forces exists. The phenomenological approach leaves room for too many choices.

5. Binding energies of H^3 and He^3. We have learned from the interpretation of the magnetic moments (see Sect. 4) that the ground state of H^3 and He^3 is nearly pure $^2S_{\frac{1}{2}}$ with an admixture of about 4% of the $^4D_{\frac{1}{2}}$ state. Arguments have also been presented that favor a ground state configuration totally symmetrical with respect to permutations of the spatial coordinates of the three bodies (cf. Sects. 4 and 18).

A value $T=\frac{3}{2}$ for the total isotopic spin would preclude such high symmetry. With $S=\frac{1}{2}$ the only isotopic spin state that allows a full symmetric spatial eigenfunction is $T=\frac{1}{2}$ (Sect. 1). Configurations with a lower degree of symmetry in space are always associated with larger values for the kinetic energy and have therefore the effect of diminishing or destroying the binding. Variational calculations strongly support these qualitative arguments showing that fully symmetrical trial-eigenfunctions yield the lowest eigenvalues.

The spatial component ψ^s, for $S=\frac{1}{2}$, obeys Eq. (2.8),

$$[\boldsymbol{T}+\tfrac{1}{2}(\lambda_{3,1}+\lambda_{1,3})\,U^s]\cdot\psi^s = E\,\psi^s - \tfrac{1}{2}(\lambda_{3,1}-\lambda_{1,3})\,(U'\,\psi' + U''\,\psi''). \tag{5.1}$$

Putting $\lambda=\frac{1}{2}(\lambda_{3,1}+\lambda_{1,3})$ and neglecting the components ψ', ψ'' on the right side, one obtains

$$(\boldsymbol{T}+\lambda\,U^s)\,\psi^s = E\,\psi^s. \tag{5.2}$$

There is no hope of obtaining from the binding energy information on the isotopic spin dependence of the potential strengths. Indeed, the only coupling constant which appears in Eq. (5.2) is $\lambda=\frac{1}{2}(\lambda_{3,1}+\lambda_{1,3})$. This is nearly equal to $\lambda_{3,1}$, as we shall see later when we consider the presence of tensor forces (Sect. 6).

Many computations of E have been carried out mostly with the aid of variational techniques[3] corresponding to several types and shapes of two-body central potentials.

An interesting attempt to improve the slow convergence of the usual variational approach has been made[4] with the so-called variation-iteration method. Starting with a variational trial function ψ_0, a set of functions $\psi_1, \psi_2 \ldots$ is

[1] F. Villars: Helv. phys. Acta **20**, 476 (1947).

[2] R. G. Sachs: Nuclear Theory, p. 253. Cambridge, Mass.: Addison-Wesley Publishing Company, Inc. 1953.

[3] E. Feenberg: Phys. Rev. **47**, 850 (1935). — E. Feenberg and S. S. Share: Phys. Rev. **50**, 253 (1936). — S. Flügge: Z. Physik **105**, 522 (1937). — H. Margenau and D. T. Warren: Phys. Rev. **52**, 790 (1937). — W. Rarita and R. D. Present: Phys. Rev. **51**, 788 (1937). — F. W. Brown: Phys. Rev. **56**, 1107 (1939). — J. Irving: Phil. Mag. **42**, 338 (1951). J. M. Blatt and V. F. Weisskopf: Theoretical Nuclear Physics. New York: J. Wiley 1952.

[4] Nils Svartholm: Thesis, Lund 1945. — P. M. Morse and H. Feshbach: Methods of theoretical Physics. New York: Mc Graw-Hill 1953.

constructed by means of an iterative procedure with

$$\psi_n = \left(-\frac{1}{T-E} \cdot U^s\right)^n \cdot \psi_0 . \tag{5.3}$$

E is kept fixed in Eq. (5.1) and equal to the experimental value. The sequence $\lambda_0, \lambda_{\frac{1}{2}}, \lambda_1 \ldots$ defined by

$$\lambda_n = \frac{\langle n \,|\, U^s \,|\, n-1 \rangle}{\langle n \,|\, U^s \,|\, n \rangle}, \quad \lambda_{n+\frac{1}{2}} = \frac{\langle n |\, U^s \,|\, n \rangle}{\langle n+1 \,|\, U^s \,|\, n \rangle} \tag{5.4}$$

where n indicates the function ψ_n given in Eq. (5.3) converges monotonically to the lowest eigenvalue. This is valid if the operators $T - E$ and $-U^s$ are positive definite, and λ is bounded from below. If $-U^s$ is not positive definite, there is a weaker condition on the convergence of the λ_n. This method is very useful whenever the integrations involved in Eq. (5.4) are actually practicable.

A different approximation to the solution of Eq. (5.2) consists in its reduction to an ordinary differential equation by restricting the three spatial degrees of freedom to a single one. One may, for instance, introduce a spherical coordinate reference in a six-dimensional space, with

$$R = \sqrt{\tfrac{1}{3}\left[|\boldsymbol{r}_1 - \boldsymbol{r}_2|^2 + |\boldsymbol{r}_1 - \boldsymbol{r}_3|^2 + |\boldsymbol{r}_2 - \boldsymbol{r}_3|^2\right]} \tag{5.5}$$

as radius[1].

Averaging Eq. (5.2) over the full five-dimensional solid angle with the assumption that ψ^s depends on R only, one obtains

$$\left\{-\frac{\hbar^2}{2M}\frac{1}{R^5}\frac{\partial}{\partial R}\left(R^5 \frac{\partial}{\partial R}\right) + \lambda V_M(R)\right\}\psi^s(R) = E\,\psi^s(R) \tag{5.6}$$

where

$$V_M(R) = \frac{1}{\pi^3}\int d\omega_5 \cdot U^s = \frac{48}{\pi}\int_0^1 U(Rx)\, x^2 \sqrt{1-x^2}\, dx$$

and U is the two-body interaction.

The reduction of Eq. (5.2) to the equivalent two-body problem Eq. (5.6) can be carried out with many other choices of the single variable R which may be taken as an arbitrary symmetric function of the mutual distances of the three particles. Thus, for example, if $R = \frac{1}{2}\{|\boldsymbol{r}_1 - \boldsymbol{r}_2| + |\boldsymbol{r}_1 - \boldsymbol{r}_3| + |\boldsymbol{r}_2 - \boldsymbol{r}_3|\}$ the result of the averaging of Eq. (5.2) in a domain Δ between R and $R + dR$, is[2]

$$\left\{-\frac{15}{14}\frac{\hbar^2}{M}\frac{1}{R^5}\frac{\partial}{\partial R}\left(R^5 \frac{\partial}{\partial R}\right) + \lambda V_F(R)\right\}\psi^s(R) = E\,\psi^s(R) \tag{5.7}$$

where

$$V_F(R) = \frac{\int_\Delta d\tau\, U^s}{\int_\Delta d\tau} = \frac{15}{14}\cdot 24 \cdot \int_0^1 U(Rx)\, x^2\left(1 - x + \frac{1}{6}x^2\right) dx .$$

From the numerous numerical results obtained with the aid of all methods discussed above we may draw the following conclusions.

With a potential strength $\lambda_{3,1}$ chosen so that the central force alone provides the deuteron binding, the predicted energy for the three-body ground state is much larger than the experimental value. The value of $\lambda_{3,1}$ compatible with the presence of the tensor force is too small for insuring an appreciable three-body binding.

---

[1] G. MORPURGO: Nuovo Cim. 9, 461 (1952).

[2] H. FESHBACH and S. I. RUBINOW: Phys. Rev. 98, 188 (1955).

Even if the three-body ground state is predominantly spherical symmetric, we cannot ignore the presence of the tensor interaction. In the triton as well as in the deuteron the tensor force plays an important role in the binding mechanism. This will be the subject of a brief discussion in the next section.

We wish to add a few more remarks about the dependence of the three-body binding energy on the range μ of the two-body interaction.

Measuring all distances in units of μ the SCHRÖDINGER equation (2.1) reads:

$$(\boldsymbol{T}+\Lambda V)\,\Psi=\mu^2 E\cdot\Psi$$

$\Lambda=\mu^2\lambda$ and λ is a two-body potential strength. A small change in Λ brings, as it is well known, a corresponding variation of the eigenvalue $\mu^2 E$ given by

$$\delta(\mu^2 E)=\delta\Lambda\cdot\langle\mu|\,V\,|\mu\rangle \tag{5.8}$$

$|\mu\rangle$ stands for the normalized eigenfunction belonging to the given strength Λ and eigenvalue $\mu^2 E$.

We only consider positive definite operators $-V$. Then from Eq. (5.8) it follows that

$$\left.\begin{aligned}\left(\frac{\partial\Lambda}{\partial\mu^2}\right)_{E=\text{const}}&=\frac{E}{\langle\mu|\,V\,|\mu\rangle}>0,\\ \left(\frac{\partial\Lambda}{\partial E}\right)_{\mu^2=\text{const}}&=\frac{\mu^2}{\langle\mu|\,V\,|\mu\rangle}<0.\end{aligned}\right\} \tag{5.9}$$

This means that on keeping E fixed, Λ is a monotonic increasing function of μ^2, and for a fixed range μ, larger binding energies correspond to increased values of Λ.

For a range zero, $\Lambda(0)$ is obviously independent of the choice of E

$$\Lambda(0)=-\frac{\langle 0|\,T\,|0\rangle}{\langle 0|\,V\,|0\rangle}>0. \tag{5.10}$$

Eqs. (5.9) and (5.10) are valid, under the restrictions imposed, for a two-body as well as for a three-body system. The value of μ^2 compatible with the same two-body potential strength for the deuteron and triton must satisfy the relation

$$\Lambda_d(\mu^2 E_d)=\Lambda_t(\mu^2 E_t).$$

If $\Lambda_d(0)<\Lambda_t(0)$ Eqs. (5.9) and (5.10) imply that for an infinitely small range μ and a two-body strength that insures the deuteron's binding an infinite large triton's binding will result.

If $\Lambda_d(0)>\Lambda_t(0)$ for small μ no binding of three bodies is possible, and in the case $\Lambda_d(0)=\Lambda_t(0)$ the binding energy for the three-body system is arbitrary. No zero range potential[1] is consistent with well-defined energies for both the ground states of H^2 and H^3.

6. The role of the tensor force in the binding. An accurate numerical analysis of the two-body data at low energies has been made[2] with a phenomenological potential of the following form:

$$\left.\begin{aligned}U_{12}(r)&=\lambda_{2s+1,\,2t+1}\left[\frac{e^{-r/r_c}}{r/r_c}+\gamma\,\frac{e^{-r/r_t}}{r/r_t}\,S_{12}(r)\right],\\ S_{12}(r)&=3\,\frac{(\boldsymbol{\sigma}^{(1)}\cdot\boldsymbol{r})\,(\boldsymbol{\sigma}^{(2)}\cdot\boldsymbol{r})}{r^2}-(\boldsymbol{\sigma}^{(1)}\cdot\boldsymbol{\sigma}^{(2)}).\end{aligned}\right\} \tag{6.1}$$

---

[1] L. H. THOMAS: Phys. Rev. **47**, 903 (1935).
[2] H. FESHBACH and J. SCHWINGER: Phys. Rev. **84**, 194 (1951).

For the singlet even parity state in which the tensor force is absent, the two constants $\lambda_{1,3}$ and r_c have the values

$$\lambda_{1,3} = -46.48 \text{ Mev}, \qquad r_c = 1.184 \times 10^{-13} \text{ cm}.$$

The three constants $\lambda_{3,1}$, γ and r_t belonging to the triplet even parity state can be determined using the experimental values of the deuteron binding energy, its electric quadrupole moment and the triplet n-p effective range at zero energy.

The triton binding energy provides a further parameter which can serve to test the potential (6.1) under the assumption of charge independence and absence of two-body odd parity states. A variational calculation[1] has shown that the potential (6.1) is compatible with the experimental value $E_T = -8.48$ Mev of the triton binding energy.

An interesting feature revealed by extensive numerical computations[1] is that E_T is a sensitive function of the range r_t of the tensor force. It is not possible to accept equal ranges for central as well as tensor interaction in Eq. (6.1), because in this case the variational eigenvalue is well below the experimental one. A range $r_t = 1.44\, r_c = 1.70 \times 10^{-13}$ cm is needed for obtaining a variational eigenvalue $E_T = -8.48$ Mev. The singlet constants $\lambda_{3,1}$ and γ have the values

$$\lambda_{3,1} = -46.96 \text{ Mev}, \qquad \gamma = 0.51.$$

They are fixed by the deuteron binding energy and its electric quadrupole moment. The triplet n-p effective range, which is insensitive to variation of r_t, turns out to be 1.74×10^{-13} cm and therefore slightly larger than its experimental value. It is worthy of note that $\lambda_{3,1} - \lambda_{1,3} = -0.48$ Mev. This small value is a justification a posteriori for neglecting in Eq. (5.1) the components ψ', ψ'' of the S state.

The variational trial function which gives the triton binding energy contains a 3.1% ${}^4D_{\frac{1}{2}}$ admixture. This is very reasonable (see Sect. 4). The expectation value of e^2/r, r being the distance between two like particles, turns out to be $E_c = 1.04$ Mev. The COULOMB energy in He^3 is 0.76 Mev (see Sect. 3). The reason for the discrepancy may be attributed to the trial eigenfunction which probably fails to describe correctly the ground state configuration for distances between two nucleons much shorter than their average separation.

7. Hard-core potential and three-body forces. No nuclear potential deduced with the aid of a meson field theory can be accepted as fully trustworthy[2]. This is in particular relevant for distances between two nucleons smaller than the π-meson COMPTON wavelength $\frac{\hbar}{\mu c} = 1.4 \times 10^{-13}$ cm. Yet, some properties of the field theoretical potential have been supported by independent arguments of phenomenological nature. One of these is the necessity of introducing a very strong repulsion at short distances between two nucleons, the so-called hard-core model of nuclear forces[3]. For this model the wave function of two interacting nucleons must vanish for any distance smaller than the hard-core radius.

Many attempts have been made[2] for an interpretation of the low energy two-body data with the static potential of the pseudoscalar meson theory with a pseudoscalar coupling. A meson-nucleon coupling constant $\frac{G^2}{4\pi} \approx 10$ and a

---

[1] R. L. PEASE and H. FESHBACH: Phys. Rev. **88**, 945 (1952). We refer to this paper for older literature on the same subject.

[2] Cf. e.g.: M. M. LÉVY and R. E. MARSHAK: The 1954 International Conference on Nuclear and Meson Physics, Glasgow, Pergamon Press, London and New York, 1955, p. 10.

[3] R. JASTROW: Phys. Rev. **81**, 165 (1951).

hard-core radius $r_c = 0.38\hbar/\mu c$ seem to account satisfactorily for the two-body low energy phenomena[1].

Some types of mesonic hard-core potential without their tensor term, have also been tested by carrying a numerical computation[2,3] of the triton binding energy E_T.

The approximation used for the three-body ground state was analogous to that already described for the phenomenological potential Eq. (5.7) in one case[2] and to the conventional variational approach in the other[3]. The boundary conditions with the hard-core, were fulfilled by choosing wave functions of the form:

$$\psi(r_{12}, r_{13}, r_{23}) = \left\{ \begin{array}{c} 0 \quad \text{for} \quad r_{ij} < r_c \\ \varphi(r_{12} + r_{13} + r_{23})\,(r_{12} - r_c)\,(r_{13} - r_c)\,(r_{23} - r_c) \text{ otherwise} \end{array} \right\} \tag{7.1}$$

r_{ij} being the distance between the pair of nucleons i and j. In full similarity with the case of the phenomenological central potential the numerical analysis allows the following conclusions.

No binding for the three-body system occurs unless a larger strength for the triplet even state is chosen. If the choice is adjusted so that the central potential alone provides the deuteron binding the theoretical eigenvalue E_T is larger than the experimental one. This lets us conjecture that here again, as for the phenomenological potential, the presence of the tensor force may produce the correct binding.

Meson field theories of nuclear interactions predict the existence of many-body forces. The strong meson nucleon coupling allows meson exchanges not only between pairs but also between three or more nucleons at a time. The total interaction energy in a many-body system is no longer the sum of the two-body interaction energies. There are in addition terms which depend on the mutual distances and spin variables of three and more nucleons.

Three-body systems should provide a first basis for learning the real significance of three-body forces. Unfortunately this basis is rather unsafe. We do not possess a well founded two-body interaction, furthermore there are uncertainties in accepting the mesonic three-body force of the same kind as for the two-body forces[4]. One may be consistent by choosing terms of the lowest order in $\lambda = \frac{G^2}{4\pi} \cdot \frac{\mu}{M}$ and $\frac{\mu}{M}$, the ratio of the π-meson to the nucleon mass.

The so-called leading term, of the order λ^3, is repulsive, spin and charge independent[4,5]. It has the form

$$V_3 = 12\lambda^3 \mu \frac{2}{\pi} \frac{K_1[\mu(r_{12} + r_{13} + r_{23})]}{\mu^3 r_{12} r_{13} r_{23}}, \tag{7.2}$$

where K_1 is a Bessel function of imaginary argument.

The expectation value of V_3 calculated with the wave function ψ [Eq. (7.1)] belonging to the two-body central part of the total interaction, has been estimated[3] to diminish the eigenvalue by about 10%. However, the expectation value of the three-body force of order λ^4 which is also spin and charge independent but attractive, is of the same order of magnitude and serves to compensate the repulsive action of V_3. Other three-body potentials of order $\lambda^2 \frac{\mu}{M}$ and $\lambda^3 \frac{\mu}{M}$, bring contributions difficult to be safely estimated because of their non-central character which involves cross terms of the $^2S_{\frac{1}{2}}$ state with the $^4D_{\frac{1}{2}}$ admixture.

---

1 See, however, J. M. Blatt and M. H. Kalos: Phys. Rev. **94**, 762 (1954).
2 H. Feshbach and S. I. Rubinow: Phys. Rev. **98**, 188 (1955).
3 E. M. Gelbard: Phys. Rev. **100**, 1530 (1955).
4 A. Klein: Phys. Rev. **89**, 1158 (1953); **90**, 1101 (1953).
5 S. D. Drell and K. Huang: Phys. Rev. **91**, 1527 (1953).

III. The continuum states of three nucleons.

8. The continuum states. Many measurements of scattering differential cross sections for collisions of neutrons and protons on deuterons have been assembled. They provide the richest and most direct informations on the continuum three-body configurations (Figs. 1 and 2).

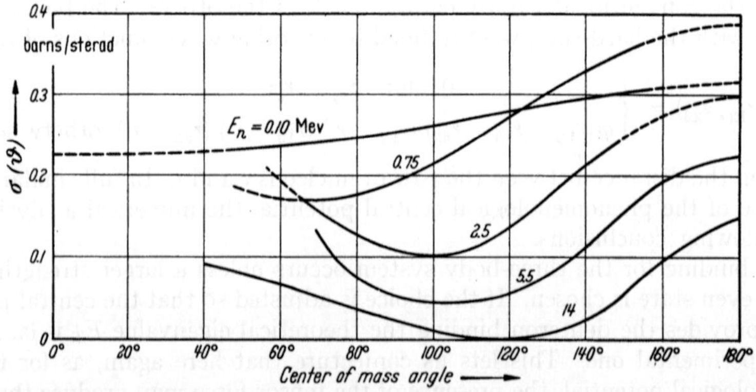

Fig. 1. Some experimental differential cross sections[1] for elastic *n-d* scattering in the center-of-mass system, for energies of the incident neutrons between 0.10 and 14 Mev.

At low energies an interesting feature of the angular distributions is the strong anisotropy, and in particular the presence of a pronounced maximum at 180°. There seems to be symmetry about 90° for energies of approximately 2 Mev. The broad minimum of the angular distribution near 120° is a characteristic behavior at somewhat higher energies which persist even at 340 Mev (see Fig. 4, p. 170). For a theoretical discussion of the angular distributions we refer to Sect. 14 for the low energies and to Sects. 9 to 11 and 13 for the high.

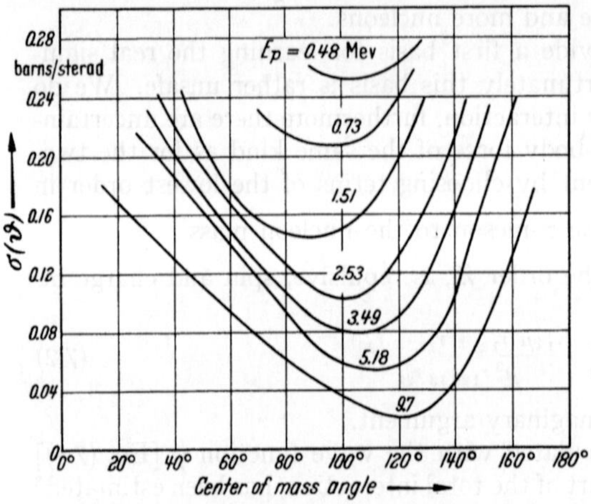

Fig. 2. Some experimental differential cross sections[2] for elastic *p-d* scattering in the center-of-mass system, for energies of the incident protons between 0.48 and 9.7 Mev.

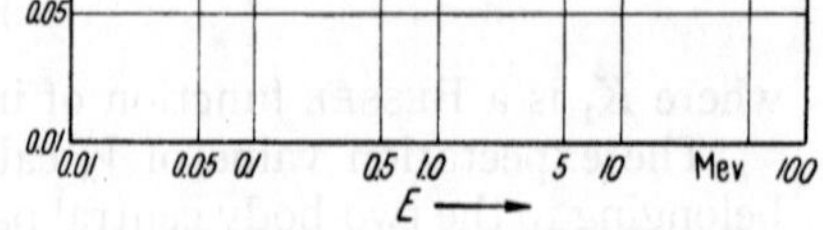

Fig. 3. Experimental total cross section for *n-d* scattering[1] as function of the energy of the incident neutrons. The scales are logaritmic for the energy as well as for the cross section.

---

[1] D. J. HUGHES and J. A. HARVEY: Neutron cross sections. Brookhaven National Laboratory, BNL 325, 1955.

[2] R. J. S. BROWN *et al.*: Phys. Rev. **88**, 253 (1952). — J. C. ALLRED *et al.*: Phys. Rev. **91**, 90 (1953) and references given in this paper.

The total cross section for scattering of neutrons on deuterons as function of the energy is given in Fig. 3.

9. Perturbation theory of the elastic n-d scattering. We consider in this section the case of central interactions for which the total spin S and isotopic spin T are good quantum numbers. The presence of the tensor force and the COULOMB effects will be treated later in Sects. 10 and 17, respectively.

Since the deuteron in its ground state has an isotopic spin zero, the only value of T which comes into play is $\frac{1}{2}$. For S, however, we have the two possibilities, $\frac{3}{2}$ and $\frac{1}{2}$. We will discuss in some detail the quartet case ($S=\frac{3}{2}$) and give the corresponding formulas for the doublet ($S=\frac{1}{2}$), which can be derived in an entirely similar way.

The scattering amplitude in the center of mass system of the incoming nucleon and the deuteron is given by the well-known matrix element:

$$F^{(\frac{3}{2})}(\vartheta) = -\frac{1}{4\pi}\int \left(e^{i\boldsymbol{k}_f\cdot\boldsymbol{q}}\,\varphi(r)\,\chi^s\zeta'\right)^* H_{\text{int}}\left(\psi'\zeta''-\psi''\zeta'\right)\chi^s\cdot d\tau. \tag{9.1}$$

We are using here the same notation for the spins and space variables as adopted in Sects. 1 and 2. It is not necessary to specify the m-quantum numbers since $F(\vartheta)$ does not depend on them. We recognize in Eq. (9.1) the final configuration of the deuteron in its ground state and a free nucleon with the final momentum $\boldsymbol{k}_f$.

The initial eigenstate $(\psi'\zeta''-\psi''\zeta')\cdot\chi^s$ belongs to $T=\frac{1}{2}$ and $S=\frac{3}{2}$, and evolves from the incident wave function:

$$e^{i\boldsymbol{k}_i\cdot\boldsymbol{q}}\,\varphi(r)\,\chi^s\zeta'. \tag{9.2}$$

The deuteron's ground state eigenfunction $\varphi(r)$ must be chosen normalized to one in the $\boldsymbol{r}$-space.

$$\int|\varphi(r)|^2\,d^3r = 1.$$

As $\boldsymbol{q}$ is the coordinate of the outgoing nucleon 1 relative to the center of mass of the nucleons 2 and 3, H_{int} is given by:

$$H_{\text{int}} = \frac{4M}{3\hbar^2}\{U_{12}\,\boldsymbol{O}(12)+U_{13}\,\boldsymbol{O}(13)\}. \tag{9.3}$$

U_{12} is the radial part of the interaction between 1 and 2 and $\boldsymbol{O}(12)$ contains the spin dependence.

Owing to the antisymmetry with respect to permutation of all variables of the body 2 with those of 3 for the initial as well as for the final states, we may substitute in Eq. (9.1) for H_{int} instead of Eq. (9.3) merely the operator $\frac{8M}{3\hbar^2}U_{12}\,\boldsymbol{O}(12)$.

Eliminating the spin variables by using the expectation values

$$\langle\chi^s\zeta'|\,\boldsymbol{O}(12)\,|\chi^s\zeta''\rangle = \frac{\sqrt{3}}{4}(\lambda_{3,3}-\lambda_{3,1}),$$

$$\langle\chi^s\zeta'|\,\boldsymbol{O}(12)\,|\chi^s\zeta'\rangle = \frac{1}{4}(3\lambda_{3,3}+\lambda_{3,1})$$

[see Eqs. (2.4), (1.3) and (1.5)], one obtains

$$\left.\begin{aligned} F^{(\frac{3}{2})}(\vartheta) = &-\frac{1}{3\pi}\frac{M}{\hbar^2}\int e^{-i\boldsymbol{k}_f\cdot\boldsymbol{q}}\,\varphi^*(r)\,U_{12}\Big[\frac{\sqrt{3}}{2}(\lambda_{3,3}-\lambda_{3,1})\,\psi' - \\ &-\frac{1}{2}(3\lambda_{3,3}+\lambda_{3,1})\,\psi''\Big]\,d\tau. \end{aligned}\right\} \tag{9.4}$$

$\lambda_{2s+1,\,2t+1}$ is the two-body potential strength for a spin s and isotopic spin t [Eq. (2.3)].

For the doublet state the corresponding formula reads

$$F^{(\frac{1}{2})}(\vartheta) = -\frac{1}{3\pi}\cdot\frac{M}{\hbar^2}\int e^{-i\boldsymbol{k}_f\cdot\boldsymbol{q}}\,\varphi^*(r)\,U_{12}\left[\Lambda_1\psi^a + \Lambda_2\psi' - \Lambda_3\psi'' - \Lambda_4\psi^s\right]d\tau, \quad (9.5)$$

where the Λ's are twice the matrix elements of $\boldsymbol{O}(12)$ between the initial and final spin states:

$$\left.\begin{aligned}
\Lambda_1 &= 2\langle\chi''\zeta'|\,\boldsymbol{O}(12)\,|\xi^s\rangle = \frac{1}{2}\sqrt{\frac{3}{2}}\,(\lambda_{3,3} - \lambda_{1,1}),\\
\Lambda_2 &= 2\langle\chi''\zeta'|\,\boldsymbol{O}(12)\,|\xi''\rangle = \frac{1}{4}\sqrt{\frac{3}{2}}\,(\lambda_{3,3} + \lambda_{3,1} - 3\,\lambda_{1,3} + \lambda_{1,1}),\\
\Lambda_3 &= 2\langle\chi''\zeta'|\,\boldsymbol{O}(12)\,|\xi'\rangle = \frac{1}{4}\,\frac{1}{\sqrt{2}}\,(3\,\lambda_{3,3} - \lambda_{3,1} + 3\,\lambda_{1,3} + 3\,\lambda_{1,1}),\\
\Lambda_4 &= 2\langle\chi''\zeta'|\,\boldsymbol{O}(12)\,|\xi^a\rangle = -\frac{1}{2}\,\frac{1}{\sqrt{2}}\,(\lambda_{3,1} + 3\,\lambda_{1,3})
\end{aligned}\right\} \quad (9.6)$$

[cf. Eqs. (2.4) and $(1.6)_{s,a,d}$].

The initial doublet state

$$\psi^a\,\xi^s + \psi'\,\xi'' - \psi''\,\xi' - \psi^s\,\xi^a$$

must evolve from the incident wave

$$e^{i\boldsymbol{k}_i\cdot\boldsymbol{q}}\,\varphi(r)\cdot\frac{1}{\sqrt{2}}\,(\xi' - \xi^a). \quad (9.7)$$

This implies that for large values of $\boldsymbol{q}$, the two components ψ^a and ψ' rapidly vanish, whereas

$$-\psi'' \approx \psi^s \approx e^{i\boldsymbol{k}_i\cdot\boldsymbol{q}}\,\frac{\varphi(r)}{\sqrt{2}}. \quad (9.8)$$

The first BORN approximation of the scattering amplitudes is obtained by substituting in Eqs. (9.4) and (9.5) the unperturbed wave functions for the initial states.

For $S=\frac{3}{2}$ the two components ψ' and ψ'' are generated by acting with the spatial symmetry operators T' and T'' [as defined in Eq. (1.1)] upon the initial state $e^{i\boldsymbol{k}_i\cdot\boldsymbol{q}}\,\varphi(r)$.

For $S=\frac{1}{2}$ the component ψ^a is identically zero, whereas ψ^s, ψ', ψ'' are analogously generated operating with T^s, T' and T'' on the state $e^{i\boldsymbol{k}_i\cdot\boldsymbol{q}}\,\frac{\varphi(r)}{\sqrt{2}}$. The wave functions thus obtained have the asymptotic behavior imposed by the conditions expressed by Eqs. (9.2) and (9.8).

The scattering amplitudes clearly become linear combinations of the following three integrals[1]

$$\left.\begin{aligned}
J_1(\vartheta) &= \frac{k}{3\pi}\,\frac{M}{\hbar^2}\int e^{-i\boldsymbol{k}_f\cdot\boldsymbol{q}}\,\varphi^*(r)\,U_{12}\,e^{i\boldsymbol{k}_i\cdot\boldsymbol{q}}\,\varphi(r)\,d\tau,\\
J_2(\vartheta) &= \frac{k}{3\pi}\,\frac{M}{\hbar^2}\int e^{-i\boldsymbol{k}_f\cdot\boldsymbol{q}}\,\varphi^*(r)\,U_{12}(12)\,[e^{i\boldsymbol{k}_i\cdot\boldsymbol{q}}\,\varphi(r)]\,d\tau,\\
J_3(\vartheta) &= \frac{k}{3\pi}\,\frac{M}{\hbar^2}\int e^{-i\boldsymbol{k}_f\cdot\boldsymbol{q}}\,\varphi^*(r)\,U_{12}(13)\,[e^{i\boldsymbol{k}_i\cdot\boldsymbol{q}}\,\varphi(r)]\,d\tau,
\end{aligned}\right\} \quad (9.9)$$

[1] T. Y. WU and J. ASHKIN: Phys. Rev. **73**, 986 (1948).

where (12) and (13) indicate a permutation of the coordinates of nucleon 1 with those of 2 and 3, respectively. From Eqs. (9.4) to (9.6) we deduce

$$k F^{(\frac{3}{2})}(\vartheta) = -\tfrac{1}{2}(3\lambda_{3,3} + \lambda_{3,1}) J_1 + \tfrac{1}{2}(3\lambda_{3,3} - \lambda_{3,1}) J_2 + \lambda_{3,1} J_3, \tag{9.10}$$

$$\left.\begin{aligned} k F^{(\frac{1}{2})}(\vartheta) = &-\tfrac{1}{8}(3\lambda_{3,3} + \lambda_{3,1} + 9\lambda_{1,3} + 3\lambda_{1,1}) J_1 + \\ &+ \tfrac{1}{8}(3\lambda_{3,3} - \lambda_{3,1} - 9\lambda_{1,3} + 3\lambda_{1,1}) J_2 - \tfrac{1}{2}\lambda_{3,1} J_3. \end{aligned}\right\} \tag{9.11}$$

The integrals of Eq. (9.9) have a simple physical interpretation. $J_1(\vartheta)$ corresponds to a collision in which the momentum $\boldsymbol{k}_i - \boldsymbol{k}_f$ is transferred to the whole deuteron. Therefore it gives a large contribution only in the direction of the impinging nucleon and becomes the dominant term at high energies. The integrals J_2 and J_3 are consequences of the PAULI principle and can be interpreted as due to processes by which the incoming neutron exchanges its position with the deuteron's neutron, or to collisions where the incident neutron picks up the deuteron's proton forming a new deuteron. In either case, after the collision the neutron is mostly left behind, and the contribution of J_2 and J_3 occurs mainly in the backward direction.

A slight transformation of Eq. (9.9) brings more clearly in evidence these features[1]. We have indeed

$$\left.\begin{aligned} J_1(\vartheta) &= \frac{k}{3\pi}\frac{M}{\hbar^2}\left(\int e^{-i(\boldsymbol{k}_i-\boldsymbol{k}_f)\cdot\frac{\boldsymbol{y}}{2}}\varphi^2(y)\,d^3y\right)\cdot\left(\int e^{i(\boldsymbol{k}_i-\boldsymbol{k}_f)\cdot\boldsymbol{x}}\,U(x)\,d^3x\right), \\ J_2(\vartheta) &= \frac{k}{3\pi}\frac{M}{\hbar^2}\int e^{i(\boldsymbol{k}_i-\boldsymbol{k}_f)\cdot\frac{\boldsymbol{y}}{2}}\varphi\left(\left|\boldsymbol{y}+\frac{\boldsymbol{x}}{2}\right|\right)\varphi\left(\left|\boldsymbol{y}-\frac{\boldsymbol{x}}{2}\right|\right)e^{-\frac{3}{4}i(\boldsymbol{k}_i+\boldsymbol{k}_f)\cdot\boldsymbol{x}}\,U(x)\,d^3x\,d^3y, \\ J_3(\vartheta) &= -\frac{k}{3\pi}\frac{M}{\hbar^2}\left[k_d^2 + \left(\boldsymbol{k}_f + \frac{\boldsymbol{k}_i}{2}\right)^2\right]\cdot\left(\int e^{-i\left(\boldsymbol{k}_f+\frac{\boldsymbol{k}_i}{2}\right)\cdot\boldsymbol{x}}\varphi(x)\,d^3x\right)\cdot \\ &\qquad\cdot\left(\int e^{i\left(\boldsymbol{k}_i+\frac{\boldsymbol{k}_f}{2}\right)\cdot\boldsymbol{y}}\varphi(y)\,d^3y\right). \end{aligned}\right\} \tag{9.12}$$

The first two expressions are obtained by simply renaming the integration variables, and for J_3 we have made use of the deuteron's equation of motion:

$$(\Delta_x - k_d^2)\,\varphi(x) = \lambda_{3,1}\,U(x)\,\varphi(x). \tag{9.13}$$

$\varphi(x)$ is chosen to be real.

For the double integral $J_2(\vartheta)$ a further simplification occurs at high energies and small ϑ. In this case we may substitute $\varphi^2(y)$ for $\varphi\left(\left|\boldsymbol{y}+\frac{\boldsymbol{x}}{2}\right|\right)\varphi\left(\left|\boldsymbol{y}-\frac{\boldsymbol{x}}{2}\right|\right)$ because of the short range of the potential $U(x)$.

$$J_2(\vartheta) = \frac{k}{3\pi}\frac{M}{\hbar^2}\left(\int e^{i(\boldsymbol{k}_i-\boldsymbol{k}_f)\cdot\frac{\boldsymbol{y}}{2}}\varphi^2(y)\,d^3y\right)\left(\int e^{-\frac{3}{4}i(\boldsymbol{k}_i+\boldsymbol{k}_f)\cdot\boldsymbol{x}}\,U(x)\,d^3x\right).$$

A direct connection with the scattering amplitudes in a two-body collision is now evident.

Introducing the form factor for the deuteron ground state

$$S^{\frac{1}{2}}(k,\vartheta) = \int e^{i(\boldsymbol{k}_i-\boldsymbol{k}_f)\cdot\frac{\boldsymbol{y}}{2}}\varphi^2(y)\,d^3y \tag{9.14}$$

and the FOURIER transform of the potential U:

$$V(p) = \int e^{i\boldsymbol{p}\cdot\boldsymbol{x}}\,U(x)\,d^3x \tag{9.15}$$

[1] G. F. CHEW: Phys. Rev. **74**, 809 (1948). — R. S. CHRISTIAN and J. L. GAMMEL: Phys. Rev. **91**, 100 (1953).

one obtains for large k^2 and $\vartheta \ll 180°$)

$$\left.\begin{aligned} J_1(\vartheta) &= \frac{k}{3\pi}\frac{M}{\hbar^2} S^{\frac{1}{2}}(k,\vartheta)\, V(|\boldsymbol{k}_i - \boldsymbol{k}_f|)\,, \\ J_2(\vartheta) &= \frac{k}{3\pi}\frac{M}{\hbar^2} S^{\frac{1}{2}}(k,\vartheta)\, V(\tfrac{3}{4}\,|\boldsymbol{k}_i + \boldsymbol{k}_f|)\,. \end{aligned}\right\} \tag{9.16}$$

$J_3(\vartheta)$ cannot be related in a similar way to a two-body elastic scattering. It represents a typical three-body phenomenon, and becomes negligible in the scattering at small angles and high energies.

The two-body scattering amplitude corresponding to a spin s and isotopic spin t, is given by

$$f^{s,t}(\omega) = -\frac{1}{4\pi}\frac{M}{\hbar^2}\lambda_{2s+1,\,2t+1}\{V(|\boldsymbol{p}_i - \boldsymbol{p}_f|) + (-1)^{s+t}\, V(|\boldsymbol{p}_i + \boldsymbol{p}_f|)\}, \tag{9.17}$$

where ω is the center-of-mass scattering angle. From Eqs. (9.10), (9.11), (9.16) and (9.17) we deduce the following simple relationship between the three-body and two-body scattering amplitudes[1]

$$\left.\begin{aligned} F^{\frac{3}{2}}(k,\vartheta) &= \tfrac{4}{3} S^{\frac{1}{2}}(k,\vartheta)\,\{f^{\text{tripl}}_{n,p}(p,\omega) + f^{\text{tripl}}_{n,n}(p,\omega)\}\,, \\ F^{\frac{1}{2}}(k,\vartheta) &= \tfrac{4}{3} S^{\frac{1}{2}}(k,\vartheta)\,\{\tfrac{1}{4}\,(f^{\text{tripl}}_{n,p}(p,\omega) + f^{\text{tripl}}_{n,n}(p,\omega)) + \\ &\qquad + \tfrac{3}{4}\,(f^{\text{sing}}_{n,p}(p,\omega) + f^{\text{sing}}_{n,n}(p,\omega))\}. \end{aligned}\right\} \tag{9.18}$$

Since we have neglected the contribution of J_3 and taken the approximate value [Eq. (9.16)] for J_2, the formulas above are valid at high energies and small angles.

The symbols used in (9.18) have an obvious meaning. $f^{\text{tripl}}_{n,p}$ indicates, for instance, the triplet neutron-proton scattering amplitude.

The angles ϑ and ω and the momenta $\boldsymbol{k}$ and $\boldsymbol{p}$ are related by the two equations:

$$\left.\begin{aligned} \tan\frac{\omega}{2} &= \frac{4}{3}\tan\frac{\vartheta}{2} \\ p^2 &= \frac{9}{16}\,k^2\left(1 + \frac{7}{9}\sin^2\vartheta\right) \end{aligned}\right\} \tag{9.19}$$

which follow from the two conditions

$$\left.\begin{aligned} \boldsymbol{k}_i - \boldsymbol{k}_f &= \boldsymbol{p}_i - \boldsymbol{p}_f\,, \\ \tfrac{3}{4}\,(\boldsymbol{k}_i + \boldsymbol{k}_f) &= \boldsymbol{p}_i + \boldsymbol{p}_f \end{aligned}\right\} \tag{9.20}$$

[cf. Eqs. (9.16) and (9.17)].

We now use the well-known theorem which relates the integrated cross section with the imaginary part of the scattering amplitude in the forward direction

$$\sigma = \frac{4\pi}{k}\,\mathrm{Im}\,F(0)\,. \tag{9.21}$$

This permits us to deduce from Eqs. (9.18) and (9.19)

$$\begin{aligned} \sigma^{(\frac{3}{2})}_{n,d} &= \sigma^{\text{tripl}}_{n,p} + \sigma^{\text{tripl}}_{n,n}\,, \\ \sigma^{(\frac{1}{2})}_{n,d} &= \tfrac{1}{4}\,(\sigma^{\text{tripl}}_{n,p} + \sigma^{\text{tripl}}_{n,n}) + \tfrac{3}{4}\,(\sigma^{\text{sing}}_{n,p} + \sigma^{\text{sing}}_{n,n})\,. \end{aligned}$$

With an unpolarized incident beam:

$$\sigma_{n,d}(k) = \tfrac{2}{3}\,\sigma^{(\frac{3}{2})}(k) + \tfrac{1}{3}\,\sigma^{(\frac{1}{2})}(k) = \sigma_{n,p}(p) + \sigma_{n,n}(p)\,. \tag{9.22}$$

[1] P. B. DAITCH and J. B. FRENCH: Phys. Rev. **85**, 695 (1952).

This is equivalent to saying as $p = \frac{3}{4} k$ that the n-d cross section is equal to the sum of the n-p and n-n cross sections taken at the same energy in the laboratory system.

For the differential cross section we may only approximately deduce from Eq. (9.18)

$$\frac{d\sigma_{n,d}}{d\Omega_\vartheta}\bigg|^{(k,\vartheta)} \approx \frac{16}{9} S(k,\vartheta)\left\{\frac{d\sigma_{n,p}}{d\Omega_\omega}\bigg|^{(p,\omega)} + \frac{d\sigma_{n,n}}{d\Omega_\omega}\bigg|^{(p,\omega)}\right\}. \tag{9.23}$$

This implies an approximate additivity of the two-body differential cross sections at the same energy and small angles in the laboratory system.

Since $S(k,\vartheta)$ is fairly well known, the importance of Eq. (9.23) lies in the possibility of obtaining direct information on the elastic neutron-neutron cross section from the experimental knowledge of $\sigma_{n,d}(k,\vartheta)$ and $\sigma_{n,p}(p,\omega)$.

In addition, one can check the domain of validity of Eq. (9.23) by measuring the proton-deuteron cross section which is expressed by the analogous formula:

$$\frac{d\sigma_{p,d}}{d\Omega_\vartheta}\bigg|^{(k,\vartheta)} \approx \frac{16}{9} S(k,\vartheta)\left\{\frac{d\sigma_{n,p}}{d\Omega_\omega}\bigg|^{(p,\omega)} + \frac{d\sigma_{p,p}}{d\Omega_\omega}\bigg|^{(p,\omega)}\right\}. \tag{9.24}$$

This is because we can independently measure $\sigma_{p,p}$ and $\sigma_{n,p}$. As we shall see later on (Sect. 13) the two cross sections $\sigma_{p,d}$ and $\sigma_{n,d}$ provide a strong argument in favor of the charge symmetry of nuclear forces at high energies.

Eqs. (9.22) to (9.24) have been established here with the aid of the first Born approximation. We have, however, good reasons to reckon on a less restrictive validity of them. Indeed, when the two-body scattering length becomes at high energies small in comparison with the deuteron size, we should expect the additivity of the cross sections for each struck nucleon. This should occur even if the two-body interaction is not so weak as to justify the perturbation assumption, owing to the relatively large distance between the two nucleons in the deuteron.

In Sect. 14 we will discuss some other approximations and see how the reduction of the three-body to the two-body cross sections can be accomplished.

There are several numerical computations of the differential cross section for particular shapes of the static potential $U(x)$ and the deuteron's eigenfunction.

The angular distributions so obtained[1] show a large peak forward due to the presence of $J_1^2(\vartheta)$ and a weaker maximum backwards due to $J_2^2(\vartheta)$ and $J_3^2(\vartheta)$.

The isotopic spin dependence of the potential strength determines the relative weights with which the three integrals enter in the differential cross section. It is noteworthy that an isotopic dependence of the type $(\boldsymbol{\tau}^{(1)}\cdot\boldsymbol{\tau}^{(2)})$ for the two-body force eliminates the forward maximum. Indeed, in this case the coefficients of J_1 in Eqs. (9.10) and (9.11) $3\lambda_{3,3}+\lambda_{3,1}$, $3\lambda_{1,3}+\lambda_{1,1}$ are both zero.

The experiments show always a maximum forward (see Figs. 1, 2 and 4) and this would seem an argument against the $(\boldsymbol{\tau}^{(1)}\cdot\boldsymbol{\tau}^{(2)})$ isotopic spin dependence of the central interaction. Yet, we have no right to draw conclusions of this kind because we are still ignoring the tensor force which brings important contributions in the forward direction (see next section).

These qualitative arguments are certainly valid beyond the domain of strict applicability of a perturbation theory because they are based upon the values that the weights of the integrals J_k in Eqs. (9.10) and (9.11) assume for different isotopic spin dependence.

---

[1] T. Y. Wu and J. Ashkin: Phys. Rev. **73**, 986 (1948). — G. C. Chew: Phys. Rev. **74**, 809 (1948). — M. Verde: Helv. phys. Acta **22**, 339 (1949).

10. Effects of the tensor force on the scattering. The scattering amplitude which rises from the tensor interaction can be evaluated using the procedure of the preceding section.

The differential cross section calculated in the presence of the tensor forces only, must be simply added to that due to the central interaction. There are indeed no interference terms between the central and tensor scattering amplitudes. This is because the central interaction conserves the magnetic quantum numbers and the interference terms in the differential cross section become proportional to the trace of the tensor force in spin space, which is zero.

The operator of the tensor force between nucleons 1 and 2 is

$$\gamma\, U_{12}\, S_{12} = \gamma\, U(r_{12}) \cdot \left\{ \frac{3\,(\boldsymbol{\sigma}^{(1)} \cdot \boldsymbol{r}_{12})\,(\boldsymbol{\sigma}^{(2)} \cdot \boldsymbol{r}_{12})}{r_{12}^2} - (\boldsymbol{\sigma}^{(1)} \cdot \boldsymbol{\sigma}^{(2)}) \right\} \tag{10.1}$$

where $\boldsymbol{r}_{12} = \boldsymbol{r}_1 - \boldsymbol{r}_2$. We assume the isotopic spin as a good quantum number. The potential strength γ has then two possible values γ_3 and γ_1 corresponding to an isotopic spin $t=1$ or $t=0$. The operator S_{12} of Eq. (10.1) may be written as the scalar product of a tensor operator $X_\mu^{(2)}$ of rank two operating in the spin space and the spherical harmonic of the second degree:

$$S_{12} = \sum_{\mu=-2}^{+2} (-1)^\mu\, P_\mu^{(2)}(\boldsymbol{r}_{12}) \cdot X_{-\mu}^{(2)}(\boldsymbol{\sigma}^{(1)}, \boldsymbol{\sigma}^{(2)}). \tag{10.2}$$

If the $P_\mu^{(2)}(\boldsymbol{r}_{12})$ are chosen normalized so that $P_0^{(2)}$ is the Legendre polynomial of the second degree, the $\mu=0$ component of $X^{(2)}$ is

$$X_0^{(2)} = 3\,\sigma_z^{(1)} \cdot \sigma_z^{(2)} - \boldsymbol{\sigma}^{(1)} \cdot \boldsymbol{\sigma}^{(2)}.$$

The total spin is no longer a good quantum number as the interaction Eq. (10.2) induces doublet-quartet transitions. There are, however, no doublet-doublet transitions since the operator $X^{(2)}$ is of rank 2.

The dependence of the scattering amplitude for a transition S, S' on the spin m-quantum numbers is given by the following formula:

$$F^{S'M',SM}(\vartheta, \varphi) = \sum_{\mu=-2}^{+2} C^{2,\mu}_{S',-M';S,M}\, A_\mu^{S',S}(\vartheta, \varphi) \tag{10.3}$$

where $C^{2,\mu}_{S',-M';S,M}$ are CLEBSCH-GORDAN coefficients. The differential cross section for a S, S' transition with an unpolarized beam is then:

$$d\sigma^{S',S} = \frac{1}{2S+1} \sum_{M,M'} |F^{S'M',SM}|^2 = \frac{1}{2S+1} \sum_{\mu=-2}^{+2} |A_\mu^{S',S}|^2$$

and the total contribution to the cross section becomes:

$$d\sigma_{\text{tensor}}(\vartheta) = \tfrac{2}{3}\left[d\sigma^{\frac{3}{2},\frac{3}{2}} + d\sigma^{\frac{1}{2},\frac{3}{2}}\right] + \tfrac{1}{3}\, d\sigma^{\frac{3}{2},\frac{1}{2}}.$$

As in the case of the central interaction the quantities $A_\mu^{S',S}$ [Eq. (10.3)] are expressible as linear combinations of three types of integrals.

$$\left.\begin{aligned} k\, A_\mu^{\frac{1}{2},\frac{3}{2}} &= (3\gamma_3 + \gamma_1)\, T_\mu^{(1)} - (3\gamma_3 - \gamma_1)\, T_\mu^{(2)} - 2\gamma_1 T_\mu^{(3)}, \\ A_\mu^{\frac{3}{2},\frac{3}{2}} &= 2 A_\mu^{\frac{1}{2},\frac{3}{2}}, \\ A_\mu^{\frac{3}{2},\frac{1}{2}} &= A_\mu^{\frac{1}{2},\frac{3}{2}} + 6\,\frac{\gamma_1}{k}\, T_\mu^{(3)}. \end{aligned}\right\} \tag{10.4}$$

The integrals $T_\mu^{(n)}$ are defined by inserting in Eq. (9.9) $U_{12} Y_\mu^{(2)}(\boldsymbol{r}_{12})$ instead of U_{12}. We neglect the small D admixture in the deuteron's ground state.

The same kind of transformation performed on the integrals of Eq. (9.9) to deduce their form of Eq. (9.12) can be carried out with the result:

$$\left.\begin{aligned} T_\mu^{(1)} &= k\, S^{\frac{1}{2}}(k,\vartheta) \sqrt{\frac{5}{4\pi}}\, P_\mu^{(2)}(\omega_1)\, \frac{4M}{3\hbar^2} \int_0^\infty j_2(\omega_1 x)\, x^2\, U(x)\, dx\,, \\ T_\mu^{(2)} &= k\, S^{\frac{1}{2}}(k,\vartheta) \sqrt{\frac{5}{4\pi}}\, P_\mu^{(2)}(\omega_2)\, \frac{4M}{3\hbar^2} \int_0^\infty j_2(\omega_2 x)\, x^2\, U(x)\, dx \end{aligned}\right\} \tag{10.5}$$

where

$$\omega_1 = \boldsymbol{k}_i - \boldsymbol{k}_f\,,$$

$$\omega_2 = \frac{3}{4}(\boldsymbol{k}_i + \boldsymbol{k}_f), \qquad j_2(\omega x) = \sqrt{\frac{\pi}{2\omega x}}\, J_{\frac{5}{2}}(\omega x)\,.$$

The value of $T_\mu^{(2)}$ is approximated as for the central case and its validity is restricted to angles $\vartheta \ll 180°$. $S^{\frac{1}{2}}(k,\vartheta)$ is the deuteron form factor as defined in Eq. (9.14).

The third type of integrals $T_\mu^{(3)}$ can also be written as product of two factors, yet as with the central forces no deuteron's form factor appears. Its importances should be negligible at high energies.

The two-body scattering amplitude due to the tensor force and for a value t of the isotopic spin, is given by:

$$\left.\begin{aligned} f_{M',M}^{t} &= -\frac{\sqrt{6}}{2\pi}\frac{M}{\hbar^2}\,\gamma_{2t+1} \sum_{\mu=-2}^{+2} C_{1,-M';1,M}^{2,\mu} \left(T_\mu(\boldsymbol{p}_i - \boldsymbol{p}_f) + (-1)^t\, T_\mu(\boldsymbol{p}_i + \boldsymbol{p}_f)\right) \\ &\text{where} \\ T_\mu(\boldsymbol{p}) &= k\,\frac{M}{\hbar^2} \sqrt{\frac{5}{4\pi}}\, P_\mu^{(2)}(\boldsymbol{p}) \int_0^\infty j_2(p\,x)\, x^2\, U(x)\, dx\,. \end{aligned}\right\} \tag{10.6}$$

Therefore, remembering Eqs. (10.4) and (10.5), we can establish a relationship between the three-body scattering amplitudes and the two-body amplitudes. This is in full analogy with the previous case of merely central forces. We may in particular reach the same conclusions as before in connection with the approximate Eq. (9.23).

Numerical calculations of the cross sections have been done corresponding to only a few choices for the shape and the isotopic spin dependence of the tensor interaction[1].

The comparison of the theoretical curves with the experimental ones has not been brought so far as to give us a clear discrimination in favor of a particular potential. For an isotopic spin dependence of the type $(\boldsymbol{\tau}^{(1)} \cdot \boldsymbol{\tau}^{(2)})$ the integral $T_\mu^{(1)}$ does not appear in the scattering amplitudes [Eq. (10.4)]. The presence of $T_\mu^{(2)}$, however, is sufficient to bring an appreciable contribution in the forward direction. With this type of interaction the tensor force alone would be responsible for the scattering at small angles and high energies.

11. Impulse approximation and diffraction theory. *α) Impulse approximation.* An answer to the problem of reducing three-body to two-body scattering amplitudes has been attempted with the so-called impulse approximation[2].

[1] B. H. Bransden: Proc. Roy. Soc. Lond., Ser. A **209**, 380 (1950). — Proc. Phys. Soc Lond. A **65**, 972 (1952). — H. Horie *et al.*: Progr. Theor. Phys. **8**, 341 (1952).

[2] G. F. Chew: Phys. Rev. **80**, 196 (1950). — G. F. Chew and G. C. Wick: Phys. Rev. **85**, 636 (1951). — G. F. Chew and M. L. Goldberger: Phys. Rev. **87**, 778 (1952).

The three-body continuum wave function which evolves from the initial state:

$$e^{i\boldsymbol{k}_i\cdot[-\boldsymbol{r}_1+\frac{1}{2}(\boldsymbol{r}_2+\boldsymbol{r}_3)]}\,\varphi(|\boldsymbol{r}_2-\boldsymbol{r}_3|)$$

is expanded in terms of two-body continuum eigenstates $\psi_{\boldsymbol{k}_i,\boldsymbol{k}}(\boldsymbol{r}_2-\boldsymbol{r}_1)$

$$\Psi(1,2,3)=\int d^3k\,g(\boldsymbol{k})\,\psi_{\boldsymbol{k}_i,\boldsymbol{k}}(\boldsymbol{r}_2-\boldsymbol{r}_1)\,e^{i\left[\left(\boldsymbol{k}-\frac{\boldsymbol{k}_i}{2}\right)\cdot\left(\frac{\boldsymbol{r}_1+\boldsymbol{r}_2}{2}-\boldsymbol{r}_3\right)\right]} \tag{11.1}$$

The normalization of $\psi_{\boldsymbol{k}_i,\boldsymbol{k}}(\boldsymbol{x})$ is chosen so that for large $\boldsymbol{x}$, $\psi_{\boldsymbol{k}_i,\boldsymbol{k}}(x)\approx e^{i(\frac{3}{4}\boldsymbol{k}_i+\frac{1}{2}\boldsymbol{k})\cdot\boldsymbol{x}}$ and $g(\boldsymbol{k})$ is assumed to be the momentum distribution in the deuteron ground state

$$\int d^3k\,g(k)\,e^{i(\boldsymbol{k}\cdot\boldsymbol{x})}=\varphi(x)\,. \tag{11.2}$$

The assumptions, Eqs. (11.1) and (11.2), have a simple underlying physical interpretation. During the collision of two bodies the presence of the third body is ignored, so that the incident particle collides with a free nucleon whose momentum distribution corresponds to the actual bound state.

These conditions are certainly fulfilled with short range interactions if the distance between the two struck nucleons can be retained large in comparison with the two-body scattering length at high energies.

The presence of spin and the requirements of the PAULI principle do not present difficulties. It is indeed possible to follow closely the procedure of the perturbation approximation. We have merely to substitute the wave function, Eq. (11.1), instead of the unperturbed wave function $\psi=e^{i(\boldsymbol{k}_i\cdot\boldsymbol{q})}\,\varphi(r)$ in all formulas of Sect. 9. There is again the possibility of factorizing the integrals $J_n(\vartheta)$, [Eq. (9.9)]. J_1, for instance, becomes:

$$J_1=\frac{8\pi^2}{3}\cdot\frac{kM}{\hbar^2}\int d^3k\,g^*\left(\boldsymbol{k}+\frac{1}{2}\left[\boldsymbol{k}_i-\boldsymbol{k}_f\right]\right)\cdot g(\boldsymbol{k})\times$$
$$\times\int e^{-i(\boldsymbol{k}_f\cdot\boldsymbol{x})}\,U(x)\,e^{i\left(\frac{\boldsymbol{k}_i}{2}-\boldsymbol{k}\right)\cdot\frac{\boldsymbol{x}}{2}}\,\psi_{\boldsymbol{k}_i,\boldsymbol{k}}(\boldsymbol{x})\,d^3x\,.$$

For incoming particles of high energy and at small angles we may further simplify:

$$J_1\approx\frac{k}{3\pi}\,\frac{M}{\hbar^2}\left(\int d^3k\,g^*\left(\boldsymbol{k}+\frac{1}{2}(\boldsymbol{k}_i+\boldsymbol{k}_f)\right)g(\boldsymbol{k})\right)\cdot\left(\int e^{-i(\boldsymbol{k}_f\cdot\boldsymbol{x})}\,U(x)\,e^{i(\boldsymbol{k}_i\cdot\boldsymbol{x})/4}\,\psi_{\boldsymbol{k}_i,0}(\boldsymbol{x})\,d^3x\right).$$

Remembering Eq. (11.2)

$$J_1\approx\frac{k}{3\pi}\,\frac{M}{\hbar^2}\,S^{\frac{1}{2}}(k,\vartheta)\cdot\int e^{-i(\boldsymbol{k}_f\cdot\boldsymbol{x})}\,U(x)\,e^{i(\boldsymbol{k}_i\cdot\boldsymbol{x})/4}\,\psi_{\boldsymbol{k}_i,0}(\boldsymbol{x})\,d^3x$$

where $S^{\frac{1}{2}}(k,\vartheta)$ is the deuteron's form factor, Eq. (9.14).

The formulas given in Eq. (9.18) maintain their validity if we consider that the two-body scattering amplitudes have now, as in the formula above, their exact value.

The assumptions, Eqs. (11.1) and (11.2), of the impulse approximation lead as the first BORN approximation of Sect. 9 to the conclusion that the three-body scattering amplitudes can be expressed as linear combination of two-body amplitudes. This yields the additivity of the integrated cross section as in the perturbation theory.

We will see later in this section that the quantities which are really additive at high energies are the phase shifts [Eq. (11.6)]. The scattering amplitudes as consequence cannot be simply added [cf. e.g. Eq. (11.8)], with the exception of weak interactions where the perturbation calculations are reliable.

The impulse approximation can be considered valid for the prediction of the approximate additivity of the differential cross section for not too large angles [Eq. (9.23)]. Yet terms in the differential cross section which contain the scattering amplitudes of both struck bodies are not accurate unless the interaction is weak.

An example which illustrates this situation is the scattering by two fixed point sources[1]. This problem can be solved exactly if one assumes that the two sources scatter waves of a fixed angular momentum, e.g. S-waves. The impulse approximation properly predicts the additivity of the two-body cross sections for large distances between the two scattering centers. However, the answer for the interference correction is the same as that given by the first BORN approximation.

β) Diffraction theory. The scattering amplitude in a two-body collision and spherically symmetric interaction

$$f(\boldsymbol{k}_i, \boldsymbol{k}_f) = \frac{1}{i k} \sum_l \left(l + \frac{1}{2}\right) (e^{2 i \delta_l} - 1) P_l(\cos\vartheta) \tag{11.3}$$

can be approximated at small angles ϑ and high energies k^2 by the following integral

$$f(\boldsymbol{k}_i, \boldsymbol{k}_f) = \frac{k}{i} \int_0^\infty b\, d b\, (e^{2 i \delta(b)} - 1) J_0(k b \vartheta). \tag{11.4}$$

This is a consequence of the asymptotic formula[2] $P_l(\cos\vartheta) \approx J_0[(l+\frac{1}{2})\vartheta]$ valid for small ϑ. l is assumed to be a continuous variable and the summation is substituted by an integration because of the large number of terms which at high energies occur in Eq. (11.3).

An equivalent form of Eq. (11.3) valid under the same assumption of small angles is:

$$f(\boldsymbol{k}_i, \boldsymbol{k}_f) = \frac{k}{\pi} \int_0^{2\pi} \int_0^\infty d\varphi\, b\, d b\, e^{i(\boldsymbol{k}_i - \boldsymbol{k}_f)\cdot \boldsymbol{b}}\, t(b) \tag{11.5}$$

where the integration is over the whole plane of $\boldsymbol{b}$ orthogonal to $\boldsymbol{k}_i + \boldsymbol{k}_f$. $t(b)$ is a short notation for the scattering amplitude $\frac{1}{2i}(e^{2 i \delta} - 1)$ which is a function of the impact parameter b.

If the kinetic energy of the incoming particle is larger than the potential $V(r)$ everywhere, the phase shift $\delta(b)$ can be approximated by the formula[3]

$$\delta(b) = \frac{1}{\hbar v} \int_{-\infty}^{+\infty} V\left(\sqrt{b^2 + z^2}\right) d z. \tag{11.6}$$

This has a simple physical interpretation. The incoming particles with a velocity v and impact parameter b follows a straight line experiencing an elementary action $\frac{1}{v} V\, d z$ in each path element $d z$.

The validity of Eq. (11.5) may be retained for many-body scattering[4]. In the particular case of deuterons the elastic scattering amplitude becomes

$$F(\boldsymbol{k}_i, \boldsymbol{k}_f) = \frac{k}{\pi} \int d^2 b\, d^3 r\, e^{i(\boldsymbol{k}_i - \boldsymbol{k}_f)\cdot \boldsymbol{b}} . t(\boldsymbol{b}, \boldsymbol{s}_1, \boldsymbol{s}_2)\, |\varphi(r)|^2. \tag{11.7}$$

---

[1] K. A. BRUECKNER: Phys. Rev. **89**, 834 (1953).

[2] Cf. e.g., W. MAGNUS and F. OBERHETTINGER: Formeln und Sätze. Berlin: Springer 1948.

[3] G. MOLIÈRE: Z. Naturforsch. **2**, 133 (1947).

[4] R. J. GLAUBER: Phys. Rev. **100**, 242 (1955).

$|\varphi(r)|^2$ is the probability of finding the two struck particles a distance r apart, $\boldsymbol{s}_1$ and $\boldsymbol{s}_2$ are the projections on the plane of $\boldsymbol{b}$ orthogonal to $\boldsymbol{k}_i+\boldsymbol{k}_f$ of the coordinates $\boldsymbol{r}_1$ and $\boldsymbol{r}_2$ of the two nucleons in the deuteron. Because of the additivity of the phase shifts [see Eq. (11.6)] we may write for the total t

$$t(\boldsymbol{b},\boldsymbol{s}_1,\boldsymbol{s}_2)=t_1\left(\boldsymbol{b}-\frac{\boldsymbol{s}}{2}\right)+t_2\left(\boldsymbol{b}+\frac{\boldsymbol{s}}{2}\right)+2i\,t_1\left(\boldsymbol{b}-\frac{\boldsymbol{s}}{2}\right)t_2\left(\boldsymbol{b}+\frac{\boldsymbol{s}}{2}\right) \tag{11.8}$$

where $\boldsymbol{s}=\boldsymbol{s}_1-\boldsymbol{s}_2$, and the origin is taken at the deuteron's center of mass. Substituting Eq. (11.8) in Eq. (11.7) and considering the forward scattering only, we have:

$$F(0)=f_1(0)+f_2(0)+\frac{2ik}{\pi}\int d^2b\,d^3r\,t_1\left(\boldsymbol{b}-\frac{\boldsymbol{s}}{2}\right)t_2\left(\boldsymbol{b}+\frac{\boldsymbol{s}}{2}\right)e^{i(\boldsymbol{k}_i-\boldsymbol{k}_f)\cdot\boldsymbol{b}}\,|\varphi|^2.$$

Using Eq. (9.21) we obtain for the deuteron cross section

$$\sigma_{n,d}=\sigma_1+\sigma_2+8\,\mathrm{Re}\int d^2b\,d^3r\,t_1\left(\boldsymbol{b}-\frac{\boldsymbol{s}}{2}\right)t_2\left(\boldsymbol{b}+\frac{\boldsymbol{s}}{2}\right)|\varphi(r)|^2. \tag{11.9}$$

The third term in this formula represents the interference effects due to the presence of the two scattering centers.

Since $t(b)$ is a function which rapidly vanishes for impact parameters outside the range of interaction, the interference term is appreciable only if the interaction regions of the two nucleons overlap. We can more clearly formulate this qualitative argument transforming further the integral in Eq. (11.9). Introducing cylindrical coordinates for $\boldsymbol{r}$ and operating a simple variable transformation we have

$$\left.\begin{aligned}&\int d^2b\,d^3r\,t_1\left(\boldsymbol{b}-\frac{\boldsymbol{s}}{2}\right)t_2\left(\boldsymbol{b}+\frac{\boldsymbol{s}}{2}\right)|\varphi(r)|^2\\&\quad=2\int d^2x\,d^2y\,t_1(\boldsymbol{x})\,t_2(\boldsymbol{y})\int_0^\infty dz\,\left|\varphi\left(\sqrt{z^2+(\boldsymbol{y}-\boldsymbol{x})^2}\right)\right|^2\approx\\&\quad\approx 2\int d^2x\,d^2y\,t_1(\boldsymbol{x})\,t_2(\boldsymbol{y})\int_0^\infty dz\,|\varphi(z)|^2.\end{aligned}\right\} \tag{11.10}$$

With $\int|\varphi(z)|^2\,dz=\frac{1}{4\pi}\left\langle\frac{1}{r^2}\right\rangle_d$. Eq. (11.9) becomes

$$\sigma_{n,d}=\sigma_1+\sigma_2+\frac{4\pi}{k^2}\,\mathrm{Re}\,[f_1(0)\,f_2(0)]\cdot\left\langle\frac{1}{r^2}\right\rangle_d. \tag{11.11}$$

This expression is valid under the restriction made in Eq. (11.10), i.e. the two scattering regions are only weakly overlapping.

Any prediction of the importance of the interference term requires the knowledge of the complex two-body scattering amplitude in the forward direction. Its numerical value decreases with increasing energy and separation of the two struck nucleons.

In the analysis made above, nuclear spin and the PAULI principle were not considered. It is straightforward to include the influence of spin, but the role of the exclusion principle is a more difficult task to be attempted.

Since the expression (11.11) can be trusted in the forward direction only, we are not able to make any prediction of the differential cross sections. Moreover, the incident energy must be high enough to justify the approximation (11.5) for the two-body scattering amplitudes.

12. Inelastic collisions. Impacts of nucleons with an energy larger than 3.3 Mev on deuterons may lead to their disintegration. This is the only type of inelastic collision which occurs until the threshold for meson production above 200 Mev is reached.

At low energy the calculation of the transition probability requires an accurate knowledge of the continuum three-body wave functions[1]. They will be considered in Sect. 14 in connection with elastic scattering.

The cross sections at low energies, as in the elastic case, are rather insensitive to the isotopic spin dependence of nuclear forces[2]. This circumstance and the uncertainties of the theoretical wave functions do not allow definite conclusions.

Inelastic collisions at high energies are in some degree more profitable. The Born approximation has received great attention and has been improved with a recipe equivalent to the impulse approximation by substituting the experimental two-body cross sections instead of the perturbation ones. The total cross section which includes elastic and inelastic collisions, is a quantity which can be approximated, better than the pure inelastic cross section, using plane waves for the final states[3]. For the total cross section indeed, a detailed knowledge of the continuum two-body wave functions is not necessary. It is possible to sum over all final states on the energy shell, using the sole condition that these states form a complete set. The total cross section is expressed then in terms of the deuteron's ground state wave function and the two-body interactions. It is a sum of two-body cross sections and an interference term.

The differential inelastic cross section requires, however, knowledge of the two-body continuum states, therefore, its numerical value is more uncertain.

As in the elastic case, the interference terms are certainly not accurate for strong interactions.

Inelastic cross sections have been computed numerically with the restriction of purely central two-body forces. Since we know that tensor forces play an important role in elastic scattering, no entirely reliable comparisons between theory and experiments can be made.

It is possible to give a qualitative picture of the inelastic process which leads to the disintegration of the deuteron[3,4]. We consider the case of neutron impacts. Apart from minor effects due to the Coulomb repulsion the same picture holds for protons.

The emitted protons form two groups, clearly distinct in energy and angular distribution. There is a group of fast protons ejected mainly in the forward direction, and a second group more-or-less isotropic of slow protons.

The fast protons come from a neutron-proton exchange collision, they are mostly emitted forward because small momentum transfers are favored. A sharp peak in the energy spectrum of these protons can be predicted owing to the virtual singlet state of the two slow neutrons left behind.

The low energy group of protons originates from neutron-neutron or non exchange neutron-proton encounters. The incident neutron will conserve its direction of motion in most of the cases and the freed protons will practically have the low kinetic energy possessed in the bound configuration[5].

It is worth while to mention that the diffraction theory discussed in Sect. 11 may be extended to take care of inelastic collisions[6]. Eq. (11.11) is still valid; the new feature being that the phase shifts are no longer real numbers. One can,

---

[1] B. H. Bransden and E. H. S. Burhop: Proc. Phys. Soc. Lond. A **63**, 1337 (1950).

[2] R. M. Franck and J. L. Gammel: Phys. Rev. **93**, 163 (1954).

[3] R. L. Gluckstern and H. A. Bethe: Phys. Rev. **81**, 761 (1951).

[4] G. F. Chew: Phys. Rev. **84**, 710 (1951).

[5] For some numerical evaluations of yields and angular distributions of the fast and slow proton groups we refer to the papers 3, 4 mentioned above.

[6] R. J. Glauber: Phys. Rev. **100**, 242 (1955).

for instance, at very high energies assume purely absorptive interactions. Then the phase shifts and the scattering amplitudes become pure imaginary, so that $f(0) = \frac{i\,k}{4\pi}\,\sigma$.

In this case the interference term of Eq. (11.11) is

$$-\frac{1}{4\pi}\left\langle\frac{1}{r^2}\right\rangle_d \sigma_1\,\sigma_2\,.$$

This formula has been used to explain the non-additivity of the total cross sections for (n, d) and (p, d) reactions at 1.4 Bev.

13. Charge symmetry of nuclear forces at high energies. Experiments[1] on elastic scattering of 340 Mev protons on deuterons have demonstrated that for certain angles practically no interference between p-n and p-p scattering exists.

The measured (p, d) cross section for center-of-mass angles between 20 and 90° shows the additivity of the two-body cross sections in agreement with Eq. (9.24). Independent experiments on (n, d) elastic scattering at 300 Mev leads to the same conclusion[2].

Additional evidence for this conclusion comes from measurements of (n, d) elastic scattering at 590 Mev[3]. From the (n, d) scattering after subtracting the known n-p cross section we have direct informations on the n-n differential cross section. Like the p-p angular distribution, the n-n differential cross section is nearly isotropic at 300 Mev, but shows a definite anisotropy at 590 Mev. The n-n angular distribution at 590 Mev energy coincides for angles between 20 and 90° with the p-p curve meassured at Brookhaven[4]. We have therefore convincing experimental evidence in favor of the charge symmetry of the nuclear interaction at high energy.

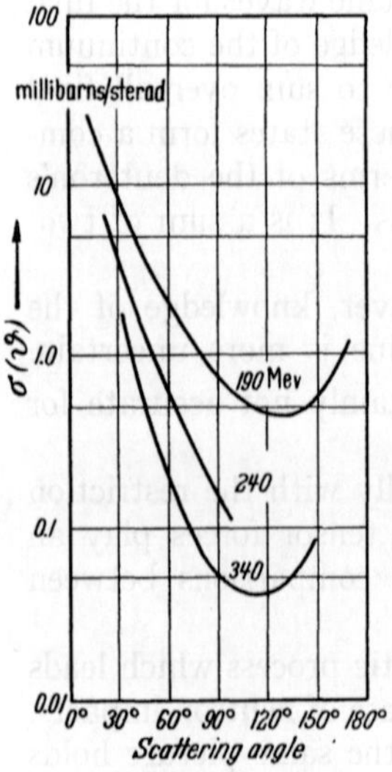

Fig. 4. Experimental elastic differential cross sections[5] for p-d scattering in the center-of-mass system, for energies of the incident protons in the laboratory as indicated on each curve.

14. Elastic n-d scattering at low energies. Many experimental data on the differential p-d and n-d elastic cross sections have been obtained for a wide range of low energies (see Figs. 1 and 2, p. 158).

Measurements of impacts of *thermal* neutrons in deuterium gas have provided informations on the two scattering lengths a_4, a_2 at very low energy. Unfortunately, the measured quantity is quadratic in the a's, and is compatible with two different sets of values[6]. They are, in units 10^{-13} cm, either

$$a_4 = 6.38 \pm 0.06, \quad a_2 = 0.7 \pm 0.3 \tag{14.1}$$

or

$$a_4 = 2.6 \pm 0.2, \quad a_2 = 8.26 \pm 0.12 \tag{14.2}$$

for incident neutrons of 0.0724 eV.

An interesting feature of the angular distribution is the presence of a maximum in the backward direction even for impinging neutrons of 10 kev. Measurements

[1] O. CHAMBERLAIN and D. D. CLARK: Phys. Rev. **102**, 473 (1955).
[2] V. P. DZELEPOV *et al.*: Dokl. Acad. Nauk. USSR. **99**, 943 (1954).
[3] V. D. DZELEPOV *et al.*: The 1956 CERN Symposium in Geneva, Switzerland.
[4] L. W. SMITH *et al.*: Phys. Rev. **97**, 1186 (1955).
[5] O. CHAMBERLAIN and D. D. CLARK: Phys. Rev. **102**, 473 (1955) and references given in this paper.
[6] D. G. HURST and N. Z. ALCOCK: Canad. J. Phys. **29**, 36 (1951).

in the forward direction are difficult. From the extrapolation of the integrated cross section it is, however, possible to recognize a trend of increasing forward scattering with increasing energy. Near 2 Mev the angular distribution seems to be symmetric about 90°. At higher energies forward scattering is predominant and the minimum of the differential cross section has shifted to 120° (see Figs. 1, 2 and 4, p. 158 and 170).

The analysis of the angular distribution in terms of phase shifts is marred by the familiar uncertainties that do not allow a unique solution. Starting at very low energy with the values (14.2) for the zero scattering lengths a set of phase shifts ^2S, ^4S, ^2P, ^4P up to an energy of 3 Mev has been suggested[1]. A different approach[2] which consists in using the ^2P, ^4P phase shifts calculated in BORN approximation, has indicated that the values (14.1) of the scattering lengths gave the best fit to the experimental cross sections.

In both analyses the orbital angular momentum is considered as a good quantum number. This is an acceptable assumption as soon as the D phase shift plays a minor role, probably for energies less than 3 Mev.

The remarkable importance of the P waves for the anisotropy at low energies could seem to yield a favorable ground for information on the odd parity nuclear potentials. Yet the anisotropy with respect to the center-of-mass of the struck deuteron does not necessarily imply that the two scattering centers are interacting in P or higher states.

An anisotropy exists even if the two centers are emitting merely S-waves. The presence of the back scattering is an exchange phenomenon, already known in the collision of electrons with hydrogen atoms. The incoming nucleon changes place with the deuteron's similar particle.

In contrast with the high energy case, the most important contribution to the P-scattering amplitude [cf. Eqs. (9.10) and (9.11)] comes from the overlap integral J_3. This can be seen most easily in BORN approximation taking the P wave components of the integrals $J_k (k=1, 2, 3)$ which appear in Eqs. (9.10) and (9.11), but has a more general validity (see Sect. 15).

Since the coefficients of J_3 in Eqs. (9.10) and (9.11) are $\lambda_{3,1}$ and $-\frac{1}{2}\lambda_{3,1}$ for quartet and doublet respectively, the ^2P phase shift at very low energy is approximately equal to half of the ^4P phase shift and has the opposite sign. This feature appears in both mentioned analyses of the experimental angular distributions. Since $\lambda_{3,1}$ is the potential strength for the deuteron ground state we can be convinced as already anticipated that the experimental anisotropy at low energies cannot lead to a knowledge of the odd parity potentials[2].

For the S-wave scattering an expansion in powers of the energy can be deduced in complete analogy with the two-body case.

Let us consider the quartet spin state. From the equation of motion, Eq. (2.7), if ψ_0', ψ_0'' are the three-body eigenfunctions at zero energy, we deduce:

$$(\psi_0' \Delta\psi' + \psi_0'' \Delta\psi'') - (\psi' \Delta\psi_0' + \psi'' \Delta\psi_0'') + k^2 (\psi' \psi_0' + \psi'' \psi_0'') = 0. \quad (14.3)$$

The asymptotic behavior for large values of q is chosen to be

$$\left.\begin{aligned} \psi'' &\to -\varphi(r) \frac{\sin(k \cdot q + \delta_4)}{q \sin \delta_4} = u \\ \psi' &\to 0, \end{aligned}\right\} \quad (14.4)$$

δ_4 is the quartet S phase shift and we always consider real eigenstates. $\varphi(r)$ is the deuteron's ground state normalized to one in the $\boldsymbol{r}$ space.

[1] R. K. ADAIR *et al.*: Phys. Rev. **89**, 1165 (1953). (See Fig. 3, p. 1168 of this paper.)
[2] R. S. CHRISTIAN and J. L. GAMMEL: Phys. Rev. **91**, 100 (1953).

From the equation

$$(\Delta_q + k^2)\, u = 4\pi\, \delta(\boldsymbol{q})\, \varphi(r)$$

follows:

$$u_0\, \Delta_q u - u\, \Delta_q u_0 + k^2\, u\, u_0 = 4\pi\, \delta(\boldsymbol{q})\, \varphi(r)\, (u_0 - u) \tag{14.5}$$

where u_0 using Eq. (14.4) is $u_0 = -\varphi(r)\left(-\frac{1}{a_4} + \frac{1}{q}\right)$, $\delta(\boldsymbol{q})$ is the DIRAC delta-function, and Δ_q refers to the kinetic energy of the incoming nucleon relative to the deuteron's center-of-mass.

We subtract Eq. (14.3) from Eq. (14.5) and integrate over all $\boldsymbol{r}$ and $\boldsymbol{q}$. Remembering that $\Delta = \frac{1}{3} T^s(\Delta_q)$ [see Eq. (2.11)] we deduce[1]:

$$k \cot \delta_4 = -\frac{1}{a_4} + \frac{k^2}{4\pi} \int [u_0\, u - (\psi_0'\, \psi' + \psi_0''\, \psi'')]\, d\tau. \tag{14.6}$$

An analogous procedure for the doublet case gives:

$$k \cot \delta_2 = -\frac{1}{a_2} + \frac{k^2}{4\pi} \int [u_0\, u - (\psi_0^s\, \psi^s + \psi_0'\, \psi' + \psi_0''\, \psi'' + \psi_0^a\, \psi^a)]\, d\tau. \tag{14.7}$$

It is possible to extend to any value of the orbital angular momentum Eqs. (14.6) and (14.7) given above.

To evaluate the scattering lengths and the effective ranges, it is necessary to know the three-body wave function at zero energy. Approximate methods for the computation of this function will form the subject of the next two sections.

15. Reduction to an equivalent two-body scattering problem. The equations of motion for the quartet spin state, Eq. (2.10),

$$(\Delta + k^2 - k_d^2)\, \psi' = \tfrac{1}{2}(\lambda_{3,3} + \lambda_{3,1})\, U^s \psi' - \tfrac{1}{2}(\lambda_{3,3} - \lambda_{3,1})\, (U' \psi'' + U'' \psi'), \tag{15.1a}$$

$$(\Delta + k^2 - k_d^2)\, \psi'' = \tfrac{1}{2}(\lambda_{3,3} + \lambda_{3,1})\, U^s \psi'' - \tfrac{1}{2}(\lambda_{3,3} - \lambda_{3,1})\, (U' \psi' - U'' \psi'') \tag{15.1b}$$

can be reduced to an integral-differential equation[2] in the single vector-variable $\boldsymbol{q}$, if an ansatz is made for the dependence of the configuration on the variable $\boldsymbol{r}$. We know that for large values of $\boldsymbol{q}$ which is the coordinate of the incoming body 1 relative to the center-of-mass of the remaining bodies 2 and 3, ψ'' becomes a product of the deuteron ground state eigenfunction $\varphi(r)$ and a free wave in the variable $\boldsymbol{q}$.

This suggests for the ψ's the following form:

$$\left.\begin{aligned}
\psi' &= T'[\varphi(23)\cdot f(1)] = \frac{\sqrt{3}}{2}[\varphi(21)\, f(3) - \varphi(13)\, f(2)],\\
\psi'' &= T''[\varphi(23)\cdot f(1)] = -\varphi(23)\, f(1) + \frac{1}{2}[\varphi(21)\, f(3) + \varphi(13)\, f(2)].
\end{aligned}\right\} \tag{15.2}$$

T' and T'' are the symmetry operators defined in Eq. (1.1). The two components ψ' and ψ'' given by Eq. (15.2) belong to the correct symmetry class with respect to permutations of spatial coordinates.

$\varphi(23)\, f(3)$ is the product of the deuteron's ground state eigenfunction formed by the bodies 2 and 1, with an unknown function $f(3)$. 3 stands for the vector coordinate of particle 3 relative to the center-of-mass of 1 and 2.

Eq. (15.2) is the so-called "no-polarization" approximation. With its use we are implying that the three-body wave functions can be expanded in terms of two-body eigenstates and that the deuteron ground state constitutes by far the most important term.

[1] M. VERDE: Atti Accad. Naz. Lincei **8**, 228 (1950). — G. BREIT: Rev. Mod. Phys. **23**, 228 (1951).

[2] L. MOTZ and J. SCHWINGER: Phys. Rev. **58**, 26 (1940).

The integro-differential equation for f follows by multiplying Eq. (15.1b) by $\varphi(23)$ and integrating over the full space of 2 and 3. One obtains in the first step:

$$\left.\begin{aligned}&\int \varphi(23)\cdot[\Delta+k^2-k_d^2-\lambda_{3,1}U(23)]\,\psi''\,d\tau_{2,3}\\&=\frac{3\lambda_{3,3}+\lambda_{3,1}}{2}\int\varphi(23)\,U(12)\,\psi''\,d\tau_{2,3}+-\frac{\sqrt{3}}{2}(\lambda_{3,3}-\lambda_{3,1})\int\varphi(23)\,U(12)\,\psi'\,d\tau_{2,3}.\end{aligned}\right\}\tag{15.3}$$

Making use of Eq. (15.2), interchanging variables 3 and 2 whenever $f(3)$ appears in the integrals and remembering that

$$[\Delta-\Delta_1-k_d^2-\lambda_{3,1}U(23)]\,\varphi(23)=0,\qquad \int\varphi^2(23)\,d\tau_{2,3}=1\tag{15.4}$$

Eq. (15.3) becomes:

$$\left.\begin{aligned}&-(\Delta_1+k^2)\,f(1)+\int\varphi(23)\,(\Delta_2+k^2)\,\varphi(13)\,f(2)\,d\tau_{2,3}\\&=-\frac{3\lambda_{3,3}+\lambda_{3,1}}{2}\int\varphi(23)\,U(12)\,\varphi(23)\,f(1)\,d\tau_{2,3}+\frac{3\lambda_{3,3}-\lambda_{3,1}}{2}\times\\&\times\int\varphi(23)\,U(12)\,\varphi(13)\,f(2)\,d\tau_{2,3}+\lambda_{3,1}\int\varphi(23)\,U(23)\,\varphi(13)\,f(2)\,d\tau_{2,3}.\end{aligned}\right\}\tag{15.5}$$

It is possible furthermore to transform the integral on the left side of Eq. (15.5) using either GREEN's theorem[1]

$$\int\varphi(23)\,\varphi(13)\,\Delta_2 f(2)\,d\tau_{2,3}=\int\Delta_2\big(\varphi(23)\,\varphi(13)\big)\,f(2)\,d\tau_{2,3}\tag{15.5'}$$

or the identity

$$\begin{aligned}&\int\varphi(23)\,\big(\Delta_2+k^2+\lambda_{3,1}U(13)\big)\,\varphi(13)\,f(2)\,d\tau_{2,3}\\&\qquad=\int\varphi(23)\,\big(\Delta_1+k^2+\lambda_{3,1}U(23)\big)\,\varphi(13)\,f(2)\,d\tau_{2,3}\end{aligned}$$

which follows from Eq. (15.4).

In this second case the integro-differential equation reads[2]:

$$\left.\begin{aligned}&-(\Delta_1+k^2)\left[f(1)-\int\varphi(23)\,\varphi(13)\,f(2)\,d\tau_{2,3}\right]\\&\qquad=-\frac{3\lambda_{3,3}+\lambda_{3,1}}{2}\int\varphi(23)\,U(12)\,\varphi(23)\,f(1)\,d\tau_{2,3}+\\&\qquad+\frac{3\lambda_{3,3}-\lambda_{3,1}}{2}\int\varphi(23)\,U(12)\,\varphi(13)\,f(2)\,d\tau_{2,3}+\\&\qquad+\lambda_{3,1}\int\varphi(23)\,U(12)\,\varphi(12)\,f(3)\,d\tau_{2,3}.\end{aligned}\right\}\tag{15.6}$$

We observe that on the right-hand side of Eq. (15.6) the constants which multiply the integrals are exactly the same as in Eq. (9.10). The structure of the three integrals is also similar to those appearing in the BORN approximation, Eq. (9.9). Here the perturbed wave f appears instead of a free-wave.

At low energy the third integral which is weighted by $\lambda_{3,1}$ is predominant so that the configuration is insensitive to the isotopic spin dependence of the potential strengths.

Eqs. (15.5) and (15.6) can be written as integral equations and solved numerically[1,2].

The answer for $f(\boldsymbol{q})$ may be slightly different in the two cases if the $\varphi(r)$ used is not an exact solution of Eq. (15.4).

---

[1] R. A. BUCKINGHAM and H. S. W. MASSEY: Proc. Roy. Soc. Lond., Ser. A **179**, 123 (1941). — R. A. BUCKINGHAM *et al.*: Proc. Roy. Soc. Lond., Ser. A **211**, 183 (1952).

[2] R. S. CHRISTIAN and J. L. GAMMEL: Phys. Rev. **91**, 100 (1953).

The doublet case can be dealt with similarly. The" no-polarization" approximation is however a more severe restriction. Indeed, the component ψ^a is entirely neglected and the component ψ^s is forced to be generated by T^s from the same state $\varphi(23)\, f(1)$ as ψ''.

At zero energy the results of numerical calculations are:

$$a_4 = 5.9 \times 10^{-13}\ \text{cm}, \qquad a_2 = 1.5 \times 10^{-13}\ \text{cm}.$$

These would favor the choice of Eq. (14.1) for the experimental scattering lengths.

Extensive numerical computations have been performed at several higher energies using conventional types of central interactions (see references, p. 173).

16. Variational method for the calculation of phase shifts.. The variational technique so largely used in two-body problems can be extended to many-body collisions in a quite natural way[1]. We shall briefly illustrate the case of *n-d* quartet scattering with central forces.

The equation of motion (15.1) will be rewritten in a more concise form as

$$(\Delta + k^2 - k_d^2)\, \boldsymbol{\psi} = \boldsymbol{U} \cdot \boldsymbol{\psi}. \tag{16.1}$$

$\boldsymbol{\psi}$ is the one-column matrix $\begin{pmatrix} \psi' \\ \psi'' \end{pmatrix}$ and $\boldsymbol{U}$ is a two-dimensional symmetrical matrix. We consider a solution of (15.1) with orbital angular momentum L and asymptotic behavior for large $\boldsymbol{q}$:

$$\psi'' \xrightarrow[q\to\infty]{} -\frac{\varphi(r)}{q}\left[\Lambda \sin\left(k q - \frac{L\pi}{2}\right) + \cos\left(k q - \frac{L\pi}{2}\right)\right] \cdot Y_L(\omega_q). \tag{16.2}$$

The integral expression

$$J = k\Lambda - \langle \boldsymbol{\psi} | \Delta + k^2 - k_a^2 - \boldsymbol{U} | \boldsymbol{\psi} \rangle \tag{16.3}$$

is stationary if $\boldsymbol{\psi}$ is a solution of Eq. (15.1) with the boundary condition [Eq. (16.2)].

The first variation of Eq. (16.3) is indeed

$$\delta J = k\,\delta\Lambda - \langle \delta\boldsymbol{\psi} | \Delta | \boldsymbol{\psi} \rangle - \langle \boldsymbol{\psi} | \Delta | \delta\boldsymbol{\psi} \rangle - 2 \langle \delta\boldsymbol{\psi} | k^2 - k_d^2 - \boldsymbol{U} | \boldsymbol{\psi} \rangle.$$

Remembering Eq. (2.11)

$$\Delta = \tfrac{1}{3}\left[1 + (12) + (13)\right] \Delta_q$$

and Eq. (16.2) we may write

$$\langle \boldsymbol{\psi} | \Delta | \delta\boldsymbol{\psi} \rangle - \langle \delta\boldsymbol{\psi} | \Delta | \boldsymbol{\psi} \rangle = \langle \boldsymbol{\psi} | \Delta_q | \delta\boldsymbol{\psi} \rangle - \langle \delta\boldsymbol{\psi} | \Delta_q | \boldsymbol{\psi} \rangle = k\,\delta\Lambda$$

and moreover

$$\delta J = -2 \langle \delta\boldsymbol{\psi} | \Delta + k^2 - k_d^2 - \boldsymbol{U} | \boldsymbol{\psi} \rangle,$$

which proves our assertion.

With a trial function $\boldsymbol{\psi}_t$, such that the first variation of J in Eq. (16.3) is zero, we have

$$J_t = k\Lambda_t - \langle \boldsymbol{\psi}_t | \Delta + k^2 - k_d^2 - \boldsymbol{U} | \boldsymbol{\psi}_t \rangle.$$

If $\boldsymbol{\psi}$ were an exact solution of Eq. (15.1) we would have

$$J = k\Lambda.$$

---

[1] W. KOHN: Phys. Rev. **74**, 1763 (1948). — M. VERDE: Helv. phys. Acta **22**, 339 (1949). — T. REGGE and M. VERDE: Nuovo Cim. **10**, 997 (1953).

The quantity

$$\Lambda_t - \frac{1}{k} \langle \psi_t | \Delta + k^2 - k_d^2 - \boldsymbol{U} | \psi_t \rangle$$

is therefore the best approximation for $\Lambda = \cot \delta_L$.

The variational principle discussed here can be easily generalized to include the presence of tensor forces[1].

The main advantage of the variational approach lies in the great freedom in choosing trial functions not restricted to a "no-polarization" approximation. Of course, the method discussed in Sect. 15 is the most powerful for the "no-polarization" case[2].

IV. Effects of the electromagnetic interaction.

17. Some remarks on COULOMB effects in p-d scattering. The presence of the COULOMB repulsion between the two protons at high energies modifies the differential cross section at *small* angles only. The effect is similar to that already known from p-p scattering.

The quartet and doublet scattering amplitudes given by Eqs. (9.10) and (9.11) in the perturbation approximation must be completed by adding to each the matrix elements $\frac{1}{2}(J_1^c - J_2^c)$ of the COULOMB interaction[3]. J_1^c and J_2^c are defined by the same Eq. (9.9) with $U(12)$ taken as the COULOMB potential e^2/r_{12}.

The COULOMB scattering amplitude at small angles

$$\frac{1}{2k} J_1^c(\vartheta) \approx \frac{2}{3} \frac{M}{\hbar^2} \frac{e^2}{k^2 \vartheta^2}$$

may become the same order of magnitude as the nuclear quartet or doublet amplitudes. From (9.12) we deduce an approximate value of J_1, which is the only term important in the forward direction,

$$J_1(0) \approx \frac{4}{3} \frac{M k}{\hbar^2} \frac{R^3}{3} U_0,$$

where R is the range of nuclear forces. Calling $\frac{1}{2}\Lambda$ the coefficient of J_1 in Eq. (9.10) or (9.11), we expect the interference effect to appear in the vicinity of an angle ϑ for which the two amplitudes are nearly equal, that is near

$$\vartheta \approx \frac{1}{kR} \left(\frac{3e^2}{U_0 \Lambda R} \right)^{\frac{1}{2}}.$$

This amounts to a few degree for an energy of 100 Mev.

It is interesting to remark that in the particular case of a $(\boldsymbol{\tau}^{(1)} \cdot \boldsymbol{\tau}^{(2)})$ isotopic spin dependence of the nuclear potential the integral J_1 disappears in Eqs. (9.10) and (9.11) as already mentioned, and a much less appreciable COULOMB interference follows.

There are indications of a small COULOMB interference in a measurement of p-d scattering at 20 Mev[4]. At these and still lower energies where an analysis

---

[1] T. REGGE and M. VERDE: Nuovo Cim. **10**, 997 (1953).

[2] Some crude numerical results with an older formulation of the variational method and no-polarization for the quartet case can be found in the paper by A. TROESCH and M. VERDE, Helv. Phys. Acta **24**, 39 (1951). The value of $a_4 = 3 \times 10^{-13}$ cm is much lower than that found in reference 2 of p. 173, with the method of Sect. 15 which must be retained a much better approximation.

[3] G. F. CHEW: Phys. Rev. **74**, 809 (1948).

[4] D. O. CALDWELL and J. R. RICHARDSON: Phys. Rev. **98**, 28 (1955).

in terms of phase shifts is unavoidable, the main effect of the COULOMB force is to bring in the familiar penetration factor $C=2\pi\eta/(e^{2\pi\eta}-1)$ with $\eta=e^2/\hbar v$. The scattering amplitude $F_{p,d}$ for the *p-d* collision is obtained from the *n-d* amplitude by the relation: $F_{p,d}=CF_{n,d}$.

A small correction must be introduced for the *S*-wave and for a better approximation of the distorted COULOMB wave[1].

There is in the *p-d* scattering a reduction of the differential cross section of about C^2 in comparison with the *n-d* case.

18. Electromagnetic transitions. An electromagnetic field may induce transitions from the bound states of H^3 and He^3 to the continuum of a deuteron and a free nucleon. Conversely a neutron or a proton can be captured by a deuteron to form the two mirror nuclei H^3 or He^3, the fusion energy being freed as γ radiation. This latter phenomenon is more convenient for experimental study, and may serve to give further informations on the three nucleon eigenfunctions.

Thermal neutrons have an exceptionally small chance to be captured by deuterons. The measured cross section (at the standard neutron velocity of 2200 m/sec) amounts to 0.57 mb[2]. This is to be compared with the thermal capture cross section of neutrons by protons which is 330 mb[2].

The $D(p,\gamma)$ He^3 reaction has been investigated in a proton energy range between 0.5 and 1.75 Mev. The emitted γ rays are distributed about the incident proton direction according to a practically pure $\sin^2\vartheta$ law. A small isotropic component of a few percent seems to be present[3].

An interesting characteristic of this reaction is the polarization of the emitted electromagnetic radiation.

This has been proved by letting the γ rays disintegrate deuterons imbedded in a photographic plate[4]. The photoprotons are mostly emitted in a direction $\boldsymbol{n}$ orthogonal to the plane formed by the lines of the incident protons and the photons. The experiment indicates a complete polarization with the electric vector along $\boldsymbol{n}$.

This radiative transition shows therefore the character of a pure electric dipole, in which the selection rule $\Delta J_z=0$ is effective.

The capture of thermal neutrons by deuterons should occur mainly through a *S-S* magnetic dipole transition. We have learned from the study of the three-body ground state (see Sects. 4 and 5) that this is predominantly a $^2S_{\frac{1}{2}}$ state and belongs to the spin eigenfunction ξ^a (Sect. 1) totally antisymmetric with respect to permutation of two particles.

The dipole magnetic moment operator [Eq. (1.4)]

$$\boldsymbol{\mu}=\tfrac{1}{2}(\mu_P+\mu_N)\sum_{k=1}^{3}\boldsymbol{\sigma}^{(k)}+\tfrac{1}{2}(\mu_P-\mu_N)\sum_{k=1}^{3}\tau_3^{(k)}\cdot\boldsymbol{\sigma}^{(k)}$$

conserves the eigenstate ξ^a. This is clearly a consequence of the complete symmetry of $\boldsymbol{\mu}$ with respect to permutations; $\boldsymbol{\mu}$ operating on ξ^a cannot give rise to spin eigenfunctions of a symmetry class different than the totally antisymmetric.

This is the cause of a strong selection rule. No *S-S* magnetic dipole transition occurs because the continuum state must belong to the same spin and

[1] R. S. CHRISTIAN and J. L. GAMMEL: Phys. Rev. **91**, 100 (1953).

[2] D. J. HUGHES and J. A. HARVEY: Neutron Cross Sections, Brookhaven National Laboratory, BNL 325, 1955.

[3] W. A. FOWLER *et al.*: Phys. Rev. **76**, 1767 (1949). — G. M. GRIFFITHS: The 1954 International Conference on Nuclear and Meson Physics, Pergamon Press, London and New York. Glasgow 1955.

[4] D. H. WILKINSON: Phil. Mag. **43**, 659 (1952).

isotopic spin quantum numbers as the ground state and therefore is orthogonal to it[1].

We must find an explanation for the very small value of the *n-d* capture cross section by a different mechanism. The electric quadrupole transition which is certainly present because of the 4% ${}^4D_{\frac{1}{2}}$ admixture in the triton ground state, is negligibly small[2].

The magnetic dipole operator, Eq. (1.4), has matrix elements different from zero for the spin eigenstates ξ', ξ'' (Sect. 1). By allowing a very small percentage of S components (ψ', ψ'') in the H^3 ground state, one could account for the entire capture cross section[3,1].

With this assumption, however, the magnetic moments of H^3 and He^3 would still remain unexplained. Their values would be practically the same as given by Eq. (4.11), with an anomaly of about $\frac{1}{4}$ nuclear magnetons.

It is suggestive to interpret the *n-d* capture[4] as a manifestation of the "exchange magnetic moments" (see Sect. 3).

A magnetic S-S transition is allowed with a phenomenological exchange moment of the form

$$\mu_{\text{exch}} = \frac{e}{2} \sum_{i<j} U(r_{ij})\, \boldsymbol{\varphi}(\sigma_i, \sigma_j)\, (\boldsymbol{\tau}_i \times \boldsymbol{\tau}_j)_3 . \tag{18.1}$$

The τ-dependence is the only compatible with the requirement of expectation values opposite in sign for mirror nuclei. $U(r_{ij})$ in a phenomenological theory is an unknown function of the mutual distance of the pair i, j. For $\boldsymbol{\varphi}(\sigma_i, \sigma_j)$ two possible forms:

$$\boldsymbol{\varphi} = \sigma_i - \sigma_j \quad \text{or} \quad \boldsymbol{\varphi} = \sigma_i \times \sigma_j$$

can be considered. This is because μ_{exch} must be an axial vector. With the choice $\sigma_i - \sigma_j$ the expectation value of Eq. (18.1) for the three-body ground state vanishes. The choice of $\boldsymbol{\varphi} = \sigma_i \times \sigma_j$ or a combination of both possibilities may account for the magnetic moment anomalies of H^3 and He^3 and explain at the same time the photomagnetic transition.

Choosing in Eq. (18.1) $U(r_{ij}) = g\,\delta(r_{ij})$, δ being the DIRAC delta function, the constant g can be determined by the knowledge of the magnetic moment anomalies.

A numerical computation[5] of the photomagnetic transition with the value of g so determined has shown that the thermal capture cross section can be interpreted as entirely due to the exchange currents.

---

[1] M. VERDE: Helv. phys. Acta **23**, 453 (1950).
[2] N. AUSTERN: Phys. Rev. **85**, 147 (1952).
[3] E. H. S. BURHOP *et al.*: Proc. Roy. Soc. Lond., Ser. A **192**, 156 (1948).
[4] N. AUSTERN: Phys. Rev. **83**, 672 (1951) and references given in this paper.
[5] N. AUSTERN: Phys. Rev. **83**, 672 (1951).

Matter and Charge Distribution within Atomic Nuclei.

By

DAVID L. HILL.

With 17 Figures.

Introduction.

An abrupt increase in our knowledge of nuclear densities has occurred in recent years. When one recalls the meagre knowledge available prior to 1953 regarding the distribution of matter and charge within atomic nuclei, it is remarkable to consider the relatively detailed information available when this article is written in January 1956. Before this recent flood of new observation and theory from many quarters, our concepts of nuclear size, shape and density distribution had changed little from the conclusions emerging as a result of the startling experiments performed by Lord RUTHERFORD[1] shortly after the turn of the century. His measurements on the scattering of alpha particles by thin foils gave the first quantitative evidence on the distribution of positive charge in the atom and demonstrated that both positive charge and mass must be condensed in a volume described by a radius of about 10^{-12} cm or, roughly, one ten-thousandth of the atomic size. The earlier proposal that the electric neutrality of the atom is provided by positive charge uniformly distributed throughout the atom, through which the (then newly discovered) electrons must therefore move in simple harmonic motions—a speculation advanced by J. J. THOMSON[2]—was abolished and the necessary foundation was laid for the understanding of atomic structure and for the beginning of the science of nuclear physics. Even today, more than forty years after Lord RUTHERFORD'S measurements, we recognize that his research, and extensions of it which soon followed [*17*], stands as the greatest single step forward in the search for detailed understanding of the distributions of mass and charge in atomic nuclei.

1. Early studies of nuclear size. In the forty years following RUTHERFORD'S work, his results were somewhat refined but there was no break-through to a qualitatively new level of understanding such as may be said to have occurred within forty-five years after his work. We may recall that the primary conclusion in Lord RUTHERFORD'S initial research was that, in order to explain by a classical impact parameter calculation the fact that alpha particles of several Mev energy could be scattered into large backward angles, it was necessary to assume that the $(1/R)^2$ COULOMB field of the scattering center was maintained down to very small distances. Only if the charge *and* the mass of the scattering center were assumed to be contained within the tiny volume so defined could the observed angular distributions of scattered alpha particles be understood. Later studies [*17*] showed that when the alpha particle and the nucleus come sufficiently close

[1] E. RUTHERFORD: Phil. Mag. **21**, 669 (1911).

[2] J. J. THOMSON: The Corpuscular Theory of Matter, p. 103. London: Archibald Constable & Co., Ltd. 1907, 172 pp.

together, as judged by the classical relation involving also angle of scattering and energy of projectile, then the alpha particles deviate from the COULOMB field scattering law and produce "anomalous" scattering, or even coalescence of the alpha particle with the scattering nucleus to produce a nuclear transformation. The distance at which such anomalous scattering begins is then a well defined experimental quantity and must represent the sum of the nuclear radius and the alpha particle radius plus a distance representing the range of nuclear forces involved.

Another method of measuring nuclear radii was soon developed. After the quantum mechanical theory of barrier penetration had been worked out it was applied to determine the radius at which nuclear forces overwhelmed the COULOMB field. Also proton scattering and barrier penetration factors for artificially induced charged particle reactions were used to measure the nuclear force radius.

A new tool was provided when the neutron was discovered, for by elastic scattering of neutrons with a few Mev of energy, measures were obtained of nuclear force radii without the complication of the repulsive COULOMB field previously involved in nuclear projectiles. None of these methods, of course, were capable of giving detailed information for nuclear density distributions. They had in common that they measured a single parameter of these distributions and that the "radius" which they measured did not depend so much on the distribution of charge or mass but rather on the spacial distribution of nuclear potential energy.

Another measure of nuclear size was obtained by assuming that the energy differences between mirror nuclei were directly attributable to the difference in COULOMB energy when a neutron was replaced by a proton with both assumed to be uniformly distributed within the same "radius". Not until much later were the special features of nuclear structure ignored by this assumption to be understood.

It was reassuring that these different methods gave similar results and led in common to the conclusion that the density of nuclear matter is about constant throughout the periodic table with nuclear "radii" given by the relationship

$$R = r_0 A^{\frac{1}{3}}, \qquad r_0 = 1.4 \text{ to } 1.5\,^1. \tag{1.1}$$

It was also reassuring that this nuclear size was consistent with the idea that nuclei are made up of neutrons and protons retaining their identity within the nuclear structure, as suggested by the systematics of the nuclear statistics and spin. Thus the wavelength of a proton or neutron of mass m as given by

$$\lambda\!\!\!^{-} = \frac{\hbar}{p} = \frac{\hbar}{\sqrt{2MT}} \tag{1.2}$$

is equal to 1.6 if the kinetic energy T (by way of an order of magnitude estimate) is equal to the observed average binding energy of 8 Mev for nucleons in atoms of medium atomic weight. As most protons or neutrons within nuclei will be in states requiring the wave function to have several nodes, the observed nuclear size was reasonably consistent with the idea that nuclei are composed of protons and neutrons.

2. Outlook and definitions. When we come to consider more recent observations and theory, we find sufficiently detailed information to be available on nuclear densities that it becomes fruitful to distinguish between the distribution of mass and the distribution of charge throughout atomic nuclei. We shall, therefore, discuss both matter and charge densities, although after reviewing

[1] Unless otherwise stated, distances in this article will be measured in units of 10^{-13} cm.

available information we shall reach the tentative conclusion that these densities are the same except for scale factors necessarily in the ratio of atomic mass and charge numbers. While in an ultimate theory of the nucleus it may be possible to describe the mass and charge density in terms of subnucleonic particles, any effort to do so at this time would be quite premature. Instead we express the mass and charge densities in terms of the neutron and proton densities according to the relations

$$\varrho^{(M)}(\boldsymbol{r}) = N(\boldsymbol{r}) + P(\boldsymbol{r}) \tag{2.1}$$

and

$$\varrho^{(c)}(\boldsymbol{r}) = P(\boldsymbol{r}). \tag{2.2}$$

Here for simplicity we have neglected the mass difference between neutrons and protons and we have expressed the mass and charge densities in units of the nucleon mass and the electron charge respectively. The neutron and proton densities are, therefore, expressed as particles per unit volume. Quantitative discussion of these proton and neutron distributions falls rather naturally, in the present state of our knowledge, into two parts:

(a) The radial variation of the distribution, which to avoid confusion with the geometrical shape, we shall term the "radial shape" of the distribution. For its description we shall require two or more parameters, the most familiar of which is of course the "radius". We shall return to the question of how the radial parameter is to be defined.

(b) The angular variation of the distribution which we shall term the "angular shape" of the distribution.

Neither of these variations of particle densities may be measured alone. Usually, experiments must be interpreted in terms of particular weighted averages over the radial or angular variation of the charge or mass distribution. Measurements of charge distribution will usually be much more clear-cut than measurements of mass distribution, for the reason that with the former the electrical force field of interaction is well known so that the distribution of charge may be deduced without ambiguity. In contrast the measurement of mass distribution —or at least of the neutron component of mass distribution—involves interpretation of the nuclear force field, as it enters the measurements, and confronts us with the difficult question of how this force field is to be connected quantitatively with the density of nucleons producing it. Hence our conclusions on mass distribution must be reached by less direct paths and will depend more strongly on the present development of nuclear theory. Even in the favorable case of charge distribution, we seldom find an experiment which can measure the radial or angular variations separately. For example, we may have a mean square radius or higher radial moment which in fact represents an average of the radial distribution over all angles. Alternatively the measurement of a predominantly angular effect such as the quadrupole moment will necessarily involve consideration of the manner in which charge is radially distributed. Nevertheless, by bringing together the various types of information, both experimental and theoretical, bearing on different combinations of radial and angular variations of particle distributions, it becomes possible to give an approximate description of how the radial distribution of particle densities deviates from the idealized uniform density and how the angular distribution deviates from spherical symmetry. At the conclusion of our survey we shall summarize in Sect. 58 the emergent picture of nuclear mass and charge distributions.

3. Functional descriptions. It is fully adequate to the present state of our knowledge—and indeed it seems likely to be a close approximation even when the mass and charge distribution is ultimately specified in complete detail—to assume that the functional form, apart from a radial scale factor, of the radial variation of particle densities is constant as the angle is varied relative to a system of coordinates fixed with respect to the intrinsic nuclear shape. We therefore describe radial density variations in the form,

$$\varrho(r) = \varrho_0 f\left(\frac{r}{r_1}\right) \tag{3.1}$$

where ϱ_0 is the central density of charge or mass, $f(0)$ equals 1, and r_1 is the "range" parameter related to the size of the nucleus. In practice the functions $f(x)$ are taken to be one or two parameter families which include some plausible shapes for nuclear densities and which are amenable to relatively simple calculations. Within this framework, then, the angular variation is described by representing the range parameter r_1 as a function of ϑ and φ, the polar and azimuthal angles measured with respect to body-fixed axes in the nucleus. We know that in some processes, such as fission and alpha decay, important effects must be attributed to angular deformations of order higher than the second (that is, described by an angular function containing more than two loops or more than one node in the polar variable). In the present survey, however, we shall be mainly concerned with effects which may be interpreted in terms of ellipsoidal deformation so that the angular function describing r_1 is given by

$$r_1 = \overline{r_1}\left[1 + \sum_{\mu} \alpha_\mu Y_{2\mu}(\vartheta, \varphi)\right] \tag{3.2}$$

in which for the surface harmonics and their coefficients we follow the notation of CONDON and SHORTLY [5]. We may choose the axes to which the polar angle ϑ and the azimuthal angle φ refer to be the major axes of the nuclear ellipsoid, so that $\alpha_2 = \alpha_{-2}$, and $\alpha_1 = \alpha_{-1} = 0$. There are thus two independent parameters in (3.2) which may be denoted as a deformation parameter, β, and a symmetry parameter, γ, which are related to the parameters of (3.2) by

$$\alpha_0 = \beta \cos\gamma, \tag{3.3}$$

$$\alpha_2 = \alpha_{-2} = \frac{\beta \sin\gamma}{\sqrt{2}}. \tag{3.4}$$

The parameters β and γ thus define the general ellipsoid[1]. Let us choose β always positive; then $\gamma = 0$ defines a prolate spheroid and $\gamma = \pi$ defines an oblate spheroid. With each increase of γ by $\pi/3$ the sequence of ellipsoidal deformations is repeated for a different orientation of the ellipsoidal forms relative to the body-fixed axes.

It is to be emphasized that both (3.1) and (3.2) define averages over time. When the particles under consideration are sufficiently energetic to have wavelengths comparable to the separation of nucleons, the difference between the continuous distribution defined by (3.1) and the more realistic distribution which provides for clustering of mass and charge at the nucleonic loci becomes significant. Likewise some applications may require that the dynamic properties of the deformation parameters α_μ be included in the treatment in addition to the properties of the average static deformation given by (3.2).

4. Equivalent radius. The range parameter r_1 entering into (3.1) has a different meaning for each different functional form for $f(r/r_1)$; hence it is not very suitable for the discussion of the nuclear radius. Moreover, the various experiments for studying charge and mass densities measure different moments of the distributions, with the second moment of the nuclear charge or mass densities occurring

[1] For further details see Fig. 13 of reference [*14*].

with particular frequency. Thus the root mean square radius is a useful measure for the extension of mass or charge densities. A still more unambiguously defined nuclear radius is given in terms of the uniform density model. In this idealization, particle densities are assumed to extend with constant value to the surface radius at which they drop abruptly to zero. The simplicity of this schematic representation of nuclear densities has led to its frequent appearance in the literature even though the first theoretical treatment [20] of nuclear density of course made it clear that the principles of quantum mechanics require the nuclear surface to be represented by a continuous function of moderate gradient instead of the discontinuity given by the uniform density model. For this simplest radial shape, the function f is described by

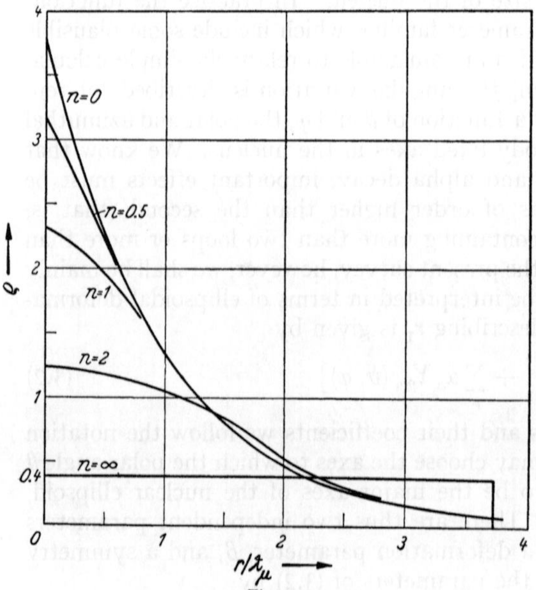

Fig. 1 a.

Fig. 1a—c. Here are shown charge distributions for Families II and III, defined in Sect. 22. For each of the curves here plotted, the radial extension has been adjusted to make them "equivalent" in the sense that when the charge distribution in lead is represented by these functions each will give the same energy (6.00 Mev) for the $2\,P_{\frac{3}{2}}\to 1\,S$ transition of the μ-meson. The horizontal scale is in units of the meson reduced COMPTON wavelength, $ƛ_\mu$. The vertical scale is in number of protons per $ƛ_\mu^3$. (These curves are taken from reference [13].)

$$\left.\begin{aligned} f(x) &= 1, \quad x \leq 1, \\ f(x) &= 0, \quad x > 1. \end{aligned}\right\} \tag{4.1}$$

It is therefore convenient to define for non-uniform shapes an "equivalent radius", R_{eq}, such that, *for a given experiment*, a uniform distribution with $R = R_{eq}$ would yield the same result as the given distribution. It necessarily follows that this definition can apply only to those experiments which measure a single parameter of the charge or mass distribution. For example, low energy electron scattering measures the charge average value of r^2:

$$\langle r^2\rangle_{Av} = \frac{1}{z}\int \varrho^{(c)}(\boldsymbol{r})\, r^2 d\,(\mathrm{vol}) \tag{4.2}$$

The equivalent radius for the distribution $f(x)$ and for this particular experiment is therefore found by equating the right side of (4.2) to the corresponding value of $\langle r^2\rangle_{Av}$ for a uniform distribution, namely, $\frac{3}{5}R_{eq}^2$. Even for those experiments which measure only a single parameter of, say, the charge distribution the deduced values of R_{eq} may be substantially different, as indicated by the following three types of measurements:

(a) Charge distributions which are equivalent for isotope shifts and x-ray fine structure have equal values of $\langle r^{2\sigma}\rangle_{Av}$ where σ is about 0.8 for nuclei in the neighborhood of Pb (see Sect. 15).

(b) Electron scattering below about 40 Mev measures approximately $\langle r^2\rangle_{Av}$, the mean square radius for the charge distribution (see Sect. 9).

(c) The COULOMB energy of a nucleus measures yet another function of the charge distribution (see Sect. 25).

If the equivalent radius is then defined for these three processes as a function of the forms for $f(x)$ such as are illustrated in Figs. 1a and b, the results are as graphed in Fig. 2. The abscissa in this figure is so chosen that one passes continuously from the uniform distribution at the left-hand side to the exponential distribution at the right-hand side. Fig. 2 serves to emphasize the disparate

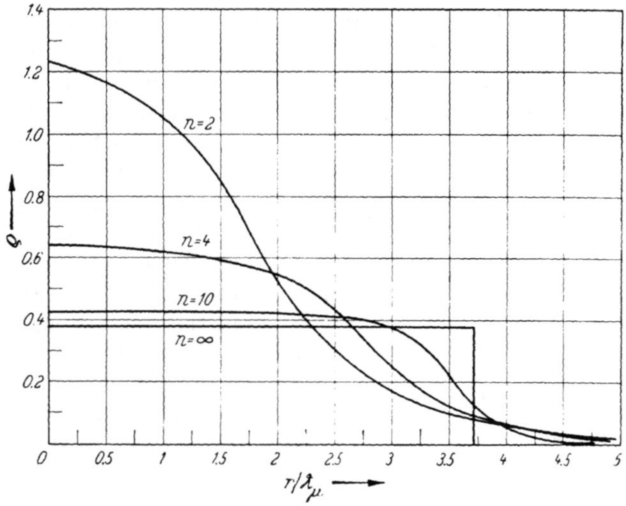

Fig. 1 b.

values of R_{eq} which will be obtained if one attempts to analyze three such different types of observations as (a), (b) and (c) in terms of the schematic uniform charge distribution. As will later be evident the charge distribution in heavy

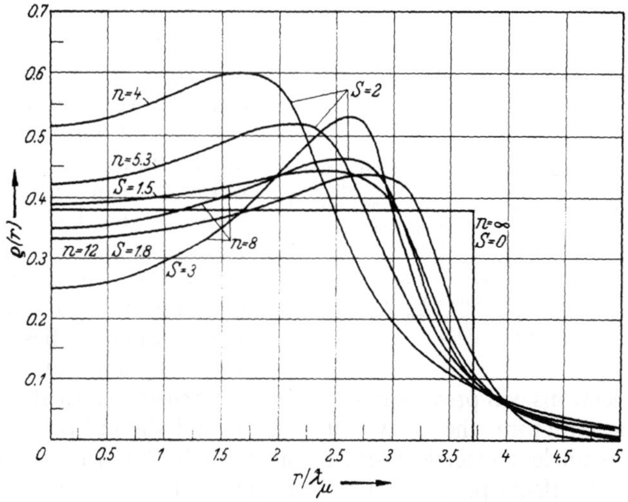

Fig. 1 c.

elements has a shape corresponding to an abscissa in Fig. 2 of about 0.2. For a shape thus proximate to the uniform distribution we note that the disparities between R_{eq} for the three illustrative measurements considered would not exceed about 3%. Thus the uniform model density of the nucleus has proved adequate prior to the analysis of the more discriminating modern experiments.

5. Plan of discussion. We have seen that in our survey of nuclear densities two dichotomies of the subject arise, and these will serve in our survey to define the four divisions A, B, C and D devoted to the review of available experimental information. First there is the division between the electrical experiments that measure the size and shape of the charge distribution and the nuclear force experiments which measure the size and shape of the nucleonic potential energy distribution. Secondly, there is the division for convenience of discussion between the radial shape and the angular shape of charge and mass distributions.

Recognizing that our most clear-cut information is derived from electric measures of nuclear size and radial shape, we review observations of this class first in Part A. Within this division we first examine electron scattering from nuclei which has given us by far the most detailed insight available from any source concerning charge densities. Particularly is this remark applicable to high energy electron scattering (100 to 200 Mev) for which many angular momentum states of the scattered electrons are involved and for which therefore many different moments of the nuclear charge distribution are measured.

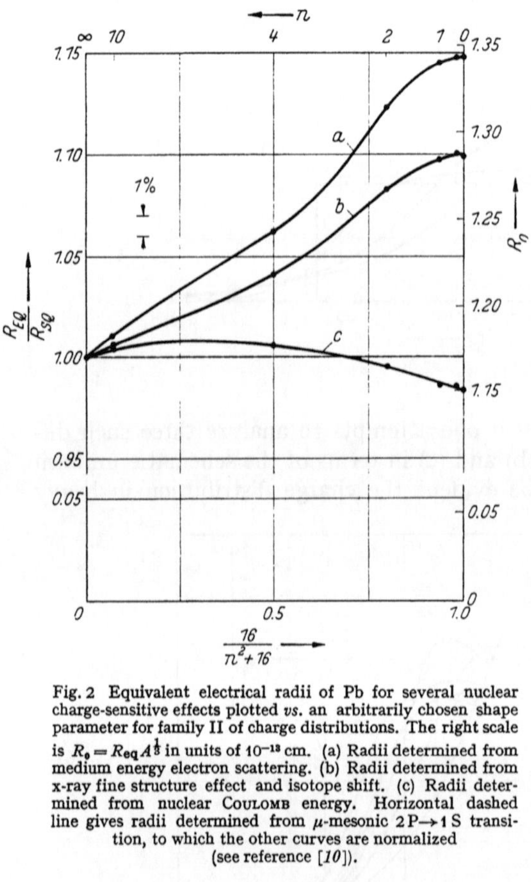

Fig. 2 Equivalent electrical radii of Pb for several nuclear charge-sensitive effects plotted *vs.* an arbitrarily chosen shape parameter for family II of charge distributions. The right scale is $R_0 = R_{eq} A^{\frac{1}{3}}$ in units of 10^{-13} cm. (a) Radii determined from medium energy electron scattering. (b) Radii determined from x-ray fine structure effect and isotope shift. (c) Radii determined from nuclear COULOMB energy. Horizontal dashed line gives radii determined from μ-mesonic $2\,P \rightarrow 1\,S$ transition, to which the other curves are normalized (see reference [*10*]).

The greatest accuracy in the measurement of a single parameter of nuclear size is probably available from the study of the x-rays indicating the energy difference between the two lowest states of μ-mesons bound about atomic nuclei. The work on μ-mesonic x-rays also gives promise of providing data on the radial shape of charge distributions when transitions between higher states have been observed. In time useful information may also be derived from studies of the hyperfine structure of the spectra as well as from studies of the μ-meson scattering and pair creation.

Bound electrons will provide information on nuclear electrical size from studies of x-ray fine structure, the x-ray isotope shift and the optical isotope shift. If, however, the nuclear electrical size is taken to be known from other sources, then the x-ray effects provide an experimental means for measuring the LAMB shift of closely bound electrons in heavy elements. Similarly, from a known electric charge distribution in the nucleus, the optical isotope shift provides a measurement of nuclear compressibility.

The COULOMB energy of the nucleus as deduced from energy differences of mirror nuclei and from the semi-empirical formula for nuclear masses also gives the measure of nuclear electrical size.

In Part B we examine nuclear force measures of nuclear size and radial shape. The experiments subject to most clear-cut interpretation are probably those involving neutron scattering which have been studied at energies ranging from a few Mev through a vast range up to one or two Bev. Likewise, proton scattering is available over a broad energy range. The important strides which have been made in analyzing neutron and proton scattering with complex potential well models permit definite conclusions regarding nuclear size and radial shape.

It is of interest that the original method of Lord RUTHERFORD is still actively employed to study nuclear size. Recent studies of high energy alpha-particle scattering give further information on the nuclear force radius for many different nuclei.

Further data on the nuclear force radii of nuclei is available from charged particle reaction cross-sections and from alpha decay in both of which the characteristic quantum mechanical barrier penetration phenomenon comes into play.

Deuteron pick-up and stripping reactions are of special interest as they tend to give not only the nuclear force size but also the relative population of neutrons and protons near the nuclear surface. Part B closes with a brief review of the properties of bound pions and of the scattering of positive energy pions.

Having completed the survey of the radial variation of nuclear densities, we turn next to the angular variation. In Part C we examine electric measures of nuclear angular shape. The most intuitively meaningful of these are nuclear quadrupole moments as deduced from optical hyperfine spectra. Similar direct information on the oblateness or prolateness of the nuclear charge density may be obtained from the study of electric quadrupole transition rates. The latter may be considered to measure the off-diagonal elements of the same operator, with the representation defined by the particle states of the given nucleus. Rather clear-cut information on the angular shape of nuclear charge is also provided by the study of isotope shifts over a succession of neighboring elements. Similar information should follow from that part of the hyperfine structure of μ-mesonic x-rays which results from the interaction of the bound meson with the quadrupole moment of the nucleus.

The angular variation of nuclear charge affects fast electron scattering in two ways. First, there are the inelastic scattering processes. In the second place is the effect on elastic scattering arising from the fact that the effective radial distribution of charge is given by an average over all angles.

In Part D we examine non-electric measures of nuclear angular shape. The nuclear moments of inertia relevant to the analysis of the prolific data on nuclear rotational states provide a rather direct measure of the deviation of the mass distribution from spherical symmetry. Nuclear deformations have an important effect on the probability of alpha decay. Such ground state deformations likewise have a significant bearing on the probability of observing angular correlations between fission fragments and the neutron or proton beam exciting the target nuclei. Likewise, the alpha emission and fission properties from spin-oriented nuclei are dependent on the ground state angular shape of the collective nuclear force field.

We have seen that a theory of nuclear matter is required to make the step from observed distributions of the nuclear potential to the distribution of nuclear matter. In Part E we examine the status of current nuclear matter theory, which includes the examination of the nuclear surface energy, the nuclear compressibility and the relative distributions of neutrons and protons. We shall examine some of the radial distributions of nuclear matter as predicted by several radial potentials for the independent particle model and for the statistical model. Bringing into evidence the interaction between individual particle states and the collective potential, we shall then be able to review the nuclear angular shapes predicted by the collective model of the nucleus.

Finally in Part F we summarize the picture of nuclear matter and charge densities following from the experimental and theoretical sources which we have reviewed.

A. Electric measures of nuclear size and radial shape.

Among experiments sensitive to the nuclear charge distribution, the recent studies by HOFSTADTER and his collaborators [*15*] on the elastic scattering of high-energy electrons by nuclei are undoubtedly of the greatest significance, since their work permits the measurement of a large number of parameters of the nuclear shape and hence leads to the delineation of the radial shape as well as of the size of the nuclear charge distribution.

We, therefore, consider electron scattering experiments first and then pass on to consider other experiments sensitive to the nuclear charge distribution, recognizing that the latter fill an important corroborative role with respect to the information provided by electron scattering.

I. Electron scattering.

6. Electron as nuclear probe. The analysis of electron scattering experiments is fortunately quite direct. High intensity electron beams of nearly constant energy are readily available from modern accelerators. The nature of the primary interaction between electron and nucleus is well known and may be represented to a high degree of approximation by the COULOMB forces between electrons and protons. The motion of the scattered electrons may therefore be described by the DIRAC equation for a static, central field generated by the COULOMB forces from the extended distribution of protons.

The RAMSAUER effects, well known in the passage of slow electrons through atoms, have their counterparts in the scattering of fast electrons by atomic nuclei. The study of these nuclear RAMSAUER effects have given us a picture of the nuclear charge distribution.

The small mass of the electron requires that it be accelerated to high energy in order to be sensitive to the finite nuclear size. Nuclear effects become significant in electron scattering when the energy of the electron exceeds 1 Mev and the detail with which the nuclear charge distribution may be measured increases with the energy of the electron. For example, a 150 Mev electron possesses a reduced wavelength, $\lambda\!\!\!^{-}$, which is about one-quarter of the radial extension of the charge of the lead nucleus.

The non-COULOMB forces between electron and nucleon have somewhat different significance for bound state electrons and for positive energy electrons. The bound-electron-free-neutron interaction has been measured[1-3] to be equivalent to an attractive square well of depth only 45 kev over a volume $\frac{4}{3}\pi r_e^3$ where $r_e = 2.82 \times 10^{-13}$ cm. This interaction has been attributed to the neutron magnetic moment[4]. The corresponding interaction between bound atomic electrons and nuclear protons should be somewhat greater but the total electron nucleus interaction from this source should be much less than A times the single

[1] W. W. HAVENS jr., L. J. RAINWATER and I. I. RABI: Phys. Rev. **82**, 345 (1951).
[2] D. J. HUGHES: Phys. Rev. **86**, 606 (1952).
[3] M. HAMERMESH, G. R. RINGO and A. WATTENBERG: Phys. Rev. **85**, 483 (1952).
[4] L. L. FOLDY: Phys. Rev. **87**, 693 (1952).

electron-proton interaction. We conclude that this interaction is negligible in comparison with the direct COULOMB interaction. The electron-nucleon beta decay interaction is further orders of magnitude smaller than the magnetic moment interaction. On the other hand, for the lightest nuclei substantial velocity dependent effects come into play from nuclear recoil in the scattering of fast electrons. The contribution of the nucleon magnetic moment to the scattering of electrons by protons has been calculated by ROSENBLUTH[1]; and has been calculated by JANKUS[2] for the scattering of electrons by deuterons. The coherent scattering resulting from magnetic effects in heavy nuclei should not be much greater than for a single nucleon, and, in the case of even-even nuclei, possibly less. However, at great energies and at large angles the elastic cross-section from a heavy nucleus is less than from a single proton; hence in this limit the non-COULOMB interaction may become important, although the fits to experimental data currently available give no indication of such an effect.

7. BORN approximation treatments. The BORN approximation [*16*] affords the simplest way to gain qualitative understanding of the effect of the radial extension and variation of the charge distribution on fast electron scattering [*18*]. While the BORN formula (7.1) gives a close description of the scattering for low Z elements, it becomes quite inaccurate for the heavier elements. The experience in non-relativistic scattering that the BORN approximation improves with energy does not hold[3] for the electron scattering of interest for study of nuclear charge distribution inasmuch as these electrons are already in the relativistic domain. Here the proportionality of energy to momentum implies that the phase shifts do not decrease with increasing energy: The phase shift of a relativistic electron in passing through a heavy nucleus is not small (compared to one radian) and is roughly proportional to the product $(R\,\Delta k)$ of nuclear radius by the shift of the wave number which occurs as the electron comes into the nuclear field. It follows that the BORN approximation is not accurate at any energy for heavy nuclei. We may nevertheless usefully employ this approximate method to gain qualitative insight for more precise calculations. Thus it is not surprising that the first discussions[4,5] of the effect of finite nuclear size on electron scattering were based on the BORN approximation.

The electron differential cross-section for scattering per unit solid angle,

$$\frac{d\sigma}{d\Omega} = \sigma(\vartheta) = \left[\frac{Z e^2 \cos\frac{\vartheta}{2}}{2E\sin^2\frac{\vartheta}{2}}\right]^2 F^2 \tag{7.1}$$

contains the form factor

$$F = \frac{1}{Z}\int\limits_0^\infty \varrho^{(c)}(r)\,\frac{\sin q r}{q r}\,4\pi r^2\,dr \tag{7.2}$$

in which q is the magnitude of the electron momentum change divided by $\hbar$, being related to the electron energy and angle of scattering by the relation

$$q = \frac{2E}{\hbar c}\sin\frac{\vartheta}{2}\,. \tag{7.3}$$

[1] M. ROSENBLUTH: Phys. Rev. **79**, 615 (1950).
[2] V. Z. JANKUS: Phys. Rev. **102**, 1586 (1956).
[3] G. PARZEN: Phys. Rev. **80**, 261 (1950).
[4] E. GUTH: Anz. Akad. Wiss. Wien, Math.-naturwiss. Kl. **24**, 299 (1934).
[5] M. E. ROSE: Phys. Rev. **73**, 279 (1948); **82**, 389 (1951).

At zero angle the form factor is unity. For a given shape of charge distribution F falls away from unity more rapidly with increasing angle as the charge becomes more extended. For a fixed size of the nucleus, the variation of F depends on the smoothness of the charge variation with the radius. Thus a uniform distribution gives rise to maxima and minima in the form factor (which are substantially greater than the maxima and minima given by an exact calculation), while a sufficiently gradual change of charge density, such as an exponential or gaussian dependence, gives rise to a monotonically decreasing form factor. These features are illustrated in Fig. 3.

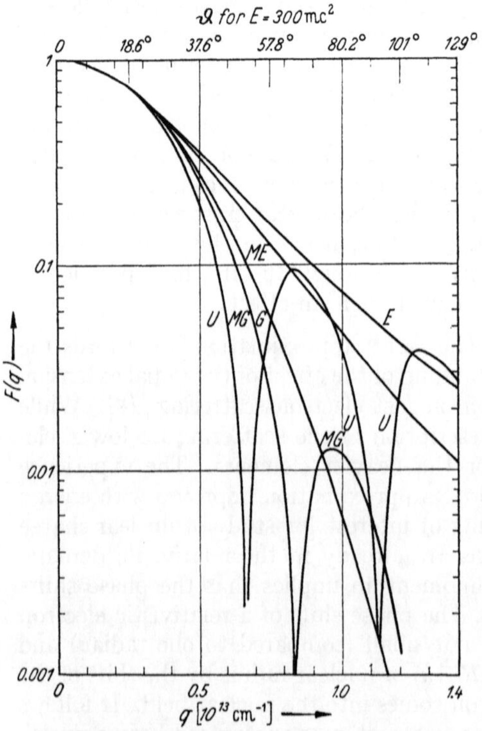

Fig. 3. BORN approximation form factors for high energy electron scattering. The absolute value of $F(q)$ is plotted *vs.* $q = (2E/\hbar c)\sin\frac{1}{2}\vartheta$. The mass number is chosen to be 200, and the radius of the uniform distribution, $1.20A^{\frac{1}{3}} = 7.02$ (this and other radii in units 10^{-13} cm). The definition, see (3.1) of the different shapes, and the sizes of each, are as follows:

U: uniform, $\varrho = \text{const.}$, $r < 7.02$.
MG: modified gaussian, $\varrho \sim (1 + x^2)\, e^{-x^2}$, $x = r/3.80$.
G: gaussian, $\varrho \sim e^{-x^2}$, $x = r/4.53$.
ME: modified exponential, $\varrho \sim (1 + x)\, e^{-x}$, $x = r/1.33$.
E: exponential, $\varrho \sim e^{-x}$, $x = r/1.64$.

The sizes are chosen such that the distributions have the same expectation value of $r^{2\sigma} \approx r^{1.6}$, and are therefore equivalent for bound electron effects. The curves are of qualitative significance only. Note that the gaussian shape gives the most rapid fall-off of cross-section at large angles; either a sharper or a less sharp fall-off of proton density produces more back scattering.

When the electron energy is sufficiently low that qr is much less than unity within the nucleus, or at small angles for any energy, $\frac{\sin qr}{qr}$ may be expanded to yield

$$F = 1 - \tfrac{1}{6}\, q^2 \langle r^2 \rangle_{\text{Av}} + \cdots. \tag{7.4}$$

This result, that scattering at low energy depends only on the mean square nuclear radius, is characteristic not only of the BORN approximation, but may be derived[1] by means of a partial wave expansion, and holds whenever the finite size of the nucleus has a substantial effect only on S-wave electrons.

a) Low energy scattering.

For purposes of the present discussion we shall speak of electron scattering below 60 Mev as low energy scattering, scattering above 60 Mev as high energy scattering. Although the early work[2] on electron scattering led to values of the radius of the nuclear charge distribution small relative to the then accepted nuclear radius, expressed by (1.1), there was sufficient uncertainty in the experiments and in their interpretation that the significance of the result was not widely appreciated. Subsequent work in the low energy region[3-6] as well as the high energy electron scattering and μ-mesonic x-ray studies, have borne out and made more precise these early indications.

---

[1] H. FESHBACH: Phys. Rev. **84**, 1206 (1951).
[2] E. LYMAN, A. HANSON and M. SCOTT: Phys. Rev. **84**, 626 (1951).
[3] C. L. HAMMER, E. C. RAKA and R. W. PIDD: Phys. Rev. **90**, 341 (1953).
[4] R. W. PIDD, C. L. HAMMER and E. C. RAKA: Phys. Rev. **92**, 436 (1953).
[5] W. PAUL and H. REICH: Z. Physik **131**, 326 (1952).
[6] R. W. PIDD and C. L. HAMMER: Phys. Rev. **99**, 1396 (1955).

8. Analysis of low energy scattering. For electrons of energy less than about 15 Mev, only the S wave phase shift deviates appreciably from its value for a point nucleus. Then for low Z, we note from (7.4) that the scattering may be described by the simple formula

$$\frac{\sigma}{\sigma_c} = \left[1 - \frac{1}{6}\left(\frac{Z e}{\hbar c}\right)^2 \langle r^2\rangle_{\text{Av}} \sin^2\frac{\vartheta}{2}\right]. \tag{8.1}$$

The evaluation of σ_c, the point nucleus cross-section requires a substantial numerical calculation. Early calculations[1,2] were made with an accuracy sufficient for electrons of a few Mev energy. More accurate evaluations of the point nucleus cross-section sufficient for treatment of scattering at several hundred Mev have since been made [*11*], [*23*]. For high Z, Eq. (8.1) is not applicable and an accurate fit to the low energy scattering data requires the numerical evaluations of the S wave phase shifts, as reported by several authors[3–5]. Table 1 summarizes the results obtained for both low and high Z. The results for the mean square radii of the nuclear charge distributions are expressed in terms of the radius of an equivalent uniform distribution:

$$R_{\text{eq}} = \left[\tfrac{5}{3}\langle r^2\rangle_{\text{Av}}\right]^{\frac{1}{2}}. \tag{8.2}$$

Table 1. *Nuclear radius measurements from low energy electron scattering.*

Results are given in terms of r_0 where $R_{\text{eq}} = r_0 A^{\frac{1}{3}}$, and the equivalent radius R_{eq} is defined by (8.2). The first clear-cut evidence for the presently accepted electrical radius of the nucleus appears to have come in the work of LYMAN, HANSON and SCOTT [*9*].

Element	Z	Energy Mev	Reference for data	Reference for analysis	Result for $10^{13}\, r_0$ (cm)
Silver . . .	47	15.7	(1)	(1)	1.15 ± 0.1
Gold . . .	79	15.7	(1)	(1)	1.15 ± 0.1
Tin	50	30—45	(2)	(2)	1.1 ± 0.1
Tungsten .	74	30—45	(2)	(2)	1.0 ± 0.1
Copper . .	29	15.7	(1)	(3)	1.0
Silver	47	15.7	(1)	(3)	1.1
Gold . . .	79	15.7	(1)	(3)	1.2
Tungsten .	74	31—60	(4)	(4)	1.18 ± 0.10

(1) E. LYMAN, A. HANSON and M. SCOTT: Phys. Rev. **84**, 626 (1951).
(2) C. HAMMER, E. RAKA and R. PIDD: Phys. Rev. **90**, 341 (1953).
(3) F. BITTER and H. FESHBACH: Phys. Rev. **92**, 837 (1953).
(4) R. W. PIDD and C. L. HAMMER: Phys. Rev. **99**, 1396 (1955).

9. Connection of phase shifts and radial moments. The results stated in (8.1) that the second moment of the charge distribution is measured by low energy scattering, when only S waves are significantly distorted by the extension of the nucleus, may be generalized[6] beyond the limits of the BORN approximation and shown to hold equally for heavy elements. The simplicity and utility of this result warrants its extension [*12*] to the higher energy region in which, say, 3 or 4 partial waves of the angular momentum decomposition are appreciably affected by the finite size of the nucleus.

[1] J. H. BARTLETT and R. E. WATSON: Proc. Amer. Acad. Arts a. Sci. **74**, 53 (1940).
[2] H. FESHBACH: Phys. Rev. **88**, 295 (1952).
[3] A. E. GLASSGOLD: Ph. D. Diss., Massachusetts Inst. of Technology 1954 (unpublished). — Phys. Rev. **94**, 757 (1954).
[4] F. BITTER and H. FESHBACH: Phys. Rev. **92**, 837 (1953).
[5] A. R. BODMER: Proc. Phys. Soc. Lond. A **66**, 1041 (1953).
[6] H. FESHBACH: Phys. Rev. **84**, 1206 (1951).

We shall see that $\langle r^{2k}\rangle_{\rm Av}$ is determined by the phase shift η_k of the partial wave denoted by the DIRAC quantum number k:

$$k = \mp (j + \tfrac{1}{2}) \quad \text{for} \quad j = l \pm \tfrac{1}{2}. \tag{9.1}$$

(The integer l is, of course, not a good quantum number in the relativistic theory of the electron, but is introduced as a convenient mode for describing the relativistic solutions in non-relativistic language. Thus we refer to the $k=-1$ solution as the "$S_{\frac{1}{2}}$" solution, the $k=+1$ solution as the "$P_{\frac{1}{2}}$" solution, etc.) We shall derive this result by a method which also makes clear the limits of its validity.

Consider the radial DIRAC equations for an electron in a central field:

$$\left.\begin{aligned} \frac{dF}{d\xi} &= \frac{k}{\xi} F - [\varepsilon - 1 - U(\xi)]\, G, \\ \frac{dG}{d\xi} &= -\frac{k}{\xi} G + [\varepsilon + 1 - U(\xi)]\, F. \end{aligned}\right\} \tag{9.2}$$

G and F are the large and small components of the wave function; ε is the total energy and $U(\xi)$ the potential energy, both in units of mc^2; ξ is the radius in units of $\lambda\!\!\!^{-}_c = \hbar/(mc)$.

The phase shift η_k for the l-th partial wave depends on $X = (F/G)$, the ratio of the DIRAC amplitudes for this component as evaluated at some radius $\xi = a$ outside the nucleus. It is, therefore, useful to reduce Eqs. (9.2) to a single equation for the ratio X. We consider alternative integral forms, depending on the sign of k:

$$\left(\frac{F}{G}\right)_\xi = -\xi^{2k} \int_0^\xi (\varepsilon - 1 - U)\left[1 + \frac{\varepsilon + 1 - U}{\varepsilon - 1 - U}\left(\frac{F}{G}\right)^2\right] \xi^{-2k}\, d\xi, \quad k < 0, \tag{9.3}$$

$$\left(\frac{F}{G}\right)_\xi = +\xi^{-2k} \int_0^\xi (\varepsilon + 1 - U)\left[1 + \frac{\varepsilon - 1 - U}{\varepsilon + 1 - U}\left(\frac{G}{F}\right)^2\right] \xi^{2k}\, d\xi, \quad k > 0. \tag{9.4}$$

We suppose, for the energy region in question, and for $k = \pm 1$ (therefore for all k), that there are no nodes of F or G for ξ smaller than the "fitting" radius a. Within the low energy region, the second terms in the square brackets of (9.3) and (9.4) contribute only a small part of the total value of the integrals. For 40 Mev electrons in Pb, for example, these terms contribute less than 5%. (Here the fitting radius $R_a = a\lambda\!\!\!^{-}_c$ must be outside the charge distribution but, subject to this restraint, may be quite close to the nuclear surface. The result just quoted applies under these conditions when the nuclear charge distribution falls off sharply at the nuclear surface as illustrated in Fig. 5.) An approximate solution for (9.3) and (9.4) is given when the point nucleus solutions for the amplitudes F and G are substituted in the right members of the equations. Meaningful results are obtained, however, even from the lowest approximation in which the terms of the integrands involving F and G are deleted. In this lowest approximation we have for positive k,

$$\left(\frac{G}{F}\right)_a \approx a^{-2k} \int_0^a [\varepsilon + 1 - U(\xi)]\, \xi^{2k}\, d\xi \tag{9.5}$$

with a similar relation for negative k. Integrating in (9.5) twice by parts the term involving $U(\xi)$ we find

$$\left(\frac{G}{F}\right)_a \approx \frac{a(\varepsilon + 1)}{2k + 1} + \frac{1}{2k}\,\frac{Ze^2}{\hbar c}\left[1 - \frac{1}{2k + 1}\,\frac{\langle r^{2k}\rangle_{\rm Av}}{R_a^{2k}}\right] \tag{9.6}$$

where $R_a = a\lambda\!\!\!^{-}_c$. The corresponding expression for negative k is given by replacing $(\varepsilon + 1)$ by $(\varepsilon - 1)$ in this formula, and by replacing k by $|k|$. We have thus demonstrated the result, stated at the beginning of this section. More detailed considerations indicate[1] the general result that only even moments of the charge distribution are needed to determine the scattering.

b) High energy scattering.

By far the greatest amount of information concerning the radial variation of nuclear charge has been derived from fast electron scattering, that is, for energies above 60 Mev. A beautiful series of measurements by HOFSTADTER and his associates at Stanford University [*15*] have provided the stimulus for a number of theoretical analyses [*11*], [*18*], and [*23*] of the data.

10. Analysis of high energy scattering. The analysis of the fast electron scattering experiments has necessarily taken the form of integrating the DIRAC equation (9.2) numerically. A series expansion, valid near the origin, of the amplitudes F and G derived from these differential equations is used to start the integration, which is carried forward to a radius a outside the charge distribution where the field corresponds to that of a point charge at the origin and where, therefore, the amplitudes for the actual electrical field may be expressed as a superposition of the regular and irregular solutions for the case of the point charge field. The amount by which the phase of the (regular) solution for the actual field is shifted from the phase of the regular solution for the pure COULOMB field is then computed for all values of the quantum number k for which the shift is appreciable. The phase shifts for the pure COULOMB field are well known for all values of k. As the method of calculation is familiar, we here summarize only the principal definitions and steps involved.

The phase shift η_k depends on the ratio,

$$X_k = \frac{F_k}{G_k} \tag{10.1}$$

evaluated at some radius, a, outside the nucleus. If we put $\eta_k = \zeta_k + \delta_k$ the difference δ_k between η_k and the phase shift ζ_k for a point charge field is given by

$$\tan \delta_k = \frac{\sin(\zeta_k^I - \zeta_k^R)}{C + \cos(\zeta_k^I + \zeta_k^R)} \tag{10.2}$$

where

$$C = -\frac{X_k G_k^I(a) - F_k^I(a)}{X_k G_k^R(a) - F_k^R(a)} \tag{10.3}$$

is expressed in terms of the regular and irregular solutions of (9.2). The difference $(\zeta_k^I - \zeta_k^R)$ between the point charge phase shifts for the irregular and regular amplitudes, respectively, is given by

$$\zeta_k^I - \zeta_k^R = \pi\left(\varrho_k + \frac{1}{2}\right) - \operatorname{arc\,tan}\left(\frac{\operatorname{Tan} \pi\gamma}{\tan \pi \varrho_k}\right) \tag{10.4}$$

in which

$$\varrho_k = (k^2 - Z^2\alpha^2)^{\frac{1}{2}}, \quad \gamma = Z\alpha\varepsilon(\varepsilon^2 - 1)^{-\frac{1}{2}}, \quad \alpha = \frac{e^2}{\hbar c}. \tag{10.5}$$

Combining these values of the phase shifts δ_k with the phase shifts for the pure COULOMB field, ζ_k, as given [*16*] by

$$e^{2i\zeta_k} = \frac{-k - i\gamma'}{\varrho_k - i\gamma} \frac{\Gamma(\varrho_k + 1 - i\gamma)}{\Gamma(\varrho_k + 1 + i\gamma)} e^{-i\pi(\varrho_k - l)} \tag{10.6}$$

---

[1] A. E. GLASSGOLD: Ph. D. Diss., Massachusetts Inst. of Technology 1954, unpublished. — Phys. Rev. **94**, 757 (1954).

with $\gamma' = \gamma \varepsilon^{-1}$, and l defined by (9.1), we obtain η_k. With the total phase shifts η_k for the actual field thus in hand for all values of k, the expressions

$$f = \frac{\lambda\!\!\!^{-}}{2i} \sum_{k \neq 0} [|k| + \operatorname{sign}(k)]\, e^{2i(\zeta_k + \delta_k)} P_k(\cos\vartheta), \tag{10.7}$$

$$g = \frac{\lambda\!\!\!^{-}}{2i} \sum_{k \neq 0} [-\operatorname{sign}(k)]\, e^{2i(\zeta_k + \delta_k)} P_k'(\cos\vartheta), \tag{10.8}$$

may be summed to yield the differential angular cross section for fast electron scattering as given by the well known formula

$$\sigma(\vartheta) = |f|^2 + |g|^2. \tag{10.9}$$

When the electron energy is large, $\varepsilon \gg 1$, the electron mass may be neglected relative to the total energy, with considerable simplifications in the analysis. The number of phase shifts entering the calculation is halved[1], but the computed scattering cross-section is but little affected[2] by this approximation.

11. Analysis of data. To determine the charge distribution which most nearly produces a theoretical cross section proportional throughout the range of observation to the observed angular distribution of scattered electrons, cross sections are calculated [*11*] as just described for a series of possible choices of parameters in (22.5) to (22.7) defining an assumed charge distribution. The cross sections so obtained are then folded into the estimated angular resolution with which the electron scattering data are taken and these folded cross sections compared with the angular distributions observed.

Because the intensity of the angular distribution varies by a factor of about 10^5 in the angular range of observation, comparison of theory and experiment is facilitated by taking the ratios of the predicted and observed distributions at the angles of observation. The constancy of this ratio at all points would indicate an exact fit. Hence the standard deviation of this ratio is taken as the measure of closeness of fit; that is, the fit is judged by the value of D where

$$D^2 = \frac{\sum_{i=1}^{I} w_i (\gamma_i - \bar{\gamma})^2}{\bar{\gamma}^2 \sum w_i} \tag{11.1}$$

and $\gamma_i = \sigma_{\exp}(\vartheta_i)/\sigma_{\text{th}}(\vartheta_i)$ is the ratio of observed to calculated cross section at the angle ϑ_i. Here $\gamma = \sum w_i \gamma_i / \sum w_i$, the weights w_i being chosen such that each value γ_i makes the same probable contribution to the error in $\bar{\gamma}$, as determined by counting statistics and other known sources of variation in the experiments. Plotted in Fig. 4 are the values of D for various choices of parameters n and R in the neighborhood of the best fit for a charge distribution in lead described by (22.6).

Both from the behavior of individual cases, such as the one illustrated in Fig. 4, and from the consistency in results obtained as the electron energy and the target atomic number are varied, we conclude that the charge distribution in the neighborhood of Pb^{208} is determined to about one unit in n and to 1%

[1] HERMAN FESHBACH: Scattering of Electrons by Nuclei, Proceedings of the U.S. National Bureau of Standards Symposium on Electron Physics, 1951. National Bureau of Standards Circular 527, 1954.

[2] L. R. B. ELTON: Proc. Phys. Soc. Lond. A **68**, 741 (1955).

in r_1. The values of D so determined for lead and gold over a range of possible charge distributions are tabulated in Table 2a. In the second part of this table are listed the parameters of the charge distributions giving the closest fit to the electron angular distributions. It has been found possible to fit the angular distributions at different energies with substantially the same charge distributions. This fact suggests that radiative corrections and other energy dependent effects do not significantly change the observed angular distributions.

The most careful study of the radial variation of charge distribution as given by electron scattering has been made for elements in the neighborhood of lead. The charge distributions so determined are plotted in Fig. 5 with indication of

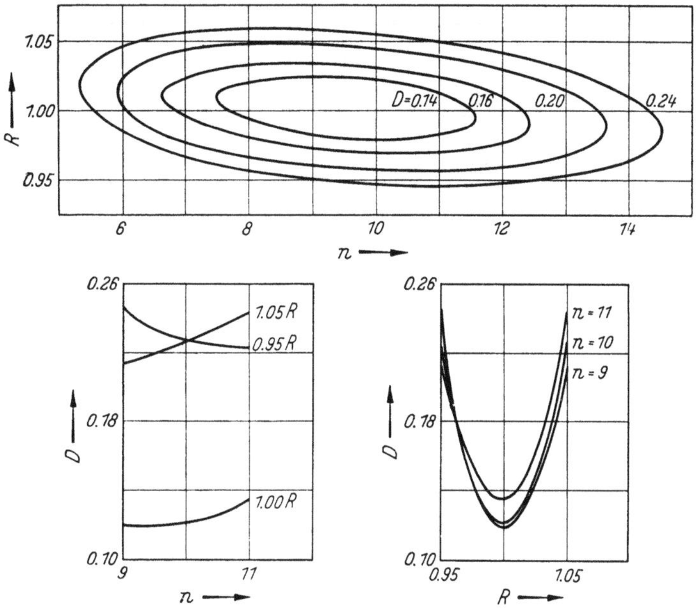

Fig. 4. Sections of the surface defined in Sect. 11 by the fit measure, D, and listed in Table 2, for high energy elastic scattering of electrons on nuclei. The particular domain of D illustrated is for gold at 183 Mev, and is obtained by fitting the expansion,

$$D(n, R, S=0) = D_0 + D_n\, n + D_R\, R + 2D_{nR}\, nR + D_{nn}\, n^2 + D_{RR}\, R^2$$

to the nine values of D corresponding to $n=9$, 10, and 11 for $R=0.95$, 1.00, and 1.05.

the uncertainties as stated above in the parameters defining these distributions. In Fig. 6 are plotted the observed angular distributions in gold at four different energies with the theoretical distribution indicated by the solid lines, as determined from the "best fit" function plotted in Fig. 5.

It should be noted that nuclei in the neighborhood of lead are expected to be nearly spherical so that the function plotted in Fig. 5 may be expected to be a close approximation to the true radial distribution of charge—not a radial distribution showing the more gradual surface gradient produced by averaging over angles in a non-spherical nucleus. When the survey of fast electron scattering is extended to other nuclei which are expected to be approximately spherical, it is found [15] that the central density remains about constant and that the surface slope or surface thickness also remains about constant over a wide range of mass numbers, as shown in Fig. 7.

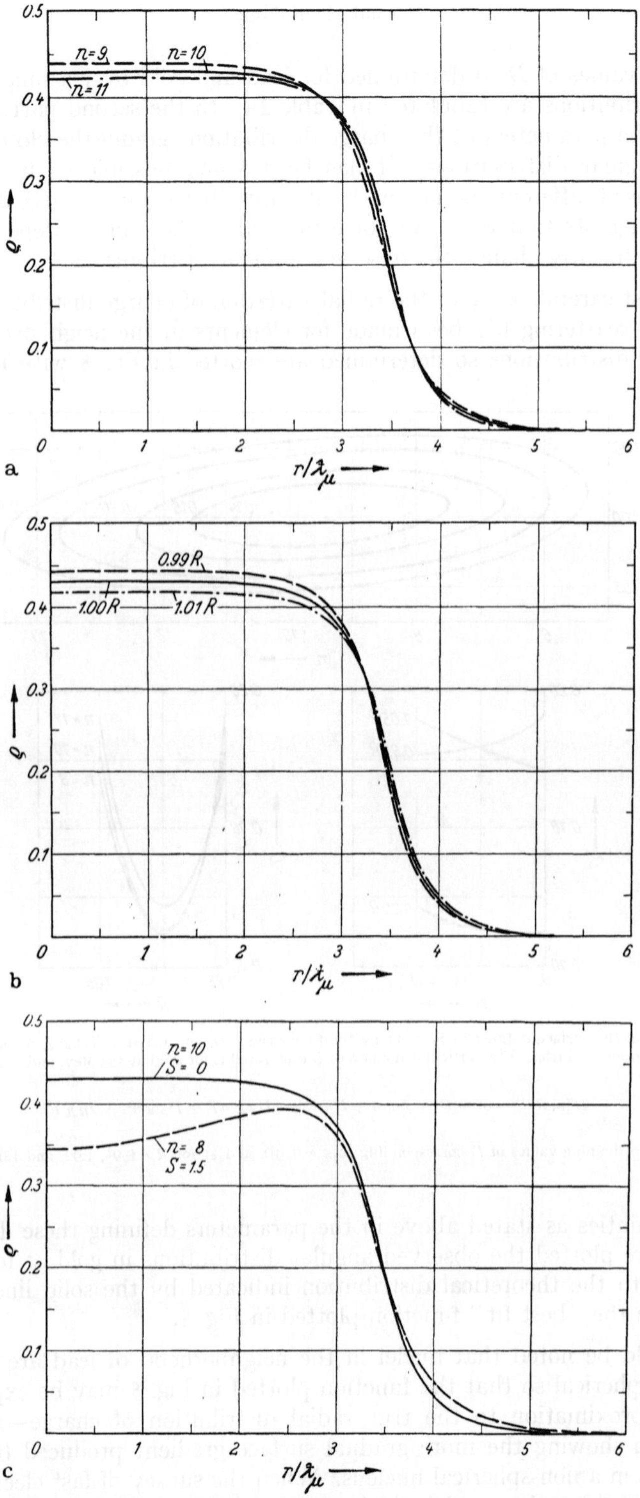

Fig. 5a–c. The nuclear charge density for Pb^{208} as a function of radius, measured in units of $ƛ_\mu = 1.867 \times 10^{-13}$ cm. Charge density units are positive electron charges per $ƛ_\mu^3$. Determinations [*11*] of $\varrho(r)$, as defined in (22.6) and sequel, for high energy electron elastic scattering [*15*] on lead and gold indicate an uncertainty (Table 2) of less than 1 unit in n and less than 1% in R. Fig. (a) and (b), respectively, illustrate these maximum uncertainties about the best fit ($n = 10$, $R = 1.00$, $S = 0$). Fig. (c) compares this best fit with the most nearly acceptable distribution ($n = 8$, $R = 1.05$, $S = 1.5$) with a central depression of charge density as suggested from elementary considerations of proton redistributions. It should be noted that the 5% discrepancy in this case with the nuclear electrical radius given by analysis [13] of μ-mesonic x-rays is in sharp contrast with the agreement of about 1% in all cases for the best fit with no central depression of charge density.

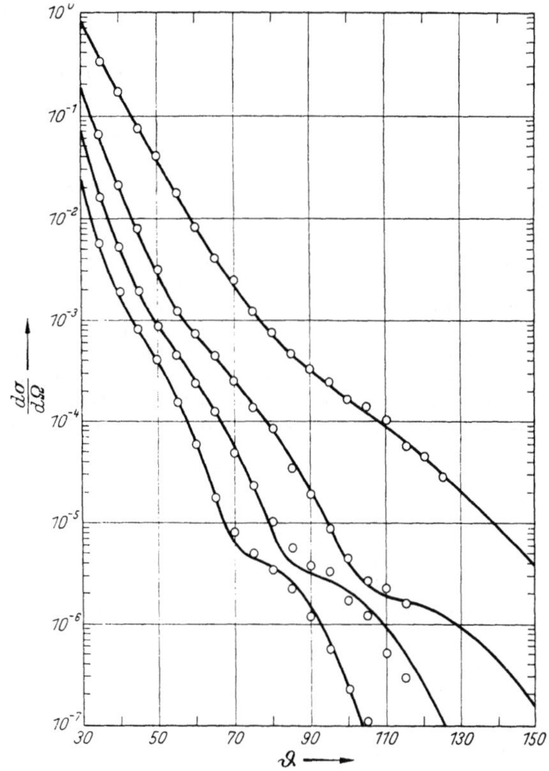

Fig. 6. Absolute differential cross-sections for fast electron elastic scattering on gold as functions of the angle ϑ. The points show the scatter of the experimental data about $d\sigma/d\Omega$ computed for $n=10$, $R=1.00$, $s=0$. The curves, proceeding from top to bottom, are for 84, 126, 154, and 183 Mev. The experimental data are normalized by $\bar{\gamma}$, defined after (11.1).

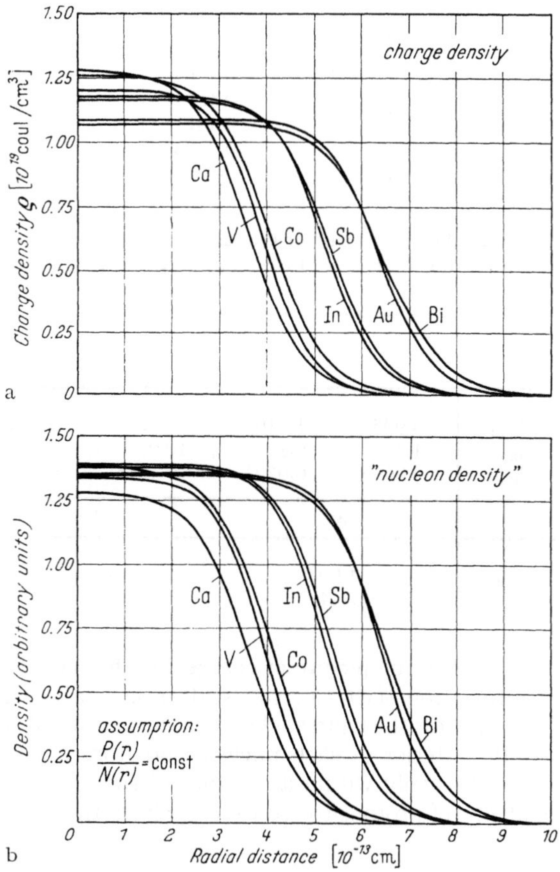

Fig. 7a and b. (a) Charge distributions $\varrho(r)$ ror Ca, V, Co, In, Sb, Au, and Bi. They are Fermi smoothed uniform shapes as determined in [*15*]. (b) A plot of $(A/Z)\varrho(r)$ for the above nuclei. On the assumption that the distribution of matter in the nucleus is the same as the distribution of charge, this function represents the "nucleon density".

Table 2.

(a) Comparison of experimental results [*15*] on high energy electron scattering with predictions [*11*] for an illustrative set of assumed charge distributions, energies, and targets. The values of D, as defined in (11.1) are tabulated for shapes of charge distribution defined by (22.6) et seq. The restriction of spherical symmetry probably involves no significant error for nuclei in the neighborhood of Pb^{208}. Also tabulated is $D_{\text{statistical}}$, the most probable value of D for an optimal theoretical fit if the only source of experimental error were the fluctuations due to a finite number of counts. The parameter labeled R is the ratio of the radial constant used to the radial constant determined by the μ-mesonic x-ray analysis [*13*]. The best fit for all energies for both lead and gold is in the neighbourhood of $n=10$, $s=0$, and $R=1.00$. (The possible charge distribution $n=8$, $s=1.5$, $R=1.05$ seems unlikely in view of the close agreement of all other good fits (small D) with the radius determined from the μ-mesonic $2\,P \to 1\,S$ transition, as indicated in part (b) of the table. Further increases in R for the cases with $s \neq 0$ do not give better fits than the best fits shown in the table.)

		Pb at 153 Mev			Au at 183 Mev		
		$D_{\text{statistical}}$					
		0.0771			0.0637		
		D_{observed}					
Shape parameters n	s	$0.95\,R$	R	$1.05\,R$	$0.95\,R$	R	$1.05\,R$
0 (exponential)	0					0.5063	
∞ (uniform)	0		0.3176			0.3206	
9	0	0.2165	0.1016	0.1450	0.2452	0.1231	0.1983
10	0	0.2018	0.1064	0.1713	0.2275	0.1214	0.2208
11	0	0.2017	0.1271	0.1957	0.2225	0.1354	0.2418
8	1.5		0.1921			0.2162	0.1198
8	3		0.2423			0.2637	0.1554
8	3		0.3583			0.3658	0.2947

(b) In the second part of the table are listed the values of n and R (for $s=0$) which locate the minimum of D in each case, as determined by the procedure described in the caption of Fig. 4.

E	Au				Pb		
Mev	84	126	154	183	84	153	186
$R_{\min}$	0.998	1.010	1.020	1.002	1.018	1.014	0.987
$n_{\min}$	11.37	9.56	10.05	9.58	10.20	8.44	10.88
$D_{\min}$	0.0656	0.0723	0.1868	0.1209	0.0575	0.0979	0.1343

	Au	Pb	Au Pb
$R_{\min\,\text{Av}}$	1.008	1.006	1.007
$n_{\min\,\text{Av}}$	10.14	9.84	10.01

12. Resulting picture of charge distribution. The picture of charge distribution resulting from fast electron scattering analysis may be summarized as follows: (a) except for the lightest elements the density of charge is about constant throughout the interior of each nucleus and from element to element. (b) The surface falls off sharply within a distance of the order of 2×10^{-13} cm. (c) The expected tendency of protons to crowd away from the nuclear center and to leave a central charge depression is not borne out. (d) Expressed in terms of the "mean square radial constant" of (1.1), the radial size is

$$\boxed{r_0 = 1.20 \pm 0.01.} \tag{12.1}$$

It is highly satisfactory that this radial size agrees within 1% with the radial size determined from the analysis of μ-mesonic x-rays.

In view of the consistency of the fit between theory and experiment for different energies and different elements and as obtained by different authors, it seems that such slight differences as may still exist between theory and experiment can be largely accounted for by the limitations of the particular two and three parameter families of charge distributions used to fit the data.

13. Corrections to the theoretical analysis. We have confined our attention thus far to nuclei which were nearly spherical, and we shall defer until Part C consideration of the effects of aspherical nuclear charge.

α) *Inelastic scattering.* The data intended to represent elastic scattering may be affected by inelastic scattering in two ways. First, because of the finite energy resolution in the observations, particles scattered inelastically with small energy loss may be counted as elastic particles. Such inelastic collisions may be attributable to bremsstrahlung (discussed below) or to the excitation of low-lying nuclear states[1,2]. The latter effect, which provides an interesting new way to study nuclear levels, is not of particular importance for the analysis of charge distributions inasmuch as there are a number of nuclei in which the first excited levels are sufficiently above the ground state that the inelastically scattered electrons may easily be separated from the elastically scattered particles. At the other extreme are the very low-lying rotational states which will be discussed in Parts C and D.

A second effect of inelastic scattering may be more important, just as it is less amenable to simple quantitative discussion. The existence of inelastic processes modifies the outgoing elastic wave. Whereas in Born approximation, elastic and inelastic processes are independent, it must be expected that this independence does not carry through in actual scattering processes, particularly in heavy elements. We do know, however, that in the true situation, as in the Born approximation, the outgoing elastic wave is only slightly distorted from a plane wave for energies well in excess of 100 Mev. Therefore, even though at large angles the elastic and inelastic cross sections are observed to be of comparable magnitude, we are led to believe that the processes are still approximately independent so that the modification of elastic scattering by the inelastic processes may be rather small.

The magnitude of the inelastic cross section may be roughly estimated by means of the Born approximation. We consider the situation in which the more probable excitation energies ΔE are small compared to the electron energy and make an approximate sum for transitions to all nuclear states. Then the result[3,4] follows,

$$\sigma(\vartheta) = \sigma_c \left[|F|^2 + \frac{1}{Z}\left(1 - |F|^2\right) \right], \tag{13.1}$$

[1] J. A. Thie, C. J. Mullin and E. Guth: Phys. Rev. **87**, 962 (1952).

[2] L. I. Schiff: Phys. Rev. **96**, 765 (1954).

[3] J. H. Smith: The Scattering of High Energy Electrons by Heavy Nuclei. Doctoral thesis, Cornell University, Ithaca, N.Y. 1951 (unpublished).

[4] K. W. Ford: Personal communication.

in which σ_c is the point-nucleus cross section and F is the form factor given in (7.2). Note that the inelastic term is of order Z^{-1}; it vanishes in the forward direction where F becomes unity, but the inelastic term may become dominant at large angles where $|F| \ll 1$. If we make the alternative assumption that the inelastic cross section arises from elastic collisions between the electrons and single protons in the nucleus, and further ignore the exclusion principle, then the second term in (13.1) becomes just Z^{-1}. This form does agree with (13.1) at large angles for which the proton recoil is large, but it disagrees at small angles, where, in fact, the exclusion principle must operate to reduce the inelastic scattering from single proton encounters. Hence (13.1) is the better approximation to the actual scattering.

Evidently if the energy resolution of the experiment is poor, the observed electron intensity in backward directions will be erroneously large, and hence the analysis will lead to the conclusion that the charge distribution drops off more sharply at the surface than it actually does. If the energy resolution is high relative to the energy interval to the first excited states, then the qualitative arguments as to the effect of inelastic processes on the analysis clearly do not apply. In much of the work that has been done [*15*], especially in the neighborhood of lead, this favorable situation for study of elastic scattering has prevailed.

β) Non-COULOMB *forces.* In Sect. 6 we have already seen that one of the principal advantages of the electron as a nuclear probe is that the non-COULOMB interactions between the electron and the nucleus are probably negligible in their effect on observed elastic scattering below 200 Mev for angles up to 120°, as illustrated in Fig. 6.

γ) Granularity of nuclear charge. The difference between the idealized smooth charge distribution employed in the theoretical analysis of electron scattering and the actual distribution of charge should make itself felt primarily in the inelastic processes, as discussed above. As noted in Sect. 14, scattering from single protons indicates that the fast electron sees a finite extension of the proton charge and magnetic moment for which the diameter is somewhat less than the average separation of nucleons in nuclei, about 2.5×10^{-13} cm. This granularity of charge should have some effect on the elastic scattering itself inasmuch as the velocity of the electron in passing through the nucleus is very nearly the velocity of light, thus being sufficiently more rapid than the nucleon velocities within the nucleus that the fast electron "sees" an almost instantaneous snapshot of the nucleons. Estimates of the size of this effect, employing wave packets for the electrons, indicate that the difference between the effect of uniform charge distribution and the clustered charge distribution actually present is not of significant size for coherent scattering.

δ) Radiative corrections. The largest radiative correction is due to bremsstrahlung, but for some time it has been known[1-3] that, in first order in the BORN approximation, electrons which have emitted soft protons are scattered with the same angular distribution as electrons which have emitted no photons. Recently it has been further shown[4] that to all orders in the BORN approximation electrons which have emitted single soft photons are scattered with very nearly the same angular distribution as non-radiating electrons. While the role of other processes and of higher order corrections (in number of photons) is yet

[1] M. E. ROSE: Phys. Rev. **73**, 279 (1948); **82**, 389 (1951).
[2] F. BLOCK and A. NORDSIECK: Phys. Rev. **52**, 54 (1937).
[3] J. SCHWINGER: Phys. Rev. **75**, 898 (1949). See also reference 1, p. 192.
[4] H. SUURA: Phys. Rev. **99**, 1020 (1955).

to be studied, information now available suggests that corrections to angular distributions may be small even at several hundred Mev and for high Z. This expectation is borne out as noted in Sect. 11 by the possibility of fitting angular distributions over a broad energy range with the same nuclear charge function.

14. Size of lightest nuclei. In addition to the work cited on nuclei of medium and high atomic number, fast electron scattering has also been applied to gain knowledge of the charge distribution in the lightest nuclei[1-3]. An important effect in scattering of fast electrons from protons is the contribution of the magnetic moment of the proton which gives greater intensity for back-angle scattering than would be expected from point charge scattering alone. The experimental points are found to fall midway between the predictions for point charge scattering and the predictions[4] for point charge scattering including the effect of the anomalous moment of the proton localized in the point. While this discrepancy may find its explanation in features of the electron-proton interaction not yet understood, it seems plausible to examine how the fit may be improved by assuming finite extension of the proton charge and magnetic moment. It is found that the scattering data may be reproduced by a calculation invoking a model of the proton in which the charge and magnetic moment are both taken to have finite extension corresponding to a root mean square radius of 0.8, or with $R_{\text{eq}} = 1.0$. It is interesting that this value for the "proton radius" is only 20% smaller than the r_0 of (12.1) which has the meaning of the radius of the volume occupied per nucleon in heavy nuclei. Accepting the present result literally then implies that the proton "size" is almost as large as the volume which the proton occupies in heavy nuclei. If we assume the neutron to have about the same radial extension as the proton, then we obtain a picture of neutrons and protons indeed closely packed together in heavy nuclei.

For the α-particle the analysis[3] of data is made directly in terms of finite extension. The resulting RMS radius of 1.5 or slightly less than twice the extension of the proton charge, suggests that nucleons are close-packed even in the lightest nuclei, but the RMS radius of 2.0 for the deuteron[2] is consistent with the low binding energy of this nucleus. Thus it appears that an equation with the form of (1.1) may, when applied to the charge distribution, have approximate validity for all nuclei.

II. Bound electrons.

The very great accuracy with which electronic energy level differences may be measured largely compensates for the smallness of effects due to the finite size of the nucleus on the energies of bound electrons. Observations on the properties of negative electronic energy states can therefore yield some information on the nuclear charge distribution. We distinguish three sources for information of this kind.

a) X-ray fine structure.

15. Energy shifts of bound states. In optical and x-ray spectra, the effects of finite nuclear size are seen in terms of differences between lines corresponding to states that are affected to different degrees by the finite extension of the nucleus. By far the largest energy displacements occur for the $S_{\frac{1}{2}}$ and the $P_{\frac{1}{2}}$ states, compared with which the energy displacements of all other electronic

---

[1] R. Hofstadter and E. E. Chambers: Bull. Amer. Phys. Soc., Ser. II **1**, 10 (1956).
[2] J. A. McIntyre and R. Hofstadter: Phys. Rev. **98**, 158 (1955).
[3] R. Blankenbecler and R. Hofstadter: Bull. Amer. Phys. Soc., Ser. II **1**, 10 (1956).
[4] M. N. Rosenbluth: Phys. Rev. **79**, 615 (1950).

states are negligible. The sensitivity of an electronic state to the nuclear size depends upon the electronic density in the neighborhood of $r=0$ for the state in question. The significance of the electronic density variation in the neighborhood of the nucleus is most easily seen by use of first order perturbation theory, which is known to be inadequate for computing the magnitude of the finite nucleus effects, yet has been shown by a number of interesting calculations to give the correct dependence of energy shifts on the size and shape of the nuclear charge distribution. According to first order perturbation theory, the energy shift of the n-th electronic level may be written

$$\Delta E_1 = 4\pi \int_0^\infty \varrho^{(c)}(r)\, V_e^{(0)}(r)\, r^2\, dr \tag{15.1}$$

where $V_e^{(0)}(r)$ is the potential due to the (unperturbed) n-th electron, normalized to be zero at the origin. The unperturbed electronic charge density is given by the relativistic expression

$$\varrho_e^{(0)} = -\frac{e}{r^2}\left(F_0^2 + G_0^2\right) \tag{15.2}$$

in which F_0 and G_0 are the small and large components of the point nucleus solution of the radial DIRAC equation, and we have used the normalization,

$$4\pi \int_0^\infty \left(F_0^2 + G_0^2\right) dr = 1\,. \tag{15.3}$$

In the vicinity of the nucleus, the electronic density is given by

$$\varrho_e^{(0)}(r) = -A\, e\, r^{2\sigma-2} \tag{15.4}$$

for the $S_{\frac{1}{2}}$ and $P_{\frac{1}{2}}$ states, the normalization constant A being larger in the former case, and with

$$\sigma = \sqrt{1-(Z\alpha)^2} \approx \sqrt{1-\left(\frac{Z}{137}\right)^2}\,. \tag{15.5}$$

Thus for the $S_{\frac{1}{2}}$ and $P_{\frac{1}{2}}$ states, there is a sharp rise in the unperturbed functions near the origin as indicated by the negative exponent in (15.4). For all other states the exponent of the density near the origin is positive and the electronic potential therefore has a much smaller variation near the origin than does the potential for these two states:

$$V_e^{(0)}(r) = \left[\frac{4\pi A e}{2\sigma(2\sigma+1)}\right] r^{2\sigma}\,. \tag{15.6}$$

Then the perturbation energy shift defined in (15.1) takes the form

$$\Delta E_1 = \left[\frac{4\pi Z e^2 A}{2\sigma(2\sigma+1)}\right] \langle r^{2\sigma}\rangle_{\mathrm{Av}}\,. \tag{15.7}$$

Thus the property of the charge distribution determined by experiment is the average value of $r^{2\sigma}$,

$$\langle r^{2\sigma}\rangle_{\mathrm{Av}} = (Z)^{-1} \int_0^\infty r^{2\sigma}\, \varrho^{(c)}(r)\, 4\pi r^2\, dr \tag{15.8}$$

and this fact is unaltered by the modifications required to convert first order perturbation theory into an accurate theory[1,2] [*10*], [*6*], [*9*], [*13*].

---

[1] W. HUMBACH: Z. Physik **133**, 589 (1952).
[2] J. A. WHEELER: Rev. Mod. Phys. **21**, 133 (1949).

16. L_{II}—L_{III} x-ray separation. The fine structure splitting of the 2P states of heavy elements has been calculated[1] and employed to measure the fine structure constant. The finite size of the nucleus and electrodynamic processes contribute significantly but with opposite sign to this splitting. The fact that a rather accurate value for the fine structure constant resulted from this early work is partly connected with the fact that both sorts of corrections were omitted. It has been more recently made clear[2] that the observed x-ray fine structure can be fit only if one adds to the theory based on the point nucleus a very small, but a very sharply Z-dependent term. The finite nucleus effect has such a strong Z-dependence because, in addition to the usual Z^3 rate of increase of electron density at the nucleus, it depends on the rapid rate of increase of the small component of the $P_{\frac{1}{2}}$ wave function with Z, as relativistic effects become more important. The nuclear charge radius thus deduced by SCHAWLOW and TOWNES corresponds to

$$r_0 \approx 2 \times 10^{-13}\ \text{cm}. \tag{16.1}$$

It should be stressed, however, as suggested by SHACKLETT[3], that much of the anomalously large value given by (16.1) probably results from inaccurate values of the $L_{\alpha 2}$ wavelengths for several high—Z elements used in the analysis. The much more precise data recently obtained by SHACKLETT yields, by the same analysis, involving theoretical values for the nuclear size effect, a value of nuclear radius corresponding to

$$r_0 = 1.07. \tag{16.2}$$

Comparable in magnitude to the effect of finite nuclear size on the fine structure splitting, is the aggregate of radiative corrections to the $P_{\frac{1}{2}}$ level[4, 5] [*10*]. The vacuum polarization contribution has been evaluated[6], and is opposite in sign to the reactive or LAMB shift which is more difficult to compute in heavy elements. As SHACKLETT has observed[3], when the theoretical splitting is calculated with corrections for vacuum polarization and a nuclear radius constant of $r_0 = 1.2 \times 10^{-13}$ cm, a comparison with experiment shows that a discrepancy remains which is then used to evaluate an empirical correction term. The sign, magnitude, and Z-dependence of this term suggest that the remaining discrepancy might arise from the LAMB shift effect.

17. X-ray isotope shift. In principle an excellent method for separating nuclear volume effect and electrodynamic effects for deeply bound electronic states in heavy nuclei is provided by the x-ray isotope shift. The electrodynamic processes are almost entirely Z-dependent and would therefore contribute almost nothing to the shift of energy levels among different isotopes of the same element. Thus the nuclear volume effect would be subject to relatively simple study. Successful measurements of the x-ray isotope shift have not yet been reported[7], but appear to be just within experimental reach. The same uncertainties which apply to nuclear radius determination by optical isotope shifts are, of course, present in the isotope shifts of the more tightly bound levels.

[1] R. F. CHRISTY and J. M. KELLER: Phys. Rev. **61**, 147 (1942).

[2] A. L. SCHAWLOW and C. H. TOWNES: Science, Lancaster, Pa. **115**, 284 (1952). — Phys. Rev. **100**, 1273 (1955).

[3] R. L. SHACKLETT: A Precision Measurement of the $L_{II}-L_{III}$ X-Ray Energy Level Difference in Some Heavy Elements and a Comparison with the Predictions of the SCHAWLOW-TOWNES Theory of the Nuclear Size Effect, Doctoral Diss., California Institute of Technology, 1956.

[4] C. H. TOWNES: Phys. Rev. **94**, 773 (1954).

[5] N. M. KROLL: Phys. Rev. **94**, 747 (1954).

[6] E. WICHMANN and N. M. KROLL: Phys. Rev. **96**, 232 (1954).

[7] G. IGO and M. S. WERTHEIM: Phys. Rev. **95**, 1097 (1954). References to experimental work are given in this paper.

b) Optical fine structure.

18. Optical isotope shift. Recent increased understanding of the regularities in atomic isotope[1–6] shifts has stimulated considerable theoretical activity in the interpretation[7–10] of the shifts [22]. Whereas for the x-ray fine structure the energy difference is measured between the $2P_{\frac{1}{2}}$ and the $2P_{\frac{3}{2}}$ states in the same isotope, in the optical (and x-ray) isotope shift the energy difference is measured between, at least in first order, the same $S_{\frac{1}{2}}$ state in two or more isotopes of the same element. As mentioned in Sect. 15, the effect of nuclear size is greatest on the $S_{\frac{1}{2}}$ level and much greater than on any other save the $P_{\frac{1}{2}}$ level. It should be emphasized that we are here discussing the isotope shift as it appears in heavy nuclei, for which the dominant effect is the finite extension of the nuclear charge, while the effect on electronic motion of adding one or more units to the nuclear mass becomes almost negligible in the heavier atoms. It therefore follows (Sect. 15) that the observed isotope shift in heavy elements measures the difference

$$\delta\langle r^{2\sigma}\rangle = \langle r^{2\sigma}\rangle_{Z,N+1} - \langle r^{2\sigma}\rangle_{ZN}. \tag{18.1}$$

For spherically symmetric nuclei which follow an $A^{\frac{1}{3}}$ law of nuclear radius, the fractional increase of radius is simply $\delta A/3A$ and

$$\delta\langle r^{2\sigma}\rangle = \left(\frac{2\sigma}{3}\right)\frac{\delta A}{A}\langle r^{2\sigma}\rangle. \tag{18.2}$$

The fact that with rare exceptions nuclei are not spherically symmetric has a large effect on isotope shifts and provides an important source of information on nuclear angular shapes (see part C). However, from the regular variation of nuclear deformations and the associated regular variation of isotope shifts, it is possible to deduce [22] the magnitude of the isotope shift corresponding to spherical symmetry. Interpreting this magnitude for an incompressible nuclear model with uniform charge distribution (that is, still assuming an $A^{\frac{1}{3}}$ law for nuclear charge extension) we find

$$R_{\text{eq}} = (0.9 \pm 0.1)\, A^{\frac{1}{3}}, \tag{18.3}$$

a radius appreciably too small to agree with the values determined from μ-mesonic atoms or from electron scattering. However, it is well known that nuclei possess a finite compressibility and when this nuclear property is introduced [22] into the theory of isotope shifts, the discrepancy indicated by (18.3) may be removed. The effect of compressibility is that the nuclear density depends somewhat on the neutron-proton ratio. Thus when a neutron is added to the nucleus, the density increases slightly due to the diminished COULOMB energy, and the resultant fractional increase of radius is less than $A/3$. The isotope shift

---

[1] P. BRIX and H. KOPFERMANN: Z. Physik **126**, 344 (1949). — Festschr. Akad. Wiss. Göttingen, Math.-physik. Kl. **17** (1951); other references here. — Phys. Rev. **85**, 1050 (1952).

[2] M. F. CRAWFORD and A. L. SCHAWLOW: Phys. Rev. **76**, 1310 (1949).

[3] P. BRIX and H. FRANK: Z. Physik **127**, 289 (1950).

[4] K. MURAKAWA and J. S. ROSS: Phys. Rev. **83**, 1272 (1951).

[5] P. BRIX, H. VON BUTTLER, F. G. HOUTERMANS and H. KOPFERMANN: Nachr. Akad. Wiss. Göttingen, Math.-physik. Kl. **7** (1951).

[6] H. ARROE: Phys. Rev. **93**, 94 (1954).

[7] E. K. BROCH: Arch. Mat. Naturvidensk. **48**, 25 (1945).

[8] G. BREIT, G. B. ARFKEN and W. W. CLENDENIN: Phys. Rev. **77**, 569 (1950); **78**, 390 (1950).

[9] G. BREIT: Phys. Rev. **86**, 254 (1952).

[10] W. HUMBACH: Z. Physik **133**, 589 (1952).

is consequently smaller than for constant density nuclei, and therefore an analysis of the actual shifts on the basis of incompressible nuclei leads to the low radius given in (18.3).

19. Determination of nuclear compressibility. The nuclear compressibility is commonly defined for a hypothetical nucleus in which COULOMB forces are absent and which contains equal even numbers of neutrons and protons. This hypothetical nucleus is taken to be sufficiently large that the surface tension does not appreciably effect nuclear density. If the energy is written in the form

$$E = E_v + E_s, \tag{19.1}$$

then E_v is proportional to volume and E_s is proportional to the surface. The nuclear compressibility coefficient has been defined as

$$E_0'' = R^2 \left(\frac{\partial^2 E_v}{\partial R^2} \right)_{R=R_0} \tag{19.2}$$

in which the radial derivative of the volume energy is taken in the neighborhood of R_0, the equilibrium radius. Early estimates[1] making use of the virial theorem and various analytic forms for the nuclear potential energy gave values for E_0'' ranging from 100 to 150 Amc^2. Recent evidence[2] from a theory of nuclear saturation suggests E_0'' about 50 Amc^2. As these estimates are uncertain to perhaps 50%, they may be said to be in approximate agreement. The finite value of the compressibility, if used to correct the analysis of the isotope shift, leads to radii larger than given by (18.3). In the approximate theory which has been used [*22*] the factor of correction increases slowly with Z, and at $Z = 80$ is 1.27 if the compressibility coefficient is chosen to be 50 Amc^2. Thus, near $Z = 80$, the revised estimate of R_{eq} from isotope shifts is, for this particular choice of E_0'',

$$R_{\text{eq}} = (1.16 \pm 0.13)\, A^{\frac{1}{3}}. \tag{19.3}$$

It is evident from the above discussion that the x-ray fine structure and isotope shift do not at present provide an accurate means for measuring nuclear charge radii. The theoretical analysis of the effect of the finite nucleus on these quantities, however, may be usefully applied for entirely different purposes. Since we now know the nuclear electric radius quite accurately from other experiments, we take the charge distribution as given from the μ-mesonic and electron scattering observations to make an accurate calculation of $\langle r^2 \rangle_{\text{Av}}$. We then may insert this quantity into the theory of finite nuclear extension effects on bound electrons and require that the equivalent radii given by the x-ray fine structure or the isotope shift agree with values they must have for the known charge distribution. We thus have a method of measuring the aggregate electrodynamic effects on the $2P_{\frac{1}{2}}$ electronic level as determined by x-ray fine structure data, and we may obtain the nuclear compressibility from the isotope shift data. Following this method we conclude that the *experimentally determined* nuclear compressibility is

$$E_0'' = (90 \pm 20)\, Amc^2. \tag{19.4}$$

Here the limits of error reflect primarily the estimated crudeness of the theory [*22*] of how compressibility affects isotope shifts.

---

[1] E. FEENBERG: Phys. Rev. **59**, 149, 593 (1941). — Rev. Mod. Phys. **19**, 239 (1947).
[2] K. BRUECKNER, C. LEVINSON and H. MAHMOUD: Phys. Rev. **95**, 217 (1954).

III. μ-mesons.

The μ-meson holds a position of particular importance in the science of nuclear charge extension for it was the quantitative study of x-rays from tightly bound μ-mesonic states [9] which gave the first unambiguous evidence that the nuclear charge radius is about 20% smaller than the previously accepted nuclear radius as given in (1.1).

20. μ-meson as nuclear probe. There is convincing evidence[1-3] that the μ-meson is a DIRAC particle which, for the purposes of the present discussion, has, like the electron, no significant interaction with nucleons apart from the COULOMB interaction. As a nuclear probe particle it has the further advantage of large mass compared to the electron (about 207 times greater[4]), and therefore even at low energies a relatively short wavelength and the consequent localizability needed for exploring the effects of nuclear charge distribution. It suffers the experimental disadvantage that intense monoenergetic beams cannot be readily produced, and such low intensity beams as are available tend to carry a considerable mixture of competing radiations. Nevertheless, considerable success has been achieved in obtaining spectra of bound μ-mesons, and experiments to study the elastic scattering of μ-mesons by nuclei have been started.

21. Nuclear charge size from μ-mesonic x-rays. A variety of data on the nuclear electrical properties are in principle available[5,6] from the study of μ-mesonic x-rays. Thus the fine structure splitting of the 2P levels is related [*13*] to the radial shape of the charge distribution, and the hyperfine structure includes quadropole moment effects[7-9] to which we shall return in Sect. 46. While information on these effects must await more precise data, certain primary features of the x-ray spectra[10] have been quantitatively observed [*9*], and to these we turn our attention.

The bound μ-meson in the lowest states is well within the electronic K-orbit. These states are therefore essentially hydrogenic for all Z, with the important difference, however, that the finite size of the nucleus is of major importance. For the 1 S or lowest bound state in Pb for example, the meson is bound with an energy of 10 Mev, whereas for a point nucleus the binding would be near 20 Mev. Thus the finite extension of the nucleus provides a deviation from the point nucleus COULOMB field which amounts to a repulsive potential term to reduce the actual binding. This large nuclear effect follows from the large mass of the meson which causes the 1 S state to describe the meson as spending a significant part of its time within the nucleus. The 2P states in the Pb nucleus on the other hand, are much less affected by the nuclear size [*13*] and hence the 2P→1 S transition which has been measured [*9*] to a precision of 1% serves to provide a measure of comparable precision for one parameter of the nuclear charge distribution. Some transitions among higher states have also been observed among lighter nuclei and are of value in providing an accurate measure

---

[1] E. AMALDI and G. FIDECARO: Phys. Rev. **81**, 339 (1951).
[2] J. H. KENNEDY: Phys. Rev. **87**, 953 (1952).
[3] F. FERRARI and C. VILLI: Phys. Rev. **96**, 1159 (1954).
[4] S. KOSLOV, V. FITCH and J. RAINWATER: Phys. Rev. **95**, 291 (1954).
[5] J. A. WHEELER: Rev. Mod. Phys. **21**, 133 (1949).
[6] S. FLÜGGE: Meson-Atome. Physikertagung Hamburg 1954, Hauptvorträge. Mosbach i. Baden: Physik-Verlag.
[7] J. A. WHEELER: Phys. Rev. **92**, 812 (1953).
[8] L. WILETS: Dan. Mat. fys. Medd. **29**, No. 3 (1954).
[9] B. A. JACOBSOHN: Phys. Rev. **96**, 1637 (1954).
[10] W. Y. CHANG: Rev. Mod. Phys. **21**, 166 (1949).

of the μ-meson mass. Such transitions among higher states, when observed in the heavy nuclei, will provide further information on the radial variation of charge distribution inasmuch as the different states measure different moments of the charge density [*13*].

Although the angular distribution of high energy electron scattering contains more parameters of the charge distribution, each of these parameters is somewhat less accurately determined than the single parameter fixed unambiguously by the $2\text{P}\to 1\text{S}$ transition of the μ-meson. This latter measurement therefore provides a useful constraint on the charge distribution and has been used [*11*] in conjunction with electron scattering distribution analysis to specify more precisely the shape of the charge distribution. The results listed in Table 2 were obtained both with and without the employment of this constraint. The fact that the radial parameters as measured by μ-mesonic x-ray spectra and by fast electron scattering, were in close agreement permitted this flexible use of the constraint to facilitate in the following manner the determination of charge distribution as given by fast electron scattering. A series of functional forms (3.1), each with the radial parameter r_1 as determined from μ-mesonic x-ray analysis, were employed to predict fast electron elastic angular distributions. By comparison with the observed distributions, the radial shape most nearly reproducing the data was selected. Then the parameters of this shape, *including* r_1 were varied to improve the fit. The value of r_1 from the x-ray analysis was found to give approximately the best fit even when variations of r_1 and of the other parameters of $f(r/r_1)$ were permitted over very great ranges. Having established this correspondence between r_1 as given by electron scattering and by μ-mesonic x-rays, the latter readily-determined values of r_1 for any shape of charge distribution were employed to reduce the number of parameters entering the fast electron scattering analysis for various energies, elements and families of charge distribution functions. For each of these additional cases when the best fit had been obtained, the constraint on r_1 was again relaxed and in each case it was again established that the radial size of charge as determined from fast electron scattering was in agreement with the analysis of the $2\text{P}\to 1\text{S}$ transition energies in μ-mesonic spectra.

Table 3. *Assuming the muon mass to be $\mu=210\ m_e$ and using a uniform nuclear model with $R_0=r_0 A^{\frac{1}{3}}$, values of r_0 shown in column 3 gave the best fit to the measured $2\,\text{P}_{\frac{3}{2}}\to 1\,\text{S}$ transition energies given in column 2 as a function of Z.* (*This table taken from* [*9*].)

Z	$E\,(2\,\text{P}_{\frac{3}{2}}\to 1\,\text{S})$ (experimental, Mev)	r_0 (10^{-13} cm)
22	0.955	1.17
29	1.55	1.21
51	3.50	1.22
82	6.02	1.17

Analyses of the μ-mesonic x-ray data and predictions of transition energies have been made [*9*], [*13*][1] on the basis of the DIRAC equation for the μ-meson in the smooth, static, spherically symmetric electric potential generated by assumed charge distributions in the nucleus. The results for a number of elements are summarized in Table 3 and give an equivalent radius for the light elements [defined by (12.1)]:

$$\boxed{R_{\text{eq}} = (1.20 \pm 0.03)\, A^{\frac{1}{3}}.} \tag{21.1}$$

For light nuclei where perturbation treatment is possible, the $1\,\text{S}$ ($k=-1$) and $2\,\text{P}_{\frac{1}{2}}$ ($k=+1$) levels are raised above their point nucleus values by an amount

[1] S. FLÜGGE and W. ZICKENDRAHT: Z. Physik **143**, 1 (1955).

proportional to $\langle r^{2\sigma}\rangle$ as defined in (15.8) and (15.5). The $2P_{\frac{1}{2}}$ shift, which is governed by the small component of the DIRAC radial solution, is small compared to the $1S_{\frac{1}{2}}$ shift. The energy shifts of higher angular momentum states measure higher moments of the charge distribution and are negligible in comparison to the shifts for the states with $|k|=1$. As the perturbation approximation is valid only for low Z, where σ is near unity, the $2P\rightarrow 1S$ transition energy for the light elements measures $\langle r^2\rangle_{\mathrm{Av}}$ which is the same quantity measured by low energy electron scattering.

22. Analysis in heavy elements. Perturbation treatment is no longer suitable for heavy elements and exact numerical calculation is required. Extensive calculations have been carried out [*13*] especially for Pb. For this work, the radial equations for the DIRAC particle in a central field are expressed in the form

$$\left.\begin{aligned}\frac{dG}{dx}&=-\frac{k}{x}G+[\gamma(\varepsilon+1)+\alpha Z J_f(x)]F,\\ \frac{dF}{dx}&=\frac{k}{x}F-[\gamma(\varepsilon-1)+\alpha Z J_f(x)]G\end{aligned}\right\}\tag{22.1}$$

where x and γ are respectively the radial coordinate, r, and the range parameter, r_1, expressed in units of the reduced COMPTON wavelength of the μ-meson, $\lambda\!\!\!^{-}_\mu=\frac{\hbar}{mc}=1.87\times10^{-13}$ cm; ε is the dimensionless energy E/mc^2, m is the μ-meson mass $=207\,m_e$, and the electrostatic potential is given by

$$V(x)=-\frac{Ze}{r_1}J_f(x)\tag{22.2}$$

in which J_f is the functional

$$J_f(x)=[I_f(\infty)]^{-1}\int_x^\infty I_f(x)\frac{dx}{x^2}\tag{22.3}$$

with

$$I_f(x)=\int_0^x f(x)\,x^2\,dx,\tag{22.4}$$

as defined in terms of the radial variation of charge distribution (3.1). The quantities F, G and k are as defined in (9.2).

The method of the calculation is to select a form for $f(x)$ and to vary the binding energy ε for each of the lowest levels until the amplitude functions F and G are brought sufficiently close to satisfying the boundary conditions for proper solutions. In order to cover the range of possible charge distributions, several families of functions were employed. One family was defined by

$$f_n^{\mathrm{I}}(x)=\frac{1}{n!}\int_x^\infty x^n e^{-x}\,dx=\sum_{k=0}^{n}\frac{x^k}{k!}e^{-x},\qquad n=0,1,2,\ldots\tag{22.5}$$

which gives a sequence of discrete forms ranging from the exponential to the uniform square distributions as n takes integral values between zero and infinity. Because of the inconvenience of this family for dealing with charge shapes in the neighborhood of the uniform square distribution, a second family was defined by

$$f_n^{\mathrm{II}}(x)=\frac{1}{1-\frac{1}{2}e^{-n}}\left\{\begin{matrix}1-\frac{1}{2}e^{-n(1-x)}, & x<1,\\ \frac{1}{2}e^{-n(x-1)} & x\geq 1\end{matrix}\right\}\tag{22.6}$$

in which the parameter n has a different significance and may vary continuously from zero to infinity to define a continuous sequence of shapes ranging from exponential to uniform square. In order to allow for the expectation[1-4] that the proton density is greater near the nuclear surface than in the center, a third family was defined

$$f_{ns}^{\mathrm{III}}(x) = \frac{\mathrm{Sin}\,(s\,x)}{s\,x} f_n^{\mathrm{II}}(x), \tag{22.7}$$

in which the parameter s controls the amount of central depression of charge density. Representative curves from families 2 and 3 are plotted in Fig. 2. The

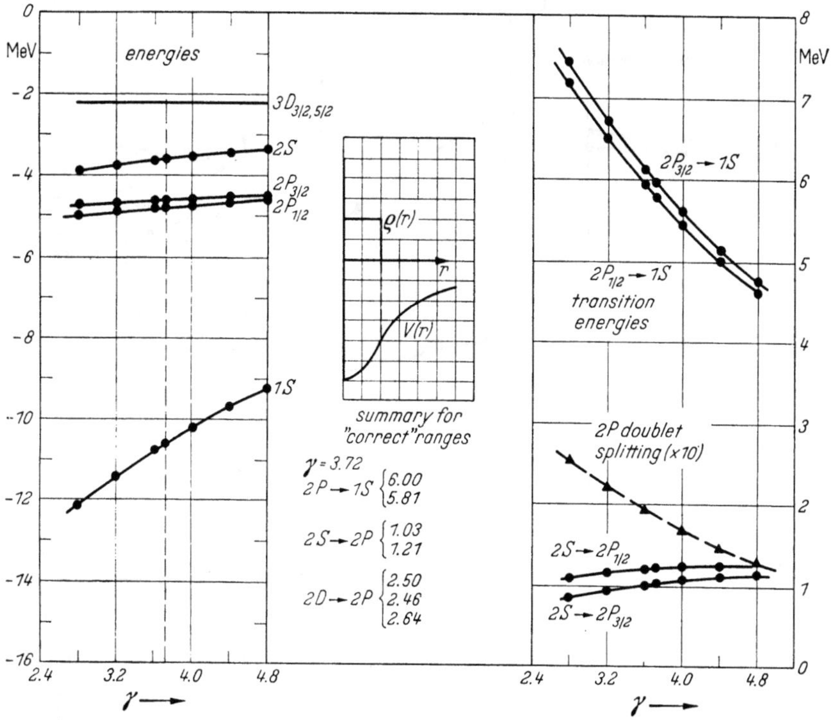

Fig. 8. Binding energies, transition energies, and 2 P doublet splitting for μ-mesons bound in orbitals about the lead nucleus, for an assumed uniform charge distribution. Energies are given in Mev, assuming a meson mass of 207.0 m_e, and are plotted vs. γ, the nuclear radius in units of the meson reduced COMPTON wavelength. The 3D levels were calculated by perturbation theory. The vertical dashed lines indicate the radius for which the $2\,P_{\frac{3}{2}} \to 1\,S$ energy equals 6.00 Mev. Taken from reference [13].

dependence of the lowest mesonic levels on the nuclear radius for a particular form of $f(r/r_1)$ is illustrated in Fig. 8. The experimental value [9] of the $2\,P_{\frac{3}{2}} \to 1\,S$ transition energy, 6.00 Mev, is chosen to select the "correct" radial size and, subject to this constraint, the variation of energy levels and transition energies, as well as fine structure splitting as a function of charge shape is illustrated in Fig. 9. This figure makes evident the sort of information on the radial shape of charge distribution which may be made available by measurements of the higher transitions in μ-mesonic atoms. It is of interest to observe that a different method of calculation[5] employing the RITZ variational procedure has been shown

[1] E. WIGNER: Bicentennial Symposium, University of Pennsylvania 1940.
[2] E. FEENBERG: Phys. Rev. 59, 149, 593 (1941). — Rev. Mod. Phys. 19, 239 (1947).
[3] W. J. SWIATECKI: Proc. Phys. Soc. Lond. A 63, 1208 (1950).
[4] S. FLÜGGE and K. WOESTE: Z. Physik 132, 384 (1952).
[5] S. FLÜGGE and W. ZICKENDRAHT: Z. Physik 143, 1 (1955).

capable of producing results similar to those available from the numerical calculation illustrated in Figs. 8 and 9. Experimentally the most easily determined energies besides the $2P\to 1S$ transition energies, are the $2P$ doublet splitting and $3D\to 2P$ energies. As a gauge of the sensitivity of the μ-meson spectra to the radial shape of the charge distribution, we observe that for Pb, and for a

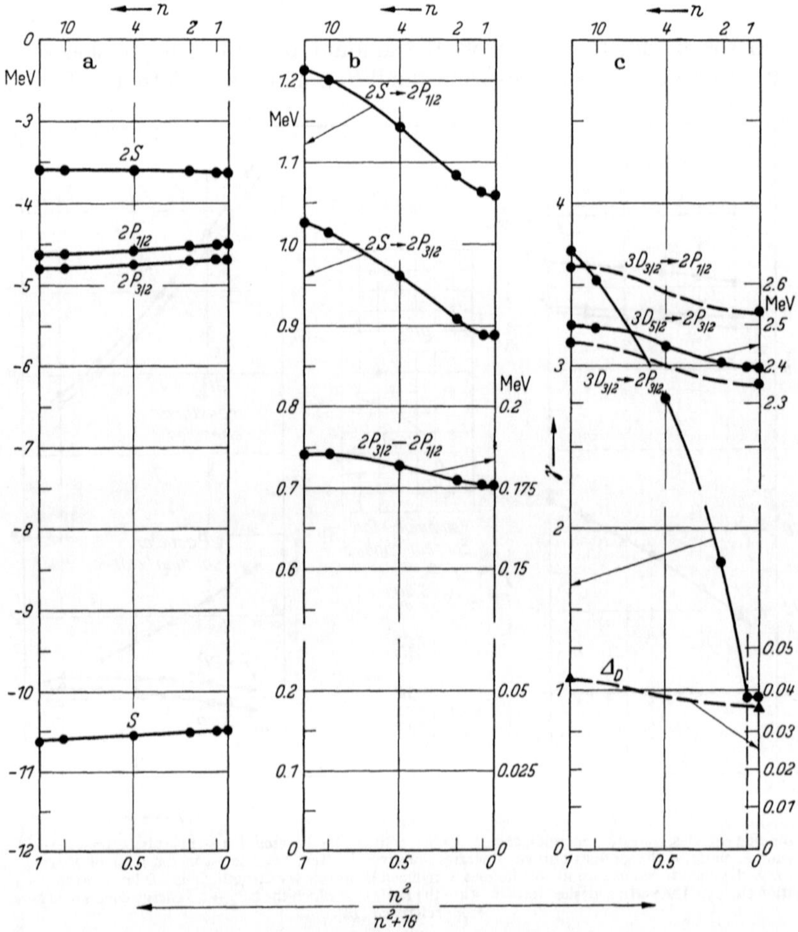

Fig. 9a–c. μ-mesonic binding energies and nuclear radii *vs.* shape of charge distribution (22.6) limited by the condition that the μ-mesonic $2P_{\frac{3}{2}}\to 1S$ energy = 6.00 Mev. (a) Binding energies of four lowest states. (b) $2S\to 2P$ transition energies and $2P$ doublet splitting. (c) $3D\to 2P$ transition energies, $3D$ doublet splitting, and range parameter γ. The quantities are plotted *vs.* an arbitrarily chosen shape parameter, $n^2/(n^2+16)$, which runs from 1 (uniform distribution) to 0 (exponential distribution). From [*13*], for the lead nucleus.

fixed $2P_{\frac{3}{2}}\to 1S$ energy, the $3D\to 2P$ energies vary by 4.5% between the uniform and the exponential charge distribution. Whereas this difference is small, it is about four times the theoretical uncertainty of the calculations. Thus it is evident that refined measurements of μ-mesonic spectra can roughly confirm the radial shapes, as well as sizes, of charge distributions as determined by fast electron scattering. The sensitivity of μ-mesonic transition energies to the radial distribution of nuclear charge may be stated in a different way as illustrated in Fig. 10 (also for the Pb nucleus). A sensitivity function, S_{mn}, may be defined for the transitions $m \to n$ by the relation [*13*],

$$\Delta E_{m\to n} = \int_0^\infty S_{mn}(r) \left[\frac{\delta \varrho(r)}{Z}\right] 4\pi r^2 dr. \tag{22.8}$$

Here ΔE is the change of transition energy resulting when the charge density differs by the amount $\delta\varrho(r)$ from a uniform square charge distribution. The functions S_{mn} were determined by using the numerically calculated μ-mesonic wave functions for the uniform square distribution as a basis for a perturbation analysis.

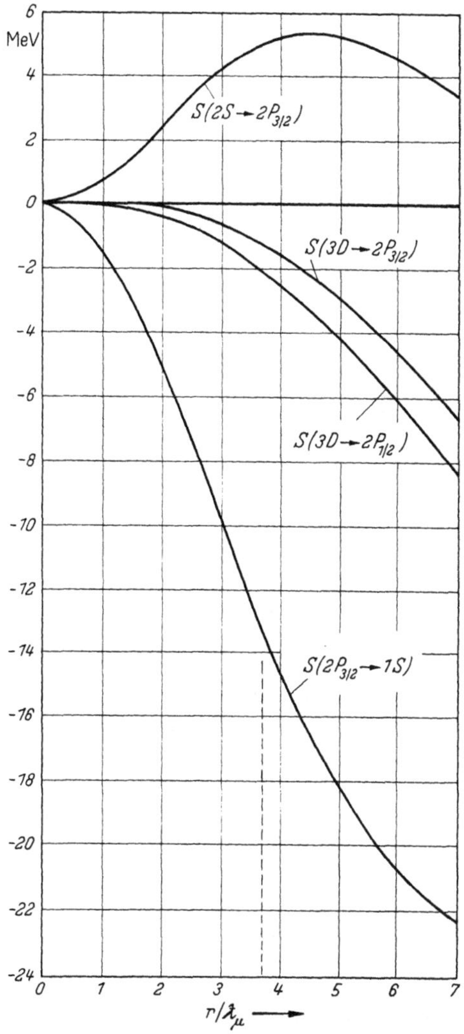

Fig. 10. The sensitivity of μ-meson transition energies to distribution of nuclear charge is defined by the functions S_{mn} of (22.8) and plotted here for four transitions. The abscissa gives the radial distance in units of λ_μ, the meson reduced COMPTON wavelength, and the dotted line at 3.7 indicates the radius in these units of the uniform charge distribution which has the correct ($2P_{\frac{3}{2}}\to 1S$) transition energy.

23. Perturbations and uncertainties of calculated μ-mesonic levels. The several perturbations to the calculated μ-mesonic energies have been considered by various authors[1–5] for spherically symmetric nuclei, and are summarized in [*13*]. There are two effects that account for most of the perturbing energy shift and one of these accounts for most of the theoretical uncertainty. We consider first the polarization of the vacuum by electrons[4] [*13*], noting that other radiative corrections involve intermediate states for which a μ-meson is created, and are probably less by order 10^{-2}. Utilizing standard results of quantum electrodynamics[6], the electron vacuum polarization may be readily calculated and turns out to have magnitude of the order of 1% of the $2P_{\frac{3}{2}}\to 1S$ transition energy in heavy elements. While this effect has thus far been calculated only approximately, a question more important than accurate calculations of the first order effect is the magnitude of higher order effects. It has been shown[5] that higher order corrections are in fact quite small in spite of the large value of αZ for the heavy elements. Thus it follows that radiative corrections account for a noticeable shift in the energy but do not give rise to any important uncertainty.

1 L. N. COOPER and E. M. HENLEY: Phys. Rev. **92**, 801 (1953).
2 H. C. CORBEN: Phys. Rev. **94**, 787 (1954).
3 W. LAKIN and W. KOHN: Phys. Rev. **94**, 787 (1954).
4 A. B. MICKELWAIT and H. C. CORBEN: Phys. Rev. **96**, 1145 (1954).
5 E. WICHMANN and N. M. KROLL: Phys. Rev. **96**, 232 (1954).
6 J. SCHWINGER: Phys. Rev. **75**, 651 (1949).

The second important perturbing effect is nuclear polarization or the admixture of excited nuclear states to the ground state configuration due to the presence of the μ-meson[1,2]. This effect for heavy nuclei is estimated to be also of the order of 1% for the $2\,P_{\frac{3}{2}}\rightarrow 1\,S$ energy but its magnitude is uncertain by about a factor of 2. The uncertainty may be minimized in two limits. (a) If there are no nuclear excited states close to the ground state as in the double-magic nucleus Pb^{208}, the whole effect and therefore its uncertainty is diminished. (b) If the nature of the low excited states and their mode of coupling to the μ-mesons are understood, the polarization effect, although not small, may be calculated to reasonable accuracy. Such is the case for nuclear rotational states (see Sect. 46).

Other perturbations have been estimated to be much smaller than these two. An interesting fact for heavy nuclei is that the $2\,P\rightarrow 1\,S$ energy is very insensitive[1] [*13*] to the value of the μ-meson mass[3,4]. This insensitivity[5] which is due to the shape of the potential is such that this transition energy for Pb varies approximately as $m^{0.2}$ for variations of the μ-meson mass in the neighborhood of the true mass. For a parabolic potential resembling the COULOMB field within the nucleus, energies vary as $m^{-\frac{1}{2}}$, while for a $1/r$ potential such as that outside the nucleus, energies vary as m. Thus the result for a μ-meson in the 1 S state which finds itself roughly half the time in each kind of potential is a weak dependence of transition energies on meson mass.

For the Pb nucleus, it has been estimated [*13*] that the combined error of theoretical transition energy for a given charge distribution is at most 1% of the energy. Hence a measurement of the $2\,P\rightarrow 1\,S$ transition energy ($\sim$6 Mev) with an error of less than 1% determines a single parameter of the charge distribution to about 1%, or more accurately than a single parameter can be determined by any other means at present.

24. Positive energy states. The scattering of μ-mesons from nuclei[6] provides in principle an excellent way to study the distribution of nuclear charge. The fact that μ-mesons are not available in intense monoenergetic beams, of course, introduces great practical difficulties but preliminary experiments have been completed to measure the angular distributions in nuclear emulsions of μ-mesons from the natural radiation. The advantage of μ-meson scattering over electron scattering is that a given small wavelength may be achieved at lower energy for the μ-meson and that the electrodynamic uncertainties resulting from the extreme relativistic energies of the electron are not present for the μ-meson.

The creation of muon pairs by protons in the field of an atomic nucleus is another phenomenon for which the cross section should depend critically upon nuclear size. The cross section has been shown[7] to be considerably smaller than that for a point charge nucleus. Experiments[8,9] have thus far provided only an upper limit for the effect, but information on nuclear size should in time be available from this source.

---

[1] See footnote 1, p. 209.
[2] See footnote 3, p. 209.
[3] F. M. SMITH, W. BIRNBAUM and W. H. BARKAS: Phys. Rev. **91**, 765 (1953).
[4] S. KOSLOV, V. FITCH and J. RAINWATER: Phys. Rev. **95**, 291 (1954).
[5] J. A. WHEELER: Rev. Mod. Phys. **21**, 133 (1949).
[6] G. D. ROCHESTER and A. W. WOLFENDALE: Phil. Mag. **45**, 980 (1954).
[7] G. H. RAWITSCHER: Phys. Rev. **101**, 423 (1956).
[8] B. T. FELD, R. JULIAN, A. C. ODIAN, L. S. OSBORNE and A. WATTENBERG: Phys. Rev. **96**, 1386 (1954).
[9] G. E. MASEK, A. J. LAZARUS and W. K. H. PANOFSKY: Phys. Rev. **99**, 650 (1950).

IV. COULOMB energy.

25. Mirror nuclei. The COULOMB energy difference between mirror nuclei is one of the oldest means to determine nuclear radii, but the difficulties of exact interpretation were recognized early [*1*], [*20*] and have been reemphasized recently[1-5] [*6*], although they have not been stressed in the interim, possibly because the simplest interpretation of the data without corrections led to radii in agreement with those obtained by alpha decay and by fast neutron total cross sections for the heavier elements. The difficulty is associated, of course, with the fact the COULOMB energy, in spite of its purely electrical nature, is dependent on certain details of the nuclear structure.

On a classical theory of uniformly distributed charge over a volume $\frac{4}{3}\pi R^3$, the nuclear COULOMB energy is

$$E_c = \frac{3}{5} Z(Z-1)\frac{e^2}{R} \tag{25.1}$$

and the COULOMB energy difference between mirror nuclei (Z and $Z+1$),

$$\Delta E_c = \frac{6}{5} Z \frac{e^2}{R}. \tag{25.2}$$

When the mirror nuclei are interpreted in this way, one finds resulting radii which follow quite closely an $A^{\frac{1}{3}}$ law [*2*] given by

$$R = 1.46\, A^{\frac{1}{3}} \tag{25.3}$$

up to $A=43$. There are two reasons why this simple interpretation is not correct. First, the anti-symmetrization of the ground state proton wave function leads to exchange terms which diminish the COULOMB energy and, therefore, diminish the calculated radius. The approximate formula [*6*] which includes the exchange energy is

$$E_c \approx \frac{3}{5} Z(Z-1)\frac{e^2}{R} - 0.46\, Z^{\frac{4}{3}}\frac{e^2}{R}. \tag{25.4}$$

This effect leads for the light nuclei to an average reduction of radius of about 10%, i.e., a change of the radial constant in (25.3) from 1.46 to 1.3. A more detailed study of the correction term in (25.4) has led to an explanation of the alternation in magnitude of the correction for very light nuclei on the basis of the symmetry properties of the ground state wave function[3].

A second and more difficult correction to (25.2) comes about from the difference between the density distribution of the proton participating in the beta decay and the average density distribution of all protons. Although (25.4) might provide a good estimate of the total nuclear COULOMB energy, the *difference* $E_c(Z+1)-E_c(Z)$ evaluated from (25.4) could be in considerable error if the disappearing proton had a very non-uniform density distribution. The nuclear shell model suggests that this is the case, and calculations of COULOMB energy differences on the basis of the shell model have been carried out[4,5] [*6*]. The results generally tend to give nuclear electrical radii smaller than (25.3) by 10 to 15%, but no longer so smoothly varying with A as the results of the

---

[1] R. R. WILSON: Phys. Rev. **88**, 350 (1952). References to experimental work here.
[2] J. B. EHRMANN: Phys. Rev. **81**, 412 (1951).
[3] E. FEENBERG and G. GOERTZEL: Phys. Rev. **70**, 597 (1946).
[4] B. G. JANCOVICI: Phys. Rev. **95**, 389 (1954).
[5] B. C. CARLSON and I. TALMI: Phys. Rev. **96**, 436 (1954).

simpler analysis. The final result of the mirror nucleus analysis is a radius given by

$$\boxed{R = (1.3 \pm 0.1)\, A^{\frac{1}{3}}.} \tag{25.5}$$

It is still uncertain whether this slightly high value is due to inadequacies of the analysis, or to the fact that the proton density in light nuclei is indeed somewhat lower than in heavy nuclei. The latter conclusion is suggested[1] by evidence for an increasing trend of nuclear density between $A=13$ and $A=27$.

Two omissions in the analysis so far which may be of significance are the following. (1) Nonuniform nuclear density. Since the COULOMB energy and electron scattering measure different properties of the charge distribution, the nuclear radii obtained should agree only for a uniform distribution. For a gaussian distribution, the radius, R_{eq}, deduced from COULOMB energy should be smaller by 5% than that deduced from low energy scattering (or any other effect measuring $\langle r^2 \rangle$), as shown in Fig. 2. Since the radius in (25.5) is already larger than that found from electron scattering, this effect would increase the discrepancy somewhat. (2) The short range correlations in proton positions within the nucleus brought about by repulsive cores in the nuclear forces would reduce the COULOMB energy. This effect would further decrease the radius deduced from the mirror nuclei energy difference, and would be in the direction to diminish the discrepancy between the radius in (25.5) and the radius deduced from the more purely electromagnetic means.

26. Nuclear radius from nuclear masses. The semi-empirical formula of nuclear masses [8] includes the term proportional to $A^{-\frac{1}{3}}$ which as suggested by (1.1) and (25.1) has been interpreted to be the contribution of COULOMB energy to nuclear mass. For the constants in the formula which were for some time accepted [8] this assumption led for a uniform square model of nuclear charge, to a value of r_0 consistent with the value given in (1.1). A recent reexamination[2] of the value of r_0 as determined from a new fit to nuclear masses, has given a different value. From a least squares analysis employing an objective criterion of best fit, the value of r_0 then deduced is

$$\boxed{r_0 = (1.216 \pm 0.012)} \tag{26.1}$$

in which the indicated probable error is given by the standard deviation of the fit, and the analysis has been carried through for the assumption of a uniform square charge distribution. From Fig. 2 it is evident that the equivalent radius for COULOMB energy is about one-half percent larger than the equivalent radius for the μ-mesonic x-ray, as read by choosing an abscissa on this graph of 0.2 corresponding to the known radial shape of the nuclear charge. Allowing for special effects of shell closing which should cause the radial constant for the Pb nucleus to be slightly smaller than the radial constant for the average nucleus, we find this value of the radial constant from COULOMB energy to be in excellent agreement with radial size as given by μ-mesonic x-rays.

B. Nuclear force measures of nuclear size and radial shape.

When methods employing nuclear forces are used to measure a nuclear radius, then the quantity measured becomes the spatial distribution of nuclear potential

[1] See footnote 5, p. 211.

[2] A. E. S. GREEN: Phys. Rev. 95, 1006 (1954).

and may, therefore, be expected to differ from the proton density itself which was the quantity observed by the measurements discussed in Part A. The nuclear force measures include both scattering and reaction processes.

I. Neutron scattering.

27. Neutron total cross sections. The experiments of AMALDI, *et al.*[1] and of SHERR[2] some years ago, provided one of the first sets of data on nuclear radii spanning the whole periodic table. For neutron energies in the range 10 to 50 Mev, one can define a reasonable nuclear radius by the simple relation,

$$\sigma_T = 2\pi R^2 \tag{27.1}$$

where σ_T is the total cross section. Below 5 Mev, the finite nuclear wavelength causes this approximate relation to break down. Above 50 Mev, nuclear transparency becomes important. A slightly better definition of radius is given[3] by

$$\sigma_T = 2\pi (R + \lambda\!\!\!^{-})^2, \tag{27.2}$$

where $\lambda\!\!\!^{-}$ is the reduced wave length of the neutron. In either case, the early experiments led to

$$R = 1.4\, A^{\frac{1}{3}}. \tag{27.3}$$

More recent experiments[4] at 14 Mev have confirmed this order of magnitude for the nuclear radius, but these more accurate experiments also show that $(\sigma_T)^{\frac{1}{2}}$ vs. $A^{\frac{1}{3}}$ is not a straight line, but has some regular and significant deviations from linearity. Although a radial constant of about 1.4 to 1.5 fits the data on the average, the derivative $d(\sigma_T/2\pi)^{\frac{1}{2}}/d(A^{\frac{1}{3}})$ shows large fluctuations away from this value. Somewhat less accurate inelastic cross section measurements are consistent with these results[5].

A nuclear transparency correction[6] made it possible to obtain nuclear radii from 90 Mev neutron scattering[7,8] in agreement with radii obtained from lower energy scattering. A fit[9] to all high energy neutron scattering data in terms of this "optical model" indicates the nuclear radii to be well approximated by:

$$\boxed{R = 0.8 + 1.23\, A^{\frac{1}{3}}.} \tag{27.4}$$

The high energy nuclear transparency comes from the weakening of the nuclear forces with increasing energy. At low energies ($E \lesssim 5$ Mev) a transparency of quite another sort seems to exist, due to the coherent motion of a nucleon through nuclear matter. The "cloudy crystal ball" model of WEISSKOPF and collaborators[10] has been successful in fitting[11,12] total average neutron cross sections

[1] E. AMALDI, D. BOCCIARELLI, B. N. CACCIAPUOTI and G. C. TRABACCHI: Nuovo Cim. **3** 15, 203 (1946).

[2] R. SHERR: Phys. Rev. **68**, 240 (1945).

[3] H. FESHBACH and V. WEISSKOPF: Phys. Rev. **76**, 1550 (1949).

[4] J. H. COON, E. R. GRAVES and H. H. BARSCHALL: Phys. Rev. **88**, 562 (1952).

[5] D. D. PHILLIPS, R. W. DAVIS and E. R. GRAVES: Phys. Rev. **88**, 600 (1952).

[6] S. FERNBACH, R. SERBER and T. B. TAYLOR: Phys. Rev. **75**, 1352 (1949).

[7] L. COOK, E. MCMILLAN, J. PETERSON and D. SEWELL: Phys. Rev. **75**, 7 (1949).

[8] A. BRATENAHL, S. FERNBACH, R. HILDEBRAND, C. LEITH and B. MOYER: Phys. Rev. **77**, 597 (1950).

[9] THEODORE B. TAYLOR: Nuclear Scattering of High Energy Neutrons and the Optical Model of the Nucleus. Thesis, Cornell University 1954.

[10] H. FESHBACH, C. PORTER and V. WEISSKOPF: Phys. Rev. **90**, 166 (1953). — M. I. T. Technical Report 62, 1953; also later unpublished work with revised nuclear parameters.

[11] H. H. BARSCHALL: Phys. Rev. **86**, 431 (1952).

[12] M. WALT and H. H. BARSCHALL: Phys. Rev. **93**, 1062 (1954).

in the energy region 0 to 3 Mev. In the low energy region, total cross sections are not sensitive either to nuclear size (since a variation in size can be nearly compensated by a variation in well depth) or to the thickness of the surface transition region. Satisfactory fits have been obtained[1,2] with square wells of (real) depth 45 Mev and radius $1.4\,A^{\frac{1}{3}}$. A recent analysis of 14 Mev elastic neutron scattering data indicates[3] a potential depth of 42 Mev and a radius $R = 1.22\,A^{\frac{1}{3}} + 0.74$.

Interesting new analyses[4,5] of high energy neutron scattering have suggested the radial shape as well as radius of the nuclear density. For the analysis of 1.4 Bev neutron scattering, WILLIAMS has assumed a shape of the nuclear density corresponding to the proton density as given by electron scattering, such as plotted in Fig. 5. He finds that the very fast neutron scattering is best fit by a radius corresponding to the radius indicated in Fig. 5 from electromagnetic measurements.

28. Neutron angular distribution. When angular distributions of low energy elastically scattered neutrons are considered as well, the best fits to the data determine radius and well depth separately. A recent value of nuclear radius based on analysis[6] of this kind of data, is

$$R \approx 1.3\,A^{\frac{1}{3}}. \tag{28.1}$$

At higher energies ($\gtrsim 10$ Mev) the analysis of elastic neutron angular distributions reveals something about the thickness of the nuclear surface. Most such information has come, however, from proton scattering, discussed below.

II. Proton scattering.

29. Advantages of proton scattering. Because detectors for protons are usually more efficient than detectors for neutrons, we may expect that proton angular distributions may, on the average, be measured at greater precision than neutron angular distributions. Observations on scattered protons have the additional advantage that the observed intensity must include interference between the COULOMB scattering and the specifically nuclear scattering which is capable of providing an absolute calibration of the nuclear scattering amplitude. Hence we should expect proton scattering to reveal more clearly the finer details of the effective nuclear potential well. Thus it is not surprising that the analysis of scattered proton angular distributions first demonstrated the inadequacy of a sharp edge on the potential well and led to the acceptance of the more realistic potential well shapes with smoothly sloping boundaries.

30. Results of proton scattering analyses. Detailed calculations of the elastic scattering of protons by nuclei have been made, using a complex well model[7–9]. Fits to the experimental data have fixed, at least roughly, four parameters of the effective nuclear potential well: the real potential depth, the imaginary potential depth, the nuclear radius, and the surface thickness. The particular form

[1] K. W. FORD and D. BOHM: Phys. Rev. **79**, 745 (1950).
[2] R. K. ADAIR: Phys. Rev. **94**, 737 (1954).
[3] G. CULLER, S. FERNBACH and N. SHERMAN: Phys. Rev. **101**, 1047 (1956).
[4] R. JASTROW and J. ROBERTS: Phys. Rev. **85**, 757 (1952).
[5] ROBERT W. WILLIAMS: Phys. Rev. **98**, 1387, 1393 (1955).
[6] W. S. EMMERICH: Phys. Rev. **98**, 1148 (1955).
[7] D. CHASE and F. ROHRLICH: Phys. Rev. **94**, 81 (1954).
[8] R. D. WOODS and D. S. SAXON: Phys. Rev. **95**, 577 (1954).
[9] D. S. SAXON: Elastic Scattering (Report presented at Brookhaven on Statistical Aspects of the Nucleus, January, 1955).

of the potential assumed by SAXON and collaborators[1] is

$$V = \frac{V_0 + i W_0}{1 + e^{\frac{r-R}{b}}}, \tag{30.1}$$

the four parameters mentioned above being V_0, W_0, R, and b. The limit $b \to 0$ corresponds to a complex square well. Calculation of scattering from such a potential is, of course, easier for neutrons than for protons, but the proton experimental data are more extensive and more accurate. Fits[2] to the proton data in the energy region around 20 Mev have led to the mean values

$$\left.\begin{aligned} R &= 1.33\,A^{\frac{1}{3}}, \\ b &= 0.5 \end{aligned}\right\} \tag{30.2}$$

(with $V_0 = 38$ Mev, $W_0 = 9$ Mev). It has been noted in this region that the elastic scattering of neutrons and protons are nearly the same when the proton energy is greater than the neutron energy by the amount of the COULOMB barrier.

Data on very high energy proton scattering[3] have also been analysed and indicate a nuclear radius of about $1.25 \times A^{\frac{1}{3}}$.

31. Spin orbit coupling. When the radial shape of the nuclear potential is obtained from analysis of proton or neutron scattering, its reliability, of course, depends upon the appropriateness of choice for the potential itself. A persistent discrepancy between calculation and observation in the work thus far mentioned is the excessive sharpness of minima in the calculated curve for the angular distribution of scattered particles. At least a partial explanation[4] may be the absence of a term to represent the spin-orbit coupling which should be present in the optical potential. We should expect that the somewhat different interaction of the nucleon with the nucleus corresponding to the two values of nucleon spin should produce minima in the diffraction curves for the two spin states at slightly different angles. The superposition of these displaced minima may account largely for the flatter minima found in the measured angular distribution.

32. Relative inelastic scattering of high-energy nucleons. An analysis[5] of the inelastic scattering from heavy nuclei of protons and of neutrons of energy near 150 Mev indicates that the scattered protons "see" a somewhat larger effective radius than the neutrons. Inasmuch as the *n-p* cross section at this energy is about twice as large as the *p-p* cross section or the *n-n* cross section, it may therefore be argued that the proton inelastic cross sections depend mainly on the radius of the neutron distribution and that the neutron inelastic cross sections depend mainly upon the radius of the proton distribution. This simple argument, which is made semi-quantitative in terms of the optical model, suggests that the protons in the nucleus have a smaller radius of extension than the neutrons, in contrast to the result cited above from high energy neutron scattering and its comparison with the electromagnetic radius given by fast electron scattering.

---

[1] See footnote 9, p. 214.

[2] See footnote 9, p. 214.

[3] R. E. RICHARDSON, W. P. BALL, C. E. LEITH jr. and B. J. MOYER: Phys. Rev. **83**, 859 (1951); **86**, 29 (1952).

[4] D. C. PEASLEE: Nuclear Reactions of Intermediate Energy Heavy Particles. Annual Review of Nuclear Science, Vol. 5. 448 pp. Stanford, California: Annual Review, Inc. 1955.

[5] R. G. P. VOSS and R. WILSON: Phys. Rev. **99**, 1056 (1955).

Although there have been some other suggestions[1,2] from experimental sources that the neutron extension exceeds the proton extension in heavy nuclei, these arguments and experimental observations must be made more conclusive before they can be accepted in preference to observations (see Sect. 27) and theoretical calculations (see Sect. 53) which support the idea that neutrons and protons have about the same radial extension.

III. Alpha particle scattering.

33. Angular distribution. The scattering of alpha particles many years ago gave the first quantitative evidence on nuclear size. Recent experiments in the range 20 to 40 Mev have been interpreted in terms of a nuclear size[3-6]. The theoretical model employed was of a cruder sort than the complex well model employed to analyze the neutron and proton scattering data. It utilized a "black", i.e., totally absorbing, nucleus corresponding, in the complex well model, to an infinite imaginary potential. The nuclear radius deduced from the angular distribution is, however, quite insensitive to the model. The elastic angular distribution is approximately a COULOMB distribution up to some angle, beyond which it drops rapidly below COULOMB. An interpretation in terms of a transparent nucleus (i.e., a real potential) also leads to about the same nuclear radius as the interpretation in terms of a black nucleus[6,7]. The radius obtained, which we denote by $R+R_\alpha$, the sum of nuclear and alpha particle radii, is given by

$$R+R_\alpha = (1.7 \text{ to } 2.0)\, A^{\frac{1}{3}}. \quad (33.1)$$

A reasonably generous allowance of about 2.5×10^{-13} cm for R_α brings the nuclear radius down to

$$R = (1.4 \text{ to } 1.5)\, A^{\frac{1}{3}}. \quad (33.2)$$

In spite of uncertainties associated with the interpretation, it seems clear that the nuclear radius determined in this way is among the largest of nuclear-force determined radii, and is surely larger than the electric radius of the nucleus.

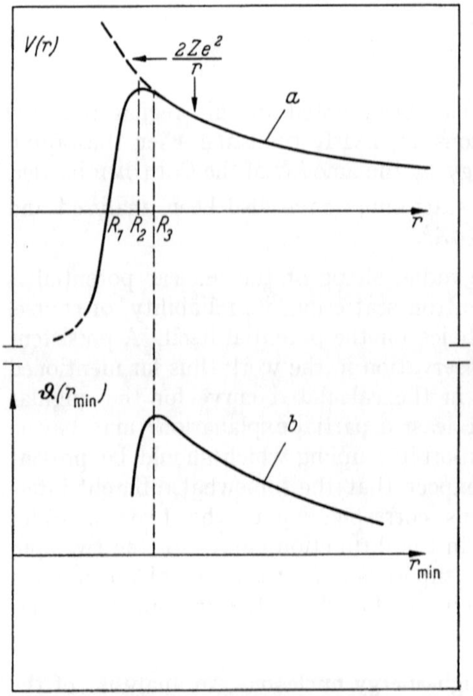

Fig. 11. Qualitative explanation of large radii deduced from elastic scattering of alpha particles. (a) Potential, $V(r)$, experienced by alpha particle if the finite thickness of the nuclear surface is taken into account. (b) Classical scattering angle as a function of $r_{\min}$, the point of closest approach of the classical trajectory. R_1 is the point where $V_{\text{nuclear}}+V_{\text{coulomb}}=0$, and will be close to the electric radius of the nucleus. R_2 is the point where $dV/dr=0$. R_3 is the point of closest approach for the maximum classical deflection angle, and will be greater than R_2. The "interaction radius" determined by the semiclassical analysis of alpha-particle scattering is approximately R_3, and should be substantially greater than the nuclear electric radius. This argument is independent of the strength of absorption in the nuclear interior.

[1] YOSHIHIRO NAKANO: Phys. Rev. **98**, 842 (1955).
[2] OSAMU MIYATAKE and CLARK GOODMAN: Phys. Rev. **99**, 1040 (1955).
[3] G. FARWELL and H. WEGNER: Phys. Rev. **93**, 356 (1954); **95**, 1212 (1954).
[4] J. S. BLAIR: Phys. Rev. **95**, 1218 (1954).
[5] N. S. WALL, J. R. REES and K. W. FORD: Phys. Rev. **97**, 726 (1955).
[6] H. WEGNER, R. EISBERG and G. IGO: Phys. Rev. **99**, 825, 1606 (1955).
[7] K. W. FORD and J. A. WHEELER: Phys. Rev. (to be published).

Since the alpha particle scattering problem is semi-classical for heavy nuclei ($Ze^2/\hbar v \gg 1$), it is reasonable to seek an explanation for the large "interaction" radius in classical terms. Such an explanation[1] is afforded by Fig. 11. The classical particle orbits in a pure COULOMB field experience a major fraction of their deflection close to their turning point (point of closest approach). Therefore only a slight weakening of the repulsive field near the point of closest approach will cause the total particle deflection to diminish appreciably. As the impact parameter is decreased, the scattering angle will increase until the scattered particle begins to feel the tail of the nuclear potential. A further decrease of impact parameter will cause the total particle deflection to pass through a maximum and decrease again. The maximum of the deflection will occur when the scattered particle still feels only the tail of the nuclear potential. The nuclear "radius" determined from the scattering cross section will be approximately the point of closest approach of the particle trajectory corresponding to the maximum deflection, ϑ_1. Therefore, the radius determined from the alpha particle data will correspond roughly to a "maximum" nuclear radius. In summary, the reason for this result is that only a weak deviation from the COULOMB force is required near the point of nearest approach of the alpha particle in order to alter substantially the deflection of the particle. This "anomalously" large radius should therefore disappear if the alpha particle scattering, like the proton scattering, were analyzed in terms of an optical model[2]. This point is made clear when we note that the constants of (30.2) inserted in (30.1), for example, cause the nuclear potential to fall to 0.1 of its central value at a radius, for heavy nuclei,

$$R_{\max} = 1.52\,A^{\frac{1}{3}}. \tag{33.3}$$

The effect of α-particle absorption in the diffuse nuclear surface has been considered[3], and provides an improvement on the cruder treatments we have mentioned. The results obtained for the nuclear radius are similar.

The appreciable tail of the nucleon density is also shown up in its effect on inelastic processes, which indicate[4] that nuclear effects extend to a radius at least as great as $1.7A^{\frac{1}{3}}$. The elastic alpha results confirm this figure.

IV. Charged particle reaction cross sections.

34. Barrier penetrability indications on nuclear radius. Reactions initiated by protons or alpha particles incident below the barrier height have a cross section with a rapid energy dependence characteristic of the barrier penetrability factor. Analysis of (p, n) reactions[5] has led to a radial constant of 1.5. The determination is so crude, however, that it is only possible to comment that this radius is consistent with the nuclear force radius measured in other ways.

A recent analysis[6] of the fission of Th^{232} by 33 Mev alpha particles gives for the radius of the interaction

$$r_0 A^{\frac{1}{3}} + 1.60 \tag{34.1}$$

with $1.14 < r_0 < 1.24$.

[1] K. W. FORD: Personal communication.

[2] *Note added in proof:* This expectation is borne out by a recent optical model analysis of alpha particle scattering. See G. IGO, R. M. THALER and D. L. HILL: Bull. Amer. Phys. Soc. II **1**, 384 (1956).

[3] C. E. PORTER: Phys. Rev. **99**, 1400 (1955).

[4] G. P. MILLBURN, W. BIRNBAUM, W. E. CRANDALL and L. SCHECHTER: Phys. Rev. **95**, 1268 (1954).

[5] J. P. BLASER, F. BOEHM, P. MARMIER and D. C. PEASLEE: Helv. phys. Acta **24**, 3 (1951). See also MATHEW M. SHAPIRO: Phys. Rev. **90**, 171 (1953).

[6] R. M. THALER: Personal communication regarding an analysis of data taken by J. JUNGERMAN, Phys. Rev. **79**, 632 (1950).

V. Alpha decay.

35. Large nuclear radii from alpha decay lifetimes. The GAMOW theory of alpha decay[1] provided one of the earliest means to determine nuclear radii. This theory has not been seriously altered up to the present time[2,3] and radii recently determined from alpha decay are very little different from those given by the original form of the theory. Alpha decay radii for $Z \gtrsim 82$ follow quite closely an $A^{\frac{1}{3}}$ law,

$$R = 1.5\, A^{\frac{1}{3}}. \tag{35.1}$$

More fluctuation is noted for elements below $Z = 82$, but with about the same mean radial constant. This may be in part due to nuclear deformations, which would generally tend to increase radii determined from alphy decay. There are factors of inhibition, however, which are known[4] to work in the opposite direction, and must be due to details of nuclear structure, and not merely to nuclear size.

General conclusions from the theory and analysis of alpha decay lifetimes are these: (1) The factor in the decay rate arising from intrinsically nuclear factors, sometimes thought of as the formation probability of an alpha particle at the nuclear surface, is very uncertain. (2) Nuclear radii deduced from alpha decay are so insensitive to the lifetime that it would take a very major alteration in alpha decay theory to change very much the nuclear radii. (3) The nuclear radii deduced from alpha decay with *any* current model are given roughly by (35.1) and are definitely larger than nuclear electric radii.

VI. Deuteron pick-up and stripping.

36. Results from deuteron reactions. The (p, d) and (n, d) pick-up reactions and their inverse stripping reactions have revealed several significant facts about nuclear structure. First, they have provided a measure of the high momentum components in the nuclear wave function[5]. Second, they have tested the shell model assignments for orbital angular momentum, and to some extent tested the validity of the angular momentum, l, of a single nucleon as a good quantum number[6].

COOPER and TOBOCMAN have pointed[7] out that any differences between the neutron and proton radii will lead to very great differences in the (d, p) and (d, n) probabilities. Such differences appear to have been observed[8] although a number of uncertainties surround the interpretations of the experiment.

VII. Pion scattering and bound pions.

37. Uncertain information from pions. Bound pion transition energies (e.g. $2P \rightarrow 1S$) have been interpreted[9] for the light elements in terms of an average repulsive potential between nucleus and pion[10]. The pion-nucleus force is not well

[1] G. GAMOW: Z. Physik **52**, 510 (1928).
[2] M. A. PRESTON: Phys. Rev. **71**, 865 (1947).
[3] J. J. DEVANEY: Phys. Rev. **91**, 587 (1953).
[4] I. PERLMAN, A. GHIORSO and G. T. SEABORG: Phys. Rev. **77**, 26 (1950). — F. ASARO and I. PERLMAN: Phys. Rev. **91**, 763 (1953).
[5] K. BRUECKNER, R. EDEN and N. FRANCIS: Phys. Rev. **98**, 1445 (1955).
[6] S. T. BUTLER: Phys. Rev. **80**, 1095 (1950). — Proc. Roy. Soc. Lond., Ser. A **208**, 559 (1951).
[7] L. N. COOPER and W. TOBOCMAN: Phys. Rev. **97**, 243 (1955).
[8] W. N. HESS and B. J. MOYER: Phys. Rev. **96**, 859 (1954).
[9] M. B. STEARNS, M. STEARNS, S. DE BENEDETTI and L. LEIPUNER: Phys. Rev. **96**, 804 (1954).
[10] S. DESER, M. L. GOLDBERGER, K. BAUMANN and W. THIRRING: Phys. Rev. **96**, 774 (1954).

enough understood to enable analysis of the bound state transitions to provide an independent measure of the nuclear radius.

The scattering of pions by nuclei is somewhat more sensitive to the nuclear radius, but again uncertainties of interpretation are still too great to permit an independent measure of nuclear radius in this way. Experiments of the elastic scattering of 60 Mev pions by carbon[1] have been used[2] to determine properties of the pion-nucleon interaction, and have been interpreted[3] by means of a refined optical model[4], with the conclusion that a diffuse nuclear boundary is required to fit the data but that the nuclear radius is not determined by the experiments. Study has also been made[5] of the interaction of 79 Mev pions on aluminum.

C. Electric measures of nuclear angular shape.

Just as with the radial variation of nuclear density the electrical measures are most clear-cut in their interpretation, so for the angular variation our most reliable information comes from electromagnetic observations. This discussion of these effects is largely taken from [*12*].

38. Deformation parameters. Simple considerations of the energetics of nuclear deformation show that the predominant mode of deformation should be ellipsoidal, as expressed by (3.2). As noted under that equation, such deformations are characterized by two parameters, a deformation parameter, β, related to the absolute magnitude of the deformation, and an asymmetry parameter, γ, measuring the deviation from cylindrical symmetry. These are classically defined quantities, however, assuming a static nucleus. For large nuclear deformations, β and γ may indeed have classical significance (amplitude of surface vibration $\ll$ deformation), but even then approximate degeneracies between different combinations of β, γ may lead to mixing of different deformational configurations. For small deformations, β and γ have significance only as expectation values.

I. Quadrupole moments.

39. Origin of quadrupole moments. The shell model of the nucleus in its simplest form—nucleons moving independently in a common central potential—predicts quadrupole moments for any number of particles which are of the same order or less than the quadrupole moment of a single particle. This result follows because closed shells have zero quadrupole moments and groups of equivalent particles have quadrupole moments less than one particle of the group. Many observed moments are many times larger than can be explained by this central-well shell model. These large moments can be explained in the same shell model language in terms of very great configuration mixing in which many partly filled shells contribute additively to the quadrupole moment. An equivalent but far simpler explanation is in terms of a collective description of deformation of the whole nucleus. The shell model together with a finite nuclear deformability offers a ready explanation of the mechanism of the deformation and accounts for the order of magnitude of the observed quadrupole moment.

Townes, Foley and Low[6] first pointed out the correlation with nuclear shell structure of the regularities in quadrupole moments as a function of

[1] H. Byfield, J. Kessler and L. M. Lederman: Phys. Rev. **86**, 17 (1952).
[2] D. C. Peaslee: Phys. Rev. **87**, 862 (1952).
[3] L. S. Kisslinger: Phys. Rev. **98**, 761 (1955).
[4] N. C. Francis and K. M. Watson: Phys. Rev. **92**, 291 (1953).
[5] A. Pevsner and J. Rainwater: Phys. Rev. **100**, 1431 (1955).
[6] C. H. Townes, H. M. Foley and W. Low: Phys. Rev. **76**, 1415 (1949).

mass number. RAINWATER[1] showed that surface coupling of extra nucleons can account for the large magnitude of some observed moments, and the theory of this surface coupling was developed by BOHR[2], HILL and WHEELER [6], and BOHR and MOTTELSON [27]. In a more recent survey of quadrupole moments, GOEPPERT-MAYER[3] has considered the combined contributions of extra-shell protons and of total nuclear deformation. She finds that in the vicinity of closed shells, the moments can be accounted for with zero deformation, but that far from closed shells quite substantial deformations are required. The general picture obtained from observed moments is therefore a regularly varying nuclear deformation with minima for either proton or neutron closed shells, and maxima between closed shells.

Since we wish to avoid detailed discussion of the shell model and nuclear structure, we merely emphasize here that for nuclei with small quadrupole moments the nuclear deformation appears to be almost negligible. This conclusion comes both from the quadrupole evidence given by GOEPPERT-MAYER, and from the success near closed shells of shell model calculations based on spherical symmetry[4-6]. There are, then, at least a few nuclei for which a spherically symmetric charge distribution is a very good approximation. It is upon these nuclei that we have concentrated in Parts A and B for the determination of the radial variation of nuclear charge and mass distributions.

40. Connection with nuclear deformation. For nuclei in the opposite limit, with very large quadrupole moments, it is reasonable to ignore the extra-particle contribution and to evaluate the angular shape of the nucleus on the assumption that all of the observed quadrupole moment is due to the deformation of a homogeneous charge distribution. The "intrinsic" quadrupole moment, or moment relative to body-fixed axes (or "classical" moment appropriate for angular momentum, $I\to\infty$), may then be defined for the charge distribution having the form given by (3.1) and (3.2):

$$Q_0 = \int_0^\infty \varrho(r')\,(3z'^2 - r'^2)\,r'^2\,dr'\,d\Omega', \tag{40.1a}$$

$$= (5/\pi)^{\frac{1}{2}} Z\langle r^2\rangle_{\text{Av}}\,\alpha_0 = (5/\pi)^{\frac{1}{2}} Z\langle r^2\rangle_{\text{Av}}\,\beta\cos\gamma. \tag{40.1b}$$

In (40.1a), the primes denote body-fixed axes such as are used in (3.2). The relation (40.1b) is only accurate to order α; higher terms are omitted. The mean square radius which appears here is averaged over angle, i.e.

$$\langle r^2\rangle_{\text{Av}} = (4\pi\,\varrho_0/Z)\,r_1^5 \int_0^\infty f(x)\,x^4\,dx. \tag{40.2}$$

(40.1) and (40.2) have the same form as the usual definitions for a uniform charge distribution, but they hold (to order α) for an arbitrary radial shape, $f(x)$. If the angular momentum is directed along the z' axis in the body, then the observed quadrupole moment is less than Q_0 by a projection factor [3]

$$P_I = \frac{I(2I-1)}{(I+1)(2I+3)}. \tag{40.3}$$

[1] J. RAINWATER: Phys. Rev. **79**, 432 (1950).
[2] A. BOHR: Dan. Mat. fys. Medd. **26**, No. 14 (1952).
[3] M. GOEPPERT-MAYER: Paper on Nuclear Shell Theory, Proc. Intern. Conf. of Theoretical Physics (Kyoto and Tokyo, Japan, 1953) (Science Council of Japan, Tokyo, Japan, 970 pp., 1954).
[4] J. P. ELLIOTT and B. H. FLOWERS: Proc. Roy. Soc. Lond., Ser. A **229**, 536 (1955).
[5] D. E. ALBURGER and M. H. L. PRYCE: Phys. Rev. **95**, 1482 (1954).
[6] C. LEVINSON and K. W. FORD: Phys. Rev. **99**, 792; **100**, 1, 13 (1955).

The simplest interpretation of observed quadrupole moments is then obtained by supposing that the deformation is cylindrically symmetric about the axis defined by the total angular momentum. One then sets $\cos\gamma = \pm 1$ according as $Q_0 \gtrless 0$, and uses (40.1 b) together with the projection factor (40.3) to obtain $\beta\langle r^2\rangle_{\mathrm{Av}}$ as the parameter of the nuclear charge distribution measured by experiment. Whether the number so determined is indeed $\beta\langle r^2\rangle_{\mathrm{Av}}$ depends on the orientation of the angular momentum in the body fixed system and on the magnitude of the deformation, both of which change the projection factor (40.3); on the direct contribution of extra-shell protons to the quadrupole moment; and on the

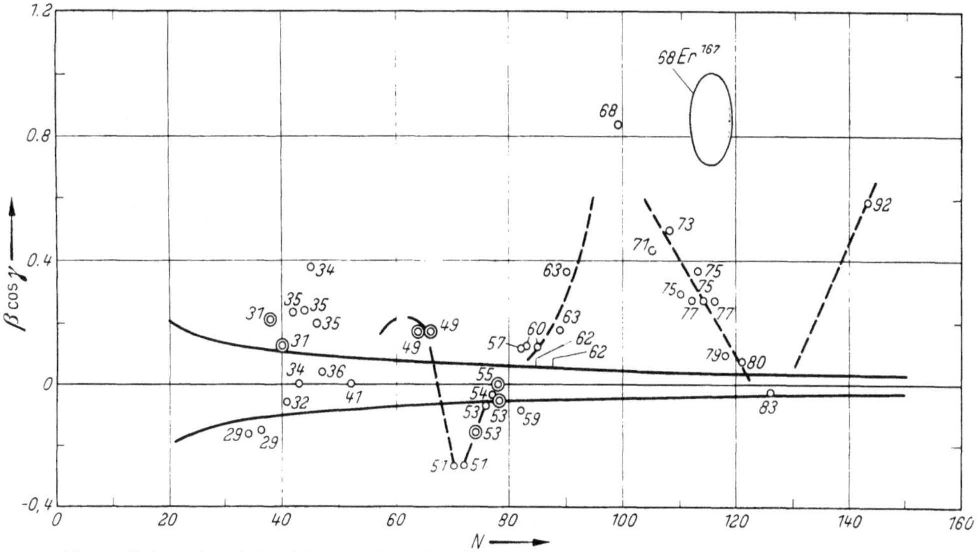

Fig. 12. Deformations deduced from static quadrupole moments. The deformation, $\beta\cos\gamma$, from (40.1), is plotted *vs.* neutron number, and Z values are indicated by each point. A dashed line shows the approximately regular behaviour of the deformation among the heavy nuclei. Most points are not accurate, errors being about 10% to a factor two. Those points which are accurate to within a few percent are doubly circled. The solid lines enclose the region the points would be expected to occupy if the moments were due to one or a few protons outside a spherically symmetric closed shell. The shape of the Er^{167} nucleus, with a measured quadrupole moment of 10 barns, is illustrated. From [*12*].

degree of symmetry of the charge distribution, which changes $\cos\gamma$. The last effect, in particular, could cause a small observed moment and a consequent underestimate of β. Near closed shells the direct particle contribution causes instead, with this interpretation, an overestimate of β. However, in the region of large deformations, the only region where the simple picture of a deformed homogeneous charge distribution is appropriate, the assumptions described above are expected to be nearly correct.

Although the quantity determined by the quadrupole moment is $\beta\langle r^2\rangle_{\mathrm{Av}}$, we can take $\langle r^2\rangle_{\mathrm{Av}}$ as a known quantity from the μ-mesonic energies or from electron scattering results, and therefore determine from the data the deformation, β, alone, by the formula

$$\beta = (\pi/5)^{-\frac{1}{2}}\,(Z\langle r^2\rangle]^{-1}\,(P_I)^{-1}\,|Q|\,. \tag{40.4}$$

Nuclear deformations determined from quadrupole moments[1] by this formula are shown in Fig. 12, except that $\beta\cos\gamma$ is plotted in order to illustrate the sign of Q. The values are expected to be meaningful measures of the actual deformation of total nuclear charge only where they are large ($\beta \gtrsim 0.1$ for the heavy nuclei,

[1] J. E. MACK: Rev. Mod. Phys. **22**, 64 (1950).

$\beta \gtrsim 0.2$ for the intermediate nuclei). Measured quadrupole moments are not of high accuracy, but Fig. 12 nonetheless gives valuable evidence as to the magnitude of nuclear charge deformation, as to the regularity of this deformation as a function of neutron number (for $N \gtrsim 70$), and as to the marked preference of large deformations for the prolate shape.

II. Electric quadrupole transition rates.

41. Transitions among rotational states. Electric quadrupole (E2) transition rates measure the off-diagonal matrix elements of the same operator whose diagonal element is the ground state quadrupole moment. As a means to measure nuclear deformation, the transition rate has one disadvantage, but several advantages, as compared to the quadrupole moment measurement. The disadvantage is that two nuclear states are involved; if the excited state deformation differs from that of the ground state, one does not measure the desired quantity, the ground state deformation. There is good theoretical reason to believe, however, that certain low excited states of nuclei with large deformation have nearly the same angular shape as the ground state [*3*]. For such nuclei, whose angular shape is in fact of the greatest interest, this feature is therefore not a serious disadvantage. The advantages of the transition rate measurements, relative to quadrupole moments are: (1) the available data are more extensive; (2) the accuracy of measurement is frequently much greater; (3) the quadrupole matrix elements may be determined even for spin zero or one-half, when quadrupole moments vanish, and therefore, in particular, the deformations of even-even nuclei may be determined.

In nuclei which are spherically symmetric or have only very small deformations the E2 transition rates will depend on shell structure effects as well as collective effects. For nuclei with large deformations, there are two simplifying features. (a) The contribution to E2 rates of extra-shell protons will be small compared to the contribution of the collective nuclear deformation. We therefore consider only the collective contribution. (b) Nuclear rotational states are expected to have deformations nearly the same as the ground state, and to be connected by large quadrupole matrix elements to the ground state (for $\Delta I \leq 2$). Moreover, the theoretical treatment of E2 rates is very easy in the limit of large deformations.

The collective quadrupole operator for a homogeneous distribution of charge, defined by (3.1) and (3.2) is

$$Q'_{2m} = \int_0^\infty \varrho(r')\, Y_{2m}(\vartheta', \varphi')\, r'^4\, dr'\, d\Omega', \tag{41.1a}$$

$$= \frac{5}{4\pi} Z \langle r^2 \rangle \alpha'_m, \tag{41.1b}$$

where $\langle r^2 \rangle$ is defined by (40.2) for an arbitrary radial shape function, $f(x)$. Note that the normalizations of (41.1b) and (40.1b) are different: $Q_0 = 4(\pi/5)^{\frac{1}{2}} Q_{20}$. The factor of difference stems from the standard practice of defining the quadrupole moment in terms of $3z^2 - r^2$, but the quadrupole transition operator in terms of $r^2\, Y_{2m}(\vartheta, \varphi)$.

Since the radial shape of the charge distribution enters only through the mean square radius, $\langle r^2 \rangle$, which is determined from other experiments, the problem is only to determine α'_0 and α'_2 ($\alpha'_{-2} = \alpha'_2$, $\alpha'_{\pm 1} = 0$), the two independent deformation parameters in the body fixed system. One supposes, on the theoretical basis of the collective model, and the experimental basis of the symmetric top rotational energies proportional to $I(I+1)$, that in fact most nuclei with large deformations are cylindrically symmetric. Therefore $\alpha_2' = 0$, and $\alpha_0' (= \pm\beta)$ is the single deformation parameter determined by experiment. There is very little experimental information so far which bears on the interesting question whether there are nuclei with large but asymmetric deformations.

The theory of the fast E2 transitions of deformed nuclei is based on the collective theory of nuclear rotational states[1, 2] [3], the particle structure being assumed the same in initial and final states. It should be stressed that an important feature of the collective theory of rotational states is just that it predicts that many nuclei have a classically definable angular shape (in general spheroidal) and provides a ready means to determine this shape from E2 transition rate measurements. Although the interpretation of the transition rates in terms of charge deformation rests on this theory of nuclear structure, it is insensitive to all details, depending only on the idea of rotational states.

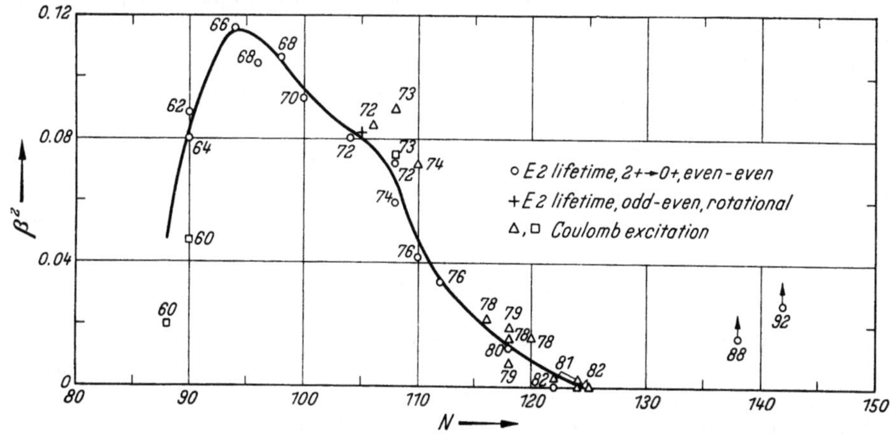

Fig. 13. Nuclear deformations determined from E2 transition rates, *vs.* neutron number N. Circles: from directly measured lifetimes, as summarized in ref. 4, p. 223 of Sect. 42. *Squares:* from COULOMB excitation cross section of ref. 4, p 224 of Sect. 43. *Triangles:* from ref. 2, p. 224 of Sect. 43. The nuclear electric radius is taken to be $1.20 \times 10^{-13} A^{\frac{1}{3}}$ cm. Errors are not indicated, but the direct lifetime measurements are, on the average, of higher accuracy. From [12].

42. Gamma decay. Lifetimes of nuclear states undergoing E2 transitions have been measured for a number of nuclei among the heavy elements[3, 4] [3]. The determination is limited to transition energies sufficiently small to make the lifetime measurably long, but this condition is satisfied for most of the low-lying rotational states of heavy nuclei. These lifetimes determine a charge deformation, β_E, within the limits set by the assumptions above (rotational theory, direct particle contribution small), and with the further assumptions of cylindrical symmetry,

$$\beta_E^2 \equiv \langle \beta \cos\gamma \rangle^2, \tag{42.1}$$

and of known mean square electric radius. From the available data, we plot in Fig. 13, β_E^2 *vs.* neutron number N, with mean square radius taken to be

$$\langle r^2 \rangle = \tfrac{3}{5} (1.20 A^{\frac{1}{3}})^2. \tag{42.2}$$

[1] A. BOHR: Ph. D. Diss., Univ. Inst. for Teoretisk Fysik, Copenhagen: Rotational States of Atomic Nuclei. 55 pp. København: Ejnar Munksgaards 1954.

[2] K. FORD: Phys. Rev. **90**, 29 (1953).

[3] M. GOLDHABER and A. W. SUNYAR: Phys. Rev. **83**, 906 (1951).

[4] A. W. SUNYAR: Phys. Rev. **98**, 653 (1955).

The data are taken from a recent compilation by SUNYAR[1] and his estimates of internal conversion coefficients are used. The deformation shows a very regular trend with neutron number in agreement with the trend from quadrupole moments (Fig. 12). The transition rates lead, however, to β^2 smaller by about a factor two than β^2 from quadrupole moments. (An exception is the quadrupole moment of Er^{167}, for which the discrepancy is greater.) According to Fig. 12, $\beta^2_{max}=0.12$ in the rare earth region, or $\beta_{max}\approx 0.35$. This value corresponds to a nuclear shape $\sim[1+0.22\,P_2(\cos\vartheta)]$, which we shall call a 22% deformation. The quadrupole moment of Er^{167}, if taken seriously, gives a 50% deformation for that nucleus.

43. COULOMB excitation. The process of electric excitation of nuclear levels has been exploited recently to measure the E2 rates of many transitions. The method is in a sense complementary to lifetime measurements, being most easily applied to fast E2 transitions (although the dependence of the excitation cross section on transition energy also favors low-lying levels). Also the COULOMB excitation process has the advantage of picking out almost exclusively the E2 rate between levels of the same parity, not being affected by the possible competition of M1 in the decay process (important in the odd-even nuclei).

COULOMB excitation cross sections have been measured by a number of workers[2-5]. Some results are included in Fig. 13, reduced to the square of the nuclear deformation, β_E^2, according to the assumed relations (42.1) and (42.2), for the symmetry of the deformation, and for the electric radius. These results are seen to be in good agreement with deformations determined by lifetime measurements.

III. Isotope shifts.

44. Anomalous shifts from nuclear deformation. The atomic isotope shift provides the most sensitive measure now available of the change of nuclear electric radius from one isotope to another. Specifically, the measured quantity is $\delta\langle r^{2\sigma}\rangle$ (18.1), where $\langle r^{2\sigma}\rangle$ is averaged over the radius and over angle. As discussed in Sect. 19, the important role of nuclear compressibility in the phenomenon invalidates the isotope shift as an accurate way to measure nuclear electric radii. Since, however, the isotope shift depends on the angular shape as well as the radial shape of the charge distribution, one may regard as known the radial shape, and seek to extract information from the isotope shift on the angular shape.

BRIX and KOPFERMANN[6] first suggested that nuclear deformation might account for certain anomalously large isotope shifts (by more than a factor two) in the neighborhood of neutron number 90. WILETS further suggested that nuclear deformation might explain all of the anomalies, and the regular variation of the reduced isotope shift with neutron number. An analysis of isotope shifts from this point of view [*22*] confirmed the idea that all of the *variation* of anomaly with mass number could be accounted for in terms of the deformation of nuclear charge. Moreover, the isotope shift provides a measure of the change of nuclear

---

[1] See Footnote 4, p. 223.

[2] T. HUUS and C. ZUPANČIČ: Dan. Mat. fys. Medd. **28**, No. 1 (1953).

[3] G. M. TEMMER and N. P. HEYDENBURG: Phys. Rev. **96**, 426 (1954); and other references there.

[4] F. MCGOWAN and P. STELSON: Phys. Rev. **99**, 112 (1955).

[5] C. L. MCCLELLAND, H. MARK and C. GOODMAN: Phys. Rev. **93**, 904 (1954); **97**, 1191 (1955).

[6] P. BRIX and H. KOPFERMANN: Z. Physik **126**, 344 (1949). — Festschr. Akad. Wiss. Göttingen, Math.-physik. Kl. **17** (1951); other references here. — Phys. Rev. **85**, 1050 (1952).

deformation which is independent of uncertainty associated with the nuclear compressibility.

Consider [12] the nuclear charge distribution defined by (3.1) and (3.2) and suppose that the deformation is spheroidal. Then $\alpha_{\mu'} = \beta \cos\gamma\, \delta_{\mu 0}$, in which $\delta_{\mu 0}$ is the KRONECKER delta, and the mean value of $r^{2\sigma}$ is given, to second order in β, by

$$\langle r^{2\sigma}\rangle = (4\pi\,\varrho_0/Z)\,\bar{r}_1^{3+2\sigma} \int\limits_0^\infty f(x)\, x^{2\sigma+2}\, dx \left[1 + \frac{1}{8\pi}(3+2\sigma)(2+2\sigma)\,\beta^2\right]. \tag{44.1}$$

Now the radial constant, $\bar{r}_1$, is independent of β only to first order [cf. Eq. (40.2) where second order terms were ignored]. To second order, the normalization of charge gives

$$1 = (4\pi\,\varrho_0/Z)\,\bar{r}_1^3 \int f(x)\, x^2\, dx\, [1 + (3/4\pi)\,\beta^2]. \tag{44.2}$$

A revised radial constant, $\bar{\bar{r}}_1$, is therefore defined such that $\varrho_0 \bar{\bar{r}}_1^3$ is a constant through second order in β:

$$\bar{\bar{r}}_1 = \bar{r}_1 \left[1 + \frac{1}{4\pi}\beta^2\right]. \tag{44.3}$$

Substitution of (44.2) and (44.3) into (44.1) gives

$$\langle r^{2\sigma}\rangle = (\bar{\bar{r}}_1)^{2\sigma} \cdot \mathfrak{R} \cdot \left[1 + \frac{1}{8\pi}\, 2\sigma(2\sigma+3)\,\beta^2\right] \tag{44.4}$$

where $\mathfrak{R}$ is the ratio of two integrals over the radial shape function, $f(x)$:

$$\mathfrak{R} = \int f(x)\, x^{2\sigma+2}\, dx \Big/ \int f(x)\, x^2\, dx. \tag{44.5}$$

Now we suppose that two isotopes of a given element differ in nuclear radius, or in central charge density ($\bar{\bar{r}}_1$ changes), and in deformation (β changes), but not in the radial shape of the charge distribution ($\mathfrak{R}$ does not change). In that case, the change in $\langle r^{2\sigma}\rangle$, which is proportional to the isotope shift, is given by,

$$\delta\langle r^{2\sigma}\rangle = \mathfrak{R}\left[\delta(\bar{\bar{r}}_1^{2\sigma}) + (\bar{\bar{r}}_1^{2\sigma})\,\frac{1}{8\pi}\, 2\sigma(2\sigma+3)\,\delta\beta^2\right]. \tag{44.6}$$

In the second term within the brackets, the change of $\bar{\bar{r}}_1$ between the isotopes is ignored, since in general the fractional change of β^2 is much greater than the fractional change of $\bar{\bar{r}}_1$. The isotope shift has been calculated for spherical incompressible nuclei—i.e. assuming $\beta^2 = 0$, and $\delta(\bar{r}_1^{2\sigma})/(\bar{r}_1^{2\sigma}) = (2\sigma/3)\,(\delta A/A)$. If this model is called the "standard", then

$$[\delta\langle r^{2\sigma}\rangle]_{\text{st'd}} = \mathfrak{R}\,[\delta(\bar{\bar{r}}_1^{2\sigma})]_{\text{st'd}}. \tag{44.7}$$

and the ratio of the observed isotope shift to the standard shift is just the ratio of the expressions in (44.6) and (44.7):

$$\frac{(\text{Shift})_{\text{exp}}}{(\text{Shift})_{\text{st'd}}} = \frac{\delta(\bar{\bar{r}}_1^{2\sigma})}{[\delta(\bar{\bar{r}}_1^{2\sigma})]_{\text{st'd}}} + \frac{3}{8\pi}\, A(2\sigma+3)\,\frac{\delta\beta^2}{\delta A}. \tag{44.8}$$

This discussion is a slight generalization of that given in [22] to include non-uniform charge distributions. The result, however, is not essentially altered. The ratio given by the left side of (44.8) has first been plotted by BRIX and KOPFERMANN[1], who showed that the deviations of the ratio from unity are large but regular.

[1] P. BRIX and H. KOPFERMANN: Z. Physik **126**, 344 (1949). — Festschr. Akad. Wiss. Göttingen, Math-physik. Kl. **17** (1951); other references here. — Phys. Rev. **85**, 1050 (1952).

45. Indicated deformations. The right side of (44.8) breaks up very conveniently into two terms, one associated only with the change of mean charge density between the two isotopes, the other only with the change of deformation. All of the uncertainty coming from the nuclear compressibility is in the first term. The ratio in the first right-hand side term in (44.8) should be less than 1. Experimentally, it appears to be about 0.64, for the heavy elements, which is in

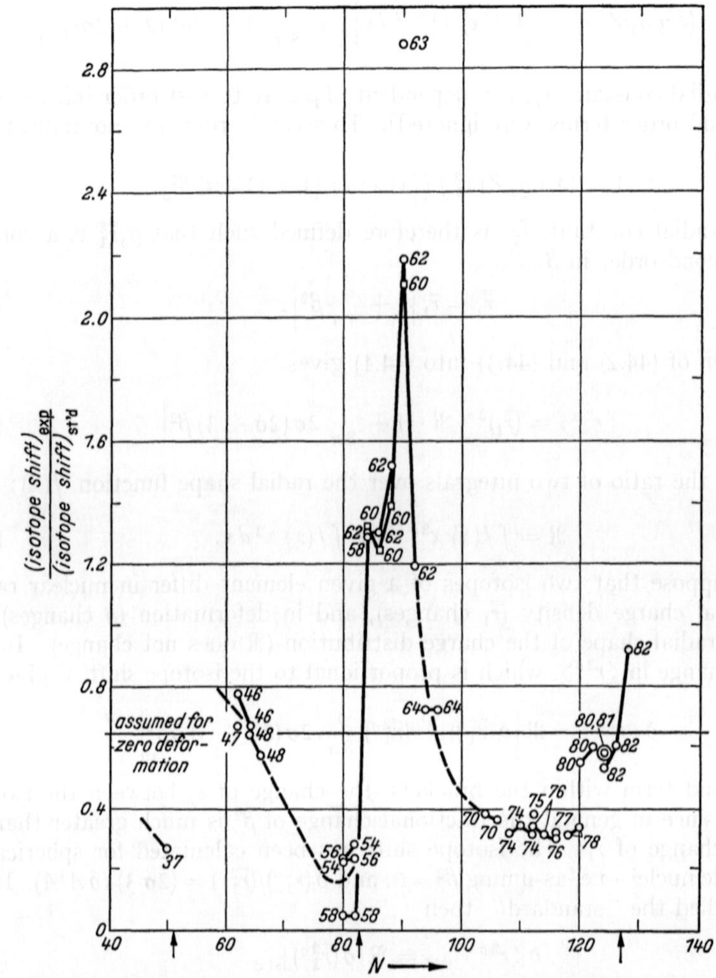

Fig. 14. Ratio of observed isotope shift to "standard" shift for uniform density nuclei with $R = 1.20 \times 10^{-13} A^{\frac{1}{3}}$, plotted *vs.* neutron number N. Z values are indicated on graph. The assumed line of zero deformation is drawn in at ordinate 1.65. [The Nd points ($Z = 60$) are only known relatively, but are included because they confirm the large anomaly at $N = 88$ to 90.] Points are placed at the higher N value of each pair of isotopes. From [12].

good agreement with what is expected for a reasonable nuclear compressibility, as pointed out in Sect. 19. The final assumption required in order to interpret the isotope shifts in terms of nuclear deformation is that the value of the ratio in the first term is nearly constant, or only slowly varying with A, among the heavy elements. This assumption would be wrong if, for example, shell structure effects caused a change in the average nuclear density among isotopes. However, it is the COULOMB energy which causes the ratio in question to be ~ 0.64 instead of 1. Shell energies are smaller than COULOMB energies in the heavy elements and should therefore change this ratio by a lesser amount.

Within the framework of the assumptions outlined above, the change of nuclear deformation, $\delta\beta^2$, between isotopes can be found as follows. The ratio of observed isotope shift to the "standard" shift is plotted *vs.* N or A. One reads off the ratio at several points where, on the basis of other evidence, $\delta\beta^2$ is expected to be nearly zero (e.g. near Pb^{208} or at the maximum of β^2 *vs.* N shown in Figs. 12 and 13). These values should be (and are) nearly the same. They determine the value of the first term on the right of Eq. (44.8). For the rest of the plotted points, the value of the first term is taken to be the same, whence the values of the second term, and therefore of $\delta\beta^2/\delta A$, are read off. Fig. 14 shows a plot of $(\text{Shift})_{\text{exp}}/(\text{Shift})_{\text{st'd}}$, which differs from that given in [22] only in that the nuclear electric radius is here taken to be $1.20\,A^{\frac{1}{3}}$ instead o

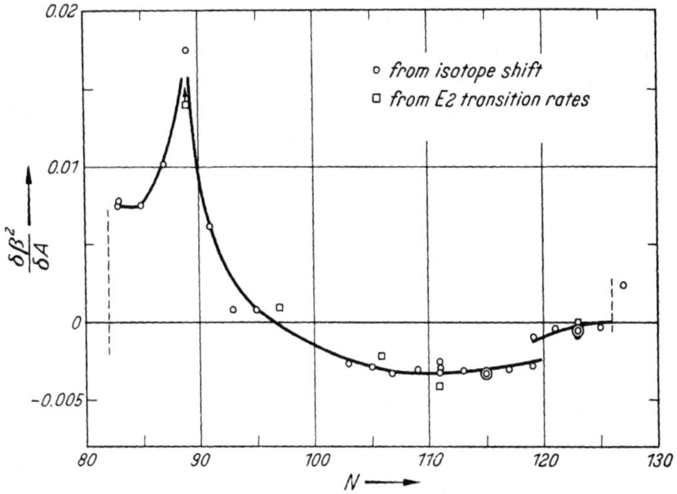

Fig. 15. Rate of change of nuclear deformation with mass number, *vs.* neutron number, N. Circles are the derivatives $\delta\beta^2/\delta A$ evaluated from isotope shift data (Fig. 14). The squares are obtained by differencing the values of β^2 for fixed Z, which are calculated from E2 transition rates and illustrated in Fig. 13. From [12].

$1.40\,A^{\frac{1}{3}}$. Values of $\delta\beta^2/\delta A$ determined from this curve are plotted in Fig. 15, together with the same quantity obtained by differencing values of β^2 plotted in Fig. 13, where these are known for more than one isotope of a given element. The agreement is excellent in view of the magnification of error inherent in taking small differences of numbers from experimental sources. The marked asymmetry of this curve is due to the fact that large deformations develop very rapidly between neutron numbers 88 and 90, while the deformation falls off gradually and smoothly between $N=100$ and 120. The small discontinuity at $N=118$ is probably real, since it is seen also in the energies of the first excited state of even-even nuclei. The isotopes of Hg and Pb appear to be nearly spherically symmetric.

IV. μ-mesonic x-rays.

46. Quadrupole effects on spectra. WHEELER[1,2] first pointed out that the effect of the quadrupole moment of the nucleus on the levels of the bound μ-meson might be quite appreciable—of the same order as the fine structure splitting in some cases, which amounts to several tenths of a Mev for heavy nuclei. He has calculated some quadrupole patterns, although the experiments have not as yet

[1] J. A. WHEELER: Rev. Mod. Phys. **21**, 133 (1949).
[2] J. A. WHEELER: Phys. Rev. **92**, 812 (1953).

resolved the quadrupole effect. WILETS[1] and JACOBSOHN[2] have shown that the dynamic interaction of the μ-meson with a deformed nucleus is as important as the static interaction, and that in addition the dynamic interaction modifies the fine structure of the μ-meson even for even-even nuclei of spin-zero, for which, of course, the static interaction vanishes. This large effect comes about from the large electric quadrupole interaction of the μ-meson with the nuclear rotational states, which causes nuclear excitation with substantial probability by the cascading μ-meson. WILETS and JACOBSOHN calculate typical fine structure patterns to be expected for $2\,P \to 1\,S$ μ-mesonic transitions. A particularly interesting feature is that this pattern depends on the sign of the intrinsic nuclear deformation (sign of $\cos\gamma$), and affords the only method known at present which permits in principle the measurement of this quantity for even-even nuclei. From these calculations it appears that, although individual lines of the fine structure pattern will not all be resolved with conventional techniques, the intensity patterns for positive and negative $\cos\gamma$ will be sufficiently different that a determination of the sign of the deformation will be possible without unreasonable demands on the experimental energy resolution of the gamma rays. Such measurements will be of great interest, since one feature of the large nuclear deformations in heavy nuclei of special significance is the marked predominance[3] of positive $\cos\gamma$ for the odd-even nuclei. The magnitude of the deformation will also be determined by the $2\,P \to 1\,S$ fine structure pattern, but not with the accuracy expected of transition rate measurements.

V. Electron scattering.

47. Elastic scattering from aspherical charge. A characteristic of the high energy electron scattering angular distributions from the highly deformed nuclei is a smoother variation with angle than shown by the distribution from nearly spherical nuclei. It is reasonable to attribute this behavior largely to the averaging over angles of the highly deformed nuclear surface so that the radial distribution of charge which is effective in the scattering is more gradually tapered at the nuclear surface than in the case of spherically symmetric nuclei. One such calculation has recently been performed[4] for the element Ta^{181}, a nucleus whose charge deformation has been rather accurately determined by other means, and is large. The quadrupole effect on the scattering has been treated as a perturbation on the exactly calculated scattering for a spherically symmetric nucleus. The large deformation of the nucleus perturbs the scattering in three ways. (1) The electrons which excite the nucleus to low lying rotational states (137 kev and 300 kev in this case) and are inelastically scattered are included experimentally as part of the elastic group. (2) The quadrupole field contributes to the elastic scattering. (This effect is averaged over orientations of the nucleus, i.e. over M values, and contributes only for $I \geq 1$.) (3) The deformation averaged classically over all angles extends and smooths the radial charge distribution. For even-even nuclei and for ideal energy resolution, this effect is the only one present and could not be distinguished from a change in the radial shape of the charge distribution. The effect of the subtraction of the inelastic electrons on the elastic amplitude is considered to be of negligible importance and is hence omitted in this discussion.

---

[1] L. WILETS: Dan. Mat. fys. Medd. **29**, No. 3 (1954).
[2] B. A. JACOBSOHN: Phys. Rev. **96**, 1637 (1954).
[3] S. A. MOSZKOWSKI and C. H. TOWNES: Phys. Rev. **93**, 306 (1954).
[4] B. W. DOWNS, D. G. RAVENHALL and D. R. YENNIE: Phys. Rev. **98**, 277 (1955).

The measured angular distribution of 180 Mev electrons from Ta, together with fits for various assumed nuclear deformations, gives a value of nuclear deformation corresponding to an intrinsic quadrupole moment of 10 barns, which is in good agreement with other values for the same nucleus.

D. Nonelectric measures of nuclear angular shape.

48. Rotational states. The collective theory of nuclear rotational states [*3*][1] has had outstanding success correlating data on many low-lying nuclear levels. Its greatest utility has involved a number of heavy nuclei well removed from closed shells for which certain facts are well established. (1) These nuclei have large deformations as deduced from the consistent evidence of several methods for measuring the deformation of charge; see Part C. (2) These nuclei possess rotational states with energies given, very accurately in many cases, by

$$E_I = [I(I+1) - I_0(I_0+1)]\,\frac{\hbar^2}{2\mathfrak{J}} \tag{48.1}$$

in which I is the total angular momentum number for a band of excited states, $\mathfrak{J}$ is the moment of inertia for the nucleus (assumed constant over the sequence of states, in first approximation). For even-even nuclei the angular momentum number of the ground state, I_0, of course, vanishes. (3) The moments of inertia, $\mathfrak{J}$, are closely proportional to the square of the charge deformation,

$$\mathfrak{J} = \text{const} \times \beta^2 \tag{48.2}$$

as well illustrated in a recent survey[2] comparing β^2 as given by (48.2) with β^2 obtained from lifetime measurements of E2 transitions. The curves for β^2 *vs.* neutron number, as evaluated from these two sources, are closely proportional.

In the early applications of the theory[1], the moment of inertia (48.2) was taken in the strong coupling limit of the collective model for which its value is the same as for a classical spheroidal liquid droplet rotating about an axis perpendicular to its symmetry axis with irrotational flow,

$$\mathfrak{J}_{\text{irrot}} = 3\,B\,\beta^2 \tag{48.3}$$

where $B = \frac{1}{2}\varrho_0 R^5$ for a uniform density droplet of mean radius R. However, when the moment of inertia as given by (48.3) was introduced into (48.1) to fit the observed bands of levels, very large nuclear deformations resulted corresponding to β^2 values about six times as large as were given by the electrical measurements (see Part C). If, on the other hand, the deformations are taken to be known from the electrical measures, then the moments of inertia given by the rotational spectra, come close to the larger values appropriate for rigid body rotation,

$$\mathfrak{J}_{\text{rigid}} = \frac{16}{15\pi}\,B. \tag{48.4}$$

This discrepancy in the early applications of the collective model to rotational states has proven highly instructive. It has been a stimulus to a careful analysis[3] of features appearing in the rotations and vibrations of the nuclear shell structure. We summarize briefly some of the conclusions emerging from this study.

[1] A. Bohr: Ph. D. Diss. Univ. Inst. for Teoretisk Fysik. Copenhagen: Rotational States of Atomic Nuclei. 55 pp. København: Ejnar Munksgaards 1954.

[2] A. W. Sunyar: Phys. Rev. **98**, 653 (1955).

[3] A. Bohr and B. Mottelson: Dan. Mat. *30*, Nr. 1 (1955).

For independent particle motion in an average nuclear field, the rotational moments of inertia for deformed nuclei would be approximately those corresponding to rigid rotation. However, the correlation in the nucleonic motion arising from residual internucleon coupling modifies this result in an important way and gives rise for slightly deformed nuclei to surface waves as a component of the rotational motion. If the residual interaction between nucleons becomes sufficiently strong, the shell structure is broken down completely and moments of inertia corresponding to irrotational flow result. In actual nuclei the residual interactions appear to remain always less than the effect of the average nuclear field, so that irrotational moments are not produced. In the limit of highly deformed nuclei, the particles are coupled more strongly to the nuclear shape and the residual interactions are less effective in breaking down the shell structure. Hence the moments of inertia for highly deformed nuclei tend to approach the moments corresponding to rigid rotation.

If the residual interactions between nucleons were well known from other sources, then the observations of rotational states would provide a measure of the angular variations of the matter distributions within nuclei. Conversely, if we accept the evidence that nuclear matter distribution is closely proportional to the nuclear charge distribution, then observations on nuclear rotational states provide valuable independent data on the residual interactions required to describe nucleonic motion in the average nuclear field of a chosen model.

49. Alpha decay. Nuclear deformation may be expected to affect alpha decay in two principal ways: (1) The probability of alpha emission through the COULOMB barrier will be enhanced by the thinning of the barrier at the positions of greatest radial extension of the nuclear surface. (2) The probability of alpha formation from the constituent nucleons will be affected by the rearrangement of these states as a result of surface deformation.

The barrier penetration effect has been computed (see Fig. 31 of [*14*]) in the simple approximation of considering the effects at the tips of a prolate spheroid to dominate the process, and also by calculation[1,2] of the probability for alpha particle penetration through a spheroidal surface. The effect of barrier penetration on deformation is probably reflected in the fluctuations of nuclear radii as derived from alpha decay[3,4], and may be associated with the exceptionally short half-life for alpha decay of Sm^{147} which is in the rare earth group characterized by exceptionally large deformations.

The effect of deformation on the probability of alpha particle formation from its constituent nucleons should be most clearly seen[5,6] in the relative intensities of various lines appearing in the fine structure of alpha decay. Such information will play an important role in the refinement of current theories of nuclear structure and hence should ultimately be associated in a quantitative way with nuclear deformation magnitudes.

Recent progress[7] in observing alpha decay from oriented nuclei promises to provide valuable information on the angular distributions of alpha processes relative to the nuclear symmetry axis. Such angular distributions should reflect the effects of both types of deformation influences mentioned above.

---

[1] R. F. CHRISTY: Phys. Rev. **98**, 1205 (1955).

[2] J. O. RASMUSSEN and B. SEGALL: Phys. Rev **103**, 1298 (1956).

[3] J. J. DEVANEY: Phys. Rev. **91**, 587 (1953).

[4] I. PERLMAN and F. ASARO: Ann. Rev. Nucl. Sci. **4**, 157 (1954).

[5] A. BOHR, P. O. FRÖMAN and B. R. MOTTELSON: Dan. Mat. fys. Medd. **29**, No. 10 (1955).

[6] L. DRESNER: Phys. Rev. (to be published).

[7] L. D. ROBERTS, J. W. T. DABBS, G. W. PARKER and R. D. ELLISON: Bull. Amer. Phys. Soc. II **1**, 207 (1956).

50. Fission. The same influences of intrinsic particle structure which lead to ground state deformations of nuclei also affect the height of the barrier which must be surmounted by a nucleus undergoing fission (see Fig. 42 of [*14*]). In spontaneous fission which involves quantum mechanical tunneling through the fission barrier, the effect of particle structure should be clearly evident. The quantitative theory of these and other fission effects, however, is yet to be developed.

Angular correlations of fission fragments with the exciting beam for both proton induced fission[1] and neutron induced fission[2] may be expected to shed some light on ground state nuclear deformations, particularly if observations may be made on the fission of oriented nuclei.

E. Theories of nuclear density.

The increased experimental information available in recent years on nuclear densities has renewed interest in their theoretical prediction. These investigations are also the natural outgrowth of recent advances in the development of more adequate theories of nuclear structure.

I. Radial shape.

51. Statistical treatment. In the first extensive treatment of nuclear densities, v. WEIZSÄCKER [*20*] applied the THOMAS-FERMI statistical method to the nucleus. An energy density was defined from which the total energy of the nucleus was given by volume integration. Along with the mean binding of particles in nuclear matter, the surface, COULOMB, and symmetry effects were identified and led to a semiempirical formula of nuclear masses [*8*] which, with subsequent refinements, has had considerable success[3] in describing nuclear masses over the entire range of isotopes.

To provide for the smearing out of the nuclear surface as required by the indeterminancy principle of quantum mechanics, v. WEIZSÄCKER assumed a component of the energy density of the form

$$\varepsilon_w = \frac{\hbar^2}{8M} \frac{(\operatorname{grad}\varrho)^2}{\varrho} \tag{51.1}$$

involving the gradient of the density. The resulting surface thicknesses were in reasonable accord with values now indicated (Parts A and B) from experimental sources. A recent study[4] examines the errors following from the use of this approximate expression and leads to the conclusion that ε_w must be multiplied by a factor less than unity to obtain agreement with the exact quantum mechanical solutions for various potential shapes of the nuclear field. Comparisons were made for the three dimensional isotropic harmonic oscillator potential and for a step potential in plane symmetry. An earlier treatment of the infinite step potential involved the calculation [*19*] of the kinetic and potential parts of the nuclear surface energy for a statistical model, taken to first order, employing a gaussian interaction between the nuclear particles. The resulting surface energy was the right order of magnitude, being slightly more than one-half of the empirical value.

---

[1] E. J. WINHOLD, P. T. DEMOS and I. HALPERN: Phys. Rev. **87**, 1139 (1952).
[2] J. E. BROLLEY jr. and W. C. DICKINSON: Phys. Rev. **94**, 640 (1954).
[3] A. E. S. GREEN: Phys. Rev. **95**, 1006 (1954).
[4] R. BERG and L. WILETS: Proc. Phys. Soc. Lond. A **68**, 229 (1955).

The COULOMB repulsion between protons should cause them to tend to crowd toward the surface of the nucleus. Treatments of this effect[1-3] indicated that both proton and neutron densities would be expected, in the simple statistical models employed, to have a central depression in their densities. The magnitude of this predicted central depression is related to the compressibility of nuclear matter. The compressibility is likewise connected with the thickness of the nuclear surface as indicated[4] in a recent phenomenological self-consistent statistical method for studying the nuclear surface effect. The smaller the gradient of nuclear density at the surface (that is, corresponding to a thicker surface) the smaller the value of nuclear compressibility as defined by (19.2). The surprisingly sharp surfaces implied by recent electron scattering observations (Fig. 5) are thus to be associated with the absence of a central depression of proton density in the picture of charge distribution given by these measurements. From a theoretical viewpoint, one of the most interesting conclusions emerging from present experimental information, is that the nuclear densities are as nearly uniform as is consistent with the finite wavelength of the nucleons within the nucleus.

A much more elaborate theory of radial shape of nuclear densities has been given by GOMBAS[5] in terms of a statistical treatment with only one adjustable parameter, the strength of the two-body interaction. The nature of the assumed interaction, however,—namely, a pure MAJORANA force with YUKAWA radial dependence—is probably not appropriate for real nuclei. Nuclear binding energies and radii were well predicted over a wide range of mass number, A, but density distributions were peaked at the center of the nucleus with very gradual tapering at the nuclear surface. More recent work[6] along these lines has indicated that the shape of the density distribution is indeed sensitive to the radial dependence of the two-body force. For two-body potentials which are finite at the origin, density distributions more in accord with observation are obtained. It is found that an exponential interaction gives a better fit with empirical mass defects than a gaussian type of interaction. When a two-body exponential interaction is employed, the corresponding nuclear densities are found to be roughly gaussian at atomic mass 16 but for heavier masses (80 and 200) a central depression in the densities develops.

52. Relative neutron and proton extensions. The simple argument given in the preceding section suggests that COULOMB energy between protons will enhance the density of protons at the nuclear surface. However, a counter argument may be given[7] to suggest that the effect of COULOMB energy may be just the opposite, namely, to reduce the density of protons at the nuclear surface. Assume that the nucleonic component of the potential in which the protons move is identical with the potential in which the neutrons move. Such an effective average potential would be expected to be fairly flat in the interior of the nucleus and then to rise to zero with sloping walls at the nuclear surface. The actual potential in which the protons move is then given by adding the COULOMB potential

[1] E. FEENBERG: Phys. Rev. **59**, 149, 593 (1941). — Rev. Mod. Phys. **19**, 239 (1947).

[2] E. WIGNER: Nuclear Masses and Binding Energies (Paper presented at Univ. of Pennsylvania Bicentennial Conf., Philadelphia, Pa. (1940); published in Nuclear Physics, 68 pp. Philadelphia, Pa.: Univ. of Pennsylvania Press 1941.

[3] W. J. SWIATECKI: Proc. Phys. Soc. Lond. A **63**, 1208 (1950).

[4] R. A. BERG and L. WILETS: Phys. Rev. **101**, 201 (1956).

[5] P. GOMBAS: Acta phys. hung. **1**, 329 (1952); **2**, 223 (1952).

[6] P. GOMBAS, E. MAGORI, B. MOLNAR and E. SZABO: Ann. Phys., Lpz. **16**, 93 (1955). Also: Acta phys. hung. **4**, 267 (1955).

[7] M. H. JOHNSON and E. TELLER: Phys. Rev. **93**, 357 (1954).

to this nucleonic potential. Beta-stability requires that the neutrons and protons shall fill their respective potential wells to approximately the same absolute energy level. This condition, in combination with the sloping walls of the potential wells, means then that the protons move in a smaller volume than the neutrons. In other words, the nuclear surface by this argument is found to be neutron rich above the value as given by the proportion of neutrons to protons in the nucleus.

While this conclusion has received some support from other theoretical considerations[1,2], it should be noted that the initial premise of the argument ignores the correlation of neutron and proton densities in generating the effective potential in which each kind of particle moves. The more extended neutron distribution, predicted by the above argument, leads one to conclude that the effective nucleonic potential seen by the proton may be substantially different from the potential seen by the neutron. Therefore, a succession of adjustments in particle densities and effective potentials is needed to complete the above argument until a self-consistent situation is reached. In this self-consistent solution it is to be expected that the radial extension of neutrons and protons will be much more nearly the same and possibly almost identical. Some quantitative support for this view is provided in the following section.

53. Semiempirical theory of nuclear densities. Information on the relative extension of neutron and proton densities may be obtained [*21*] from a semiempirical self-consistent statistical treatment of the particle densities, $N(r)$ and $P(r)$, in nuclei. Experimental values are taken for the nuclear radius, the binding energy of the last nucleon, the surface energy, the surface thickness, and symmetry energy. In line with the arguments given in the preceding sections, the total energy of the nucleus may be written in the form:

$$\mathscr{E} = \int \left\{ \varepsilon(P,N) + \zeta \frac{\hbar^2}{8M} \left[\frac{(\operatorname{grad} P)^2}{P} + \frac{(\operatorname{grad} N)^2}{N} \right] \right\} d\,(\mathrm{vol})\,. \tag{53.1}$$

Here the coefficient ζ is less than unity. Functions P and N which make the energy a minimum, are sought by a variational procedure subject to the constraint that

$$A = \int (P+N)\, d\,(\mathrm{vol}) \tag{53.2}$$

remains a constant. Apart from surface effects, the energy density is given by

$$\varepsilon(P,N) = f(P+N) + (N-P)^2 k + V_c P \tag{53.3}$$

in which the total nucleonic density is represented by $\varrho = P + N$. The function f/ϱ (energy per nucleon) must have a minimum at the observed nuclear density and at this density must take a value corresponding to the coefficient of the term linear in A in the semiempirical mass formula [*8*]. The curvature at this minimum is related to the nuclear compressibility. The second term on the right-hand side of (53.3) represents the symmetry energy, with the constant k again determined from the semiempirical mass formula. The last term in (53.3) is the Coulomb energy density, V_c being the Coulomb potential energy function. The differential equations determining the densities as derived from the variational procedure applied to (53.1) and (53.2) are solved by iteration, with the Lagrangian multiplier serving as an eigenvalue. Trial solutions are successively improved until solutions are determined which, as found by numerical integration of the

[1] P. Mittelstaedt: Z. Naturforsch. **10**a, 379 (1955).
[2] W. J. Swiatecki: Phys. Rev. **98**, 203, 204 (1955).

differential equations, satisfy with sufficient accuracy the required boundary conditions on the densities.

The constant ζ plays the role of a scale factor in the differential equation. Both the unit of length and the surface energy vary as $\zeta^{\frac{1}{2}}$ which is adjusted to give the proton surface thickness illustrated in Figs. 5 and 7. Typical results for the radial variation of potential and particle densities in a spherical nucleus corresponding to atomic mass 225 and atomic number 93 are shown in Fig. 16. The result indicates that the neutron and proton densities extend to about the same radius, as indicated by the approximate coincidence of the radial values at which the two densities have dropped to one-half their central values.

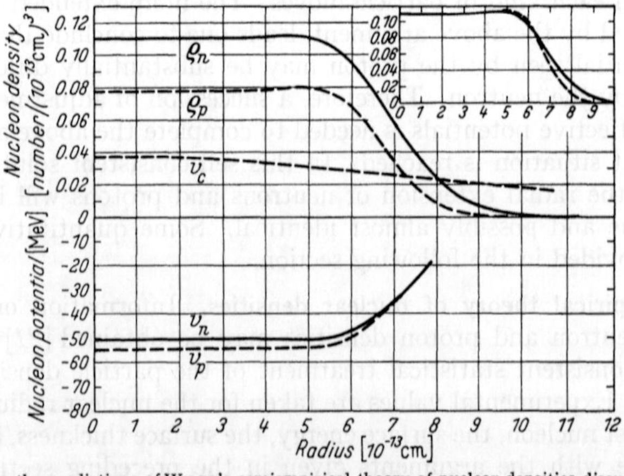

Fig. 16. Illustrated are the predictions of WILET's semiempirical theory of nuclear densities discussed in Sect. 53. The neutron and proton densities plotted are for a hypothetical nucleus of mass number $A = 225$ and atomic number $Z = 93$. Plotted in the lower half of the figure for the same radial scale are shown the nucleonic potentials in which the neutrons and protons move while the COULOMB potential affecting the protons is plotted to the same energy scale. Thus the total potential in which the protons move is given by $V_p + V_c$. Note that the midpoints of the potential well extend to a radius about 0.7 units greater than the radius to which the midpoints of the densities extend. The insert in the upper right-hand corner of the figure shows the proton (— —) and the neutron densities scaled to the same central value, in order to make more evident the approximate proportionality of the two densities throughout the nucleus. This figure is taken from [21].

As measured by the midpoint values, the potentials are seen to extend about 0.7×10^{-13} cm further than the densities. This result may be understood[1] both as depending upon the saturation of nuclear forces which causes the potential to fall off less rapidly than the densities near maximum density, and also upon the general relationship between potential density given in the THOMAS-FERMI theory,

$$V \propto \varrho^{\frac{2}{3}} \tag{53.4}$$

which, at intermediate densities, is responsible for a less rapid fall-off of the potential relative to the density.

The approximate equality of the nuclear force potentials acting on protons and neutrons, as shown in Fig. 16, is associated with the introduction [4] of an effective mass, $M^* = 0.6\,M$, to account for the velocity dependence of the interaction resulting from interparticle correlations.

A recent computation[2] of the relationship between nuclear density and potential, taking account of the saturating character and the finite range of the two-body forces, reproduces the features empirically derived [21].

54. Densities from individual particle states. More complete information on nuclear densities may in principle be obtained by summing the densities

[1] R. A. BERG and L. WILETS: Phys. Rev. **101**, 201 (1956).
[2] K. A. BRUECKNER: Phys. Rev. **103**, 1121 (1956).

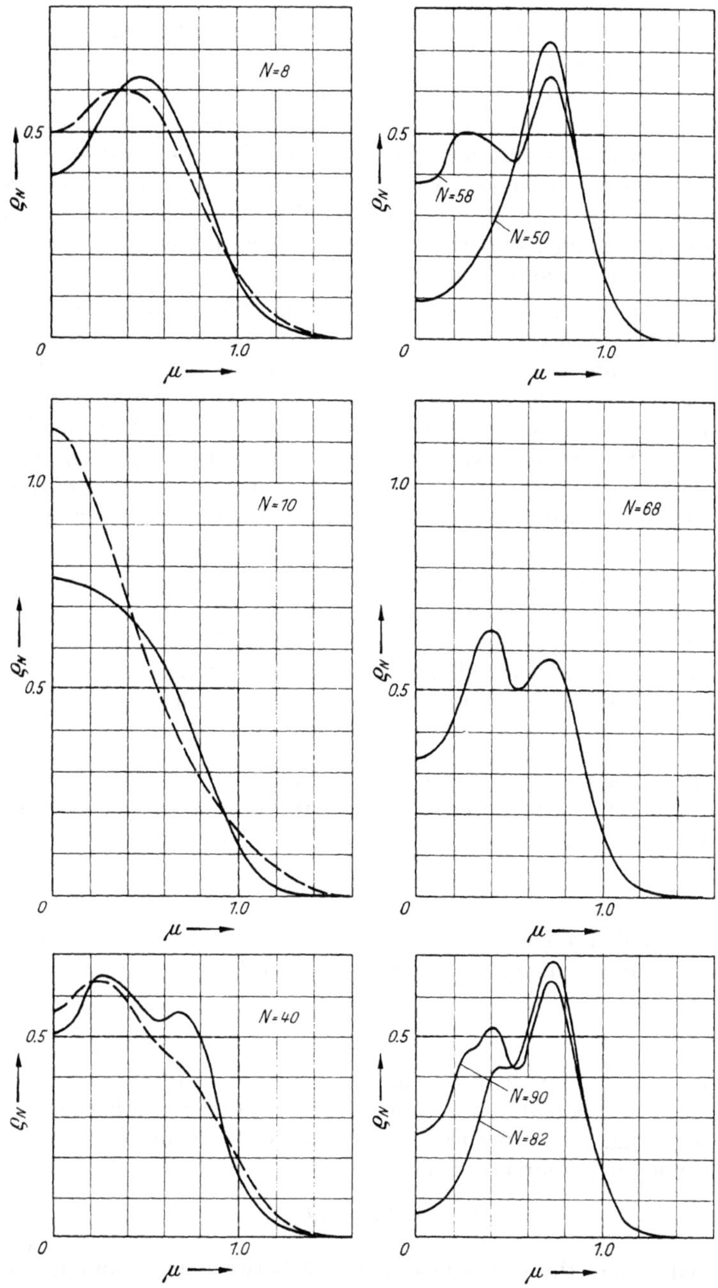

Fig. 17a and b. Total densities for independent particle nuclear models. N is the number of nucleons. Two free-particle models are used: (a) a potential well of constant depth and radius $R = r_0 A^{\frac{1}{3}}$, where $r_0 = \frac{e^2}{2mc^2} = 1.4 \times 10^{-13}$ cm, and (b) a harmonic oscillator potential for $N = 8, 10$, and 40. Densities for (b) are shown by a dashed line. Nucleon densities $\varrho_n(\mu R)$ are in particles per volume $(4\pi/3)\, r_0^3$. The dashed line in each graph represents a constant particle density out to the radius R, and is included for comparison. The results shown are taken from ref. 1, p. 236 of Sect. 54.

corresponding to individual particle states defined in an average nuclear potential with or without assumed interparticle forces. A number of results are available from this approach although, because of its greater mathematical complexity, relatively less progress has been made than with the statistical approach discussed in the preceding sections.

The simplest model of this type is the independent particle model in which all particles see the same common potential and move independently of each other. Radial density distributions computed by this model, both for an assumed square well and for a harmonic oscillator potential, have been computed[1] and are illustrated in Fig. 17. One would expect the sharp dips and rises appearing in this model to be much smoothed out in a calculation which allows for correlations between the motions of different particles.

A number of interesting results have been obtained relating shell model structure to nuclear densities. It has been shown[2] that shell closings are obtained for reasonable assumptions on nuclear density. In terms of a neutral two-body spin-orbit potential, the level splittings in heavy nuclei have been satisfactorily explained in terms of a nuclear density distribution with finite transition at the surface[3], and it has been shown[4] that particle binding energies in conjunction with neutron cross sections in heavy elements require a diffuseness of the nuclear boundary consistent with that given by the electron scattering experiments (Figs. 5 and 7).

More recently, self consistent treatments of the nuclear individual particle model have been given[5,6] by the HARTREE-FOCK method for central nucleon-nucleon interactions, including several combinations of exchange character and radial dependence. Here, as in all other independent particle calculations, results indicate the necessity of including appropriate inter-particle correlation in order to approach the properties of actual nuclei.

55. Nuclear force radius relative to matter radius. Of the various nuclear force measures of nuclear size reviewed in Part B, the proton and neutron scattering may be most directly interpreted, and are consistent with a nuclear force radius

$$R_F = 0.8 + 1.23\,A^{\frac{1}{3}}. \tag{55.1}$$

On the other hand, the electron scattering [*15*] over a broad range of assumed nearly spherical nuclei, indicates a radius to the *midpoint* of the charge distribution given by

$$R_c^{(\frac{1}{2})} = 1.07\,A^{\frac{1}{3}}. \tag{55.2}$$

However, the force radius in (55.1) has been estimated [*7*] to correspond to a point on the nuclear potential about one-third of the distance between zero and maximum potential. Adjusting the constant in (55.1) to correspond to the midpoint of the nuclear potential, we obtain

$$R_F^{(\frac{1}{2})} = 0.8 + 1.18\,A^{\frac{1}{3}}. \tag{55.3}$$

We note finally that the displacement of 0.7 between force and matter radii discussed in Sect. 53 follows from general features of the relationship between potential and density, as included in the FERMI-THOMAS statistical theory, and does *not* include special effects due to the finite range of nuclear forces. The

[1] K. C. HAMMACK: Topics in Nuclear Structure. Diss. Washington University, St. Louis 1951.
[2] M. BORN and L. M. YANG: Nature, Lond. **166**, 399 (1950).
[3] R. J. BLIN-STOYLE: Phil. Mag. **46**, 973 (1955).
[4] A. E. S. GREEN: Phys. Rev. **99**, 1410 (1955).
[5] G. E. TAUBER and T. Y. WU: Phys. Rev. **100**, 1240 (1955).
[6] M. ROTENBERG: Phys. Rev. **100**, 439 (1955).

difference between (55.2) and (55.3) in the heaviest nuclei is about 1.5, leaving a discrepancy of 0.8. This remaining discrepancy between the force and charge radii may be interpreted either by assuming that the neutrons extend about 10% farther than the protons in heavy nuclei or by attributing the difference to the range of nuclear forces. The former explanation conflicts in particular with the analysis of 1.4 Bev neutron and 0.86 Bev proton reaction cross sections [7], which indicate matter distributions for both neutrons and protons in close agreement with the charge distributions obtained from the electron scattering observations. This agreement with the charge distribution radius, obtained by analysis of very high energy nucleon scattering, contrasts with the lack of agreement at lower energies, indicated by (55.2) and (55.3), and is to be associated[1] with [7] the decrease in the effective range of the nucleon-nucleon interaction as one passes from the Mev to the Bev energy range. If we accept the alternative explanation that (55.2) and (55.3) indicate an effect of the nuclear forces, we are led [7] to the instructive conclusion that these forces must contain long-range (i. e., $\sim r_0$) direct central interactions, and short-range repulsive interactions in order to account both for the difference between charge and force radii and for the nuclear saturation.

II. Angular shape.

56. Resistance to deformation near closed shells. The deformation of nuclei is of great significance for nuclear structure and has received theoretical attention from many quarters. The first quantitative treatment of nuclear deformation was given in the classic paper by BOHR and WHEELER[2] on nuclear fission. The interaction between individual particle states and the average nuclear potential in producing deformations associated with ground state quadrupole moments, was pointed out by RAINWATER[3] and theories of nuclear deformation emphasizing the dynamic aspects of nuclear vibration and rotation have subsequently been given [*14*], [*3*].

For closed shells the effects of different individual particle states are counterbalancing so that the resultant shape is spherical if both neutron and proton shells are closed in the same nucleus (e.g., Pb^{208}). There is moreover a greater rigidity or resistance to deformation near closed shells. The direct electric evidence from quadrupole moments favors this view. The nucleus O^{17}, for example, has an extremely small quadrupole moment and at the other end of the periodic table, the nucleus Bi^{209} has[4] a moment of the same order as a single proton state. GOEPPERT-MAYER[5] has pointed out that throughout the periodic table quadrupole moments indicate a varying nuclear deformability which has minima near closed shells.

The property of nearly spherical symmetry near closed shells has been employed in several recent shell model calculations: for example, the work of ELLIOTT and FLOWERS[6] and of REDLICH[7], near ${}_8O^{16}$; the work of LEVINSON

---

[1] R. W. WILLIAMS: Phys. Rev. **98**, 1387, 1393 (1955).
[2] N. BOHR and J. A. WHEELER: Phys. Rev. **56**, 426 (1939).
[3] J. RAINWATER: Phys. Rev. **79**, 432 (1950).
[4] J. E. MACK: Rev. Mod. Phys. **22**, 64 (1950).
[5] M. GOEPPERT-MAYER: Paper on Nuclear Shell Theory, Proc. Intern. Conf. of Theoretical Physics (Kyoto and Tokyo, Japan, 1953) (Science Council of Japan, Tokyo, 790 pp., 1954).
[6] J. P. ELLIOTT and B. H. FLOWERS: Proc. Roy. Soc. Lond., Ser. A **229**, 536 (1955).
[7] M. REDLICH: Phys. Rev. **95**, 448 (1954); **99**, 1427 (1955).

and FORD[1], near $_{20}Ca^{40}$; and the work of ALBURGER and PRYCE[2] and of TRUE[3], near $_{82}Pb^{208}$ bears out this view.

57. Origin of large deformations. Deformations may be expected to arise both from the pressure of particles outside closed shells on the average nuclear potential, and from tensor forces between particles. In the present state of nuclear theory, the latter source of deformations has not been quantitatively explored, but is expected to produce small effects in comparison with those arising by the pressure of particles on the nuclear surface.

In regions of the Periodic Table in which large numbers of nucleons are required to fill a given angular momentum shell, as in the rare earths, then in the process of shell filling a large number of particles may be present with angular momenta approximately parallel and anti-parallel. Under this condition of cumulative coupling between particles and the nuclear surface, remarkably large deformations of the nucleus may appear as illustrated in Figs. 12 to 15. Midway in the process of filling an angular momentum shell, prolate and oblate forms may both be strongly preferred over the spherical form and lead to "mutational" oscillations through the ellipsoidal forms connecting the two limiting spheroidal forms [*14*].

Prediction of the equilibrium shapes in terms of the collective model requires the calculation of individual particle energies as a function of deformation, and the summation of these energies for all the particles in a given nucleus. The minimum in the resulting deformation energy surface then defines the equilibrium shape about which the nucleus will oscillate to an amplitude determined by the curvature of the energy surface in the neighborhood of the minimum. Such calculations have been performed by a number of authors[4-7] with results in approximate agreement with observations of ground state shapes.

F. Summary.

58. Emergent picture of charge and matter distributions. From the many sources of experimental and theoretical information, there results a rather well defined description of nuclear charge and matter distributions. Most sharply defined is the nuclear charge distribution with the radial shape as depicted in Figs. 5 and 7. Minor variations from this shape are, of course, to be expected for individual nuclei but the general features apparently remain constant for medium and heavy nuclei. There is very little change of the density over the interior of the nucleus and the surface thickness remains about constant in the neighborhood of 2.4×10^{-13} cm. The distance to the midpoint of the surface varies as given in (55.2). This relation even holds roughly for the lightest elements (including the proton!) with the shape of the charge distribution probably resembling a gaussian in the lightest nuclei.

The angular variation of the nuclear charge shape is indicated by ground state quadrupole moments, Fig. 12, and is consistent with information from other experimental sources, as well as with the general features expected from current theories of nuclear structure, which predict a regular change of deformation with neutron and proton numbers.

---

[1] C. LEVINSON and K. W. FORD: Phys. Rev. **99**, 792 (1955); **100**, 1, 13 (1955).
[2] D. E. ALBURGER and M. H. L. PRYCE: Phys. Rev. **95**, 1482 (1954).
[3] W. TRUE: Phys. Rev. **101**, 1342 (1956).
[4] S. G. NILSSON: Dan. Mat. fys. Medd. **29**, 16 (1955).
[5] S. A. MOSZKOWSKI: Phys. Rev. **99**, 803 (1955).
[6] B. MOTTELSON and S. G. NILSSON: Phys. Rev. **99**, 1615 (1955).
[7] D. L. HILL and J. A. WHEELER: Phys. Rev. (to be published).

Although in our survey we have encountered some apparently conflicting experimental information and theoretical suggestions regarding the relationship between the matter and the charge distribution, the balance of evidence appears to favor the view *that matter and charge distributions in nuclei are substantially the same.*

A clearly delineated picture emerges from the recent flood of new observations and theory which leads one to feel that the atomic nucleus, long dependent on RUTHERFORD's classic work for its description, is now shaping up nicely[1].

General references.

[*1*] (a) BETHE, H. A., and R. F. BACHER: Nuclear Physics. A. Stationary States of Nuclei. Rev. Mod. Phys. **8**, 82—229 (1936). — (b) BETHE, H. A.: Nuclear Physics. B. Nuclear Dynamics, Theoretical. Rev. Mod. Phys. **9**, 69—244 (1937). — These definitive treatments of nuclear physics were for years the standard reference and still retain great utility.

[*2*] BLATT, J. M., and V. F. WEISSKOPF: Theoretical Nuclear Physics. New York: John Wiley & Sons 1952. — A modern lucidly written one volume encyclopedia of nuclear physics.

[*3*] (a) BOHR, A.: The Coupling of Nuclear Surface Oscillations to the Motion of Individual Nuclei. Dan. Mat. fys. Medd. **26**, No. 14 (1952). — (b) BOHR, A., and BEN R. MOTTELSON: Collective and Individual-Particle Aspects of Nuclear Structure. Dan. Mat. fys. Medd. **27**, 16 (1953). — Comprehensive pioneering treatments of nuclear collective properties with special emphasis on rotation of the nucleus.

[*4*] (a) BRUECKNER, K. A., C. A. LEVINSON and H. M. MAHMOUD: Two-Body Forces and Nuclear Saturation. I. Central Forces. Phys. Rev. **95**, 217—228 (1954). — (b) BRUECKNER K. A.: Nuclear Saturation and Two-Body Forces. II. Tensor Forces. Phys. Rev. **96**, 508—516 (1954). — (c) BRUECKNER, K. A.: Two-Body Forces and Nuclear Saturation. III. Details of the Structure of the Nucleus. Phys. Rev. **97**, 1353—1366 (1955). — An important series of papers connecting nuclear properties with the characteristics of the interaction between constituent nucleons.

[*5*] CONDON, E. U., and G. H. SHORTLEY: The Theory of Atomic Spectra. Cambridge: University Press 1953. — Many of the mathematical procedures employed in this fundamental treatise on atomic spectra have found renewed application in the shell theory of atomic nuclei.

[*6*] COOPER, L. N., and E. M. HENLEY: μ-mesonic Atoms and the Electromagnetic Radius of the Nucleus. Phys. Rev. **92**, 801—811 (1953). — A theoretical treatment of the several corrections to be applied to the determination of nuclear charge radii as measured by μ-mesonic x-rays.

[*7*] DRELL, S. D.: Nuclear Radius and Nuclear Forces. Phys. Rev. **100**, 97—111 (1955). — An instructive treatment of the implications for nuclear forces of the difference between nuclear force radius and nuclear charge radius.

[*8*] FERMI, E.: Nuclear Physics. Revised edition of notes on a lecture course compiled by J. OREAR, A. H. ROSENFELD and R. A. SCHLUTER. Chicago: The University of Chicago Press 1950. — The genius for simplicity of presentation and the illuminating physical intuition of the late ENRICO FERMI are to some extent preserved in these notes.

[*9*] FITCH, V. L., and J. RAINWATER: Studies of X-rays from μ-mesonic Atoms. Phys. Rev. **92**, 789—800 (1953). — Reported first unambiguous evidence for presently accepted electro-magnetic radius of the nucleus. References to earlier work are given here.

[*10*] FORD, K. W., and D. L. HILL: Nonuniform Nuclear Charge Distributions and Measurements of Nuclear Electrical Radius. Phys. Rev. **94**, 1630—1637 (1954). — An examination of the effect of the radial shape of nuclear charge on the properties of electrons in bound states and of electrons scattered at medium energies.

[*11*] FORD, K. W., B. E. FREEMAN and D. L. HILL: Fast Electron Scattering Determination of Radial Charge Distribution in Heavy Nuclei. Phys. Rev. (to be published). — A careful analysis of the radial charge distribution for nuclei in the neighborhood of lead, as deduced from the scattering of electrons between 80 and 200 Mev.

[*12*] FORD, K. W., and D. L. HILL: The Distribution of Charge in the Nucleus. Annual Review of Nuclear Science, Vol. 5. Stanford: Annual Reviews, Inc. 1955. — This earlier review has provided much of the organization for the present article; in addition, certain

[1] For a comparable viewpoint on a different subject, see p. 79 of [*5*].

sections have been taken over into the present treatment with only minor revision. It is a pleasure for the author to express his gratitude to Dr. KENNETH FORD for his invaluable collaboration on this earlier survey and for his consequent important contributions to the present work.

[13] HILL, D. L., and K. W. FORD: μ-mesonic X-rays and the Shape of the Nuclear Charge Distribution. Phys. Rev. **94**, 1617—1629 (1954). — An extensive numerical investigation of μ-mesons bound in the field of the lead nucleus with the charge distribution specified by the different members of four functional families.

[14] HILL, D. L., and J. A. WHEELER: Nuclear Constitution and the Interpretation of Fission Phenomena. Phys. Rev. **89**, 1102—1145 (1953). — A fundamental study of the relation between individual particle states and the collective properties of nuclei, with special emphasis on the causes and consequences of nuclear deformation and vibration.

[15] (a) HOFSTADTER, R., H. R. FECHTER and J. A. McINTYRE: High-Energy Electron Scattering and Nuclear Structure Determinations. Phys. Rev. **92**, 978—987 (1953). — (b) HOFSTADTER, R., B. HAHN, A. W. KNUDSEN and J. A. McINTYRE: High-Energy Electron Scattering and Nuclear Structure Determinations. II. Phys. Rev. **95**, 512—515 (1954). — (c) HAHN, B., D. G. RAVENHALL, and R. HOFSTADTER: High-Energy Electron Scattering and the Charge Distribution of Selected Nuclei. Phys. Rev. **101**, 1131 (1956). — This group of papers summarizes the most significant information now available on the distribution of charge in medium and heavy nuclei.

[16] MOTT, N. F., and H. S. W. MASSEY: The Theory of Atomic Collisions, 2nd edn. Oxford: The Clarenden Press 1949. — The standard treatise on collision problems.

[17] RUTHERFORD, E., J. CHADWICK and C. D. ELLIS: Radiations from Radioactive Substances. Cambridge: University Press 1930. — The classic treatise on early work with radioactive material.

[18] SCHIFF, L. I.: Interpretation of Electron Scattering Experiments. Phys. Rev. **92**, 988—993 (1953). — The inadequacy of the BORN approximation for analysis of high-energy scattering in heavy elements was demonstrated by this early treatment.

[19] SWIATECKI, W. J.: The Nuclear Surface Energy. Proc. Phys. Soc. Lond. A **64**, 226—238 (1950). — The kinetic and potential parts of the nuclear surface energy are calculated by a statistical model with gaussian interaction between particles.

[20] (a) VON WEIZSÄCKER, C. F.: Zur Theorie der Kernmassen. Z. Physik **96**, 431—458 (1935). — (b) FLÜGGE, S.: Zum Aufbau der leichten Atomkerne. Z. Physik **96**, 459—472 (1935). — (c) HEISENBERG, W.: Die Struktur der leichten Atomkerne. Z. Physik **96**, 473—484 (1935). — This group of papers provides the earliest comprehensive theoretical treatment of nuclear densities.

[21] WILETS, L.: Neutron and Proton Densities in Nuclei. Phys. Rev. **101**, 1805 (1956). — A highly instructive semi-empirical investigation by numerical methods of neutron and proton densities in nuclei.

[22] WILETS, L., D. L. HILL and K. W. FORD: Isotope Shift Anomalies and Nuclear Structure. Phys. Rev. **91**, 1488—1500 (1953). — Both the average magnitude and the variations of isotope shifts are correlated with features of nuclear structure.

[23] (a) YENNIE, D. R., D. G. RAVENHALL and R. N. WILSON: Phase-Shift Calculations of High-Energy Electron Scattering. Phys. Rev. **95**, 500—512 (1954). — (b) RAVENHALL, D. G., and D. R. YENNIE: Results of a Phase Shift Calculation of High-Energy Electron Scattering. Phys. Rev. **96**, 239—240 (1954). — (c) BRENNER, S., G. E. BROWN and L. R. B. ELTON: Elastic Scattering of 125 Mev Electrons by Mercury. Phil. Mag. **45**, 524—532 (1954). — These papers provide the first examples of accurate numerical analysis of the high-energy electron scattering data.

The Nuclear Shell-Model.

By

J. P. Elliott and A. M. Lane.

With 28 Figures.

I. Introduction.

1. The central physical feature of a shell-model of a closed system of particles is that the motions of the various particles are largely uncorrelated, so that each particle moves essentially undisturbed in its own closed orbit. Another popular way of expressing the same thing is to say that the mean free path of a particle against collision with other particles is long compared to the linear dimensions of the system.

The broad validity of such a model for describing the motion of electrons in an atom is a universally accepted fact. In the first place, the model has been exhaustively checked experimentally. In the second place, the applicability of the model is believed to be well understood theoretically. The idea that electron motion is governed mainly by the spherical potential of the nuclear charge with a small modification from the average interactions with other electrons seems so reasonable that no opposing model for the atom has ever been seriously proposed.

It would be far from the truth to say that the shell-model of the nucleus were so widely valid, either theoretically or experimentally. It has rarely if ever been suggested that the model is as valid for the nucleus as for the atom. The great problem has always been to decide just how applicable the model is to the nucleus, i.e. to decide whether nuclear motion is "10% shell model" or "90% shell-model". The last twenty years have seen many different opinions expressed on the answer to this problem, opinions varying from one extreme to the other. In fact, it is probable that few models in physics have had such a persistently violent and chequered history as the nuclear shell-model. Striking evidence in its favour has often been followed by equally strong evidence against it, and vice-versa. It is only quite recently that the situation has reached a point when one can say confidently that the extent of the model's applicability is fairly well known.

There are two main reasons why the nuclear shell-model has taken so long to become mature and respectable. The first is that the model has no apparent theoretical basis. In fact, until recently, arguments why the model should *not* be valid were more numerous and convincing than those in favour of the model. These former arguments almost invariably commence with the observations that the nucleus, in contrast to the atom, has no overall central potential and that, furthermore, the short range of nuclear forces means that one cannot use a smooth average potential to represent the actual potential felt by a nucleon, which will have strong local fluctuations. These facts clearly throw strong doubt on the validity of the nuclear shell-model. In the absence of any further feature, they would surely lead to its rejection. However such a feature does exist in the form of the PAULI Principle which prescribes that only one particle at a time can be found in any given state. This means that, in spite of feeling a strongly fluctuating potential, a nucleon will often have a smooth path because there are not many empty states for it to be scattered into. The vital role that the PAULI Principle thus plays in "ironing" out nuclear motion has only recently been appreciated, although the first suggestion along these lines was already made a few years ago[1]. We will not pursue this topic further here, but reserve it for the concluding chapter. Suffice it for the moment to return to our original observation that, until recently, there was no apparent theoretical justification at all for a shell-model of the nucleus. This state of affairs naturally caused many physicists to be very wary in accepting the usefulness of the model, and contributed to the slowness of the model's development.

We now turn to the second main reason for this slowness. This stems from the fact that the amounts of the various kinds of experimental data have not all increased steadily together in the last twenty years. Certain types of data were measured extensively from the beginning. Others have only begun to be measured recently. When this is coupled to the fact that the shell-model has very different degrees of success in predicting different kinds of data, the model's strange history becomes understandable. For example, one of the earliest types of data to be available was binding energies (from nuclear masses). It so happens that the shell-model's worst failure is the prediction of binding energies. This fact, which was soon realised, caused a sudden arrest in the development of the model in the mid-1930's, and it was a number of years before the resulting prejudices against the model were counteracted by successes in predicting other kinds of data.

We will now outline, in chronological order, the most significant events in the history of the nuclear shell-model.

1. Pre-1935. The model was originally suggested as a possible explanation of the fluctuations in the relative abundance and relative masses of nuclei in the periodic table. Such fluctuations were associated with shell-filling and shell-closures at "magic numbers"[2].

2. 1935–1945. These years witnessed the failure of the model to fit the binding energies of nuclei[3], especially light nuclei[4]. At the same time, the success of the compound nucleus model of unbound states was also used to argue against the shell-model, although there was no obvious reason why a model applying to excited states of nuclei should have any relevance for the low-lying states.

3. 1945–1950. This was the critical period in the development of the nuclear shell-model, both in theory and applications. It was during this time that the first serious suggestion was made that the PAULI Principle could strongly inhibit collisions between nucleons and thereby help to justify the model[1]. Also during this time, the model was modified by including a strong spin-orbit force implying that each nucleon orbit was not only to be associated with a principal quantum

[1] E. FERMI: Nuclear Physics. Chicago: University of Chicago Press 1950. — V. F. WEISSKOPF: Helv. phys. Acta **23**, 187 (1950).

[2] F. W. ASTON: Isotopes. London: Edward Arnold & Sons 1924. — W. D. HARKINS: Z. Physik **50**, 1927 (1928). — G. BECK: Z. Physik **47**, 407 (1928); **50**, 548, 1927 (1928); **61**, 615 (1930). — J. H. BARTLETT: Phys. Rev. **41**, 370 (1932). — G. GAMOW: Z. Physik **89**, 572 (1934). — W. ELSASSER: J. Phys. Radium **5**, 625 (1934). — K. GUGGENHEMER: J. Phys. Radium **5**, 253 (1934).

[3] H. A. BETHE and R. F. BACHER: Rev. Mod. Phys. **8**, 82 (1936).

[4] D. R. INGLIS: Phys. Rev. **51**, 531 (1937). — E. FEENBERG and E. P. WIGNER: Phys. Rev. **51**, 95 (1937). — E. FEENBERG and M. L. PHILLIPS: Phys. Rev. **51**, 597 (1937).

number n and orbital angular momentum l, but also a total angular momentum $j = l \pm \frac{1}{2}$. This modification[1] lead to striking successes for the model. Almost without exception, the ground state spins of odd nuclei were found to be correctly predicted. Furthermore several other features of nuclei such as the occurrence of isomeric states and the values of magnetic dipole moments were explained, at least qualitatively. However the model completely failed to explain the large values of observed electric quadrupole moments and certain regularities in nuclear spectra, especially of rare earth nuclei.

4. 1950—1953. The emphatic success of the shell-model modified by a spin-orbit force gave the necessary confidence and incentive to physicists to apply the model in detail to individual nuclei. Guided by parallel calculations in atomic spectroscopy, considerable effort was devoted to computing spectra of levels of nuclear systems with the so-called "Intermediate Coupling Model" in which the independent particle motion is considered to be perturbed by central particle-particle interactions and spin-orbit forces. Computational labour restricts such calculations to nuclei near closed shells, say within four particles or holes of closed shells. This explains why only light nuclei ($A < 20$) and isolated groups of nuclei higher in the Periodic Table were thus treated. Usually such calculations were rewarded by agreement with experiment especially those for light nuclei[2] and nuclei near the double closed shell at Pb^{208} [3].

Another development during the same period was the removal of the previously imposed restriction that the nucleus should be spherical[4]. The extension of the model to allow for spheroidal shapes immediately lead to some remarkable agreement with experiment[5]. Not only were two failures of the spherical version of the model (quadrupole moments and spectra regularities) removed, but several other features of nuclei were more adequately described. A fortunate occurrence is that the spheroidal shell-model seems to be most applicable to nuclei removed from closed shells. Thus, this version of the shell-model, taken with the above Individual Particle version which is applicable near closed shells, enables one to give a detailed treatment of most nuclei. In some respects, the two versions overlap to give similar results. (This is so near F^{19}, for example[6].)

5. Post 1953. Once the shell-model became well established by its successes in predicting experimental properties of nuclei, it was natural that attention should be turned to the theoretical basis of the model. Such a trend was encouraged by the discovery that even nuclear reactions involving states of considerable excitations were influenced by shell-model effects[7]. The data from neutron and proton reactions have been analysed to obtain values for the mean free paths of nucleons against collision in a nucleus, and also a value for the mean depth of the potential felt by a nucleon in a nucleus. It appears that both quantities are independent of the particular nucleus and so they can be said to characterise "nuclear matter". This fact enables one to reduce the rather vague problem of "explaining the shell-model" to a succinct quantitative form in which the problem is to predict the observed mean potential and observed mean free path for a nucleon of given energy inside nuclear matter. Since the specification of

[1] M. G. MAYER: Phys. Rev. **75**, 1969 (1949); **78**, 16 (1950). — HAXEL, JENSEN and SUESS: Phys. Rev. **75**, 1766 (1949). — Z. Physik **128**, 295 (1950).

[2] D. R. INGLIS: Rev. Mod. Phys. **25**, 390 (1953).

[3] M. H. L. PRYCE: Proc. Phys. Soc. Lond. A **65**, 773 (1952).

[4] J. RAINWATER: Phys. Rev. **79**, 432 (1950). — A. BOHR: Dan. Mat. fys. Medd. **26**, No. 14 (1952).

[5] A. BOHR and B. R. MOTTELSEN: Dan. Mat. fys. Medd. **27**, No. 16 (1953).

[6] E. B. PAUL: To be published.

[7] FESHBACH, PORTER and WEISSKOPF: Phys. Rev. **96**, 448 (1954); **90**, 166 (1953).

mean potential as a function of nuclear energy immediately enables one to compute the total binding of a system simply by summing the kinetic and mean potential energies of the particles, the failure of the early shell-model work to give the observed binding energies reflects the failure to predict the correct mean potential as a function of energy. According to the modern work, it appears that the root cause of the paradox that the shell-model can explain so many observed qualities and yet fail so badly to give the observed binding is that there exists some small but vital correlations between nucleons due to their interactions[1]. These correlations are supposed to account for only a small impurity in the total states but nevertheless to have a strong influence on the binding. Work to develop the full implications of this idea is still in progress.

It is clear, even from this sketchy outline, that a vast amount of work has been devoted to examinations of the shell-model from theoretical and experimental points of view. In this article, we wish to present a review of most of this work. In order to preserve smooth reading and the logical development of the subject, we have paid no respect to historical order in our presentation. Rather our plan has been to describe the various versions of the model in order of increasing complexity and sophistication. After describing a version of the model, we compare it with experiment, note its successes and failures, then pass on to a modified model which attempts to retain the former while discarding the latter.

In the following Chap. 2, we begin by introducing what might be called a "primitive" version of the shell-model. Now, quite generally, the basic idea of the shell-model is that one builds up a nucleus by filling up the successive quantum levels of the mean potential well with neutrons and protons until one finally has a number of closed shells of neutrons and protons and some "loose" neutrons and protons in unfilled shells. The special feature of the "primitive" version of the shell-model, the so-called "*Single Particle Model*", is that it is further assumed that all pairs of particles including pairs of loose particles form an "inert core" of spin zero contributing nothing to most observed properties. Thus a nucleus with an even number of neutrons and an odd number of protons is predicted to be characterised by the last odd proton. In particular, its spin and magnetic moment are said to be those of the last odd proton. The same model predicts that even-even nuclei have spin zero, and says that odd-odd nuclei can be represented by the last odd proton and the last odd neutron.

Chap. 3 is concerned with comparing the Single Particle Model with *experiment* and emphasises the manner in which the model systematises vast amounts of apparently unrelated data. At the same time, various shortcomings of the primitive model emerge such as its inability to make any sensible prediction about nuclear spectra.

In Chap. 4, we begin to develop the shell-model beyond the primitive Single Particle Model to the more correct and consistent "*Individual Particle Model*". We do not introduce the latter model in its full generality but keep this for Chap. 5. The reason is that, in Chap. 4, we concern ourselves only with medium and heavy nuclei for which the spin-orbit coupling appears to be strong enough to make the total angular momentum j of individual particles a good quantum number. This means that we can treat systems of several loose particles in the convenient approximation of "j-j coupling". We do this for groups of particles with the same j and also for groups with different j, in both cases comparing with experiment. Also in this chapter, we make some remarks on configuration mixing,

[1] BRUECKNER, LEVINSON and MAHMOUD: Phys. Rev. **95**, 217 (1954).

and we end by introducing the *spheroidal shell-model.* This is done for purposes of completeness and the discussion is restricted to the essential minimum. (This generalised version of the shell-model is described in an accompanying article in this volume[1].)

Chap. 5 is devoted to the introduction of the *complete* Individual Particle Model with a view to the analysis of data on light nuclei in Chap. 6. For these nuclei, the approximation of j-j coupling has no success. The opposite extreme of "RUSSELL-SAUNDERS" or "L-S Coupling" does not work much better. The actual coupling mode seems to be intermediate between the two extremes and is called "*Intermediate Coupling*". To make calculations with Intermediate Coupling, one can either work in a representation of L-S wave-functions or a representation of j-j wave-functions. We prefer to use L-S coupling wave-functions and have framed our discussion accordingly. This is largely due to prejudice on our part. Other authors[2] have worked with j-j coupling wave-functions with equal facility.

In Chap. 7, we have collected together various methods that have been suggested for *improving* the Individual Particle Model. These methods have little in common with each other, but each has its own special interest which we consider make it worthy of note.

The last chapter discusses the main empirical features of the shell-model and reviews the recent attempts that have been made to understand the model. We do not describe these attempts in detail because much of the work is still in progress so that any results are only tentative.

Finally in an appendix we give details of the mathematics, such as the RACAH algebra, that is needed for calculations with the Individual Particle Model. This appendix is mainly intended as a supplement to Chap. 5 where inclusion of all these details would have detracted from the continuity of the development.

II. The single particle model.

In this chapter, we introduce the nuclear shell-model by specifying the form of the mean potential well in which nucleons are supposed to move. After doing this, we are then able to specify the spectrum of quantum levels available to any one nucleon. This so-called "single particle spectrum" is at the basis of any version of the nuclear shell-model, no matter how sophisticated. However, in the most developed versions of the model, it is necessary, as we shall see in Chaps. IV and V, to specify other things besides the single particle spectrum. On the other hand, it is possible to formulate some rather naive versions of the model in which the single particle spectrum is the only quantitative feature. One such model, which we will apply to experimental data in Chap. III, is the "Single Particle Model". The essential assumption of this model is that the nucleons in any odd nucleus can be regarded as filling the single-particle levels in such a way that all of them except the last (top-most) odd nucleon pair-off to form an inert "core", which does not contribute at all to the angular momentum or electromagnetic moments of the nucleus. Thus the prediction of these quantities involves only the last odd particle. In Sect. 4 of the present chapter, we evaluate the electromagnetic moments and also transition probabilities for a single particle in preparation for the applications of the Single Particle Model in Chap. III.

[1] S. A. MOSZKOWSKI: Models of Nuclear Structure. (This volume.)
[2] D. R. INGLIS: Rev. Mod. Phys- **25**, 390 (1953). — D. KURATH: Phys. Rev. **101**, 216 (1956).

2. Level sequences in spherically symmetric potentials. It is to be expected that the sequences of single particle quantum levels for various shapes of spherically symmetric potential well may be quite different. For instance, the hyperbolic COULOMB potential well, of the hydrogen atom with its divergence at the origin, has the sequence: $1s$; $2s, 2p$; $3s, 3p, 3d$; $4s \ldots$ (in the special nomenclature for the COULOMB well.[1]) In contrast, the square shaped potential well, which is constant everywhere except for a discontinuity at some radial distance, has the quite different sequence: $1s, 1p, 1d, 2s, 1f, 2p \ldots$ It is clear that, when setting up the nuclear shell-model, it is essential to know at least the general features of the shape of the well. Otherwise the predicted sequences of levels will be very ambiguous.

The necessary information has fortunately become available recently from a careful analysis[2] of the data on the scattering of nucleons by nuclei. The well-shape determined in this way is flat over the bulk of the nucleus, and falls to zero steadily at the edge of the nucleus. One can represent the shape rather well with the form

$$V(r) = V_0\left[1 + \exp\left(\frac{r-R}{a}\right)\right]^{-1}$$

with the value 0.5×10^{-13} cm for a and with $R = 1.33 A^{\frac{1}{3}} \times 10^{-13}$ cm, where A is the mass-number of the nucleus. The value of V_0 depends on nucleon energy, but is roughly 50 or 60 MeV.

There is no known well-shape representing these features for which the eigenfunction and eigenvalues are simply obtainable. Of course, long numerical computation would give the required information, but, fortunately, such labour seems unnecessary because the features of a well of the above type are expected to be closely bracketed by those of two idealised wells for which the spectra are well-known. These wells are the "Square Well" (with constant depth over the nucleus, but infinitely sharp fall-off at the edge) and the "Harmonic Oscillator Well" (with steady fall off at the edge, but not constant and still increasing inside the nucleus). Strictly speaking, such wells only have simple solutions when their walls extend up to infinity without cutting off at a finite depth and range. However we are only interested in bound states for which it makes very little difference whether the wells are cut off or not (in classical mechanics, of course, it makes no difference at all).

Before interpolating between these two idealised wells to discover the features expected of the proper potential well, we will consider the two separately.

α) *Infinite harmonic oscillator well.* The form of this well can be written:

$$V(r) = \tfrac{1}{2} M \omega^2 r^2$$

where M is nucleon mass and ω [which clearly has the dimension $(\text{time})^{-1}$] is the frequency of the oscillation of the classical oscillator.

Before discussing the particular case of the oscillator-type potential well, we note that, for any spherically symmetric well $V(r)$, the acceptable solutions of the SCHRÖDINGER equation:

$$\left(-\frac{\hbar^2}{2M}\Delta^2 + V(r) - E\right)\psi = 0$$

---

[1] In the usual nomenclature for a general well, this sequence would be written: $1s$; $2s, 1p$; $3s, 2p, 1d$; $4s \ldots$

[2] R. D. WOODS and D. S. SAXON: Phys. Rev. **95**, 577 (1954). — MELKANOFF, MOSZKOWSKI, NODVIK and SAXON: Phys. Rev. **101**, 507 (1956).

have the form of products:

$$\psi_{nlm} = u_{nl}(r)\, Y_m^{(l)}(\vartheta, \varphi)$$

where $Y_m^{(l)}(\vartheta, \varphi)$ is a spherical harmonic whose presence expresses the fact that angular momentum l and its z-component m are constants of motion. The function $u_{nl}(r)$ contains the radial dependence of the wave-function and satisfies:

$$\left(-\frac{\hbar^2}{2M}\frac{d^2}{dr^2} + V(r) + \frac{\hbar^2}{2M}\frac{l(l+1)}{r^2} - E_{nl}\right)(r\,u_{nl}) = 0\,.$$

We can refer to $u_{nl}(r)$ as the "radial part of the wave-function" to distinguish it from the so-called "radial wave-function", which is usually understood to mean the product $r u_{nl}(r)$.

Returning now to the case of the harmonic oscillator well, we first introduce the "size-parameter" b, which determines the size of the well and has dimensions of length, being related to ω by: $\hbar\omega = \frac{\hbar^2}{M b^2}$. In discussing the well and its wave-functions, it is very convenient to work with the dimensionless quantity: $q = r/b$, rather than r itself. It is easy to see that the SCHRÖDINGER equation may be written in terms of q as follows.

$$\left(-\frac{d^2}{dq^2} + q^2 + \frac{l(l+1)}{q^2} - 2\left(\frac{E}{\hbar\omega}\right)\right)(q\,u_{nl}) = 0\,.$$

The only eigenvalues of such an equation can be shown to be

$$E = E_\Lambda = (\Lambda + \tfrac{3}{2})\,\hbar\omega \quad (\Lambda = 0, 1, 2, \ldots)\,.$$

Corresponding to any given eigenvalue E_Λ, there are a number of (degenerate) eigenfunctions classified by their l-values:

$$\text{for } \Lambda \text{ even: } l = 0, 2, 4 \ldots \Lambda,$$
$$\text{for } \Lambda \text{ odd: } l = 1, 3, 5 \ldots \Lambda$$

and such that for given Λ:

$$u_{nl}(q) \sim q^l e^{-\frac{1}{2}q^2} f_{nl}(q^2)$$

where $f_{nl}(q)$ is a polynomial in q^2 beginning with a constant term.

Thus the sequence of single particle levels for the harmonic oscillator well consists of bands of degenerate levels, each band separated by energy $\hbar\omega$ from the next. If we label levels by the pairs of quantum numbers (Λ, l) and use the usual spectroscopic notation s, p, d, etc, for l-values 0, 1, 2 ... then the sequence of levels is:

$$0s;\ 1p;\ 2s,\ 2d;\ 3p,\ 3f;\ 4s,\ 4d,\ 4g;\ \ldots$$

In practice however one labels the states as follows:

$$1s;\ 1p;\ 2s,\ 1d;\ 2p,\ 1f;\ 3s,\ 2d,\ 1g;\ \ldots$$

i.e. by the pairs (n, l) where n simply denotes that it is the n-th time that the l-value appears in the sequence. (More formally, n is the number of nodes in the radial wave-function $r u_{nl}(r)$, including those at the origin, minus l.) n and Λ are simply related:

$$\Lambda = 2n + l - 2$$

In Table 1, we list the bands of levels along with their energies and the number $N_\Lambda = (\Lambda+1)(\Lambda+2)$ of particles (neutrons or protons) in each band and the accumulating total number of particles $\sum_\Lambda N_\Lambda = \frac{1}{3}(\Lambda+1)(\Lambda+2)(\Lambda+3)$. In

Table 1. *The Single Particle States of the Infinite Harmonic Oscillator Well.*

The fourth column gives the number of particles N_Λ that the PAULI principle allows in each degenerate Λ-group of orbitals and the fifth column gives the accumulating number of particles.

Λ	$E_\Lambda/\hbar\omega$	Orbitals (n, l)	$N_\Lambda = \sum\limits_{nl} 2(2l+1)$	$\sum\limits_{\Lambda} N_\Lambda$
0	$\frac{3}{2}$	$1s$	2	2
1	$\frac{5}{2}$	$1p$	6	8
2	$\frac{7}{2}$	$2s\ 1d$	12	20
3	$\frac{9}{2}$	$2p\ 1f$	20	40
4	$\frac{11}{2}$	$3s\ 2d\ 1g$	30	70
5	$\frac{13}{2}$	$3p\ 2f\ 1h$	42	112
6	$\frac{15}{2}$	$4s\ 3d\ 2g\ 1i$	56	168

Table 2. *Wave-Functions for the Lowest States of the Harmonic Oscillator in terms of the dimensionless quantity* $q = (M\omega/\hbar)^{\frac{1}{2}} r$.

To obtain the wave-functions $u_{nl}(q)$ that are properly normalised to unity $\int u_{nl}^2(q)\, q^2\, dq = 1$, the functions of the second column should be multiplied by $N_{nl}^{\frac{1}{2}}/\pi^{\frac{1}{4}}$ where N_{nl} is listed in the third column. Notice that the values of N_{nl} are given by $2^{n+l+1}/(2n+2l-1)!!$.

Orbital nl	q-dependence	Normalising factor N_{nl}
$1s$	$e^{-\frac{1}{2}q^2}$	4
$1p$	$q\, e^{-\frac{1}{2}q^2}$	$\frac{8}{3}$
$2s$	$(q^2 - \frac{3}{2})\, e^{-\frac{1}{2}q^2}$	$\frac{8}{3}$
$1d$	$q^2 e^{-\frac{1}{2}q^2}$	$\frac{16}{15}$
$2p$	$\left(q^3 - \frac{5q}{2}\right) e^{-\frac{1}{2}q^2}$	$\frac{16}{15}$
$1f$	$q^3 e^{-\frac{1}{2}q^2}$	$\frac{32}{105}$

Table 2 we list a few of the wave-functions $u_{nl}(q)$. From such wave-functions, one can straightforwardly compute the radial density distributions $\varrho(r) = \sum\limits_{nl} u_{nl}^2(r)$ for the various Λ-bands[1] and verify that, as expected, the total radial density has a strong peak at $r = 0$ reflecting the shape of the oscillator potential.

β) Infinite square well. The regular radial wave functions in this case are expressible in terms of BESSEL-functions of half integral order:

$$u \sim r^{-\frac{1}{2}} J_{l+\frac{1}{2}}(k r) \sim j_l(k r)$$

where the wave-number k is defined by

$$E = \frac{\hbar^2 k^2}{2M}.$$

The eigenfunctions and values are selected by the condition that $u_{nl}(kR) = 0$, where R is the radius of the well. Thus the eigenvalues E_{nl} are immediately found from knowledge of the zeros X_{nl} of $J_{l+\frac{1}{2}}(X)$. The energies E_{nl} in units of $\hbar^2/2MR^2$ are simply equal to the X_{nl}^2. The sequence of levels for the square well along with the values of X_{nl} and the numbers of particles in each orbit, $2(2l+1)$, and the accumulating numbers, $\sum\limits_{nl} 2(2l+1)$, are given in Table 3.

We note here that, unlike the case of the harmonic oscillator well, the eigenvalues for the square well are still found fairly easily when the well is given a finite depth. Instead of using the boundary condition above, one finds the eigenvalues by joining the above radial wave-functions across $r = R$ on to HANKEL wavefunctions that fall exponentially to zero outside $r = R$. The ordering of levels is preserved in the case of a finite depth, but the relative spacings are somewhat changed.

The radial density distribution $\varrho(r) = \sum\limits_{nl} u_{nl}^2(r)$ has been computed for various total numbers of particles in the cases of the infinite and finite wells[2]. Again, as

[1] E. FEENBERG: Shell Theory of the Nucleus, p. 20. Princeton: Princeton University Press 1955.

R [2] B. J. MALENKA: Phys. Rev. **86**, 68 (1952). — J. S. LEVINGER and D. C. KENT: Phys. ev. **95**, 418 (1954).

Table 3. *The Single Particle States of the Infinite Square Well.*

The first column lists the states and the second column gives the values of the appropriate zeros X_{nl} of the corresponding BESSEL functions. The energy eigenvalues, E_{nl}, are simply X_{nl}^2 in units of $\hbar^2/2MR^2$. The third column gives the number of particles, $N_{nl} = 2(2l+1)$, in each orbit and the fourth column gives the accumulating number, $\sum_{nl} N_{nl}$. (From E. FEENBERG: Shell Theory of the Nucleus. Princeton: Princeton University Press 1955.)

Orbital $n\,l$	X_{nl}	$N_{nl}=2(2l+1)$	$\sum_{nl} N_{nl}$
$1s$	3.142	2	2
$1p$	4.493	6	8
$1d$	5.763	10	18
$2s$	6.283	2	20
$1f$	6.988	14	34
$2p$	7.725	6	40
$1g$	8.183	18	58
$2d$	9.095	10	68
$3s$	9.425	2	70
$1h$	9.356	22	92
$2f$	10.417	14	106
$3p$	10.904	6	112
$1i$	10.513	26	138
$2g$	11.705	18	156

for the oscillator well, the resulting density distribution is found to mirror the shape of the potential well except just at the edge, where there is a smooth fall to zero instead of a sharp one. This fall is considerably steeper than the experimental fall-off, and so the square well is too extreme a shape for representing the actual nuclear well, just as the oscillator well is too extreme in the opposite direction.

γ) *Interpolation between the square and harmonic oscillator wells.* Upon examination of Tables 1 and 2 one can see the differences in the spectra of the two wells. The most obvious difference is that the degeneracies of the oscillator well have been removed in the square well. The degenerate states split up in the order of the l-values with the highest l-value lowest. This is the direction of splitting that one expects on going from the oscillator to the square well, since one essentially deepens the well at the edges where the particles of higher l-values are to be found.

Besides the removal of degeneracies there are some instances in the higher states of actual inversion of level order. The only two cases in the range of our interest are the $(1h, 3s)$ and $(1i, 3p)$ pairs. In Fig. 1 we plot the spectra of the two wells alongside each other and tentatively interpolate between them. The order of levels for any intermediate stage is quite unambiguous except in the case of the two inversions. For our standard single particle spectrum, given in the middle of Fig. 1, we choose these two pairs to have the order of the oscillator levels, because their inversion arises from the anomalously extended constant region of the square well.

On the Figure we have indicated the numbers of particles (neutrons or protons) that are assembled when the various shells are filled. We may anticipate Chap. 3 at this point to note that the so-called "magic numbers", i.e. the major shell-closures, occur experimentally at N or $Z = 8$, 20, 28, 50, 82 and 126. Only the first two of these coincide with the numbers on the Figure. Consequently we must look for some refinement or modification of our simple single-particle model that will reproduce the other numbers as well. One suggestion that had some success in this direction proposed a radical change in well-shape to include a central hump which would raise the orbits of lower l-values[1]. This "wine-bottle" potential was shown to be capable of giving the magic numbers at 50 and 82, but it is not generally accepted nowadays because no corresponding dip appears in the experimental density distributions in the nucleus. A second suggestion proposed that, in addition to the central spherical symmetric potential a spin-orbit force should be included in the single particle model[2]. We consider this generally accepted suggestion in the following.

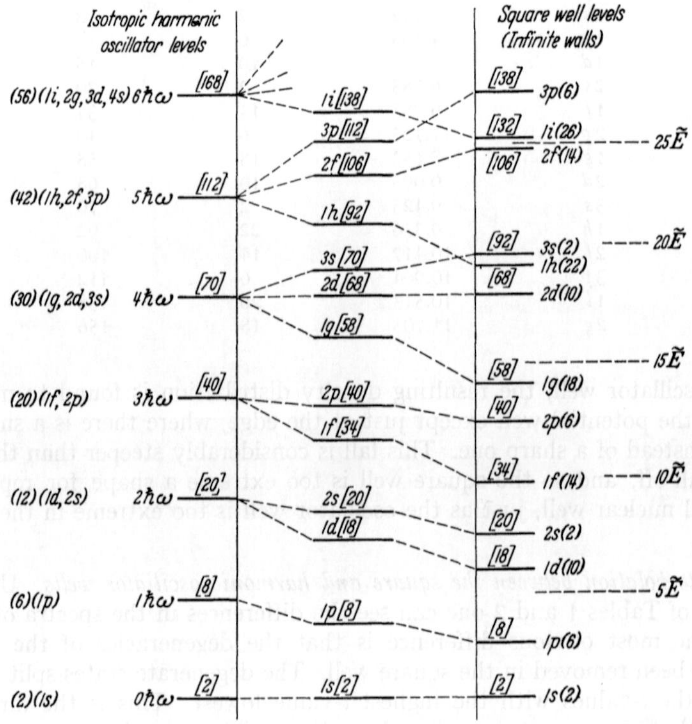

Fig. 1. Level Systems of the Isotropic Harmonic Oscillator Well and the Square Well with infinitely high walls. The Oscillator Well spectrum is shown on the left in units of the oscillator energy quantum $\hbar\omega$. The Square Well spectrum is on the right in units of $\tilde{E}$, defined as $\frac{2\hbar^2}{MR^2}$. A qualitative spectrum for an intermediate shape is interposed between the two. The numbers of particles of one kind (neutrons or protons) in the various shells are given in round brackets. The accumulating numbers are given in square brackets. (From M. G. Mayer and J. H. D. Jensen: The Elementary Theory of Nuclear Structure, p. 53. New York: John Wiley & Sons 1955.)

3. Effects of spin-orbit forces. From Fig. 1, it can be seen that one way to obtain the magic numbers 28, 50, 82 and 126 would be to split each of the $1f$, $1g$, $1h$ and $1i$ single particle levels into two parts so that a number $2(l+1)$ of the total number of $2(2l+1)$ particles is lowered into the region of the next oscillator band of levels below. The simplest way to bring about such a splitting is to hypothesize the existence of an attractive spin-orbit force of the type:

$$U(r)\,(\mathbf{s}\cdot\mathbf{l})$$

where $\mathbf{s}$ is the intrinsic particle spin and where $U(r)$ is a function of radial distance and is mainly negative. This type of force removes the degeneracy in each

[1] W. Elsasser; J. Phys. Radium **5**. 625 (1934). — E. Feenberg and K. C. Hammack: Phys. Rev. **75**, 1877 (1849).

[2] M. G. Mayer: Phys. Rev. **75**, 1969 (1949). — Haxel, Jensen and Suess: Phys. Rev. **75**, 1766 (1949).

single particle level arising from the equality in energy of the two values of total particle spin: $j = l \pm \frac{1}{2}$. When $\boldsymbol{l}$ is coupled to $\mathbf{s}$ in some way, the two component states are split up. For the above spin-orbit force, the state with $\boldsymbol{l}$ and $\mathbf{s}$

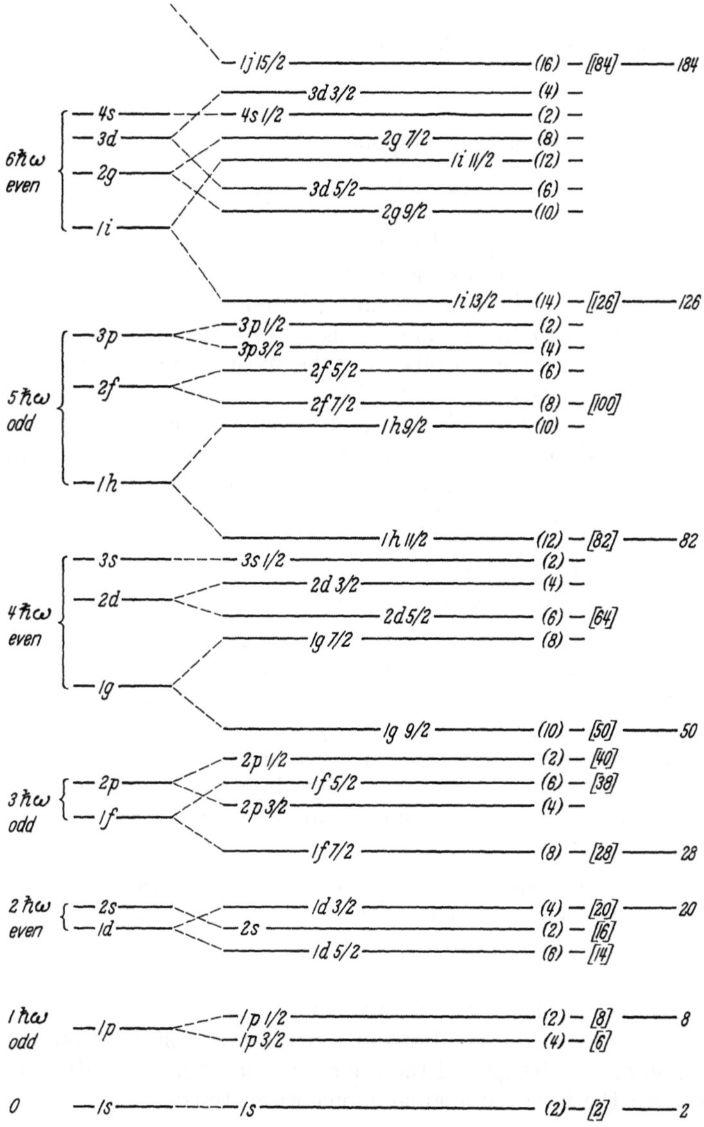

Fig. 2. Effect of Spin-Orbit Coupling on the Level System of a Well of a shape intermediate between the Square and Oscillator Wells. The number of particles of one kind (neutrons or protons) in the various shells are indicated in round brackets. The accumulating numbers are given in square brackets. (From M. G. Mayer and J. H. D. Jensen: The Elementary Theory of Nuclear Structure, p. 58. New York: John Wiley & Sons 1955.)

parallel ($j = l + \frac{1}{2}$) is depressed below that with $\boldsymbol{l}$ and $\mathbf{s}$ antiparallel ($j = l - \frac{1}{2}$) Using the fact that

$$\boldsymbol{j}^2 = (\boldsymbol{l} + \mathbf{s})^2 = \boldsymbol{l}^2 + \mathbf{s}^2 + 2(\mathbf{s} \cdot \boldsymbol{l})$$

we find that:

$$\mathbf{s} \cdot \boldsymbol{l} = \tfrac{1}{2}\left[\boldsymbol{j}^2 - \boldsymbol{l}^2 - \mathbf{s}^2\right] = \begin{cases} \tfrac{1}{2} l & \text{for } j = l + \tfrac{1}{2}, \\ -\tfrac{1}{2}(l+1) & \text{for } j = l - \tfrac{1}{2}. \end{cases}$$

It follows that, if the spin-orbit force is small so that it can be assumed not to disturb the radial wave-function, then it leads to a depression of the $j=l+\frac{1}{2}$ component of an l-orbit by:

$$-\tfrac{1}{2}l\,\xi_{nl}$$

and an elevation of the $j=l-\frac{1}{2}$ component by

$$-\tfrac{1}{2}(l+1)\,\xi_{nl}$$

and a total splitting of the two components by:

$$-\tfrac{1}{2}(2l+1)\,\xi_{nl}.$$

In these expressions ξ_{nl} is the average of $U(r)$ taken over the state (n, l): $\xi_{nl}=\langle U(r)\rangle_{nl}$. From the properties of $U(r)$, ξ_{nl} is evidently negative. Provided that ξ_{nl} does not vary with l significantly, it follows that orbits are split by an amount increasing with the l-value. This is just the trend needed to account for the fact that the high l-orbits must be most strongly split. Fig. 2 shows qualitatively the resulting spectrum of levels when the splitting is taken to be roughly proportional to $(2l+1)$. It can be seen that all of the observed magic numbers are reproduced. The ordering of levels in the so-called "major shells" between magic numbers is somewhat ambiguous, depending on the precise magnitude of the spin-orbit force. For instance, the order of the $2p_{\frac{3}{2}}$ and $1f_{\frac{5}{2}}$ levels may be as illustrated, or the reverse, depending on the magnitude of the spin-orbit splitting of levels relative to the $1f-2p$ separation. In such cases, the ordering has been fixed by the empirical evidence.

One very important feature of the new spectrum that should be noted is that levels of opposite parity and very different spin are brought close together. In the absence of spin-orbit effects, levels are banded in groups of levels of the same parity in which there are no large spin differences between adjacent levels. The spin-orbit splitting of the type we have described leads to the presence of an "intruder" in each band of levels. This intruder comes from the next oscillator band above, and so is of opposite parity to the levels of the band and has a high spin.

4. Electromagnetic moments and transition probabilities. We now derive expressions for the electromagnetic moments and the radiative transition probabilities of a single particle moving in a spherically symmetric potential with spin-orbit coupling.

α) *Magnetic dipole moment.* The magnetic dipole vector of a moving charged particle consists of two contributions in general, one from the current generated by the motion of the charge and the other from intrinsic currents in the particle associated with the intrinsic angular momentum, thus:

$$\boldsymbol{\mu}=\boldsymbol{\mu}_{\text{orbital}}+\boldsymbol{\mu}_{\text{spin}}$$

where:

$$\boldsymbol{\mu}_{\text{orbital}}=\boldsymbol{l}\,g_l\left(\frac{e\hbar}{2Mc}\right),$$

$$\boldsymbol{\mu}_{\text{spin}}=\boldsymbol{s}\,g_s\left(\frac{e\hbar}{2Mc}\right).$$

Here, $\boldsymbol{l}$ and $\boldsymbol{s}$ are the orbital angular momentum and intrinsic angular momentum operators for the particle, and g_l and g_s are the orbital and spin gyromagnetic ratios. $\frac{e\hbar}{2Mc}$ is the nuclear magneton (n.m.) in conventional notation. Supposing

$\boldsymbol{l}$ and $\boldsymbol{s}$ to be coupled by a spin-orbit force to give total angular momentum $\boldsymbol{s}+\boldsymbol{l}=\boldsymbol{j}$, the magnetic dipole moment μ is conventionally defined as the expectation value of the z-component of $\boldsymbol{\mu}$ taken in the nuclear substate with the z-component, m, of j equal to j:

$$\mu = (j\, m = j \,|\, \mu_z \,|\, j\, m = j)\,.$$

To evaluate this matrix element in terms of l, s and j, we might use the general methods of the RACAH algebra, described in Appendix 4. In fact, we shall use such methods later to evaluate the matrix element of a general magnetic λ-pole operator $M_0^{(\lambda)}$. [As defined later, $M_0^{(\lambda)}$ in the special case $\lambda = 1$ is related to μ_z by: $M_0^{(1)} = \left(\frac{3}{4\pi}\right)^{\frac{1}{2}} \mu_z$.]

However, in the simple case of the magnetic dipole, there is no need to use such powerful methods. Instead we can use the general theorem that, for an isolated system of spin J, component M, the expectation value of any vector operator $\boldsymbol{V}$ can be written:

$$(J\,M \,|\, \boldsymbol{V} \,|\, J\,M) = (J\,M \,|\, \boldsymbol{J} \,|\, J\,M)\,\frac{(JM \,|\, \boldsymbol{V}\cdot\boldsymbol{J} \,|\, JM)}{(JM \,|\, \boldsymbol{J}^2 \,|\, JM)}\,.$$

Taking the z component of this equation, in the case $M = J$

$$(J,\, M = J \,|\, V_z \,|\, J,\, M = J) = (J+1)^{-1}\,(J,\, M = J \,|\, \boldsymbol{V}\cdot\boldsymbol{J} \,|\, J,\, M = J)\,.$$

Replacing $\boldsymbol{V}$, $\boldsymbol{J}$ by $\boldsymbol{\mu}$, $\boldsymbol{j}$, we find:

$$\mu = (j,\, m = j \,|\, \mu_z \,|\, j,\, m = j) = (j+1)^{-1}\,(j,\, m = j \,|\, \boldsymbol{\mu}\cdot\boldsymbol{j} \,|\, j,\, m = j)\,.$$

Using the simple identities:

$$2\boldsymbol{j}\cdot\boldsymbol{l} = \boldsymbol{j}^2 + \boldsymbol{l}^2 - \boldsymbol{s}^2\,, \qquad 2\boldsymbol{j}\cdot\boldsymbol{s} = \boldsymbol{j}^2 - \boldsymbol{l}^2 + \boldsymbol{s}^2$$

and our definition of μ, it follows that:

$$\mu = \left(j - \tfrac{1}{2}\right) g_l + \tfrac{1}{2}\, g_s \qquad \text{for} \quad j = l + \tfrac{1}{2}$$

$$\mu = \frac{j}{j+1}\left[\left(j + \tfrac{3}{2}\right) g_l - \tfrac{1}{2}\, g_s\right] \quad \text{for} \quad j = l - \tfrac{1}{2}\,,$$

in units of $\frac{e\hbar}{2Mc}$, the nuclear magneton (n.m.).

To conclude this discussion of the single particle magnetic dipole moment, we draw attention to the existence of a small correction arising from the presence of the spin orbit force[1]. In principle, any velocity dependent force, such as a spin-orbit force of the type:

$$U(r)\, \boldsymbol{s}\cdot\boldsymbol{l} = U(r)\, \boldsymbol{s}\cdot\left(\frac{\boldsymbol{r}\times\boldsymbol{p}}{\hbar}\right)$$

gives rise to a contribution to the effective magnetic moment of a charged particle. (For an uncharged particle, there is no effect.) In general, for a charged particle in the presence of an electromagnetic field described by the vector potential $\boldsymbol{A}(\boldsymbol{r})$, one ensures the fulfilment of the gauge invariance condition by replacing $\boldsymbol{p}$ in the usual Hamiltonian equation by $\boldsymbol{p} - \frac{e}{c}\boldsymbol{A}(\boldsymbol{r})$. This not only affects the kinetic energy term, but also any potential energy term that depends

[1] J. H. D. JENSEN and M. G. MAYER: Phys. Rev. **85**, 1040 (1952).

on $\boldsymbol{p}$ such as the one above. In this case, the additional interaction energy thereby introduced is:

$$-\frac{e}{\hbar c}\, U(r)\, \mathbf{s}\cdot(\boldsymbol{r}\times\boldsymbol{A}).$$

For a constant magnetic field $\boldsymbol{H}$, $\boldsymbol{A}$ has the form $\frac{1}{2}(\boldsymbol{H}\wedge\boldsymbol{r})$, and the interaction energy has the form $-\boldsymbol{\mu}'\cdot\boldsymbol{H}$ where:

$$\boldsymbol{\mu}' = -\frac{e}{2\hbar c}\, U(r)\,[-r^2\,\mathbf{s} + (\boldsymbol{r}\cdot\mathbf{s})\,\boldsymbol{r}].$$

The contribution to the single particle magnetic dipole moment as defined above is easily shown to be:

$$\mu' = \pm\frac{\langle r^2\,U(r)\rangle_{nl}}{\hbar^2/M}\,\frac{2j+1}{2j+2} \quad \text{for} \quad j = l \pm \frac{1}{2}$$

in units of $\frac{e\hbar}{2Mc}$, the nuclear magneton (n.m.), where $\langle r^2\,U(r)\rangle_{nl}$ is the average value of $r^2\,U(r)$ over the particle orbit, $(n\,l)$ say, i.e. if $r\,u_{nl}(r)$ is the radial wave-function of the orbit:

$$\langle r^2\,U(r)\rangle_{nl} = \int r^2\,U(r)\,r^2\,u_{nl}^2(r)\,dr.$$

This has been roughly estimated[1] with the result that:

$$\mu' \sim \frac{1}{4}\,\frac{2j+1}{2j+2} \quad \text{for} \quad j = l \pm \frac{1}{2}.$$

β) Electric quadrupole moment. The electric quadrupole tensor of a particle of charge e is expressible in terms of the five basic components:

$$Q_q^{(2)} = e\,r^2\,Y_q^{(2)}(\vartheta,\varphi).$$

The electric quadrupole moment, Q, is conventionally defined as the expectation value of $\left(\frac{16\pi}{5}\right)^{\frac{1}{2}} Q_0^{(2)}$ taken in the nuclear substate with $m=j$,

$$Q = (j, m=j|\left(\frac{16\pi}{5}\right)^{\frac{1}{2}} Q_0^{(2)}\,|j, m=j).$$

Now a state $|jm)$ involves the vector coupling of angular (spherical harmonic) wave functions $|l m_l) \equiv Y_m^{(l)}(\vartheta,\varphi)$ to the intrinsic spin wave-functions $|s m_s)$:

$$|j\,m) = u_{nl}(r) \sum_{m_s+m_l=m} (s\,l\,m_s\,m_l\,|j\,m)\,|l\,m_l)\,|s\,m_s)$$

where $(s l m_s m_l|j m)$ are the CLEBSCH-GORDAN vector-coupling coefficients.

Inserting such an expansion on both sides of the matrix element of $Q_0^{(2)}$, we obtain:

$$Q = e\left(\frac{16\pi}{5}\right)^{\frac{1}{2}}\langle r^2\rangle_{nl} \sum_{m_s+m_l=m} (s\,l\,m_s\,m_l\,|j, m=j)^2\,(l\,m_l|\,Y_0^{(2)}\,|l\,m_l)$$

where $\langle r^2\rangle_{nl}$ is the mean square radius:

$$\langle r^2\rangle_{nl} = \int_0^\infty r^2\,u_{nl}^2(r)\,r^2\,dr.$$

It only remains to specify the matrix elements of the spherical harmonic and evaluate the sum over m_s+m_l. It can be shown that the matrix elements are

[1] J. H. D. JENSEN and M. G. MAYER: Phys. Rev. **85**, 1040 (1952).

(see Appendix 8):

$$(l\, m_l |\, Y_0^{(2)}\, | l\, m_l) = \left(\frac{5}{4\pi}\right)^{\frac{1}{2}} (2 l\, 0\, 0\, | l\, 0)\, (2 l\, 0\, m_l\, | l\, m_l)\,.$$

On using the known algebraic expressions for the various vector-coupling coefficients, the two terms in the sum over $m_s + m_l$ can be combined to give the result:

$$Q = -\frac{2j-1}{2j+2}\, e \langle r^2 \rangle_{nl} \quad \text{for both cases} \quad j = l \pm \frac{1}{2}\,.$$

If we assume the radial wave-functions to be constant inside the point $r = R$ and zero outside:

$$\langle r^2 \rangle_{nl} = \tfrac{3}{5} R^2\,.$$

For rough estimates of Q we can use this relation putting R equal to the nuclear radius.

We note that a single neutron does not generate any electric quadrupole moment because it does not posses a charge. However, for an actual nucleus of A particles visualized as a single neutron and a "core" of all the other $(A-1)$ particles, some quadrupole moment will be generated by the recoil motion of the core that is needed to keep the centre-of-mass fixed. If the charge on the core is Ze, the fact that the radius of recoil motion of the core is only a fraction $1/A$ of that of the odd neutron implies that this moment will be smaller than the above single proton moment by a factor Z/A^2. For medium and heavy nuclei, this factor is very small. (For a nucleus visualised as a set of single particles moving in orbits, it can be shown that proper antisymmetrisation leads to zero recoil effect, at least in the case of an oscillator potential and a nucleus with one particle outside closed shells. See Sect. 22.)

γ) Magnetic octupole moment. Like the magnetic dipole moment, the octupole moment of a single particle arises from the currents associated with orbital and intrinsic angular momentum of the particle. The conventional definition of magnetic octupole moment Ω is:

$$\Omega = (j\, m = j\, |\, M_0^{(3)} \left(\frac{4\pi}{7}\right)^{\frac{1}{2}} |\, j\, m = j)$$

where the operator $M_0^{(3)}$ is one of the seven basic components of the magnetic octupole tensor:

$$M_q^{(3)} = [\nabla r^3\, Y_q^{(3)}(\vartheta, \varphi)] \cdot \left[\frac{1}{2} g_l\, \boldsymbol{l} + g_s\, \mathbf{s}\right] \left(\frac{e\hbar}{2Mc}\right).$$

Following the same standard procedure as described above in the case of the electric quadrupole moment, one can show that[1]:

$$\Omega = \frac{3(2j-1)(j+2)}{2(2j+4)(2j+2)} \left[\left(j - \frac{3}{2}\right) g_l + g_s\right] \langle r^2 \rangle_{nl} \left(\frac{e\hbar}{2Mc}\right) \quad \text{for} \quad j = l + \frac{1}{2}\,,$$

$$\Omega = \frac{3(2j-1)(j-1)}{2(2j+4)(2j+2)} \left[\left(j + \frac{5}{2}\right) g_l - g_s\right] \langle r^2 \rangle \left(\frac{e\hbar}{2Mc}\right) \quad \text{for} \quad j = l - \frac{1}{2}\,.$$

As in the case of the magnetic dipole moment, there may be small corrections to these expressions arising from the velocity dependence of the spin-orbit force that couples $\boldsymbol{l}$ to $\mathbf{s}$.

δ) Transition probabilities. The general formula for the transition probability by λ-pole type radiation from an upper state of spin j to a lower state of spin j'

[1] C. Schwarz: Phys. Rev. **97**, 380 (1955).

has often been given[1]. It is

$$T(\lambda) = \frac{1}{\hbar} \frac{8\pi(\lambda+1)}{\lambda[2\lambda+1)!!]^2} \left(\frac{E_\gamma}{\hbar c}\right)^{2\lambda+1} \sum_{m'+q=m} |(j\,m|\,O_q^{(\lambda)}\,|j'\,m')|^2 .$$

In this formula, E_γ is the energy removed by the photon, and λ and q are the angular momentum and its z-component that are removed by the photon. m and m' are the z-components of the spins of the initial and final states. The sum is over all pairs $m'+q$ that add up to a given m. (It does not matter which m, since the sum is independent of m.) The $O_q^{(\lambda)}$ are the appropriate operators for the particular type of radiation involved in the transition. Their forms are natural generalisations of the expressions for electric quadrupole ($\lambda=2$) and magnetic octupole ($\lambda=3$) operators given above:

$$\text{electric } \lambda\text{-pole:} \quad O_q^{(\lambda)} \to Q_q^{(\lambda)} \equiv e\, r^\lambda Y_q^{(\lambda)}(\vartheta, \varphi),$$

$$\text{magnetic } \lambda\text{-pole:} \quad O_q^{(\lambda)} \to M_q^{(\lambda)} \equiv [\nabla r^\lambda Y_q^{(\lambda)}(\vartheta, \varphi)] \cdot \left[\frac{2 g_l}{\lambda+1} \boldsymbol{l} + g_s\, \mathbf{s}\right] \left(\frac{e\hbar}{2Mc}\right).$$

In the expression for the electric λ-pole operator, the (small) contribution from spin magnetic moments has been dropped. Using the standard methods of RACAH algebra described in the appendix the matrix elements and the sum over $m'+q$ can be evaluated.

For electric radiation, one finds[2]:

$$\sum_{m'+q=m} |(j\,m|\,Q_q^{(\lambda)}\,|j'\,m')|^2 = \frac{e^2}{4\pi} \langle r^\lambda \rangle_{nl}^2 \frac{(Z^0(l\,j\,l'\,j', \tfrac{1}{2}\lambda))^2}{2j+1}$$

where the Z^0-function is defined in terms of the familiar RACAH W-function by:

$$Z^0(l\,j\,l'\,j', s\,\lambda) = [(2j+1)(2l+1)(2j'+1)(2l'+1)]^{\frac{1}{2}} (l\,l'\,0\,0\,|\lambda\,0)\, W(l\,j\,l'\,j', s\,\lambda).$$

For the purpose of estimating a *typical* single particle transition probability[3], as opposed to that for a *particular* transition with particular $l\,j\,l'\,j'$ and potential well shape, we can replace the spinfactor $(Z^0)^2/(2j+1)$ in the above expression by unity and $\langle r^\lambda \rangle_{nl}$ by $\left(\frac{3R^\lambda}{\lambda+3}\right)$, where R is the nuclear radius. (The latter estimate follows on assuming constant radial wave-functions.) This gives the typical single particle transition probability:

$$T_E(\lambda) = \frac{4.4(\lambda+1)}{\lambda[(2\lambda+1)!!]^2} \left(\frac{3}{\lambda+3}\right)^2 \left(\frac{E_\gamma}{197\,\text{MeV}}\right)^{2\lambda+1} R^{2\lambda} \quad (10^{21}\,\text{sec})$$

where E_γ is in MeV, R in 10^{-13} cm.

For magnetic radiation, the sum over $m'+q$ can be reduced[1] to:

$$\sum_{m'+q=m} |(j\,m|\,M_q^{(\lambda)}\,|j'\,m')|^2 = \frac{e^2}{4\pi} \langle r^{\lambda-1} \rangle_{nl}^2 \left(\frac{\hbar}{mc}\right)^2 \frac{[Z^0(l\pm 1, j, l', j', \tfrac{1}{2}\lambda)]^2}{2j+1} S^2$$

where the sign $\pm$ in the Z^0-function is chosen according as $j = l \pm \frac{1}{2}$, and where the dimensionless quantity S is tabulated in Table 4. Notice that, for a transition with $|j-j'| = \lambda$:

$$S = \frac{\lambda g_s}{2} - \frac{\lambda g_l}{\lambda+1}.$$

---

[1] J. M. BLATT and V. F. WEISSKOPF: Theoretical Nuclear Physics, p. 595. New York: John Wiley & Sons 1952.

[2] J. M. KENNEDY and W. T. SHARP: Chalk River Report CRT 580, 1954.

[3] M. GOLDHABER and J. WENESER: Ann. Rev. Nucl. Sci. 5, 1 (1955).

Table 4. *Quantities F an G for the Calculation of the Statistical Factor S in Magnetic Multipole Matrix Elements*: $S=\frac{F}{2}g_s+\frac{FG}{\lambda+1}g_l$. (From J. M. KENNEDY and W. T. SHARP: Statistical Factors for Electromagnetic Transitions, CRT 580, Chalk River, 1954.)

j	j'	F	G
$l+\frac{1}{2}$	$l'+\frac{1}{2}$	$\frac{1}{2}(j+j'+\lambda+1)$	$j+j'-\lambda$
$l+\frac{1}{2}$	$l'-\frac{1}{2}$	$\frac{1}{2}(j-j'+\lambda)$	$j-j'-\lambda-1$
$l-\frac{1}{2}$	$l'+\frac{1}{2}$	$\frac{1}{2}(j'-j+\lambda)$	$j'-j-\lambda-1$
$l-\frac{1}{2}$	$l'-\frac{1}{2}$	$\frac{1}{2}(j+j'-\lambda+1)$	$-j-j'-\lambda-2$

In order to estimate a typical single particle transition probability, we again replace the spin factor by unity and $\langle r^{\lambda-1}\rangle_{nl}$ by $\frac{3R^{\lambda-1}}{\lambda+2}$. For S, we choose the value for a single proton transition in the case $|j-j'|=\lambda$, with g_s equal to the gyromagnetic ratio for an isolated proton i.e. $S=\lambda\mu(P)-\frac{\lambda}{\lambda+1}$, where $\mu(P)$ is the observed proton magnetic dipole moment. The transition probability is then:

$$T_M(\lambda)=\frac{0.19(\lambda+1)}{\lambda[(2\lambda+1)!!]^2}\left(\frac{3}{\lambda+2}\right)^2\left(\frac{E_\gamma}{197\,\mathrm{MeV}}\right)^{2\lambda+1}\left(\lambda\mu(P)-\frac{\lambda}{\lambda+1}\right)^2R^{2\lambda-2}\quad(10^{21}\,\mathrm{sec})$$

where E_γ is in MeV, R in 10^{-13} cm and $\mu(P)$ in nuclear magnetons.

Finally we note the trivial relations between the width Γ, the lifetime τ and the transition probability T for any given transition:

$$\tau=\frac{1}{T};\quad \Gamma=\frac{\hbar}{\tau}=\hbar T.$$

III. Applications of the single particle model to medium and heavy nuclei.

Having defined the Single Particle Model in the last chapter, we will now compare it with some of the vast amount of experimental data that exists on nuclei, especially that on low-lying states. We have seen in the Introduction that the Single Particle Model is only a very primitive version of the complete shell-model, so it is not surprising that there are certain types of data which the model, as formulated so far, is intrinsically incapable of describing or predicting. Evidently the model only applies to odd-mass nuclei, and it cannot describe even-mass nuclei. Furthermore, its predictions of nuclear spectra in both even and odd nuclei are extremely incomplete. This is a result of the assumption that all particles, except the last odd one, are paired off in an inert "core", so that the only degrees of freedom available for generating excited states of the system are those of the odd particle. Thus the only spectrum that the model predicts is a single-particle one, with energy spacings of the order of 1 MeV. Such a prediction cannot sensibly account for many observed spectra, whose small energy spacings indicate a multi-particle structure for the states. In the next chapter we will see how multi-particle states can be constructed according to a more sophisticated shell-model. For the present however we will take the extreme single particle version of the shell-model and discuss only those data for which this naive model can make predictions. Some of the comparisons with the data will be repeated in the next chapter using the extended model.

5. Magic number anomalies. It has been known for some time that nuclei with Z or N equal or near to the numbers 8, 20, 28, 50, 82 and 126 exhibit many anomalous features. As a consequence these numbers have been distinguished by the term "magic". Before reviewing these anomalies, we mention some general features of the shell-model interpretation of the magic numbers. These numbers are, from a shell-model point of view, to be identified with major shell-closures, and we have already seen in the previous chapter how such closures can be made to occur at the above numbers by including a spin-orbit coupling in the single particle potential. The most evident manifestation of such a shell-closure would be a sudden decrease in binding energy above the shell-closure due to particles having to enter higher orbits. For a similar reason, the binding energies of excited states of the closed shell nuclei themselves would be expected to be especially low, i.e. the states would be especially high above the ground state. Such an effect could be exaggerated by the existence of extra binding energy in the closed shell ground state (from the "pairing" effect as in atoms in which pairs of particles in the same orbit couple more effectively than those in different orbits).

We now review the various experimental anomalies and discuss them in terms of the interpretation of the magic numbers as shell-closures. The most significant anomalies are:

α) *Unusual binding energies* giving rise to discontinuities in the Mass Surface. These discontinuities are well-defined and quite strong (∼2 MeV), and come from the anomalously low binding of neutrons above magic N numbers and protons above magic Z numbers. Such effects have been found to some extent at all the magic N and Z numbers, and they are often to be noticed in a direct fashion in the energy releases in β and α decay, which are related to the binding energies. The best developed effects are perhaps those at the double closed shell Pb^{208} ($Z=82$, $N=126$). From Table 5, it can be seen that the nucleon binding energies have large discontinuities, both at $N=126$ and $Z=82$.

It is difficult to see from a study of the individual nucleon binding energies whether or not there are anomalies above and below the closed shells, besides

Table 5. *The Nucleon Binding Energies of Nuclei in the Region of* Pb^{208}.

The sharp breaks at the magic numbers $Z=82$, $N=126$ are evident. (Binding energies are given in mMU: 1 mMU ≡ 0.931 MeV.)

Z \ N		124		125		126		127		128	
				(7.73)							
81	(8.03)	Tl^{205}	(6.69)	Tl^{206}	(7.30)	Tl^{207}	(4.11)	Tl^{208}	(5.30)	Tl^{209}	(4.05)
		(7.47)		(8.01)		(9.63)		(7.67)		(9.00)	
82	(8.70)	Pb^{206}	(7.23)	Pb^{207}	(7.92)	Pb^{208}	(4.15)	Pb^{209}	(5.63)	Pb^{210}	(4.06)
		(3.88)		(3.97)		(4.04)		(4.86)		(4.72)	
83		Bi^{207}	(7.32)	Bi^{208}	(7.99)	Bi^{209}	(4.97)	Bi^{210}	(5.48)	Bi^{211}	(4.70)
		(5.41)		(5.15)		(5.38)		(5.31)		(6.27)	
84		Po^{208}	(7.06)	Po^{209}	(8.22)	Po^{210}	(4.90)	Po^{211}	(6.45)	Po^{212}	(4.63)

actually at the closed shells. However such anomalies are easily seen if one deals with the total nuclear binding energies, which are effectively the integrated nucleon binding energies. For instance, one can plot[1] the total observed energies together with a smoothed standard energy such as that given by the semi-empirical mass formula. The anomalously high binding associated with the Pb^{208} closed shell can be discerned in nuclei which are 10 or more mass units above or below the shell-closure. The extra binding below the shell closure may be due to pairing energy, just as in atoms, where the effect is much more evident. There, the individual electron binding energies can be seen to increase steadily as a shell is filled as a result of pairing energy; the fall in binding at the shell-closures is about 50%, which is much the same as with the nucleus.

An interesting consequence of the low binding of the 9th, 51st and 83rd neutron in O^{17}, Kr^{87} and Xe^{137} respectively is that the β-decay of N^{17}, Br^{87} and I^{137} can lead to states of these nuclei above their neutron thresholds. Thus the phenomenon of delayed neutron emission occurs.

β) Large total abundances. In a plot of abundance against mass number[2], one finds anomalies at $A \sim 56$, 90, 135 and 200. These anomalies are generally describable as increases and they are separated by shallow troughs. One notes however that the anomalies are not sharp peaks at the magic numbers of Z and N. In the first place there is no real evidence for increased abundances associated with magic numbers in Z at all, because the above four mass numbers can be correlated with $N=28$, 50, 82 and 126. Furthermore the anomalies at 135 and 200 are *double* peaks (at 128, 138 and 194, 208) and the magic N number coincides with the upper peak. Some light is thrown on why there are not just simple sharp peaks at the magic N numbers by the case of Fe^{56} which accounts for the sharp (single) peaked anomaly at $A \sim 56$. The nucleus Fe^{56} itself has no magic number in either N or Z. However Ni^{56} which has both N and Z magic, decays by β-emission into Fe^{56} in a matter of days. Thus a tendency for Ni^{56} to be produced in element-forming reactions is expected to manifest itself as an increased abundance of the stable isobar Fe^{56}. Now let us consider the magic number $N=82$. This number gives rise to a β-stable isotope in each of seven elements from $Z=54$ to $Z=62$. In addition, this number is expected to occur in many β-unstable isotopes with $Z<54$ and $Z>62$. Such isotopes will tend to decay into stable isotopes with $N<82$ and $N>82$ respectively. An essential difference between the two cases is that the isotopes with $N>82$ are anomously weakly bound against neutron emission so that β-unstable isotopes with $Z>62$ may actually lead to heavy particle emission from some of the daughters. (This has been established experimentally; see the mention of "delayed neutron emission" above.) Thus a tendency for the magic number $N=82$ to be formed in element-forming reactions will manifest itself not as a sharp peak in the abundance plot, but as a broad peak centred somewhat below the mass numbers where $N=82$ occurs in β-stable nuclei. Similar considerations apply to the case of $N=126$. Thus one can understand the general breadth and position of the anomalies in the plot of abundance against mass number in terms of a tendency for magic N numbers to be formed in heavy particle reactions. However no argument has yet been advanced to explain the double-peak shape of the anomalies at $N=82$ and 126.

It should be emphasised that we have assumed that magic number nuclei are formed preferentially in the nuclear reactions responsible for the natural

[1] A. E. S. Green: Phys. Rev. **95**, 1006 (1954).
[2] H. E. Suess and H. C. Urey: Rev. Mod. Phys. **28**, 56 (1956).

abundances without giving any explanation for this suggestion. Such an explanation will presumably relate features of such reactions to the extra stability (closed shell aspects) of the magic number nuclei.

γ) Large relative abundances of individual isotopes and isotones. We now describe anomalies that emerge when the total abundances of particular elements are analysed into the relative abundances of component isotopes. The origin of the anomalies can, at least in part, be traced back to the sudden decrease in binding above magic numbers. The anomalies are[1]:

(a) the numbers of stable and long lived isotopes for $Z=20$, 28, 50 and 82 are larger than for other Z-values (elements) nearby. The total numbers of isotopes, including the β-unstable ones, are also larger.

(b) the numbers of stable and long lived isotones for $N=20$, 28, 50, 82 and 126 are larger than for other N-values nearby. The total numbers of isotones including the β-unstable ones, are also larger.

(c) the abundance of individual even-even isotopes relative to that of the whole element is considerably larger when N falls at 50 or 82. This relative abundance is always $<60\%$, except for Sr^{88} ($N=50$, 82%), Ba^{138} ($N=82$, 72%) and Ce^{140} ($N=82$, 90%). These large values are explained by the fact that the abundances of these elements are shared among less stable isotopes than usual due to the absence of stable isotopes above the magic numbers.

(d) the occurrence of more than one odd stable isotone for even N when $N=20$, 50 and 82. For all other N, there is only one such isotone, but, at these N-values there are two:

$$N=20:\quad Cl^{37},\ K^{39},$$
$$N=50:\quad Rb^{87},\ Y^{89},$$
$$N=82:\quad La^{139},\ Pm^{141}.$$

One can express this in another way: each odd A has only one stable isobar. To make the isobar of $A+2$ nucleons from that of A nucleons, one normally must add either two neutrons, or a neutron and a proton. In only three cases does one add two protons. These are the three pairs we have cited.

(e) the occurrence of the same N value for adjacent stable even-even nuclei. Normally when the Z value is increased by two, the N value of the heaviest isotope increases by 2 or more. The exceptions are:

$$Ca^{48},\ Ti^{50}\quad N=28,$$
$$Kr^{86},\ Sr^{88}\quad N=50,$$
$$Xe^{136},\ Ba^{138}\quad N=82.$$

Similarly for the lightest isotope. The exceptions here are:

$$Zr^{90},\ Mo^{93}\quad N=50,$$
$$Nd^{142},\ Sm^{144}\quad N=82.$$

δ) Large excitation energies of first excited states, especially of even-even nuclei. A rather different, but equally striking, type of anomaly associated with the magic numbers is the sharp rise in the excitation of first excited states. This is especially apparent for even-even nuclei as is shown in Fig. 3[2]. It can be seen, for instance, that the excitation energy in the region $154<A<188$ is of the

[1] M. G. MAYER and J. H. D. JENSEN: Elementary Theory of Nuclear Shell Structure, p. 16. New York: John Wiley & Sons 1955. — E. FEENBERG: Shell Theory of the Nucleus, p. 5. Princeton: Princeton University Press 1955.

[2] G. SCHARFF-GOLDHABER: Phys. Rev. 90, 587 (1953).

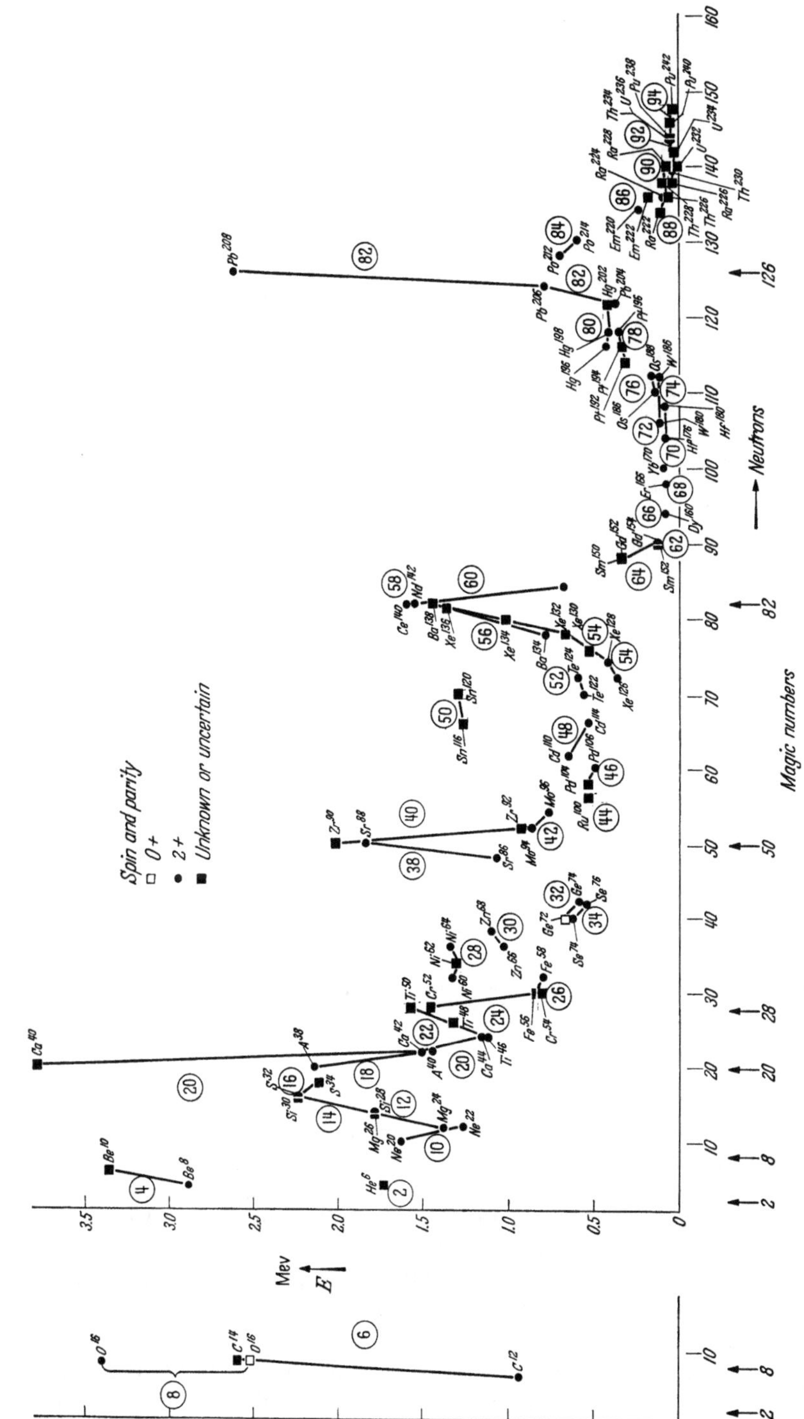

Fig. 3. Energies of First Excited States of Even-Even Nuclei plotted against the Number of Neutrons. Points for isotopes of the same element are connected by straight lines. The corresponding proton number, encircled, appears next to the connecting line. [From G. Scharff-Goldhaber: Phys. Rev. **90**, 587 (1953) as presented by M. G. Mayer and J. H. D. Jensen: The Elementary Theory of Nuclear Structure, p. 191. New York: John Wiley & Sons 1955.]

order of 100 keV, but there is a sharp rise below this region to a peak of ~1.5 MeV at Ce^{140} and a sharp rise above to a peak of ~2.5 MeV at Pb^{208}. These two nuclei occur, of course, at magic number $N=82$ and double magic numbers $Z=82$,

$N=126$ respectively. In a qualitative manner, the naive shell-model predicts such peaks because, whereas in a normal nucleus the first excited state arises from re-orientation of the particles, in a magic nucleus it can only arise from the actual elevation of a particle into a higher orbit. This elevation requires considerable energy in itself; in addition, more energy is needed to compensate for the closed-shell pairing energy. We note that the spin of the first excited state is not expected to be the normal 2^+, but may be quite different depending on the spin and parity of the higher orbit. This checks with observation; for instance, the first excited state of Pb^{208} is 3^-. Before concluding that the behaviour of first excited states at magic numbers is in total keeping with a closed shell interpretation of the magic numbers, it should be emphasised that the shell-model, strictly speaking, only predicts an anomaly *at* the closed shell. Experimentally, the anomalously high excitation of the first excited state occurs over a fair range of mass numbers, although the effect is much stronger actually at the closed shell. It follows that, in any attempt to refine the naive shell-model, one should look for a means of "smearing out" the closed shell effects to adjoining nuclei to some extent.

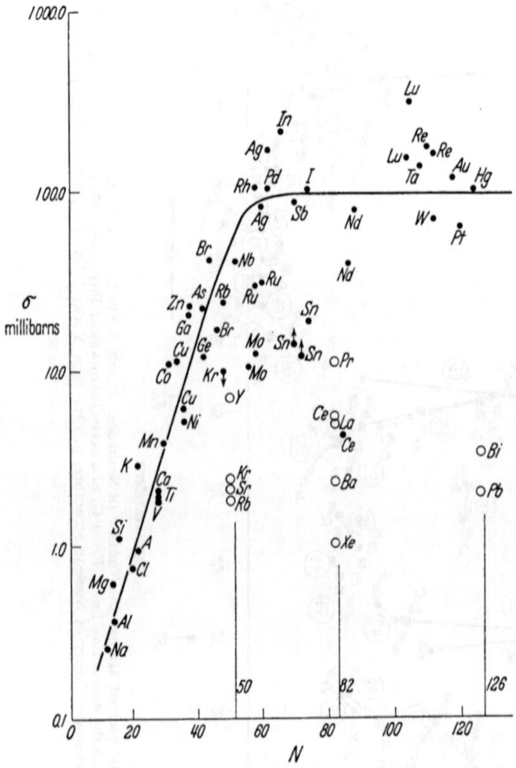

Fig. 4. Neutron Capture Cross Sections at Effective Energy 1 MeV as a Function of the Number of Neutrons in the Target Nuclide. Magic values of N are denoted by open circles. [From Hughes, Garth and Levin: Phys. Rev. **91**, 1423 (1953) as presented by E. Feenberg: Shell Theory of the Nucleus, p. 7. Princeton: Princeton University Press 1955.]

ε) Small level densities and neutron capture cross-sections. Level densities at the neutron binding energies are obtainable by observation of neutron resonances. Such densities are dependent on the amount of neutron binding. Nuclei in the neighbourhood of magic numbers are found to have anomalously low level densities (after allowance is made for neutron binding energy according to a standard law of level density, *vs.* excitation energy). For instance, the level spacing in the case of $Pb^{206}+n$, which gives Pb^{207} at an excitation of ~ 7.0 Mev, is several tens of kilovolts which indicates a level density at least 10^3 below normal. On the other side of the double closed shell Pb^{208}, the level spacing of Pb^{209} at 4.2 MeV from $Pb^{208}+n$ is of the order of hundreds of kilovolts which again indicates a level density several orders of magnitude below normal.

These anomalously low level densities are also revealed by the very low capture cross-sections at thermal energies and at ~ 1 MeV which are expected to vary as some positive power of the level density. The thermal cross-sections can be occasionally influenced by the fortuitously close presence of a particular capture resonance, but the 1 MeV cross-sections[1] do not suffer this drawback

[1] Hughes, Garth and Levin: Phys. Rev. **91**, 1423 (1953). — H. Rose: Harwell Report AERE R/R 1911, 1956.

and exhibit very striking closed shell effects as shown in Fig. 4. It is interesting that the low cross-sections do not only occur *at* shell-closures, but also, to some extent, above and below the closures. For instance, the cross-sections of Au, Hg and especially Tl are low, although these elements occur before the closed shell at Pb^{208}. The implied low level density of these nuclei is parallelled by their observed excess total binding. One might be able to relate these facts quantitatively by using the suggestion that level densities ought to be measured from a standard energy, such as is given by the semi-empirical formula[1], and not from nuclear ground states. It is evident that the special (shell) nature of ground states make them unsuitable as reference states for statistical properties of nuclei.

6. Ground state spins of stable nuclei of odd mass-number. The most immediate and simple prediction of any shell-model is the occurrence of those magic numbers effects at the shell-closures that we have discussed in the preceding section. Perhaps the next most simple prediction[2] is that of the spins J of individual odd-mass nuclei. The most primitive version of the shell-model, i.e. the Single Particle Model of Chap. II, regards all even numbers of nucleons to be paired off to form an inert core, so that most properties of an odd-mass nucleus arise from the last odd nucleon. This model immediately predicts the ground state spin of an odd nucleus to be that of the last nucleon orbital, but it is intrinsically incapable of predicting spins of even-mass nuclei. In the next chapter, we will develop a more self-consistent and sophisticated version of the shell-model in which a general nucleus is pictured as a number of "loose" neutrons and protons in unfilled shells outside closed shells. We will see that the number of linearly independent states that can be constructed from these loose particles is often considerable, even after taking account of anti-symmetrisation requirements. These states have many different spins, so that the predicted spin is indeterminate unless one assumes the presence of some inter-particle forces in order to remove the energy degeneracy of the states. For a fairly general family of such forces that are weak enough to preserve the predominance of the spin-orbit coupling, and are of rather short range relative to the size of the nucleus, it will be shown in the next chapter that the spin of the lowest state of a group of like particles (neutrons or protons) in the j-shell is $J=0$ if the number of particles is even, and $J=j$, if it is odd. Of course, the prediction of $J=0$ for an even group of particles is a feature of almost all nuclear models, and this prediction is checked by the observation that even-even nuclei always have zero spin. Thus even-even nuclei are not useful for our present purpose of checking the shell-model, so we will restrict the discussion to odd-mass nuclei (odd N, even Z or odd Z, even N). For such nuclei then, the model now predicts that the ground state spin should be equal to the spin of the last odd nucleon. This prediction is just the same as that made by the primitive model.

If this last rule is applied, using the single particle spectrum of Fig. 2 for the order of filling of levels, it will be found that the rule has many successes but also a few failures of a rather systematic type[3]. These failures mostly arise from the observation of certain low spins in regions just above the predicted shell-closures of such spins, where high spins are expected. Thus it seems that there

---

[1] H. A. Bethe and H. Hurwitz: Phys. Rev. **81**, 898 (1951). — T. D. Newton: Canad. J. Phys. **34**, 804 (1956).

[2] M. G. Mayer: Phys. Rev. **75**, 1969 (1949); **78**, 16 (1950).

[3] For an exhaustive analysis of ground state spins and the ordering of shell filling, see N. Zeldes: Nuclear Phys. **2**, 1 (1956).

Table 6. *Spins, Moments and Shell-Model*

Odd neutron nuclei											
N	Atom	A	J	μ	l_j	Q	Expected configuration				
								$1s$ $\frac{1}{2}$ (2)	$1p$ $\frac{3}{2}$ (4)	$1p$ $\frac{1}{2}$ (2)	
1	n	1	$\frac{1}{2}$	-1.913	$s_{\frac{1}{2}}$			1			
	He	3	$\frac{1}{2}$	-2.127	$s_{\frac{1}{2}}$			1			
3								2	1		
5	Be	9	$\frac{3}{2}$	-1.177	$p_{\frac{3}{2}}$	$+0.02$		2	3		
7	C	13	$\frac{1}{2}$	$+0.702$	$p_{\frac{1}{2}}$			2	4	1	
8								2	4	2	
							8	$1d$ $\frac{5}{2}$ (6)	$2s$ $\frac{1}{2}$ (2)	$1d$ $\frac{3}{2}$ (4)	
9	O	17	$\frac{5}{2}$	-1.894	$d_{\frac{5}{2}}$	-0.005	8	1			
11	Ne	21	$\frac{3}{2}$	<0	$(d_{\frac{5}{2}})^{3}_{\frac{3}{2}}$?		8	3			
13	Mg	25	$\frac{5}{2}$	-0.855	$d_{\frac{5}{2}}$		8	5			
15	Si	29	$\frac{1}{2}$	-0.555	$s_{\frac{1}{2}}$		8	6	1		
17	S	33	$\frac{3}{2}$	$+0.643$	$d_{\frac{3}{2}}$	-0.055	8	6	2	1	
19	S	35	$\frac{3}{2}$	$+1.0$	$d_{\frac{3}{2}}$	$+0.038$	8	6	2	3	
20							8	6	2	4	
							20	$1f$ $\frac{7}{2}$ (8)			
21							20	1			
23	Ca	43	$\frac{7}{2}$	-1.315	$f_{\frac{7}{2}}$		20	3			
25	Ti	47	$\frac{5}{2}$	-0.79	$(f_{\frac{7}{2}})^{5}_{\frac{5}{2}}$		20	5			
27	Ti	49	$\frac{7}{2}$	-1.104	$f_{\frac{7}{2}}$		20	7			
28							20	8			
							28	$2p$ $\frac{3}{2}$ (4)	$1f$ $\frac{5}{2}$ (6)	$2p$ $\frac{1}{2}$ (2)	$1g$ $\frac{9}{2}$ (10)
29	Cr	53	$\frac{3}{2}$	-0.475	$p_{\frac{3}{2}}$		28	1			
31	Fe	57	$\frac{3}{2}$	$\lvert\leq 0.05\rvert$	$p_{\frac{3}{2}}$		28	3			
33	Ni	61		~ 0	$p_{\frac{3}{2}}$		28	3	2		
35							28	3	4		
37	Zn	67	$\frac{5}{2}$	$+0.876$	$f_{\frac{5}{2}}$		28	4	5		

Configurations of the Stable Odd Nuclei.

Odd Proton nuclei											
Z	Atom	A	J	μ	l_j	Q	Expected configuration				
								$1s$	$1p$	$1p$	
								$\frac{1}{2}$	$\frac{3}{2}$	$\frac{1}{2}$	
								(2)	(4)	(2)	
1	H	1	$\frac{1}{2}$	$+2.793$	$s_{\frac{1}{2}}$			1			
	H	3	$\frac{1}{2}$	$+2.979$	$s_{\frac{1}{2}}$			1			
3	Li	7	$\frac{3}{2}$	$+3.256$	$p_{\frac{3}{2}}$	-0.02 ± 0.02		2	1		
5	B	11	$\frac{3}{2}$	$+2.689$	$p_{\frac{3}{2}}$	$+0.05$		2	3		
7	N	15	$\frac{1}{2}$	-0.283	$p_{\frac{1}{2}}$			2	4	1	
8								2	4	2	
								$1d$	$2s$	$1d$	
							8	$\frac{5}{2}$	$\frac{1}{2}$	$\frac{3}{2}$	
								(6)	(2)	(4)	
9	F	17	$\frac{5}{2}$		$d_{\frac{5}{2}}$		8	1			
	F	19	$\frac{1}{2}$	$+2.629$	$s_{\frac{1}{2}}$		8		1		
11	Na	23	$\frac{3}{2}$	$+2.218$	$(d_{\frac{5}{2}})^3_{\frac{3}{2}}$	$+0.1$	8	3			
13	Al	27	$\frac{5}{2}$	$+3.641$	$d_{\frac{5}{2}}$	$+0.149$	8	5			
15	P	31	$\frac{1}{2}$	$+1.132$	$s_{\frac{1}{2}}$		8	6	1		
17	Cl	35	$\frac{3}{2}$	$+0.822$	$d_{\frac{3}{2}}$	-0.078	8	6	2	1	
	Cl	37	$\frac{3}{2}$	$+0.684$	$d_{\frac{3}{2}}$	-0.061	8	6	2	1	
19	K	39	$\frac{3}{2}$	$+0.391$	$d_{\frac{3}{2}}$		8	6	2	3	
	K	41	$\frac{3}{2}$	$+0.215$	$d_{\frac{3}{2}}$		8	6	2	3	
20							8	6	2	4	
								$1f$			
							20	$\frac{7}{2}$			
								(8)			
21	Sc	45	$\frac{7}{2}$	$+4.756$	$f_{\frac{7}{2}}$		20	1			
23	V	51	$\frac{7}{2}$	$+5.147$	$f_{\frac{7}{2}}$	$+0.3$	20	3			
25	Mn	53	$\frac{7}{2}$	5.050	$f_{\frac{7}{2}}$		20	5			
25	Mn	55	$\frac{5}{2}$	$+3.468$	$(f_{\frac{7}{2}})^3_{\frac{5}{2}}$	$+0.5$	20	5			
27	Co	57	$\frac{7}{2}$	$+4.65$	$f_{\frac{7}{2}}$		20	7			
	Co	59	$\frac{7}{2}$	$+4.649$	$f_{\frac{7}{2}}$	$+0.5$	20	7			
28							20	8			
								$2p$	$1f$	$2p$	$1g$
							28	$\frac{3}{2}$	$\frac{5}{2}$	$\frac{1}{2}$	$\frac{9}{2}$
								(4)	(6)	(2)	(10)
29	Cu	63	$\frac{3}{2}$	$+2.227$	$p_{\frac{3}{2}}$	-0.16	28	1			
	Cu	65	$\frac{3}{2}$	$+2.385$	$p_{\frac{3}{2}}$	-0.15	28	1			
31	Ga	69	$\frac{3}{2}$	$+2.016$	$p_{\frac{3}{2}}$	$+0.23$	28	3			
	Ga	71	$\frac{3}{2}$	$+2.561$	$p_{\frac{3}{2}}$	$+0.15$	28	3			
33	As	75	$\frac{3}{2}$	$+1.439$	$p_{\frac{3}{2}}$	$+0.3$	28	3	2		
35	Br	79	$\frac{3}{2}$	$+2.106$	$p_{\frac{3}{2}}$	$+0.335$	28	3	4		
	Br	81	$\frac{3}{2}$	$+2.270$	$p_{\frac{3}{2}}$	$+0.280$	28	3	4		
37	Rb	81	$\frac{3}{2}$	$+1.0$	$p_{\frac{3}{2}}$		28	3	6		
	Rb*	81	$\frac{9}{2}$		$g_{\frac{9}{2}}$		28	4	4		1

Table 6.

Odd neutron nuclei

N	Atom	A	J	μ	l_j	Q	Expected configuration					
39	Zn	69	$\frac{1}{2}$		$p_{\frac{1}{2}}$		28	4	6	1		
41	Ge	73	$\frac{9}{2}$		$g_{\frac{9}{2}}$	-0.2	28	4	6	2	1	
43	Se	77	$\frac{1}{2}$	$+0.534$	$p_{\frac{1}{2}}$		28	4	6	1	4	
45	Se	79	$\frac{7}{2}$	-1.0	$(g_{\frac{9}{2}})_{\frac{7}{2}}$	$(+0.7)$	28	4	6		7	
47	Kr	83	$\frac{9}{2}$	-0.969	$g_{\frac{9}{2}}$	$+0.15$	28	4	6	2	7	
49	Sr	87	$\frac{9}{2}$	-1.1	$g_{\frac{9}{2}}$		28	4	6	2	9	
50							28	4	6	2	10	
								$2d$	$1g$	$3s$	$1h$	$2d$
							50	$\frac{5}{2}$	$\frac{7}{2}$	$\frac{1}{2}$	$\frac{11}{2}$	$\frac{3}{2}$
								(6)	(8)	(2)	(12)	(4)
51	Zr	91	$\frac{5}{2}$	-1.3	$d_{\frac{5}{2}}$		50	1				
53	Mo	95	$\frac{5}{2}$	-0.933	$d_{\frac{5}{2}}$		50	3				
55	Mo	97	$\frac{5}{2}$	-0.952	$d_{\frac{5}{2}}$		50	5				
	Ru	99	$\frac{5}{2}$				50	5				
57	Ru	101	$\frac{5}{2}$		$d_{\frac{5}{2}}$		50	5	2			
59	Pd	105	$\frac{5}{2}$	(-0.6)	$d_{\frac{5}{2}}$		50	5	4			
61							50	5	6			
63	Cd	111	$\frac{1}{2}$	-0.595	$s_{\frac{1}{2}}$		50	6	6	1		
	Cd*	111	$\frac{5}{2}$	-0.78	$d_{\frac{5}{2}}$		50	5	8			
65	Cd	113	$\frac{1}{2}$	-0.622	$s_{\frac{1}{2}}$		50	6	8	1		
	Sn	115	$\frac{1}{2}$	-0.918	$s_{\frac{1}{2}}$		50	6	8	1		
67	Sn	117	$\frac{1}{2}$	-1.000	$s_{\frac{1}{2}}$		50	6	8	1	2	
69	Sn	119	$\frac{1}{2}$	-1.046	$s_{\frac{1}{2}}$		50	6	8	1	4	
71	Te	123	$\frac{1}{2}$	-0.736	$s_{\frac{1}{2}}$		50	6	8	1	6	
73	Te	125	$\frac{1}{2}$	-0.887	$s_{\frac{1}{2}}$		50	6	8	1	8	
75	Xe	129	$\frac{1}{2}$	-0.777	$s_{\frac{1}{2}}$		50	6	8	1	10	
77	Xe	131	$\frac{3}{2}$	$+0.708$	$d_{\frac{3}{2}}$	-0.12	50	6	8	2	10	1
79	Ba	135	$\frac{3}{2}$	$+0.836$	$d_{\frac{3}{2}}$		50	6	8	2	12	1
81	Ba	137	$\frac{3}{2}$	$+0.932$	$d_{\frac{3}{2}}$		50	6	8	2	12	3
82							50	6	8	2	12	4

(Continued.)

Odd proton nuclei												
Z	Atom	A	J	μ	l_j	Q	Expected configuration					
	Rb	83	$\frac{5}{2}$		$f_{\frac{5}{2}}$		28	4	5			
	Rb	85	$\frac{5}{2}$	$+1.353$	$f_{\frac{5}{2}}$	$+0.31$	28	4	5			
	Rb	87	$\frac{3}{2}$	$+2.751$	$p_{\frac{3}{2}}$	$+0.14$	28	3	6			
39	Y	89	$\frac{1}{2}$	-0.137	$p_{\frac{1}{2}}$		28	4	6	1		
41	Nb	93	$\frac{9}{2}$	$+6.167$	$g_{\frac{9}{2}}$		28	4	6	2	1	
43	Tc	99	$\frac{9}{2}$	$+5.680$	$g_{\frac{9}{2}}$		28	4	6	2	3	
45	Rh	103	$\frac{1}{2}$	-0.11	$p_{\frac{1}{2}}$		28	4	6	1	6	
47	Ag	107	$\frac{1}{2}$	-0.113	$p_{\frac{1}{2}}$		28	4	6	1	8	
	Ag	109	$\frac{1}{2}$	-0.129	$p_{\frac{1}{2}}$		28	4	6	1	8	
	Ag	111	$\frac{1}{2}$	-0.1	$p_{\frac{1}{2}}$		28	4	6	1	8	
49	In	113	$\frac{9}{2}$	$+5.523$	$g_{\frac{9}{2}}$	$+1.14$	28	4	6	2	9	
	In	115	$\frac{9}{2}$	$+5.534$	$g_{\frac{9}{2}}$	$+1.16$	28	4	6	2	9	
50							28	4	6	2	10	
								$1g$	$2d$	$1h$	$2d$	$3s$
							50	$\frac{7}{2}$	$\frac{5}{2}$	$\frac{11}{2}$	$\frac{3}{2}$	$\frac{1}{2}$
								(8)	(6)	(12)	(4)	(2)
51	Sb	121	$\frac{5}{2}$	$+3.359$	$d_{\frac{5}{2}}$	-0.5	50		1			
	Sb	123	$\frac{7}{2}$	$+2.547$	$g_{\frac{7}{2}}$	-0.7	50	1				
53	I	127	$\frac{5}{2}$	$+2.809$	$d_{\frac{5}{2}}$	-0.75	50	2	1			
	I	129	$\frac{7}{2}$	$+2.617$	$g_{\frac{7}{2}}$	-0.53	50	3				
	I	131	$\frac{7}{2}$		$g_{\frac{7}{2}}$	-0.4		3				
55	Cs	131	$\frac{5}{2}$	$+3.48$	$d_{\frac{5}{2}}$		50	4	1			
	Cs	133	$\frac{7}{2}$	$+2.579$	$g_{\frac{7}{2}}$		50	5				
	Cs	135	$\frac{7}{2}$	$+2.727$	$g_{\frac{7}{2}}$	-0.003	50	5				
	Cs	137	$\frac{7}{2}$	$+2.850$	$g_{\frac{7}{2}}$		50	5				
57	La	139	$\frac{7}{2}$	$+2.778$	$g_{\frac{7}{2}}$	$\sim+0.9$	50	7				
59	Pr	141	$\frac{5}{2}$	$+3.9$	$d_{\frac{5}{2}}$	-0.05	50	8	1			
61	Pm						50	8	3			
63	Eu	151	$\frac{5}{2}$	$+3.4$	$d_{\frac{5}{2}}$	$+1.2$	50	8	5			
	Eu	153	$\frac{5}{2}$	$+1.5$	$d_{\frac{5}{2}}$	$+2.5$	50	8	5			
65	Tb	159	$\frac{3}{2}$	$\sim+1.5$			50	Region of Large Spheroidal Deformation				
67	Ho	165	$\frac{7}{2}$				50					
69	Tm	169	$\frac{1}{2}$	-0.2			50					
71	Lu	175	$\frac{7}{2}$	$+2.9$		$+5.7$	50					
73	Ta	181	$\frac{7}{2}$	$+2.1$		$+4.3$	50					
75	Re	185	$\frac{5}{2}$	$+3.172$	$d_{\frac{5}{2}}$	$+2.8$	50	8	5	12		
	Re	187	$\frac{5}{2}$	$+3.176$	$d_{\frac{5}{2}}$	$+2.6$	50	8	5	12		
77	Ir	191	$\frac{3}{2}$	$+0.16$	$d_{\frac{3}{2}}$	$+1.5$	50	8	6	10	3	
	Ir	193	$\frac{3}{2}$	$+0.17$	$d_{\frac{3}{2}}$	$+1.5$	50	8	6	10	3	
79	Au	197	$\frac{3}{2}$	$+0.16$	$d_{\frac{3}{2}}$	$+0.6$	50	8	6	12	3	
81	Tl	203	$\frac{1}{2}$	$+1.612$	$s_{\frac{1}{2}}$		50	8	6	12	4	1
	Tl	205	$\frac{1}{2}$	$+1.627$	$s_{\frac{1}{2}}$		50	8	6	12	4	1
82							50	8	6	12	4	2

Table 6.

Odd neutron nuclei

N	Atom	A	J	μ	l_j	Q		Expected Configuration						
							82	$2f\ \frac{7}{2}$ (8)	$1h\ \frac{9}{2}$ (10)	$3p\ \frac{3}{2}$ (4)	$2f\ \frac{5}{2}$ (6)	$3p\ \frac{1}{2}$ (2)	$1i\ \frac{13}{2}$ (14)	
83	Nd	143	$\frac{7}{2}$	−1.0	$f_{\frac{7}{2}}$	$\lvert<1\rvert$	82	1						
85	Nd	145	$\frac{7}{2}$	−0.62	$f_{\frac{7}{2}}$	$\lvert<1\rvert$	82	3						
	Sm	147	$\frac{7}{2}$	−0.68	$f_{\frac{7}{2}}$	$\lvert\leq 0.72\rvert$	82	3						
87	Sm	149	$\frac{7}{2}$	−0.55	$f_{\frac{7}{2}}$	$\lvert\leq 0.72\rvert$	82	5						
89							82	7						
91	Gd	155	$\geq\frac{3}{2}$	−0.31			82	Region of Large Spheroidal Deformation (N = 91–107)						
93	Gd	157	$\geq\frac{3}{2}$	−0.38			82							
95	Dy	161	$\frac{7}{2}$	+0.38			82							
97	Dy	163	$\frac{7}{2}$	+0.53			82							
99	Er	167	$\frac{7}{2}$	−0.5		+10.2	82							
101	Yb	171	$\frac{1}{2}$	+0.45			82							
103	Yb	173	$\frac{5}{2}$	−0.66		+3.9	82							
105	Hf	177	$\frac{7}{2}$	+0.61		∼3.0	82							
107	Hf	179	$\frac{9}{2}$	−0.47		∼3.0	82							
109	W	183	$\frac{1}{2}$	+0.087	$p_{\frac{1}{2}}$		82	8	10	4	4	1		
111	Os	187					82	8	10	4	6	1		
113	Os	189	$\frac{1}{2}(\frac{3}{2})$	+0.71	$p_{\frac{1}{2}}$		82	8	10	4	6	1	2	
115							82	8	10	4	6	1	4	
117	Pt	195	$\frac{1}{2}$	+0.606	$p_{\frac{1}{2}}$		82	8	10	4	6	1	6	
	Hg	197	$\frac{1}{2}$	+0.5	$p_{\frac{1}{2}}$		82	8	10	4	6	1	6	
119	Hg	199	$\frac{1}{2}$	+0.504	$p_{\frac{1}{2}}$		82	8	10	4	6	1	8	
121	Hg	201	$\frac{3}{2}$	−0.613	$p_{\frac{3}{2}}$	+0.65	82	8	10	3	6	0	12	
123							82	8	10	4	6	1	12	
125	Pb	207	$\frac{1}{2}$	+0.589	$p_{\frac{1}{2}}$		82	8	10	4	6	1	14	
126							82	8	10	4	6	2	14	
							126	$2g\ \frac{9}{2}$ (10)	$3d\ \frac{5}{2}$ (6)	$1i\ \frac{11}{2}$ (12)	$2g\ \frac{7}{2}$ (8)	$4s\ \frac{1}{2}$ (2)	$3d\ \frac{3}{2}$ (4)	$1j\ \frac{15}{2}$ (16)
141	U	233	$\frac{5}{2}$				126	Region of Large Spheroidal Deformation (N = 141–147)						
143	U	235	$\frac{7}{2}$	−0.8			126							
145	Pu	239	$\frac{1}{2}$	±0.5			126							
147	Pu	241	$\frac{5}{2}$	±1.4			126							

is an occasional inversion in the order of shell filling that occurs at the expense of the high spins. One can formulate the empirical rule that when a high spin shell comes after a low spin shell in the single particle sequence, the high spin shell has a tendency to begin to fill, two particles at a time, before the low spin shell fills completely. From a theoretical point of view (see Sect. 11), it can be shown that the "pairing energy" between two high spin particles arising from

(Continued.)

Odd proton nuclei													
Z	Atom	A	J	μ	l_j	Q		Expected configuration					
								$1h$	$2f$	$3p$	$2f$	$3p$	$1i$
							82	$\frac{9}{2}$	$\frac{7}{2}$	$\frac{3}{2}$	$\frac{5}{2}$	$\frac{1}{2}$	$\frac{13}{2}$
								(10)	(8)	(4)	(6)	(2)	(14)
83	Bi	209	$\frac{9}{2}$	$+4.080$	$h_{\frac{9}{2}}$	-0.4	82	1					
85	At												
87	Fr												
89	Ac	227	$\frac{3}{2}$	$+0.6$		-1	82	Region of Large Spheroidal Deformation					
91	Pa	231	$\frac{3}{2}$				82						
93	Np	237	$\frac{5}{2}$	~ 6			82						
	Np	239	$\frac{1}{2}$	~ 0.7									
95	Am	241	$\frac{5}{2}$	$+1.4$		$+4.9$	82						
	Am	243	$\frac{5}{2}$	$+1.4$		$+4.9$							
97	Bk												
99													

their interaction is considerably larger than for two particles of low spin. This fact gives the empirical rule a possible theoretical basis[1].

In Table 6, the odd-mass nuclei of known spin (mostly stable) are listed along with their spins and the shell-model configurations predicted by the single particle model, revised in accordance with the empirical rule just mentioned where necessary. The places where the rule is involved are:

(i) $28 < (N \text{ or } Z) < 38$. Normally the $p_{\frac{3}{2}}$ shell would fill at 32 and be followed by the $f_{\frac{5}{2}}$ shell. The fact that almost all spins in the region are $\frac{3}{2}$ implies that the $f_{\frac{5}{2}}$ shell fills in pairs.

(ii) $38 < (N \text{ or } Z) < 50$. Normally the $p_{\frac{1}{2}}$ shell would fill at 40 and the $g_{\frac{9}{2}}$ shell at 50. The occurrence of several spins of $\frac{1}{2}$ implies that the $g_{\frac{9}{2}}$ shell tends to fill sin pair.

(iii) $50 < Z < 82$. The spins of these odd proton nuclei exhibit serious anomalies which are only partly removed by assuming that the normal sequence of $g_{\frac{7}{2}}$ (50 to 58), $d_{\frac{5}{2}}$ (58 to 64) and $h_{\frac{11}{2}}$ (64 to 76) is disturbed by filling in pairs which leads to the complete absence of spin $\frac{11}{2}$. The remainder of this region seems to fill normally with the $d_{\frac{3}{2}}$ (76 to 80) and $s_{\frac{1}{2}}$ (80 to 82) shells.

(iv) $50 < N < 82$. Again the spins indicate that the normal sequence $g_{\frac{7}{2}}$, $d_{\frac{5}{2}}$, $h_{\frac{11}{2}}$ is disturbed by pair filling. It is noteworthy that the effect is considerably stronger for the odd neutron nuclei than the odd proton ones of (iii). For instance, from 63 to 75, all of eight odd neutron nuclei have spin $\frac{1}{2}$. The seven odd proton nuclei have spins up to $\frac{7}{2}$ and only one has spin $\frac{1}{2}$. One might expect that the single particle level order is different for protons and neutrons because, for the former, the Coulomb potential will lower the higher orbits relatively.

(v) $82 < N < 126$. Normally the $f_{\frac{7}{2}}$ shell would fill at 90 followed by the $h_{\frac{9}{2}}$ at 100. As with region (iii), there are anomalies which are only partly removed by allowing filling in pairs. Similar remarks apply to the next levels $p_{\frac{3}{2}}$, $f_{\frac{5}{2}}$. Above 110, there are the final levels, $p_{\frac{1}{2}}$ and $i_{\frac{13}{2}}$, that fill the 126 major shell. As expected, the $i_{\frac{13}{2}}$ shell fills in pairs, so that no spin of $\frac{13}{2}$ is observed.

[1] M. G. Mayer: Phys. Rev. **78**, 16 (1950).

Summarising these remarks and the entries of Table 6, we see that the shell-model gives a very good account of the ground state spins when supplemented by the empirical rule about the high orbits filling in pairs. We note that, in nuclei whose ground state is changed by this effect, one expects the normal ground state to appear as a low-lying excited state. Usually it will be an isomeric state because of the big spin change caused by the effect, and we will see in the later discussion that many isomeric states can be interpreted in this way.

Finally we note certain nuclei whose spins cannot be plausibly explained on our present model, either with or without the pairing rule. Amongst the lighter nuclei, there are Ne^{21}, Na^{23} ($d_{\frac{5}{2}}$ shell; $J=\frac{3}{2}$), Ti^{47} ($f_{\frac{7}{2}}$ shell; $J=\frac{5}{2}$), Mn^{55} ($f_{\frac{7}{2}}$ shell; $J=\frac{5}{2}$), Se^{79} ($g_{\frac{9}{2}}$ shell; $J=\frac{7}{2}$). All these anomalies are characterised by the observed spin being $J=j-1$ rather than j. Some heavy nuclei like Tb^{159} ($J=\frac{3}{2}$), Tm^{169} ($J=\frac{1}{2}$) and Yb^{171} ($J=\frac{1}{2}$) are anomalous, although these spins can be accounted for by assuming capricious preliminary filling of the next higher shells.

Amongst the heavier nuclei, it is very significant that all the anomalous spins are concentrated in the region from $A=158$ to 180. All five odd-proton nuclei in this region in Table 6 are anomalous, and seem to imply capricious filling of higher orbits or emptying of filled orbits. The same remarks apply to all four odd-neutron nuclei from $A=170$ to 180 in Table 6. It so happens that the region around mass 170 is just where the shell-model fails to give the observed large electric quadrupole moments (see Sect. 8). The believed explanation for these failures is a breakdown in the assumption of a spherical nucleus. There is now considerable evidence that the nuclei near mass 170 have strong quadrupole deformations. The single particle spectra for potential wells of various eccentricities have recently been evaluated[1] and they show considerable departures from Fig. 2. Thus we expect to find anomalous spins in regions of appreciable deformation.

7. Ground state spins of unstable nuclei of odd mass number and β-decays[2]. We have seen that the shell-model gives a good account of the ground state spins of stable odd nuclei. Experimentally, these spins are measured directly. Now we turn to the spins of ground states of unstable odd nuclei (and also the spins of certain excited states). Although the spins of such nuclei cannot be directly measured, one can infer important evidence from the study of their β-decay which usually leads to stable daughter nuclei of known spin. From the nature of the β-decay (the degree and type of forbiddeness), one can find the difference in spins and parities between the two nuclei. The Single Particle Model can then be used to interpret these differences in terms of the spins of the last odd neutron in one nucleus and the last odd proton in the other. As we shall see, the model has considerable success in this connection.

Occasionally, the actual spin of the unstable nucleus has been found from angular correlation work or other sources. In such a case, of course, one interprets the spin directly in terms of the Single Particle Model as for stable nuclei. In general, however one only knows the spectrum and lifetime of the β-transition. In such cases, one cannot usually find the ground state spin of the parent nucleus, but one can find the spin and parity differences between the parent and daughter nuclei by appealing to the systematics of β-decay or possibly, if available, to the shape of the β spectrum. We will now outline the essentials of β-decay theory and its systematics that are necessary for following the subsequent discussion.

[1] S. G. Nilsson: Dan. Mat. Fys. Medd. **29**, No. 16 (1955). — S. A. Moszkowski: Phys. Rev. **99**, 803 (1955). — K. Gottfried: Phys. Rev. **103**, 1017 (1956).

[2] For this section cf. also the articles on β-decay in Vol. XLI of this Encyclopedia.

From β-decay theory, the half-life against β-decay, $t_{\frac{1}{2}}$, can be expressed by:

$$\frac{\log_e 2}{t_{\frac{1}{2}}} = \frac{f(W_0)}{2J_i+1} \sum_{M_f M_i} |(J_f M_f|\, g m\, |J_i M_i)|^2 .$$

In the matrix elements $(J_f M_f|\, g m\, |J_i M_i)$, the initial and final states are specified by their spins and z components $|J_i M_i)$ and $(J_f M_f|$. g is the β coupling constant and m is an operator whose form depends on the nature of transition and the type of β interaction. (For the allowed type transitions, and for each nuclear particle, $m = t$ for the FERMI interaction, $2\mathbf{s}t$ for the GAMOW-TELLER interaction, where t transforms neutron into proton or vice-versa, and $\mathbf{s}$ is the intrinsic nucleon spin.)

In this expression, $f(W_0)$ contains all the phase-space factors and depends on W_0, the maximum β energy. The quantity $f(W_0)$ is tabulated and so, given $t_{\frac{1}{2}}$ for a transition, one can calculate the so-called ft-value which is a convenient measure of the nuclear matrix element with the irrelevant phase-space factors removed. It is also desirable to remove the dependence of the half-life on spin. This can be done by working with f' instead of f where[1]:

$$f' = f \qquad \text{for} \quad J_i \geqq J_f,$$

$$f' = \left(\frac{2J_f+1}{2J_i+1}\right) f \quad \text{for} \quad J_i \leqq J_f .$$

From a theoretical point of view, the value of this matrix element is very sensitive to the "nature of the transition" by which is meant the change of angular momentum ($\Delta J = |J_i - J_f|$) and of parity in the transition. The various transitions have names assigned to them and these are given in Table 7 along with the

Table 7. *Nomenclature and Representative $f't$-Values for β-Transitions of various changes in spin ($\Delta J = |J_i - J_f|$) and parity (+ or −).* (From E. FEENBERG: Shell Theory of the Nucleus, p. 82. Princeton: Princeton University Press 1955.)

Nature of transition		Name	$\log f't$	$\log f_n't$
$\Delta J = 0, 1$;	+	Allowed, favoured	3−3.6	
		Allowed, unfavoured	4.3−6	
$\Delta J = 0, 1$;	−	1st Forbidden, ordinary	6−8	
$\Delta J = 2$;	−	1st Forbidden, Unique Shape		8−9
$\Delta J = 2$;	+	2nd Forbidden, ordinary	10.3−13.3	
$\Delta J = 3$;	+	2nd Forbidden, Unique Shape		12
$\Delta J = 3$;	−	3rd Forbidden, ordinary	18	
$\Delta J = 4$;	−	3rd Forbidden, Unique Shape		15.6
$\Delta J = 4$;	+	4th Forbidden, ordinary	23	
$\Delta J = 5$;	+	4th Forbidden, Unique Shape		19
$\Delta J = n$,	$(-)^n$	n-th Forbidden, ordinary		
$\Delta J = n+1$,	$(-)^n$	n-th Forbidden, Unique Shape		

representative $\log f't$-values as known from a study of β-decay systematics. For the particular transitions with $\Delta J = n+1$, we actually list a quantity $\log f_n't$,

[1] E. FEENBERG: Shell Theory of the Nucleus, p. 81. Princeton: Princeton University Press 1955.

where f'_n is defined for these transitions as the product of f' and the factors:

$$\frac{1}{20}(W_0^2-1) \quad \text{for} \quad n=1,$$

$$\frac{1}{2520}(W_0^2-1)^2 \quad \text{for} \quad n=2,$$

$$\frac{1}{210\times 72^2}(W_0^2-1)^3 \quad \text{for} \quad n=3.$$

These factors can be extracted from the squares of the nuclear matrix elements for this special class of transitions. They represent an extra dependence on β-decay energy which is unambiguous for these transitions which are correspondingly called "Unique Shape" transitions. It is found that the spread in $\log f'_n t$-values is considerably smaller than the spread in $\log f' t$-values for these transitions, so that the systematics of $\log f'_n t$-values are more precise and meaningful than those of $\log f' t$-values. Unfortunately, it is not possible to extract a similar factor unambiguously for other types of transition and so, for these, one must make do with the usual $f' t$ values.

From the representative values of $\log f' t$ and $\log f'_n t$ given in Table 7, and with some analysis of spectrum shape, it is usually possible to make a fairly definite assignment to the nature of any given transition. In Tables 8a—h, we list the experimental β-transitions according to their nature. We also list the shell-model assignments to the ground states. It can be seen that these complement the nature of the β-transitions very well. However, in contrast to this success, the Single Particle Model does not provide a good account of the actual $f' t$-values of individual transitions. In the first place, except for the favoured

Table 8a. *ft-Values for Super-Allowed β-Transitions in Odd Nuclei.*
Bracketed entries are assumed, not experimental. (From E. FEENBERG, Shell Theory of the Nucleus, p. 117. Princeton: Princeton University Press 1955.) In cases where other transitions affect the half life, the proportion of the listed transition is given.

Transition	Spin	Half life	E (MeV)	ft
${}_0n_1 \to {}_1H_0$	$1/2$	12.5 m	0,783	1280
${}_1H_2 \to {}_2He_1$	$1/2$	12.5 y	0.0185	1120
${}_4Be_3 \to {}_3Li_4$	$3/2$	52.9 d/0.89	0.863	2300
$\to {}_3Li_4^*$	$3/2 \to 1/2$	52.9 d/0.11	0.385	3600
${}_6C_5 \to {}_5B_6$	$3/2$	20.4 m	0.958	4150
${}_7N_6 \to {}_6C_7$	$1/2$	10.1 m	1.202	4800
${}_8O_7 \to {}_7N_8$	$1/2$	118 s	1.683	3700
${}_9F_8 \to {}_8O_9$	$5/2$	66 s	1.72	2250
${}_{10}Ne_9 \to {}_9F_{10}$	$1/2$	18.5 s	2.18	1750
${}_{11}Na_{10} \to {}_{10}Ne_{11}$	$(3/2)$	22.8 s	2.50	3900
${}_{12}Mg_{11} \to {}_{11}Na_{12}$	$3/2$	12.3 s	2.99	4600
${}_{13}Al_{12} \to {}_{12}Mg_{13}$	$5/2$	7.3 s	(3.10)	3000
${}_{14}Si_{13} \to {}_{13}Al_{14}$	$5/2$	5.4 s	3.48	3700
${}_{15}P_{14} \to {}_{14}Si_{15}$	$(1/2)$	4.6 s	3.94	5500
${}_{16}S_{15} \to {}_{15}P_{16}$	$1/2$	3.2 s	4.06	4300
${}_{17}Cl_{16} \to {}_{16}S_{17}$	$3/2$	1.8 s	4.43	3500
${}_{18}A_{17} \to {}_{17}Cl_{18}$	$3/2$	1.84 s	4.4	3400
${}_{19}K_{18} \to {}_{18}A_{19}$	$3/2$	1.2 s	4.57	2700
${}_{20}Ca_{19} \to {}_{19}K_{20}$	$3/2$	1.1 s	5.13	4000
${}_{21}Sc_{20} \to {}_{20}Ca_{21}$	$(7/2)$	0.87 s	4.94	2550

Table 8b. *$f't$-Values for Probable $\Delta J = 1\,(+)$ Allowed β-Transitions in Odd Nuclei.*

Directly measured spins are underlined. The labels *g*, *m*, *e* and *u* mean ground, metastable (isomeric), excited and uncertain. (From E. FEENBERG: Shell Theory of the Nucleus, p. 84. Princeton: Princeton University Press 1955.)

Parent → Daughter	States	Energy (MeV)	Spins and parities	$\log f't$
${}_{8}O_{11} \to {}_{9}F_{10}$	$g \to e$	4.5	$5/2^+ \to \underline{3/2}^+$	4.35
${}_{10}Ne_{13} \to {}_{11}Na_{12}$	$g \to g$	4.21	$5/2^+ \to \underline{3/2}^+$	5.10
${}_{13}Al_{16} \to {}_{14}Si_{15}$	$g \to e$	2.5	$5/2^+ \to 3/2^+$	5.25
${}_{13}Al_{16} \to {}_{14}Si_{15}$	$g \to e$	1.4	$5/2^+ \to 3/2^+$	4.61
${}_{15}P_{14} \to {}_{14}Si_{15}$	$g \to e$	2.68	$1/2^+ \to 3/2^+$	5.0
${}_{15}P_{14} \to {}_{14}Si_{15}$	$g \to e$	1.51	$1/2^+ \to 3/2^+$	4.7
${}_{14}Si_{17} \to {}_{15}P_{16}$	$g \to g$	1.49	$3/2^+ \to \underline{1/2}^+$	5.60
${}_{15}P_{18} \to {}_{16}S_{17}$	$g \to g$	0.26	$1/2^+ \to \underline{3/2}^+$	5.38
${}_{21}Sc_{26} \to {}_{22}Ti_{25}$	$g \to g$	0.490	$7/2^- \to \underline{5/2}^-$	5.4
${}_{22}Ti_{29} \to {}_{23}V_{28}$	$g \to g$	2.24	$5/2^- \to \underline{7/2}^-$	5.70
${}_{24}Cr_{25} \to {}_{23}V_{26}$	$g \to g$	1.54	$5/2^- \to 7/2^-$	5.05
${}_{24}Cr_{31} \to {}_{25}Mn_{30}$	$g \to g$	2.85	$3/2^- \to \underline{5/2}^-$	5.35
${}_{27}Co_{34} \to {}_{28}Ni_{33}$	$g \to g$	1.42	$7/2^- \to 5/2^-$	5.66
${}_{28}Ni_{37} \to {}_{29}Cu_{36}$	$g \to g$	2.10	$5/2^- \to 3/2^-$	6.59
${}_{29}Cu_{32} \to {}_{28}Ni_{33}$	$g \to g$	1.20	$3/2^- \to 5/2^-$	5.14
${}_{29}Cu_{38} \to {}_{30}Zn_{37}$	$g \to g$	0.58	$3/2^- \to 5/2^-$	6.25
${}_{30}Zn_{35} \to {}_{29}Cu_{36}$	$g \to g$	0.32	$5/2^- \to 3/2^-$	7.38
${}_{30}Zn_{39} \to {}_{31}Ga_{38}$	$g \to g$	0.86	$1/2^- \to 3/2^-$	4.64
${}_{30}Zn_{41} \to {}_{31}Ga_{40}$	$g \to g$	2.1	$1/2^- \to 3/2^-$	$\geqq 4.81$
${}_{33}As_{38} \to {}_{32}Ge_{39}$	$g \to g$	0.82	$3/2^- \to 1/2^-$	5.70
${}_{31}Ga_{42} \to {}_{32}Ge_{41}$	$g \to (m)$	1.4	$3/2^- \to 1/2^-$	5.95
${}_{32}Ge_{43} \to {}_{33}As_{42}$	$g \to g$	1.3	$1/2^- \to 3/2^-$	5.54
${}_{35}Br_{40} \to {}_{34}Se_{41}$	$g \to g$	1.70	$3/2^- \to 1/2^-$	6.76
${}_{33}As_{44} \to {}_{34}Se_{43}$	$g \to g$	0.7	$3/2^- \to 1/2^-$	5.75
${}_{35}Br_{42} \to {}_{34}Se_{43}$	$g \to g$	0.336	$3/2^- \to 1/2^-$	$\geqq 5.36$
${}_{36}Kr_{41} \to {}_{35}Br_{42}$	$g \to g$	1.7	$1/2^- \to 3/2^-$	5.71
${}_{34}Se_{47} \to {}_{35}Br_{46}$	$g \to g$	1.38	$1/2^- \to 3/2^-$	5.01
${}_{37}Rb_{44} \to {}_{36}Kr_{45}$	$g \to m$	0.990	$3/2^- \to 1/2^-$	5.35
${}_{34}Se_{49} \to {}_{35}Br_{48}$	$m \to g$	3.4	$1/2^- \to 3/2^-$	$\geqq 5.50$
${}_{35}Br_{48} \to {}_{36}Kr_{47}$	$g \to m$	0.94	$3/2^- \to 1/2^-$	5.05
${}_{35}Br_{50} \to {}_{36}Kr_{49}$	$g \to m$	2.5	$3/2^- \to 1/2^-$	5.08
${}_{36}Kr_{49} \to {}_{37}Rb_{48}$	$m \to e$	0.817	$1/2^- \to 3/2^-$	5.46
${}_{40}Zr_{55} \to {}_{41}Nb_{54}$	$g \to e$	0.371	$5/2^+ \to 7/2^+$	6.62
${}_{41}Nb_{56} \to {}_{42}Mo_{55}$	$g \to e$	1.267	$9/2^+ \to 7/2^+$	5.34
${}_{42}Mo_{59} \to {}_{43}Tc_{58}$	$g \to e$	2.2	$5/2^+ \to 7/2^+$	6.2
${}_{43}Tc_{58} \to {}_{44}Ru_{57}$	$g \to e$	1.4	$9/2^+ \to 7/2^+$	4.9
${}_{45}Rh_{60} \to {}_{46}Pd_{59}$	$g \to g$	0.570	$7/2^+ \to 5/2^+$	5.67
${}_{50}Sn_{61} \to {}_{49}In_{62}$	$g \to g$	1.51	$7/2^+ \to 9/2^+$	4.83
${}_{50}Sn_{71} \to {}_{51}Sb_{70}$	$g \to g$	0.383	$3/2^+ \to 5/2^+$	5.22
${}_{50}Sn_{73} \to {}_{51}Sb_{72}$	$g \to e$	1.26	$3/2^+ \to 5/2^+$	5.42
${}_{53}I_{68} \to {}_{52}Te_{69}$	$g \to e$	1.2	$5/2^+ \to 3/2^+$	5.04
${}_{52}Te_{75} \to {}_{53}I_{74}$	$g \to g$	0.7	$3/2^+ \to \underline{5/2}^+$	5.67
${}_{54}Xe_{79} \to {}_{55}Cs_{78}$	$g \to e$	0.345	$3/2^+ \to 5/2^+$	5.80
${}_{54}Xe_{81} \to {}_{55}Cs_{80}$	$g \to e$	0.90	$3/2^+ \to 5/2^+$	6.11
${}_{60}Nd_{81} \to {}_{59}Pr_{82}$	$g \to g$	0.7	$3/2^+ \to 5/2^+$	5.20

Table 8c. *ft-Values of Probable $\Delta J = 0(+)$ Allowed β-Transitions in Odd Nuclei.* (From E. FEENBERG: Shell Theory of the Nucleus, p. 86. Princeton: Princeton University Press 1955.)

Parent → Daughter	States	Energy (MeV)	Spins and parities	log *ft*
${}_{16}S_{19} \rightarrow {}_{17}Cl_{18}$	$g \rightarrow g$	0.168	$3/2^+ \rightarrow 3/2^+$	5.01
${}_{18}A_{23} \rightarrow {}_{19}K_{22}$	$g \rightarrow e$	1.245	$7/2^- \rightarrow 7/2^-$	5.11
${}_{20}Ca_{27} \rightarrow {}_{21}Sc_{26}$	$g \rightarrow g$	2.060	$7/2^- \rightarrow 7/2^-$	8.5
${}_{21}Sc_{22} \rightarrow {}_{20}Ca_{23}$	$g \rightarrow g$	1.18	$7/2^- \rightarrow 7/2^-$	5.00
${}_{22}Ti_{23} \rightarrow {}_{21}Sc_{24}$	$g \rightarrow g$	1.022	$7/2^- \rightarrow 7/2^-$	4.59
${}_{21}Sc_{28} \rightarrow {}_{22}Ti_{27}$	$g \rightarrow g$	2.4	$7/2^- \rightarrow 7/2^-$	6.08
${}_{36}Kr_{43} \rightarrow {}_{35}Br_{44}$	$g \rightarrow e$	0.595	$1/2^- \rightarrow 1/2^-$	5.40
${}_{36}Kr_{49} \rightarrow {}_{37}Rb_{48}$	$g \rightarrow m$	0.15	$9/2^+ \rightarrow 9/2^+$	9.05
${}_{39}Y_{48} \rightarrow {}_{38}Sr_{49}$	$g \rightarrow m$	0.7	$1/2^- \rightarrow 1/2^-$	7.56
${}_{40}Zr_{49} \rightarrow {}_{39}Y_{50}$	$g \rightarrow m$	0.905	$9/2^+ \rightarrow 9/2^+$	6.14
${}_{40}Zr_{49} \rightarrow {}_{39}Y_{50}$	$m \rightarrow g$	2.5	$1/2^- \rightarrow 1/2^-$	6.82
${}_{42}Mo_{49} \rightarrow {}_{41}Nb_{50}$	$m \rightarrow g$	3.7	$1/2^- \rightarrow 1/2^-$	$\gtrsim 5.72$
${}_{42}Mo_{49} \rightarrow {}_{41}Nb_{50}$	$g \rightarrow u$	2.6	$9/2^+ \rightarrow 9/2^+$	$\gtrsim 4.00$
${}_{45}Rh_{60} \rightarrow {}_{46}Pd_{59}$	$g \rightarrow e$	0.25	$7/2^+ \rightarrow 7/2^+$	~ 5.79

Table 8d. *f't-Values of Probable $\Delta J = 1(-)$ First Forbidden β-Transitions in Odd Nuclei.* (From E. FEENBERG: Shell Theory of the Nucleus, p. 86. Princeton: Princeton University Press 1955.)

Parent → Daughter	States	Energy (MeV)	Spins and Parities	log *f't*
${}_{35}Br_{52} \rightarrow {}_{36}Kr_{51}$	$g \rightarrow g$	8.0	$3/2^- \rightarrow 5/2^+$	7.49
${}_{36}Kr_{51} \rightarrow {}_{37}Rb_{50}$	$g \rightarrow g$	3.63	$5/2^+ \rightarrow \underline{3/2}^-$	7.30
${}_{47}Ag_{64} \rightarrow {}_{48}Cd_{63}$	$g \rightarrow e$	0.70	$1/2^- \rightarrow 3/2^+$	8.03
${}_{48}Cd_{65} \rightarrow {}_{49}In_{64}$	$m \rightarrow g$	0.59	$(11/2^-) \rightarrow 9/2^+$	8.75
${}_{48}Cd_{67} \rightarrow {}_{49}In_{66}$	$m \rightarrow g$	1.61	$11/2^- \rightarrow \underline{9/2}^+$	8.75
${}_{49}In_{68} \rightarrow {}_{50}Sn_{67}$	$m \rightarrow e$	1.61	$1/2^- \rightarrow 3/2^+$	7.22
${}_{58}Ce_{83} \rightarrow {}_{59}Pr_{82}$	$g \rightarrow g$	0.581	$7/2^- \rightarrow \underline{5/2}^+$	7.72
${}_{58}Ce_{85} \rightarrow {}_{59}Pr_{84}$	$g \rightarrow g$	1.37	$7/2^- \rightarrow 5/2^+$	7.72
${}_{59}Pr_{84} \rightarrow {}_{60}Nd_{83}$	$g \rightarrow g$	0.932	$5/2^+ \rightarrow \underline{7/2}^-$	7.72
${}_{59}Pr_{86} \rightarrow {}_{60}Nd_{85}$	$g \rightarrow g$	3.2	$5/2^+ \rightarrow \underline{7/2}^-$	7.96
${}_{61}Pm_{86} \rightarrow {}_{62}Sm_{85}$	$g \rightarrow g$	0.229	$5/2^+ \rightarrow \underline{7/2}^-$	7.54
${}_{62}Sm_{91} \rightarrow {}_{63}Eu_{90}$	$g \rightarrow g$	0.81	$7/2^- \rightarrow 5/2^+$	7.25
${}_{71}Lu_{106} \rightarrow {}_{72}Hf_{105}$	$g \rightarrow g$	0.495	$5/2^+ \rightarrow 3/2^-$	6.78
${}_{71}Lu_{106} \rightarrow {}_{72}Hf_{105}$	$g \rightarrow e$	0.366	$5/2^+ \rightarrow 7/2^-$	7.08
${}_{74}W_{111} \rightarrow {}_{75}Re_{110}$	$g \rightarrow g$	0.428	$3/2^- \rightarrow \underline{5/2}^+$	7.62
${}_{79}Au_{120} \rightarrow {}_{80}Hg_{119}$	$g \rightarrow g$	0.460	$3/2^+ \rightarrow \underline{1/2}^-$	7.75

Table 8e. *ft-Values of Probable $\Delta J = 0(-)$ First Forbidden β-Transitions in Odd Nuclei.* The shell-model assignments are unambiguous in the cases marked ‡, but the *ft* values are anomalously low. (From E. FEENBERG: Shell Theory of the Nucleus, p. 87. Princeton: Princeton University Press 1955.)

Parent → Daughter	States	Energy (MeV)	Spins and parities	log *ft*
${}_{47}Ag_{64} \rightarrow {}_{48}Cd_{63}$	$g \rightarrow g$	1.04	$1/2^- \rightarrow \underline{1/2}^+$	7.32
${}_{47}Ag_{66} \rightarrow {}_{48}Cd_{65}$	$g \rightarrow g$	2.1	$1/2^- \rightarrow 1/2^+$	6.96
${}_{47}Ag_{68} \rightarrow {}_{48}Cd_{67}$	$g \rightarrow g$	3.0	$1/2^- \rightarrow 1/2^+$	6.40
${}_{48}Cd_{67} \rightarrow {}_{49}In_{66}$	$g \rightarrow m$	1.12	$1/2^+ \rightarrow 1/2^-$	7.13
${}_{49}In_{66} \rightarrow {}_{50}Sn_{65}$	$m \rightarrow g$	0.830	$1/2^- \rightarrow \underline{1/2}^+$	6.66

Table 8e. (Continued.)

Parent → Daughter	States	Energy (Mev)	Spins and parities	log ft
${}_{49}In_{68} \to {}_{50}Sn_{67}$	$(m) \to g$	1.73	$1/2^- \to \underline{1/2^+}$	6.21
${}_{49}In_{70} \to {}_{50}Sn_{69}$	$(m) \to g$	2.7	$1/2^- \to \underline{1/2^+}$	6.19
${}_{54}Xe_{83} \to {}_{55}Cs_{82}$	$g \to g$	4	$7/2^- \to \underline{7/2^+}$	$\gtrsim$6.33
${}_{56}Ba_{83} \to {}_{57}La_{82}$	$g \to g$	2.27	$7/2^- \to 7/2^+$	$\gtrsim$6.67
${}_{58}Ce_{83} \to {}_{59}Pr_{82}$	$g \to e$	0.44	$7/2^- \to 7/2^+$	7.00
${}_{66}Dy_{99} \to {}_{67}Ho_{98}$	$g \to g$	1.25	$7/2^- \to \underline{7/2^+}$	6.21
${}_{68}Er_{101} \to {}_{69}Tm_{100}$	$g \to g$	0.33	$1/2^- \to \underline{1/2^+}$	6.10
${}_{69}Tm_{102} \to {}_{70}Yb_{101}$	$g \to g$	0.10	$1/2^+ \to \underline{1/2^-}$	6.36
${}_{78}Pt_{121} \to {}_{79}Au_{120}$	$g \to g$	1.8	$3/2^- \to 3/2^+$	6.22
${}_{79}Au_{120} \to {}_{80}Hg_{119}$	$g \to e$	0.250	$3/2^+ \to 3/2^-$	6.17
${}_{80}Hg_{125} \to {}_{81}Tl_{124}$	$g \to g$	1.8	$1/2^- \to 1/2^+$	5.60#
${}_{81}Tl_{126} \to {}_{82}Pb_{125}$	$g \to (g)$	1.44	$1/2^+ \to \underline{1/2^-}$	5.13#
${}_{82}Pb_{127} \to {}_{83}Bi_{126}$	$g \to g$	0.620	$9/2^+ \to \underline{9/2^-}$	5.60#
${}_{82}AcB_{129} \to {}_{83}Bi_{128}$	$g \to g$	1.39	$9/2^+ \to 9/2^-$	6.08
${}_{83}Bi_{130} \to {}_{84}Po_{129}$	$g \to g$	1.39	$9/2^- \to 9/2^+$	6.34

Table 8f. *ft-Values of Probable First Forbidden β-Transitions in Odd Nuclei for cases where the spin change ΔJ is uncertain.* [From E. FEENBERG: Shell Theory of the Nucleus, p. 88. Princeton: Princeton University Press 1955.)

Parent → Daughter	States	Energy (MeV)	Spins and parities	log ft
${}_{27}Co_{30} \to {}_{26}Fe_{31}$	$g \to e$	0.26	$7/2^- \to$	6.96
${}_{32}Ge_{45} \to {}_{31}Ga_{46}$	$g \to e$	2.196	$7/2^+ \to$	7.55
${}_{32}Ge_{45} \to {}_{31}Ga_{46}$	$g \to e'$	1.379	$7/2^+ \to$	6.75
${}_{36}Kr_{53} \to {}_{37}Rb_{52}$	$g \to g$	4.00	$5/2,\ 7/2^+ \to 5/2,\ 3/2^-$	6.17
${}_{37}Rb_{52} \to {}_{38}Sr_{51}$	$g \to g$	4.5	$5/2,\ 3/2^- \to 5/2^+$	$\gtrsim$6.89
${}_{40}Zr_{57} \to {}_{41}Nb_{56}$	$g \to m$	1.91	$\to 1/2^-$	7.16
${}_{42}Mo_{57} \to {}_{43}Tc_{56}$	$g \to m$	1.225		7.13
${}_{56}Ba_{85} \to {}_{57}La_{84}$	$g \to (g)$	2.8	$7/2,\ 9/2^- \to 5/2,\ 7/2^+$	$\gtrsim$6.38
${}_{57}La_{84} \to {}_{58}Ce_{83}$	$g \to g$	2.43	$5/2,\ 7/2^+ \to 7/2^-$	7.26
${}_{57}La_{84} \to {}_{58}Ce_{83}$	$g \to e$	0.9	$5/2,\ 7/2^+ \to 9/2,\ 5/2^-$	6.86
${}_{58}Ce_{83} \to {}_{59}Pr_{82}$	$g \to g$	0.581	$7/2^- \to 5/2,\ 7/2^+$	7.70
${}_{58}Ce_{83} \to {}_{59}Pr_{82}$	$g \to e$	0.442	$7/2^- \to 5/2,\ 7/2^+$	6.99
${}_{61}Pm_{88} \to {}_{62}Sm_{87}$	$g \to (g)$	1.05	$(5/2^+) \to \underline{7/2^-}$	7.00
${}_{62}Sm_{89} \to {}_{63}Eu_{88}$	$g \to e$	0.0755	$9/2^- \to$	7.63
${}_{63}Eu_{92} \to {}_{64}Gd_{91}$	$g \to u$	0.243		8.08
${}_{70}Yb_{105} \to {}_{71}Lu_{104}$	$g \to g$	0.50	$\to \underline{7/2^+}$	> 6.37
${}_{72}Hf_{109} \to {}_{73}Ta_{108}$	$g \to m$	0.42	$1/2,\ 3/2^- \to \underline{1/2^+}$	7.20
${}_{80}Hg_{123} \to {}_{81}Tl_{122}$	$g \to e$	0.208	$1/2,\ 3/2^- \to 3/2^+$	6.42

Table 8g. *ft-Values of Probable $\Delta J = 2(-)$ First Forbidden β-Transitions in Odd Nuclei.* In the cases marked *, the measured shape is consistent with the unique first forbidden assignment. (From E. FEENBERG: Shell Theory of the Nucleus, p. 89. Princeton: Princeton University Press 1955.)

Parent → Daughter	States	Energy (MeV)	Odd group spins and parities: Odd Z	Odd N	log $f_1' t$
${}_{38}Sr_{51} \to {}_{39}Y_{50}$	$g \to g$	1.463	$\underline{1/2^-}$	$5/2^+$	8.34*
${}_{38}Sr_{52} \to {}_{39}Y_{51}$	$g \to g$	0.54	$1/2^-$	$5/2^+$	8.89*
${}_{38}Sr_{53} \to {}_{39}Y_{52}$	$g \to g$	2.665	$1/2^-$	$5/2^+$	8.24*

Table 8g. (Continued.)

Parent Daughter	Sates	Energy (MeV)	Odd group spins and parities Odd Z	Odd N	log $f_1't$
${}_{39}Y_{52} \to {}_{40}Zr_{51}$	$g \to g$	1.54	$1/2^-$	$5/2^+$	8.98*
${}_{39}Y_{54} \to {}_{40}Zr_{53}$	$g \to (g)$	3.1	$1/2^-$	$5/2^+$	$\geq$8.17
${}_{40}Zr_{53} \to {}_{41}Nb_{52}$	$g \to m$	0.034	$1/2^-$	$5/2^+$	8.1
${}_{40}Zr_{55} \to {}_{41}Nb_{54}$	$g \to m$	0.887	$1/2^-$	$5/2^+$	8.93
${}_{40}Zr_{57} \to {}_{41}Nb_{56}$	$g \to m$	1.91	$1/2^-$	$5/2^+$	7.2
${}_{46}Pd_{61} \to {}_{47}Ag_{60}$	$g \to g$	0.04	$1/2^-$	$5/2^+$	8.5
${}_{47}Ag_{64} \to {}_{48}Cd_{63}$	$g \to m$	0.80	$1/2^-$	$5/2^+$	8.64
${}_{74}W_{113} \to {}_{75}Re_{112}$	$g \to g$	1.33	$5/2^+$	$1/2^-$	8.14
${}_{16}S_{21} \to {}_{17}Cl_{20}$	$g \to g$	4.3	$3/2^+$	$7/2^-$	7.67
${}_{18}A_{21} \to {}_{19}K_{20}$	$g \to g$	0.565	$3/2^+$	$7/2^-$	8.78*
${}_{18}A_{23} \to {}_{19}K_{22}$	$g \to g$	2.61	$3/2^+$	$7/2^-$	8.75
${}_{36}Kr_{49} \to {}_{37}Rb_{48}$	$g \to g$	0.695	$5/2^-$	$9/2^+$	8.24
${}_{34}Se_{45} \to {}_{35}Br_{44}$	$g \to g$	0.15	$3/2^-$	$7/2^+$	9.1
${}_{48}Cd_{67} \to {}_{49}In_{66}$	$m \to e$	0.7	$7/2^+$	$11/2^-$	$\sim$8.37
${}_{50}Sn_{73} \to {}_{51}Sb_{72}$	$g \to g$	1.42	$7/2^+$	$11/2^-$	8.82*
${}_{50}Sn_{75} \to {}_{51}Sb_{74}$	$g \to g$	2.33	$7/2^+$	$11/2^-$	8.96*
${}_{51}Sb_{74} \to {}_{52}Te_{73}$	$g \to m$	0.616	$7/2^+$	$11/2^-$	8.73
${}_{53}I_{78} \to {}_{54}Xe_{77}$	$g \to m$	0.815	$7/2^+$	$11/2^-$	8.80
${}_{55}Cs_{82} \to {}_{56}Ba_{81}$	$g \to m$	0.51	$7/2^+$	$11/2^-$	9.03*

Table 8h. *$f't$-Values of Probable $\Delta J = \pm 2(+)$ Second Forbidden β-Transitions in Odd Nuclei.* (From E. FEENBERG: Shell Theory of the Nucleus, p. 90. Princeton: Princeton University Press 1955.)

Parent Daughter	Spins and parities	Energy (MeV)	Half life	log $f't$
${}_{26}Fe_{33} \to {}_{27}Co_{32}$	$3/2^- \to 7/2^-$	1.56	46 d/0.003	10.8
${}_{40}Zr_{53} \to {}_{41}Nb_{52}$	$5/2^+ \to 9/2^+$	0.063	9.5×10^5 y/0.75	11.2
${}_{43}Tc_{56} \to {}_{44}Ru_{55}$	$9/2^+ \to 5/2^+$	0.292	2.16×10^5 y	12.3
${}_{53}I_{76} \to {}_{54}Xe_{75}$	$7/2^+ \to 3/2^+$	0.12	1.72×10^7 y	13.2
${}_{55}Cs_{80} \to {}_{56}Ba_{79}$	$7/2^+ \to 3/2^+$	0.20	2.5×10^6 y	13.2
${}_{55}Cs_{82} \to {}_{56}Ba_{81}$	$7/2^+ \to 3/2^+$	1.17	33 y/0.08	11.8

allowed transitions, the representative $f't$-values are larger than those expected for allowed single particle transitions. It has been suggested that, for allowed transitions, this may be due to the fact that certain single particle transitions are l-forbidden because of the spatial properties of the β operators (for instance, the transition $d_{\frac{3}{2}} \to s_{\frac{1}{2}}$ is allowed by total spin, but is l-forbidden by the l values since the β operator for allowed transitions cannot change the l-value). The existence of l-forbiddeness on the Single Particle Model would lead to two groups of $f't$-values, for each type of transition, corresponding to non l-forbidden (with single particle $f't$-values) and l-forbidden (with $f't$-values $\gg$ single particle $f't$-values). For allowed transitions, the $f't$-values *do* fall into two distinct groups (favoured and unfavoured), but the l-forbiddeness cannot give an adequate account of this splitting. Although all the favoured transitions are l-allowed by the shell-model assignments, some of the unfavoured transitions are l-allowed and some l-forbidden with little distinguishable effect on the $f't$-values. Thus it appears that, although the single particle components in nuclear ground states are decisive in determining the spins of these states, there are important additional components in the states that modify the transition probabilities.

For ordinary first forbidden transitions, the $f't$-values have a tendency to fall into two groups differing by an order of magnitude[1]. However, again this splitting is not a reflection of any l-forbiddeness, which does not exist for first forbidden transitions, but rather according as $\Delta J = 0$ or 1 (the former have the smaller $f't$-values).

8. Electromagnetic moments of odd mass nuclei. In the preceding discussion, we have seen that the single particle model reproduces the observed ground-state spins of odd-mass nuclei rather well. We now turn to a comparison of the values of the observed electromagnetic moments of such nuclei with those predicted by the Single Particle Model. For this comparison, we have made extensive use of a recent review article on this subject[2] and we refer the reader to this article for additional details. The observed moments are listed in Table 6.

α) Magnetic dipole moments. In the previous chapter, we have given formulae for the single particle magnetic dipole moments in terms of the gyromagnetic ratios g_l and g_s. In applications, g_l is set equal to unity for an odd proton and zero for an odd neutron. There has been some debate about the values to be used for g_s. One's first guess would naturally be the values that reproduce the observed moments of isolated particles, viz: $g_s = 5.587$ for the proton $[\mu(P) = 2.793]$ and $g_s = -3.826$ for the neutron $[\mu(N) = -1.913]$. However certain authors[3] have suggested various other choices, all of which fall between these values and those of "bare" particles without virtual mesons: $g_s = 2$ for the proton, $g_s = 0$ for the neutron. Such suggestions have been mainly motivated by the attempts to obtain the best fit of the Single Particle Model to the data. In a qualitative way, the partial "quenching" of the anomalous contributions to the nucleon moments inside nuclear matter is ascribed to the effects of exchange and other interactions, which are required to effectively reduce the number of virtual mesons carried by a nucleon. There is no convincing evidence that such a quenching effect actually exists and we will ignore it for the present.

Accepting then the values of g_s for isolated nucleons, we may write the previous formulae in terms of the observed moments of isolated nucleons:

$$\text{odd neutron:} \quad \mu = \mu(N) \qquad \text{for} \quad j = l + \tfrac{1}{2},$$

$$\mu = -\frac{j}{j+1}\,\mu(N) \qquad \text{for} \quad j = l - \tfrac{1}{2},$$

$$\text{odd proton:} \quad \mu = \left(j - \tfrac{1}{2}\right) + \mu(P) \qquad \text{for} \quad j = l + \tfrac{1}{2},$$

$$\mu = \frac{j}{j+1}\left[\left(j + \tfrac{3}{2}\right) - \mu(P)\right] \qquad \text{for} \quad j = l - \tfrac{1}{2}.$$

To compare these formulae with experimental values, it is usual to plot curves of μ against j, such curves being referred to as SCHMIDT diagrams and the resulting lines as SCHMIDT lines. The SCHMIDT diagrams are given in Fig. 5a and b on which are plotted the measured magnetic moments of odd A nuclei. Of course the SCHMIDT lines only have any meaning at points corresponding to allowed values (i.e. half integral values) of j.

---

[1] R. W. KING and D. C. PEASLEE: Phys. Rev. **94**, 1284 (1954).
[2] R. J. BLIN-STOYLE: Rev. Mod. Phys. **28**, 75 (1956).
[3] H. MIYAZAWA: Progr. Theor. Phys. **6**, 263 (1951). — A. DE SHALIT: Helv. phys. Acta **24**, 296 (1951). — F. BLOCH: Phys. Rev. **83**, 839 (1951).

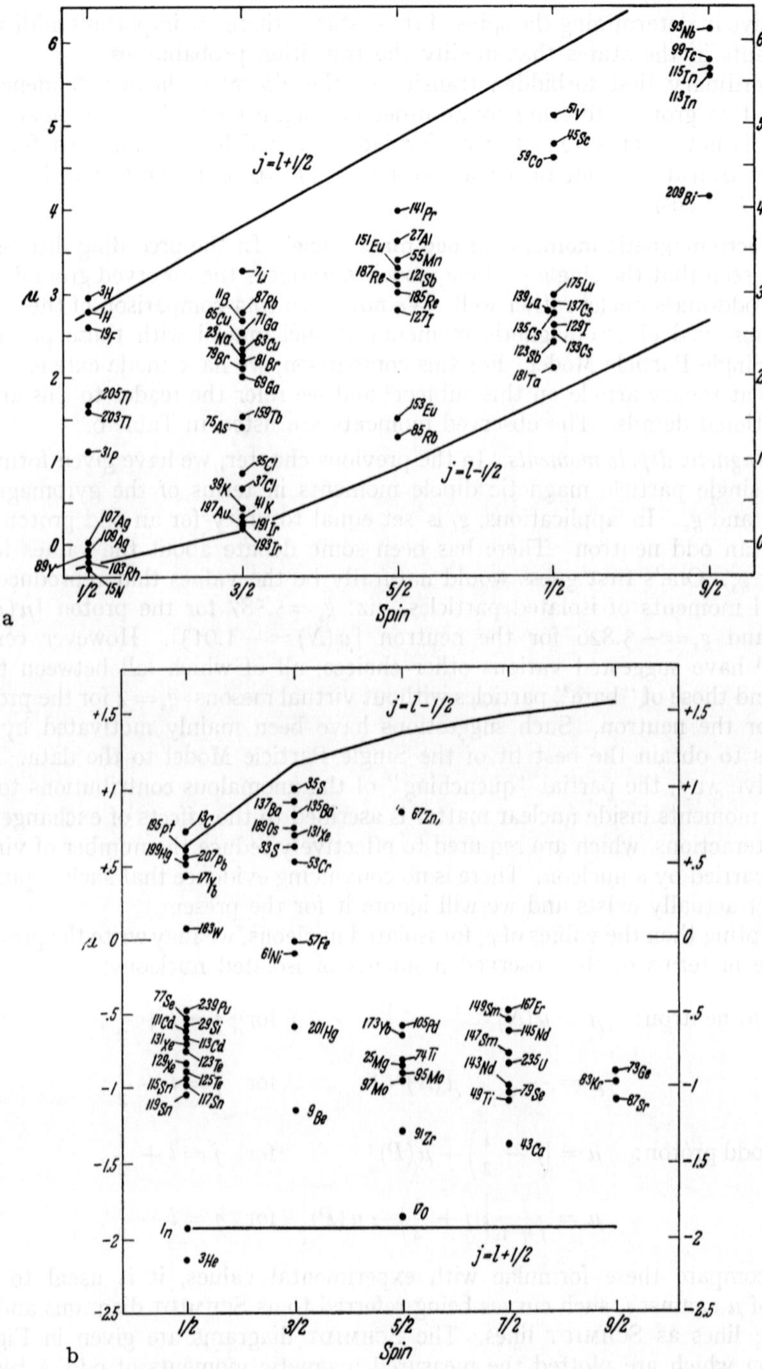

Fig. 5a and b. (a) Schmidt Diagram for Odd-Proton Nuclei. [From R. J. Blin-Stoyle: Rev. Mod. Phys. 28, 75 (1956).] (b) Schmidt Diagram for Odd-Neutron Nuclei. [From R. J. Blin-Stoyle: Rev. Mod. Phys. 28, 75 (1956).]

There are several points to be noticed in the Schmidt diagrams:

(1) The moments of most nuclei deviate from the Schmidt lines by amounts varying between about $\frac{1}{2}-1\frac{1}{2}$ n.m.

(2) Apart from the light nuclei H^3, He^3, N^{15}, C^{13}, the deviations of the magnetic moments are always inwards from the SCHMIDT lines.

(3) Lines drawn to represent average deviations are roughly parallel to the SCHMIDT lines.

(4) The average deviation of odd proton nuclei is a little larger (about 20%) than the average deviation of odd neutron nuclei.

(5) The only nuclei which do not deviate by more than about 0.2 n.m. are (apart from those of $A=1$ and $A=3$) O^{17}, K^{39}, K^{41} and all $p_{\frac{1}{2}}$ nuclei.

In spite of the deviations, there is clearly a good qualitative correlation between the single particle predictions and the observed moments. It is possible to associate most moments with one line or the other, and this enables parity allocations to be assigned to nuclear ground states. Such assignments are almost invariably correct. The only really bad moments are those of As^{75}, which is mid-way between the SCHMIDT lines, and of most nuclei between masses 158 and 180. We have already seen in Sect. 6 that even the spins of nuclei in this latter region are not correctly given by the Single Particle Model, so the model cannot be expected to give correct moments.

Perhaps the most striking of the above observations is that of the tendency for deviations always to be inside the SCHMIDT lines, by about 1 n.m. on the average. This systematic feature is very useful for guiding one in attempts to refine the simple Single Particle Model.

We have already mentioned two factors which give effects in the right direction, namely the velocity dependence of the spin-orbit force giving an additional moment $\sim \pm\frac{1}{4}\frac{2j+1}{2j+2}$, and the "quenching" of the intrinsic moments. However, the first of these factors apparently gives effects that are too small and only apply to odd protons and the second has no sound theoretical basis at present. We remark that such considerations typify attempts to improve the SCHMIDT lines by alteration of the effective gyromagnetic ratios. A second, quite different, type of approach seeks the source of improvement in modifications to the single particle wave-functions, and only appeals to possible changes in the gyromagnetic ratios as a last resort. Since the extreme single particle model is an idealisation and *must* be somewhat modified when applied to actual nuclei, the second type of approach is more logical. A common feature of all the suggested modifications to the extreme Single Particle Model is the taking into account of the presence of other particles which are coupled to the single particle and thereby share its angular momentum. This partial transfer of angular momentum means that the other particles can contribute to the magnetic moment. Depending on the value of the effective gyromagnetic ratio for the motion of the other particles, the particle moments may be slightly or severely modified. In the next chapter, we will review the attempts to improve the Single Particle Model so as to lead to more correct magnetic moments.

We close this discussion of magnetic moments on the Single Particle Model by mentioning two interesting features of the experimental magnetic moments. This first feature is the tendency for the deviations of the magnetic moments from the SCHMIDT line to be dependent only on the number of odd particles present. Fig. 6 shows the dimensionless quantity $\frac{\mu_s-\mu}{\mu_s-\mu_d}$ plotted against N or Z depending on which is odd. μ is the observed moment and μ_s is the SCHMIDT moment. The quantity $\mu_s-\mu_d$ (where μ_d is the DIRAC moment: $\mu_d=1$ for proton, 0 for neutron) is divided into the deviation $\mu_s-\mu$ merely to give a dimensionless measure. One can see that the two curves, one for odd N nuclei, one for odd Z nuclei, follow each other closely over the whole range $N, Z < 50$. This is in spite of the fact that nuclei of quite different masses are being compared. For instance, at 25, the two nuclei in question are Ti^{47} ($Z=22$, $N=25$) and Mn^{55} ($Z=25$, $N=30$). Both have the same spin ($\frac{5}{2}$) because, of course, this is always found to be determined by the number of odd particles.

The second feature of the observed moments is concerned with the change in moments when the number of odd or even particles is increased by two. On the Single Particle Model one predicts the moment to be unchanged, provided the addition of the two nucleons does not involve the filling of a shell. Experimentally (Table 6, pp. 264–269), one finds that the pairs of values of moments are close, in 27 of the 28 cases of two-neutron addition H (0, 2), Cl (18, 20), K (20, 22), Co (30, 32), Cu (34, 36), Ga (38, 40), Br (44, 46), Ag (60, 62, 64), In (64, 66), Cs (78, 80, 82), Eu (88, 90), Re (110, 112), Ir (114, 116), Tl (122, 124), Am (146, 148), Mo (53, 55), Cd (63, 65), Sn (65, 67, 69), Te (71, 73), Ba (79, 81), Nd (83, 85), Sm (85, 87), Gd (91, 93), Dy (95, 97), Hg (117, 119) and in all three cases of two-proton addition: (Cd, Sn), (Nd, Sm), (Cl, K). The one exception is Eu and it has been suggested that the large difference in moment in this case is due to an abrupt change in nuclear shape between $N=88$ and $N=90$[1] (see Sect. 15). In general, the values are some way inside the Schmidt lines and it is found that, of 23 pairs, the heavier nucleus has the smaller deviation in 17 cases. This empirical rule holds for addition to either the even or odd group of particles in a nucleus.

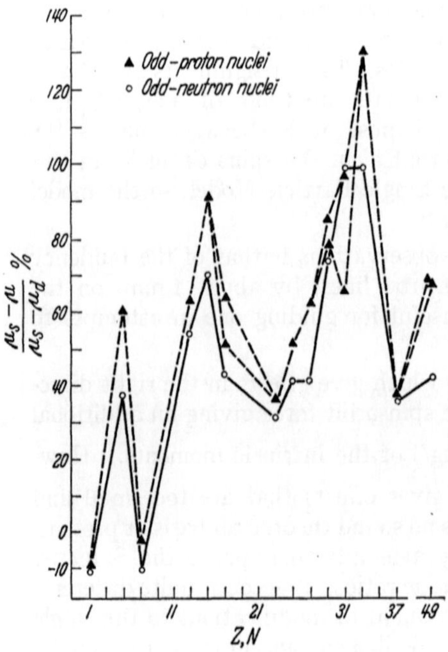

Fig. 6. Comparison of the Magnetic Moment Deviations for nuclei of odd Z and odd N. (From M. G. Mayer and J. H. D. Jensen: The Elementary Theory of Nuclear Structure, p. 90. New York: John Wiley & Sons 1955.)

β) Electric quadrupole moments. Unlike the case of the magnetic dipole moments, the electric quadrupole moments of nuclei (Tables 6) are often not even qualitatively similar to the predictions of the Single Particle Model. For this reason, there is no point in constructing a Schmidt type diagram which emphasises the comparison with the single particle picture. Instead it is more instructive to plot the observed quadrupole moments against Z (for odd proton nuclei) and N (for odd neutron nuclei). In Fig. 7, the quantity actually plotted is the ratio $|Q|/R^2$. R is the nuclear radius, taken as $1.40 A^{\frac{1}{3}} \times 10^{-13}$ cm, and R^2 can be regarded as a rough measure of the single particle moment Q_j of Chap. II. Some remarkable features of the observed quadrupole moments are:

(1) The only nuclei for which the quadrupole moments are roughly equal to Q_j in magnitude and sign are those immediately above major closed shells. Notice that one of these (Ge^{73}, $Z=32$, $N=41$) is an odd *neutron* nucleus.

---

[1] G. Scharff-Goldhaber and J. Weneser: Phys. Rev. **98**, 212 (1955). — B. R. Mottelson and S. G. Nilson: Phys. Rev. **99**, 1615 (1955).

(2) Almost all other nuclei have positive quadrupole moments and so do not agree with the single proton value even in sign.

(3) Most of these positive quadrupole moments do not agree with the single proton value in magnitude either, being much too large.

(4) The large moments occur in odd neutron as well as odd proton nuclei.

(5) The largest moments occur in the middle of the major shells, and the values fall off considerably on either side to approach roughly single particle magnitudes at the closed shells.

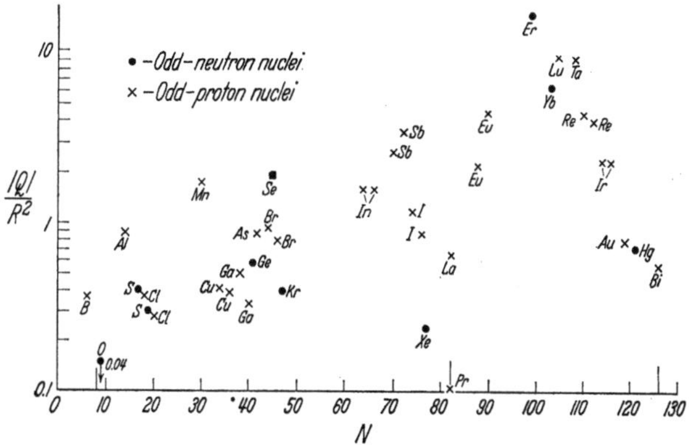

Fig. 7. Plot of the Quantity $|Q|/R^2$ against Neutron Number N. Q is electric quadrupole moment and R is nuclear radius taken as $R = 1.40 A^{\frac{1}{3}} \times 10^{-13}$ cm. The dimensionless $|Q|/R^2$ can be regarded as a measure of quadrupole moment in units of the single particle quadrupole moment. (From M. G. MAYER and J. H. D. JENSEN: The Elementary Theory of Nuclear Structure, p. 108. New York: John Wiley & Sons 1955.)

Thus we find that, although the quadrupole moments of nuclei are strongly related to shell-structure (i.e. to the proximity of closed shells), the Single Particle Model fails almost completely to predict the observed values. The only exceptions are those nuclei immediately above closed shells. For these nuclei one might indeed expect the model to succeed, if it succeeds at all. For other nuclei, it is apparent that one must allow for the fact that the single nucleon circulating about the "core" of other nucleons, must polarise this core and induce strong quadrupole moments in it. The size of some quadrupole moments correspond to really strong quadrupole distortions, which can be considered to be classically permanent in the sense that they are much larger than the expected zero-point fluctuations. It follows that, in attempting to refine the Single Particle Model, one must be prepared to relax the condition of spherical symmetry which is one of its main simplifying features. We consider such refinements in Sect. 15.

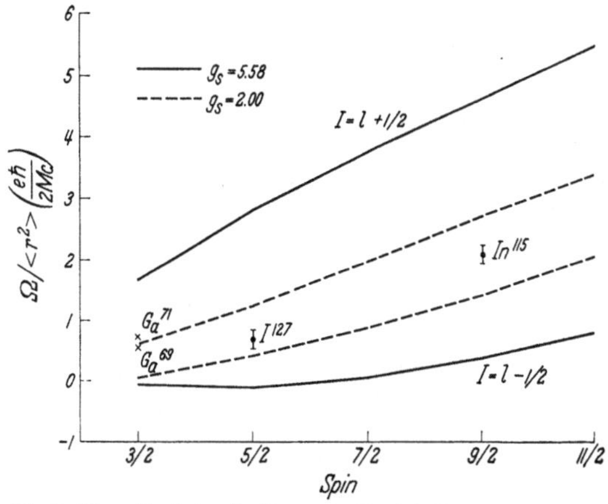

Fig. 8. Magnetic Octupole Moments, Ω, of Odd-Proton Nuclei in units of $\langle r^2\rangle\left(\frac{e\hbar}{2Mc}\right)$. The mean square radius $\langle r^2\rangle$ is evaluated as $\frac{3}{5}R^2$ with $R = 1.35 A^{\frac{1}{3}} \times 10^{-13}$ cm. The continuous and broken lines are the predictions of the single particle model taking $g_s = 5.58$ and 2.00 respectively. [From C. SCHWARTZ: Phys. Rev. 97, 380 (1955).]

γ) Magnetic octupole moments. One can construct a SCHMIDT-type diagram for magnetic octupole moments exactly analogous to that for magnetic dipole moments[1]. In Fig. 8 the quantity $\Omega/\langle r^2\rangle$ has been plotted for odd protons as a function of j for the two cases $j = l \pm \frac{1}{2}$. There are only four experimentally

[1] C. SCHWARTZ: Phys. Rev. 97, 380 (1955).

determined octupole moments, all for odd proton nuclei, and these are included in the Figure after estimating $\langle r^2 \rangle$ by $\frac{3}{5} R^2$ with $R = 1.35 A^{\frac{1}{3}} \times 10^{-13}$ cm. It can be seen that the observed values are inside the SCHMIDT lines. As with the dipole moments, one may try to improve the SCHMIDT lines from two rather independent points of view. One of these emphasises that the extreme Single Particle Model must be subject to some refinement when applied to actual nuclei. No work has been reported in this connection. The other point of view is that the gyromagnetic ratios are altered in nuclear matter from their values for isolated particles. The effect of changing g_s from the usual value of 5.58 to 2.00 is shown by the dotted lines in the Figure.

9. Isomeric states in odd-mass nuclei[1]. Isomeric states of nuclei are defined as excited states of nuclei that have such long lifetimes that they can be considered as stable for many purposes. The lifetime of a state is, of course, inversely proportional to the probability of its decay by the emission of radiation. Thus the longevity of a state implies that its radiative decay is improbable. This may be due to the softness of the radiation, to its high multipole order, or to both. The widespread occurrence of isomeric states is a very significant experimental fact, expecially as it shows that large differences in angular momentum can occur between the ground states of nuclei and the nearby first excited states. Such differences are hard to understand from the standpoint of certain nuclear models, such as the liquid drop model, where the excitation of the first few states of a nucleus might be thought to involve the consecutive addition of only single units of angular momentum. However, before we speculate on the implications of isomeric states, we will review the experimental situation.

First we must set up a standard lifetime to distinguish isomeric states from other states. This lifetimes is necessarily somewhat arbitrary. For the present discussion, we choose it as 10^{-15} sec and only consider states of lifetime longer than this. In Tables 9a to e, the known isomeric states are listed according to the type of radiation by which they decay. All the transitions listed are found in odd nuclei. Of course there exist many known transitions in even-even nuclei, almost all of which are $E2$ because this is the only type allowed by the spins (0^+ and 2^+ for the ground and first excited states). We have not included such

Table 9a. *Isomeric M1 Transitions in Odd Nuclei: Lifetimes, Energies and Shell-Model Interpretation.* (From E. FEENBERG: Shell Theory of the Nucleus, p. 71. Princeton: Princeton University Press 1955.)

Nucleus	log τ_γ	W (keV)	Initial state	Final state
$_3Li_4$	$\overline{14}.87$	478	$(p^3)_{\frac{1}{2}}$	$(p^3)_{\frac{3}{2}}$
$_{26}Fe_{31}$	$\bar{6}.24$	14	$p_{\frac{1}{2}}$	$p_{\frac{1}{2}}$
$_{52}Te_{71}$	$\overline{10}.52$	159	$d_{\frac{3}{2}}$	$s_{\frac{1}{2}}$
$_{52}Te_{73}$	$\bar{8}.64$	35.4	$d_{\frac{3}{2}}$	$s_{\frac{1}{2}}$
$_{54}Xe_{77}$	$\bar{9}.32$	80	$s_{\frac{1}{2}}$	$d_{\frac{3}{2}}$
$_{55}Cs_{78}$	$\bar{8}.43$	81	$d_{\frac{5}{2}}$	$g_{\frac{7}{2}}$
$_{55}Cs_{80}$	$\overline{10}.64$	248	$d_{\frac{5}{2}}$	$g_{\frac{7}{2}}$
$_{61}Pm_{86}$	$\bar{8}.05$	91.5	—	$d_{\frac{5}{2}}$
$_{80}Hg_{119}$	$< \bar{8}.20$	50	$p_{\frac{3}{2}}$	$f_{\frac{5}{2}}$
$_{80}Hg_{119}$	$< \overline{10}.76$	209	$p_{\frac{3}{2}}$	$p_{\frac{1}{2}}$

Table 9b. *Isomeric E2 and M2 Transitions in Odd Nuclei: Lifetimes, Energies and Shell-Model Interpretation.* (From E. FEENBERG: Shell Theory of the Nucleus, p. 71. Princeton: Princeton University Press 1955.)

Nucleus	log τ_γ	W (keV)	Type	Initial state	Final state
$_{19}K_{22}$	$\bar{9}.98$	1300	$M2$	$f_{\frac{7}{2}}$	$d_{\frac{3}{2}}$
$_{37}Rb_{48}$	$\bar{6}.12$	513	$M2$	$g_{\frac{9}{2}}$	$f_{\frac{5}{2}}$
$_{48}Cd_{63}$	$\bar{7}.09$	243	$E2$	$d_{\frac{5}{2}}$	$s_{\frac{1}{2}}$
$_{71}Lu_{106}$	$\bar{6}.27$	150	$(M2)$		
$_{73}Ta_{108}$	$\bar{5}.89$	134	$E2$	$1/2^+$	$3/2^+$
$_{73}Ta_{108}$	$\bar{7}.18$	345	$E2$	$3/2^+$	$7/2^+$
$_{73}Ta_{108}$	$\bar{8}.26$	481	$E2$	$3/2^+$	$g_{\frac{7}{2}}$
$_{75}Re_{112}$	$\bar{5}.13$	133	$M1 + E2$		
$_{80}Hg_{117}$	$\bar{8}.53$	133	$E3$	$f_{\frac{5}{2}}$	$p_{\frac{1}{2}}$
$_{80}Hg_{119}$	$\bar{9}.78$	159	$E2$	$f_{\frac{5}{2}}$	$p_{\frac{1}{2}}$

[1] For details on isomers cf. ALBURGER's article in Vol. XLII of this Encyclopedia.

Table 9c. *Isomeric E3 Transitions in Odd Nuclei: Lifetimes, Energies and Shell-Model Interpretation.* (From E. FEENBERG: Shell Theory of the Nucleus, p. 72. Princeton: Princeton University Press 1955.)

Nucleus	$\log \tau_\gamma$	W (keV)	Initial state	Final state	Nucleus	$\log \tau_\gamma$	W (keV)	Initial state	Final states
$_{32}Ge_{45}$	> 1.94	380	$p_{\frac{1}{2}}$	$7/2^+$	$_{45}Rh_{60}$	2.55	130	$p_{\frac{1}{2}}$	$7/2^+$
$_{34}Se_{43}$	1.68	160	$7/2^+$	$p_{\frac{1}{2}}$	$_{47}Ag_{60}$	3.03	94	$7/2^+$	$p_{\frac{1}{2}}$
$_{34}Se_{45}$	3.89	80	$p_{\frac{1}{2}}$	$7/2^+$	$_{47}Ag_{62}$	3.04	87	$7/2^+$	$p_{\frac{1}{2}}$
$_{34}Se_{47}$	4.64	98	$7/2^+$	$p_{\frac{1}{2}}$	$_{46}Pd_{63}$	2.78	170	$(h_{\frac{11}{2}})$	$(d_{\frac{5}{2}})$
$_{36}Kr_{43}$	2.47	127	$(7/2^+)$	$(p_{\frac{1}{2}})$	$_{48}Cd_{63}$	4.12	149	$h_{\frac{11}{2}}$	$d_{\frac{5}{2}}$
$_{36}Kr_{45}$	1.45	187	$(p_{\frac{1}{2}})$	$(7/2^+)$	$_{54}Xe_{73}$	2.41	175	$h_{\frac{11}{2}}$	$d_{\frac{5}{2}}$
$_{36}Kr_{47}$	7.4	32	$p_{\frac{1}{2}}$	$7/2^+$	$_{79}Au_{118}$	1.89	130	$h_{\frac{11}{2}}$	$d_{\frac{5}{2}}$
$_{38}Sr_{47}$		7.5	$p_{\frac{1}{2}}$	$7/2^+$	$_{66}Dy_{99}$	3.72	109	$s_{\frac{1}{2}}$ or $i_{\frac{13}{2}}$	$f_{\frac{7}{2}}$
$_{43}Tc_{56}$		2.0	$p_{\frac{1}{2}}$	$7/2^+$	$_{74}W_{109}$	~3.0	80	$7/2^+$	$p_{\frac{1}{2}}$
$_{45}Rh_{58}$	6.95	40	$7/2^+$	$p_{\frac{1}{2}}$					

Table 9d. *Isomeric M3 Transitions in Odd Nuclei: Lifetimes, Energies and Shell-Model Interpretation.* (From E. FEENBERG: Shell Theory of the Nucleus, p. 73. Princeton: Princeton University Press 1955.)

Nucleus	$\log \tau_\gamma$	W (keV)	Initial state	Final state
$_{72}Hf_{107}$	2.95	160	$(h_{\frac{9}{2}})$	$(p_{\frac{3}{2}})$
$_{73}Ta_{108}$	$\bar{3}.36$	610	$1/2^+$	$g_{\frac{7}{2}}$
$_{76}Os_{115}$		74	$i_{\frac{13}{2}}$	$7/2^+$

Table 9e. *Isomeric M4 Transitions in Odd Nuclei: Lifetimes, Energies and Shell-Model Interpretation.* (From E. FEENBERG: Shell Theory of the Nucleus, p. 73. Princeton: Princeton University Press 1955.)

Nucleus	$\log \tau_\gamma$	W (keV)	Initial state	Final state	Nucleus	$\log \tau_\gamma$	W (keV)	Initial state	Final state
$_{30}Zn_{39}$	4.88	439	$g_{\frac{9}{2}}$	$p_{\frac{1}{2}}$	$_{52}Te_{71}$	10.35	88	$h_{\frac{11}{2}}$	$d_{\frac{3}{2}}$
$_{36}Kr_{49}$	5.23	305	$p_{\frac{1}{2}}$	$g_{\frac{9}{2}}$	$_{52}Te_{73}$	9.43	109	$h_{\frac{11}{2}}$	$d_{\frac{3}{2}}$
$_{38}Sr_{47}$	6.17	232	$p_{\frac{1}{2}}$	$g_{\frac{9}{2}}$	$_{52}Te_{75}$	10.26	88	$h_{\frac{11}{2}}$	$d_{\frac{3}{2}}$
$_{38}Sr_{49}$	4.27	390	$p_{\frac{1}{2}}$	$g_{\frac{9}{2}}$	$_{52}Te_{77}$	9.34	106	$h_{\frac{11}{2}}$	$d_{\frac{3}{2}}$
$_{39}Y_{48}$	4.97	384	$g_{\frac{9}{2}}$	$p_{\frac{1}{2}}$	$_{52}Te_{79}$	6.98	183	$h_{\frac{11}{2}}$	$d_{\frac{3}{2}}$
$_{39}Y_{50}$	1.31	913	$g_{\frac{9}{2}}$	$p_{\frac{1}{2}}$	$_{52}Te_{81}$	~4.13	~300	$h_{\frac{11}{2}}$	$d_{\frac{3}{2}}$
$_{39}Y_{52}$	3.68	555	$g_{\frac{9}{2}}$	$p_{\frac{1}{2}}$	$_{54}Xe_{75}$	7.39	196	$h_{\frac{11}{2}}$	$d_{\frac{3}{2}}$
$_{40}Zr_{49}$	2.62	588	$p_{\frac{1}{2}}$	$g_{\frac{9}{2}}$	$_{54}Xe_{77}$	7.91	163	$h_{\frac{11}{2}}$	$d_{\frac{3}{2}}$
$_{41}Nb_{50}$	8.90	104	$(g_{\frac{9}{2}})$	$(p_{\frac{1}{2}})$	$_{54}Xe_{79}$	6.42	232	$h_{\frac{11}{2}}$	$d_{\frac{3}{2}}$
$_{41}Nb_{54}$	6.39	216	$p_{\frac{1}{2}}$	$g_{\frac{9}{2}}$	$_{54}Xe_{81}$	3.22	520	$h_{\frac{11}{2}}$	$d_{\frac{3}{2}}$
$_{41}Nb_{56}$	1.94	749	$p_{\frac{1}{2}}$	$g_{\frac{9}{2}}$	$_{56}Ba_{77}$	6.05	276	$h_{\frac{11}{2}}$	$d_{\frac{3}{2}}$
$_{43}Tc_{52}$	13.2	39	$p_{\frac{1}{2}}$	$g_{\frac{9}{2}}$	$_{56}Ba_{79}$	5.77	269	$h_{\frac{11}{2}}$	$d_{\frac{3}{2}}$
$_{43}Tc_{54}$	9.67	97	$(p_{\frac{1}{2}})$	$(g_{\frac{9}{2}})$	$_{56}Be_{81}$	2.40	661	$h_{\frac{11}{2}}$	$d_{\frac{3}{2}}$
$_{43}Tc_{56}$	8.03	142	$p_{\frac{1}{2}}$	$g_{\frac{9}{2}}$	$_{78}Pt_{117}$	~8.45	129	$i_{\frac{13}{2}}$	$f_{\frac{5}{2}}$
$_{49}In_{64}$	4.14	390	$p_{\frac{1}{2}}$	$g_{\frac{9}{2}}$	$_{78}Pt_{119}$	4.86	337	$(i_{\frac{13}{2}})$	$(f_{\frac{5}{2}})$
$_{49}In_{66}$	4.70	335	$p_{\frac{1}{2}}$	$g_{\frac{9}{2}}$	$_{80}Hg_{117}$	7.65	165	$i_{\frac{13}{2}}$	$f_{\frac{5}{2}}$
$_{50}Sn_{67}$	7.95	159	$h_{\frac{11}{2}}$	$d_{\frac{3}{2}}$	$_{80}Hg_{119}$	4.50	368	$i_{\frac{13}{2}}$	$f_{\frac{5}{2}}$
$_{50}Sn_{69}$	11.12	65	$h_{\frac{11}{2}}$	$d_{\frac{3}{2}}$	$_{82}Pb_{125}$	0.164	1063	$i_{\frac{13}{2}}$	$f_{\frac{5}{2}}$
$_{52}Te_{69}$	10.67	82	$h_{\frac{11}{2}}$	$d_{\frac{3}{2}}$					

transitions, partly because the Single Particle Model as we have developed it so far has no application to such transitions, and partly because the lifetimes have not been measured directly. (Notice however that the method of COULOMB excitation gives the transition matrix elements from which the lifetimes can be deduced.)

In order to analyse the large number of observed lifetimes, we first wish to extract from the data such irrelevant factors as the energies of the transitions. The most convenient way to do this is to set up a quantity representing the ratio of observed lifetime to single-particle (proton) lifetime of the previous chapter. (The irrelevant factors cancel in such a ratio.) Such a quantity can be called a "reduced" lifetime and may be defined as[1]:

$$\tau_E(\lambda)\,\{E_\gamma^{2\lambda+1} A^{2\lambda/3}\}$$

for electric transitions and:

$$\tau_M(\lambda)\,\{E_\gamma^{2\lambda+1} A^{2\lambda-2/3}\}$$

for magnetic transitions. The dependence of the single particle lifetimes on nuclear radius R is taken into account in these quantities by the presence of A, the mass-number (assuming $R\propto A^{\frac{1}{3}}$). The logarithms of the reduced lifetimes for all the observed isomeric electric and magnetic transitions are plotted in Figs. 9a and b. Also plotted, as horizontal lines, are the single particles estimates of these quantities with the assumption: $R=1.2A^{\frac{1}{3}}\times 10^{-13}$ cm.

We draw attention to the following significant features of the Figures and Tables:

(1) The isomeric states tend to occur in well defined mass-ranges in the *upper* halves of major shells. These so called "islands of isomerism" are bracketed by the odd numbers of particles: 39→49, 65→81, 101→125.

(2) Most of the isomeric transitions are $E2$, $E3$ and $M4$; very few are $M2$, $M3$ or $E4$.

(3) The $M4$ reduced lifetimes are extraordinarily constant in comparison with the others. Most points are within a factor of 2 of the average value (which is rather larger than the single particle value).

(4) The $E2$ reduced lifetimes are not as constant as the $M4$ ones but, nevertheless, there is a remarkable systematic feature of the plotted values, viz. the decrease from the single particle lifetime near the double closed shell at $Z=82$, $N=126$ to a much smaller value for nuclei with $A\sim 160$.

(5) The many $E3$ and the few $M2$, $M3$ and $E4$ transitions each exhibit a fairly wide scatter of points on the figures. However one may discern a possible trend of the $E3$ transitions to increase from the single particle lifetime at $Z=82$, $N=126$ to a much larger value for nuclei with $A\sim 160$.

We will now examine the predictions that the single particle shell-model makes about the above features. To anticipate the results, we will find that the model has considerable success in predicting the occurrence and nature of isomeric states [features (1) and (2)], but fails to give adequate account of the magnitudes of the lifetimes [(3), (4) and (5)]. We note in passing that this is the first occasion on which the model has been applied to anything other than ground states.

From the single particle level sequences given in Fig. 2, we see that, occasionally, there are big spin changes between adjacent levels. This provides the necessary conditions for the occurrence of isomeric states, which are essentially long lived states whose longevity is due to their only being able to decay by soft

[1] M. GOLDHABER and J. WENESER: Ann. Rev. Nucl. Sci. 5, 1 (1955).

radiation of high multiple order. The most important single feature of the single particle level sequences, as far as isomerism is concerned, is the large depression

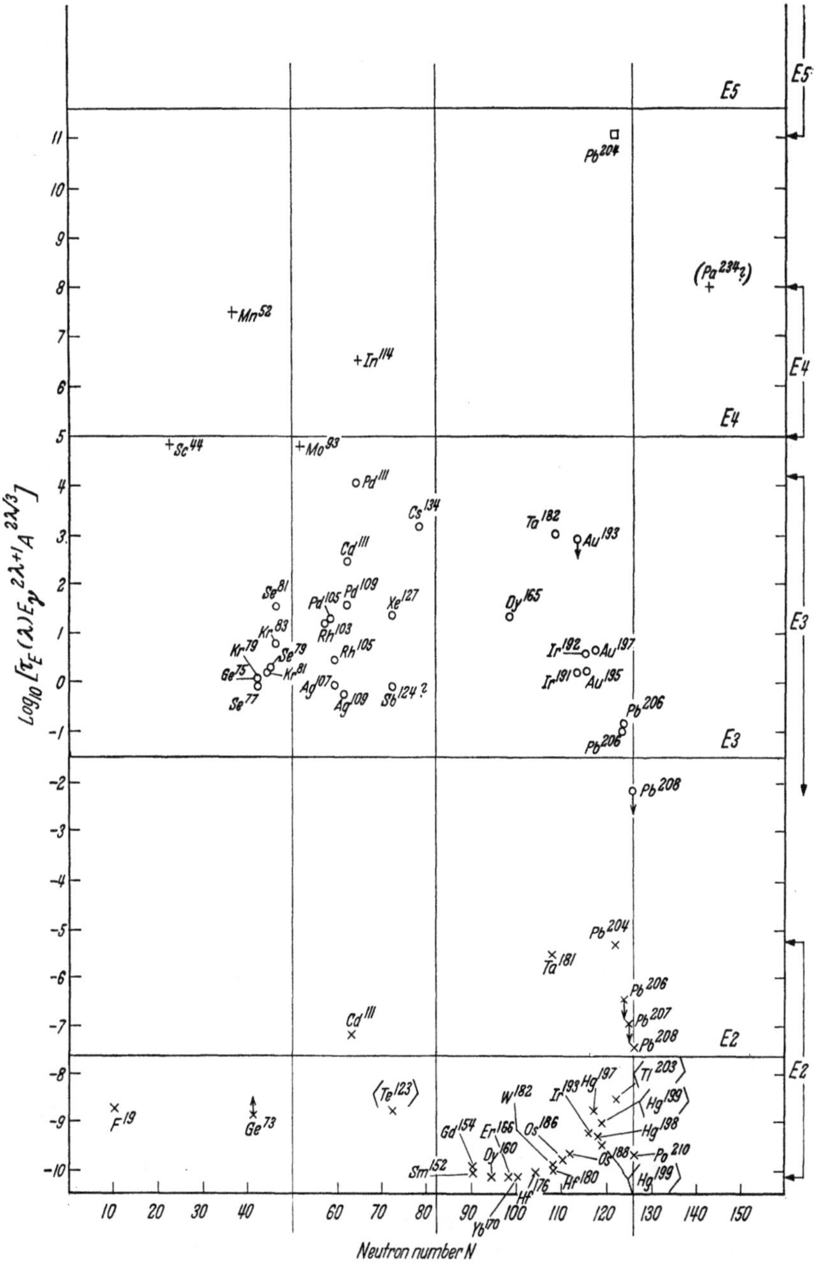

Fig. 9a. Reduced Lifetimes for $E2$, $E3$, $E4$ and $E5$ Transitions plotted against Neutron Number. The horizontal lines are the values for a single proton calculated with the spin factor $\frac{(Z^0)^2}{2j+1}$ put equal to unity and with a radius $R=1.20A^{\frac{1}{3}}\times 10^{-13}$ cm. [From M. Goldhaber and J. Weneser: Ann. Rev. Nucl. Sci. 5, 1 (1955), p. 13.]

of the $j=l+\frac{1}{2}$ member for the orbits of high l-value. These are expected to be depressed into the low spin levels of the next major shell of opposite parity to become part of that shell. (This has been confirmed by the observed ground state spins as we have already seen.) For instance, just below the magic number Z or $N=50$, we find the $1g_{\frac{9}{2}}$ level, which has been depressed to the region of the

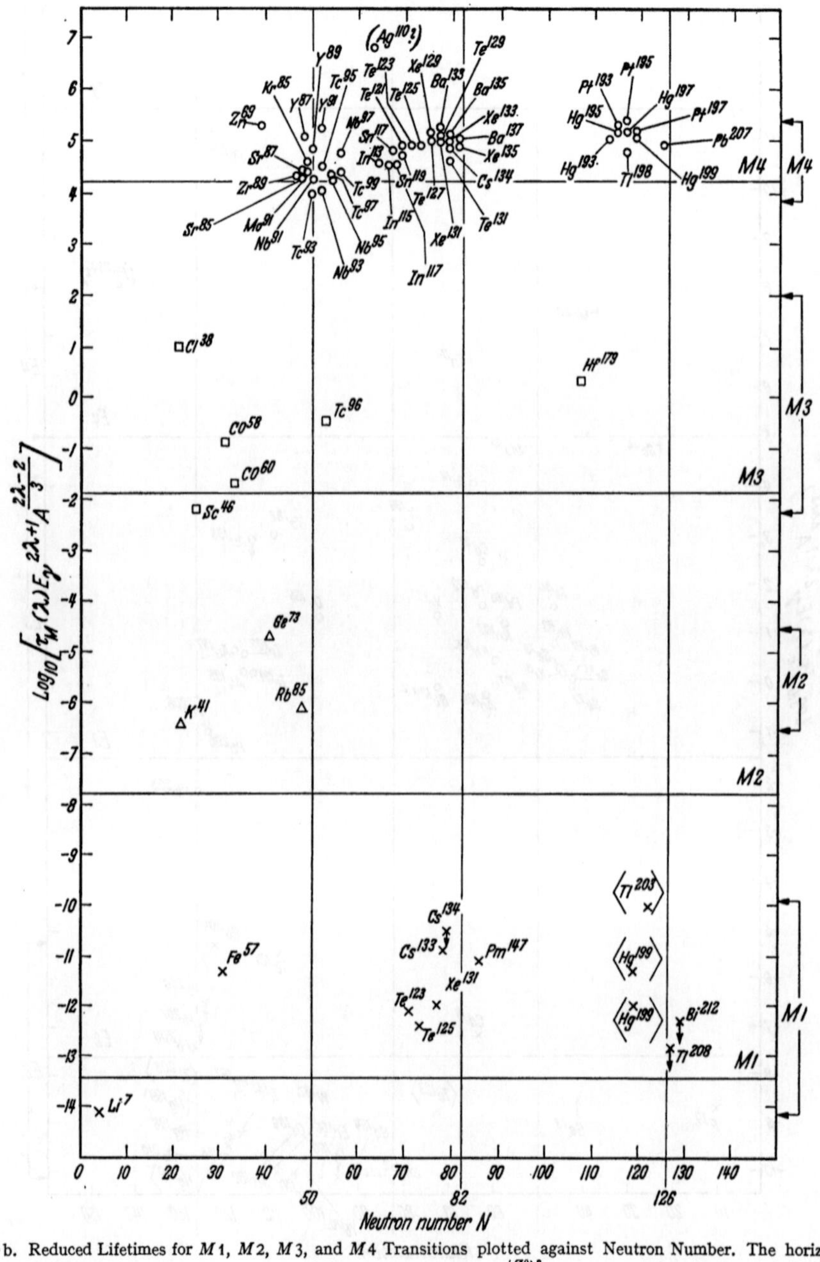

Fig. 9b. Reduced Lifetimes for $M1$, $M2$, $M3$, and $M4$ Transitions plotted against Neutron Number. The horizontal lines are the values for a single proton calculated with the spin factor $\frac{(Z^0)^2}{2j+1}$ put equal to unity and with a radius $R=1.20A^{\frac{1}{3}}\times10^{-13}$ cm. [From M. Goldhaber and J. Weneser: Ann. Rev. Nucl. Sci. 5, 1 (1955), p. 12.]

$2p_{\frac{1}{2}}$ shell and $1f_{\frac{5}{2}}$ shells. Radiation from the $1g_{\frac{9}{2}}$ state to these levels will be $M4$ and $E3$ respectively. Just below magic number Z or $N=82$, the $1h_{\frac{11}{2}}$ level occurs near the $3s_{\frac{1}{2}}$, $2d_{\frac{3}{2}}$ and $2d_{\frac{5}{2}}$ levels. Radiation from the $1h_{\frac{11}{2}}$ level to these levels will be $E5$, $M4$ and $E3$ in type. Just below magic number Z or $N=126$, the $1i_{\frac{13}{2}}$ level occurs near the $3p_{\frac{1}{2}}$, $3p_{\frac{3}{2}}$, $2f_{\frac{5}{2}}$ and $2f_{\frac{7}{2}}$ levels with corresponding radiation of the $M6$, $E5$, $M4$ and $E3$ types.

Thus, one expects to find isomers concentrated just below magic numbers and we expect the transitions to involve a parity change i.e. to be $E3$ or $M4$. Of course, this is just what is found experimentally. In more detail, it is found that individual isomeric states can usually be identified as single particle levels as we have done in the tables, and so the Single Particle Model can be said to predict the existence and nature of isomeric states very successfully.

The model has less success in the matter of predicting the lifetimes of isomeric states. As we have seen, the predicted lifetimes are too short (by as much as a factor 100) for $E2$ transitions and too long for all other types. Thus, as we found in discussing moments and β-decay lifetimes, we have a situation suggesting that states contain strong single particle components which determine many general features of the states, but there are also present admixtures of many-particle (collective type) components which can strongly modify the moments and transition probabilities. These admixtures will be examined in the next chapter with especial reference to those types that can *increase* the $E2$ transition probabilities to the observed values away from the closed shells.

IV. Refinements of the single particle model.

So far we have used an extreme version of the shell-model, viz. the Single Particle Model, which can be applied to odd nuclei on the assumption that all nucleons except the last odd one form an inert core. Although it is attractive by virtue of its simplicity, this model has no logical basis and, as soon as one attempts to apply it to detailed data such as nuclear spectra, one discovers its inconsistencies. The shortcomings of the extreme model come from its complete neglect of particle-particle forces. The manner in which such forces remove the degeneracies of the extreme model by splitting up the states of a configuration will be discussed in the following four sections. Such states must be properly constructed in accordance with the Exclusion Principle. If this is done, the model can be said to satisfy the formal quantum-mechanical requirements—unlike the extreme model. However this does not establish the correctness of the model in physical applications. Even though the shell-model states are properly constructed, there is still the possibility that the particle-particle forces are strong enough to cause strong correlations between nucleons and thereby invalidate the whole basis of the shell-model. We will see that, although the model has many successes, it also has some significant failures. In the last two sections, these failures will be discussed in terms of configuration mixing and departures from sphericity in nuclear shape.

10. States of several particles in the same shell: Theory. In the preceding two chapters we have discussed the "Single Particle Model" and its application to experimental data. This model rests on the shell-model picture that ascribes to any given nucleus a configuration in which there are a number of filled shells for neutrons and protons and some additional "loose" particles in unfilled shells. Such a configuration for the protons (say) would be:

$$(n_1 l_1 j_1)^{2j_1+1} (n_2 l_2 j_2)^{2j_2+1} \dots (n_i l_i j_i)^{2j_i+1} (n\, l\, j)^k.$$

In this configuration there are several filled shells in each of which there is the full quota of protons allowed by the Pauli principle. There are also k "loose" protons in an unfilled shell. Now closed shells never contribute to the spin or electromagnetic moments of the nucleus since they are spherically symmetric. Therefore these features of the nucleus may be discussed solely in terms of the configuration of the loose particles, let us say $(n\,l\,j)^k$ for the protons and $(n'\,l'\,j')^{k'}$ for the neutrons.

The extreme assumption that characterises the Single Particle Model is that, for odd nuclei, not only can one ignore the closed shells, but also all even numbers of neutrons and protons, even including the loose ones. This means that the spins, moments, etc. of odd nuclei are considered to be completely determined by the odd particle, neutron or proton as it happens to be. As we have hinted previously, this extreme picture is not implied by any of the features of the shell-model that have been assumed so far. In fact, it is not at all evident that it could be justified even after making extra assumptions. It is the purpose of the present and the following sections to discuss the replacement of the extreme Single Particle Model by a more correct model in which all the loose particles are taken into account. In particular we will examine what changes, if any, must be made in the previous predictions based on the extreme Single Particle Model.

With the shell-model features assumed so far, a nucleus can be represented by the configuration of its loose particles:

$$(n\,l\,j)^k\,(n'\,l'\,j')^{k'}.$$

Let us now consider the group of protons only and imagine that small forces act between pairs of particles to couple their angular momenta together to form a total angular momentum $\boldsymbol{J}$:

$$\boldsymbol{j}+\boldsymbol{j}+\cdots(k\text{ terms})=\boldsymbol{J}.$$

The presence of the forces implies that the states of the configuration $(n\,l\,j)^k$, which are all degenerate in the absence of coupling forces, are split up in energy from each other, i.e. the degeneracy is removed. It is very desirable to be able to classify all these states in some way. Such a classification must label all those linearly independent combinations of one-particle states that have total spin as a quantum number, and satisfy the PAULI principle requirement of being anti-symmetric in all particles. We will give more details of such classifications below in the more general case when neutrons and protons are considered together. For the present, we list (Table 10) the number of states of each J value that are allowed for various configurations of protons (or neutrons) by the requirement of anti-symmetry. These allowed states are most directly obtained by a simple but tedious chain-calculation from the allowed states of $(j)^{k-1}$, which are obtained in turn from those of $(j)^{k-2}$, and so on[1].

The requirement that the wave-function of a state should be anti-symmetric in all the k particles which are coupled up to some total spin means that, even in the cases $J=j$, the actual nuclear wave-functions cannot be written in the form of a simple product of a single particle wave-function for a particular particle times a wave-function for an "inert core" made up by the other particles. Rather they are multi-particle wave-functions that, in general, cannot be resolved in this simple fashion. However we will see later, that, in spite of this, the predictions of the correct multi-particle wave-functions coincide with those of the Single Particle Model in several respects.

We now consider in more detail the classification of the states of a general configuration $(n\,l\,j)^k$. We will, in fact, be more general than hitherto and allow

---

[1] M. G. MAYER and J. H. D. JENSEN: Elementary Theory of Nuclear Shell Structure, p. 64. New York: John Wiley & Sons. 1955.

Table 10. *Possible Total Spins J for various Configurations* $(j)^k$. (From M. G. MAYER and J. H. D. JENSEN: Elementary Theory of Nuclear Shell Structure. New York: John Wiley & Sons 1955.)

$j=\frac{3}{2}$		$j=\frac{5}{2}$		$j=\frac{7}{2}$	
$k=1$	$\frac{3}{2}$.	$k=1$	$\frac{5}{2}$.	$k=1$	$\frac{7}{2}$.
$=2$	0, 2.	$=2$	0, 2, 4.	$=2$	0, 2, 4, 6.
		$=3$	$\frac{3}{2}, \frac{5}{2}, \frac{9}{2}$.	$=3$	$\frac{3}{2}, \frac{5}{2}, \frac{7}{2}, \frac{9}{2}, \frac{11}{2}, \frac{15}{2}$.
				$=4$	0, 2 (twice), 4 (twice), 5, 6, 8.

$j=\frac{9}{2}$

$k=1$	$\frac{9}{2}$.
$=2$	0, 2, 4, 6, 8.
$=3$	$\frac{3}{2}, \frac{5}{2}, \frac{7}{2}, \frac{9}{2}$ (twice, $\frac{11}{2}, \frac{15}{2}, \frac{17}{2}, \frac{21}{2}$.
$=4$	0 (twice), 2 (twice), 3, 4 (3 times), 5, 6 (3 times), 7, 8, 9, 10, 12.
$=5$	$\frac{1}{2}, \frac{3}{2}, \frac{5}{2}$ (twice), $\frac{7}{2}$ (twice), $\frac{9}{2}$ (3 times), $\frac{11}{2}$ (twice), $\frac{13}{2}$ (twice), $\frac{15}{2}$ (twice), $\frac{17}{2}$ (twice), $\frac{19}{2}, \frac{21}{2}, \frac{25}{2}$.

$j=\frac{11}{2}$

$k=1$	$\frac{11}{2}$.
$=2$	0, 2, 4, 6, 8, 10.
$=3$	$\frac{3}{2}, \frac{5}{2}, \frac{7}{2}, \frac{9}{2}$ (twice), $\frac{11}{2}$ (twice), $\frac{13}{2}, \frac{15}{2}$ (twice), $\frac{17}{2}, \frac{19}{2}, \frac{21}{2}, \frac{23}{2}, \frac{27}{2}$.
$=4$	0 (twice), 2 (3 times), 3, 4 (4 times), 5 (twice), 6 (4 times), 8 (4 times), 9 (twice), 10 (3 times), 11, 12 (twice), 13, 14, 16.
$=5$	$\frac{1}{2}, \frac{3}{2}$ (twice), $\frac{5}{2}$ (3 times), $\frac{7}{2}$ (4 times), $\frac{9}{2}$ (4 times), $\frac{11}{2}$ (5 times), $\frac{13}{2}$ (4 times), $\frac{15}{2}$ (5 times), $\frac{17}{2}$ (4 times), $\frac{19}{2}$ (4 times), $\frac{21}{2}$ (3 times), $\frac{23}{2}$ (3 times), $\frac{25}{2}$ (twice), $\frac{27}{2}$ (twice), $\frac{29}{2}, \frac{31}{2}, \frac{35}{2}$.
$=6$	0 (3 times), 2 (4 times), 3 (3 times), 4 (6 times), 5 (3 times), 6 (7 times), 7 (4 times), 8 (6 times), 9 (4 times), 10 (5 times), 11 (twice), 12 (4 times), 13 (twice), 14 (twice), 15, 16, 18.

the k nucleons to be composed of both neutrons and protons. When dealing with both types of particles together on the same footing, it is convenient to use the isotopic spin formalism in which each nucleon is assigned a so-called "isotopic spin" $\boldsymbol{t}$. This vector is such that its third component t_3 is $-\frac{1}{2}$ for protons, $+\frac{1}{2}$ for neutrons. The real advantage of this formalism is that, in most practical cases, T, the total isotopic spin, is a good quantum number and so is useful for classifying the states. (In nuclei with $A<50$, the effect of COULOMB forces is still relatively weak so that one finds neutrons and protons filling the same orbits simultaneously, and one can assume T is a good quantum number. In nuclei with $A>50$, the COULOMB forces repel protons to such an extent that neutrons and protons are not found to fill the same shells together. Thus, for such nuclei, a shell that is filling will contain all neutrons or all protons so that for that shell, T is trivially a good quantum number, viz. $T=k/2$ in spite of the COULOMB effects. Of course, this does not imply that T is a good quantum number for the whole nucleus.)

In addition to T, of course, total angular momentum J is a good quantum number as always. In general, J and T alone are not sufficient to classify the states of $(n\,l\,j)^k$ completely. Therefore one tries to find extra quantum numbers that will help to do this. Such quantum numbers are the so-called *seniority number* s and *reduced isotopic spin* t[1], which, although not universally good

[1] B. H. FLOWERS: Proc. Roy. Soc. Lond., Ser. A **212**, 248 1952).

quantum numbers like J, are expected to be reasonably good quantum numbers in the situations of practical interest.

In order to give some tangible meaning to these new quantum numbers, we first consider the relation of the ordinary isotopic spin T to the symmetry properties of a state. When we use the isotopic spin formalism, we are regarding all particles as equivalent and so the wave-function must be anti-symmetrised in all particles. The fact that each particle has only two isotopic spin states means that at most two particles at a time can be in the same space state i.e. symmetric (in space). Thus the wave-function can be characterised by the number of symmetrically coupled pairs S (anti-symmetric in isotopic spin) and the number of anti-symmetrically coupled particles A (symmetric in isotopic spin). We have, of course, that the total number of particles $k=2S+A$. Since all the symmetric pairs have $T=0$, the isotopic spin arises purely from the A, anti-symmetrically coupled, particles and so: $2T=A$. We can express this alternatively by saying that $2T$ is defined as the number of particles left after any symmetric pairs ($T=0$) have been removed from the state. The seniority number s of a state is defined as the number of particles left after any *anti-symmetric pairs* ($T=1$) *with* $J=0$ have been removed from the state. The reduced isotopic spin t is the isotopic spin of the state of the remaining particles. In practice, it is especially simple to build up states with quantum numbers s and t. (Of course, an arbitrary classification of states will not have these quantum numbers.)

States characterised by s and t have the property of being nearly diagonal for ordinary and HEISENBERG (charge-exchange) central forces. This may not seem very advantageous at first sight as the nuclear force mixture is believed to have a strong MAJORANA (space exchange) component and possibly a BARTLETT (spin-exchange) component. However such forces reduce to ordinary and HEISENBERG forces in the short-range limit and so, provided forces of not too long a range are used, s and t are still approximately good quantum numbers. In practice, the relevant quantity for deciding whether a range is long or short for these purposes is the ratio of the range of forces to the nuclear radius. If this is $\ll 1$, the proposed classification is useful.

We have considered the classification of states of a configuration $(n\,l\,j)^k$. This will suffice for light nuclei ($A<50$), where the nucleus in general can be regarded as a number of closed shells plus k nucleons (both neutrons and protons in general) in the $(n\,l\,j)$ shell. In heavier nuclei, as we have mentioned, neutrons and protons in a given nucleus are found to be filling different shells so that the nucleus in general consists of closed shells plus (say) k protons in the $(n\,l\,j)$ shell and k' neutrons in the $(n'\,l'\,j')$ shell. Although our previous considerations can be applied to the group of neutrons or the group of protons separately, none of our remarks so far indicate a procedure for classifying states of groups taken together. In general, this is a much more complex situation, but one can simplify matters considerably by assuming that each total state consists of a simple product (appropriately vector-coupled to some total angular momentum) of a proper state of the neutron group and a proper state of the proton group. Of course, the usefulness of this classification depends critically on whether proper total nuclear states have approximately this structure or not. One can argue that states do tend to have this simple structure, because the loose neutrons and protons in nuclei with $A>60$ are in such different shells that there is only small overlap between their spatial distributions. The short range of nuclear forces then implies that the effective interaction between the loose neutrons and protons is weak. In Sect. 12, we will give more attention to states consisting of a group of neutrons in one orbit and a group of protons in another orbit.

11. States of several particles in the same shell: Applications. In this Section, the comparisons of theory and experiment given in Chap. III will be revised in the light of the more consistent shell-model in which all loose particles are properly taken into account. There can be significant differences between the predictions of this model and the extreme Single Particle Model (see, for example, the subsequent discussion of magnetic dipole moments). Agreement with experiment is found to be generally improved on using the more correct model. However there remain certain types of data which are still very poorly predicted (see the subsequent discussion of electric quadrupole moments).

α) *Spectra of states.* The extreme Single Particle Model is intrinsically incapable of predicting reasonable spectra of states because of the very large degeneracies that it implies. According to the present improved shell-model, these degeneracies are removed by the interparticle forces. The resulting discrete sets of states are diagonal in these forces, and each one is in general a multi-particle state with all loose particles coupled together in some definite manner. In general, one can only find the spectrum generated from a given configuration by some given type of central force by following the usual methods of atomic spectroscopy, i.e. setting up the Hamiltonian matrix in some representation and diagonalising. It is straightforward to do this for an individual nucleus, given the necessary fractional parentage expansions of the representation. However this procedure is of no use for discussing systematics of nuclear spectra. For this we need closed formulae for the energies of eigenstates. It is fortunate that such formulae have been established for most cases of practical interest. As we have already pointed out, for WIGNER (i.e. ordinary) forces, proper eigenstates are approximately those characterised by the quantum numbers s and t (provided, as always, that the effective strength of the forces is weak compared with the spin-orbit coupling). It can be shown, by some use of group theoretical methods[1], that the diagonal energy of ordinary forces $\sum_{i<j} v_{ij}$ taken in a state of j^k characterised by quantum numbers J, T, s, t and α is:

$$E(j^k; J, T, s, t, \alpha) = \frac{1}{2} k(k-1) E_0 + \frac{1}{2}\left\{\frac{3k}{2} - 2T(T+1) + k(j+1) - \frac{1}{2} g(s,t)\right\} E_1 + \cdots.$$

In this expression $E_0, E_1, E_2 \ldots E_{j-\frac{1}{2}}$ are certain linear combinations of the well-known SLATER integrals, $F^0, F^2, \ldots F^{2l}_r$, (see Appendix 8). In the short-range limit, where we can write $v(r_{ij}) = V\delta(\mathbf{r}_{ij})$, we have:

$$F^p = (2p+1) F^0, \quad \text{where} \quad F^0 = V \int (r u_l)^4 \frac{dr}{4\pi r^2}.$$

In the same limit, E_0 and E_1 are much larger than the other $E_2 \ldots E_{j-\frac{1}{2}}$, and they satisfy:

$$E_1 = 2E_0 = \left(\frac{j+\frac{1}{2}}{j+1}\right) F^0,$$

$$E_0, E_1 \gtrsim 100\, E_2.$$

We remark that the coefficients of E_0 and E_1 do not contain total spin J and so the main ordering in the spectrum is determined by s, t and T, with J-splitting as fine structure. The quantity $g(s, t)$ is the so-called "eigenvalue of the CASIMIR operator" and can be expressed in terms of s and t by:

$$g(s,t) = (2j+2)\, s - \tfrac{1}{2} s(s-1) + \tfrac{3}{2} s - 2t(t+1).$$

---

[1] A. R. EDMONDS and B. H. FLOWERS: Proc. Roy. Soc. Lond., Ser. A **214**, 515 (1952).

The non-diagonal elements of a WIGNER force in the present states do not contain E_0 and E_1 but only $E_2, E_3 \ldots$ etc, so these elements are small and the interaction matrix is approximately diagonal in these states.

We now extend the discussion to types of forces other than the WIGNER force. No closed formulae like the one above have been given for other types of forces, but we can deduce appropriate expressions in special cases. For instance, in the short-range limit, MAJORANA forces effectively reduce to WIGNER forces and so the above formula can be used for MAJORANA forces. The other two types of force, HEISENBERG and BARTLETT, become equivalent to each other in the short-range limit, but do not become equivalent to WIGNER forces in general. Thus we cannot cope with these forces using the above formula. However, in the special case when all particles are of the same type, neutrons or protons, (i.e. maximum isotopic spin, $T=k/2$), the spin exchange forces become equivalent to WIGNER forces because the short range limit only allows the singlet interaction to be effective. To be more explicit, let us suppose that the central force between any two particles i and j is:

$$v_{ij}=(W+M\,P_{ij}^{r}-H\,P_{ij}^{\tau}+B\,P_{ij}^{\sigma})\,V_0(r_{ij})\,.$$

We can assume that the coefficients W, M, H and B satisfy $W+M+H+B=1$, so that $V_0(r_{ij})$ is the attractive (negative) potential in the deuteron ground state. The various permutation operators P are defined[1] as:

$$P_{ij}^{\sigma}=\tfrac{1}{2}(1+\boldsymbol{\sigma}_i\cdot\boldsymbol{\sigma}_j)\,,\qquad P_{ij}^{\tau}=\tfrac{1}{2}(1+\boldsymbol{\tau}_i\cdot\boldsymbol{\tau}_j)\,,\qquad P_{ij}^{r}=-P_{ij}^{\tau}P_{ij}^{\sigma}.$$

If all particles are of the same type (i.e. $T=k/2$, $t=s/2$), then, in the short-range limit where $E_1=2E_0$, we have:

$$E(j^k;J,s,\alpha)=(W+M-H-B)\left[\left\{(j+1)\,k-(2j-s)\,\frac{s}{2}\right\}E_0+\text{small terms}\right].$$

The potential $(W+M-H-B)\,V_0(r_{ij})$ is the interaction in the singlet state of the deuteron, which is believed to be attractive. Accordingly, we can assume $(W+M-H-B)$ to be positive and thereby make predictions about the ordering of states in the spectrum of j^k. It is clear that, since $s<2j$, the ordering is that of increasing s with the state of lowest s being the ground state. Since for even k, the lowest s-value is zero, and since $s=0$ always specifies a unique state which has $J=0$, one predicts that the ground state of an even group of particles has spin zero. For odd k, the lowest s-value is 1 and this again specifies a unique state, namely $J=j$, so the spin of the ground state of an odd group of particles is predicted to be that of a single particle. We note that the above formula for $E(j^k;J,s,\alpha)$ shows that the binding energy of the ground state $(s=0)$ of an even number of like particles contains the factor $(2j+1)\,F^0$. If F^0 is assumed be independent of l, this result gives a theoretical basis to the empirical rule of Sect. 6 that pairing energy increases with j-value.

Before applying these results to actual medium and heavy nuclei, we must take some account of the fact that such nuclei generally contain *two* groups of "loose" particles, one of protons and one of neutrons, in different orbits. In order to apply our results, it follows that we must invoke the extra assumptions that the two groups do not seriously perturb each other. This means that the ground state of an odd nucleus is thought to consist of an even number of one type of nucleons with spin $J=0$, and an odd number of the other type with spin $J=j$

[1] $\boldsymbol{\sigma}$ and $\boldsymbol{\tau}$ are related to the intrinsic spin vector $\boldsymbol{s}$ and the isotopic spin vector $\boldsymbol{t}$ by: $\boldsymbol{\sigma}=2\boldsymbol{s}$, $\boldsymbol{\tau}=2\boldsymbol{t}$.

where j is the spin characterising the shell that the odd particles are filling. The spin of the total system, neutrons and protons is thus predicted unambiguously to be $J=j$. For even-even nuclei, the prediction is also unambiguous, viz. $J=0$. For odd-odd nuclei however, the ground state spin of the total system is not determined by the present considerations, but can only be said to fall between the limits $J=|j_n-j_p|$ and $J=j_n+j_p$, where j_n, j_p are the spins of the last neutron and proton shells respectively. The ground state spins and the spectra in such cases can only be obtained after further study. The methods to be used in this connection are described below in Sects. 12 and 13 which are devoted to consideration of states containing inequivalent particles. Consequently we will withhold discussion of odd-odd nuclei with the loose neutrons and protons in different shells until Sect. 13. This excludes all medium and heavy odd-odd nuclei from the discussions of the present section. There are some light odd-odd nuclei for which the loose neutrons and protons are in the same shell. Such nuclei can of course be treated straightforwardly with the methods for equivalent particles. Notice however that specification of s and t for the states of lowest T in an odd-odd nucleus does not lead to a unique J. Thus, to find the ground state spin, it is necessary to specify the higher (smaller) terms in the above formula that depend on J. We will not discuss the light odd-odd nuclei here however, because it appears that the basis of j-j coupling is a poor one for such nuclei.

To conclude the discussion of spectra, we call attention to the striking behaviour of the observed 0^+ to 2^+ separation energies of even-even nuclei as a function of mass number that was noted in Chap. 2. On the present model, this separation is easily found from the above formula using the fact that the 0^+ and 2^+ states belong to $s=0$, 2 respectively. It is simply $(W+M-H-B)(j-1)|E_1|$ and is independent of k, the number of particles in the shell[1]. An attempt to reconcile this result with the observed steady decrease of the separation with mass number distance from major closed shells has been made in terms of the interaction between the neutron and proton groups[2]. Without interaction, the spectrum of the combined groups is clearly 0^+, then two 2^+ states close together, then many other states higher up. Interaction between the groups will tend to push one of the 2^+ states downwards towards the ground state. However, it is not clear why this effect should be stronger in between major shells and fall off smoothly in the observed fashion. Consequently this explanation cannot be regarded as satisfactory although it may contain some truth. An alternative approach will be discussed in Sect. 15.

β) Binding energies. The energy with which a group of k "loose" particles $(j)^k$ are bound to closed shells can generally be written as a sum of the single particle energy, say $k\mathcal{B}$ where $\mathcal{B}$ represents the binding of a single j-nucleon, and the interaction energy $E(j^k; J, T, s, t, \alpha)$ where J, T, s, t, α are the quantum numbers of the lowest state of $(j)^k$. As we have seen in the preceding discussion, the theoretical expression for a general interaction energy involves the SLATER integrals and the coefficients of the interaction force mixture. However these integrals and coefficients enter in definite combinations so that the number of parameters available in fitting is reduced. On specialising the formulae for $E(j^k; J, T, s, t, \alpha)$ to the case of ground states, we obtain expressions which can be used in attempts to fit binding energies. We remark here that the appropriate expressions for a somewhat more general case have been derived directly in an elegant fashion[3].

---

[1] C. SCHWARTZ and A. DE SHALIT: Phys. Rev. **94**, 1257 (1954).

[2] A. DE SHALIT and M. GOLDHABER: Phys. Rev. **92**, 1211 (1953).

[3] G. RACAH: L. Farkas Memorial Volume. Research Council of Israel **1**, 29 (1952).

One first notes that there are always only three types of states of j^2, namely those with $(s, t) = (0, 0)$, $(2, 0)$, $(2, 1)$ with the first containing only state $J=0$, $T=1$, the second containing all states of odd J values and $T=0$ and the third containing all states of even J values (except $J=0$) and $T=1$. Defining the barycentric energy of a group of levels of given (s, t) as:

$$E(j^2; s, t) = \sum_J (2J+1)\, E(j^2; J, T, s, t, \alpha) \Big/ \sum_J (2J+1)$$

the existence of only three (s, t) values means that one can always write $E(j^2; s, t)$ in terms of three parameters a, b, and c as follows:

$$E(j^2; s, t) = a + \frac{b}{2}\left\{T(T+1) - \frac{3}{2}\right\} + \frac{c}{2}\left\{g(s, t) - 4(j+1)\right\}.$$

It can now immediately be shown that for the case of k particles:

$$E(j^k; s, t) = \tfrac{1}{2}k(k-1)\, a + \tfrac{1}{2}b\left\{T(T+1) - \frac{3k}{4}\right\} + \frac{c}{2}\left\{g(s, t) - 2k(j+1)\right\}.$$

Now for the case of an even number of neutrons and an even number of protons, the lowest energy is given by $(s, t) = (0, 0)$ and this has only one J-state, namely $J=0$. Using $g(0, 0) = 0$, we have immediately an expression for the binding of the loose nucleons in ground states of even-even nuclei in terms of the four constants $\mathcal{B}, a, b, c$. Similarly for odd nuclei, the lowest $(s, t) = (1, \tfrac{1}{2})$, and this has only one J-state, namely $J=j$, so that the binding of the loose nucleons in the ground states of odd nuclei can also be written in terms of $\mathcal{B}, a, b, c$.

Observed binding energies have been analysed with this approach for nuclei with $A < 50$[1]. Because one is comparing nuclei with different numbers of neutrons and protons, it is necessary to introduce a fifth term for the COULOMB energy. The observed bindings in each shell are fitted with the least squares method, and it is found that, even when there are of the order of thirty observed energies, the fitting can be made with a root mean-square deviation of only a fraction of one percent. The COULOMB energy term emerges in each case with a very reasonable value. The four constants $\mathcal{B}, a, b, c$ are assigned values in the fitting but, as yet, no attempt to interpret these values in terms of interaction forces has been reported. At first sight, the fitting of so many observed binding energies with only a few parameters is a striking success for the j-j coupling model. In contrast, it is known that for certain spectra of nuclei of $A < 50$, the j-j model fails to give agreement (for example, F^{19}, see Sect. 27). Thus one must be cautious in interpreting the fitting of the binding energies.

γ) Magnetic dipole moments. The magnetic dipole moment of the lowest state of given T of the configuration j^k (k odd) can be written in closed form[2]. The k particles consist of an odd number of one kind (neutrons or protons) and an even number of the other kind. If the odd number exceeds the even number:

$$\mu = \mu_0 + (\mu_e - \mu_0)\left(\frac{k/2 - T}{(2j+2)(2T+2)}\right),$$

and if the even number exceeds the odd number:

$$\mu = \mu_0 + (\mu_e - \mu_0)\left(\frac{2j + 1 - T - k/2}{(2j+2)(2T+2)}\right).$$

[1] R. THIEBERGER and I. TALMI: Phys. Rev. **102**, 923 (1956).

[2] M. MIZUSHIMA and M. UMEZAWA: Phys. Rev. **83**, 463 (1951); **85**, 37 (1952). — B. H. FLOWERS: Phil. Mag. **43**, 1330 (1952). — Report of Birmingham Conference on Nuclear Physics, p. 8. 1952.

The quantities μ_0 and μ_e are the single particle moments of the particles in the odd and even groups respectively. It is evident that, in general, the magnetic moment is not simply the single particle moment μ_0. A noteworthy point about the above expressions is that the coefficients of $(\mu_e - \mu_0)$ are both always positive. Furthermore when μ_0 is obtained from the upper SCHMIDT line, $(\mu_e - \mu_0)$ is negative, and when it is obtained from the lower SCHMIDT line, $(\mu_e - \mu_0)$ is positive. Thus, when the moments of the above formulae deviate from the SCHMIDT line values they always deviate inwards, i.e. towards the observed values. However such deviations are only possible in applications to the lighter nuclei ($A < 50$), where the j-j coupling model is believed to be inaccurate (see Chap. VI). For $A > 50$, the loose neutrons and loose protons occur in different shells. If the neutrons and protons can be treated independently in such nuclei as we have previously suggested, the moment of an odd nucleus arises solely from the odd group, since the moment of the even group with $J = 0$ vanishes. Thus we are concerned with the moment of an odd group of particles of the same type ($T = k/2$). From the first of the above formulae, the moment of the ground state of such a group is simply the single particle moment μ_0. Consequently the more correct treatment of nuclei with $A > 50$ in which all the loose particles are taken into account gives the same moment as that of the extreme Single Particle Model. It follows that the observed deviations from the SCHMIDT line for these nuclei cannot be explained in this way. Instead, one must develop the model further and either take into account the interaction between the states of the loose neutrons and protons or else consider configuration mixing in some way.

δ) Electric quadrupole moments. No general formula has been given for the electric quadrupole moment Q of a general state of a configuration j^k of neutrons and protons. However, in the special case where all particles are of the same type, a simple expression for Q is obtainable. For a group of neutrons Q is zero, of course, apart from a small recoil contribution from the protons in the rest of the nucleus. For a group of protons, Q is zero for the ground state of an even number ($J = 0$) and, for the ground state of an odd number[1] with $s = 1$, $J = j$:

$$Q = Q_j\left(1 - \frac{2k-2}{2j-1}\right)$$

where Q_j is the single particle value. It is disappointing from the viewpoint of the comparison of the present model with experiment that the factor multiplying Q_j never rises above unity for any allowed value of k. Thus the model is not capable of explaining the large observed values of Q shown in Fig. 7. Neither does it appear capable of explaining the predominance of positive Q values, because the factor changes sign for k in the middle of the shell and leads to as many negative as positive moments. Notice however that some agreement with observation is achieved just at the beginning of a shell ($Q \sim Q_j$) and at the end ($Q \sim -Q_j$). Apart from this, the predictions of the more correct shell-model are just as bad as those of the extreme single particle model.

To conclude this discussion of moments, we note a very simple derivation of the moment of any state of a configuration j^k of the same type of particle, neutron or proton[2]. The essential simplifying feature is that all particles are equivalent and so have the same individual moment μ_j which we can write as

[1] R. G. SACHS: Nuclear Theory, Appendiy 4. Cambridge: Addison Wesley Publishing Co. Ltd. 1953.

[2] M. G. MAYER and J. D. H. JENSEN: Elementary Theory of Nuclear Shell Structure, Appendix 4. New York: John Wiley & Sons 1953.

$\boldsymbol{\mu}_j = g_j \boldsymbol{j}$ where g_j is thus defined as the effective gyromagnetic ratio. The moment, μ, of a state of k such particles with spin J is, by definition, the z-component of:

$$(J, M=J \mid \sum \boldsymbol{\mu}_j \mid J, M=J) = g_j (J, M=J \mid \sum \boldsymbol{j} \mid J, M=J).$$

Thus:

$$\mu = g_j J = \mu_j \left(\frac{J}{j}\right).$$

Notice that this formula is independent of whether k is even or odd and independent of whether states are classified by seniority (s, t) or not.

ε) The β-transitions and isomeric transitions. Perhaps the most serious deficiency of the Single Particle Model in applications to radiative and β-transitions is that it predicts too large values for the matrix elements. This is true for all types of transition except $E2$ radiative transitions (usually too small values) and super-allowed β-transitions (correct values). Let us take allowed β-transitions as an example. The Single Particle Model always predicts roughly the same log ft-value for such transitions ($\log ft \sim 3$). In practice however the values fall roughly into two groups; the super-allowed transitions with $\log ft \sim 3$, and the ordinary-allowed transitions with $\log ft \sim 5$. The super-allowed transitions mostly take place between states in light nuclei which, from charge independence, are identical except for the occurrence of a neutron in one state where the other state has a proton. The ft-values for such transitions are not strongly dependent of specific models such as the shell-model, and need not be considered further. The other transitions do not take place between states that are simply related by charge independence. For these, the matrix elements are at least an order of magnitude smaller than the single particle predictions in general. Qualitatively such reductions are understandable when the Single Particle Model is extended to take into account all loose particles. On the one hand, this change leads to increased matrix elements since there are now several particles that can make the transitions. On the other hand, the change leads to reductions because of the incomplete overlap of the states of particles not taking part in the transition. In practical cases, it is usually found that the latter (reducing) factor more than compensates the former (enhancing) factor, so that matrix elements are expected to be less than the single particle values on the whole. This will be so except when the overlap between the initial and final states is anomalously good (as in the case of super-allowed transitions in mirror nuclei, where charge independence implies very good overlap).

Thus, in a qualitative sense, the improved shell-model predicts results in accordance with observation except in the case of $E2$ transitions, where the prediction of lifetimes longer than the single particle one is completely opposite to the observed short lifetimes.

12. States of several particles in different shells: Theory. So far in this Chapter, only the coupling of particles in the same orbit has been considered. We will now consider the coupling of particles in different orbits in some simple situations. The simplest situation is that of one particle (say a neutron) in the (l, j) orbit and one particle (say a proton) in the (l', j') orbit. In the absence of interactions between the two particles, they form a group of degenerate states with all spins from $J = |j - j'|$ to $J = j + j'$. If a small central interaction is introduced between the particles, these states are split up in energy. In addition to its zero-order energy, each state of given J acquires an energy which is just the diagonal element of the interaction:

$$E_J = (J(j j') M \mid \boldsymbol{v}_{12} \mid J(j j') M)$$

where $\boldsymbol{v}_{12}$ is the interaction.

In the short-range limit, any central force like that of the previous section can be written in the form:

$$v_{12} = [1 - \alpha + \alpha(\boldsymbol{\sigma}_1 \cdot \boldsymbol{\sigma}_2)]\, V_0(r_{12}).$$

where $V_0(r_{12}) = V\delta(\boldsymbol{r}_{12})$ and where the constant α is related to our previous constants W, M, H, B by the relation:

$$\alpha = \tfrac{1}{2}(H + B) = \tfrac{1}{2}(1 - W - M).$$

The contribution of the ordinary force, $(1-\alpha)\, V_0(r_{12})$, to E_J can be shown by straightforward application of standard RACAH algebra to be[1] (see Appendix 4, 8):

$$E_0(jj', J) = \frac{(1-\alpha)\, F^0(ll')}{2} (2j+1)(2j'+1)(j\tfrac{1}{2}j' - \tfrac{1}{2}|J0)^2 \times \\ \times \left[\frac{4J(J+1) + [(2j+1) + (-)^{j+j'+J}(2j'+1)]^2}{4J(J+1)(2J+1)}\right]$$

and the contribution of the spin force $\alpha(\boldsymbol{\sigma}_1 \cdot \boldsymbol{\sigma}_2)\, V_0(r_{12})$ is:

$$E_\sigma(jj', J) = \frac{\alpha F^0(ll')}{2} (2j+1)(2j'+1)(j\tfrac{1}{2}j' - \tfrac{1}{2}|J0)^2 \times \\ \times \left[\frac{[(2j+1) + (-)^{j+j'+J}(2j'+1)]^2 - [1 + 2(-)^{l+l'+J}]\, 4J(J+1)}{4J(J+1)(2J+1)}\right].$$

In these expressions $F^0(ll')$ is a SLATER integral (see Appendix 8), which, for zero-range forces, has the form of an overlap integral:

$$F^0(ll') = \frac{V}{4\pi} \int (r u_l)^2 (r u_{l'})^2 \frac{dr}{r^2}$$

where $r u_l(r)$, $r u_{l'}(r)$ are the radial wave-functions of the l and l' orbits.

Having considered the simple case of only one particle in each of the two orbits, we will now mention the more general case of k particles in the (l, j) orbit and k' particles in the (l', j') orbit, where k and k' are both odd numbers. In general the problem of diagonalising the interaction energy completely for the $(k+k')$ particles is lengthy. In practice, when treating odd-odd nuclei, one usually makes the following assumption: not only is the effective interaction strength assumed to be weak compared to the spin-orbit force (thereby permitting a j-j coupling approach), but also the effective interaction strength between two particles in the orbits l and l' is assumed to be weak compared with that between two particles in the same orbit. Such an assumption should be justified if the orbits l and l' have small overlap in view of the short-range nature of nuclear forces. The essential simplification consequent on the assumption is that each total state wave-function can be regarded as a simple vector-coupled product of the internal wave-functions of the two groups. The lowest states of the total system on this picture will contain the two groups in their lowest states. These component states have $s=1$, $t=\frac{1}{2}$ and the single particle spin. The interaction energy for short range forces between the two groups can be written as a sum of contributions from the ordinary and spin forces. These contributions are simply expressible

[1] A. DE SHALIT and M. GOLDHABER: Phys. Rev. **92**, 1211 (1953). — A. DE SHALIT: Phys. Rev. **91**, 1479 (1953). — C. SCHWARTZ: Phys. Rev. **94**, 95 (1954).

in terms of those for the two-particle case[1]:

$$E_0\big((j)_j^k\,(j')_{j'}^{k'},J\big)=\left(\frac{2j+1-2k}{2j-1}\right)\left(\frac{2j'+1-2k'}{2j'-1}\right)\times$$
$$\times\left[E_0(jj',J)-(1-\alpha)\,F^0(ll')\right]+(1-\alpha)\,k k' F^0(ll'),$$
$$E_\sigma\big((j)_j^k\,(j')_{j'}^{k'},J\big)=E_\sigma(jj',J).$$

Notice that the spin force contribution does not depend on k and k' and so it remains the same if the k particles are changed into $2j+1-k$ particles, i.e. k holes. The ordinary force contribution depends on k. When k particles are changed into $2j+1-k$ particles (k holes), the sign of E_0 changes since $\frac{2j+1-2k}{2j-1}$ changes sign. When both groups of particles are changed into holes E_0 is unchanged.

We conclude with a few remarks about the theoretical values of electromagnetic moments of odd-odd nuclei. Assuming as usual that the lowest state of an odd-odd system can be written as a vector coupled product of the lowest state of the proton group (spin $J_p=j_p$, moment $g_{j_p}\boldsymbol{j}_p$) and the lowest state of the neutron group (spin $J_n=j_n$, moment $g_{j_n}\boldsymbol{j}_n$), the magnetic dipole vector of the total system is $\boldsymbol{\mu}=g_{j_p}\boldsymbol{j}_p+g_{j_n}\boldsymbol{j}_n$. The magnetic dipole moment of the state of the total system of spin $\boldsymbol{j}_n+\boldsymbol{j}_p=\boldsymbol{J}$ can be expressed in the following way on using the theorem that was quoted in Sect. 4:

$$\mu=\frac{(J,M=J\,|\,\boldsymbol{\mu}\cdot\boldsymbol{J}\,|\,J,M=J)}{J+1}.$$

A little easy algebra gives finally:

$$\mu=\frac{J(J+1)\,(g_{j_n}+g_{j_p})+\big(j_n(j_n+1)-j_p(j_p+1)\big)\,(g_{j_n}-g_{j_p})}{2(J+1)}.$$

The corresponding formula for the electric quadrupole moment can also be obtained straigtforwardly[2]:

$$\varphi=\frac{(2J+1)!}{(2j_p)!}\left\{\frac{(2j_p-2)!\,(2j_p+3)!}{(2J-2)!\,(2J+3)!}\right\}^{\frac{1}{2}}W(j_pJj_pJ,j_n2)\,(-)^{j_n-j_p-J}\,Q_{j_p}$$

where Q_{j_p} is the single particle moment of Chap. II.

13. States of several particles in different shells: Applications. The considerations of the last section have been applied in practice to a variety of nuclear data. Here we shall review the work on (α) odd-odd nuclei with $A\geq 50$, (β) the nuclei near Pb^{208}, (γ) Zr^{90}.

α) Odd-odd nuclei with $A>50$. In Tables 11a—c, we list the odd-odd nuclei of $A\geq 50$ whose spins are established as 0^-, 1^+ or 1^- from the β-decay associating these nuclei with the lowest 0^+ and 2^+ states of even-even nuclei. The Table includes some relevant β-decay data, along with shell-model orbits for the last odd neutron and odd proton. In Table 12, we list those odd-odd nuclei whose spins are suggested to be 2^- from β-decay data.

The ground state spins J are found to obey certain empirical rules which can be stated in terms of the so-called NORDHEIM number[3] $N=(j_p-l_p)+(j_n-l_n)$:

Strong Rule $\quad J=|j_n-j_p|\quad$ when N is even,

Weak Rule $\quad |j_n-j_p|<J<j_n+j_p$ when N is odd.

[1] C. SCHWARTZ: Phys. Rev. **94**, 95 (1954).
[2] R. J. BLIN-STOYLE: Rev. Mod. Phys. **28**, 75 (1956).
[3] L. NORDHEIM: Phys. Rev. **78**, 294 (1950).

One can supplement the Weak Rule by saying that, for odd N, J is always $>|j_n-j_p|$ and it tends towards the maximum allowed value of j_n+j_p. Both rules can be expressed compactly by saying that there is a tendency for the two intrinsic spins to be parallel in the lowest state. It is not surprising that one

Table 11a. *Some β-Transitions of the $\Delta J=\pm 1\,(+)$ Type between Ground States of Even-Mass Nuclei.* (From E. Feenberg: Shell Theory of the Nucleus, p. 104. Princeton: Princeton University Press 1955.)

Parent Daughter	Energy (MeV)	Odd group spins and parities		log $f't$
		Odd Z	Odd N	
$_{29}Cu_{31} \to {}_{28}Ni_{32}$	3.3	$3/2^-$	$3/2, 5/2^-$	>7.12
$_{29}Cu_{35} \to {}_{30}Zn_{34}$	0.571	$3/2^-$	$3/2, 5/2^-$	5.31
$_{29}Cu_{35} \to {}_{28}Ni_{36}$	0.657	$3/2^-$	$3/2, 5/2^-$	4.98
$_{29}Cu_{37} \to {}_{30}Zn_{36}$	2.63	$3/2^-$	$3/2, 5/2^-$	5.34
$_{31}Ga_{35} \to {}_{30}Zn_{36}$	4.144	$3/2^-$	$3/2, 5/2^-$	7.81
$_{31}Ga_{37} \to {}_{30}Zn_{38}$	1.88	$3/2^-$	$3/2, 5/2^-$	5.24
$_{31}Ga_{39} \to {}_{32}Ge_{38}$	1.65	$3/2^-$	$1/2^-$	$\gtrsim 5.04$
$_{35}Br_{45} \to {}_{36}Kr_{44}$	1.99	$3/2^-$	$1/2^-$	5.56
$_{35}Br_{45} \to {}_{34}Se_{46}$	0.868	$3/2^-$	$1/2^-$	4.54
$_{45}Rh_{59} \to {}_{46}Pd_{58}$	2.6	$9/2^+$	$7/2^+$	4.69
$_{44}Ru_{62} \to {}_{45}Rh_{61}$	0.0392	$9/2^+$	$7/2^+$	4.75
$_{45}Rh_{61} \to {}_{46}Pd_{60}$	3.53	$9/2^+$	$7/2^+$	5.26
$_{47}Ag_{57} \to {}_{46}Pd_{58}$	2.70	$9/2^+$	$7/2^+$	5.40
$_{47}Ag_{59} \to {}_{46}Pd_{60}$	1.945	$9/2^+$	$7/2^+$	~ 4.97
$_{47}Ag_{61} \to {}_{48}Cd_{60}$	1.5	$9/2^+$	$7/2^+$	~ 4.42
$_{47}Ag_{63} \to {}_{48}Cd_{62}$	2.82	$9/2^+$	$7/2^+$	5.03
$_{46}Pd_{66} \to {}_{47}Ag_{65}$	0.2	$9/2^+$	$(7/2^+)$	4.39
$_{47}Ag_{65} \to {}_{48}Cd_{64}$	4.1	$9/2^+$	$(7/2^+)$	8.60
$_{49}In_{63} \to {}_{50}Sn_{62}$	0.67	$9/2^+$	$7/2^+$	4.00
$_{49}In_{63} \to {}_{48}Cd_{64}$	1.52	$9/2^+$	$7/2^+$	4.40
$_{49}In_{65} \to {}_{50}Sn_{64}$	1.98	$9/2^+$	$(7/2^+)$	4.47
$_{49}In_{67} \to {}_{50}Sn_{66}$	2.95	$9/2^+$	$(7/2^+)$	4.45
$_{51}Sb_{69} \to {}_{50}Sn_{70}$	1.70	$5/2^+$	$7/2^+$	4.63
$_{53}I_{75} \to {}_{54}Xe_{74}$	2.02	$5/2^+$	$(3/2^+)$	5.94
$_{55}Cs_{75} \to {}_{56}Ba_{74}$	0.442	$5/2^+$	$3/2^+$	4.96
$_{55}Cs_{75} \to {}_{54}Xe_{76}$	1.97	$5/2^+$	$3/2^+$	5.01
$_{57}La_{77} \to {}_{56}Ba_{78}$	2.7	$5/2^+$	$3/2^+$	4.81
$_{57}La_{79} \to {}_{56}Ba_{80}$	2.1	$5/2^+$	$3/2^+$	4.49
$_{59}Pr_{81} \to {}_{58}Ce_{82}$	2.23	$5/2^+$	$3/2^+$	4.30

Table 11b. *Some β-Transitions of the $\Delta J=0\,(-)$ Type between Ground States of Even-Mass Nuclei.* (From E. Feenberg: Shell Theory of the Nucleus, p. 107. Princeton: Princeton University Press 1955.)

Parent Daughter	Energy (MeV)	Odd group spins and parities		log ft
		Odd Z	Odd N	
$_{58}Ce_{86} \to {}_{59}Pr_{85}$	0.304	$7/2^+$	$7/2^-$	7.40
$_{59}Pr_{85} \to {}_{60}Nd_{84}$	2.97	$7/2^+$	$7/2^-$	6.53
$_{59}Pr_{87} \to {}_{60}Nd_{86}$	3.8	$7/2^+$	$7/2^-$	7.17
$_{72}Hf_{98} \to {}_{71}Lu_{99}$	2.4	$7/2^+$	$(7/2^-)$	6.15
$_{81}Tl_{125} \to {}_{82}Pb_{124}$	1.51	$1/2^+$	$1/2^-$	5.18

Table 11c. *Some β-Transitions of the $\Delta J = \pm 1\,(-)$ Type between Ground States of Even-Mass Nuclei.* (From E. FEENBERG: Shell Theory of the Nucleus, p. 107. Princeton: Princeton University Press 1955.)

Parent Daughter	Energy (MeV)	Odd group spins and parities		log f't
		Odd Z	Odd N	
$_{67}Ho_{99} \to {}_{68}Er_{98}$	1.9			8.14
$_{69}Tm_{101} \to {}_{70}Yb_{100}$	0.968			8.97
$_{73}Ta_{107} \to {}_{74}W_{106}$	0.71			6.85
$_{75}Re_{111} \to {}_{76}Os_{110}$	1.07			7.71
$_{77}Ir_{117} \to {}_{78}Pt_{116}$	2.18			≥ 8.16
$_{82}Pb_{128} \to {}_{83}Bi_{127}$	0.065	$7/2^-$, $9/2^-$	$9/2^+$	7.5
$_{83}Bi_{127} \to {}_{84}Po_{126}$	1.17	$7/2$, $9/2^-$	$9/2^+$	8.03
$_{82}Pb_{130} \to {}_{83}Bi_{129}$	0.589	$7/2$, $9/2^-$	$9/2^+$	7.29
$_{83}Bi_{129} \to {}_{84}Po_{128}$	2.25	$7/2$, $9/2^-$	$9/2^+$	7.23

Table 12. *Some β-Transitions of the $2^- \to 2^+$ Type.* (From E. FEENBERG: Shell Theory of the Nucleus, p. 108. Princeton: Princeton University Press 1955.)

Parent Daughter	Energy (MeV)	Half life	lg f't
$_{17}Cl_{21} \to {}_{18}A_{20}$	2.77	37.3m/0.16	6.88
$_{19}K_{23} \to {}_{20}Ca_{22}$	2.04	12.5h/0.20	7.58
$_{33}As_{39} \to {}_{32}Ge_{38}$	2.50	26h/0.175	7.77
$_{33}As_{41} \to {}_{32}Ge_{42}$	0.92	19d/0.622	6.93
$_{33}As_{41} \to {}_{34}Se_{40}$	0.69	19d/p.166	7.55
$_{33}As_{43} \to {}_{34}Se_{42}$	2.57	26.1h/0.21	8.52
$_{37}Rb_{47} \to {}_{36}Kr_{48}$	0.82	34d/0.04	7.75
$_{37}Rb_{49} \to {}_{38}Sr_{48}$	0.72	19.5d/0.20	7.53
$_{37}Rb_{51} \to {}_{38}Sr_{50}$	3.6	17.7m/0.13	7.42
$_{45}Rh_{55} \to {}_{44}Ru_{56}$	2.07	20.8h/0.0188	8.2
$_{51}Sb_{71} \to {}_{52}Te_{70}$	1.36	> 2.8d	> 7.40
$_{53}I_{71} \to {}_{52}Te_{72}$	1.50	4.5d/0.10	7.50
$_{53}I_{73} \to {}_{54}Xe_{72}$	0.85	13d/0.75×0.40	7.86
$_{79}Au_{117} \to {}_{80}Hg_{116}$	0.27	5.55d/0.045	7.30

cannot explain the rule in terms of ordinary interaction forces alone. Instead one must allow for spin dependent forces which depress the triplet state relative to the singlet state. Now, for a general force of the type $[1-\alpha+\alpha(\boldsymbol{\sigma}_1\cdot\boldsymbol{\sigma}_2)]\,V_0$, the triplet interaction is V_0 and the singlet is $(1-4\alpha)\,V_0$. Thus we require that α should be positive. In actual fact, this is not quite sufficient and one requires not only that α is positive, but that it exceeds $\frac{1}{10}$. In such a case, one can see from our previous spectrum formulae that the above rules are predicted for any simple system of one neutron and one proton[1]. In more detail, it can be shown that, for N even, the state $J=|j_n-j_p|$ is well below the others, which are in a bunch and that for N odd, the state $J=j_n+j_p$ is lowest with other states not far away and evenly spaced. It is interesting to see that the prediction of the rules has been tested theoretically in the case when the requirement of zero range is relaxed[2]. It is found that the rules are still predicted in such a case subject to the same condition as above, i.e. $\alpha>\frac{1}{10}$.

[1] A. DE SHALIT: Phys. Rev. **91**, 1479 (1953).
[2] D. M. BRINK: Proc. Phys. Soc. Lond. A **67**, 757 (1954).

Many of the experimentally found odd-odd nuclei have more complicated configurations than simply one neutron plus one proton, and so it is desirable to extend the derivation of the rules to more general cases. Certainly the above comments apply equally well to the case of one proton hole and one neutron hole. In the case of k neutrons and k' protons, the spectrum formula of Sect. 12 shows that the Strong Rule is still valid provided that both k and k' are less than half a filled shell or provided both are greater than half a filled shell. It does not appear possible, in general, to remove this restriction on k and k' and give a general derivation of the two rules. However one can consider actual experimental cases and enquire what extra conditions, if any, are required to give the observed ground state spins. One piece of work reports that the spins of Co^{60} (5^+), Rb^{88} (2^-) and Sb^{122} (2^-) can be fitted provided α lies in the range $\frac{1}{6}$ to $\frac{1}{4}$[1].

Finally we note that ten spins and magnetic dipole moments are known for odd-odd nuclei of $A \geqq 50$ (Table 13). The calculated moment for each case is obtainable from the simple formula of the previous section, once the neutron and proton shells are specified. The moments of seven cases are rather well accounted for, but those of Co^{58} and Co^{60} require that the $f_{\frac{5}{2}}$ and $p_{\frac{3}{2}}$ fill in a rather special fashion with the 31st neutron in the $f_{\frac{5}{2}}$ and the 33rd in the $p_{\frac{3}{2}}$ level. An analysis[2] of the Lu^{176} data suggests that the most consistent fits are those with $J = 10$ and configuration $h_{\frac{11}{2}}\, i_{\frac{13}{2}}$ or $J = 8$ and $g_{\frac{7}{2}}\, h_{\frac{9}{2}}$. Both give dipole moments close to the observed value.

Table 13. *The Calculated and Experimental Values of the Magnetic Dipole Moments of Odd-Odd Nuclei with* $A \geqq 50$.

The shell-model configurations on which the calculations are based are also given.

Nucleus	J	Protons Z	Protons orbits	Neutrons N	Neutrons orbits	μ_{calc}	μ_{exp}
V^{50}	6	23	$(f_{\frac{7}{2}})^3$	27	$(f_{\frac{7}{2}})^7$	+3.33	+3.34
Mn^{54}	2	25	$(f_{\frac{7}{2}})^5$	29	$(p_{\frac{3}{2}})^1$	+6.23	+5.1
Co^{56}	4	27	$(f_{\frac{7}{2}})^7$	29	$(p_{\frac{3}{2}})^1$	+4.28	+3.85
Co^{58}	2	27	$(f_{\frac{7}{2}})^7$	31	$\{(p_{\frac{3}{2}})^3$	+6.23	+3.5 ± 0.3
					$(f_{\frac{5}{2}})^1$	+3.49	
Co^{60}	5	27	$(f_{\frac{7}{2}})^7$	33	$\{(p_{\frac{3}{2}})^3$	+3.88	+3.8
					$(f_{\frac{5}{2}})^1$	+6.15	
Cu^{64}	1	29	$(p_{\frac{3}{2}})^1$	35	$(f_{\frac{5}{2}})^3$	+0.94	0 ± 0.4
Rb^{86}	2	37	$(f_{\frac{5}{2}})^5$	49	$(g_{\frac{9}{2}})^9$	−2.05	−1.67
Cs^{134}	4	55	$(g_{\frac{7}{2}})^5$	79	$(d_{\frac{3}{2}})^1$	+1.83	+2.95
La^{138}	5	57	$(g_{\frac{7}{2}})^7$	81	$(d_{\frac{3}{2}})^3$	+2.48	3.7
Lu^{176}	≧7	71	?	105	?	?	+4.2 ± 0.8

β) Nuclei near Pb^{208}. Several nuclei near the double closed shell Pb^{208} have been rather well investigated experimentally. The levels that are known at present along with their spins are given in Table 14.

In a shell-model analysis of this data, one first tries to assign shell-model labels to the single particle states of Pb^{209} and Bi^{209} and to the single hole states of Pb^{207} and Tl^{207}. This can be done especially well in Pb^{207} where there are at

[1] C. Schwartz,: Phys. Rev. **94**, 95 (1954).

[2] Feenberg: Shell Theory of the Nucleus, p. 58. Princeton: Princeton University Press 1955.

Table 14. *The States of nuclei in the Region of Pb^{208}.*

The first row of each tableau gives the observed energies of states in kilovolts, the second row the experimental spin values, and the third row the theoretical shell-model configurations of the states.

Pb^{202}

0	961	1383	2040	2169
0^+	2^+	4^+	5^-	9^-
$p_{\frac{1}{2}}^{-2} f_{\frac{5}{2}}^{-4}$				$p_{\frac{1}{2}}^{-2} f_{\frac{5}{2}}^{-4} i_{\frac{13}{2}}^{-1}$

Pb^{204}

0	890	1264	2169
0^+	2^+	4^+	9^-
$p_{\frac{1}{2}}^{-2} f_{\frac{5}{2}}^{-2}$			$p_{\frac{1}{2}}^{-2} f_{\frac{5}{2}}^{-1} i_{\frac{13}{2}}^{-1}$

Pb^{205}

0	282	703	988	1044	1614	1705	1766	2055	2566
$5/2^-$	$3/2^-$	$9/2^-(7/2^-)$	$7/2^-(9/2^-)$	$5/2^-$	$7/2^-$	$1/2^{(+)}$	$1/2^-$	$11/2^{(+)}$	$9/2^+$
$p_{\frac{1}{2}}^{-2} f_{\frac{5}{2}}^{-1}$	$p_{\frac{1}{2}}^{-2} p_{\frac{3}{2}}^{-1}$								

Bi^{205}

0
$9/2^-$
$(p_{\frac{1}{2}}^{-2} f_{\frac{5}{2}}^{-2})_N (h_{\frac{9}{2}})_P$

Pb^{206}

0	803	1341	1684	1998	2200	2385	2525	2783	3017	3280	3403
0^+	2^+	3^+	4^+	4^+	7^-	6^-	3^- (?)	5^-	6^-	5^-	5^-
$p_{\frac{1}{2}}^{-2}$	$p_{\frac{1}{2}}^{-1} f_{\frac{5}{2}}^{-1}$	$p_{\frac{1}{2}}^{-1} f_{\frac{5}{2}}^{-1}$	$f_{\frac{5}{2}}^{-2}$	$f_{\frac{5}{2}}^{-1} p_{\frac{3}{2}}^{-1}$	$p_{\frac{1}{2}}^{-1} i_{\frac{13}{2}}^{-1}$	$p_{\frac{1}{2}}^{-1} i_{\frac{13}{2}}$	*	$f_{\frac{5}{2}}^{-1} i_{\frac{13}{2}}^{-1}$	$f_{\frac{5}{2}}^{-1} i_{\frac{13}{2}}^{-1}$	$p_{\frac{3}{2}}^{-1} i_{\frac{13}{2}}^{-1}$	*

* Probably proton excitation.

Bi^{206}

0
6^+ (?)
$h_{\frac{9}{2}} f_{\frac{5}{2}}^{-1}$

Tl^{207}

0	350
?	?
$s_{\frac{1}{2}}^{-1}$	$d_{\frac{3}{2}}^{-1}$

Pb^{207}

0	569	900	1633	2350
$1/2^-$	$5/2^-$	$3/2^-$	$13/2^+$	$7/2^-$
$p_{\frac{1}{2}}^{-1}$	$f_{\frac{5}{2}}^{-1}$	$p_{\frac{3}{2}}^{-1}$	$i_{\frac{13}{2}}^{-1}$	$f_{\frac{7}{2}}^{-1}$

Bi^{207}

0
$9/2$
$h_{\frac{9}{2}}$

Pb^{208}

0	2615	3198	3475	3709	3961
0	3^-	5^-	4^-	5^-	4, 5, 6^-
closed shell	$d_{\frac{3}{2}}^{-1} h_{\frac{9}{2}}$	$s_{\frac{1}{2}}^{-1} h_{\frac{9}{2}}$	$s_{\frac{1}{2}}^{-1} h_{\frac{9}{2}}$	$d_{\frac{3}{2}}^{-1} h_{\frac{9}{2}}$	$d_{\frac{3}{2}}^{-1} h_{\frac{9}{2}}$

Tl^{208}

0	40	327	471	492	617
5	4	(4, 5)	(4, 5)	(3)	?
$s_{\frac{1}{2}}^{-1} g_{\frac{9}{2}}$	$s_{\frac{1}{2}}^{-1} g_{\frac{9}{2}}$	$d_{\frac{3}{2}}^{-1} g_{\frac{9}{2}}$	$d_{\frac{3}{2}}^{-1} g_{\frac{9}{2}}$	$d_{\frac{3}{2}}^{-1} g_{\frac{9}{2}}$	$d_{\frac{3}{2}}^{-1} g_{\frac{9}{2}}$

Pb^{209}

0	750	1560	2030	2540
?	?	?	?	?
$g_{\frac{9}{2}}$	$j_{\frac{15}{2}}$	$i_{\frac{11}{2}}$		

Bi^{209}

0	?	?
$9/2$	?	?
$h_{\frac{9}{2}}$	$i_{\frac{13}{2}}$	$f_{\frac{7}{2}}$

Table 14. (Continued.)

Bi^{210}

0
1^-
$h_{\frac{9}{2}} i_{\frac{11}{2}}$

Po^{210}

0	1185	1431	1478	2918	3035
0	2	4	(4)	5^-	4, 5^-
$(h_{\frac{9}{2}})^2$	$(h_{\frac{9}{2}})^2$	$(h_{\frac{9}{2}})^2$			

least five states for which spins are suggested by experiment, and all of these spins correspond to expected shell-model states. In the other three nuclei, data are more scanty and most states are given shell-model labels on little or no experimental evidence.

Once the single-particle and single-hole spectra have been decided in this way, one can proceed to consider the two-particle nuclei Bi^{210}, Po^{210}, the two-hole nucleus Pb^{206} and the one-particle-plus-one-hole nuclei Tl^{208}, Pb^{208}. (In the case

Table 15. *The Configurations of* Pb^{206} *in order of ascending zero-order energies.* These energies, given in the third column, are all relative to the zero-order energy of the lowest $(p_{\frac{1}{2}})^2$ configuration which is fixed as zero. The last column lists the semi-empirically determined values of the interaction parameters $\varepsilon_s(l l')$.

	Configuration	Zero order energy	Allowed spins	Parity	Interaction parameter $\varepsilon_s(l l')$
1	$p_{\frac{1}{2}}^2$	0	0	+	0.4
2	$p_{\frac{1}{2}} f_{\frac{5}{2}}$	0.57	2, 3	+	0.3
3	$p_{\frac{1}{2}} p_{\frac{3}{2}}$	0.87	1, 2	+	0.4
4	$f_{\frac{5}{2}}^2$	1.14	0, 2, 4	+	0.4
5	$f_{\frac{5}{2}} p_{\frac{3}{2}}$	1.44	1, 2, 3, 4	+	0.3
6	$p_{\frac{1}{2}} i_{\frac{13}{2}}$	1.63	6, 7	—	0.2
7	$p_{\frac{3}{2}}^2$	1.74	0, 2	+	0.4
8	$f_{\frac{5}{2}} i_{\frac{13}{2}}$	2.20	4, 5, 6, 7, 8, 9	—	0.2
9	$p_{\frac{1}{2}} f_{\frac{7}{2}}$	2.35	3, 4	+	
10	$p_{\frac{3}{2}} i_{\frac{13}{2}}$	2.50	5, 6, 7, 8	—	0.2
11	$f_{\frac{5}{2}} f_{\frac{7}{2}}$	2.92	1, 2, 3, 4, 5, 6	+	
12	$p_{\frac{3}{2}} f_{\frac{7}{2}}$	3.22	2, 3, 4, 5	+	
13	$i_{\frac{13}{2}}^2$	3.27	0, 2, 4, 6, 8, 10, 12	+	0.5

of Pb^{208}, we mean the excited states of course.) Let us, as our main example, take the case of Pb^{206} where a dozen or so states have been found experimentally from the β-decay of Bi^{206}.

Pb^{206}. In the absence of interactions between the particles the predicted levels of Pb^{206} will fall into degenerate groups, one group for each two-hole configuration. The energy of each group is determined simply by the sum of the two single-hole energies as given by the spectrum of Pb^{207}. The groups of states are listed in Table 15 in order of ascending energies. It can be seen that there are many more states predicted below 3.4 MeV than are observed. This is simply a consequence of the highly selective nature of the means of observation, i.e. the β-decay process and the gamma cascade process.

Now we consider the interaction between the two particles. The fact that there are two particles of the same type (neutrons) present instead of a neutron and a proton means that we have to modify the results of the last section to take account of this. Normally this would imply the tedious evaluation of exchange terms arising from the anti-symmetrisation. However, in the case of very short-range forces, this can be avoided by using the observation that only the singlet state interaction can contribute. (Two neutrons or two protons can only be at the same point in space in the singlet state.) Generally one can write the force that we have been using as a sum of singlet and triplet parts:

$$1-\alpha+\alpha(\boldsymbol{\sigma}_1\cdot\boldsymbol{\sigma}_2)=(1-4\alpha)\left[\frac{1-\boldsymbol{\sigma}_1\cdot\boldsymbol{\sigma}_2}{4}\right]+\left[\frac{3+\boldsymbol{\sigma}_1\cdot\boldsymbol{\sigma}_2}{4}\right].$$

In a triplet component of a state of a neutron-proton system, the first term is zero and the second gives the only contribution. Since the triplet contribution must be zero for a system of two neutrons or two protons, we conclude that the exchange integral of the second term must exactly cancel the usual direct integral of this term as computed from the formula of the last section. Thus, for a system of two neutrons or two protons, the effective short range interaction is simply the singlet part:

$$(1-4\alpha)\left[\frac{1-\boldsymbol{\sigma}_1\cdot\boldsymbol{\sigma}_2}{4}\right].$$

The matrix elements of this are, from the previous formulae:

$$E_J=\frac{(1-4\alpha)}{2}F^0(l l')\,(2j+1)\,(2j'+1)\,(j\tfrac{1}{2}j'-\tfrac{1}{2}\,|\,J\,0)^2\,\frac{\left[1+(-)^{l+l'+J}\right]}{2J+1}.$$

An extra factor of 2 in this formula enters because the exchange contribution is exactly equal to and has the same sign as the direct contribution. It can be seen that this energy vanishes for alternate J values.

In the case of Pb^{206}, not only are we dealing with two particles of the same type (neutrons), but, in addition, some of the configurations are such that both particles are in the same shell. In such cases, the PAULI principle disallows the existence of states of odd J. Also the last formula for E_J should be divided by two.

In general for a neutron-proton system, the interaction energies can be expressed from our formulae in terms of two parameters, α and $F^0(l l')$. In published work[1], two alternative parameters are used. These are the effective strengths of the singlet and triplet interactions $\varepsilon_s(l l')$ and $\varepsilon_t(l l')$ and they are simply related to our parameters by:

$$\varepsilon_s(l l')=(1-4\alpha)\,F^0(l l'),\qquad \varepsilon_t(l l')=F^0(l l').$$

From what we have said above, the spectrum of a configuration $(l l')$ in the case of a two-neutron or two-proton system depends on only one parameter, viz. $\varepsilon_s(l l')$. This parameter depends essentially on the strength V of the interaction, and the overlap between the orbits l and l'. In Table 15, values of $\varepsilon_s(l l')$ are listed for the various configurations. These values are determined semi-empirically. It can be seen that the values depend on how close l and l' are. The predicted spectrum is not very sensitive to the values of ε_s because the coefficients of ε_s are not too large in general (except for 0^+ states).

The evaluation of the spectrum of Pb^{206} is now perfectly straightforward with the above formula for E_J, and with the zero-order energies and the suggested values of ε_s in Table 15. When this is done (Table 16) it follows that all but

[1] M. H. L. PRYCE: Proc. Phys. Soc. Lond. A **65**, 773 (1952). — D. E. ALBURGER and M. H. L. PRYCE: Phys. Rev. **95**, 1482 (1954).

Table 16. *The Total Energies (Zero-Order plus Interaction Energies) of the Predicted States of Pb^{206} in terms of the singlet interaction parameters.*

The subscripts on ε denote the configuration $(l l')$ as numbered in Table 15. The configuration for each state given above can be found immediately by comparing the Zero-Order Energy Contribution to the entries of Table 15.

Spin and parity	Predicted energies	Spin and parity	Predicted energies
0^+	$0-\varepsilon_1$, $1.14-3\varepsilon_4$, $1.74-2\varepsilon_7$, $3.27-7\varepsilon_{13}$	5^+	2.92
		5^-	$2.20-0.3\varepsilon_8$, $2.50-1.8\varepsilon_{10}$,
1^+	0.87, 1.44, 2.92		
2^+	$0.57-1.2\varepsilon_2$, $0.87-0.8\varepsilon_8$, $1.14-0.7\varepsilon_4$, $1.44-0.35\varepsilon_5$, $1.74-0.4\varepsilon_7$	6^+	3.2 (proton excitation)
		6^-	1.63, 2.20, 2.50
3^+	0.57, 1.44, 2.35, 2.92	7^+	2.92
3^-	2.6 (proton excitation)	7^-	$1.63-0.9\varepsilon_6$, $2.20-0.6\varepsilon_8$, $2.50-0.6\varepsilon_{10}$
4^+	$1.14-0.3\varepsilon_4$, $1.44-1.1\varepsilon_5$		
4^-	2.20	8^-	2.20, 2.50
		9^-	$2.20-1.6\varepsilon_8$

two of the observed levels can be identified in the spectrum. The two exceptions are the 3^- level at 2525 keV and the 5^- level at 3403 keV. It is remarkable that two levels of such spins and about the same excitation form the first two excited states of Pb^{208}. It thus seems natural to regard both pairs of levels as arising from proton core excitation[1].

We will now make a few remarks about the reasons why the levels found experimentally should be the particular ones listed. The β-decay of Bi^{206} appears to lead to the two highest levels listed at 3280 and 3403 MeV. Now the ground state of Bi^{206} is expected to be a single $h_{\frac{9}{2}}$ proton coupled to the three neutron hole group $p_{\frac{1}{2}}^{-2} f_{\frac{5}{2}}^{-1}$. Since $p_{\frac{1}{2}}^{-2}$ is a closed neutron shell, the ground state of Bi^{206} is expected to be the ground state of $h_{\frac{9}{2}} f_{\frac{5}{2}}^{-1}$. The spin of the ground state will depend on the ratio $\varepsilon_t/\varepsilon_s$ or alternatively, on α. For α not too large, the expected value is 6^+ (as can be simply shown from the spectrum formulae of the last section). This agrees nicely with the decay taking place to states of spin 5^- in Pb^{206}.

If now one assumes these two 5^- states to be formed and examines the probable decay modes of such states through the large number of shell-model states below, one finds that the observed cascade is the expected one except for the absence of $E1$ transitions from the 3^- state of 2525 keV to the five 2^+ states below it. However, this state is rather uncertain anyway. One feature of the decay that is not in accord with the two-neutron-hole model is that electric radiation occurs as strongly as if protons were involved. There are both $E2$ and $E3$ type transitions observed that are much too intense for neutron transitions (via the recoil effect). Thus it appears that, although the shell-model gives a remarkably good account of the spectra of states, there must be certain small admixtures in the wave-function in which the proton core is excited so that electric transitions can occur.

Finally we will briefly review similar work[2] to the above that has been carried through for the nuclei Bi^{210} and Po^{210}, Pb^{208*} and Tl^{208}, and Pb^{204}:

Bi^{210} and Po^{210}. Po^{210} consists of two protons outside the double closed shell Pb^{208}. We know from Bi^{209} that the lowest proton orbit is $h_{\frac{9}{2}}$, so one tries to fit the observed states with the configuration $(h_{\frac{9}{2}})^2$. The 2^+ level at 1185 keV

[1] D. E. Alburger: and M. H. L. Pryce: Phys. Rev. **95**, 1482 (1954).

[2] The nuclei Pb^{205} and Bi^{205} have been discussed by M. H. L. Pryce: Nuclear Phys. **2**, 226 (1956/57).

and 4^+ level at 1431 keV are well fitted with this basis on taking $\varepsilon_s = 0.3$ MeV. The same value of ε_s also gives the correct binding energy of Po^{210} to Bi^{209}. The same theory predicts a 6^+ state just at 1480 keV which is where a state has been found, although it has been tentatively assigned spin 4.

The ground state of Bi^{210} is believed to have spin 1^-. The lowest configurations are $h_{\frac{9}{2}}\, g_{\frac{9}{2}}$ and $h_{\frac{9}{2}}\, i_{\frac{11}{2}}$. Only the latter is capable of giving a 1^- ground state on the present theory. The former does not do this for any values of the parameters ε_s and ε_t. It is interesting to note[1] that the ground state of Pb^{209} appears, from an analysis of the ft-value of its β-decay, to be a $g_{\frac{9}{2}}$ neutron state and not an $i_{\frac{11}{2}}$ state. Presumably the $i_{\frac{11}{2}}$ state occurs at a low excitation so that the extra pairing energy of an $i_{\frac{11}{2}}$-neutron with an $h_{\frac{9}{2}}$-proton is sufficient to make the ground state of Bi^{210} $h_{\frac{9}{2}}\, i_{\frac{11}{2}}$ rather than $h_{\frac{9}{2}}\, g_{\frac{9}{2}}$.

Attempts have been made to relax the requirement of zero-range forces in these calculations[2]. An interesting result of these efforts is that the zero-range approximation is always good for states of high spin (> 4) for ranges up to and possibly larger than 20% of the nuclear radius. For states of lower spin however it is found that, for a range of about 20% of the nuclear radius, the results for MAJORANA and HEISENBERG forces can be seriously changed from the zero-range results, although results for WIGNER and BARTLETT forces are essentially the same. For two particles of the same type, MAJORANA and HEISENBERG forces reduce to BARTLETT and WIGNER forces independently of range and so, for such cases, one can say that the zero-range results give a good approximation for ranges up to 20% of the nuclear radius without further qualification. For a system of neutrons and protons however, there may be serious deficiencies in the zero range results for states of low spin in the presence of strong MAJORANA and HEISENBERG forces. We note that the ROSENFELD type force mixture (see Chap. VI) contains a strong MAJORANA component. Calculations with this mixture show sharp differences between the predictions of zero-range forces and forces with a range of 20% of the nuclear radius for states of spin <4, especially $J=0$ and 1. To take an example, for a zero-range MAJORANA force the configuration $i_{\frac{11}{2}}\, i_{\frac{13}{2}}$ has the state spin 1^+ lowest and well below the next state which has spin 12^+. (Notice that this order is in contradiction to the Weak Rule for spin coupling. This is due to the absence of spin-exchange forces. Such forces would reverse the order of the 1^+ and 12^+ states.) When the force is given a range of 20% of the nuclear radius, the 1^+ state is raised so much that it comes just above the 12^+ state. In the case of Bi^{210}, the raising of the lowest spin state to a point where it is near a state of large spin brings about an improvement in the predictions because, besides having a 1^- ground state, Bi^{210} is believed to have a long lived alpha decaying state of high spin close by. If spin exchange forces are introduced into this calculation in the proportion present in the ROSENFELD mixture, the general picture is unchanged.

Tl^{208} and Pb^{208}. Tl^{208} is predicted[3] to have a ground state doublet of spins 5^+, 4^+ arising from the configuration $s_{\frac{1}{2}}^{-1}\, g_{\frac{9}{2}}$, and then, higher up, a quartet of states from $d_{\frac{3}{2}}^{-1}\, g_{\frac{9}{2}}$ with spins 5^+, 4^+, 6^+, 3^+ in ascending order. The observed spectrum of six states (Table 14) does indeed fall into two such groups and the data in the ground state doublet are consistent with spins of 5^+ and 4^+. The data on the higher quartet of states supports 5^+ and 4^+ for the first two but prefers 3^+ instead of 6^+ for the third. Thus there is agreement between experiment and

---

[1] G. E. LEE-WHITING: Chalk River Report TPI-76, 1954.
[2] D. M. BRINK: Proc. Phys. Soc. Lond. A **67**, 757 (1954).
[3] M. H. L. PRYCE: Proc. Phys. Soc. Lond. A **65**, 773 (1952).

theory except for the inversion of the uppermost pairs of states. Fitting the doublet splitting gives $\varepsilon_t - \varepsilon_s \sim 72$ keV for the $(s\,g)$ pair of orbits. The quartet could be fitted by the choice $\varepsilon_s : \varepsilon_t \sim 4$ but this seems unreasonable. The separation between the doublet and quartet indicates that the $d_{\frac{3}{2}} - s_{\frac{1}{2}}$ separation is rather less than the 350 keV excitation of the reported first excited state in Tl^{207}. However there is no evidence that this state is the $d_{\frac{3}{2}}^{-1}$ state.

An attempt to explain the inversion of the 3^+ and 6^+ states has been made on the basis of allowing a finite range for the interaction forces[1]. It is found that the observed order can be achieved, although this result is sensitive to the particular parameters used. It is not found to be possible to explain the closeness of the 3^+ and 4^+ levels relative to the separation of the 5^+ and 4^+.

The excited states of Pb^{208} can arise from configurations in which a neutron is excited $(p_{\frac{1}{2}}^{-1} g_{\frac{9}{2}}, f_{\frac{5}{2}}^{-1} g_{\frac{9}{2}})$ or those in which a proton is excited $(s_{\frac{1}{2}}^{-1} h_{\frac{9}{2}}, s_{\frac{1}{2}}^{-1} i_{\frac{13}{2}}, s_{\frac{1}{2}}^{-1} f_{\frac{7}{2}}, d_{\frac{3}{2}}^{-1} h_{\frac{9}{2}})$. These configurations give many states altogether and one can account for each of the spins of the observed five excited states several times over. One detailed analysis[2] concludes that the $d_{\frac{3}{2}}^{-1} h_{\frac{9}{2}}$ configuration of proton excitation accounts for the first, fourth and fifth excited states (assuming the latter to be 6^-) and that the $s_{\frac{1}{2}}^{-1} h_{\frac{9}{2}}$ configuration of proton excitation can account for the second and third. The latter assignment is not unique however because the $p_{\frac{1}{2}}^{-1} g_{\frac{9}{2}}$ state of neutron excitation would also give the second and third excited states. Other features of this analysis are:

(i) Values of α of $\frac{1}{6}$, $\frac{1}{4}$ and $\frac{1}{3}$ were used. Much the same results were obtained with either choice, except that the order of the fourth (5^-) and fifth (6^-) levels are wrong for the first two choices. These three choices correspond to attractive, zero and repulsive singlet interaction respectively $\left(\frac{\varepsilon_s}{\varepsilon_t} = +\frac{1}{3}, 0 \text{ and } -\frac{1}{3}\right)$.

(ii) The calculations were extended to the case of finite force range using a Gaussian type shape for the interaction potential. Paradoxically, the changes that occur in the predicted spectrum as a result of this apparently more realistic treatment are such as to worsen comparison with experiment. Possibly the Gaussian shape is the culprit.

(iii) No use has apparently been made of the normally accepted relative zero-order positions of the various configurations. Instead, the separations between the various groups are regarded in the fitting as free parameters. When the spectrum of Tl^{207} is established experimentally, most of these parameters will be fixed.

Pb^{204}. This nucleus consists of four neutron holes in the closed shell Pb^{208}. From the single-hole spectrum of Pb^{207}, one can immediately draw up a sequence of lowest configurations observed by their zero-order energy (Table 17). Account can then be taken of the interaction energies in the usual way[3], the only difference being that, because of the presence of four particles, there will now be several such energies for each configuration. For instance, for $p_{\frac{1}{2}}^{-2} p_{\frac{3}{2}}^{-1} f_{\frac{5}{2}}^{-1}$, the total interaction energy is:

$$E(p_{\frac{1}{2}} p_{\frac{1}{2}}; 0) + E(p_{\frac{1}{2}}^2 p_{\frac{3}{2}}; \tfrac{3}{2}) + E(p_{\frac{1}{2}}^2 f_{\frac{5}{2}}; \tfrac{5}{2}) + E(p_{\frac{3}{2}} f_{\frac{5}{2}}; J).$$

Of course, only the 0^+ state of $p_{\frac{1}{2}}^2$ is allowed because all particles are of the same type (neutrons). For the same reason, only the singlet interaction is relevant

[1] D. M. Brink: Thesis, Oxford 1955 (unpublished).

[2] G. Tauber: Phys. Rev. **99**, 176 (1955).

[3] W. W. True: Phys. Rev. **101**, 1342 (1956). — D. M. Brink: Thesis, Oxford 1955 (unpublished).

Table 17. *The Configurations of* Pb^{204} *in order of ascending zero-order energies.*

These energies, given in the second column, are all relative to the zero-order energy of the lowest $(p_{\frac{1}{2}}^{2} f_{\frac{5}{2}}^{2})^{2}$ configuration which is fixed as zero.

Configuration	Zero order energy	Possible J	Parity
$p_{\frac{1}{2}}^{2} p_{\frac{5}{2}}^{2}$	0	0, 2, 4	+
$p_{\frac{1}{2}}^{2} p_{\frac{3}{2}} f_{\frac{5}{2}}$	0.30	1, 2, 3, 4	+
$f_{\frac{5}{2}}^{3} p_{\frac{1}{2}}$	0.57	1, 2, 3, 4, 5	+
$p_{\frac{1}{2}}^{2} p_{\frac{3}{2}}^{2}$	0.60	0, 2	+
$f_{\frac{5}{2}}^{2} p_{\frac{1}{2}} p_{\frac{3}{2}}$	0.87	0, 1, 2, 3, 4, 5, 6	+
$p_{\frac{1}{2}}^{2} f_{\frac{5}{2}} i_{\frac{13}{2}}$	1.06	4, 5, 6, 7, 8, 9	−
$f_{\frac{5}{2}}^{4}$	1.14	0, 2, 4	+

as in the case of Pb^{206} and so the only arbitrariness in the predicted spectrum is in the scale factor in the interaction energies which determined their magnitude relative to the zero-order separations of configurations. The values for ε_s given for Pb^{206} in Table 15 are used. The total interaction energies are considerably larger than in the case of Pb^{206} and so the magnitudes of ε_s are rather more critical in fixing the spectrum.

Experimentally only two excited states have been found below 2.0 MeV and one at 2.17 MeV. The former two are determined from the simple gamma ray cascade from the upper most state of spin 9^- at 2.17 MeV. Theoretically one predicts about 30 levels below this energy but, of course, there is no contradiction with experiment because the latter is so selective. The theoretical problem is to identify the three observed states amongst the 30 predicted levels, and then to explain why only these levels should be observed. The first striking achievement of the theory is to predict a unique 9^- state at about the correct place (from $p_{\frac{1}{2}}^{-2} f_{\frac{5}{2}}^{-1} i_{\frac{13}{2}}^{-1}$). Next, there are no other levels of spin $J > 4$ predicted below this state so the cascade must go to a 4^+ state (as is observed). There are several such 4^+ levels, but two are much lower than the others and occur near the observed level. These arise from the configurations $p_{\frac{1}{2}}^{-2} f_{\frac{5}{2}}^{-2}$ and $p_{\frac{1}{2}}^{-2} p_{\frac{3}{2}}^{-1} f_{\frac{5}{2}}^{-1}$. Below these levels there are only 2^+ and 0^+ levels and so the decay must be to a 2^+ level, again agreeing with observation. There are three such 2^+ levels predicted to be quite close together $(f_{\frac{5}{2}}^{-3})_{\frac{3}{2}} p_{\frac{1}{2}}^{-1}$, $p_{\frac{1}{2}}^{-2} p_{\frac{3}{2}}^{-1} f_{\frac{5}{2}}^{-1}$ and $p_{\frac{1}{2}}^{-2} f_{\frac{5}{2}}^{-2}$). At present it is not clear why the cascade picks out a particular 4^+ level from the predicted doublet of 4^+ levels, and a particular 2^+ level from the predicted triplet of 2^+ levels. The observed $E2$ transition from the 4^+ to the 2^+ state is isomeric and the gyromagnetic ratio of the 4^+ state is known to be $g = 0.054$. This indicates that the state is mainly from the $p_{\frac{1}{2}}^{-2} p_{\frac{3}{2}}^{-1} f_{\frac{5}{2}}^{-1}$ configuration ($g = -0.137$) with some admixture (28%) of the $p_{\frac{1}{2}}^{-2} f_{\frac{5}{2}}^{-2}$ configuration ($g = 0.55$).

Unlike the $E2$ transitions in Pb^{206} $(0.57 \to 0)$ and Pb^{206} $(0.803 \to 0)$, the $E2$ transition in Pb^{204} has a lifetime considerably longer than that expected for a single proton transition. It is thus much more in accord with the shell-model prediction that the lifetimes of electric transitions should be very long for nuclei with neutron or neutron hole configurations. This is strange because Pb^{204} is further from the shell closure than Pb^{206} and Pb^{207}. One possible explanation is that the phenomenon that makes the $E2$ transition take place so readily in Pb^{206} and Pb^{207} (surface coupling—see later) is neutralised by something in Pb^{204}. This something could be the strong orthogonality of the neutron wave-functions in the initial and final states. Such orthogonality would arise if the 4^+ state were mainly $p_{\frac{1}{2}}^{-2}\, f_{\frac{5}{2}}^{-1}\, p_{\frac{3}{2}}^{-1}$ and the 2^+ state were mainly $f_{\frac{5}{2}}^{-3}\, p_{\frac{1}{2}}^{-1}$ because these configurations differ by two j-values.

Pb[202]. The configurations that are relevant for the discussion of Pb^{202}, which has six holes in the closed neutron shell, are just like those for Pb^{204} as listed in Table 17 except that there are two extra $f_{\frac{5}{2}}$ holes[1]. In those configurations for Pb^{204} that have two $f_{\frac{5}{2}}$ holes (like $p_{\frac{1}{2}}^{-2}\, f_{\frac{5}{2}}^{-2}$), the only modification for Pb^{202} is that the two holes are changed to two particles. This change preserves the spectrum and so one expects the lowest states of $p_{\frac{1}{2}}^{-2}\, f_{\frac{5}{2}}^{2}$ of Pb^{202} to resemble those of $p_{\frac{1}{2}}^{-2}\, f_{\frac{5}{2}}^{-2}$ of Pb^{204}. In fact, the observed spectra of Pb^{202} and Pb^{204} are similar not only in the states of 0^+, 2^+, 4^+, but in the presence of a low-lying 9^- state. The main difference is that the 9^- state occurs with a 5^- state in Pb^{202}. There are considerably more predicted states in Pb^{202} than Pb^{204} and so any quantitative fitting is more laborious and less rewarding and has not been attempted.

γ) The nucleus Zr^{90}. This nucleus with $N=50$, $Z=40$ has a ground state consisting of a major closed shell of neutrons and a minor closed shell of protons[2]. Presumably the first few excited states arise from proton excitation and the ensuing discussion only consider this possibility. The last proton shell to fill in the ground state is a $p_{\frac{1}{2}}$ shell. The next higher orbit is the $g_{\frac{9}{2}}$ and so the relevant configurations are $(p_{\frac{1}{2}})^2$ for the ground state and $p_{\frac{1}{2}}\, g_{\frac{9}{2}}$, $g_{\frac{9}{2}}^{2}$ for the lowest excited states. One knows from the spectrum of Y^{89}, the nucleus with one proton hole, that the $p_{\frac{1}{2}}-g_{\frac{9}{2}}$ splitting is 0.9 MeV. Thus, in zero-order, the states of the $p_{\frac{1}{2}}\, g_{\frac{9}{2}}$ configuration are 0.9 MeV above the ground state of $p_{\frac{1}{2}}^{2}$ and the states of $g_{\frac{9}{2}}^{2}$ are 0.9 MeV higher still. When interactions are taken into account, the $p_{\frac{1}{2}}^{2}$ and $g_{\frac{9}{2}}^{2}$ levels are depressed much more than the $p_{\frac{1}{2}}\, g_{\frac{9}{2}}$ because of the poor overlap in the latter case. The $g_{\frac{9}{2}}^{2}$ levels are, of course, ordered in the usual way: 0^+, 2^+, 4^+, 6^+ and 8^+ in ascending energy. By choosing the interaction parameter ε_s large enough, the level of spin 0^+ may be depressed below the levels of spins 4^- and 5^- from the $p_{\frac{1}{2}}\, g_{\frac{9}{2}}$ configuration (Fig. 10). In this way, very good accord[2] is obtained with the experimental spectrum in which a 0^+ level occurs at 1.75 MeV and an

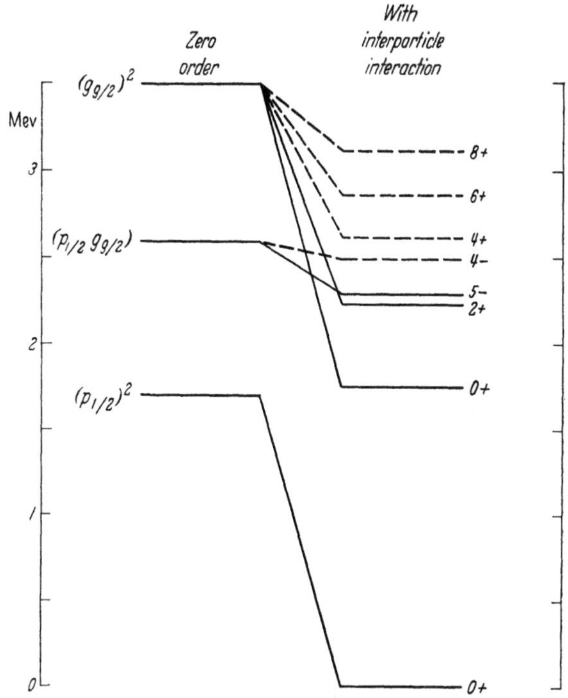

Fig. 10. Predicted Level Diagram of Zr^{90}. On the left are the relevant zero-order states as obtained from the nucleus Y^{89}. The diagram on the right is a qualitative one of the resultant states after allowance for the interaction of the last two protons. The first four states marked with solid lines agree with observed states of Zr^{90}. [From K. W. FORD: Phys. Rev. 98, 1516 (1955).]

[1] D. M. BRINK: Thesis, Oxford 1955 (unpublished).

[2] K. W. FORD: Phys. Rev. **98**, 1516 (1955).

isomeric 5^- level at 2.30 MeV. Furthermore there is evidence[1] for a level at 2.23 MeV which decays quickly to the ground state. This may well be the predicted 2^+ state.

14. Configuration mixing. In all the considerations of this Chapter so far, configuration mixing has been ignored. In other words, each state has been supposed to have a single "pure" configuration and not to contain mixtures of configurations. Of course, strictly speaking, it is only in the absence of inter-particle forces, that states have really pure configurations. However in the complete absence of interactions, states of the same configuration will be degenerate. The experimental evidence does not support this extreme picture, so one introduces interactions to split up the states. The implicit assumption that we have made in this chapter is that, although the interactions are strong enough to split up the states appreciably, they are not strong enough to give rise to severe configuration mixing. In principle, this is a perfectly permissible situation and, indeed, the success found in the comparison of theory and experiment indicates that this situation applies approximately in practice. None-the-less, we have seen that the assumption of pure shell-model configurations can by no means explain all the experimental facts, and it is of definite interest to try to improve the model to explain more data. Broadly speaking, all deviations from pure configurations could be called "configuration mixing". However it seems that one must distinguish two quite distinct types of deviation from the assumption of pure configurations. One of these is configuration mixing in the usually understood meaning of the term from atomic physics, i.e. the mixing in of some higher configurations mainly of the loose particles by virtue of the interactions amongst the loose particles, and also including excitation of individual particles from the closed-shell "core". The second type of deviation is one in which the interaction of the loose particles with the core is considered not to excite individual core particles, but to excite many core particles in a collective manner so that, although the orbits of individual particles may not be too seriously disturbed, there is an appreciable overall effect. In practice, analysis of the second type of deviation attempts to represent this special form of configuration mixing as a deformation of shape of the core from a sphere to a spheroid. We will discuss this type in the next section, and confine ourselves in this section to the discussion to the first (normal) type of configuration mixing.

Even this first type can be split up into two sub-types:

(i) Mixing between two quite different configurations that happen to lie close together in energy. This type will introduce admixtures not only of different j-values but different l-values in general.

(ii) Mixing in which only the j-values of the states are changed, i.e. in which the l-orbits are considered to be unchanged, the mixing being caused by the decoupling of individual orbits to spins by the action of the central orbit-orbit forces. Such a mixing is equivalent to the mixing of so-called "intermediate coupling", which is used to name any mode of coupling between pure j-j and pure L-S or RUSSELL-SAUNDERS coupling (see Chap. 6).

Although the separation between configurations with the same l-orbits but differing in j-values may be much greater than the spacings of adjoining configurations, the mixing will often be larger in the former case because of the excellent overlap of the space wave-functions which makes the interaction matrix elements much larger. [Quantitative estimates of the value of $\varepsilon(ll')$, the interaction energy parameter, have been made for a short-range interaction and nodeless

[1] Several new higher states have been found in Zr^{90} (N. LAZAR, to be published). Qualitatively these agree with the expected states of $(g_{\frac{9}{2}})^2$.

($n=1$) oscillator wave functions[1]. Typical results are: $\frac{\varepsilon(05)}{\varepsilon(00)} \sim 0.03$, $\frac{\varepsilon(05)}{\varepsilon(55)} \sim 0.3$.] Often adjoining configurations involve quite different l-orbits and so have small matrix elements. Thus, apart from freak cases, one expects an appreciable fraction of the total configuration mixing to be of type (ii), i.e. of the "intermediate coupling" type.

We will now investigate the effects of the usual type of configuration mixing on the predictions of electromagnetic moments.

Magnetic dipole moments. Whether configuration mixing of type (i) is important or not, there are certain data which are sensitive only to type (ii) mixing provided that the total mixing is not too large. The data enable one to examine type (ii) mixing by itself, independently of whether type (i) is a serious competitor.

The data we have in mind are the magnetic dipole moments, and their special property follows from the nature of the magnetic dipole operator which is a one-particle operator (i.e. it acts on one particle at a time) and cannot change the orbits, either in n-values or l-values. Consider a zero-order wave-function $\psi_0(n\,l\,j;\,J)$, where $n\,l\,j$ symbolises the set of individual particles orbits. The total wave-function may be written[2]:

$$\psi_{JM} = \psi_0(n\,l\,j;\,JM) + \sum_{\substack{p\\(j_p\neq j)}} \alpha_p \psi_p(n\,l\,j_p;\,JM) + \sum_{\substack{q\\(n_q l_q \neq nl)}} \alpha_q \psi_q(n_q\,l_q\,j_q;\,JM)\,.$$

α_p is the coefficient of the admixed state p, and, from first order perturbation theory, it is equal to the matrix element $(\psi_0|\sum_{i<j} v_{ij}|\psi_p)$ divided by the zero-order energy separation $E_p - E_0$. The states ψ_p correspond to mixing of type (ii). The states ψ_q correspond to mixing of type (i), and the α_q are the appropriate coefficients like the α_p. To first order in the α's, the moment of the above state is:

$$\begin{aligned}\mu &= (\psi_{JM=J}|\,\mu_z\,|\psi_{JM=J})\\ &= \mu_{00} + 2\sum_p \alpha_p \mu_{0p}\end{aligned}$$

where, for an odd nucleus, μ_{00} is the usual SCHMIDT moment $(\psi_0(J,M=J)|\,\mu_z \times |\psi_0(J,M=J))$ and μ_{0p} is the cross term: $(\psi_0(J,M=J)|\,\mu_z\,|\psi_p(J,M=J))$. Of course, μ_{0p} is only non-zero for states ψ_p in which only one j-value differs from those of ψ_0. The pairs of states ψ_0, ψ_p can be classified into three types for odd nuclei:

(a) $\psi_0 \equiv (l+\frac{1}{2})^k$; $\quad\psi_p \equiv (l+\frac{1}{2})^{k-1}(l-\frac{1}{2})^1 \quad (J = l+\frac{1}{2})$,

(b) $\psi_0 \equiv (l+\frac{1}{2})^{2l+2}(l-\frac{1}{2})^k$; $\quad\psi_p \equiv (l+\frac{1}{2})^{2l+1}(l-\frac{1}{2})^{k+1} \quad (J = l-\frac{1}{2})$,

(c) $\psi_0 \equiv (l'+\frac{1}{2})^{k_1}(l\pm\frac{1}{2})^k$; $\quad\psi_p \equiv (l'+\frac{1}{2})^{k_1-1}(l'-\frac{1}{2})^1(l'\pm\frac{1}{2})^k \quad (J = l\pm\frac{1}{2})$.

The bracketed quantities are j-values; k and k_1 are odd and even integers respectively. For cases (a) and (b), using the methods of Sect. 12 to find the interaction matrix element, one finds that the addition $\Delta\mu$ to the SCHMIDT moment has the form:

$$\Delta\mu = (g_s - g_l)\,A(l,J)\,\frac{\varepsilon_s(ll)}{\Delta E_l}\,.$$

In this expression, g_s and g_l are the usual spin and orbital g-factors for the odd nucleon, ε_s is the usual singlet interaction strength parameter, $\Delta E_l = E(j=l-\frac{1}{2}) -$

---

[1] A. DE SHALIT and M. GOLDHABER: Phys. Rev. **92**, 1211 (1953).

[2] R. J. BLIN-STOYLE and M. A. PERKS: Proc. Phys. Soc. Lond. A **67**, 885 (1954). — A. ARIMA and H. HORIE: Progr. Theor. Phys. **11**, 509 (1954).

$E(j=l+\frac{1}{2})$ is the spin-orbit splitting of the l-orbit and $A(l, J)$ is a rather involved but perfectly definite function of l and J.

Let us take the example of Bi^{209} which has an odd $h_{\frac{9}{2}}$ proton. The admixture caused by configuration mixing will consist in the elevation of an $h_{\frac{11}{2}}$ proton into the $h_{\frac{9}{2}}$ shell, i.e. type (b). Estimates of the ratio $\varepsilon_s/\Delta E$ vary between 0.25 and 0.5. The value needed to give the large observed deviation $\Delta\mu=1.04$ n.m. is 0.43 and this corresponds to an admixture α_p of about 5%. Thus it is apparent that even small admixtures in the zero-order state can cause serious deviations from the SCHMIDT lines. The case of Bi^{209} may be considered a considerable success for the configuration mixing theory because the observed deviation is accounted for in order of magnitude and direction without any need for fixing arbitrary parameters.

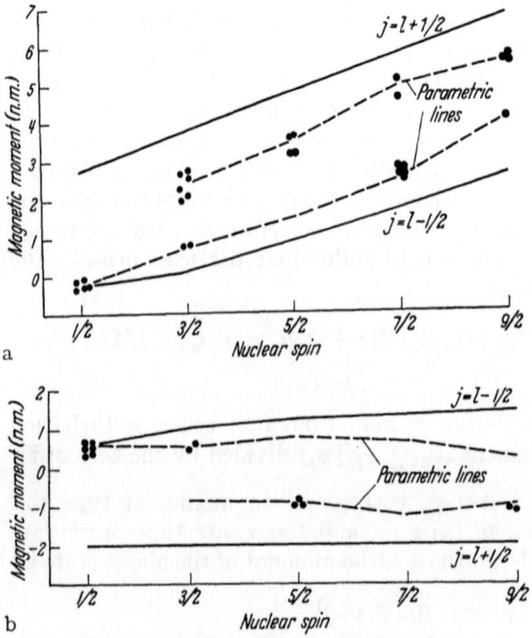

Fig. 11 a and b. (a) Predicted Deviation of the Magnetic Moments of Odd Proton Nuclei from the SCHMIDT Lines due to inter configurational mixing. [From R. J. BLIN-STOYLE and M. A. PERKS: Proc. Phys. Soc. Lond. A 67, 885 (1954).] (b) Predicted Deviation of the Magnetic Moments of Odd Neutron Nuclei from the SCHMIDT Lines due to inter configurational mixing. [From R. J. BLIN-STOYLE, and M. A. PERKS: Proc. Phys. Soc. Lond. A 67, 885 (1954).]

One very important feature of the above formula for $\Delta\mu$ is that it always gives values which bring the predicted moments inside the SCHMIDT lines towards the observed values. One can, in fact, by choosing the ratios $\varepsilon_s/\Delta\mu$ appropriately for each l-value, construct new lines on the SCHMIDT diagram that run close to all the observed moments to which the present treatment can be applied (Figs. 11 a and b). The appropriate values of the ratio are, for $l=1, 2, 3, 4$ and 5: -0.38, -0.26, -0.20, -0.24 and -0.43. These are all reasonable values.

For case (c), the addition to the SCHMIDT moment can be shown to be:

$$\Delta\mu = (g_s - g_l)\, B(l, l'; J)\, \frac{\varepsilon_s(l l')}{\Delta E_{l'}}$$

where $(g_s - g_l)$ is evaluated for the l' nucleon, and where B is, like $A(l, J)$, a rather complicated, but quite definite function of the various angular momenta involved. The values of B are larger on the whole by about a factor 2 or more for $J=l+\frac{1}{2}$ relative to the values for $J=l-\frac{1}{2}$. This agrees with the observed fact that deviations from the SCHMIDT lines are larger in the former case. One special result is that B is zero for $(l, j)=p_{\frac{1}{2}}$, and so $p_{\frac{1}{2}}$ nuclei are predicted to have small deviations. In contrast $s_{\frac{1}{2}}$ nuclei are predicted to have large deviations (especially when a $d_{\frac{5}{2}}$ nucleon is transferred to the $d_{\frac{3}{2}}$ shell). Both of these predictions are in agreement with observation as mentioned in Chap. III.

We also mentioned in Chap. III that the magnetic moments for pairs of odd-neutron and odd-proton nuclei with the same odd number of particles are remarkably correlated in the sense that the percentage deviations from the SCHMIDT lines are always about the same. This fact can also be given some theoretical backing in terms of the present model of configuration mixing. In the first place, deviations of type (c) are smaller than those of types (a) and (b). This means that the effects of configuration mixing from amongst the odd particles themselves are more important than those from mixing induced in the even group of particles. Furthermore $\Delta\mu$ for types (a) and (b) only depends on the type of odd nucleon (neutron or proton) through the factor $(g_s - g_l)$. Thus one expects the deviation in moment of an odd proton nucleus to be related to that of an odd neutron nucleus with the same odd number by[1]:

$$\frac{|\Delta\mu|_p}{|\Delta\mu|_n} = \frac{|g_s - g_l|_p}{|g_s - g_l|_n} \sim 1.20 .$$

Thus the deviations in moments for pairs of nuclei are predicted to follow each other in the observed fashion. It may be significant that the deviations in odd-proton nuclei are systematically a little greater than those in odd-neutron nuclei as is predicted by the factor 1.20.

Quadrupole moments. In view of the considerable success that the configuration mixing calculations have with magnetic moments, it is natural to do similar calculations for quadrupole moments[2]. The electric quadrupole operator does not preserve particle angular momentum like the $M1$ operator but may change it by two units, so it is convenient to express the total state simply as:

$$\psi_{JM} = \psi_0(n\,l\,j; JM) + \sum_p \alpha_p \psi_p(JM).$$

Those admixed states ψ_p which differ from ψ_0 in one proton may contribute to the quadrupole moment in general in the following way

$$Q = Q_0 + \sum_p \Delta Q_{0p}$$

where

$$Q_0 = \left(\psi_0(J, M=J)\left|\left(\frac{16\pi}{5}\right)^{\frac{1}{2}} Q_0^{(2)}\right|\psi_0(J, M=J)\right),$$

$$\Delta Q_{0p} = \alpha_p\left(\psi_0(J, M=J)\left|\left(\frac{16\pi}{5}\right)^{\frac{1}{2}} Q_0^{(2)}\right|\psi_p(J, M=J)\right).$$

Let us suppose that the state ψ_p is obtained from state ψ_0 by changing an l' orbit to an l'' orbit. When $l'' = l'$ or $l' \pm 2$, the admixture gives a contribution to the quadrupole moment. This can be written:

$$\Delta Q_{0p} = A(j, j', j''; l, l', l'')\,\frac{\varepsilon_s(l, l'l'')}{E_{j''} - E_{j'}}\,\langle r^2\rangle_{l'l''}$$

where A can be specified explicitly and where $\varepsilon(l, l'\,l'')$ is like the usual $\varepsilon(ll')$ but the square of the radial wave function $r u_{l'}$ in the SLATER integral is replaced by the product $(r u_{l'})\,(r u_{l''})$.

It is found that ΔQ_{0p} for pairs of states of types (a) (b) (c) considered for magnetic moments is smaller in general than for the pairs with $l'' = l' \pm 2$ because $E_{j''} - E_{j'}$ can be much smaller for the latter. The net effect of these latter pairs is always to give an increase in Q and so bring about closer accord with experiment. However, although the direction of the corrections is right, we cannot expect their magnitudes to be capable of explaining the very large observed quadrupole moments, and, indeed, this is found to be so. Thus one must look

[1] A. DE SHALIT: Helv. phys. Acta **24**, 296 (1951). — Phys. Rev. **90**, 83 (1953).
[2] H. HORIE and A. ARIMA: Phys. Rev. **99**, 778 (1955).

for a new source for these large moments. Such a source is provided by the alternative type of configuration mixing which we mentioned at the beginning of this section in which the core of the nucleus is excited, not through its individual nucleons, but in a collective manner. We will discuss this possibility in the next section.

15. Departures from sphericity in nuclear shape. We now consider departures from the shell-model which might be described as "coherent configuration mixing". The loose particles are imagined to be capable of exciting the "core" of a nucleus not only through individual core nucleons, but also in a collective fashion. This latter type of excitation is to be described, not in terms of individual particle co-ordinates, but with collective co-ordinates of the type encountered in the hydrodynamical study of the liquid drop[1,2]. Small forces between the loose nucleons and the core will, on this picture, be capable of inducing shape oscillations in the core. The simplest type of oscillation will be the axially symmetrical quadrupole oscillation in which the shape of the nucleus at any moment can be written as:

$$R(\vartheta) = R_0\left(1 + \beta\, Y_0^{(2)}(\cos\vartheta)\right)$$

where R_0 is the normal nuclear radius, and ϑ is the angle that the direction of R makes with the axis of cylindrical symmetry. This description of the interaction has been given in classical terms, but a quantum mechanical interpretation can be made by using the standard methods of quantum field theory in which oscillations are quantised. To be more precise, the classical quadrupole oscillations are characterised by a certain quadrupole frequency ω_2 which is related to the effective mass of the oscillation and the surface tension (restoring force) in the well-known manner. The energy of the oscillation depends continuously on the amplitude of oscillation. In the quantum approach, the oscillation frequency is retained, but the energy is restricted to integral multiples of $\hbar\omega_2$ and the oscillation amplitude is correspondingly restricted. $\hbar\omega_2$ itself is the energy of one "surface phonon" and an angular momentum of two units is associated with each phonon.

This approach to the collective form of configuration mixing due to the particle-core forces is useful as long as these forces are weak[3]. For instance, using the surface tension value from the semi-empirical mass formula and a particle-core potential of 40 MeV, one finds that the system of just one particle and a core (a closed shell) should be describable with this so-called "Weak-Coupling" approach because the interaction matrix elements are less than $\hbar\omega_2$. In contrast, when there are several loose particles outside major closed shells their individual effects on the core are additive so that the weak coupling approach is no longer valid. In this case, very large numbers of phonons can be excited, so that a kind of classical treatment becomes possible according to the Correspondence Principle which says that quantum effects vanish for large quantum numbers.

Before going further we should say here that all the striking successes of the Collective Model have been obtained in this latter approximation of "Strong Coupling" and so we will not discuss the Weak Coupling approximation further. The classical description that is permissible in the Strong Coupling case brings the notion of permanent deformation in the nuclear ground state. Clearly, a

[1] J. RAINWATER: Phys. Rev. **79**, 432 (1950). — A. BOHR and B. R. MOTTELSEN: Dan. Mat. fys. Medd. **27**, No. 16 (1953).

[2] More details on the collective model will be found in S. MOSZKOWSKI's companion article, in this volume.

[3] A. BOHR and B. R. MOTTELSEN: Dan. Mat. fys. Medd. **27**, No. 16 (1953).

condition for a deformation to be described as classically permanent is that its amplitude β (say) be greater than the amplitude of the quantum phonon oscillations β_0 (say). This condition is equivalent to the interaction matrix elements being greater than $\hbar\omega_2$.

The presence of a permanent deformation in the shape of the core of a nucleus inevitably leads to a modification of the single particle motion of the loose particles which no longer move in a spherically symmetric potential, but rather in a spheroidal one. In general, the j and l values of the particle motion will not remain good quantum numbers in such a potential but instead there will be a sharing of the total angular momentum $\boldsymbol{J}$ (say) between the loose particles and the core. If the core angular momentum is written as $\boldsymbol{R}$:

$$\boldsymbol{J} = \boldsymbol{R} + \sum_{\substack{\text{(loose}\\ \text{particles)}}} \boldsymbol{j}.$$

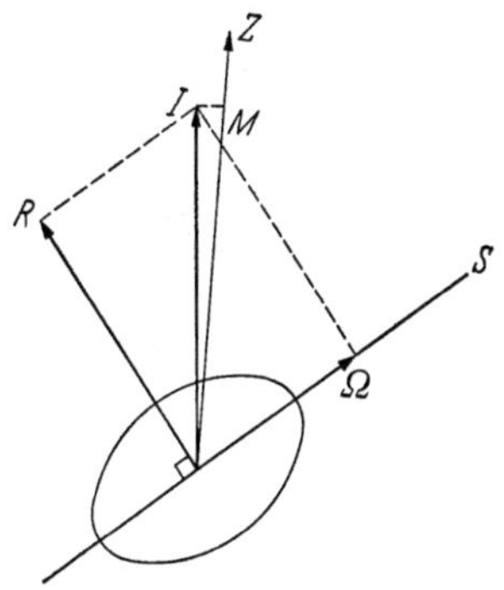

Fig. 12. Angular Momentum Coupling Scheme for a Strongly Deformed Odd Nucleus. S is the axis of symmetry of the nucleus and Z is a space-fixed axis. Ω is the component of particle angular momentum along S, and $\boldsymbol{R}$ is the rotational angular momentum, which combines with Ω to give total spin $\boldsymbol{I}$. (From A. Bohr: Rotational States of Atomic Nuclei, p. 22. Copenhagen 1954.)

The particle motion will be coupled to the axis of symmetry, S, of the core in such a way that the components of particle angular momentum along this axis (Ω_i, say) are closely constant. The total angular momentum also has constant projection on this axis (K, say). For the lowest energy states, one expects the direction of $\boldsymbol{R}$ to be perpendicular to this axis, so $K = \sum_i \Omega_i = \Omega$ (say) (see Fig. 12).

The spectrum of single particle states for various deformations have been calculated approximately for the square and oscillator type wells[1], along with the corresponding eigenfunctions which reveal the extent to which l and j values are mixed.

We now consider applications of the "Strong-Coupling" model to some of the most important types of experimental data.

Spectra. The simplest excitations of a deformed nucleus are those of simple rotation of shape[2]. Such excitations will have in general a spectrum of energies:

$$E_J = \frac{\hbar^2}{2\mathcal{J}} \{J(J+1) - J_0(J_0+1)\}$$

where J_0 is the ground state spin and $\mathcal{J}$ is the effective moment of inertia. (This is not to be evaluated for solid body rotation but rather for the irrotational motion of the surface shape.) This rotational type of spectrum has been founld in many nuclei experimentally. The spacing between levels has been previously mentioned as having a minimum towards the middle of major shells, especial y near $A \sim 170$ between the major shell closures at $A = 140$ and $A = 208$. On either side of this minimum, the spacing rises monotonically to maxima at the shell closures. The moment of inertia varies in an inverse fashion according to the above formula and so has a maximum at mid-shell. For irrotational motion, $\mathcal{J}$ is expected to be proportional to the square of the deformation parameter β and so this has a similar behaviour. Such behaviour is to be expected from the special stability of the major closed shells, which implies high surface tension and small deformations unless there are many loose particles present.

[1] S. A. Moszkowski: Phys. Rev. **99**, 803 (1955). — S. G. Nilsson: Dan. Mat. fys. Medd. **29**, No. 16 (1955). — K. Gottfried: Phys. Rev. **103**, 1017 (1956).

[2] A. Bohr and B. R. Mottelsen: Dan. Mat. fys. Medd. **27**, No. 16 (1953).

Magnetic dipole moments. Let us consider a simplified model of a single particle j and a core. If the deformation is small enough to keep j constant but large enough to allow the strong coupling approximation, the expectation value of the magnetic dipole operator:

$$\mu = \frac{e\hbar}{2Mc}(g_j \boldsymbol{j} + g_R \boldsymbol{R})$$

can be straightforwardly shown to be[1]:

$$\begin{aligned}\mu &= (J, M=J|\,\mu_z\,|J, M=J)_{J=j}\\ &= (g_j J + g_R)\frac{J}{J+1}\\ &= g_j J - (g_j - g_R)\frac{J}{J+1}\end{aligned}$$

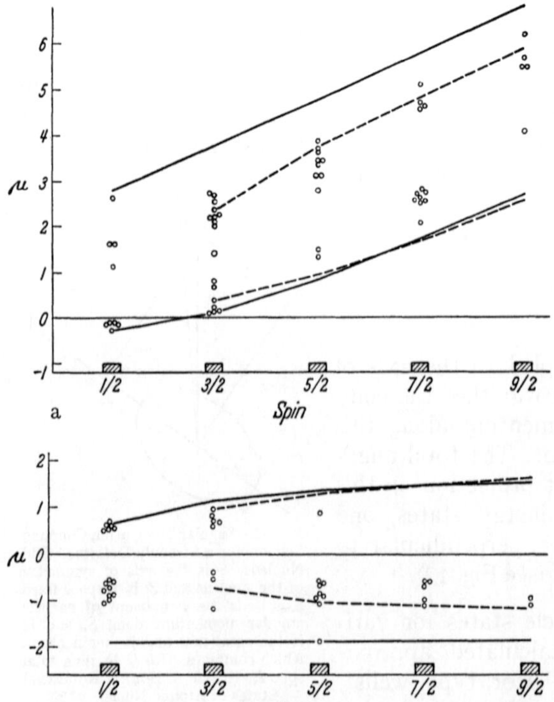

Fig. 13a and b. (a) Predicted Deviation of Magnetic Moments of Odd Proton Nuclei from the SCHMIDT Lines on the Strong Coupling Model. [From A. BOHR and B. R. MOTTELSON: Dan. Mat. fys. Medd. 27, No. 16 (1953).] (b) Predicted Deviation of Magnetic Moments of Odd Neutron Nuclei from the SCHMIDT Lines on the Strong Coupling Model. [From A. BOHR and B. R. MOTTELSON: Dans. Mat. fys. Medd. 27, No. 16 (1954).]

provided that Ω does not equal $\frac{3}{2}$ or $\frac{1}{2}$ where certain complications arise. In these expressions, g_R is the gyromagnetic ratio associated with the angular momentum $\boldsymbol{R}$ of the core. Taking g_R at the value Z/A for a uniformly charged body, one can construct revised SCHMIDT lines as in Figs. 13a and b. One sees that the new lines for $j = l + \frac{1}{2}$ are inside the old lines in agreement with experiment. The lines for $j = l - \frac{1}{2}$ are almost unchanged however. These results are only to be taken qualitatively because, in practice, j is not a very good quantum number for the particle, and it may not be permissible to consider a single particle only outside the core. The former deficiency can be removed by using the proper spheroidal well eigenfunctions which contain mixtures of j-values. This entails the replacement of g_j in the above formula by g_Ω which can be calculated from the known mixtures of j-values.

Qualitatively, the collective model type of configuration mixing has the same effect on the prediction of magnetic dipole moments as the usual, individual particle, type of configuration mixing discussed in Sect. 14. However it is possible to find individual cases where only one type of mixing is relevant. We would expect the collective type to be negligible for Bi^{209} which is only one particle outside a major closed shell. The large deviation of the moment from the SCHMIDT line value is thus to be understood solely in terms of mixing of the individual particle type as described in Sect. 14. On the other hand, this type of mixing is for instance incapable of explaining the large difference between the observed moments of Eu^{151} and Eu^{153}, but the collective type may do this because the deformation β, which controls the value of g_Ω, seems to be different for the two

[1] A. BOHR and B. R. MOTTELSEN: Dan. Mat. fys. Medd. 27, No. 16 (1953).

isotopes[1]. (Some evidence for this difference comes from the quadrupole moments which, as we point out below, give the value of β in a direct manner.)

Electric quadrupole moments and transition probabilities. The static classical quadrupole moment of a uniformly charged spheroid of deformation β is easily shown to be[2]:

$$Q_0 = \frac{3}{(5\pi)^{\frac{1}{2}}} e Z R_0^2 \beta .$$

If the total spin of the system is J, the corresponding contribution to the observed quadrupole moment is:

$$Q_{\text{core}} = \frac{J(2J-1)}{(J+1)(2J+3)} Q_0 .$$

For values of β large enough to justify the Strong Coupling approximation, the value of Q_{core} can be seen to be much larger than the single particle quadrupole moments given in Chap. 2. Thus, as we expect, the Collective Model with quadrupole distortions is capable of giving large quadrupole moments thereby agreeing with experiment. By equating the observed Q to Q_{core}, one can extract values of β for various nuclei. These values show just the same trend as the values from spectra, viz. small values near major shell closures becoming much larger towards mid-shell. At present, it is not fully understood why most observed moments are positive[3], because the Strong Coupling Model based on a proper single particle spectrum for a spheroidal well appears to give in this respect the same prediction as the usual shell-model, viz. as many negative as positive moments.

Permanent quadrupole distortions of the nuclear shape will not only give large observed moments but also give strong quadrupole transitions between states of the rotational spectrum associated with the deformation. For instance, for a transition between two adjacent states $J+2 \rightarrow J$ of the 0, 2, 4 ... spectrum of an even-even nucleus[2]:

$$\sum_{M'+q=M} |(J+2, M|\, Q_q^{(2)}\, |J, M')|^2 \equiv \frac{|(J+2\,||\boldsymbol{Q}^{(2)}||\,J)|^2}{2J+5} = \frac{15}{32\pi} Q_0^2 \frac{(J+1)(J+2)}{(2J+5)(2J+3)} .$$

The experimental matrix elements for $E2$ transitions give values of Q_0 on using this relation, and then values of β can be derived immediately. The values of Q_0 and β obtained in this way agree with those obtained from the static quadrupole moments.

V. The individual particle model.

The "Individual Particle Model" is a rather arbitrary title for the general version of the shell-model in which *all* particles are taken into account, i.e. one works with total wave-functions that are properly antisymmetrised combinations of the single particle wave-functions for all particles, and one takes some account of all the interactions between the particles. We have already introduced a special version of this model in the last chapter when we discussed various possible refinements of the extreme single particle model. In that version, which was formulated for heavy nuclei, we only considered cases of a few (two or three)

---

[1] B. R. Mottelson and S. G. Nilsson: Phys. Rev. **99**, 1615 (1955). — G. Scharff-Goldhaber and J. Weneser: Phys. Rev. **98**, 212 (1955).

[2] A. Bohr and B. R. Mottelsen: Dan. Mat. fys. Medd. **27**, No. 16 (1953).

[3] S. A. Moszkowski and C. H. Townes: Phys. Rev. **93**, 306 (1953). — K. Gottfried: Phys. Rev. **103**, 1017 (1956).

particles, and also we assumed the situation of j-j coupling with interaction forces of zero-range. In the next chapter, we shall be concerned with the application of the individual particle model to light nuclei for which these special assumptions are much too restrictive. Consequently we propose in the present chapter to re-introduce the model *ab initio* on a more general basis than previously.

The sole practical object of the present chapter is to show how one can most easily set up the total energy matrix for some nucleus in an Individual Particle Model representation. In general, to set up such a matrix, one must first prescribe a complete set of states and the interactions. Next one must write down the fractional parentage properties of the complete set, thus enabling one to reduce any matrix element to a sum over two-particle matrix elements. Finally one must evaluate these latter matrix elements. In the present chapter, after Sects. 16 and 17 on general ideas and formalism, Sects. 18 to 22 will be devoted to the definition of complete sets of states. Sect. 23 will define the various possible types of interaction, and Sect. 24 will discuss the use of fractional parentage in setting up the energy matrix. (We give details of the two-particle matrix elements in Appendix 8.) Finally Sect. 25 describes the diagonalisation of the energy matrix and the calculation of experimental quantities.

16. The role of the individual particle model in calculations. We do not intend to discuss here the rather profound questions about the theoretical basis of the individual particle version of the shell-model. Rather we will outline the basic method of practical calculation associated with the use of the model. This method is essentially a variational calculation for which the individual particle model provides a set of zero-order wave-functions[1]. In the usual way, one considers the total energy

$$E = \int \Psi^* (T + \mathscr{V})\, \Psi$$

where T is the total kinetic energy, $\mathscr{V}$ the total potential energy and Ψ is a linear combination of the set of zero-order wave-functions. Minimising the energy with respect to the coefficients of the combination, is equivalent to diagonalising the energy matrix taken in the set. Of course, if one uses a very large set of zero-order wave-functions, the source of these wave-functions (i.e. the potential well) becomes quite irrelevant. In practice, one cannot deal with very large sets because of computational tedium, so one restricts the size of the sets. This then implies that one is effectively attributing some physical reality to the ideas of a potential well and single particle motion. However we note that there is still no need to introduce the well explicitly; in particular, no reference need be made at all to its depth (although its size is reflected in the size of the zero-order wave-function). The total energy is still described by $(T + \mathscr{V})$, with

$$\mathscr{V} = \sum_{i<j} v_{ij},$$

the sum running over all pairs of particles. One may however separate the potential energy arbitrarily into

$$\mathscr{V} = V + (\mathscr{V} - V)$$

where $V = \sum_i V_i$ is some single particle potential energy arising from part of the summation. Although such a separation helps to complement ones idea of the physical role of the single particle potential well, it clearly does not introduce a well *in addition to* the inter-particle forces. The sole practical purpose of the

[1] M. G. REDLICH: Phys. Rev. **99**, 1421 (1955).

well is to guide our choice of a first set of wave-functions, all subsequent calculations, including any improvement of the functions, being carried out with the interparticle potential $\mathscr{V}$.

Finally we remark that variational calculations based on the Individual Particle Model badly underestimate the observed binding energies of nuclei[1]. As we mentioned in the Introduction to this article, it was this fact, which has been known for a long time, that arrested the development of the model. However, it is now established that the model is still very useful in spite of this failure. It has had striking successes in predicting other data such as spectra. For this reason we will present the model in unmodified form in this and the next chapter. In Chap. VII, we will describe some modifications of the model suitable for getting better binding energies.

17. Isotopic spin. It is convenient, in view of the almost certain equality of nuclear forces between nucleons whether they be neutrons or protons, to treat the neutron and proton as different states of the same particle. In analogy with the intrinsic spin s of a nucleon which has the two eigenvalues $\pm\frac{1}{2}$ (spin up or spin down) we introduce an isotopic spin t with eigenvalues $\pm\frac{1}{2}$ corresponding to neutron or proton respectively. Although attempts have been made recently, in the realm of fundamental particle theory, to give a physical meaning to the isotopic spin and its associated three-dimensional space, for the purposes of this review it will be treated simply as a mathematical device. Just as the total intrinsic spin S of a system of particles describes the symmetry of the wave function with respect to permutations of the spin co-ordinates of the particles, so the total isotopic spin T describes the symmetry[2] of that function with respect to permutations of the charge co-ordinates of the particles.

As described in Chap. I, the Pauli exclusion principle restricts the number of particles having orbital quantum number l and radial quantum number n to $4(2l+1)$, so that if we take the individual particle wave functions appropriate to an oscillator potential (Fig. 1) we shall expect closed shells at mass numbers 4, 16, 40 etc. For more than about 40 nucleons however, the Coulomb interaction between protons, because of its non-saturating character, builds up to a strength comparable with that of the nuclear interactions, so that the stable nuclei have more neutrons than protons. It is then convenient to treat neutrons and protons as different kinds of particle, abandoning the isotopic spin formalism.

18. Definition of the complete set of equivalent particle states in Russell-Saunders coupling. The individual particle wave functions will be denoted by:

$$\varphi(n\,l\,m_l\,m_s\,m_t) = u_{n\,l}(r)\,Y_{m_l}^{(l)}(\vartheta,\varphi)\,\delta(s,m_s)\,\delta(t,m_t). \tag{18.1}$$

Here, n is the radial quantum number, l is the orbital angular momentum and m_l, m_s, m_t are respectively the magnetic quantum numbers in orbital, intrinsic spin and isotopic spin spaces. The spatial co-ordinates are denoted by r, ϑ, φ

[1] H. A. Bethe and R. F. Bacher: Rev. Mod. Phys. **8**, 82 (1936). — D. R. Inglis: Phys. Rev. **51**, 531 (1937). — E. Feenberg and E. P. Wigner: Phys. Rev. **51**, 95 (1937). — E. Feenberg and M. L. Phillips: Phys. Rev. **51**, 597 (1937).

[2] In group-theoretical language, the value of T denotes the irreducible representation $\left[\frac{A}{2}+T,\ \frac{A}{2}-T\right]$ of the symmetric group S_A of the charge co-ordinates of the A nucleons according to which the function transforms. At the same time, T can be considered to label the representation of the unitary group U_2 in the two-dimensional space spanned by the charge wave functions of a single particle or to label the representation of the rotation group R_3 in the 3-dimensional isotopic spin space constructed in analogy with the physical 3-dimensional space of the intrinsic spin.

while s and t represent the spin and isotopic spin co-ordinates which can take only the values $\pm\frac{1}{2}$. The wave functions φ are orthonormal in the sense that:

$$\left.\begin{aligned} \iiint \varphi^* \varphi' r^2 dr\, d\Omega &= \delta(n, n')\, \delta(l, l')\, \delta(m_l, m_{l'}), \\ \sum_s \varphi^* \varphi' &= \delta(m_s, m_s'), \\ \sum_t \varphi^* \varphi' &= \delta(m_t, m_t'). \end{aligned}\right\} \tag{18.2}$$

The surface[1] harmonic $Y_{m_l}^{(l)}$ in (18.1) is defined with the phase convention of Condon and Shortley[2] described in Appendix 1.

In a system of A individual particles moving in a well the lowest configuration will consist of so many closed shells $n_i\, l_i$ containing $\sum_i 4(2l_i+1)$ particles and one unfilled shell nl. In considering the relative positions of levels of that configuration one can neglect the closed shells although when considering interaction between different configurations it is necessary to antisymmetrise completely. For the former case then, we must define a complete set of antisymmetric wave-functions of the k particles in an nl shell. There are of course a number of ways of defining equivalent complete sets and the one of most use will be that which is closest to the actual set of eigenstates of the energy. Clearly then, our choice will depend on the interaction to be used. For a "central" interaction, i.e. one which depends on the particle positions only through the absolute magnitudes of their separation distances (or in other words is a scalar with respect to rotations in the orbital space), both the total orbital angular momentum L and the total intrinsic spin S will be good quantum numbers. This mode of coupling is called Russell-Saunders or L-S coupling[3] and will be described in this section. The most important alternative coupling mode, the j-j scheme, will be described in Sect. 20.

Consider the $(2l+1)^2$ states of the orbital configuration l^2. They may be written

$$\varphi(l^2\, m_1\, m_2) \tag{18.3}$$

where $m_1\, m_2$, referring to the two particles, may take integral values within $-l \leq m \leq l$. These states are not however eigenstates of the angular momentum $\boldsymbol{L} = \boldsymbol{l}_1 + \boldsymbol{l}_2$. We can construct a complete set of states equivalent to (18.3) and which are eigenstates of $\boldsymbol{L}^2$ by making the transformation[4] given by the Wigner coefficients, viz:

$$\varphi(l^2\, L\, M) = \sum_{m_1(m_2)} (l_1\, l_2\, m_1\, m_2 \,|\, l_1\, l_2\, L\, M)\, \varphi(l^2\, m_1\, m_2). \tag{18.4}$$

Since M is fixed and the Wigner coefficient vanishes unless $m_1 + m_2 = M$, the value of m_2 is determined by m_1. This is indicated by the bracket round m_2. The properties of the Wigner coefficients will be discussed in Appendix 2; let it suffice for the moment to quote the result that

$$P_{12}^r\, \varphi(l^2\, L\, M) = (-1)^{2l-L}\, \varphi(l^2\, L\, M) = (-1)^L\, \varphi(l^2\, L\, M)$$

---

[1] The function $Y_m^{(l)}$ is sometimes called a spherical harmonic but strictly the spherical harmonic contains a radial factor, the angular part alone being called a surface harmonic.

[2] E. U. Condon and G. H. Shortley: Theory of Atomic Spectra. Cambridge: Cambridge University Press 1935.

[3] Note that the expression L-S coupling indicates that L and S are good quantum numbers, *not* that there is coupling between the orbital angular momentum L and the intrinsic spin S.

[4] In group theoretical language, this is the transformation which reduces the product representation $D_l \times D_l$ of R_3 to the irreducible representations D_L.

where P^r_{12} is the permutation operator of the spatial co-ordinates of particles 1 and 2. Hence the states with even L value are symmetric and those of odd L are antisymmetric.

In the same way we can construct spin states of a pair of particles which are eigenstates of the total intrinsic spin S, and then

$$P^\sigma_{12}\,\chi\{(\tfrac{1}{2})^2\,S\,M_s\} = (-1)^{1-S}\,\chi\{(\tfrac{1}{2})^2\,S\,M_s\},$$

where now P^σ_{12} is the permutation operator of the spin co-ordinates. Hence the states with even S are now antisymmetric while the odd S are symmetric. In exact analogy we may construct isotopic spin states.

The problem of constructing total wave-functions in all three spaces of orbit, spin and isotopic spin is then quite trivial for the two-body system, since they can only be the products of three antisymmetric functions or of one antisymmetric function and two symmetric functions. Hence for the configuration l^2 in question we have the totally antisymmetric states:

$$^{13}L\ (\text{even})\ ^{31}L\ (\text{even}) \quad \text{i.e.}\quad L = 0, 2 \ldots 2l,$$

$$^{11}L\ (\text{odd})\ ^{33}L\ (\text{odd}) \quad \text{i.e.}\quad L = 1, 3 \ldots 2l-1.$$

In this notation, the indices denote the multiplicities $(2T+1)$ and $(2S+1)$ in that order.

When we have more than two particles, the problem of constructing antisymmetric wave-functions is not so simple, although the principle is the same. Consider the orbital part. With two particles any function may be written as a linear combination of a symmetric and an antisymmetric function so that we had only to consider the symmetric and the antisymmetric functions to form a basis. For a number k of particles it is necessary to consider not only those functions which are symmetric in all particles and those which are antisymmetric in all particles to form a basis, but also a number of functions of intermediate symmetry. It is possible to classify these functions of intermediate symmetry into sets[1] which are invariant under any permutation of the particles[2]. These sets may be labelled by a partition $[\lambda_1\,\lambda_2\ldots\lambda_r]$ of k denoted by $[\lambda]$, where

$$\lambda_1 \geq \lambda_2 \geq \lambda_3 \ldots \geq \lambda_r \quad \text{and} \quad \sum_i \lambda_i = k.$$

A set of totally symmetric functions has only one member, expressing the fact that any permutation on such a function reproduces the same function. The same can be said for the totally antisymmetric functions except that a factor $(-1)^p$ is introduced where p denotes the parity of the permutation in question. For an intermediate symmetry, the set will contain more than one member, and any permutation on one member will produce a linear combination of the members of that set but will not involve any other set. The totally symmetric function is denoted by the partition $[k]$ while the totally antisymmetric function is denoted by $[1111\ldots1]$. The partitions $[\lambda]$ may be represented by a YOUNG tableau[2], with λ_i blocks in the i-th row, e.g.

$[5\,3\,2] =$ 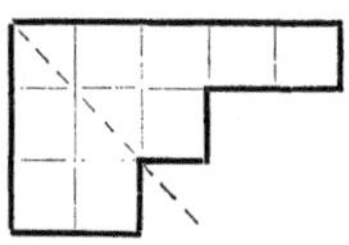

[1] The set $[\lambda]$ are the basis functions which spread out the irreducible representation $[\lambda]$ at the symmetric group S_k.

[2] D. E. RUTHERFORD: Substitutional Analysis. Edinburgh: Edinburgh University Press 1948. — T. YAMANOUCHI: Proc. Phys. Mat. Soc. Japan (3) **19**, 436 (1937).

In such a diagram the longer the rows, the greater the symmetry of the functions labelled by that $[\lambda]$. To each $[\lambda]$ we can associate an adjoint $[\tilde{\lambda}]$ obtained from $[\lambda]$ by interchanging rows and columns or in other words by a reflection of the YOUNG diagram in the diagonal indicated; e.g. with $[\lambda]=[532]$ we have $[\tilde{\lambda}]=[33211]$. It is natural therefore, and can be proved rigorously, that a totally antisymmetric function may be formed only by taking a suitable combination of products of functions of a set $[\lambda]$ with those of the corresponding $[\tilde{\lambda}]$. Let us for a moment regard the isotopic spin and spin parts of the k-particle wave-function as a single charge-spin[1] function. It then follows that the antisymmetric wave-functions of a k-particle system may be classified according to the set $[\lambda]$ to which the orbital part belongs, the charge-spin part having the symmetry of the adjoint set $[\tilde{\lambda}]$ with respect to simultaneous permutations of the charge and spin co-ordinates.

There are certain restrictions on the partitions $[\lambda]$. In the orbital space of the configuration l^k we are dealing with functions of the $(2l+1)$ states of each particle and of the k sets of particle co-ordinates. From the reciprocity between permutations of the state labels and the particle labels it follows that we cannot construct a function antisymmetric in more than $(2l+1)$ particles. This restricts the number of rows of $[\lambda]$ to $(2l+1)$. A similar argument in the charge-spin space restricts the number of rows of $[\tilde{\lambda}]$ to 4 and hence the number of *columns* of $[\lambda]$ to 4.

This reduces the problem to a consideration of the orbital and charge-spin parts of the wave-function separately. For the charge-spin part we have to find which combinations of T and S in the k-particle system form a state of symmetry $[\tilde{\lambda}]$; such a collection of TS values being the supermultiplet of WIGNER[2]. For the orbital part, we have to find which L values of a configuration l^k have symmetry $[\lambda]$; For $l>1$, where there will be many states of the same L in each $[\lambda]$, we shall have to introduce some further means of classification. RACAH[3] introduced a very elegant group theoretical method to separate such states, applying it to the atomic f-shell. His idea was interpreted for the nuclear d-shell by JAHN[4] and extended to the nuclear f-shell by FLOWERS[5]. The principles of the method will be described in Appendix 5.

To illustrate the classification, we give in Table 18 the complete list of states in the p-shell. This was originally given by HUND[6] and was derived by JAHN[3] as an example of the powerful group theoretic method. It is of course only necessary to give the states for the first half of the shell in view of the equivalence between particles and holes[7]. It will be seen from Table 18 that the quantum numbers $[\lambda]\,TSL$ serve to separate all states of the p^k configuration except when $k=6$, the middle of the shell, when there are two $[42]D$ states and two $[321]\,(33)$ states.

This classification is simply an extension of the RUSSELL-SAUNDERS coupling scheme for atoms. However, the nuclear inter-particle force is mainly attractive,

---

[1] The expressions charge function and isotopic spin function are synonymous, the precise relation between charge and isotopic spin being $T_z=\frac{1}{2}(N-Z)$ where N is the number of neutrons, Z the number of protons and T_z the z component of the isotopic spin.

[2] E. P. WIGNER: Phys. Rev. **51**, 946 (1937).

[3] G. RACAH: Phys. Rev. **76**, 1352 (1949).

[4] H. A. JAHN: Proc. Roy. Soc. Lond., Ser. A **201**, 516 (1950).

[5] B. H. FLOWERS: Proc. Roy. Soc. Lond., Ser. A **210**, 497 (1952).

[6] F. HUND: Z. Physik **105**, 202 (1937).

[7] G. RACAH: Phys. Rev. **62**, 438 (1942).

Table 18. *L-S Coupled Wave Functions for the p-shell.*

n	$[\lambda]$	L	$(2T+1, 2S+1)$
0	[0]	S	(1 1)
1	[1]	P	(2 2)
2	[2]	S, D	(1 3) (3 1)
	[1 1]	P	(1 1) (3 3)
3	[3]	P, F	(2 2)
	[2 1]	P, D	(2 2) (2 4) (4 2)
	[1 1 1]	S	(2 2) (4 4)
4	[4]	SDG	(1 1)
	[3 1]	PDF	(1 3) (3 1) (3 3)
	[2 2]	SD	(1 1) (1 5) (5 1) (3 3)
	[2 1 1]	P	(1 3) (3 1) (3 3) (3 5) (5 3)
5	[4 1]	$PDFG$	(2 2)
	[3 2]	PDF	(2 2) (2 4) (4 2)
	[3 1 1]	SD	(2 2) (2 4) (4 2) (4 4)
	[2 2 1]	P	(2 2) (2 4) (4 2) (2 6) (6 2) (4 4)
6	[4 2]	SD^2FG	(1 3) (3 1)
	[4 1 1]	PF	(1 1) (3 3)
	[3 3]	PF	(1 1) (3 3)
	[3 2 1]	PD	(1 3) (3 1) $(3\,3)^2$ (1 5) (5 1) (3 5) (5 3)
	[2 2 2]	S	(1 3) (3 1) (3 5) (5 3) (1 7) (7 1)

whereas the atomic force is repulsive, so that instead of having the states of low orbital symmetry (high spin) and high L value lowest as in the atom, the nucleus has states at high orbital symmetry and low L value lowest. In the atom there is a unique correspondence between the spin S and the symmetry $[\lambda]$ so that, for any central interaction, $[\lambda]$ is a good quantum number simply because S is one. In the nucleus this is no longer true because there are states with the same T and S values but with different $[\lambda]$, e.g. there are (22) states of all three $[\lambda]$ values in p^3. However, if we consider only central interactions independent of spin and isotopic spin, then $[\lambda]$ must be a good quantum number because of the orthogonality of the charge-spin functions for different $[\lambda]$. This approximation, introduced by WIGNER[1], allows the interaction to contain any mixture of ordinary (WIGNER) and space-exchange (MAJORANA) forces,

$$W + M P_{12}^r \tag{18.5}$$

We shall see later that such an approximation is remarkably good for certain light nuclei.

19. Definition of L-S coupled states of inequivalent particles. By inequivalent particles we mean nucleons which have different n or l values in their individual particle quantum numbers, see Eq. (18.1). Let us consider the configuration

$$(n_1 l_1)^{k_1} (n_2 l_2)^{k_2}$$

where $(n_1 l_1) \neq (n_2 l_2)$, containing two inequivalent groups of nucleons. There are two quite distinct ways of setting up a complete set of orthogonal, normalised states Ψ, antisymmetric in all particles.

The simplest way is to define the states genealogically. In this we take a definite state χ_1 of the first group with a definite state χ_2 of the second group,

[1] E. P. WIGNER: Phys. Rev. **51**, 946 (1937).

vector-couple them together in all three spaces (isotopic spin, spin and orbit) to form ψ and then antisymmetrise between the two groups to form Ψ. In more detail

$$\left.\begin{aligned}\chi_1 &\equiv \chi_1\{(n_1 l_1)^{k_1}\alpha_1 T_1 S_1 L_1 M_1^T M_1^S M_1^L\},\\ \chi_2 &\equiv \chi_2\{(n_2 l_2)^{k_2}\alpha_2 T_2 S_2 L_2 M_2^T M_2^S M_2^L\}\end{aligned}\right\} \tag{19.1}$$

where α denotes the $[\lambda]$ together with any other quantum numbers needed to specify the state of the equivalent group. The vector-coupled state ψ is then given in full by

$$\left.\begin{aligned}\psi &= \psi\{(n_1 l_1)^{k_1}(n_2 l_2)^{k_2}\alpha_1 T_1 S_1 L_1 \alpha_2 T_2 S_2 L_2;\, TSL\, M^T M^S M^L\}\\ &= \sum_{M_1^T M_1^S M_1^L (M_2^T M_2^S M_2^L)} (T_1 T_2 M_1^T M_2^T \mid T_1 T_2 T M^T)(S_1 S_2 M_1^S M_2^S \mid S_1 S_2 S M^S)\times\\ &\quad\times (L_1 L_2 M_1^L M_2^L \mid L_1 L_2 L M^L)\,\chi_1\big((n_1 l_1)^{k_1}\alpha_1 T_1 S_1 L_1 M_1^T M_1^S M_1^L\big)\times\\ &\quad\times \chi_2\big((n_2 l_2)^{k_2}\alpha_2 T_2 S_2 L_2 M_2^T M_2^S M_2^L\big)\end{aligned}\right\} \tag{19.2a}$$

which will in future be written in the concise curly bracket notation as

$$\psi = \{\chi_1, \chi_2\}\,\psi \tag{19.3}$$

where the symbol ψ on the right hand side of the equation indicates that χ_1 and χ_2 are vector-coupled to the resultant quantum numbers TSL of ψ.

The final antisymmetrising is given simply by

$$\Psi = \left(\frac{k_1!\,k_2!}{(k_1+k_2)!}\right)^{\frac{1}{2}} \sum_P (-1)^P P\psi \tag{19.2b}$$

in which ψ is defined to have particle numbers 1 to k_1, in χ_1, and particle numbers k_1+1 to k_1+k_2 in χ_2, and P runs over all permutations which preserve the natural ordering within χ_1 and χ_2 separately. The question of ordering of particle numbers is important in fixing the phase of an antisymmetric wave function. The usual convention is to take them in ascending order of particle number which we call natural ordering.

The process described above provides the required complete set of states without going into the complications of constructing functions of definite symmetry properties, that were necessary for equivalent particles. This is possible because the Pauli principle does not restrict the way in which we couple inequivalent particles. Although the states Ψ constructed above have definite symmetry properties $[\lambda_1]$, $[\lambda_2]$ with respect to permutations in the orbital space of the two groups separately, they do not in general have a definite symmetry property $[\lambda]$ with respect to permutations of all (k_1+k_2) particles. This brings us to the second way of constructing a complete set, which is simply an extension of the method described in Sect. 18 for equivalent particles.

Exactly as in Sect. 18 we break the totally antisymmetric wave function into a sum of products of a charge-spin part and an orbital part belonging to symmetry sets $[\tilde{\lambda}]$ and $[\lambda]$ with respect to permutations of all particles. Now the charge-spin parts are exactly the same as those for a group of (k_1+k_2) equivalent particles. The problem remaining is to construct orbital states of the symmetry $[\lambda]$ from the configuration $(n_1 l_1)^{k_1}(n_2 l_2)^{k_2}$ in a manner similar to that used for a group of equivalent particles. Using this approach, one constructs a complete set of states Ψ_e which have quantum numbers

$$\Psi_e \equiv \Psi_e\big((n_1 l_1)^{k_1}(n_2 l_2)^{k_2}\alpha_1 L_1 \alpha_2 L_2;\, [\lambda]\, TSL\, M^T M^S M^L\big) \tag{19.4}$$

with the additional quantum number $[\lambda]$ instead of the parental $T_1 S_2\, T_2 S_2$ values in (19.2).

Comparing these two approaches, the first has the advantage of simplicity if one is working with the complete set. If however, we are working close to the WIGNER approximation (18.5), then as we have seen, the label $[\lambda]$ will be nearly a good quantum number. Consequently it would reduce the problem considerably to neglect all but the most symmetrical $[\lambda]$. In such a calculation it would only be necessary to construct that state of the complete set (19.4) with $[\lambda]$ as a quantum number. Unfortunately however such an approximation is rarely reliable in practice so that we shall use the set (19.2) defined in the first way. One should always remember that since (19.2) and (19.4) define equivalent complete sets, there is a transformation between them, which in fact is usually very simple, enabling one to switch from one to the other for convenience and for checking. The transformation may be deduced simply from the fractional parentage coefficients of the charge-spin part alone for the reduction of $(\gamma)^{k_1+k_2}$ to $(\gamma)^{k_1}$ and $(\gamma)^{k_2}$ where γ represents the charge-spin state fo a single particle. When either k_1 or k_2 is less than three, these coefficients are tabulated[1,2].

20. Alternative approach in terms of j-j coupling. In both the preceding sections we have considered the complete set of states which have the total intrinsic spin S and the total orbital angular momentum L as quantum numbers as well as the total isotopic spin T. Such a scheme is appropriate to any charge independent central two-body interaction, for which T, S and L must necessarily be good quantum numbers. As described in earlier chapters however, the experimental evidence indicates the presence of a strong spin-orbit force

$$M = \xi \sum_i (\boldsymbol{s}_i \cdot \boldsymbol{l}_i) \tag{20.1}$$

coupling the spin and orbital angular momenta of each particle and therefore splitting the energy levels of a single particle with different values of the total angular momentum

$$\boldsymbol{j} = \boldsymbol{l} + \boldsymbol{s}.$$

Just as the L-S coupling scheme was natural for a pure central interaction let us now suppose that we have a pure spin-orbit interaction (20.1) and ask what coupling scheme is natural in that case. Since (20.1) is a single particle operator, scalar in the j space, it follows that the j value of *each* particle of a k-particle system will be a good quantum number as well as the total angular momentum J and the isotopic spin T. The coupling scheme which preserves these quantum numbers is referred to as the j-j coupling scheme. Since the intrinsic spin s of each particle has the value $\frac{1}{2}$, each l value can have two j-values viz. $(l+\frac{1}{2})$ and $(l-\frac{1}{2})$. In this pure coupling scheme we therefore have two smaller shells labelled by nlj in place of the L-S coupling shell labelled by nl. The physical significance of these j-j coupling shells has been discussed in the earlier chapters. The object of this section is to set up a classification for the complete set of states of a number of particles in an nlj shell and this will be done in quite close analogy with Sect. 18. The main difference is that now the intrinsic spin and orbital angular momenta of each particle are coupled together so that we have only to consider the resultant total particle angular momentum j. We can therefore first classify the wave-functions for k particles by their symmetry $[\lambda]$ with respect to permutations of the co-ordinates in j-space. From the antisymmetry of the total wave function the symmetry of the isotopic spin part

[1] H. A. JAHN and H. VAN WIERINGEN: Proc. Roy. Soc. Lond., Ser. A **209**, 592 (1951).
[2] ELLIOTT, HOPE and JAHN: Phil. Trans. Roy. Soc. Lond., Ser. A **206**, 912 (1953).

must be $[\tilde{\lambda}]$, the adjoint of $[\lambda]$, as described in Sect. 18. However, since the isotopic spin of a single particle is only two-valued, so that $[\tilde{\lambda}]$ can have only two rows and $[\lambda]$ only two columns, there is a direct correspondence between $[\lambda]$ and the isotopic spin T, viz. T is equal to half the number of times that 1 appears in $[\lambda]$. In other words the value of T for a given number of particles determines the symmetry of the isotopic spin part of the wave-function and hence indirectly, through the PAULI principle, the symmetry of the spin-orbital part. This is similar to the situation in atomic spectra with T and J replacing S and L, expect that the individual j are half integral whereas the individual l are integral. The problem is now reduced to a determination of the spin orbital wave functions of a given symmetry $[\lambda]$ in a number k of equivalent particles nlj, and this is best done by resorting to the group theoretical methods of RACAH. This has been done for the j-j coupling problem by FLOWERS[1].

The essence of this method is to consider the transformation properties of the wave-functions under certain groups of transformations, just as for example the $[\lambda]$ labels their properties under the group of permutations in spin-orbital space, and the J value under the rotations in spin-orbital space. From this approach it transpires that for a number k of particles there appear states with the same transformation properties as states of a fewer number of particles. This gives rise to the concept of the seniority s of a state ψ, defined as the smallest number of particles for which a state with the same transformation properties as ψ can be constructed. In particular, for an even value of $(T+\frac{1}{2}k)$ and k, i.e. for the lowest T value of an even number of neutrons and protons, there occur states of seniority zero, while for odd k there occur states of seniority unity. This result ties in well with the early shell model postulations of MAYER[2] and of HAXEL, JENSEN and SUESS[3] for the pairing of like particles to resultant spin $J=0$, with the odd particles having $J=j$. As we saw in Chap. IV, although such states of seniority unity resemble single particle states in some respect, they are nevertheless states of k particles and for example, their magnetic moments[4] are not those of a single particle in general. The group of transformations which one uses in the j-j coupling problem is called the *Symplectic Group* and the transformation properties of the states which can occur may be denoted by two labels (just as for three-dimensional rotations one has the single label J). One of these may be taken as the seniority s and the other as the "reduced isotopic spin", t, in the terminology of FLOWERS[1]. Physically the value of t in a state ψ of seniority s may be interpreted as the value of the isotopic spin T of that state of the smallest number s of particles which has the same transformation properties as ψ.

This classification according to (s, t) is valuable because in the limit of short range forces there is near degeneracy of levels with the same (s, t) so that the level ordering can be roughly estimated from (s, t) alone. For maximum isotopic spin, i.e. for neutrons only or protons only, this level ordering persists for physically reasonable values of the range, although for lower isotopic spin, corresponding to configurations of neutrons and protons, there is considerable mixing of levels as the range is increased.

To illustrate the classification, we give in Table 19 the complete set of j-j coupled states for the $p_{\frac{3}{2}}$ shell. Although $[\lambda]$ and T are equivalent quantum

[1] B. H. FLOWERS: Proc. Roy. Soc. Lond., Ser. A **212**, 248 (1952).
[2] M. G. MAYER: Phys. Rev. **75**, 1969 (1949); **78**, 16 (1950).
[3] HAXEL, JENSEN and SUESS: Phys. Rev. **75**, 1766 (1950). — Z. Physik **128**, 295 (1950).
[4] B. H. FLOWERS: Phil. Mag. **43**, 1330 (1952).

Table 19. *j-j Coupled Wave Functions for the $p_{\frac{3}{2}}$ Shell.*

n	$[\lambda]$	T	(σ)	(s, t)	J
0	[0]	0	(00)	(0, 0)	0
1	[1]	$\frac{1}{2}$	(10)	(1, $\frac{1}{2}$)	$\frac{3}{2}$
2	[2]	0	(20)	(2, 0)	1, 3
	[11]	1	(00)	(0, 0)	0
			(11)	(2, 1)	2
3	[21]	$\frac{1}{2}$	(10)	(1, $\frac{1}{2}$)	$\frac{3}{2}$
			(21)	(3, $\frac{1}{2}$)	$\frac{1}{2}, \frac{5}{2}, \frac{7}{2}$
	[111]	$\frac{3}{2}$	(10)	(1, $\frac{1}{2}$)	$\frac{3}{2}$
4	[22]	0	(00)	(0, 0)	0
			(11)	(2, 1)	2
			(22)	(4, 0)	2, 4
	[211]	1	(20)	(2, 0)	1, 3
			(11)	(2, 1)	2
	[1111]	2	(00)	(0, 0)	0

numbers we list them both. Likewise, in addition to s and t we give the equivalent quantum number (σ), which is the standard group theoretical label for states with certain transformation properties, that is, belonging to a certain representation (σ) of the group in question.

21. Transformation between the two coupling schemes. The *L-S* and *j-j* coupling schemes described in the preceding sections provide equivalent complete sets of wave functions for the k-particle system. There must therefore be a transformation between the two sets; that is, any state of one scheme may be expressed as a linear combination of states of the other.

For a pair of particles we have to couple together the four vectors $\boldsymbol{s}_1, \boldsymbol{s}_2, \boldsymbol{l}_1, \boldsymbol{l}_2$ and the two schemes differ simply in the order of coupling. In *L-S* coupling we first form

$$\boldsymbol{S} = \boldsymbol{s}_1 + \boldsymbol{s}_2 \quad \text{and} \quad \boldsymbol{L} = \boldsymbol{l}_1 + \boldsymbol{l}_2$$

and the combine the resultants to give

$$\boldsymbol{J} = \boldsymbol{S} + \boldsymbol{L}.$$

In *j-j* coupling on the other hand we first form

$$\boldsymbol{j}_1 = \boldsymbol{s}_1 + \boldsymbol{l}_1 \quad \text{and} \quad \boldsymbol{j}_2 = \boldsymbol{s}_2 + \boldsymbol{l}_2$$

and then combine the resultants to give

$$\boldsymbol{J} = \boldsymbol{j}_1 + \boldsymbol{j}_2.$$

To signify the *L-S* coupled wave function in detail we shall use the notation

$$\big((s_1 s_2)\, S\, (l_1 l_2)\, L, JM\,\big|.$$

The transformation coefficient

$$\alpha = \big((s_1 s_2)\, S\, (l_1 l_2)\, L, JM\big|\, (s_1 l_1)\, j_1\, (s_2 l_2)\, j_2, JM\big)$$

is then simply obtained from a rearrangement of the order of vector coupling. This will be described in Appendix 3, the result being

$$\alpha = \{(2S+1)(2L+1)(2j_1+1)(2j_2+1)\}^{\frac{1}{2}} \times$$
$$\times \sum_x (2x+1)\, W(S l_1 J l_2, xL)\, W(S l_1 \tfrac{1}{2} j_1, x \tfrac{1}{2})\, W(j_1 \tfrac{1}{2} J l_2, x j_2)$$

or

$$\alpha = \{(2S+1)(2L+1)(2j_1+1)(2j_2+1)\}^{\frac{1}{2}} \begin{Bmatrix} S & \frac{1}{2} & \frac{1}{2} \\ L & l_1 & l_2 \\ J & j_1 & j_2 \end{Bmatrix}$$

in terms of the RACAH W-functions or the WIGNER 9-j symbol described in Appendix 3. These coefficients have been tabulated[1] for l values less than 6.

By way of illustration, we given in Table 20 the transformation coefficients for the $T=1$ states of d^2. The transformation is clearly orthogonal.

For more than two particles, the transformation is not so simple, since one must use the additional quantum numbers specifying the states in both schemes. One would have to use a chain calculation involving fractional parentage coefficients for both schemes (see Sect. 24 and Appendix 6). This would be very laborious. However it should never be necessary to transform for more than two particles in the usual shell model calculations.

Table 20. *Transformation Coefficients between the L-S and j-j Coupling Schemes for d^2, with $T=1$.*

$J=0$

	$(\frac{5}{2})^2$	$(\frac{3}{2})^2$
^{31}S	$(\frac{3}{5})^{\frac{1}{2}}$	$(\frac{2}{5})^{\frac{1}{2}}$
^{33}P	$-(\frac{2}{5})^{\frac{1}{2}}$	$(\frac{3}{5})^{\frac{1}{2}}$

$J=1$

	$(\frac{5}{2}\,\frac{3}{2})$
^{33}P	1

$J=2$

	$(\frac{5}{2})^2$	$(\frac{5}{2}\,\frac{3}{2})$	$(\frac{3}{2})^2$
^{31}D	$(\frac{12}{25})^{\frac{1}{2}}$	$-(\frac{6}{25})^{\frac{1}{2}}$	$(\frac{7}{25})^{\frac{1}{2}}$
^{33}P	$(\frac{56}{125})^{\frac{1}{2}}$	$(\frac{63}{125})^{\frac{1}{2}}$	$-(\frac{6}{125})^{\frac{1}{2}}$
^{33}F	$-(\frac{9}{125})^{\frac{1}{2}}$	$(\frac{32}{125})^{\frac{1}{2}}$	$(\frac{84}{125})^{\frac{1}{2}}$

$J=3$

	$(\frac{5}{2}\,\frac{3}{2})$
^{33}F	1

$J=4$

	$(\frac{5}{2})^2$	$(\frac{5}{2}\,\frac{3}{2})$
^{31}G	$(\frac{1}{5})^{\frac{1}{2}}$	$-(\frac{4}{5})^{\frac{1}{2}}$
^{33}F	$(\frac{4}{5})^{\frac{1}{2}}$	$(\frac{1}{5})^{\frac{1}{2}}$

In an intermediate coupling calculation it is immaterial which scheme one uses, but since the writers prefer to work to the L-S scheme the formulae in the following sections will mostly be expressed in that scheme. The j-j coupling scheme is of most use when one restricts the set of states by retaining only the lowest one or two configurations in the j-j sense. It is useful however in interpreting results obtained in one scheme to obtain a rough estimate of the composition of those states in the other scheme. This is important to do, because the fact that a particular state is pure in one scheme does not imply that it is impure in the other. In fact in many cases there is a considerable overlap between the lowest states in the two schemes. We can obtain such an estimate of the j-j composition of an L-S coupled state by evaluating the operator $\sum_i \boldsymbol{j}_i^2$ which gives the mean value of $\boldsymbol{j}^2$ in that state. Following an intermediate coupling calculation, this can be done very simply since

$$\sum_i \boldsymbol{j}_i^2 = \sum_i \boldsymbol{s}_i^2 + \sum_i \boldsymbol{l}_i^2 + 2\sum_i (\boldsymbol{s}_i \cdot \boldsymbol{l}_i) = \tfrac{3}{4}n + n l(l+1) + 2\sum_i (\boldsymbol{s}_i \cdot \boldsymbol{l}_i)$$

and $\sum_i (\boldsymbol{s}_i \cdot \boldsymbol{l}_i)$ will already have been evaluated. If one assumes that in the j-j scheme the state may be approximated by the lowest two configurations

$$(l+\tfrac{1}{2})^k + \lambda (l+\tfrac{1}{2})^{k-1}(l-\tfrac{1}{2})$$

[1] J. M. KENNEDY and M. J. CLIFF: Chalk River Report CRT 609, 1955.

then from $\langle j^2 \rangle$ one can evaluate λ. To estimate the L-S composition in a j-j coupled state one could in the same way evaluate the HUND operator $\sum_{i<j} P_{ij}^r$ where P_{ij}^r is the space exchange operator. This would give an estimate of the mixing of different $[\lambda]$ in the L-S scheme since the value of that operator in such a state is known[1].

22. Correction for centre-of-mass motion. The shell model of the nucleus supposes that all nucleons move about a fixed centre. This results in a motion of the centre-of-mass of the nucleons about that centre and our using a model in which the centre of mass is not at rest. One must therefore be on the lookout for spurious effects due to this motion and such effects have been described by ELLIOTT and SKYRME[2]. It is possible to study them explicitly only for the case of an oscillator potential. Then the centre-of-mass motion may be separated out in the Hamiltonian and one can speak of states of the k particles in which the centre-of-mass is in an oscillator orbit. It is clear that wave functions which have the same function of the relative particle co-ordinates and differ only in the function for the centre-of-mass motion must describe the same (internal) state of the nucleus. This fact leads to the appearance of shell model states which have the same internal wave function as a shell model state of some lower configuration, but a different centre-of-mass motion. Such states, which we call spurious, must be discarded. We can generate all the internal states once only if we specify one particular state for the centre-of-mass.

In all states of a configuration, in the L-S sense, in which all shells with energy lower than the topmost is filled, it is shown that, for the oscillator potential, the centre-of-mass is in the $1s$ state. It is therefore by far the simplest to specify that the centre-of-mass be taken in the $1s$ state. Then all states of the lowest configuration will be good (not spurious) and it is only for excited configurations, with more than one shell unfilled, that spurious states will arise. One may determine the spurious states either by evaluating the matrix of R^2, where $\boldsymbol{R} = \sum_i \boldsymbol{r}_i$ is the centre-of-mass co-ordinate, and diagonalising, or by multiplying the good states of the lower configurations by $\boldsymbol{R}$ and expanding in terms of the states of the higher configurations. This procedure can become lengthy for complicated configurations.

An example is provided by the states of single excitation of O^{16}. The ground configuration is $(1s)^4\,(1p)^{12}$ so that the singly excited configurations in the oscillator will be

$$(1s)^4\,(1p)^{11}\,(2s)$$
$$(1s)^4\,(1p)^{11}\,(1d).$$

Apparently, one could construct a ${}^{11}P$ state of symmetry [4444] from both configurations, but it turns out that a particular combination of those two states is simply the ground state, $(1s)^4(1p)^{12}\ {}^{11}S$ [4444], multiplied by the centre-of-mass vector $\boldsymbol{R}$. In other words, that combination has its centre-of-mass in a $1p$ state, whereas that for the ground state is in a $1s$ state. The orthogonal combination is a good state with a $1s$ centre-of-mass motion. The good combination is found to be

$$\sqrt{\tfrac{5}{6}}\,(1s)^4\,(1p)^{11}\,(2s) + \sqrt{\tfrac{1}{6}}\,(1s)^4\,(1p)^{11}\,(1d)$$

---

[1] F. HUND: Z. Physik **105**, 202 (1937).

[2] J. P. ELLIOTT and T. H. R. SKYRME: Proc. Roy. Soc. Lond., Ser. A **232**, 561 (1955). — Nuovo Cim. **10**, 4, 164 (1956).

In an intermediate coupling calculation[1], it was found impossible to obtain even a qualitative fit to the observed spectrum until taking account of the removal of the spurious state.

If one were to use a two-body interaction rigorously taking into account consistently all the interactions between shells and removing the kinetic energy of the centre-of-mass, then the spurious states would be degenerate with the corresponding state of some lower configuration which has the same internal structure. However, in the usual semi-empirical approach of the individual particle model when this is not carried out, the spurious states could seriously perturb the calculated spectrum and so must be removed before diagonalising the energy matrix.

So far as multipole moments are concerned, there is no correction to be applied to the value calculated for a "good" state simply because the centre of mass is in an S-state; it would make no difference whether we used the multipole operator referred to the well centre or the centre of mass. In deriving the electric dipole sum rule[2] however, one sums over *all* states, good and spurious, so that there one must use the corrected operator, i.e. referred to the centre of mass. Such an operator picks out the good states in the sum, giving zero contribution from the spurious ones. One would obtain the same answer by using the uncorrected operator, i.e. referred to the well centre, and summing only over good states. In this instance, it is clearly preferable to use the former procedure.

23. Types of nuclear interaction. Although it is not rigorously valid to represent the interaction of nucleons through the meson field by an interaction potential between pairs of nucleons[3], such an approach is justifiable if $(v/c)^2 \ll 1$, where v is the mean nucleon velocity. Taking the mean kinetic energy of a nucleon in a nucleus to be 20 MeV, the ratio $(v/c)^2$ is about $\frac{1}{20}$, so the condition is satisfied. Following from this result it may be shown that terms in the interaction involving higher powers of the momenta than the first must be very small from LORENTZ invariance arguments. In discussing the interaction we may therefore limit ourselves to static and first order terms. The exact form of the interaction which best fits all the various experimental data is by no means certain, although many attempts have been made to find such a form. In this section we merely introduce the different types of interaction and classify them according to their transformation properties following WIGNER[4], and EISENBUD and WIGNER[5]. In Appendix 8 we shall give formulae for their matrix elements.

The interaction must obey the following invariance requirements:

(i) It must be Hermitian, i.e. energy must be conserved.

(ii) It must depend only on the relative co-ordinates and momenta, i.e. invariance with respect to translation and Galilean transformations.

(iii) It must be scalar in J space, i.e. conservation of total angular momentum or invariance with respect to rotations of the physical axes.

(iv) It must have even parity, i.e. invariance under spatial reflections, or conservation of parity.

---

[1] J. P. ELLIOTT: To be published.

[2] J. S. LEVINGER and H. A. BETHE: Phys. Rev. **78**, 115 (1950).

[3] We ignore many-body forces from the start, since there is no convincing theoretical or experimental evidence for them and they are much more difficult to handle, than two-body forces [S. D. DRELL and K. HUANG: Phys. Rev. **91**, 1527 (1951). — L. E. H. TRAINOR: Phys. Rev. **95**, 801 (1954)].

[4] E. P. WIGNER: Phys. Rev. **50** (1937).

[5] L. EISENBUD and E. P. WIGNER: Proc. Nat. Acad. Sci. USA. **27**, 281 (1941).

(v) Invariance under time reversal, i.e. for $\boldsymbol{p} \to -\boldsymbol{p}$, $\boldsymbol{\sigma} \to -\boldsymbol{\sigma}$, $\tau_y \to -\tau_y$.

(vi) It must be scalar in isotopic spin space, i.e. it must not depend on the z-component of the isotopic spin, i.e. it must be charge independent.

(vii) It must be symmetric with respect to permutation of the particles, taking all co-ordinates into account.

The first five of these are absolute requirements, the sixth being the result of experimental observation. The sixth condition is violated by the Coulomb interaction but here we are considering the purely nuclear forces.

The interaction $\boldsymbol{v}_{ij}$ between particles i and j may be factorised into separate isotopic spin, intrinsic spin and orbital parts:

$$\boldsymbol{v}_{ij} = \left(\mathcal{S}^{(\lambda)}(ij) \cdot \mathcal{L}^{(\lambda)}(ij)\right) \mathcal{I}(ij). \tag{23.1}$$

$\mathcal{I}$ must be scalar from (vi) and is restricted to the forms 1 or $(\boldsymbol{\tau}_i \cdot \boldsymbol{\tau}_j)$. In (23.1) we have written the spatial part, which must be scalar with respect to simultaneous rotations in spin and orbital spaces from (iii), as a scalar product of tensors of degree λ in the separate spaces. We shall speak more of tensors in Appendix 4. Put simply, they are functions which transform under rotations like a wave function with angular momentum λ and therefore obey the same laws of combination (vector coupling). Since the maximum intrinsic spin of a two-body state is unity, we have that $\lambda \leq 2$. When $\lambda = 0$, $\boldsymbol{v}_{ij}$ is scalar in both spin and orbital spaces and is called a central force. $\mathcal{S}$ can then only take the forms 1 or $(\boldsymbol{\sigma}_i \cdot \boldsymbol{\sigma}_j)$ while $\mathcal{L}$ can only be a function of the distance between the particles. $V_0(r_{ij})$. If $\lambda = 1$, the invariance requirements give uniquely

$$\mathcal{S}^{(1)} = (\boldsymbol{\sigma}_i + \boldsymbol{\sigma}_j), \qquad \mathcal{L}^{(1)} = [(\boldsymbol{r}_i - \boldsymbol{r}_j) \times (\boldsymbol{p}_i - \boldsymbol{p}_j)]\, V_1(r_{ij}) \tag{23.2}$$

where V_1 is an arbitrary function of the distance. When $\lambda = 2$, $\mathcal{S}^{(2)}$ is the tensor product of $\boldsymbol{\sigma}_i$ and $\boldsymbol{\sigma}_j$, while $\mathcal{L}^{(2)}$ is the tensor product of $(\boldsymbol{r}_i - \boldsymbol{r}_j)$ with itself. The component of $\mathcal{S}^{(2)}$ which is invariant to rotations about the z-axis is simply

$$\mathcal{S}_0^{(2)} = -\sqrt{\tfrac{1}{6}}\, \{(\boldsymbol{\sigma}_i \cdot \boldsymbol{\sigma}_j) - 3\, (\sigma_i)_z (\sigma_j)_z\}, \tag{23.3}$$

while $\mathcal{L}_0^{(2)}$ has a similar form, with an arbitrary distance dependence $V_2(r_{ij})$. In this case, the actual tensor product in (23.1) has the simple form

$$(\mathcal{S}^{(2)} \cdot \mathcal{L}^{(2)}) = \left[\frac{\{\boldsymbol{\sigma}_i \cdot (\boldsymbol{r}_i - \boldsymbol{r}_j)\}\{\boldsymbol{\sigma}_j \cdot (\boldsymbol{r}_i - \boldsymbol{r}_j)\}}{(\boldsymbol{r}_i - \boldsymbol{r}_j)^2} - \frac{1}{3}\, (\boldsymbol{\sigma}_i \cdot \boldsymbol{\sigma}_j)\right] V_2(r_{ij}), \tag{23.4}$$

although for evaluating the matrix elements of $\boldsymbol{v}_{ij}$ it is more convenient to use $\mathcal{S}^{(2)}$ and $\mathcal{L}^{(2)}$ separately, see Appendix 8.

The most general two-body interaction may therefore be written

$$\left.\begin{aligned} \boldsymbol{v}_{ij} = {} & \{(a_0 + a_\sigma (\boldsymbol{\sigma}_i \cdot \boldsymbol{\sigma}_j) + a_\tau (\boldsymbol{\tau}_i \cdot \boldsymbol{\tau}_j) + a_{\sigma\tau} (\boldsymbol{\sigma}_i \cdot \boldsymbol{\sigma}_j)(\boldsymbol{\tau}_i \cdot \boldsymbol{\tau}_j)\}\, V_0(r_{ij}) + \\ & + \{b_0 + b_\tau (\boldsymbol{\tau}_i \cdot \boldsymbol{\tau}_j)\}\, (\boldsymbol{\sigma}_i + \boldsymbol{\sigma}_j) \cdot [\boldsymbol{r}_{ij} \times \boldsymbol{p}_{ij}]\, V_1(r_{ij}) + \\ & + \{c_0 + c_\tau (\boldsymbol{\tau}_i \cdot \boldsymbol{\tau}_j)\} \left\{\frac{(\boldsymbol{\sigma}_i \cdot \boldsymbol{r}_{ij})(\boldsymbol{\sigma}_j \cdot \boldsymbol{r}_{ij})}{r_{ij}^2} - \frac{1}{3}\, (\boldsymbol{\sigma}_i \cdot \boldsymbol{\sigma}_j)\right\} V_2(r_{ij}), \end{aligned}\right\} \tag{23.5}$$

where $\boldsymbol{r}_{ij} = \boldsymbol{r}_i - \boldsymbol{r}_j$ and $\boldsymbol{p}_{ij} = \boldsymbol{p}_i - \boldsymbol{p}_j$. Following from the relations

$$P_{ij}^\sigma = \tfrac{1}{2}\left(1 + (\boldsymbol{\sigma}_i \cdot \boldsymbol{\sigma}_j)\right) \quad \text{and} \quad P_{ij}^\tau = \tfrac{1}{2}\left(1 + (\boldsymbol{\tau}_i \cdot \boldsymbol{\tau}_j)\right) \tag{23.6}$$

where P_{ij}^σ, P_{ij}^τ are the permutation or exchange operators in spin and isotopic spin spaces respectively, and from the property

$$P_{ij}^\tau\, P_{ij}^\sigma\, P_{ij}^\tau = -1 \tag{23.7}$$

for an antisymmetric state, where P^r_{ij} is the exchange operator in orbital space, it follows that the central force may be written in the alternative form

$$(W + M P^r_{ij} - H P^\tau_{ij} + B P^\sigma_{ij}) V_0(r_{ij}) \tag{23.8}$$

where

$$\begin{aligned} W &= a_0 - a_\sigma - a_\tau + a_{\sigma\tau}, \\ M &= -4a_{\sigma\tau}, \\ H &= -2a_\tau + 2a_{\sigma\tau}, \\ B &= 2a_\sigma - 2a_{\sigma\tau}, \end{aligned}$$

and

$$\begin{aligned} a_0 &= W - \tfrac{1}{4}M - \tfrac{1}{2}H + \tfrac{1}{2}B, \\ a_\sigma &= \tfrac{1}{2}B - \tfrac{1}{4}M, \\ a_\tau &= -\tfrac{1}{2}H - \tfrac{1}{4}M, \\ a_{\sigma\tau} &= -\tfrac{1}{4}M. \end{aligned}$$

For the central force, T, S and L are good quantum numbers, and if we set $H = B = 0$, then $[\lambda]$, introduced in Sect. 18 is also good. For the non-central forces, neither S, L, nor $[\lambda]$ is good. The form (23.2) is a two-body spin-orbit force, sometimes called the vector force, which can couple a state $\psi\ (SLJM)$ to another $\psi(S'L'JM)$ so long as $S' = S \pm 1$, S and $L' = L \pm 1$, L. Such a force produces the familiar single-body spin-orbit force of Eq. (20.1),

$$M = \xi \sum_i (\mathbf{s}_i \cdot \mathbf{l}_i)$$

of the JENSEN-MAYER model when one calculates[1-3] the interaction of a particle with a closed shell, although for more than one hole or one particle there will be an additional two-body contribution from the particles outside closed shells. Experience shows however that the effect of the two-body spin orbit force (23.2) may be quite well represented by the form (20.1). The form (23.4), variously called the axial-dipole or tensor force gives no interaction with closed shells but can couple states with S or L values differing by not more than two units. Few calculations[4,5] have been made with this type of force but there is evidence that it has a larger effect in second order then in first and that in second order it produces doublet splittings similar to the first order effect of a spin-orbit force (20.1) (see Sect. 31).

24. Fractional parentage coefficients (c.f.p.) and the construction of the energy matrix. Having defined our wave-functions for the many particle system and our interaction, we must now evaluate the matrix elements of that interaction in those states. In the early days of atomic spectra, the corresponding problem would have been solved using the diagonal sum method of SLATER, described by CONDON and SHORTLEY[6]. This has been superceded by the method of fractional parentage originated by GOUDSMIT and BACHER[7] and developed by RACAH[8].

---

[1] J. HUGHES and K. T. LE COUTEUR: Proc. Phys. Soc. Lond. A **63**, 1219 (1956).
[2] C. H. BLANCHARD and R. AVERY: Phys. Rev. **81**, 35 (1951).
[3] J. P. ELLIOTT and A. M. LANE: Phys. Rev. **96**, 1660 (1954).
[4] A. M. FEINGOLD: Phys. Rev. **107**, 258 (1956).
[5] J. P. ELLIOTT: Proc. Roy. Soc. Lond., Ser. A **218**, 345 (1953).
[6] E. V. CONDON and G. H. SHORTLEY: Theory of Atomic Spectra. Cambridge: Cambridge University Press 1935.
[7] S. GOUDSMIT and R. F. BACHER: Phys. Rev. **46**, 948 (1934).
[8] G. RACAH: Phys. Rev. **63**, 367 (1943).

Let ψ and $\overline{\psi}$ be antisymmetric states of the configurations $(l)^k$ and $(l)^{k-1}$ respectively, then, although it is not in general possible to write

$$\psi = \{\overline{\psi}, \varphi_k\} \psi$$

in the notation of (19.3), where φ_k is the wave-function for the k-th particle, it must be possible to write

$$\psi = \sum_{\overline{\psi}} (\psi\{|\overline{\psi}) \{\overline{\psi}, \varphi_k\} \psi \tag{24.1}$$

where the summation extends over all states $\overline{\psi}$ of the complete set for the configuration $(l)^{k-1}$. The $(\psi\{|\overline{\psi})$ are numerical coefficients, called the fractional parentage coefficients (c.f.p.) and clearly they are orthogonal in the sense that

$$\sum_{\overline{\psi}} (\psi\{|\overline{\psi}) (\psi'\{|\overline{\psi}) = \delta(\psi, \psi'). \tag{24.2}$$

The c.f.p. describe how the state ψ is built up from its possible parents $\overline{\psi}$ and can be looked upon as defining the state. Their calculation therefore goes hand in hand with their classification and we shall describe in Appendices 5 and 6 some of the methods used. Tables of c.f.p. have been prepared for $l=1$[1], for $l=2$, $k\leq 4$[2], for $l=3$ (atomic)[3] and for jj coupling[4]. For the moment let us suppose that they have been determined and let us see what use they are. Consider the matrix element

$$(\psi|\mathscr{V}|\psi')$$

where $\mathscr{V} = \sum_{i<i} v_{ij}$ is a central force and ψ is an antisymmetric state of the configuration l^k. Then, from the antisymmetry, we have

$$(\psi|\mathscr{V}|\psi') = \frac{k(k-1)}{2} (\psi|v_{12}|\psi')$$

and, using the c.f.p. expansion (24.1)

$$(\psi|\mathscr{V}|\psi') = \frac{k(k-1)}{2} \sum_{\overline{\psi}, \overline{\psi}'} (\psi\{|\overline{\psi}) (\psi'\{|\overline{\psi}') (\overline{\psi}|v_{12}|\overline{\psi}').$$

Then, from the antisymmetry of the $\overline{\psi}$, we have finally

$$(\psi|\mathscr{V}|\psi') = \frac{k}{k-2} \sum_{\overline{\psi}, \overline{\psi}'} (\psi\{|\overline{\psi}) (\psi'\{|\overline{\psi}') \quad (\overline{\psi}|\mathscr{V}|\overline{\psi}'), \tag{24.3}$$

expressing the k-particle energy matrix element in terms of that for $(k-1)$ particles. Since the two-particle matrix elements can be evaluated in a straightforward manner, this provides the basis for a chain calculation.

Such a chain calculation can be very laborious however and can be avoided by defining a new set of c.f.p. Just as (24.1) expressed the k-particle wave function ψ as a linear combination of products of states $\overline{\psi}$ of $(k-1)$ particles and φ of one particle, so it is possible to express ψ as

$$\psi = \sum_{\tilde{\psi}, \vartheta} (\psi\{|\tilde{\psi}\vartheta) \quad \{\tilde{\psi}, \vartheta_{k-1,k}\} \psi \tag{24.4}$$

where the summation runs over all states $\tilde{\psi}$ of the complete set for $(l)^{k-2}$ with all states ϑ of the last two particles $(k-1)$ and k. This new set of c.f.p. will be

---

[1] H. A. Jahn and H. van Wieringen: Proc. Roy. Soc. Lond., Ser. A **209**, 502 (1951).

[2] H. A. Jahn: Proc. Roy. Soc. Lond., Ser. A **205**, 192 (1951).

[3] G. Racah: Phys. Rev. **76**, 1352 (1949).

[4] A. R. Edmonds and B. H. Flowers: Proc. Roy. Soc. Lond., Ser. A **214**, 515 (1952).

referred to as the $(k|k-2, 2)$ set in contrast with the original set $(k|k-1, 1)$. The two sets may be related by the equation

$$(\psi\{|\tilde{\psi}\vartheta) = \sum_{\overline{\psi}} \{(2\overline{\psi}+1)(2\vartheta+1)\}^{\frac{1}{2}} W(\tilde{\psi} l \psi l, \overline{\psi}\vartheta)(\psi\{|\overline{\psi})(\overline{\psi}\{|\tilde{\psi}) \tag{24.5}$$

using the rearrangement of vector coupling described in Appendix 3. In (24.5) we have used an abbreviated notation in which the RACAH function containing wave-function symbols ψ represents a product of three RACAH functions containing the quantum numbers of those states in the T, S and L spaces respectively. The same abbreviation is used in the square root factor, and the symbol l implies values $\frac{1}{2}$, $\frac{1}{2}$, l in these spaces. As will be described in Appendix 6, it simplifies calculations to factorise the c.f.p. into charge-spin and orbital factors and it must be remembered that (24.5) is only true of the total c.f.p., more complicated expressions holding for the separate factors[1]. These $(k|k-2, 2)$ c.f.p. are very useful in evaluating the matrix elements of a two body operator as we shall see below in the case of the central force considered earlier in this section.

As before, we have

$$(\psi|\mathscr{V}|\psi') = \frac{k(k-1)}{2}(\psi|v_{k-1,k}|\psi')$$

and now, expanding with $(k|k-2, 2)$ c.f.p.,

$$(\psi|\mathscr{V}|\psi') = \frac{k(k-1)}{2}\sum_{\tilde{\psi},\vartheta,\vartheta'}(\psi\{|\tilde{\psi}\vartheta)(\psi'\{|\tilde{\psi}\vartheta')(\vartheta|v_{k-1,k}|\vartheta') \tag{24.6}$$

This Eq. (24.6) expresses $(\psi|\mathscr{V}|\psi')$ directly in terms of the two-body energy matrix elements and hence a chain calculation is avoided. In (24.6) the $\tilde{\psi}$ in the expansions of ψ and ψ' must be the same for all non-zero contributions since the operator $v_{k-1,k}$ does not involve the co-ordinates of particles 1 to $k-2$ from which the $\tilde{\psi}$ are constructed.

In this section we have only spoken about central forces, but the calculation of the matrix elements of non-central operators goes through in a similar way using the c.f.p., with the complication that one has additional factors involving RACAH functions. We describe this in detail in Appendix 7, and also give formulae for the interactions with closed shells (Appendix 9) and for the various basic two-body matrix elements (Appendix 8). The matrix elements of other operators pertaining to the various static and dynamic properties of the levels are discussed in Appendix 11.

Finally we note that there is no difficulty in writing down the kinetic energy part of the total energy matrix. The kinetic energy for a configuration is simply obtained as a sum of the single particle kinetic energies, with some correction for the centre-of-mass effect[2] if necessary. For the oscillator well, for example, the kinetic energy $\left\langle -\frac{\hbar^2}{2M}\Delta^2\right\rangle_{nl}$ for a given orbit nl is simply equal to one half of the total energy of the orbit, i.e. in the notation of Sect. 2, $\frac{1}{2}(\Lambda+\frac{3}{2})\hbar\omega$. The centre-of-mass effect is taken into account for the lowest configurations by subtracting $\frac{3}{4}\hbar\omega$ from the total kinetic energy.

25. Calculation of the spectra and other experimental quantities. The last section has described in general terms how the energy matrix H for any chosen interaction may be set up in the complete set of states of a given configuration.

---

[1] ELLIOTT, HOPE and JAHN: Phil. Trans. Roy. Soc. Lond., Ser. A **246**, 912 (1953).
[2] A. H. BETHE and R. F. BACHER: Rev. Mod. Phys. **8**, 82 (1936).

The energy levels and their corresponding wave functions are then simply obtained by solving the matrix equation

$$H\Psi_\nu = E_\nu \Psi_\nu \tag{25.1}$$

for the eigenvalues E_ν and eigenfunctions Ψ_ν. The solutions E_ν and Ψ_ν describe the spectrum with its wave functions. Eq. (25.1) is very familiar in mathematical physics and when the complete set includes more then 3 states one uses one of a number of numerical methods for its solution. Because J and T are good quantum numbers, we can of course consider different J and T values as separate problems of smaller size so far as (25.1) is concerned, and this may be extended to any other quantum numbers which are known to be good for the particular interaction being used. If a certain quantum number is known to be "nearly good", i.e. there is only a small coupling between states having different values for it, then (25.1) may be solved using a perturbation technique but one must be careful that the coupling terms are small enough compared with the difference between the corresponding diagonal elements, or else the wave function may be badly approximated.

From the solution of (25.1) we shall have wave functions

$$\Psi_\nu = \sum_\mu a_{\nu\mu} \psi_\mu$$

where the ψ_μ are the wave functions in the classification scheme used for setting up the energy matrix. To evaluate the expectation value of an operator Q, such as a multiple moment, transition probability or reduced width, we first set up the matrix $Q_{\mu'\mu}$ of that operator in the complete set of states ψ_μ using the c.f.p. and tensor operator methods described in Appendices 4 and 7. The required value is then

$$(\Psi_{\nu'} | Q | \Psi_\nu) = \sum_{\mu'\mu} a_{\nu'\mu'} a_{\nu\mu} Q_{\mu'\mu}$$

or in matrix notation, calling the set of coefficients $a_{\nu\mu}$ the row vector $\boldsymbol{a}_\nu$ we have

$$(\Psi_{\nu'} | Q | \Psi_\nu) = \boldsymbol{a}_{\nu'} \boldsymbol{Q} \tilde{\boldsymbol{a}}_\nu \tag{25.2}$$

where $\boldsymbol{Q}$ is the matrix $Q_{\mu'\mu}$ and $\tilde{\boldsymbol{a}}_\nu$ denotes the transpose of $\boldsymbol{a}_{\nu'}$, a column vector. The right-hand side of Eq. (25.2) is therefore the matrix product of a row vector, a matrix and a column vector illustrated below,

$$\boxed{a_{\nu 1}\; a_{\nu 2} \cdots} \quad \boxed{\begin{matrix} Q_{11} & Q_{22} \cdots \\ Q_{21} & Q_{22} \cdots \\ \vdots & \vdots \end{matrix}} \quad \boxed{\begin{matrix} a_{\nu 1} \\ a_{\nu 2} \\ \vdots \end{matrix}}\,.$$

The matrix elements $Q_{\mu'\mu}$ for the various physical operators of interest will be described in detail in Appendix 11.

VI. Application of the individual particle model to light nuclei.

The whole of the present chapter is devoted to applications of the Individual Particle Model described in Chap. V.

In its most extensive applications, this model has been used in a particular form called the "Intermediate Coupling Model". We introduce the latter model in Sect. 26, and apply it to nuclei of mass $A < 50$ in Sects. 27 and 28. Finally, in Sect. 29, we drop the restriction to Intermediate Coupling and mention calculations of a more general kind.

Almost all of the experimental data that is relevant for the present chapter is to be found in the excellent compilations of F. AJZENBERG and T. LAURITSEN[1] (for $A < 20$) and of P. M. ENDT and J. C. KLUYVER[2] (for $20 < A < 50$).

26. Specialisation to intermediate coupling. Any application of the individual particle model in the general form discussed in the last chapter demands in the first place a knowledge of the two-body interaction. Here is the first stumbling block because not only is the form of the interaction between free nucleons unknown, but even if it were known, it would still be questionable whether the effective interaction between nucleons in a nucleus is the same as that between free nucleons. In the very light nuclei, $A \leq 6$, attempts[3] have been made to use the information about the interaction deduced from data on the two- and three-body systems to calculate the binding energies of He^4 and Li^6. For such applications the restricted variational treatment described in Chap. V is inadequate. To obtain the correct binding energies one must have rather accurate wave functions and this surely implies taking account of mixing from many excited configurations.

Alternatively, one can approach the application of the model from an empirical viewpoint, the philosophy being to use only the lowest configuration of the oscillator wave functions and, within that limitation, to try to find an effective interaction which reproduces the properties of the light nuclei. There are, after all, many nuclei in the region $4 \leq A \leq 40$ each of which has many measured properties which can be calculated from the model. In this search for an interaction, one obviously starts from the most simple kinds; the surprising thing is that, with a very simple form, a mass of agreement is found with experiment.

To give any binding for a closed shell, the interaction must contain a central force. To produce the observed doublet splitting for one particle outside a closed shell, as in He^5, O^{17} there must be a spin-orbit (vector) force in the interaction. (Within the lowest configuration the tensor force cannot produce this splitting.) Calculations in which the interaction contains these two necessary ingredients are called "Intermediate Coupling". This name is rather vague but has arisen because a pure central force produces L-S coupling while a pure spin-orbit force produces j-j coupling, so that a mixed interaction may be said to produce an "intermediate" coupling. It should be pointed out however that there is no rule for constructing an "intermediate coupled" state other than by the full diagonalisation of the energy matrix described in Chap. V.

In most intermediate coupling calculations the single-body spin-orbit force $\xi \sum_i (\boldsymbol{s}_i \cdot \boldsymbol{l}_i)$ is used, although this term on its own cannot arise from a nucleon-nucleon interaction. Nevertheless, such a term can arise from the two-body interaction[4-7] through interaction of a particle with a closed shell. Furthermore, particular calculations have shown that in many cases the matrix elements of the two-body vector interaction in an unfilled shell are very similar to those of

[1] F. AJZENBERG and T. LAURITSEN: Rev. Mod. Phys. **27**, 77 (1955).

[2] P. M. ENDT and J. C. KLUYVER: Rev. Mod. Phys. **26**, 95 (1954).

[3] D. R. INGLIS: Phys. Rev, **51**, 531 (1937). — M. MORITA and T. TAMURA: Progr. Theor. Phys. **12**, 653 (1954). — J. IRVING and D. S. SCHONLAND: Phys. Rev. **97**, 446 (1955).

[4] J. HUGHES and K. J. LE COUTEUR: Proc. Phys. Soc., Lond. A **63**, 1219 (1950).

[5] C. H. BLANCHARD and R. AVERY: Phys. Rev. **81**, 35 (1951).

[6] J. P. ELLIOTT and A. M. LANE: Phys. Rev. **96**, 1660 (1954).

[7] R. J. BLIN STOYLE: Phil. Mag. **46**, 973 (1955).

the above one-body force, and give essentially the same result. It should be mentioned that calculations with the two-body spin-orbit force are much more complicated than for the one-body force and that the radial integrals involved depend sensitively on the radial dependence of the force and on the wave functions which are not too well known.

Although the quadrupole moment of the deuteron demands the existence of a tensor force in the nucleon-nucleon interaction there is no evidence that such a force plays any important part in the first order calculations[1] mentioned above. In Sect. 29 we shall describe an instance, the β-decay at C^{14}, where the tensor force must be included. However, the tensor force necessary for explaining this β-decay is so weak that, except for this one case which is particularly sensitive to the admixture, it will not noticeably affect the intermediate coupling results in general.

We have said that the intermediate coupling assumption includes a central force in the interaction, but as we saw in Sects. 11 and 21, this involves, besides the strength and the radial shape, the ratios of the exchange parameters W, M, H and B. What assumptions must be made about these? Experience has shown that the radial shape is not important in determining the properties of low lying levels, although the strength and range are related. For example, one can obtain essentially similar results with the YUKAWA or GAUSS radial shape, and for the YUKAWA shape a change in the range a may be approximately compensated by a change in the strength V_c by keeping $V_c a^2$ constant. These remarks refer only to central forces. For non-central forces one finds a much greater sensitivity to the radial shape. So far as the exchange character is concerned, we can say that for some nuclei the levels are to a very large degree insensitive to the exchange character, whereas for others there is only one relation between the exchange parameters which must be satisfied. This explains the fact that a number of different interactions have been used to explain the same data in some cases. From the saturation requirements, together with the triplet-singlet strength ratio deduced from the deuteron, ROSENFELD[2] determined the values[3]

$$W = -0.13, \qquad M = +0.93, \qquad H = -0.26, \qquad B = +0.46$$

$$(\text{i.e. } a_0 = a_\sigma = 0, \qquad a_\tau = -0.10, \qquad a_{\sigma\tau} = -0.23).$$

These values, or very similar values, have been extensively used in intermediate coupling calculations, although this does not mean to say that no other values would give equally good agreement. The essential characteristics of the ROSENFELD interaction in these calculations are that it has (a) a strong attractive MAJORANA (space exchange) component; (b) the value 0.6 for the singlet to triplet strength ratio $(W+M-H-B)/(W+M+H+B)$. The former is essential in giving for example, the large (15 MeV) energy difference between the lowest $T=0$ and $T=1$ levels of C^{12}. The latter is essential in giving the correct $T=0$ to $T=1$ difference in such nuclei as Li^6, F^{18}. The SERBER force ($W=M=+0.5$, $H=B=0$), which is sometimes used, has the advantage of simplicity, but does not satisfy this latter requirement.

27. Application of intermediate coupling to states of normal parity. *α) Nuclei with* $4<A\leq 16$). The first attempt to explain the spectra of light nuclei with

[1] The question whether a tensor force in higher order perturbation calculations gives rise to the doublet splitting here ascribed to a spin-orbit force in first order is mentioned in Sects. 29, 31 and 36.

[2] L. ROSENFELD: Nuclear Forces, p. 233. Amsterdam: North Holland Publishing Co. 1948.

[3] Normalised to $W+M+H+B=1$, with an attractive (negative) strength constant.

the intermediate coupling model was made by INGLIS[1]. For simplicity he chose the $1p$ shell in which case the radial wave functions and radial dependence of the interaction enter the problem only through two integrals[2] L and K. The exchange dependence assumed for the central force was a simplified version of the ROSENFELD mixture with $W=H=0$ and $M=0.8$, $B=0.2$. The problem then involves just three parameters, L, K, and ξ, the spin-orbit force strength[3]. Treating K as the unit of energy, the pattern of levels depends on the two ratios L/K and $\zeta=\xi/K$. The first ratio depends simply on the radial integrals and varies from the value 3 for zero-range forces to ∞ for very long range forces, but for reasonable values of the range and nuclear size parameter it has a value ~ 6. Since further the spectra are not very sensitive to this ratio, INGLIS assumed $L/K=6$ in all his calculations. The other parameter ζ, called the intermediate coupling parameter, measures the ratio of strengths of the two kinds of force, spin-orbit and central. INGLIS plots the spectra as functions of ζ and looks for the value which gives best fit with the experimental spectrum.

INGLIS made an exact calculation for the intermediate coupling region only for two particles or two holes, $A=6$ or 14. In the more complicated cases he merely calculated the spectra at the two extremes, L-S and j-j, together with the asymptotes to the curves for each level at those extremes and then drew smooth curves joining the asymptotes. In this way he obtained a rough indication of the levels to be expected in a given energy region and in some cases even the level order.

In spite of this approximation the results for all the $1p$ shell nuclei showed quite a good correlation with the experimental spectra, enough in fact to justify a more precise and exact calculation.

This has been done recently by KURATH[4], using a high speed electronic computer. He used the same exchange character of the central force as INGLIS but took $L/K=6.8$, a slight change from INGLIS. He also considered the variation of his results with changes in this ratio, confirming their insensitivity.

In Figs. 14 to 22, we show KURATH's results. We shall not discuss each nucleus in detail, but only mention the most interesting features, in particular the points of apparent disagreement, since these are much fewer than the points of agreement.

Masses 6 and 14. Fig. 14a shows the calculated spectra for $A=6$ and 14. These can be shown on a single diagram since, in the calculations, one can change from two particles to two holes simply by a change in sign of the spin-orbit matrix elements. For compactness the energy scale on the left has been shifted by plotting $(E+\frac{1}{2}\xi)/K$ instead of E/K. For $A=14$ one finds reasonable agreement for $\zeta\sim 5$ whereas for $A=6$ one needs $\zeta\sim 1.5$. This difference suggests an increase in ζ from beginning to end of the $1p$ shell, and this is confirmed by the other nuclei. We shall see that this increase is expected if the spin-orbit term arises from a neutral two-body interaction[5]. The comparisons with experiment, using the best values of ζ and K are given in Figs. 14b, c. In comparing theory with experiment we must remember that only the normal parity levels are being calculated and also that levels from higher configurations will begin to appear above about 6 MeV excitation.

[1] D. R. INGLIS: Rev. Mod. Phys. **25**, 390 (1953).

[2] In the notation of Appendix 8, $L=F^0+\frac{4}{25}F^2$, $K=\frac{3}{25}F^2$ where the SLATER integrals F^0, F^2 contain the negative strength constant.

[3] INGLIS uses the symbol a for ξ.

[4] D. KURATH: Phys. Rev. **101**, 216 (1956).

[5] J. P. ELLIOTT and A. M. LANE: Phys. Rev. **96**, 1660 (1954).

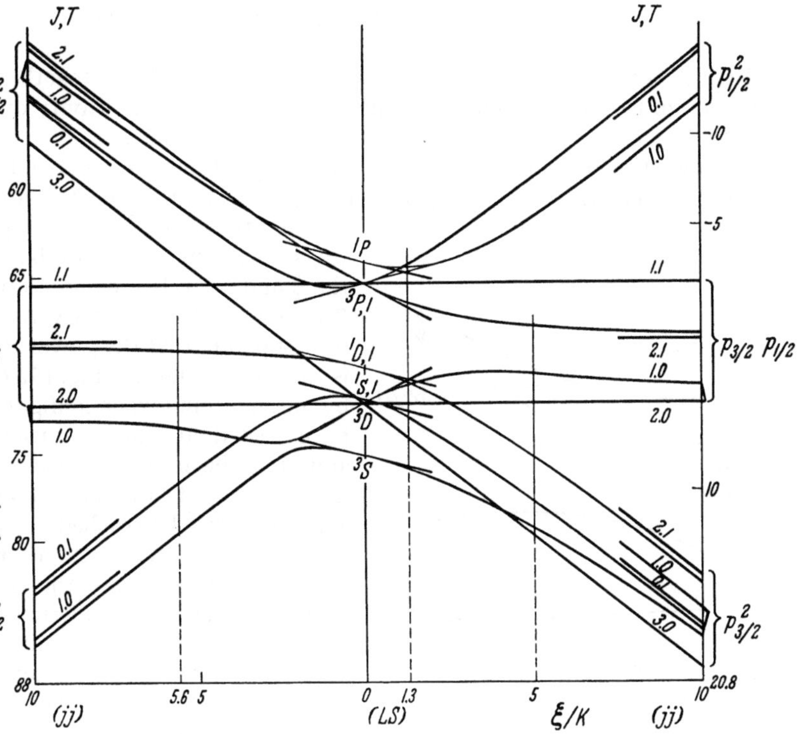

a

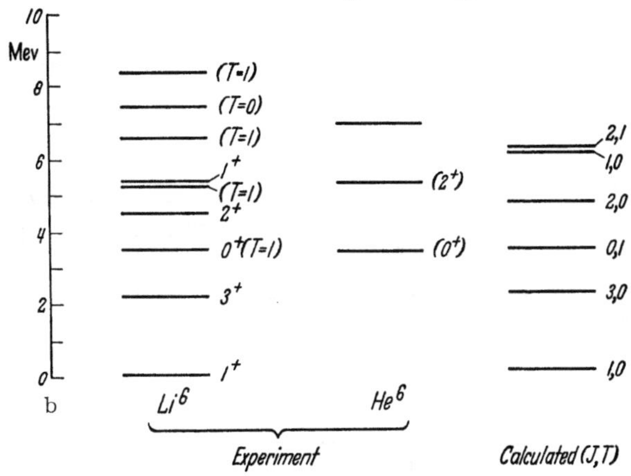

b

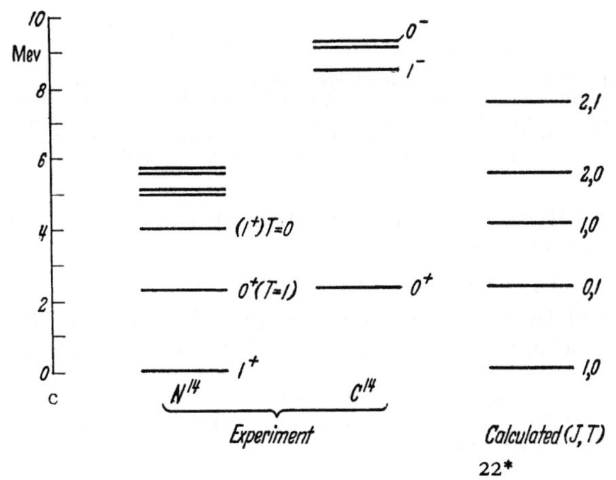

c

Fig. 14a–c. (a) Calculated Spectra for $A=6$ and 14 plotted against the Intermediate Coupling Parameter ξ/K with the choice $L/K=6.0$. The spectrum of $(1p)^2$ for $A=6$ is on the right and the spectrum of $(1p)^{10}$ for $A=14$ is on the left. [From D. R. Inglis: Phys. Rev. **87**, 915 (1952).] (b) Comparison of the Experimental Spectrum of $A=6$ with Theory taking $\xi/K=1.3$ and $K=-1.23$ MeV. (c) Comparison of the Experimental Spectrum of $A=14$ with Theory taking $\xi/K=5.0$, $K=-0.8$ MeV.

Mass 7. In Fig. 15a is shown Kurath's result for $A=7$ together with his illustration of the effect of changing the ratio L/K. The insensitivity of the lower levels to this change results from their proximity to the L-S extreme, where, for this lowest supermultiplet [3], the relative positions are independent of the integral L. Comparison with experiment is given in Fig. 15b. Of interest here is the ratio of the $\frac{1}{2}-\frac{3}{2}$ and $\frac{5}{2}-\frac{7}{2}$ splittings. Agreementc annot be reached if the $\frac{5}{2}$ and $\frac{7}{2}$ levels are those observed at 7.46 and 4.61 MeV respectively. However, if we suppose[1] that the $\frac{5}{2}$ level at 7.46 MeV is the second $\frac{5}{2}$ level in Fig. 15a depres sed by mixing from higher configurations (to be expected at this excitation energy) and that the lowest $\frac{5}{2}$ level is that believed to exist around 5.4 MeV, then agreement can be reached.

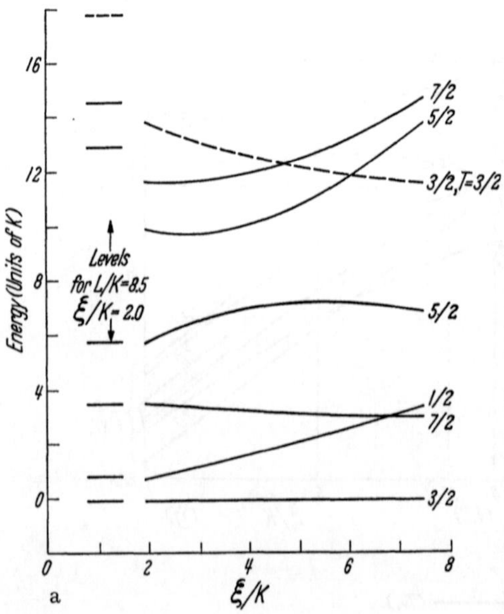

Mass 13. Fig. 16 shows Kurath's result for $A=13$ but the scarcity of levels and lack of experimental data makes a comparison unfruitful. There are only two experimental level data to use, the position of the $\frac{3}{2}$ level at 3.68 MeV and the position of the $T=\frac{3}{2}$ level from the binding energy of B^{13} with Coulomb correction. As is seen in Table 21, the values of K and ζ deduced are consistent with those found in neighbouring nuclei. We can however predict that a low $\frac{5}{2}$ level which has not yet been observed should lie around 5 MeV.

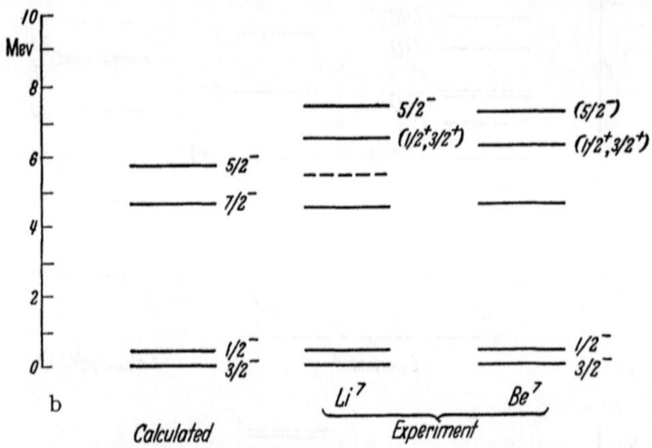

Fig. 15a and b. (a) Calculated Spectrum for $A=7$ plotted against the Intermediate Coupling Parameter ξ/K with the choice $L/K=6.8$. On the left is the spectrum for $\xi/K=2.0$, $L/K=8.5$. [From D. Kurath: Phys. Rev. **101**, 216 (1956).] (b) Comparison of the Experimental Spectrum of $A=7$ with Theory taking $\xi/K=1.0$, $K=-1.2$ MeV.

Mass 8. In Be^8 (Figs. 17a, b) the results show that the 4.2 and 5.4 MeV levels[2], if they exist, cannot be explained by the model. The 7.55 MeV level[2] is high enough to belong to an excited configuration and a spin 0^+ would be expected for the lowest state of such a configuration.

Mass 9. Of Be^9 (Figs. 18a, b) little can be said because of the lack of experimental data.

Mass 10. The nuclei with $A=10$ (Figs. 19a—c) at the middle of the shell and therefore the most complex to calculate, give, perhaps surprisingly, very good agreement with the well studied experimental spectrum. In the observed

[1] J. Soper: To be published.

[2] None at those levels are now believed to exist. See e.g. Russell, Phillips and Reich: Phys. Rev. **104**, 135 (1956).

spectrum, the first level which is not accounted for and therefore presumably contains an appreciable amount of doubly excited configurations appears to be the 1^+ at 4.7 MeV. The slow variation of the spectrum with L/K is shown in Fig. 19b.

Mass 11. Lack of experimental data prevents a detailed comparison for $A=11$ (Fig. 20a—c), but the theory predicts the 2.14 MeV level to be $^1/_2{}^-$ and the 4.46 and 5.03 MeV levels to be $^5/_2{}^-$ and $^7/_2{}^-$, although the order of these two is not reliable from the theory. The variation of the spectrum with L/K is shown in Fig. 20b.

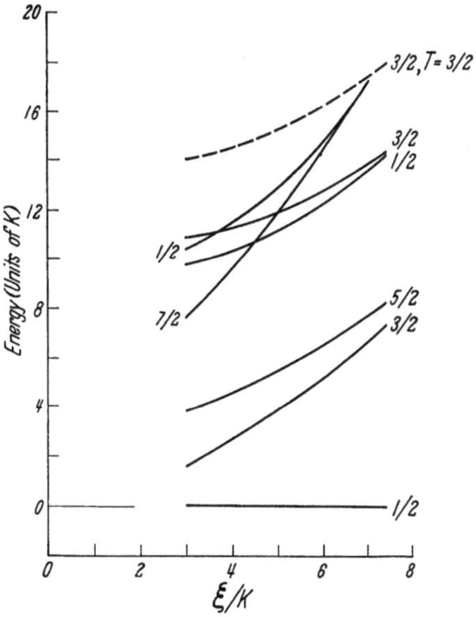

Fig. 16. Calculated Spectrum for $A = 13$ plotted against the Intermediate Coupling Parameter ξ/K with the choice $L/K = 6.8$. [From D. KURATH: Phys. Rev. 101, 216 (1956).]

Mass 12. The scarcity of low levels of C^{12} (Figs. 21a—c) prevents much comparison. Considerable interest centres on the observed (0^+) level at 7.7 MeV, because the lowest excited 0^+ state that is predicted occurs at least 3 MeV higher. However, the low position of this observed level is just what one would expect from strong interaction with the doubly excited configurations. (In this connection, we note that the 0^+ level at 6.06 MeV in O^{16} must come from such excited configurations. Again, the variation with L/K (Fig. 21b) is small, expecially for the low levels.

In these brief comments on KURATH'S work we have not discussed the values of ζ and K which he chooses for each nucleus separately to fit the data. These values are collected together in Table 21, together with the spin-orbit strength $\xi=\zeta K$. It is at once striking that K is very nearly constant as A varies, implying that the same central interaction and wave functions give consistent results throughout the shell. It would have been most unsatisfactory if a very different central interaction has to be assumed for each nucleus. There is however a variation in ζ, which from the constancy of K, is essentially a variation in ξ.

Table 21. *Intermediate coupling parameters K, ζ and ξ for the (1p) shell.*

A	5	6*	7*	8*	9*	10	11	12	13	14*	15
K (MeV)	—	−1.2		−1.2	−1.2	−0.9	−0.9	−0.9	−0.9	−0.8	—
ζ	—	1.3		2.0	1.5	4.7	6.0	5.0	5.3	5.0	—
ξ (MeV)	∼−2	−1.6		−2.4	−1.8	−4.2	−5.4	−4.5	−4.8	−4.0	−4.2

* Calculated for $L/K = 5.8$, those unstarred for $L/K = 6.8$.

The only significant feature of this variation in ξ seems to be the preference for larger values at the end of the shell. There appears to be a sharp change from $A=9$ to $A=10$ but, as KURATH points out, the ξ value for Be^9 is far from certain because of the lack of experimental data. A value of 3 for ζ which is not unreasonable from the present state of the observed spectrum would largely remove the apparent discontinuity. Nevertheless we must not expect too much regularity in ξ because the one-body spin-orbit force can only be a caricature of the true

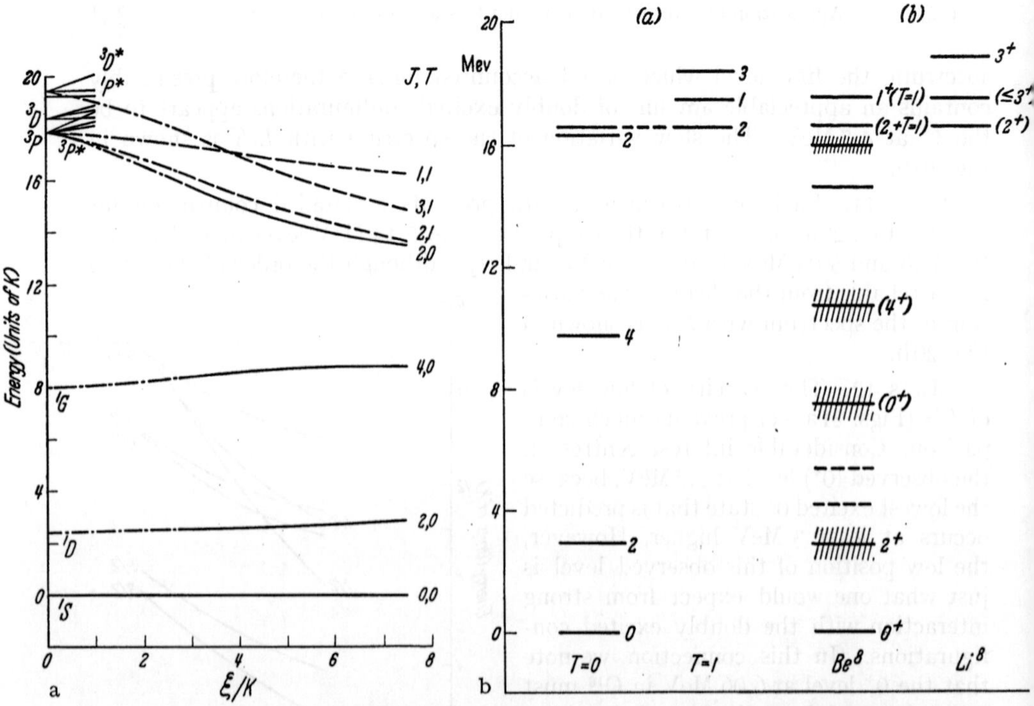

Fig. 17a and b. (a) Calculated Spectrum for $A = 8$ plotted against the Intermediate Coupling Parameter ξ/K with the choice $L/K = 6.8$. [From D. KURATH: Phys. Rev. **101**, 216 (1956).] (b) Comparison of the Experimental Spectrum (b) of $A = 8$ with Theory (a) taking $\xi/K = 2.0$, $L/K = 5.8$, $K = -1.18$ MeV. [From D. KURATH: Phys. Rev. **101**, 216 (1956).]

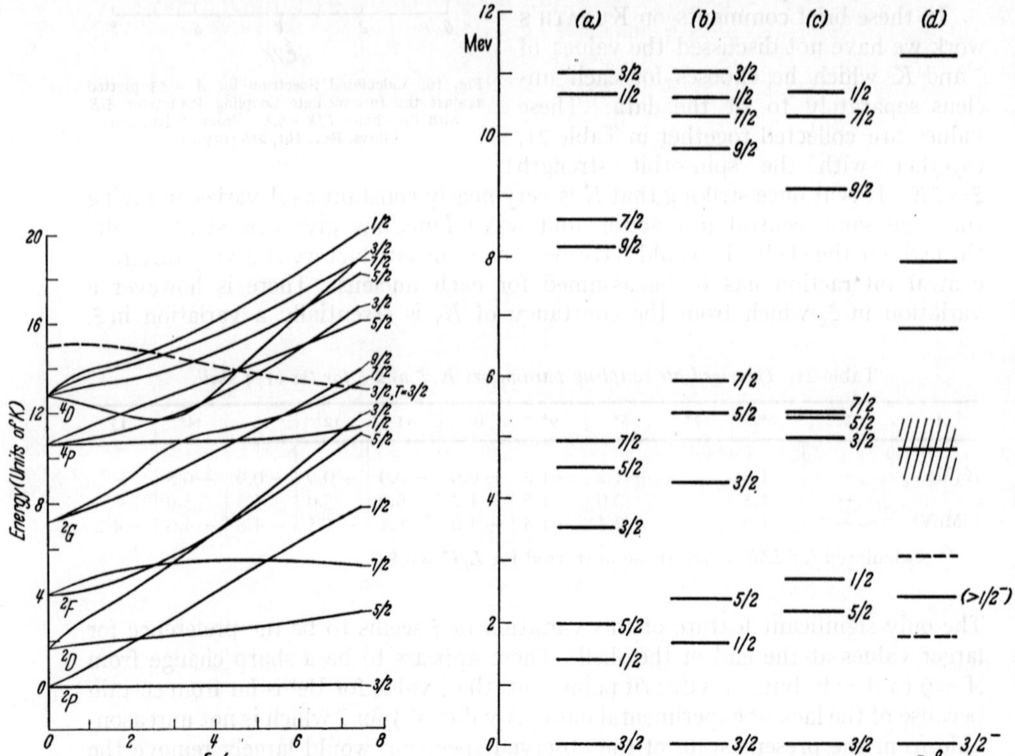

Fig. 18a and b. (a) Calculated Spectrum for $A = 9$ plotted against the Intermediate Coupling Parameter ξ/K with the choice $L/K = 6.8$. [From D. KURATH: Phys. Rev. **101**, 216 (1956).] (b) Comparison of the Experimental Spectrum (d) of $A = 9$ with Theory taking (a) $\xi/K = 1.5$, $L/K = 6.8$, $K = -1.00$ MeV; (b) $\xi/K = 1.5$, $L/K = 5.8$, $K = -1.20$ MeV; (c) $\xi/K = 2.75$, $L/K = 6.8$, $K = -1.00$ MeV. [From D. KURATH: Phys. Rev. **101**, 216 (1956).]

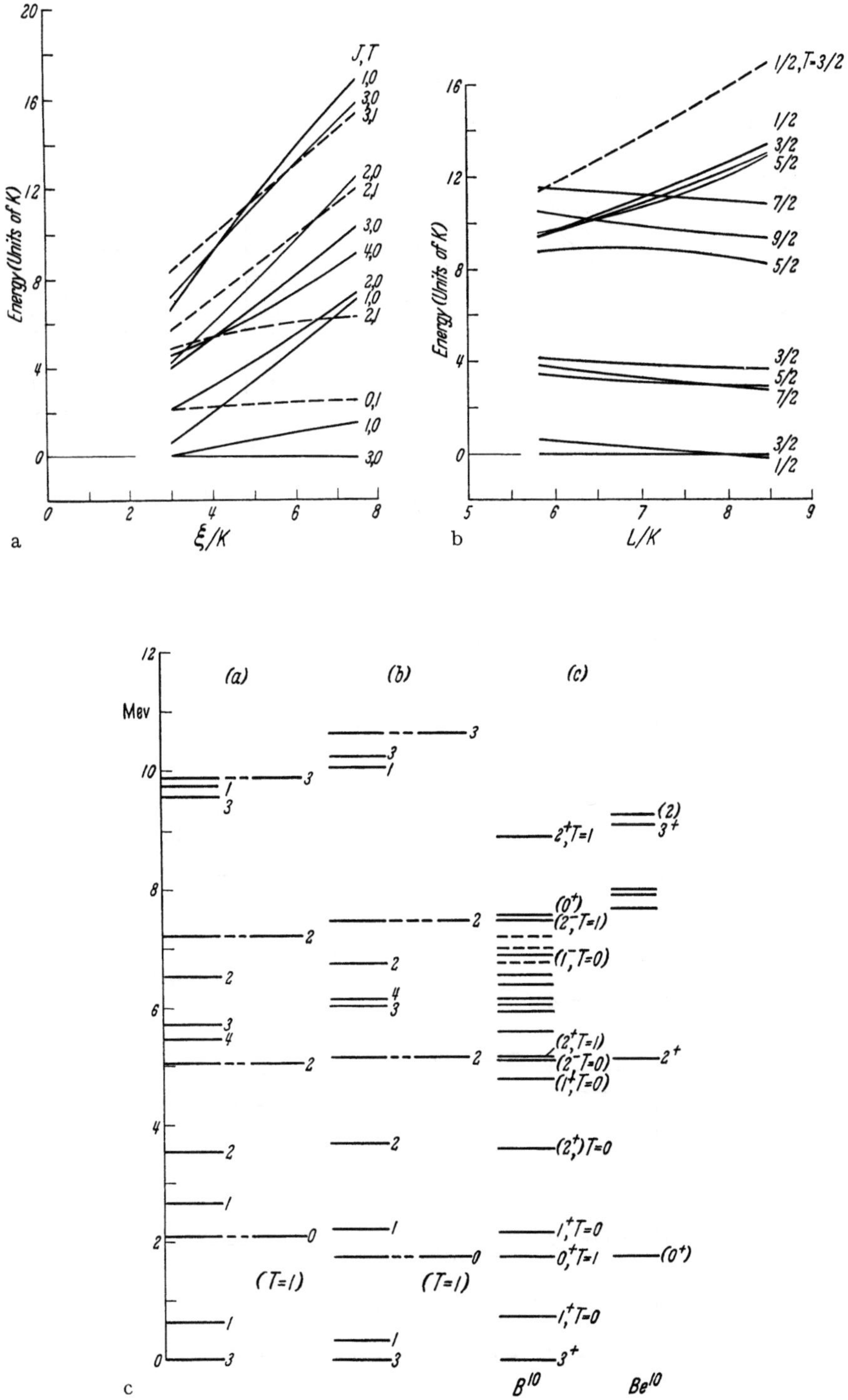

Fig. 19a–c. (a) Calculated Spectrum for $A = 10$ plotted against the Intermediate Coupling Parameter ξ/K with the choice $L/K = 6.8$. [From D. KURATH: Phys. Rev. **101**, 216 (1956).] (b) Calculated Spectrum for $A = 10$ plotted agains L/K with the choice $\xi/K = 4.0$. [From D. KURATH: Phys. Rev. **101**, 216 (1956).] (c) Comparison of the Experimental Spectrum (c) with Theory taking (a) $\xi/K = 4.75$, $L/K = 6.8$, $K = -0.90$ MeV; (b) $\xi/K = 4.0$, $L/K = 5.8$, $K = -1.13$ MeV. [From D. KURATH: Phys. Rev. **101**, 216 (1956).]

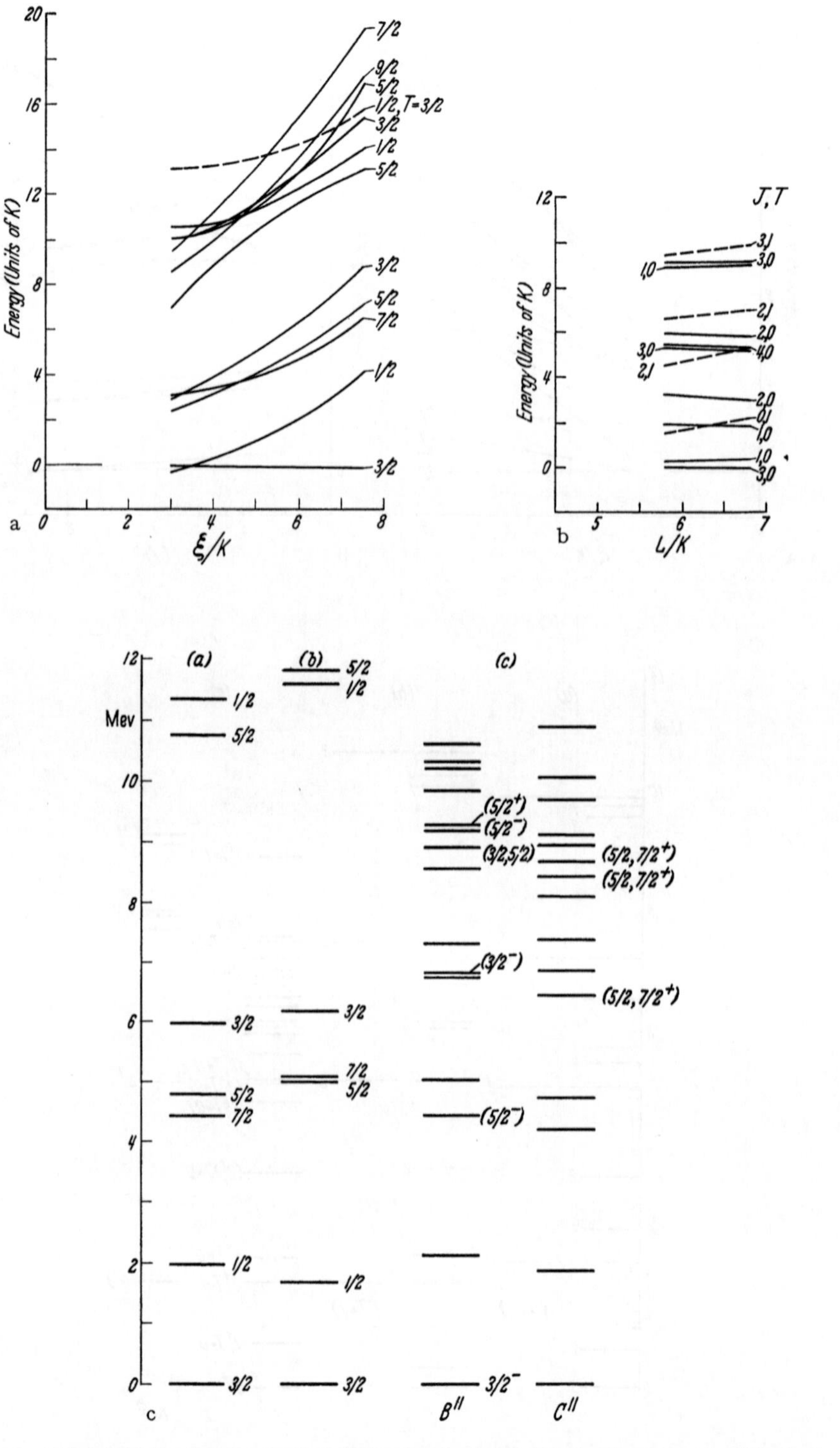

Fig. 20a–c. (a) Calculated Spectrum for $A=11$ plotted against the Intermediate Coupling Parameter ξ/K with the choice $L/K=6.8$. [From D. KURATH: Phys. Rev. **101**, 216 (1956).] (b) Calculated Spectrum for $A=11$ plotted against L/K with the choice $\xi/K=4.0$. [From D. KURATH: Phys. Rev. **101**, 216 (1956).] (c) Comparison of the Experimental Spectrum (c) with Theory taking (a) $\xi/K=6.0$, $L/K=6.8$, $K=-0.92$ MeV; (b) $\xi/K=5.0$, $L/K=5.8$, $K=-1.15$ MeV. [From D. KURATH: Phys. Rev. **101**, 216 (1956).]

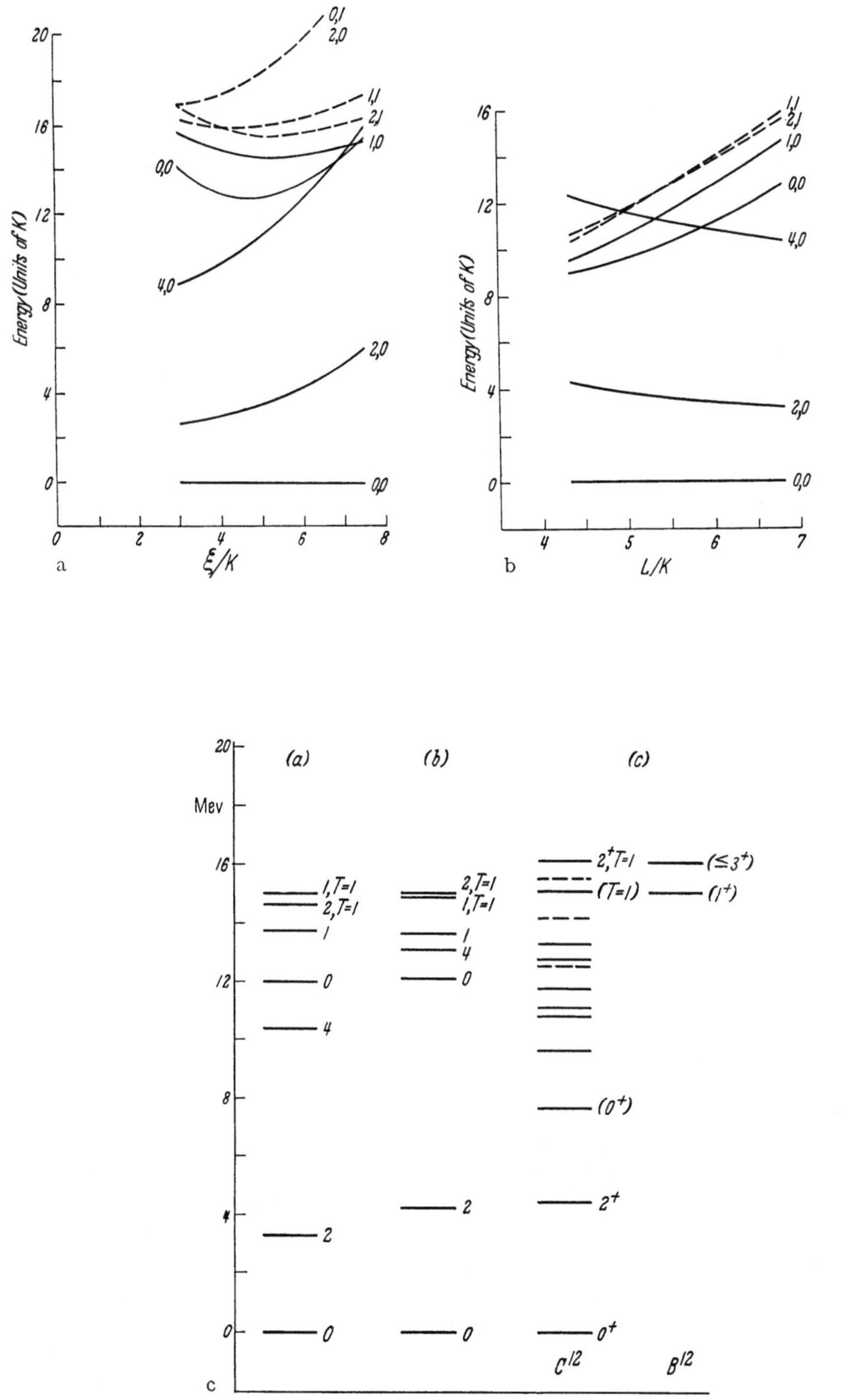

Fig. 21a—c. (a) Calculated Spectrum for $A=12$ plotted against the Intermediate Coupling Parameter ξ/K with the choice $L/K=6.8$. [From D. KURATH: Phys. Rev. **101**, 216 (1956).] (b) Calculated Spectrum for $A=12$ plotted against L/K with the choice $\xi/K=4.5$. [From D. KURATH: Phys. Rev. **101**, 216 (1956).] (c) Comparison of the Experimental Spectrum (c) of $A=12$ with Theory taking (a) $\xi/K=5.0$, $L/K=6.8$, $K=-0{,}94$ MeV; (b) $\xi/K=4.5$, $L/K=5.5$, $K=-1.17$ MeV. [From D. KURATH: Phys. Rev. **101**, 216 (1956).]

anteraction. If it results from a two-body vector interaction, then the interaction irom an unfilled shell is not simply equivalent to the one-body force, although fpproximately so, and will vary from nucleus to nucleus. In the one-hole nucleus $(1s)^4\,(1p)^{11}$ for example a neutral vector interaction among the p nucleons produces a splitting approximately equal to, and in addition to, the splitting derived from interaction with the closed $(1s)$ shell[1]. This implies a doubling in the effective ξ from $A=5$ to $A=15$ as is needed in Table 21 [2]. To explain the apparent discontinuity in ξ at $A=10$, KURATH considered the variation in nuclear radii deduced from the COULOMB energy differences. He found here also a discontinuity at $A=10$ and, tempted to relate the two phenomena, deduced that the

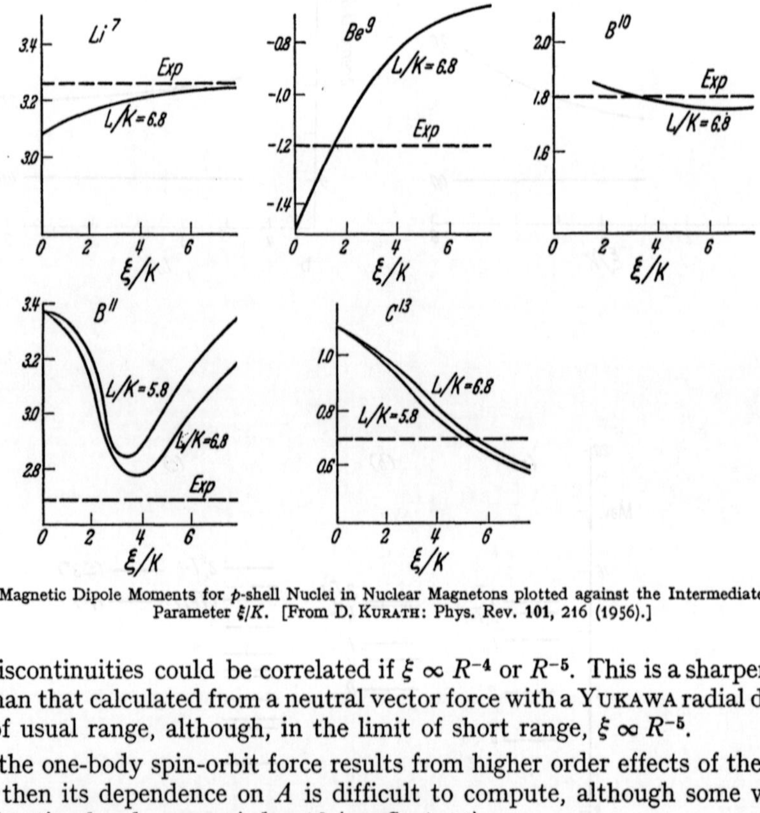

Fig. 22. Magnetic Dipole Moments for p-shell Nuclei in Nuclear Magnetons plotted against the Intermediate Coupling Parameter ξ/K. [From D. KURATH: Phys. Rev. 101, 216 (1956).]

two discontinuities could be correlated if $\xi \propto R^{-4}$ or R^{-5}. This is a sharper variation than that calculated from a neutral vector force with a YUKAWA radial dependence of usual range, although, in the limit of short range, $\xi \propto R^{-5}$.

If the one-body spin-orbit force results from higher order effects of the tensor force, then its dependence on A is difficult to compute, although some work in this direction has been carried out[3] (see Sect. 31).

KURATH also calculated the static moments of the ground states, again finding agreement for substantially the same value of the intermediate coupling parameter as was needed for the spectrum. His results are given in Fig. 22.

Using the same model, one can look for further support by calculating the "dynamic" properties at levels. This has been done[4] for the 3-particle and 3-hole problems $A=7$ and 13, which are manageable without an electronic computer. Comparison with experimental values for the nucleon reduced widths, γ-transition strengths, and β-decay ft-values gave good general agreement. In

[1] This is no longer true if the vector force is taken to be symmetric rather than neutral.
[2] J. P. ELLIOTT and A. M. LANE: Phys. Rev. 96, 1660 (1954).
[3] A. M. FEINGOLD: Phys. Rev. 101, 258 (1956).
[4] A. M. LANE: Proc. Phys. Soc. Lond. A 66, 977 (1953); A 68, 189, 197 (1955). — A. M. LANE and L. A. RADICATI: Proc. Phys. Soc. Lond. A 67, 167 (1954).

the cases studied, essentially the same results were obtained with the SERBER force ($W = M = 0.5$) as with the ROSENFELD force.

The anomalously long lifetime of C^{14} in its β-decay to N^{14} cannot be understood on the pure intermediate coupling model but the introduction of a tensor force which has negligible effect on other data brings agreement. We will discuss this in Sect. 29.

β) Nuclei with $16 < A \leq 40$. We can reasonably assume that the ground state of O^{16} can be described by the closed shell $(1s)^4 (1p)^{12}$. Beyond O^{16}, we expect, from the oscillator potential, to find the $(1d)$ and $(2s)$ shells filling together. This is confirmed on inspecting the O^{17} spectrum (Fig. 23) which has lowest levels $\frac{5}{2}$ and $\frac{1}{2}$ close together. Interpreting these as single particle $1d_{\frac{5}{2}}$, and $2s_{\frac{1}{2}}$ levels, one looks for the $1d_{\frac{3}{2}}$ level and finds that the next even parity level, at 5.08 MeV is in fact $^3/_2{}^+$ and has a large nucleon width suggestive of a single particle state. It will no doubt have some contamination from excited configurations but probably little error is caused by assuming it to be the pure single particle $1d_{\frac{3}{2}}$ level. The odd parity levels, appearing at only 3 MeV excitation, must come from excited configurations formed either by core excitation or by exciting the last particle to the $(2p, 1f)$ shell. It is disturbing to have these levels so low since although, being of opposite parity from the lowest levels, they cannot mix with them, their existence implies that states of double excitation, which *can* mix with the lowest levels, are not very high. Nevertheless, the philosophy of the intermediate coupling approach is to ignore excited configurations and try to fit the data with only the lowest major configuration. We have just seen that for nuclei with $4 < A < 16$ this approach meets with considerable success.

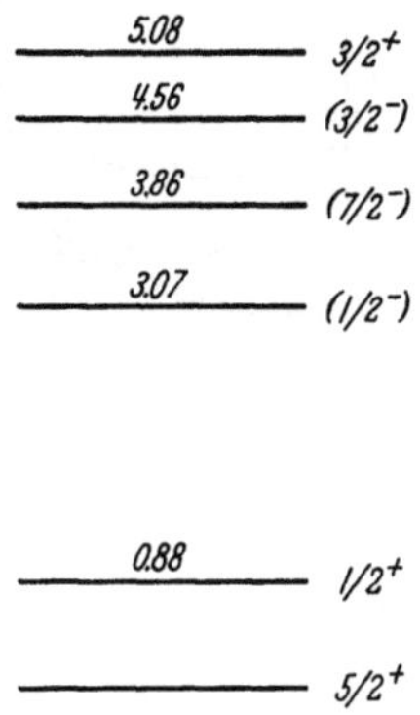

Fig. 23. The Experimental Spectrum of O^{17}.

The difficulties in extending the methods of the $1p$ shell calculations to the $(1d, 2s)$ shell[1] are several. In the first place they become more complicated because we have two different orbits, the l-value of one of them is greater than for the p-shell, and the shell is bigger. However, these difficulties are merely technical! A more important difficulty is that there are more parameters in the problem, which have either to be deduced empirically or calculated from some fundamental assumptions. In the $(1p)$ shell there were only two parameters ξ/K and L/K, apart from the scale parameter K. In the $(1d, 2s)$ shell there are eight radial integrals in place of L and K and, in addition to the spin orbit parameter ξ for the d-particles, the single particle energy difference has to be taken into account. With the interpretation of the O^{17} spectrum given in the last paragraph, one can deduce these last two parameters. For the radial integrals one has to assume a definite radial dependence of the interaction, but again the results seem to be largely insensitive to this dependence.

So far, calculations have only been made[2,3] for the nuclei of mass 18 and 19. Allowing for the mixing of the $1d$ and $2s$ particles implies taking into account for a nucleus of mass $16 + n$, all configurations $(1d)^{n-x} (2s)^x$ for $0 \leq x \leq \min(x, 4)$. From the O^{17} spectrum, one easily deduces that $\xi = -2$ MeV and that the mean $(1d)$ position (without spin-orbit splitting) is at 1.125 MeV above the $2s$ position. One set

[1] We shall see later that the $1d - 2s$ mixing is so great that they must be treated as a single shell.

[2] J. P. ELLIOTT and B. H. FLOWERS: Proc. Roy. Soc. Lond., Ser. A **229**, 536 (1955).

[3] M. G. REDLICH: Phys. Rev. **99**, 1427 (1955).

of calculations used a Rosenfeld force with Yukawa shape $V_0(r_{ij}) = V_c \frac{e^{-\left(\frac{r_{ij}}{a}\right)}}{\left(\frac{r_{ij}}{a}\right)}$ having a range $a = 1.37 \times 10^{-13}$ cm and oscillator wave functions with size parameter, (see Sect. 2) $b = 1.64 \times 10^{-13}$ cm. In Figs. 24a, b and 25a, b we show the resulting spectra for masses 18 and 19 respectively, presented as a function of the central force strength V_c which is effectively the intermediate coupling parameter since ξ is fixed.

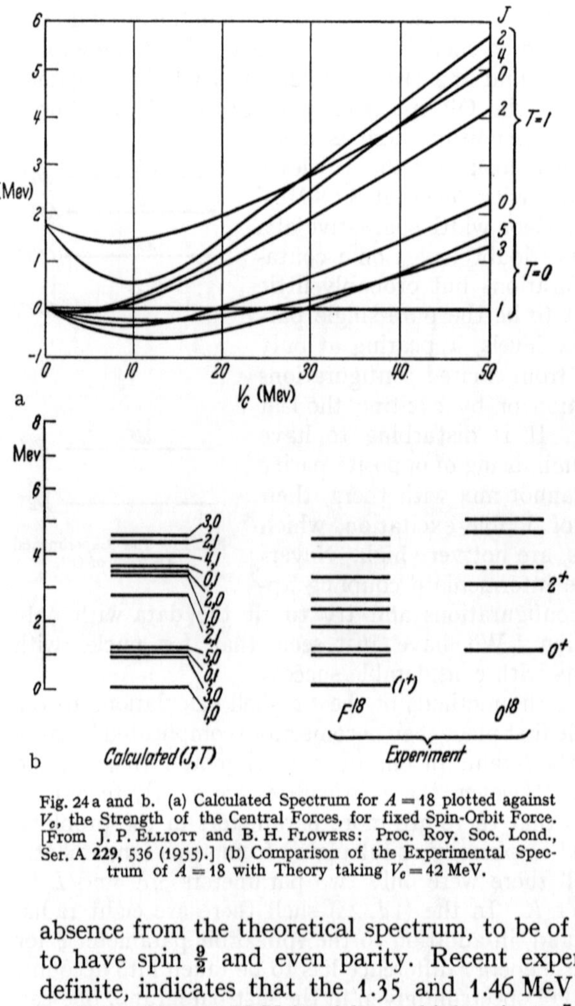

Fig. 24a and b. (a) Calculated Spectrum for $A = 18$ plotted against V_c, the Strength of the Central Forces, for fixed Spin-Orbit Force. [From J. P. Elliott and B. H. Flowers: Proc. Roy. Soc. Lond., Ser. A **229**, 536 (1955).] (b) Comparison of the Experimental Spectrum of $A = 18$ with Theory taking $V_c = 42$ MeV.

In Fig. 24a the position of the $T = 1$ group of levels relative to those with $T = 0$ is very sensitive to the exchange character, although relative positions within these groups are insensitive. (In fact, the Serber force would give[1] a $T = 1$ level as ground state, which is clearly unacceptable.) Fig. 24b shows the comparison with experiment for $A = 18$, although the experimental data is still meagre. For the comparison we have chosen $V_c = 42$ MeV making a very small change in the exchange character to bring the lowest $T = 1$ state down to 1 MeV above the F^{18} ground state. Fig. 25b shows comparison of theory and experiment for the spectrum of F^{19} taken at $V_c = 40$ MeV. This nucleus has been well studied experimentally and the agreement is very good. The theory interprets the 1.35 and 1.46 MeV levels, by their absence from the theoretical spectrum, to be of odd parity and the 2.8 MeV level to have spin $\frac{9}{2}$ and even parity. Recent experimental work although not yet definite, indicates that the 1.35 and 1.46 MeV levels are in fact of odd partity and that the 2.8 MeV level has a high spin $\geq \frac{7}{2}$.

The occurrence of the odd-parity level at so low an energy as 112 keV, with its implication of the importance of core excitation, makes the success of the calculation a little surprising. However, theory and experiment have been compared for the magnetic moment of the ground and ${}^5/_2{}^+$ level, for the γ-decay of the ${}^3/_2{}^+$ level and for the β-decay from O^{19} to F^{19} with the same success (Table 24). The calculated lifetime of the ${}^5/_2{}^+ \rightarrow {}^1/_2{}^+$ $E2$ transition is however about five times too long. This result is not so surprising if one remembers that the ${}^1/_2{}^+ \rightarrow {}^5/_2{}^+$ $E2$ transition in O^{17} has a strength approximately half of the single proton value, although the shell model describes this nucleus as a single neutron outside a closed shell and consequently forbids this transition. There is therefore clear

[1] See footnote 3, p. 347.

indication of some collective effect involving the O^{16} core. If we describe this by the weak coupling of our external particles to some surface oscillations of the core in the manner of the Unified Model (see Sect. 32) we find that in fact the $E2$ transition in O^{17} corresponds to a very weak coupling strength and that, introducing this same strength of coupling into F^{19}, the $E2$ transition for that

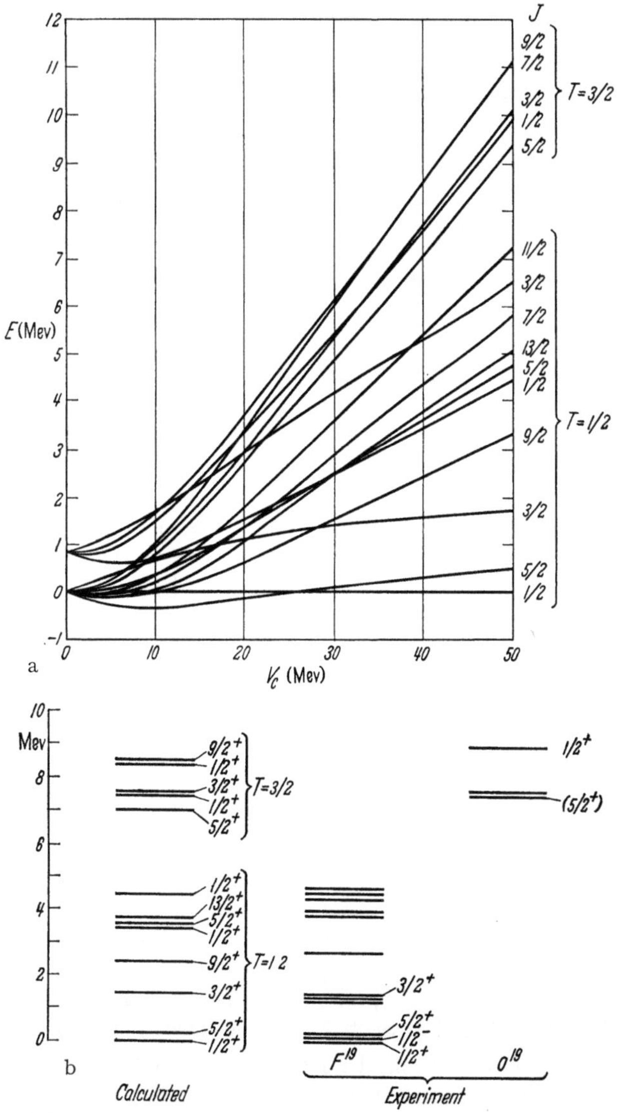

Fig. 25a and b. (a) Calculated Spectrum for $A = 19$ plotted against V_c, the Strength of the Central Force, for fixed Spin-Orbit Force. [From J. P. Elliott and B. H. Flowers: Proc. Roy. Soc. Lond., Ser. A **229**, 536 (1955).] (b) Comparison of the Experimental Spectrum of $A = 19$ with Theory taking $V_c = 40$ MeV.

nucleus also agrees. It can be shown that the effect of the weak coupling on $E2$ transitions may be represented mathematically by giving all particles an additional charge; in this instance of $\frac{1}{2}e$. The effect[1] of this strength of coupling on the level positions is very small, of the order of 100 keV, and therefore negligible in these calculations.

[1] F. C. Barker: Phil. Mag., Ser. VIII **1**, 329 (1956).

The composition of the low levels for $A=19$ is shown in Tables 22 and 23 to illustrate the mode of coupling. Table 22 shows that the symmetry of the orbital part of the wave function is close to that of the L-S limit and also illustrates the degree of configurational mixing. Table 23 shows the estimated j-j coupling structure of the wave functions indicating that although for the $T=\frac{1}{2}$ states one is far from j-j coupling, the $T=\frac{3}{2}$ states are quite well described by that extreme. We believe this to be a general result, that j-j coupling is a better approximation for higher isotopic spin. In the L-S extreme, for high orbital symmetry, i.e. low isotopic spin, there are generally a few widely spaced levels

Table 22. *Percentage Analysis of* F^{19} *and* O^{19} *Wave Functions into Orbital Symmetry and Configuration.*

T	J	Orbital symmetry			Configuration			
		[3]	[21]	[111]	d^3	$d^2 s$	$d s^2$	s^3
$\frac{1}{2}$	$\frac{1}{2}$	92	8	0	12	59	0	29
$\frac{1}{2}$	$\frac{3}{2}$	84	16	0	63	12	25	0
$\frac{1}{2}$	$\frac{5}{2}$	86	14	0	65	13	22	0
$\frac{3}{2}$	$\frac{1}{2}$	0	90	10	1	99	0	0
$\frac{3}{2}$	$\frac{3}{2}$	0	90	10	65	35	0	0
$\frac{3}{2}$	$\frac{5}{2}$	0	90	10	89	1	10	0

Table 23. *Percentage Analysis of the* F^{19} *and* O^{19} *Wave Functions into jj-coupled Configuration.*

T	J	d^3		$d^2 s$		$d s^2$		s^3
		$(\frac{5}{2})^3$	$(\frac{5}{2})^2(\frac{3}{2})$	$(\frac{5}{2})^2(\frac{1}{2})$	$(\frac{5}{2})(\frac{3}{2})(\frac{1}{2})$	$(\frac{5}{2})(\frac{1}{2})^2$	$(\frac{3}{2})(\frac{1}{2})^2$	$(\frac{1}{2})^3$
$\frac{1}{2}$	$\frac{1}{2}$	7	5	38	21	—	—	29
$\frac{1}{2}$	$\frac{3}{2}$	16	47	7	5	19	6	—
$\frac{1}{2}$	$\frac{5}{2}$	39	26	10	3	20	2	—
$\frac{3}{2}$	$\frac{1}{2}$	—	—	92	8	—	—	—
$\frac{3}{2}$	$\frac{3}{2}$	62	3	33	2	—	—	—
$\frac{3}{2}$	$\frac{5}{2}$	82	7	—	—	11	0	—

so that one needs a large spin-orbit force to mix them, while for lower orbital symmetry, i.e. higher isotipic spin, there are more levels closer together so that the spin orbit force can more easily couple them together into the j-j mode. The value $V_c \sim 40$ MeV chosen for best agreement here corresponds to the values $K \sim -1$ MeV and $\zeta \sim 2$ for the $1p$ shell parameters which are very similar to the values (Table 21) deduced in Sect. 27 for the lower part of the $1p$ shell.

The nucleus F^{19} has also been studied[1] from the viewpoint of the collective model using the spectra calculated by NILSSON[2] for the motion of a particle in a non-spherical well. It is well known[3] that the nucleus Mg^{24} has levels with $J=0, 2$ and 4 which have energy differences closely proportional to the predicted $[J(J+1)-J_0(J_0+1)]$ of the rotational model. Furthermore, the Al^{25} spectrum may be fitted[3] very well by a system of rotational bands corresponding to the states of a single particle in a distorted well. The application of the model to F^{19} meets with only very limited success so long as one keeps to the pure rotational

[1] E. B. PAUL: Phil. Mag. Ser VIII. **2**, 311 (1957).
[2] S. G. NILSSON: Dan. Mat. fys. Medd. **29**, No. 16 (1955).
[3] LITHERLAND, BARTHOLOMEW, PAUL and GOVE: Phys. Rev. **102**, 208 (1956).

bands. However, allowing a mixing[1] of the lowest $K=\frac{1}{2}$ band with the nearby $K=\frac{3}{2}$ band produces results in good agreement with experiment and very close to those obtained from the shell model. A comparison of these results is presented in Table 24. The agreement is reached for a value of the deformation parameter $\beta=0.3$ which is consistent with the values deduced from the energies of the first excited 2^+ states in the neighbouring nuclei O^{18} and Ne^{20}, assuming that they are the first excited states of a $K=0$ rotational band.

Table 24. *Comparison of the Experimental Magnetic Moment and Decay Data for the $A=19$ Nuclei with the Calculations of the Shell Model and the Rotational Model.*

Property	Rotational model	Shell model	Experimental
γ-ray branching $\frac{\Gamma(\frac{3}{2}\to\frac{1}{2})}{\Gamma(\frac{3}{2}\to\frac{5}{2})}$	0.8%	0.6%	<4%
Magnetic Moments (*n. m.*)			
$^1/_2{}^+$ state	2.75	2.80	2.64
$^5/_2{}^+$ state	3.70	3.30	3.51
Lifetime of $^5/_2{}^+$ state	0.9×10^{-7} sec	5×10^{-7} sec*	1.0×10^{-7} sec
Log ft for Ne^{19} decay to $^1/_2{}^+$ state	3.52	3.2	3.3

* Excluding the weak surface coupling contribution which improves this figure to 0.9×10^{-7} sec.

This situation, in which the two models produce very similar results, demands an explanation in terms of some fundamental relation between the models. Although one may say that the shell model approach, in allowing configurational mixing of the $2s$ and $1d$ shells, has taken a step toward the collective description and that the collective model, in allowing band-mixing, has taken a step toward the particle approach, it is unsatisfactory that the two models appear to meet so quickly unless one can find some concrete relation between the two apparently quite different models.

Beyond $A=19$ no full calculation has been made, but some tentative calculations by one of us (J.P.E.) have indicated the following generalities. Both the L-S and j-j extremes for the pure d^4 configuration fail to give the first excited state of Ne^{20} as an isolated 2^+ level, both giving a pair of levels. However, when one includes all configurations d^4, d^3s, d^2s^2, ds^3, s^4 one finds that the ground and second excited states come predominantly from configurations d^4, d^2s^2, s^4 of even numbers of d-particles while there is an isolated first excited 2^+ state which comes largely from configurations d^3s, ds^3 of an odd number of d-particles. This indicates again the importance of including all configurations.

Toward the end of the $(1d, 2s)$ shell, the one-hole nuclei, $A=39$, are not yet well understood experimentally so that similar calculations toward the close of the shell have not been carried out. However, a very pretty piece of work has been done relating[2] the levels of the nuclei K^{40} and Cl^{38}. In the j-j coupling

[1] A. K. Kerman: Dan. Mat. fys. Medd. **30**, No. 15 (1956). — E. B. Paul: To be published.
[2] S. Goldstein and I. Talmi: Phys. Rev. **102**, 589 (1956). — S. P. Pandya: Phys. Rev. **103**, 956 (1956).

limit the low levels of K^{40} are described by the configuration $(d_{\frac{3}{2}})^{-1}(f_{\frac{7}{2}})$ while for Cl^{38} we have the configuration $(d_{\frac{3}{2}})^{-3}(f_{\frac{7}{2}})$ which is equivalent to $(d_{\frac{3}{2}})(f_{\frac{7}{2}})$. The authors[1] have derived analytical relations between the levels of these two configurations which differ only in that the $d_{\frac{3}{2}}$ hole is replaced by a particle. Then, knowing the experimental level scheme for K^{40} they can predict the scheme for Cl^{38}. The agreement is remarkable and is reproduced in Table 25. This striking result

Table 25. *Energy Levels of* Cl^{38} *in* Mev *above the Ground State.*
At present, only the spins of the lowest two states have been established experimentally.

Calculated from K^{40} . . .	0 (2^-)	0.70 (5^-)	0.75 (3^-)	1.32 (4^-)
Experimental	0 (2^-)	0.67 (5^-)	0.76	1.31

would appear to imply that the *j-j* extreme is a very good approximation in this case and would also seem to indicate that three-body forces may be neglected since the formulae used were derived from the assumption of two-body forces. The success of the extreme *j-j* model in this case is not altogether surprising since the central interaction acts between inequivalent particles, i.e. different orbits, and is consequently somewhat weaker in its effects. Furthermore, the nuclei K^{40} and Cl^{38} have isotopic spin $T=1$ and $T=2$ respectively, and we have already seen that *j-j* coupling appears to be better for states of these higher isotopic spin values.

γ) Nuclei with $A>40$. The calculations of subsections α) and β) were based on the assumptions that the ground states of He^4 and O^{16} repsectively could be described as closed shells. These assumptions were supported by the single particle and single hole nature of the neighbouring nuclei. Since the neighbouring nuclei $A=39$ and 41 of Ca^{40} have not yet been established to be well described by a single hole and single particle, the assumption that the ground state of Ca^{40} may be described by a closed $(1s)^4(1p)^{12}(1d, 2s)^{24}$ shell is on less sure footing. In any calculation beyond Ca^{40}, along the lines of the calculations for nuclei beyond O^{16}, there is also the disadvantage of having many more SLATER integrals and other parameters because the appropriate shells are those of the $(2p, 1f)$ orbits.

LEVINSON and FORD[2] have approached this problem, making it manageable by a number of assumptions, and they find remarkably good agreement. They consider only the calcium isotopes, so that they are only concerned with states of maximum isotopic spin. Further more, they assume that there is no interaction in the $S=1$, $T=1$ states of a pair of particles, which implies the relation $W-M-H+B=0$ between the coefficients of the exchange mixture[3]. Their philosophy is not to calculate the levels of Ca^{42} and Ca^{43} using the single particle levels from Ca^{41} with some interaction, but to calculate the three-body spectrum of Ca^{43} from the experimentally known spectra of Ca^{41} and Ca^{42}. To do this they have to make a number of assumptions, otherwise they would have insufficient information. Firstly, they assume that the observed levels of Ca^{41} and Ca^{42} may be interpreted in the manner given in the first two columns of Fig. 26, although we stress that these assignments are not well established experimentally. Then they assume that the overlap of the $2p$ and $1f$ wave functions is zero and that the $1f$ and $1g$ wave functions have 100% overlap. This leaves them with

[1] See footnote 2, p. 351.
[2] C. LEVINSON and K. W. FORD: Phys. Rev. **100**, 13 (1955).
[3] C. LEVINSON and K. W. FORD: Phys. Rev. **99**, 792 (1955).

only one radial wave function for which the radial integrals F^k have to be determined. Since there are only four level positions in Ca^{42} to use as input data and there are still nine integrals F^k, the authors assume F^k to be a smooth function of k and by interpolation and extrapolation express the F^1, F^3, F^5, F^7 and F^8 in terms of the F^0, F^2, F^4, F^6. These four parameters may now be decuced from the four known levels in Ca^{42} and then used to calculate the Ca^{43} levels. Levinson and Ford find agreement with the four lowest level positions in the third column of Fig. 26 within 1%, and with the magnetic moment, also within 1%. In the last column of Fig. 26 is shown the Ca^{43} spectrum derived from the pure $(\frac{7}{2})^3$ configuration. The authors admit that the accuracy of the agreement is probably fortuitous but claim that the various approximations made introduce only a small error. Nevertheless the agreement to 0.01 n.m. for the magnetic moment is meaningless since exchange moments are expected to contribute corrections of the order of 0.2 n.m. The agreement to 0.01 MeV in the level positions for Ca^{43} is similarly suspect since the experimental data on which the comparison was based is in some doubt. Recently the third excited state[1] of Ca^{43} has been reported at 0.99 MeV rather than the 0.81 MeV assumed by Levinson and Ford, a change of 0.18 MeV. Certainly, the approach here which supposes that the major part of the configuration mixing arises from the $1g$ particles rather than the $2p$ particles is very different from that used in the $(1d, 2s)$ shell. One would furthermore expect configurations of the type $(1d, 2s)^{-1}(1f, 2p)^2$ to be as important as the $1g$ configuration in Ca^{41}, since they both involve a single excitation. The configurational composition of the four Ca^{42} levels deduced in the above empirical way is shown in Table 26 where the mixing from the $1g$ shell is seen to be considerable as well as the departure from pure j-j coupling. As the experimental data for these nuclei accumulates it will be interesting to see how reliable were the level assignments made by

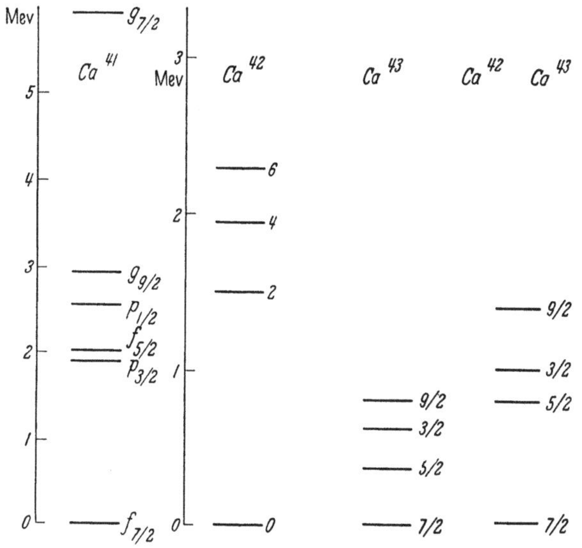

Fig. 26. The first three Spectra on the left are those of Ca^{41}, Ca^{42}, and Ca^{43} deduced from the Empirical Data. The last diagram is the spectrum of Ca^{43} predicted from the empirical levels of Ca^{42} in the absence of configuration interaction. [From C. Levinson and K. W. Ford: Phys. Rev. **100**, 13 (1955).]

Table 26. *Mixture Parameters of the States of* Ca^{42}, *based on a semi-empirical analysis of the levels of* Ca^{41} *and* Ca^{42}.

J	$(f_{\frac{7}{2}})^2$	$f_{\frac{7}{2}} f_{\frac{5}{2}}$	$(f_{\frac{5}{2}})^2$	$(g_{\frac{9}{2}})^2$	$(g_{\frac{7}{2}})^2$
0	0.87		0.34	−0.31	−0.15
2	0.95	−0.21	0.24	0	0
4	0.92	−0.37	0.17	0	0
6	0.79	−0.61		0	0

[1] C. Brahms: Thesis, Utrecht 1956 (unpublished).

Levinson and Ford and how sensitive are their results to such changes as may appear.

28. Application of intermediate coupling to states of non-normal parity. In this chapter so far, the lowest configurations obtained by putting particles into oscillator levels, have been used to describe the low lying levels of the light nuclei with some success. At quite low excitation in many of these nuclei there appear experimentally, levels with "non-normal" parity (i.e. opposite to that of the lowest configuration). In the philosophy of the individual particle model, one would expect to understand these levels in terms of mixtures of all possible configurations obtained by a single excitation in the oscillator level scheme.

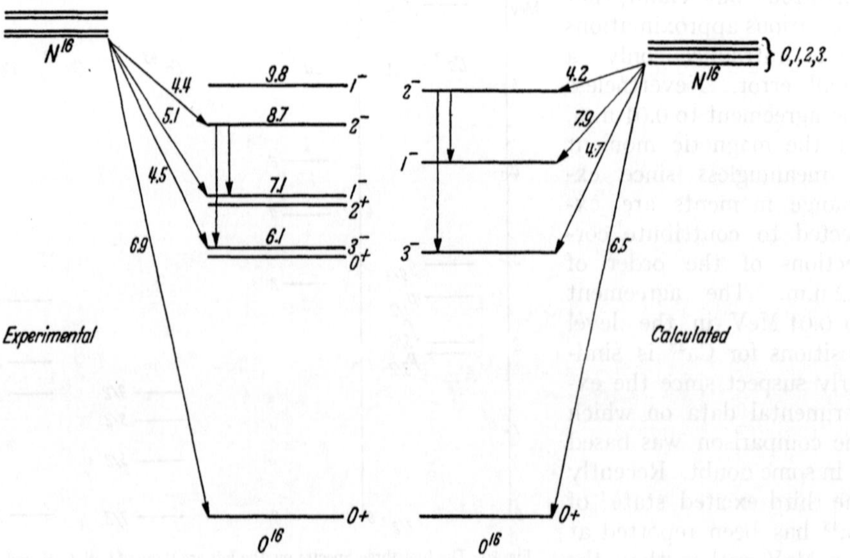

Fig. 27. Comparison of the Experimental Spectrum of Non-Normal Parity States of $A=16$ with Theory.

The simplest situation in which such levels can arise is for a closed shell, the configurations in question being those of a single hole and a single particle. Such a calculation has been carried out[1] for the O^{16} nucleus as part of a programme to understand these non-normal parity levels in that mass region. Of particular interest is the odd-parity level at 112 keV excitation in F^{19}, but to treat it rigorously on this model would involve such complicated configurations as $(1p)^{-1}\,(1d,2s)^4$. It is possible that the results for the simpler configurations may suggest some valid approximations by which such a complicated problem may be simplified.

The calculations for the levels of these non-normal parity configurations is essentially the same as for the ground configurations. The technique is a little more complicated but still within the scope of the methods described in Chap. V and the Appendix. One must remember the corrections for centre-of-mass motion discussed in Sect. 22.

The comparison of theory with experiment for O^{16} is given in Fig. 27. The even parity levels are not calculated since they must arise from doubly excited configurations and constitute a separate problem. The position of the lowest odd-parity level relative to the ground state depends on the kinetic energy difference between the excited and ground configurations, which is sensitive to the wave function parameter b and not well determined. Consequently, the

[1] J. P. Elliott: To be published.

position of the 3^- level has been fitted to the experimental value, giving the wave function parameter $b = 2 \times 10^{-13}$ cm, a reasonable value. To make these calculations one has to deduce the value of ξ_l empirically for the $1p$ and $1d$ particles from the observed N^{15} and O^{17} spectrum respectively. Then the results shown are those for the mixing of configurations $(1p)^{-1}$ $(1d, 2s)$ using the same two-body interaction as in the mass 19 calculations, in Sect. 27β. There is general agreement for the lowest three levels although the appearance of the second 1^- state at 9.8 MeV implies that it must belong to some triply excited configuration since no such level appears in the calculated spectra. The branching ratio of the 2^- level agrees closely with experiment. The theory predicts four low lying levels for N^{16} as are observed, but these are too close for the calculated level order to be reliable. The measured spins 2, 0, 3, 1 of these N^{16} levels, the lifetime of the 0^- level and the branching ratio for γ decay of the 1^- level all agree with the theory. To explain the two allowed β-decays to the 1^- and 3^- levels in O^{16} one must assume a spin 2^- for the ground state of N^{16}. With this assumption, the calculated log ft values are compared with experiment in Fig. 27 and although the decay to the 2^- and 3^- levels agree quite well, the decay to the 1^- level at 7.1 MeV is in serious disagreement. However, this decay is particularly sensitive to the ratio ξ_p/ξ_d, a small change in which reduces log ft from 1.9 to 6.1.

With this one exception, which is still being investigated, the intermediate coupling model gives a good account of the properties of the non-normal parity levels considered. A similar calculation of the levels of the mass 15 nuclei gives preliminary results which again appear to resemble the experimental levels quite closely although the experimental data are not yet very complete.

29. Departures from intermediate coupling. In the preceding sections of this chapter we have used the restricted form of the interaction referred to as Intermediate Coupling and have found that a mass of experimental data may be reproduced. This does not imply that we are using the best form of interaction, but does suggest that any more complicated interactions do not have an appreciable affect on the results described. Nevertheless one must be prepared to find particular cases in which more complicated interactions manifest themselves, and these cases will be interesting in that they should tell us something more about these interactions. There is only one instance in which it is well established that the simple intermediate coupling assumption fails. We refer to the long β-lifetime of C^{14} in its decay to N^{14}. The very long lifetime with log $ft = 9.03$ for an allowed transition can only be understood if there is a strong cancellation in the matrix element for the transition. It may be shown rigorously that, within the lowest configuration $(1s)^4$ $(1p)^{10}$ and with the intermediate coupling assumption, such a cancellation is impossible[1]. However, the introduction of a tensor force into the interaction allows this cancellation. The first reported calculation[2] along these lines used a YUKAWA radial shape for the interaction, and found that the strength of tensor force required was so large that the resulting spectrum was completely changed from the intermediate coupling result and was nothing like the observed spectrum. This was clearly an unsatisfactory state of affairs. More recent calculations set[3] out to see whether it was possible to have the best of both worlds, i.e. to introduce a tensor force sufficient to explain the β-matrix cancellation but small enough to preserve the intermediate coupling

[1] D. R. INGLIS: Rev. Mod. Phys. 25, 390 (1953).

[2] B. JANCOVICI and I. TALMI: Phys. Rev. 98, 209 (1954).

[3] J. P. ELLIOTT: Phil. Mag. Ser. VIII 1, 503 (1956). — W. M. VISSCHER and R. A FERRELL: Phys. Rev. 99, 649 (1955).

spectrum. It was found that this could be done in a very reasonable way as we describe below.

The tensor force matrix element, C, which is operative in producing the required β-decay lifetime is expressible as the difference between two radial integrals. With the Yukawa shape of the earlier work these two integrals are nearly equal, so that the difference is small and a large tensor force strength has to be assumed to make the matrix element large enough to produce the cancellation in the β-decay matrix element. This results in the other tensor force matrix element, A, being very large and thus causing a large change in the spectrum. However, the small difference in radial integrals is a property peculiar to the Yukawa shape. With a different shape, there is no need to use such a strong tensor force with the result that the spectrum remains very similar to the intermediate coupling result. Fig. 28 shows the ratio of A to C as a function of the ratio a/b for three different shapes, where a is the force range and b the well size parameter. It shows clearly that, for values of $a/b \sim 1$, the Yukawa shape gives a value different by a factor 4 from for that the Gauss or Field Theoretic shapes. These latter shapes give small values (~ 1) for A/C so that one can produce the desired β-matrix cancellation and still preserve the spectrum. This is impossible for values of $A/C \sim 4$, which result from the Yukawa shape, unless a very short range is assumed. The reader may be a little surprised that it is possible to change the log ft value from the intermediate coupling value of ~ 5 to the observed value of ~ 9 without changing the spectrum, and he may ask whether the introduction of the tensor force may not affect the decay data of other nuclei to a similar extent. The answer is that even without the tensor force the transition is quite strongly hindered. The N^{14} ground state is $\sim 90\%$ D state whereas the C^{14} ground state can only be a mixture of S and P states so that for the large part of the wave function the decay is L-forbidden; hence the unfavoured value log $ft \sim 5$ arises in intermediate coupling from the odd 10% of the wave function. To increase this log ft value to the observed 9.03 therefore calls for a change in only that small part of the wave function and this may be accomplished with a small tensor force. It is clear then that the decay in question is particularly sensitive to the tensor force and it would be interesting to find other instances of such sensitivity although they may be expected to be few. When we refer to the tensor force as being small we mean that the matrix elements appearing in this calculation are small compared with the other interaction matrix elements. The absolute strength is of the same order of magnitude as that found in fitting the low energy two-body data and it has the same sign.

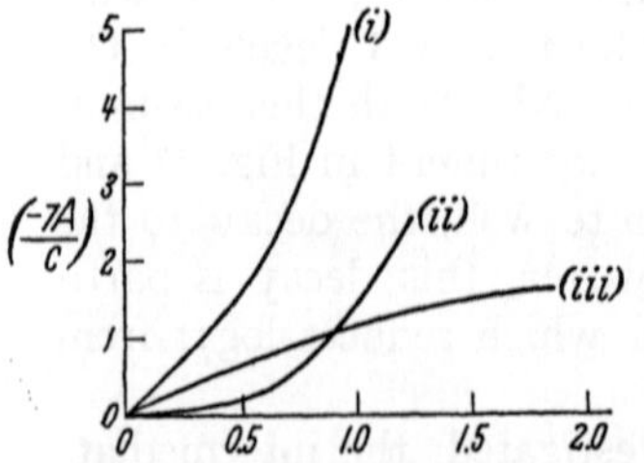

Fig. 28. The Ratio $-\frac{7A}{C}$ of Tensor Force Integrals for the Interactions:

(i) Yukawa: $\frac{e^{-r/a}}{r/a}$.

(ii) Gaussian: e^{-r^2/a^2}.

(iii) Field Theoretic:

$$\frac{e^{-r/a}}{r/a}\left[1+3\left(\frac{a}{r}\right)+3\left(\frac{a}{r}\right)^2\right].$$

Finally we mention calculations[1] in which the tensor force has been used, although the intermediate coupling model gives satisfactory agreement. These calculations were for the prediction of the spectra of Li^6 and Li^7, and they were done without assuming a vector force of any kind. All of the spin-orbit coupling in the nucleus was supposed to arise from the tensor forces, and this force was treated in second order, i.e. the effects of configuration interaction were included.

[1] A. M. Feingold: Phys. Rev. **101**, 258 (1956).

The resulting spectra were remarkably similar to those predicted by the intermediate coupling model. More than any other work, these results have suggested that the observed nuclear spin-orbit coupling may be due to second-order effects of a tensor force. Finally we note that the above remarks on C^{14} show that, even if this is true, the tensor force cannot be wholly represented as a one-body vector force.

VII. Configuration mixing and other methods for improving the individual particle model.

In spite of its striking successes, the unmodified independent particle model by no means gives a universal explanation of the properties of low lying nuclear states. This is not surprising in view of the certain fact that the model is an over-idealisation, in that it does not take account of correlations between pairs of nucleons due to their strong interaction. What is surprising at first sight is that the model can be so successful in fitting spectra and yet fail so badly to fit total nuclear binding energies. This latter failure of the model has been known for a long time[1]. In fact the model was originally used to predict binding energies rather than spectra, and it earned itself a bad name in consequence. One example[2] is the binding of the ground state of O^{16}. The shell-model expression for the total binding energy can be maximised with respect to the well parameter b for ROSENFELD central forces of a strength needed to fit spectra of nuclei near O^{16}. The result of such a calculation is that there is a flat maximum centred at a value of b rather larger than the range of forces. The corresponding energy of about -15 MeV is to be compared with the observed binding of 127 MeV. Serious discrepancies like this imply that, at least for the purposes of binding energy work, one must revise the usual shell-model calculations in order to try to cater for correlation effects, i.e. configuration mixing. Of course, when one allows in some way for the presence of correlations in the shell-model wave-function, a variational calculation is bound to give a better binding energy than in the absence of the admixture. Presumably the fact that spectra are well predicted in the absence of such admixtures implies that such admixtures are fairly small and that they tend to produce roughly the same energy depression in all the low lying states.

In the first section of this chapter we will mention a few particular calculations in which one tries to represent correlations by configuration mixing from *immediately* higher configurations. The net effect of such mixing on the binding energy is, of course, always in the right direction but is not large enough to resolve the discrepancies we have mentioned. In the next section, we describe methods that can, in principle, give the energy correction due to mixing with *all* higher configurations (at least, to second order in the interactions). Unfortunately such methods have not been applied to total binding energy calculations at present because of the computational labour. In the final section, we review suggested methods for the improvement of shell-model wave-functions by the use of collective model ideas.

30. Configuration mixing from immediately higher configurations. We can cite as a typical calculation under this heading the case of Li^7 [3]. If one evaluates the

---

[1] A. H. BETHE and R. F. BACHER: Rev. Mod. Phys. **8**, 82 (1936). — E. FEENBERG and E. P. WIGNER: Phys. Rev. **51**, 95 (1937). — E. FEENBERG and M. L. PHILLIPS: Phys. Rev. **51**, 597 (1937).

[2] J. P. ELLIOTT: Unpublished. — M. G. REDLICH: Phys. Rev. **99**, 142 (1955).

[3] J. P. ELLIOTT: Unpublished.

potential energy with central forces of the ROSENFELD type and of a strength given by fitting spectra, the total energy of the ^{22}P [43] state of $(1s)^4$ $(1p)^3$ is found to be about +9 MeV. This is a long way from the observed value of −39 MeV. Two of the next configurations above $(1s)^4$ $(1p)^3$ are $(1s)^4$ $(1p^2\,2p)$ and $(1s)^4$ $(1p^2\,1f)$. One can calculate the 4×4 energy matrix involving the three ^{22}P [43] states of these two configurations and the previous state. The result[1] of diagonalisation is that the eigenstate contains about 3% in (intensity) of each of the three new states and that the new energy is +6 MeV. This is an improvement of 3 MeV on the previous figure but the discrepancy is still large (45 MeV).

Calculations have been made for O^{17} and F^{18} but using a SERBER instead of a ROSENFELD central force mixture[1]. For the ground state of O^{17} the calculations showed that the single-particle $1d$-state is contaminated by about 5% (in intensity) of each of the single particle $2d$-state and the lowest state of $(1p)^{-2}$ $(1d)^3$. In the case of F^{18} the amount of mixing from the f^2 configuration into the low-lying states is calculated to be ~4%. These calculations were not used to find the improvement in binding energy as a result of the mixing.

In general, this approach to the problem of binding energy is rather unsatisfactory. It seems probable that the nearest configurations only contribute a fraction of the total correction to binding energies arising from configuration mixing. If this is true, then the present method is of little practical use for binding energies because the amount of labour needed to take into account many individual higher configurations is prohibitive. However we note that the present method may have some use for predicting corrections to data other than binding energies such as magnetic moments. This would be so if the admixture to the zero-order wave-function came mainly from nearby configurations, which is quite possible. It still allows the correction to the binding from other configurations to preponderate, because all the small contributing corrections from the many admixed configurations are additive (at least, in second order). For other experimental quantities, the corrections may have both signs and so tend to cancel; furthermore most operators other than the binding energy operator tend to have matrix elements that drop off much more quickly with the energy of the mixed-in state than the interaction matrix elements.

31. Evaluation of second order energy shifts due to mixing with all higher configurations. Let us suppose that we take for our zero-order wave-function the shell-model state ψ_0 corresponding to the Hamiltonian $H_0 \equiv \sum_i T_i + \frac{1}{2}\sum_i V(r_i)$ where $V(r_i)$ is the potential well for the i-th particle and T_i is its kinetic energy. Now if the actual Hamiltonian is H, we can conveniently choose the well V such that:

$$(\psi_0|\, H - H_0\, |\psi_0) \equiv \langle H - H_0 \rangle_{00} = 0,$$

thereby setting equal to zero the first order shift in energy due to the perturbation $H - H_0 = W$ (say). The second order shift in energy can be expressed as a sum over the states n of H_0 with energies E_n in the well-known manner:

$$E - E_0 = \sum_n \frac{|W_{0n}|^2}{E - E_n}.$$

Explicit reference to the states n can be immediately eliminated if desired to give:

$$E - E_0 = (\psi_0|\, W \frac{1}{E - H_0} W\, |\psi_0) \equiv \left\langle W \frac{1}{E - H_0} W \right\rangle_{00}.$$

[1] M. G. REDLICH: Phys. Rev. **95**, 448 (1954); **99**, 1421 (1955).

The evaluation of ΔE from either of these expressions is not usually a simple matter. Two methods have been proposed to help in this connection. One is approximate. The other is exact if $V(r)$ is chosen to have the form of an oscillator potential. We now discuss these methods.

First method (exact). The central problem in the evaluation of the second expression above is clearly to find a means of performing in closed form the operations of differentiation implicit in H_0. If we take for V:

$$V = \tfrac{1}{2} m \omega^2 r^2 + V_0,$$

then we can master this problem by using the following two transformations[1]:

$$\frac{1}{E - H_0} = -\int_0^\infty e^{\lambda(E-H_0)}\, d\lambda = -\int_0^\infty d\lambda\, e^{\lambda E} \cdot \prod_i e^{-\lambda h_i} \tag{31.1}$$

where

$$h_i \equiv -\frac{\hbar^2}{2M} \nabla_i^2 + \frac{1}{2} m \omega^2 r_i^2 + V_0$$

and

$$\left.\begin{aligned} e^{-\lambda h_i} f(\boldsymbol{r}_i) &= e^{-\lambda V_0} \left(\frac{m\omega}{2\pi\hbar \sinh(\hbar\omega\lambda)}\right)^{\frac{3}{2}} \int d\boldsymbol{r}_i' f(\boldsymbol{r}_i') \times \\ &\times \exp\left\{-\frac{m\omega}{2\hbar \tanh(\hbar\omega\lambda)} \left(r_i^2 + r_i'^2 - 2\boldsymbol{r}_i \cdot \boldsymbol{r}_i' \operatorname{sech}(\hbar\omega\lambda)\right)\right\}. \end{aligned}\right\} \tag{31.2}$$

Although we may thus eliminate the awkward processes of differentiation, it is clear that, for several particles, the evaluation of the resulting expression for $E - E_0$ will be very tedious. In fact the power of the present method is limited by the computational labour involved when there are more than a few particles. At present the method has only been applied to the very light nuclei of masses 2, 3 and 4. In view of the success of the method in these cases, it may well be profitable to try to use it for heavier nuclei in spite of the labour. If all the integrals are performed, one ends up with an implicit equation for E of the form:

$$E = E_0(\omega) + \int d\lambda\, e^{\lambda E} g(\lambda, \omega).$$

The final step of the calculation is then to minimise E with respect to ω, the parameter of the oscillator well, and any other free parameters.

Second method (approximate). We may always write the above second-order expression for $E - E_0$ in the form:

$$E - E_0 = \frac{\langle W^2 \rangle_{00}}{E - \tilde{E}}$$

where $\tilde{E}$ is defined as a kind of mean energy of the spectrum of states E_n that makes this last expression for $E - E_0$ equal to the previous one. The whole difficulty now lies in the evaluation of this energy. Even without such an evaluation, we can see that its lower bound is the energy of the first excited unperturbed state. This gives an upper bound to the value of $E - E_0$ which is often useful. We now outline an approximate method[2] for deriving $\tilde{E}$. For our starting point, we will consider what appears to be a rather different problem, viz. variation of the parameter λ in the trial wave-function $\psi_0 + \lambda W \psi_0$ so as to minimsie the energy. Minimisation immediately gives a second order equation for λ:

$$\lambda^2 \left(k \langle W^2 \rangle_{00}^{\frac{1}{2}}\right) - \lambda - k \langle W^2 \rangle_{00}^{-\frac{1}{2}} = 0$$

[1] M. Bolsterli and E. Feenberg: Phys. Rev. **101**, 1349 (1956).
[2] A. M. Feingold: Phys. Rev. **101**, 258 (1956).

where we define k as the dimensionless quantity:

$$k = \frac{\langle W^2\rangle_{00}^{\frac{3}{2}}}{\langle W H_0 W\rangle_{00} + \langle W^3\rangle_{00} - E_0 \langle W^2\rangle_{00}}.$$

According as the sign of k is $\pm$, so the solution that minimises the energy is:

$$(\lambda^2 \langle W^2\rangle_{00})^{\frac{1}{2}} = \frac{1}{2k}\left[1 \mp \sqrt{1+4k^2}\right].$$

Now $\lambda^2 \langle W^2\rangle_{00}$ is a measure of the admixture of $W\psi_0$ to the original zero-order wave-function ψ_0. Clearly it is always ≤ 1. Inserting this value in the expression for the energy:

$$E = \frac{E_0 + 2\lambda \langle W^2\rangle_{00} + \lambda^2 \left[E_0 \langle W^2\rangle_{00} + \langle W^2\rangle_{00}^{\frac{3}{2}} k^{-1}\right]}{1 + \lambda^2 \langle W^2\rangle_{00}}$$

it follows that:

$$E - E_0 = \lambda \langle W^2\rangle_{00} = \langle W^2\rangle_{00}^{\frac{1}{2}} \left(\frac{1 \mp \sqrt{1+4k^2}}{2k}\right)$$

If k happens to be positive and $\ll 1$, then, to order k:

$$E - E_0 = -k \langle W^2\rangle_{00}^{\frac{1}{2}} = \frac{\langle W^2\rangle_{00}}{E_0 - \tilde{E}}$$

where $\tilde{E} = [\langle W H_0 W\rangle_{00} + \langle W^3\rangle_{00}] \langle W^2\rangle_{00}^{-1}$ is the diagonal energy of $H = H_0 + W$ taken in the "state" $W\psi_0$.

Thus we see that we have derived a result for $E - E_0$ in a form like that given by second-order perturbation theory. Furthermore the energy $\tilde{E}$ can now be calculated. We make three further remarks about this last formula:

(1) The occurrence of E_0 instead of E in the denominator is unimportant because the difference between them only affects the evaluation of $E - E_0$ in higher order than the one to which we are working.

(2) In general the term in $\langle W^2\rangle_{00}$ in $\tilde{E}$ should be small compared with the leading term in $\langle W H W\rangle_{00}$. If it is ignored, this makes $\tilde{E}$ equal to the diagonal energy of H_0 in the "state" $W\psi_0$.

(3) The variational problem which we formulated above can be generalised by using a trial wave-function $\psi_0 + \lambda W'\psi_0$ where W' can differ from the perturbing potential W. In this case, the variational calculation leads to:

$$E - E_0 = \frac{\left\langle \frac{W W' + W' W}{2} \right\rangle_{00}^2 \langle W'^2\rangle_{00}^{-1}}{E_0 - \tilde{E}'}$$

where $\tilde{E}'$ is the diagonal energy of the "state" $W'\psi_0$. It is difficult to assess the accuracy of this formula compared with the exact second-order expression. We can only say that, in view of the freedom in choice of W', this formula can give a better estimate than our above approximate estimate of the second-order expression.

The present method has often been used in particular problems. A good example is an attempt to assess the importance of the tensor force in changing total binding energies of nuclear states[1]. The nuclei that were studied in this manner were Li^5, Li^6 and Li^7. Taking the zero-order states ψ_0 to be the states of $(1s)^4\,(1p)^n$ in RUSSELL-SAUNDERS coupling, the diagonal matrix elements (i.e. the first order effects) of tensor forces are zero. It is thus essential to work in

[1] A. M. FEINGOLD: Phys. Rev. **101**, 258 (1956).

second-order. The reported calculation first estimated the second-order contributions from within the ground configuration and then it was generalised to include the contributions from configuration mixing. For any given state, it is found that the contribution to the second-order energy shift from interaction with higher configurations is much larger than the contribution from interaction between the states of the ground configuration. This is not especially significant in itself, because the ratio of the two contributions depends on the choice of zero-order state. (For instance, we could have chosen the states of the ground configuration to diagonalise the tensor force, then the contribution from the ground configuration will be zero, not only in second order but to all orders.) What is really significant is the extent to which configuration mixing due to tensor forces affects the ultimate relative separation of levels. The calculations showed that configuration mixing is very important in this respect. In other words, if one first ignores configuration mixing and diagonalises the tensor force, one obtains a certain sequence of levels, which is then radically changed when the configuration mixing is included. This implies that it is quite senseless to simply diagonalise tensor forces in the ground configuration without taking into account the effects of mixing. Iu this connection, an interesting result of the calculations on Li^6 and Li^7 is that the spectra resulting from mixing due to the tensor force are very much like those due to the simple diagonalisation of a vector force in the ground configuration. This result is a purely numerical one, but has given rise to speculation on the possibility that the effects on the energies of states of the ground configuration due to mixing from tensor forces can be rather generally reproduced by introducing a one-body vector force. Conversely it has been suggested that the well-known one-body force felt by a nucleon in a nucleus has its origin in the higher order effects of tensor forces (see Sect. 36).

Finally we comment on the possible importance of mixing effects for central forces, which would be disastrous from the shell-model point of view. It seems certain that, even for central forces, the contribution to the energy of a level from configuration mixing is very important (if only because neglect of such interaction always leads to an underestimate of binding energies). The only way to reconcile this fact with the success of the spectra calculations that ignore the mixing is to assume that the energy from the mixing is roughly the same for all levels, so that the internal structure of a spectrum is not much affected by the mixing. This implies that the effects of mixing by central forces is radically different from the effects of mixing by tensor forces as far as the relative spacings of levels is concerned.

32. Improvement of wave functions through the introduction of collective modes. Although the general validity and applicability of the shell-model has been well confirmed in practice, one must bear in mind that the model is an over-idealisation especially in its starting point, which is the complete neglect of correlation between nucleons. As we have seen in the preceding sections, one can attempt to take into account correlations within the terms of reference of the shell-model by introducing higher order effects in the form of configuration mixing from nearby configurations. Apart from being laborious, this is not a very satisfactory approach, because there are no real grounds for assuming that most mixing comes from the lowest few states. Rather, the contrary might be expected because of the strong short-range nature of nuclear forces which implies that, when two nucleons do interact, they can scatter into states with very different energy. Consequently one looks for some way of expressing compactly the correlation effects between the freely moving nucleons of the shell-model.

Perhaps the most direct and obvious way to do this would be to multiply the shell-model wave-function by a "correlation function" of the form: $\prod_{(i<j)} C(|r_i - r_j|)$. Each function $C(|r_i - r_j|)$ is allowed to deviate from unity only for small $|r_i - r_j|$. In principle, it should be possible to vary the total energy and minimise with respect to the form of $C|r_i - r_j|)$. Although such a technique has been employed in the calculation of the gross properties of nuclear matter[1], no work has been reported on the application of the technique to individual closed nuclear systems.

Alternatively, one can try to express the effects of correlations by introducing the concepts of some collective model into the shell-model wave-function. There are several types of collective model that have enjoyed some mild success in particular applications. In general however it is not believed that they have any real validity in their pure form because, far from under-emphasising correlations like the shell-model, they *over*-emphasise them. Nevertheless these models may have real use in guiding one in attempts to introduce correlation effects into shell-model wave-functions. In the following we outline how this may be done for two kinds of collective model, the alpha-particle model and the liquid drop model.

α) The liquid drop model. The ideas of the liquid drop model have been introduced into the shell-model in two distinct ways. In one of these, the so-called Unified Model[2], one splits the nucleons of a nucleus into two groups, conveniently taken as the loose nucleons and the nucleons in closed shells. It is now assumed that the dynamical behaviour of the closed shell group is like that of an incompressible liquid drop so that the excitation of this group can be described in terms of surface oscillations. In this treatment then, one picks out part of the nucleus and treats it just like a liquid drop. In the second treatment one works throughout with shell-model states of the whole nucleus but uses the liquid drop model to guide one in the choice of linear combinations of these states to be taken to represent proper states. (We will see afterwards that the alpha particle model has also been used in this manner.)

First treatment. The interaction between the k loose nucleons (labelled by i) and the closed shell nucleons takes on the following form[3] when the latter are regarded as forming a liquid drop whose excitation is describable in terms of quantised quadrupole oscillations (phonons):

$$H' = -\sum_{i=1}^{k} \varkappa(r_i - R)\left(\frac{\hbar\omega}{2C}\right)^{\frac{1}{2}} \sum_{\mu} \left(b_\mu + (-)^\mu b_\mu^*\right) Y_\mu^{(2)}(\Omega_i)$$

$$= -\sum_{i=1}^{k} \varkappa(r_i - R)\, h_i \text{ (say)}$$

where $\varkappa(r_i - R)$ is a narrow function concentrated about $r_i = R$ and such that $\int \varkappa(r_i - R)\, d\boldsymbol{r}_i$ is of the order of twice the mean kinetic energy of a nucleon, $\hbar\omega = \hbar\,(C/B)^{\frac{1}{2}}$ is the phonon energy, B is the mass constant in the surface kinetic energy, C is the surface tension constant in the surface potential energy, b_μ and b_μ^* are destruction and creation operators for phonons of spin 2 and z-component μ, and the index i labels the loose particles.

---

[1] S. D. DRELL and K. HUANG: Phys. Rev. **91**, 1527 (1953).

[2] A. BOHR and B. R. MOTTELSEN: Dan. Mat. fys. Medd. **27**, No. 16 (1953).

[3] A. BOHR and B. R. MOTTELSEN: Dan. Mat. fys. Medd. **27**, No. 16 (1953). — K. W. FORD and C. LEVINSON: Phys. Rev. **100**, 1 (1955). — E. FEENBERG: Shell Theory of the Nucleus, p. 148. Princeton: Princeton University Press 1955.

The set of unperturbed states are characterised by a particle state α with principal quantum numbers n and spin I and a number N of phonons coupled to spin R to give total spin J, component M for the whole system, thus: $|n\alpha I; NR; JM)$.

A typical matrix element of the interaction H' in this representation can be reduced to the following form on using a little RACAH algebra (see Appendix 4):

$$(n\alpha I; NR; JM|\, H'\, |n'\alpha' I'; N'R'; JM) = k(n|\varkappa(r_i - R)\,|n') \times$$
$$\times(\alpha I; NR; JM|\, h_i\, |\alpha' I'; N'R', JM) = k\left(\frac{\hbar\omega}{2C}\right)^{\frac{1}{2}}(-)^{2I'+I+J+R+1}\times$$
$$\times W(RR'II', 2J)\,(NR||\boldsymbol{b}||N'R')\,(\alpha I||\boldsymbol{Y}^{(2)}(\Omega_i)||\alpha' I')\,(n|\,\varkappa(r_i - R)\,|n')$$

where we have assumed states αI, $\alpha' I'$ to be antisymmetric in all k loose nucleons.

The reduced matrix element $(NR\,||\boldsymbol{b}||\,N'R')$ is easily evaluated and has been tabulated. From the nature of the operators b, we have the selection rules $\Delta N = \pm 1, |\Delta \boldsymbol{R}| = 2$. The other reduced matrix element $(\alpha I||\boldsymbol{Y}^{(2)}(\Omega_i)\,||\alpha' I')$ is also easily evaluated by the straightforward methods for quadrupole operators. The selection rules in this case are that α and α' can only differ in the state of one particle and that this difference must involve a spin change of 2 units.

Since H' contains creation and destruction operators, all the diagonal elements of H' are zero. This means that the energy shift of a state begins with a second order contribution. For a state $(n\alpha J; 00; JM|$, the shift is

$$\Delta E\,(n\,\alpha\,J) = -\sum_{\alpha' n' J'} \frac{|(n\alpha J, 00, JM|\,H'\,|\,n'\alpha' I', 12, JM)|^2}{\hbar\omega + E(n'\alpha' I') - E(n\alpha J)}$$
$$= -k^2\sum_{n'}|(n|\,\varkappa(r-R)\,|\,n')|^2\sum_{\alpha' J'}\frac{|(n\alpha J, 00, JM|\,h_i|\,n'\alpha' I', 12, JM)|^2}{\hbar\omega + E(n'\alpha' I') - E(n\alpha J)}\,.$$

Provided that the variation in $E(n'\alpha' I')$ is small for the complete set of angular states of given (n'), one can now apply closure over the angular wave-functions to obtain:

$$\Delta E\,(n\,\alpha\,J) = -k^2(n\,\alpha\,J, 00, J\,M|\,h_i^2\,|\,n\,\alpha\,J, 00, JM)\sum_{n'}\frac{|(n|\,\varkappa(r_i - R)\,|\,n')|^2}{\hbar\omega + \Delta(n n')}$$

where $\Delta(nn')$ is a suitable mean value of $E(n'\alpha' I') - E(n\alpha J)$ averaged over $\alpha' I'$ and αJ. On performing the integration over the phonon variables, we can write:

$$\Delta E\,(n\,\alpha\,J) = (n\,\alpha\,J\,|\,V\,|\,n\,\alpha\,J)$$

where V is an effective scalar potential of the particle co-ordinates:

$$V = -\sum_{n'}\frac{|(n|\,\varkappa(r_i - R)\,|\,n')|^2}{\hbar\omega + \Delta(n n')}\,\frac{\hbar k^2}{2B\omega}\sum_{ij}\left(\sum_{\mu} Y_\mu^{(2)*}(\Omega_i)\,Y_\mu^{(2)}(\Omega_j)\right)$$

The terms of $i=j$ in the sum over i and j can be shown to give a contribution to $\Delta E(n\alpha J)$ that is independent of J. Consequently these terms can be ignored as far as the J-splitting of levels is concerned. The remaining terms with $i \neq j$ reduce to $\sum_{i\neq j}\left(\frac{4\pi}{5}\right)^{\frac{1}{2}} P_2(\cos\Omega_{ij})$ where Ω_{ij} is the angle between Ω_i and Ω_j. Thus we see that the particle-surface coupling gives energy shifts in the particle levels like those produced by an effective particle-particle force. Furthermore this has only one non-zero SLATER integral, viz. F_2. Thus it is simple to modify a standard shell-model calculation of a spectrum to take into account particle-surface coupling: one simple alters the size of F_2.

Before closing this outline of the theory of particle-surface coupling, we note the interesting fact[1] that, when $(n\alpha J|$ is a state of equivalent particles the only J-dependence of $\Delta E(n\alpha J)$ arises from the states with the same angular quantum numbers as $(n\alpha J|$. Thus, in the case of a state of equivalent particles, the J-splitting from the particle-surface coupling is not dependent upon the mixing from other configurations.

One reported application of the present treatment has been to the nuclei just above O^{16}, regarding this nucleus as a closed-shell liquid-drop[2]. (In this connection we note that recent evidence[3] supports the existence of a collective state of quadrupole oscillation at 15 MeV in O^{16}.) The object of this work is to account for the large observed strengths of electric quadrupole transitions in O^{17} and F^{19}. In spite of O^{17} being a single neutron nucleus, the transition strength of the $s_{\frac{1}{2}}-d_{\frac{5}{2}}$ jump is about that of a single proton. This implies unequivocally that the single particle states are not pure ones, but contain a fraction of core excitation. If one tries to describe this excitation in terms of the lowest shell-model excited states of O^{17}, one finds that improbably large admixtures are needed to obtain the observed transition strength. If however one uses the present approach, only a small admixture is needed because of the collective nature of the excited states and the consequent large quadrupole matrix elements. In contrast to the $E2$ transition strength the $E2$ moment of O^{17} is just about as small as that expected from the shell model. It is not at all clear how one can simultaneously fit both the large value of the quadrupole transition strength and the small value of the quadrupole moment. If the difficulty is ignored, one can estimate the strength of particle-surface coupling from the transition strength, and then consider the effect of this coupling on the shell-model predictions for F^{19}. These predictions are as we have seen, in agreement with experiment except that the $E2$ transition strength between the second excited and ground states is underestimated. Inclusion of the weak surface coupling remedies this defect whilst preserving the existing agreement.

Another application[4] has been made to nuclei between Si^{28} and Ca^{40}. Using j-j coupling for the basic particle wave-functions, some agreement has been found with magnetic moments, spectra and the ft-values of β-decay.

The same treatment has been applied to the cases of Ca^{42} and Ca^{43}, regarding Ca^{40} as the closed-shell liquid drop[1]. It was found that the introduction of particle-surface coupling was not necessary to fit the spectrum. In fact, unless the coupling was taken to be weak, the introduction disturbed existing agreement with experiment. Unfortunately no $E2$ transition strengths are known experimentally for these nuclei.

Second treatment. In general, when one attempts to assign a shell-model configuration to an excited state of a nucleus, one finds that there are several eligible configurations each with the same zero-order energy. Each configuration generally gives several independent states and the problem of diagonalising the Hamiltonian to find the correct linear combination becomes very tedious. In such a case, one may try to avoid the labour of diagonalising by appealing to the concepts of some collective model for some guide as to the correct linear

---

[1] K. W. FORD and C. LEVINSON: Phys. Rev. **100**, 1 (1955).

[2] J. P. ELLIOTT and B. H. FLOWERS: Proc. Roy. Soc. Lond., Ser. A **229**, 536 (1955). — F. C. BARKER: Phil. Mag., Ser. VIII (8) **1**, 329 (1956).

[3] D. H. WILKINSON: Phys. Rev. **99**, 1347 (1955).

[4] E. FEENBERG: Shell Theory of the Nucleus, p. 148. Princeton: Princeton University Press 1955. — P. GOLDHAMMER: Phys. Rev. **101**, 1375 (1956).

combination. We illustrate these remarks by an example to which the method has been applied[1]. This example is the sequence of states of $J=0^+$, $T=0$ in O^{16}. According to the shell-model, the first 0^+ state, i.e. the ground state, will be the closed shell $(1s)^4\,(1p)^{12}$. On the same model, the next state of $J=0^+$, $T=0$ (at 6.06 MeV) can in principle be a linear combination of all the many states of $J=0^+$, $T=0$ from the configurations $(1s)^{-1}\,2s$, $(1p)^{-1}\,2p$, $(1p)^{-2}\,(2s)^2$, $(1p)^{-2}\,2s\,1d$, $(1p)^{-2}\,(1d)^2$. To avoid the labour of setting up a large Hamiltonian matrix and diagonalising it, one tries to use the often-made suggestion that the 6.06 MeV state in O^{16} is a compressive mode collective oscillation of the ground state. This implies that the internal spatial structure of the excited state must be the same as the closed shell ground state (ψ_0, say) except for a radial, spherically symmetric, oscillation. There are only two states of the larger total number that have this property, viz. the two ${}^{11}S\,[4444]$ states of the configurations $1s^{-1}\,2s$, $1p^{-1}\,2p$, and so the problem reduces to finding the correct linear combination of these two states that has zero coupling with the closed shell ground state ψ_0. It so happens in this case that the required combination, which is $\sqrt{\frac{1}{6}}\,\psi(1s^{-1}\,2s)+\sqrt{\frac{5}{6}}\,\psi(1p^{-1}\,2p)$, can be verified directly without the labour of evaluating interaction matrix elements. This is a consequence of the fact that the above combination identically equals $\frac{b}{6}\,\frac{\partial\psi_0}{\partial b}$, where b is the usual size parameter of the oscillator well (see Sect. 2). It is evident that there is no coupling between $\partial\psi_0/\partial b$ and ψ_0 provided b is given the value, b_0 say, that minimises the energy E_0 of ψ_0:

$$0=\frac{\partial E_0}{\partial b}=\frac{\partial}{\partial b}\,(\psi_0|\,H\,|\psi_0)=2\left(\frac{\partial\psi_0}{\partial b}\right|H\left|\psi_0\right).$$

It is found that the extra single particle energy of the state $\partial\psi_0/\partial b$ relative to ψ_0 is largely compensated by extra interaction energy in the former state, so that one obtains qualitative success in predicting the lowness of the 0^+ state at 6.06 MeV. The actual theoretical figure is rather larger—about 10 MeV—but this may be reduced by allowing the s and p shells to have different size parameters, say b_s and b_p. If the ground state closed shell energy is minimised simultaneously with respect to both of these parameters, then ψ_0 will have no coupling with either $\frac{\partial\psi_0}{\partial b_s}\sim\psi(1s^{-1}\,2s)$ or $\frac{\partial\psi_0}{\partial b_p}\sim\psi(1p^{-1}\,2p)$ separately. This extension of the formulation brings double benefit. First, one will get a value closer to the observed 6.06 MeV separation energy. Second, one removes a problem of consistency that is implicit in the above approach. This problem arises from the presence of the third state $\sqrt{\frac{5}{6}}\,\psi(1s^{-1}\,2s)-\sqrt{\frac{1}{6}}\,\psi(1p^{-1}\,2p)$ that is orthogonal to ψ_0 and $\partial\psi_0/\partial b$. In general this third state couples with both the previous states and so its presence should be taken into account. However one finds qualitatively that its presence probably can be ignored as a result of the fact that its diagonal energy turns out to be very different from (much higher than) the diagonal energies of ψ_0 and $\partial\psi_0/\partial b$. In the case when one allows the size constants for s and p orbits to differ, the coupling of any combination $\alpha\psi(1s^{-1}\,2s)+\beta\psi(1p^{-1}\,2p)$ to ψ_0 must vanish, so it only remains to choose α and β so that this combination has no coupling with $\beta\psi(1s^{-1}\,2s)-\alpha\psi(1p^{-1}\,2p)$, the third state. From the fact that the states of the more restricted approach gives a roughly diagonal interaction matrix, we can assume that the more general approach finds $b_s\approx b_p\approx b_0$ and $\alpha\approx\sqrt{\frac{1}{6}}$, $\beta\approx\sqrt{\frac{5}{6}}$.

[1] R. A. Ferrell and W. M. Visscher: Phys. Rev. **102**, 450 (1956).

Although the present approach gives some agreement with experiment, it should be stressed that the neglect of configurations of two-particle excitation and of states of non-maximum spatial symmetry is not at all evidently justified from the usual shell-model point of view. It would be very nice to see if the standard shell-model approach to the 0^+ states of O^{16} corroborated this neglect or not. This would be especially interesting in view of the power and wide field of applicability of the present method which can be formulated much more generally than we have indicated hitherto.

Originally the method was suggested for improving wave-functions in binding energy calculations. Suppose we are given a shell-model state $\psi(\boldsymbol{r}, b)$ where $\boldsymbol{r}$ represents all the particle co-ordinates and b certain parameters of the well (size, deformation, etc.). Then we can consider the generalised state function

$$\Psi_0 = \int \psi(\boldsymbol{r}, b)\, f_0(b)\, db$$

and minimise the energy with respect to the functional form of $f_0(b)$. Certain excited states Ψ_i can be built up in the same way:

$$\Psi_i = \int \psi(\boldsymbol{r}, b)\, f_i(b)\, db$$

subject to the condition that the states i are orthogonal. In practice, it may be justified to suppose that the lowest state Ψ_0 is well-approximated by a pure shell-model state for some value of b, say b_0. This implies that $f_0(b)$ is peaked about $b=b_0$ and zero otherwise i.e. $f_0(b) \equiv \delta(b-b_0)$. This assumption may enable one to circumvent variational calculations for some excited states. For instance, to return to our example of O^{16}, if we assume that the expansion of

$$\Psi_1 = \int \psi(\boldsymbol{r}, b)\, f_1(b)\, db$$

around b_0 only contains states with at most two quanta of energy more than the ground state $\psi(\boldsymbol{r}, b_0)$, then the orthogonality condition determines $f_1(b)$ as $\delta^{(1)}(b-b_0)$. We stress again here that the method is more general than its specific application to compressive modes in O^{16} would suggest. For instance, b may be taken to represent the quadrupole deformation parameter β of the well and the orientation ϑ of the well in space[1]. Assuming the ground state to be spherical implies taking $f_0(b) \equiv f_0(\beta, \vartheta) = \delta(\beta)\, \delta(\vartheta)$. Proceeding as above, one would hope to find a form for $f_1(\beta, \vartheta)$ that could be associated with the excitation of a surface phonon. We note that, for heavier nuclei, where the ground states are non-spherical with a permanent deformation $\beta=\beta_0$, one can take $f_i(\beta,\vartheta) \equiv \delta(\beta-\beta_0)\, f_i(\vartheta)$, and use the present method to find the rotational spectrum and effective moment of inertia of the deformed nucleus[2].

β) The alpha-particle model. It has recently been shown that the wave-functions of the S states of maximum spatial symmetry of the shell model configurations $(1s)^4(1p)^n$ with $n=0, 4, 8, 12$ can be rewritten in quasi-alpha particle form[3]. This suggests that one tries to represent configuration mixing in the shell-model wave-function by allowing a partial "condensation" of the system into alpha-particles. As an illustration, we will consider the configuration $(1s)^4(1p)^4$ which represents Be^8. The properly anti-symmetrised shell model wave-function

[1] J. A. WHEELER: Proceedings of Tokyo International Physics Conference, 1953. — J. YUCCOZ: Private communication.

[2] J. YUCCOZ: Private communication.

[3] J. PERRING and T. H. R. SKYRME: Proc. Phys. Soc. Lond. A **69**, 600 (1956).

^{11}S [44] based on the oscillator potential can be transformed identically into the following form:

$$\sum_P (-)^P P(1,2)\, \psi_\alpha(1)\, \psi_\alpha(2)\, (\boldsymbol{R}_1 - \boldsymbol{R}_2)^4\, \mathrm{e}^{-(\boldsymbol{R}_1-\boldsymbol{R}_2)^2/b^2}$$

where $\psi_\alpha(1)$ is a alpha-particle-like wave function of four particles, say $i=1, 2, 3, 4$, with space part:

$$\mathrm{e}^{-\frac{1}{2b^2}\sum_{i=1}^{4}|\boldsymbol{r}_i-\boldsymbol{R}_1|^2}$$

and $\boldsymbol{R}_1$ is the centre of mass of four particles with co-ordinate vectors $\boldsymbol{r}_i$ $(i=1, 2, 3, 4)$: $\boldsymbol{R}_1 = \frac{1}{4}\sum_{i=1}^{4} \boldsymbol{r}_i$. $\psi_\alpha(2)$ and $\boldsymbol{R}_2$ are defined similarly for the other four particles, $i=5, 6, 7, 8$. The anti-symmetrisation between the two groups of particles is denoted by $\sum_P (-)^P P(1,2)$, where P runs over all permutations of the particles in the two groups.

We notice that the mean size of the two alpha-particles and their mean separation are of the same order of magnitude, namely b, the usual constant of the oscillator well (see Sect. 2). This means that the two alpha-particles overlap severely in space and that the energy needed to "excite" an alpha particle is of the order of that needed to excite the relative motion. Thus, although the above wave-function has the *form* of an alpha-particle model wave-function, it cannot be said to imply the existence of rotational and vibrational excited states of the type expected on the usual alpha-particle model. The necessary condition for these states is that the mean separation of the alpha particle groups be considerably greater than the size of the alpha particle. This can be arranged by replacing b in the above wave function of relative motion by bx where x is allowed to be $\gg 1$. In principle one can now carry through a variational calculation with x as the variable. If the minimum is achieved for x different from (larger than) unity, this implies the existence of alpha-particle type clustering in the shell-model wave-function. As far as we know, this programme has not been carried through in precisely this form. Related calculations have been done but these have not been based on the shell-model wave-function of Be^8 in the way we have outlined. In the case of O^{16}, however, some calculations of just the type we have discussed have been done[1].

It was found that the plots of total energy against quantities analagous to b and x were rather sensitive to the force mixture used. Perhaps significantly, the most reasonable plot (i.e. one that gave a minimum for x rather larger than unity) was obtained in the case of a force mixture that just satisfied the saturation requirement.

Finally we mention that the above identical transformation between the alpha-particle and shell-model wave functions may be exploited in a rather different way[2]. Instead of using it to build an alpha-particle type admixture into the shell-model state with a view to improving the calculation of binding energy, one can use it to try to calculate the structure and spectrum of excited states. In the case of Be^8, one first notices that the alpha particle wave-function $\left(\frac{R}{b}\right)^4 \mathrm{e}^{-R^2/b^2}$ (where $\boldsymbol{R}=\boldsymbol{R}_1-\boldsymbol{R}_2$), is the solution of the potential: $\left(\frac{R}{b}\right)^2 + 5\left(\frac{b}{R}\right)^2$. Now if we assume that states of relative motion of the two alphas can be excited

[1] C. Bloch: Private communication.

[2] J. Perring and T. H. R. Skyrme: Proc. Phys. Soc. Lond. A **69**, 600 (1956).

without disturbing the internal motion, we can say that the next state is the first excited state of this potential. This is the state with wave-function:

$$R^{\frac{1}{2}(\sqrt{105}-1)}\,\mathrm{e}^{-R^2/b^2}\,Y_2(\vartheta,\varphi)\,.$$

The corresponding total wave-function of all eight particles can now be transformed back into shell-model form. One finds that the main component is the ^{11}D [44] state of $(1s)^4\,(1p)^4$; the rest is accounted for by ^{11}D [44] states of higher configurations.

A similar calculation has been made for O^{16} with a view to finding what kind of configuration structure is implied for the first excited state of spin 0^+ at 6.06 MeV. Although more complicated, the calculation is, in principle, just the same as for Be^8. The intensities of the various configurations are:

$1s^{-2}\,2s$	$1p^{-1}\,2p$	$1p^{-2}\,1d^2$	$1p^{-2}\,1d\,2s$	$1p^{-2}\,2s^2$
29%	52%	17%	1.9%	0.15%

It is noteworthy that the two configurations of one-particle excitation account for 80% of the total state. This might have been expected as a consequence of the nature of the alpha-particle state which is a spherically symmetric vibrational (breathing) mode. As we have seen in the preceding discussion of the liquid drop model, such a mode is most naturally represented in terms of configurations of one-particle excitation. However the above figures differ from those of the liquid drop model in that the ratio of s to p excitation is about 3:5 instead of 1:5.

VIII. The theoretical basis of the shell-model.

Almost all of this article has been concerned with establishing the success of the nuclear shell-model in applications. Apart from a few hints and qualitative remarks, we have not touched the problem of providing a respectable theoretical basis for the model. In other words, we have been more concerned with *how* the shell-model works in practice, rather than *why* its works in theory. This attitude has been exemplified in nearly all the calculations we have reviewed by the implicit assumption from the start of pure configurations. Only occasionally have we mentioned the possibility of configuration mixing, and, even then, there has always been the assumption that such mixing is small.

Most models in physics are introduced on an *ad hoc* basis in order to describe certain phenomena when it is impossible to proceed directly from basic theory. It may happen, as in the case of the nuclear shell-model, that the model enjoys striking successes in detailed applications as well as in the general applications for which it was formulated in the first place. When enough successes are established to ensure the validity of the model, interest tends to shift to the problem of finding a fundamental understanding of the model. This has happened to the shell-model in the last years which have witnessed several initial attacks made on aspects on the very important and far-reaching question "Why does the shell-model work?" These efforts have not met with conclusive success; in fact, some reported results[1] almost seem to demonstrate that the shell-model ought *not* to work! It is apparent that much work is still to be done in this field.

In this chapter, the first two sections summarise the empirical features of the shell-model. The theoretical attempts to predict these features starting from the nucleon-nucleon interaction are described in the remaining sections.

[1] E. P. WIGNER: L. Farkas Memorial Volume. Research Council of Israel 1, 45 (1951).

33. Summary of the main empirical features of the shell-model. At the basis of all the shell-model calculations given in this article is the idea that nucleons move fairly independently of each other, each in its own potential well. For a nucleon i with energy E_i we can write this well as a sum of three terms:

$$V(E_i, r_i) + U(E_i, r_i)(\boldsymbol{l}_i \cdot \mathbf{s}_i) + i\,W(E_i, r_i)$$

where V, U and iW are the real (refractive) potential, the spin-orbit potential and the imaginary (absorbing) potential respectively. Each of these is a function of radial distance r_i and also of the energy E_i. We have not spoken before about the imaginary potential iW. We introduce it here as a purely phenomenological device[1] to represent the fact that a nucleon does not in general move for ever in its orbit but eventually has a collision with another nucleon and loses energy, i.e. it is "absorbed". One can easily check that the probability of absorption per unit time for a nucleon in the presence of an absorbing potential W is $2W/\hbar$ for W not too large. Since, by definition of mean free path λ, this same probability is v/λ where v is the nucleon velocity, we immediately have the simple relation:

$$W = \frac{\hbar v}{2\lambda}$$

or, numerically:

$$W\,(\text{MeV}) \sim \frac{4.5\sqrt{[E + V(E)]\,(\text{MeV})}}{\lambda\,(10^{-13}\,\text{cm})}.$$

Evidently W is a vital quantity in that it specifies the degree of purity of the shell-model states. It represents the amount of configuration mixing and we can say that the success of the usual version of the shell-model, which ignores configuration mixing, gives us an empirical limit for W applying to bound nucleon orbits, i.e. for negative energies:

$$W \ll 0.5\ \text{MeV}.$$

This estimate follows from the above equation for W under the condition that the mean free path λ is large compared with the length of a nucleon orbit $2\pi R$, and taking $E + V$ as 50 MeV.

An empirical value of $\langle U \rangle$, the spatial average of $U(r)$, for the most energetic particles in a nucleus is immediately obtained from the observed spin-orbit splitting of single-particle levels. Such splittings do not reveal anything significant[2] about the spatial distribution of $U(r)$ which, according to most theoretical arguments[3], should be concentrated on the surface. Also, nothing is revealed about the energy dependence, since the observed levels are always those of particles at the top of the nucleon energy distribution. However, no one has advanced a theoretical reason why U should be strongly energy dependent. The empirical values of $\langle U \rangle$ decrease smoothly from ~ -2 MeV near O^{16} in light nuclei to ~ -0.5 MeV near Pb^{208} in heavy nuclei[3].

Information about the refractive potential V is obtainable from observed single particle levels[2] and binding energies. The former shows that the well-shape is close to that described in Chap. II. From the latter we may obtain the value of $\langle V \rangle$ averaged first over r_i and then over particles i, on applying the relation:

$$\text{Total energy} = \sum_i \langle T_i \rangle + \tfrac{1}{2} \sum_i \langle V(E_i, r_i) \rangle.$$

---

[1] FESHBACH, PORTER and WEISSKOPF: Phys. Rev. **96**, 448 (1954).
[2] ROSS, MARK and LAWSON: Phys. Rev. **102**. 1613 (1956); **104**, 401 (1956).
[3] R. J. BLIN-STOYLE: Phil. Mag. **46**, 973 (1955). — J. S. BELL and T. H. R. SKYRME: Phil. Mag., Ser. VIII **1**, 1055 (1956).

The factor $\frac{1}{2}$ arises from the fact that V is assumed to arise from two-body forces. This expression does not consider the splitting of the states of a configuration due to residual forces and spin orbit forces so, strictly speaking, it must be applied to the centroid of the observed group of such states and not simply to the ground state. Usually however, this distinction is unimportant because the splitting is negligible compared with the total binding energy for all but the lightest nuclei.

In principle, the above formula can be applied to observed total binding energies of individual nuclei. Even if practicable, this is not a satisfactory approach because we are now interested in systematic features of nuclei and not individual properties. A better method is to use the semi-empirical mass formula as a representation of the observed binding energies. Guided by this formula, the theoretical binding energy for a general nucleus of N neutrons and Z protons can be broken down into component parts such as the volume energy, COULOMB energy, symmetry energy and surface energy. One can then equate the experimental and theoretical values of the four separate quantities. Now the first three can be computed without reference to any surface, i.e. for a very large system. It is here that a vital simplification enters in that one can use a FERMI gas of plane waves to represent the particle motion. With such a situation the total kinetic energy in a volume Ω containing equal numbers of nucleons in each of the four intrinsic nucleon states can be expressed as $\frac{\hbar^2}{2M}\tau$, where τ is given by:

$$\frac{\tau}{4} = \frac{3}{5}(6\pi^2)^{\frac{2}{3}}\left(\frac{\varrho}{4}\right)^{\frac{5}{3}}$$

and is the total density of square momentum, $\tau = \Omega^{-1}\sum_i k_i^2$ and ϱ is the total ordinary density of nucleons: $\varrho = \Omega^{-1}\sum_i 1$. Taking ϱ as $\left(\frac{4\pi}{3}r_0^3\right)^{-1}$, with $r_0 = 1.20\times 10^{-13}$ cm, the mean kinetic energy per nucleon, which is $\frac{\hbar^2\tau}{2M\varrho}$, has the value 21 MeV. We will now assume that the energy dependence of $V(E)$ can be represented as a linear function over the relevant energy range

$$V(E) = \alpha + \beta E.$$

It follows that, for the i-th particle

$$E_i = T_i + V(E_i) = T_i + \alpha + \beta E_i$$

whence:

$$E_i = \frac{1}{1-\beta}(T_i + \alpha)$$

and:

$$V(E_i) = \alpha + \frac{\beta}{1-\beta}(T_i + \alpha) = \frac{1}{1-\beta}(\beta T_i + \alpha).$$

Assuming the four spin states to be equally populated, the volume binding energy per particle, B (say), is given by:

$$-B = \frac{\hbar^2}{2M}\frac{\tau}{\varrho} + \frac{1}{2(1-\beta)}\left(\beta\frac{\hbar^2}{2M}\frac{\tau}{\varrho} + \alpha\right)$$

$$= \frac{1}{2-2\beta}\left[\left(\frac{\hbar^2}{2M}\frac{\tau}{\varrho}\right)(2-\beta) + \alpha\right].$$

Putting in the experimental equilibrium values, i.e. $B = 15$ MeV and $\frac{\hbar^2}{2M}\frac{\tau}{\varrho} = 21$ MeV, we find:

$$51\beta - \alpha = 72$$

where α is in MeV. We note that we cannot use the fact that B is a maximum (i.e. $dB/d\varrho=0$) at the equilibrium density to obtain another linear relation involving α and β because we do not know the dependence of α and β on ϱ. In Sect. 35 we will assume a model which implies definite dependences, and then we will use this condition.

Now we consider the symmetry energy. This enters the semi-empirical mass-formula as a consequence of the COULOMB repulsion between protons in a nucleus, which leads to the neutron excess. The COULOMB repulsion itself gives rise to a term in the semi-empirical formula of the well-known form $\frac{3}{5}\frac{e^2}{R}Z(Z-1)$ where R is the nuclear radius. On the shell-model, this repulsion can be represented as a rise of amount $\frac{6}{5}\frac{e^2}{R}(Z-1)$ in the position of the bottom of each proton well. Since β-stable nuclei must have the top-most proton and neutron energies approximately equal, this rise implies that, for the stable isobar of a given total number of particles, there are more neutrons than protons, so that there is more kinetic energy in the neutron well and less in the proton well than in the absence of the COULOMB force. There is a net excess of such energy that constitutes the kinetic energy contribution to the symmetry energy. The total symmetry energy is the change in energy due to having N not equal to Z, and it arises partly from this increase in kinetic energy, and partly from a change in potential energy. The former is straight-forwardly shown to be:

$$\frac{1}{2}(3\pi^2)^{\frac{2}{3}}\left(\frac{\hbar^2}{M r_0^2}\right)\frac{(N-Z)^2}{4A}\approx 45\ \text{MeV}\,\frac{(N-Z)^2}{4A}\,.$$

The contribution from potential energy arises in part from the energy dependent part of the single particle potential and equals $\frac{\beta}{2(1-\beta)}$ times the kinetic energy contribution. Combining these two contributions gives $\left(\frac{2-\beta}{2-2\beta}\right)$ 45 MeV in units of $\frac{(N-Z)^2}{4A}$. If this is directly equated to the observed energy of ~ 95 MeV, the quantity β is determined as $\beta\sim 0.7$. The above relation from fitting the volume energy then gives $\alpha\sim -36$ MeV.

A weak point in this analysis is that it ignores a second source of change in potential energy, viz. that from interaction forces of the $(\boldsymbol{\tau}_i\cdot\boldsymbol{\tau}_j)$ type. When all intrinsic spin states are equally populated the total non-exchange potential energy from these forces is zero. When there are more neutrons than protons, this is no longer correct. It may be that, in an indirect way, the change in potential energy from this source is partly taken into account by the change in single particle potential energy as above, but this is not evident. If the $(\boldsymbol{\tau}_i\cdot\boldsymbol{\tau}_j)$ forces are repulsive in $T=1$, attractive in $T=0$ states, the symmetry energy contribution would be positive like the previous ones. This means that the value of 0.7 must be taken as an upper limit for the value of β. To get some idea of magnitude, if half of the non-kinetic part of the symmetry energy is ascribed to the interaction potentials, this leaves 25 MeV for the contribution of the single particle potentials, and this implies $\beta\sim 0.5$. The change in β from 0.7 to 0.5 only changes the value of α obtained from fitting the volume energy from -36 to -46 MeV.

34. Evidence from nucleon scattering. It is a fortunate fact that the recent successes of the shell-model in fitting properties of low-lying states have been parallelled by successes of a very similar model applied to nucleon scattering data.

This model is the so-called Optical Model which analyses the data with exactly the same type of single particle potential that we introduced in the last section. The magnitudes and, to some extent, the spatial dependence of the three types of potential, V, U and W are used as parameters in fitting the data at each energy. Of course, the nucleon energies involved in reactions are positive in contrast to the negative energies involved in the discussion of bound state data in the last section.

Analysis of the low energy neutron scattering data[1] gives a value for α in the expression $V(E)=\alpha+\beta E$ of about -45 MeV. Analysis of the proton scattering data[2] at energies up to 30 MeV gives a value of about 0.6 for β. This is about the same value as that quoted in the last section, and suggests that the linear approximation to the energy dependence of $V(E)$ is valid over a considerable energy range. If quadratic and higher terms were important, the two values of β from the bound state data and scattering data would be appreciably different. We note that the value of V deduced from proton scattering data is found, after correcting for the COULOMB potential, to be larger than that from neutron scattering data by about 10 MeV. Such a difference is expected if we assume that the potentials felt by a neutron and proton of the same *kinetic* energy differ only in the mean proton COULOMB potential, say V_{Coul}. This implies that the potential felt by a proton is

$$V=\frac{1}{1-\beta}(\beta T+\alpha)+V_{\text{Coul}}$$

or, in terms of total energy $E=T+V$:

$$V=\alpha+\beta E+(1-\beta)\,V_{\text{Coul}}.$$

Thus the potentials felt by a neutron and proton of the same *total* energies are expected to differ, not only by the COULOMB energy V_c but also by an opposing term βV_c. For a medium heavy nucleus and $\beta\approx 0.6$, this term is about equal to the observed 10 MeV difference.

Analysis of neutron and proton scattering data establishes that $W(E)$ decreases from $W\sim 10$ MeV at $E\sim 20$ MeV to $W\sim 2$ MeV at $E\sim 0$ MeV. We recall that the bound state data implies a small value of W ($\lesssim 0.5$ MeV) for the negative energies $E\lesssim -10$ MeV. It is natural to conclude that the decrease of $W(E)$ with decreasing positive energy is continued smoothly through zero energy to negative energies to where $W\to 0$ for the lowest bound orbits.

Evidence for the existence of a non-zero spin-orbit potential U comes from the polarisation of scattered nucleons. The magnitude of the averaged potential $\langle U\rangle$ is not very well determined[3], although it is of the magnitude (~ -1 MeV) inferred from bound state data as described in the previous section. No reported analysis has examined the problem of the spatial distribution of U, at least not for the low bombarding energies we are considering here. For high bombarding energies[4], there is some evidence that the potential is concentrated at the surface.

35. Prediction of the refractive potential *V*. We will now discuss the problem of predicting the mean refractive potential $V(E_i)$ felt by a nucleon i with energy E_i inside a typical nucleus. The essentials of this problem are best discussed with

---

[1] FESHBACH, PORTER and WEISSKOPF: Phys. Rev. **96**, 448 (1954).

[2] R. D. WOODS and D. S. SAXON: — Phys. Rev. **95**, 577 (1954). MELKANOFF, MOSZKOWSKI, NODVIK and SAXON: Phys. Rev. **101**, 507 (1956).

[3] ADAIR, DARDEN and FIELDS: Phys. Rev. **96**, 503 (1954). — A. OKAZAKI: Phys. Rev. **99**, 55 (1955).

[4] R. M. STERNHEIMER: Phys. Rev. **100**, 886 (1955).

reference to a very large system, thereby eliminating irrelevant surface effects. Therefore the major part of the ensuing discussion will deal with such systems. At the end we will add a few remarks on calculations for finite systems.

α) Infinitely large system. The simplest way to obtain a value for $V(E_i)$ is to perform what has been called a BORN approximation type calculation. This means evaluating the expectation value of the actual potential felt by nucleon i with a FERMI gas of uncorrelated plane wave states: $V(E_i) = \langle \sum_j v_{ij} \rangle$. From the remarks in Sect. 33 it is clear that, once one has obtained $V(E_i)$ in this way, one can go on and compute the nuclear volume binding energy per particle B by using:

$$-BA = \left\langle \sum_i \left(T_i + \tfrac{1}{2} V(E_i)\right) \right\rangle .$$

Now we have seen in Sect. 33 that the observed values of α and β in the expression $V(E_i) = \alpha + \beta E_i$ lead to roughly the correct value of B if the nuclear radius parameter r_0 is taken as 1.20×10^{-13} cm. Provided that the theoretical expression for $V(E_i)$ can be stated in this form, it follows that, if α and β are predicted correctly, then the correct value of B is implied. Conversely if B is incorrectly predicted, either α or β or both must be incorrectly predicted. We have mentioned in previous chapters, especially Chap. VII, that the shell model estimates of the total binding energy for individual nuclei are much less than the observed values if the interaction v_{ij} is restricted by the usual conditions: (i) it saturates, (ii) it qualitatively fits the two-body data. It is not surprising to find that the FERMI gas calculation also badly underestimates B under the same conditions. In one calculation[1], v_{ij}, was chosen to be

$$v_{ij} = (\boldsymbol{\tau}_i \cdot \boldsymbol{\tau}_j)\,(0.10 + 0.23\,\boldsymbol{\sigma}_i \cdot \boldsymbol{\sigma}_j)\, 2.6 \left(\frac{\hbar^2}{M a^2}\right) \frac{e^{-r/a}}{r/a}$$

where a, the range of the YUKAWA interaction, was taken as 1.74×10^{-13} cm. The energy minimum was found to occur at $r_0 = 1.56 \times 10^{-13}$ cm and the best value of B was only ~ 2 MeV instead of the observed 15 MeV. From what we have said, this failure indicates that any similar calculation of V_i will also fail. This has been checked directly[2] using the same expression for v_{ij} and evaluating $V(E_i)$ with $r_0 = 1.43 \times 10^{-13}$ cm. For all energies up to $+300$ MeV. $V(E_i)$ is underestimated[3] by a factor of about 4.

From these remarks it is clear that the BORN approximation calculations are not good enough and must be improved. A possible way to make this improvement has been suggested in Chap. VII. There (Sect. 32) we noted the paradox that shell-model wave-functions make good predictions about many observed nuclear features, but fail badly to give enough total binding energy, and it was proposed that this paradox is due to the fact that the binding energy is especially sensitive to the correlations in the nuclear wave-function. These correlations certainly exist but are ignored in the shell-model wave-functions, and also in BORN approximation calculations.

With this last fact in mind, one is encouraged to try to introduce small correlation effects into the calculation. The direct way to do this is to perform the

---

[1] R. HUBY: Proc. Phys. Soc. Lond. A **62**, 62 (1949).

[2] A. KIND and C. VILLI: Nuovo Cim., Ser. X **1**, 749 (1955).

[3] The apparent agreement with experiment in the published paper is due to an error in the theoretical expression for V_i which is four times too large. We note also that, if r_0 were taken as 1.20×10^{-13} instead of 1.45×10^{-13} cm, then V_i would be increased, but not sufficiently to compensate the factor 4.

averaging of $\sum_j v_{ij}$ using a modified FERMI gas in which the product of plane waves is multiplied by a product of two-particle correlation functions. However, this kind of approach has only been applied to the present problem on a limited scale[1]. The main trouble is that, in any realistic approach with saturating (exchange) forces, one must introduce, not one, but several correlation functions corresponding to the various spin states of a pair of nucleons. Furthermore, each function ought to be allowed to depend on relative momentum $\boldsymbol{k}_{ij} = \boldsymbol{k}_i - \boldsymbol{k}_j$ besides $\boldsymbol{r}_{ij} = \boldsymbol{r}_i - \boldsymbol{r}_j$. This implies almost prohibitive computations. Because of this trouble, most attention has been focussed on a quite distinct theory in which, instead of altering the wave-functions, one alters the potentials v_{ij}.

This new approach, the so-called BRUECKNER Theory[2], was stimulated by the failure of the BORN approximation type calculation for the nuclear binding energy[2]. In a calculation of the later type for an infinite medium, one uses a FERMI gas to evaluate the expectation value of $\sum_{i<j} v_{ij}$, with the result:

$$\left\langle \sum_{i<j} v_{ij} \right\rangle = \sum_{i<j} (\boldsymbol{k}_i \boldsymbol{k}_j |\, v_{ij} \,| \boldsymbol{k}_i \boldsymbol{k}_j) - (\boldsymbol{k}_i \boldsymbol{k}_j |\, v_{ij} \,| \boldsymbol{k}_j \boldsymbol{k}_i)$$

where $\boldsymbol{k}_i$ and $\boldsymbol{k}_j$ label the normalised plane wave states. The two matrix elements are simply proportional to the BORN approximation to the direct and exchange forward scattering amplitudes of nucleon-nucleon scattering. The suggestion of the original BRUECKNER Theory[3] was that one might obtain a better answer by replacing these amplitudes by the actual observed scattering amplitudes. However this step leads to some inconsistencies because the scattering of two nucleons inside nuclear matter must differ from that between free nucleons. (For instance, the presence of other nucleons affects the scattering because of the PAULI principle.) Thus the original version of the theory has been somewhat changed to cater for this difficulty[4]. At present as far as practical applications are concerned, the theory can be summarised thus: for each pair of particles the potential $v_{ij} \equiv v(\boldsymbol{r}_{ij}, \boldsymbol{p}_{ij})$ is replaced by an operator $t_{ij} \equiv t(\boldsymbol{r}_{ij}, \boldsymbol{p}_{ij})$ chosen so that the matrix element $(\boldsymbol{k}_i \boldsymbol{k}_j |\, t_{ij} \,| \boldsymbol{k}'_i \boldsymbol{k}'_j)$ is essentially the *self-consistently determined* scattering amplitude for particles i and j scattering from the state of relative momentum $\boldsymbol{k}_{ij} = \boldsymbol{k}_i - \boldsymbol{k}_j$ to the state of relative momentum $\boldsymbol{k}'_{ij} = \boldsymbol{k}'_i - \boldsymbol{k}'_j$ *inside nuclear matter*. (For scattering on the energy shell: $|\boldsymbol{k}_{ij}|^2 = |\boldsymbol{k}'_{ij}|^2$.) Since it is not yet known how to determine these self-consistent amplitudes, $t(\boldsymbol{r}_{ij}, \boldsymbol{p}_{ij})$ is a very ill-known quantity and can be assigned rather arbitrary properties. In practice, in order to reduce the number of parameters, one can assume that the matrix elements of t_{ij} have the form[5]

$$(\boldsymbol{k}_i \boldsymbol{k}_j |\, t_{ij} \,| \boldsymbol{k}'_i \boldsymbol{k}'_j) = \frac{1}{\Omega} (t_0 + t_{1e} k_{ij}^2 + t_{1o} \boldsymbol{k}_{ij} \cdot \boldsymbol{k}'_{ij})$$

where Ω is the volume of the large FERMI box in which the states are normalised. The subscripts e and o identify the contributions of even and odd state scattering respectively, t_0, t_{1e}, t_{1o} are (unknown) parameters that are supposed to be independent of $\boldsymbol{k}_{ij}$ and $\boldsymbol{k}'_{ij}$. Now the expectation value of any two-body operator

[1] S. D. DRELL and K. HUANG: Phys. Rev. **91**, 1527 (1953).

[2] BRUECKNER, LEVINSON and MAHMOUD: Phys. Rev. **95**, 217 (1954).

[3] BRUECKNER, LEVINSON and MAHMOUD: Phys. Rev. **95**, 217 (1954). — K. A. BRUECKNER: Phys. Rev. **96**, 508 (1954).

[4] K. A. BRUECKNER: Phys. Rev. **97**, 1353 (1955).

[5] We can assume that t_{ij} is defined to take account of exchange scattering, so that there is no need to introduce any exchange matrix elements. In addition, t_{ij} may be assumed to contain the result of averaging over the possible spin states of a pair of particles, so no explicit reference to particle spins is necessary.

taken in a FERMI gas reduces to a sum of two-particle matrix elements diagonal in $\boldsymbol{k}_i$, $\boldsymbol{k}_j$. Thus in calculations with an unexcited FERMI gas, it is only necessary to assume a form for these diagonal matrix elements. On specialising the above form to the diagonal (i.e. forward scattering) part, we have

$$(\boldsymbol{k}_i\,\boldsymbol{k}_j|\,t_{ij}\,|\boldsymbol{k}_i'\,\boldsymbol{k}_j') = \frac{1}{\Omega}\,(t_0 + t_1\,k_{ij}^2)$$

where we write t_1 for $(t_{1e}+t_{1o})$.

We now apply the BRUECKNER Theory to the calculation of the nuclear volume binding energy. In the prescribed way, we replace v_{ij} by t_{ij} so that, with the above form for the matrix elements of t_{ij}, we have:

$$\begin{aligned}\text{Energy per unit volume} &= \frac{1}{\Omega}\left\langle\sum_i T_i + \frac{1}{2}\sum_{i\neq j} t_{ij}\right\rangle,\\ &= \frac{1}{\Omega}\sum_i \frac{\hbar^2 k_i^2}{2M} + \frac{t_0}{2\Omega^2}\left(\sum_i 1\right)^2 + \frac{t_1}{\Omega^2}\left(\sum_i k_i^2\right)\left(\sum_i 1\right),\\ &= \left(\frac{\hbar^2}{2M} + t_1\,\varrho\right)\tau + \frac{1}{2}\,t_0\,\varrho^2\end{aligned}$$

where we define ϱ and τ as the volume density of particles and of square wave-number as in Sect. 33:

$$\varrho = \frac{1}{\Omega}\sum_i 1;\quad \tau = \frac{1}{\Omega}\sum_i k_i^2.$$

We note that the label i runs over all particles, i.e. those in all four intrinsic spin states. The binding energy per particle, B say, follows immediately from the energy density:

$$-\,B(\tau,\varrho) = \left(\frac{\hbar^2}{2M} + t_1\,\varrho\right)\left(\frac{\tau}{\varrho}\right) + \frac{1}{2}\,t_0\,\varrho\,.$$

This expression is very similar to that given in Sect. 33 in terms of α, β. It is more specific in the sense that, if t_0 and t_1 are assumed independent of ϱ, the dependence on ϱ is quite explicit, whereas in Sect. 33 it was not, because the dependence of α and β on ϱ was not determined. This means that one can now state a new condition on the parameters, viz. that the minimum value of B is achieved at the observed equilibrium density, $\varrho=\varrho_0$ (say). Using the THOMAS-FERMI relation: $\frac{\tau}{4} = \frac{3}{5}\,(6\pi^2)^{\frac{2}{3}}\left(\frac{\varrho}{4}\right)^{\frac{5}{3}}$ and putting $\frac{dB}{d\varrho}=0$ gives:

$$0 = -\,\varrho\,\frac{dB}{d\varrho} = \left(\frac{2}{3}\,\frac{\hbar^2}{2M} + \frac{5}{3}\,t_1\,\varrho\right)\left(\frac{\tau}{\varrho}\right) + \frac{1}{2}\,t_0\,\varrho\,.$$

Taking ϱ_0 as $\left(\frac{4}{3}\,\pi\,r_0^3\right)^{-1}$ with $r_0=1.20\times10^{-13}$ cm, the value of $\frac{\hbar^2}{2M}\,\frac{\tau}{\varrho}$ is 21 MeV; on inserting this number along with $B=15$ MeV in the last two equations, we find:

$$21\,t_1\,\varrho\left(\frac{2M}{\hbar^2}\right) + \frac{1}{2}\,t_0\,\varrho = -\,36$$

$$35\,t_1\,\varrho\left(\frac{2M}{\hbar^2}\right) + \frac{1}{2}\,t_0\,\varrho = -\,14$$

with solutions:

$$t_0\,\varrho = -\,138,\quad t_1\,\varrho = \frac{11}{7}\,\frac{\hbar^2}{2M}\,.$$

The mean potential $\sum_j t_{ij}$ felt by the i-th nucleon is evidently:

$$V(k_i^2) = \varrho\, t_0 + t_1 \tau + \varrho\, t_1 k_i^2$$

or, in terms of E_i, on using $E = T + V$:

$$V(E_i) = \left(1 + \varrho\, t_1 \left(\frac{2M}{\hbar^2}\right)\right)^{-1} \left(\varrho\, t_0 + t_1 \tau + \varrho\, t_1 \left(\frac{2M}{\hbar^2}\right) E_i\right).$$

Comparing this with the equation $V(E_i) = \alpha + \beta E_i$ of Sect. 33, we have:

$$\alpha = \frac{\varrho t_0 + t_1 \tau}{1 + \varrho t_1 \left(\frac{2M}{\hbar^2}\right)}; \qquad \beta = \frac{\varrho t_1 \left(\frac{2M}{\hbar^2}\right)}{1 + \varrho t_1 \left(\frac{2M}{\hbar^2}\right)}$$

as the relation between the old pair of parameters, α, β and the new pair t_0, t_1. The above numerical values for t_0 and t_1 imply that $\alpha = -40\frac{5}{6}$ MeV, $\beta = \frac{11}{18}$.

These values agree closely with those deduced in Sect. 33 and so we have the rather striking result that, with two parameters, one can fit the nuclear volume energy, the equilibrium density and the constant and linear terms in the observed refractive potential. Also one can qualitatively account for the observed difference in neutron and proton well depths mentioned in Sect. 34, and also the observed symmetry energy subject to the uncertainties mentioned in Sect. 33. In spite of this agreement, one cannot claim at present that the BRUECKNER Theory gives any fundamental explanation of the various quantities that are fitted. This will only be possible when t_0 and t_1, which we have treated as parameters, are calculated from the nucleon-nucleon interaction in a properly self-consistent way[1].

β) Finite systems. The calculation of $V(E_i, r_i)$ and nuclear binding energy using independent particle wave functions for a finite system implies use of the HARTREE-FOCK self-consistent field method. This method enables one to establish the self-consistent mean potential by variation of the total energy with respect to the form of the single particle wave-functions. The minimal condition has the form of a SCHRÖDINGER equation for these wave-functions and gives an explicit form for $V(E_i, r_i)$ in terms of the v_{ij}. This type of calculation has been done, both in the usual way in which some form for v_{ij} is assumed, and also in the spirit of the BRUECKNER Theory where the v_{ij} are replaced by the t_{ij}.

In one reported calculation[2] on a hypothetical mass 184 system of 92 protons and 92 neutrons, the potential v_{ij} was taken in two forms:

$$v_{ij} = (1 + x P_{ij}^r)\,(-55.6\ \text{MeV})\, e^{-(r_{ij}/\sqrt{2})^2},$$

$$v_{ij} = (1 + x P_{ij}^r)\,(-25.4\ \text{MeV})\, \frac{e^{-(r_{ij}/1.4)}}{(r_{ij}/1.4)},$$

where the radial separation r_{ij} is in units of 10^{-13} cm. The usual saturation conditions in this case reduce to $x \geq 4$. The self-consistent method was found to rapidly converge at $x = 4$ when the zero-order mean potential was taken to be a square well of depth 32 MeV, radius $1.40\, A^{\frac{1}{3}} \times 10^{-13}$ cm. Values of $x < 4$ were found to lead to collapse and values > 4 to a less attractive mean potential.

[1] H. A. BETHE: Phys. Rev. **103**, 1353 (1956).
[2] M. ROTENBERG: Phys. Rev. **100**, 439 (1955).

Another calculation[1] was for a mass 150 system of 75 neutrons and 75 protons. This attempt began with a rectangular density distribution and two cycles were performed. There was no evident convergence but a tendency towards collapse as was expected from the use of ordinary (i.e. non-saturating) forces.

The important feature of these results for our purposes is that the implied values of $V(E_i) = \alpha + \beta E_i$ and the total nuclear energy are too small. In the first calculation, no attempt was made to obtain α and β separately but only the mean value of $V(E_i)$ averaged over particles i. The observed mean value computed from $\alpha = -41$ MeV, $\beta = \frac{11}{18}$ is about 75 MeV, whereas the self-consistent calculation became stable at about 33 MeV. Thus the usual independent particle type calculations badly underestimate potential energies for the case of finite systems as well as infinite ones.

The remedy that has been suggested for the infinite case, i.e. the BRUECKNER Theory, has also been applied to the finite case[2]. As might be expected, in the finite case, we can no longer work only with the diagonal matrix elements of $t(\boldsymbol{r}, \boldsymbol{p})$, but we must now give some explicit spatial dependence to $t(\boldsymbol{r}, \boldsymbol{p})$. One can choose this to be the simplest expression involving δ-functions of position that gives the previous expression for $(\boldsymbol{k}_i \boldsymbol{k}_j|\, t_{ij}\, |\boldsymbol{k}'_i \boldsymbol{k}'_j)$ viz. $t_0 + t_{1e} k_{ij}^2 + t_{1o} \boldsymbol{k}_{ij} \cdot \boldsymbol{k}'_{ij}$. Such an expression is

$$t(\boldsymbol{r}, \boldsymbol{p}) = t_0\, \delta(\boldsymbol{r}) + \frac{1}{2} t_{1e} \left[\delta(\boldsymbol{r}) \left(\frac{\boldsymbol{p}}{h}\right)^2 + \left(\frac{\boldsymbol{p}}{h}\right)^2 \delta(\boldsymbol{r})\right] + t_{1o} \left[\left(\frac{\boldsymbol{p}}{h}\right) \cdot \delta(\boldsymbol{r}) \left(\frac{\boldsymbol{p}}{h}\right)\right]$$

where, as usual, the momentum $\boldsymbol{p} = i\hbar \nabla$. Unlike the prediction of volume quantities, the discussion of surface effects does not always involve t_{1e} and t_{1o} in the combination $t_1 = t_{1e} + t_{1o}$ and so we effectively have a new parameter. If both of t_{1e} and t_{1o} are regarded as parameters, it is found to be possible to fit both the observed nuclear surface thickness and surface energy as well as the volume density and energy. (The prediction of the surface quantities involves t_{1e} and t_{1o} in the combination $t_{1e} - \frac{1}{3} t_{1o}$, and the fitting implies that $\varrho_0 (t_{1e} - \frac{1}{3} t_{1o}) = 0.5 \left(\frac{\hbar^2}{2M}\right)$.

If one is guided by the weakness of the observed p-wave contribution to the free nucleon-nucleon scattering and puts $t_{1o} = 0$, it follows that it is not possible to fit both the volume and surface magnitudes. Thus one must introduce some new freedom into the theory. This has been done by allowing t_0 to be a function of density. Such a step has radical implications and destroys many simple features of the theory. For instance, let us again consider an infinite system. The new freedom does not alter the fact that the total energy is $\left\langle \sum_i T_i + \frac{1}{2} \sum_{i \neq j} t_{ij} \right\rangle$ and it follows that the above expression for B as a function of τ and ϱ is still correct. In the self-consistent field method, τ and ϱ are both regarded as being functions of the single particle wave functions φ_i:

$$\tau = \frac{1}{\Omega} \sum_i |\nabla \varphi_i|^2, \quad \varrho = \frac{1}{\Omega} \sum_i |\varphi_i|^2$$

and so the total energy and B are expressible in terms of the φ_i. Minimisation of B with regard to the form of the $\varphi_i(r)$ gives the following condition:

$$\left(\frac{\hbar^2}{2M} + \varrho\, t_1\right) \nabla^2 \varphi_i + V' \varphi_i = E\, \varphi_i$$

---

[1] J. D. TALMAN: Phys. Rev. **102**, 455 (1956).

[2] T. H. R. SKYRME: Phil. Mag., Ser. VIII **1**, 1043 (1956).

where V' is thereby derived as the best self-consistent potential equal to $-\frac{\partial(\varrho B)}{\partial\varrho}$. (The partial differentiation with respect to ϱ regards τ as an independent variable.) We see that, when t_0 and t_1 are regarded as independent of ϱ we immediately obtain the same expression for $V(E_i)$ as was obtained from $V(E_i) = \frac{1}{\Omega}\langle\sum_j t_{ij}\rangle$,

$$V(E_i) = \varrho t_1 k_i^2 + V' = \varrho t_1 k_i^2 + \varrho t_0 + t_1 \tau.$$

When t_0 is allowed to depend on ϱ, there is an extra term $+\frac{\varrho^2}{2}\frac{\partial t_0}{\partial\varrho}$. This implies that, in this case, the best self-consistent potential is no longer simply $\frac{1}{\Omega}\langle\sum_j t_{ij}\rangle$. It also implies that the total energy can no longer be written $\langle\sum_i T_i + \frac{1}{2}\sum_i V(E_i)\rangle$. These changes run quite contrary to one's usual ideas, but they are a direct consequence of allowing t_0 to depend on density.

It is interesting to see that, in spite of these complications, the new freedom does not change a simple corollary of the HARTREE-FOCK Theory, viz. the highest nucleon energy is equal to minus the ionization energy. Still considering the infinite system we have:

$$V' = -\frac{\partial(\varrho B)}{\partial\varrho} = -B - \varrho\frac{\partial B}{\partial\varrho} = -B - \varrho\left(\frac{dB}{d\varrho} - \frac{\partial B}{\partial\tau}\frac{\partial\tau}{\partial\varrho}\right).$$

Now:

$$\frac{dB}{d\varrho} = 0$$

from the equilibrium condition, and:

$$\varrho\frac{\partial B}{\partial\tau} = -\left(\frac{\hbar^2}{2M} + t_1\varrho\right)$$

and

$$\left(\frac{\hbar^2}{2M}\right)\frac{\partial\tau}{\partial\varrho} = \text{maximum kinetic energy in the FERMI gas, } T_F.$$

Thus:

$$V' = -B - \left[1 + t_1\varrho\left(\frac{2M}{\hbar^2}\right)\right] T_F.$$

Since total particle energy $= T + V(T) = T\left[1 + t_1\varrho\left(\frac{2M}{\hbar^2}\right)\right] + V'$, the required result follows immediately. An analagous proof can be given for finite systems[1].

36. Prediction of the spin-orbit potential, *U*. It is natural that one should first try to explain the nuclear spin-orbit coupling by analogy with the atomic case. In the atom, a large part of the observed coupling is explained as a relativistic effect, the so-called "THOMAS coupling", given by the equation:

$$(\boldsymbol{l}\cdot\boldsymbol{s})\, U(r) = (\boldsymbol{l}\cdot\boldsymbol{s})\left(\frac{\hbar^2}{2M^2c^2}\right)\frac{1}{r}\frac{\partial V(r)}{\partial r}$$

where $V(r)$ is the single particle potential in which the electron of mass M moves. In the case of the nucleus, the same expression evaluated with M as the nucleon mass underestimates the observed coupling by an order of magnitude[2]. Thus one must look for some further source of the coupling. This source is presumably

[1] T. H. R. SKYRME: Phil. Mag., Ser. VIII 1, 1043 (1956).
[2] D. R. INGLIS: Rev. Mod. Phys. 25, 390 (1953).

a spin-orbit coupling in the nucleon-nucleon interaction. Such a coupling may either be vector or tensor in type. The easiest way to make a prediction of the potential $(\boldsymbol{l}\cdot\boldsymbol{s})\,U(r)$ felt by a "test" nucleon i in a nucleus is simply to evaluate the expectation value of all the two-body spin-orbit forces that the nucleon feels:

$$(\boldsymbol{l}\cdot\boldsymbol{s})\,U(r_i) = \langle \sum_j \boldsymbol{v}_{ij} \rangle.$$

If we assume $\boldsymbol{v}_{ij}$ to be a vector force:

$$\boldsymbol{v}_{ij} = J(r_{ij})\,(\boldsymbol{r}_{ij}\times\boldsymbol{p}_{ij})\cdot(\boldsymbol{s}_i+\boldsymbol{s}_j),$$

where $\boldsymbol{p}_{ij}$ is the relative momentum of particles i and j, then the above expectaction value for a closed shell nucleus can be easily shown to give[1]:

$$U(r_j) = -2\sum_{nlm}\left\langle J(r_{ij})\left[\frac{\boldsymbol{r}_i\cdot\boldsymbol{r}_j}{\boldsymbol{r}_j^2}-1\right]\right\rangle_{nlm}$$

where n, l, m are the quantum numbers of the single particle orbits in the closed shells. Here exchange integrals are ignored on the grounds that they are small for large nuclei. In the same situation, one can assume the range of J to be small compared with the size of the nucleus, and this leads to the final simple expression:

$$U = \left(\frac{4\pi r_0^5}{3}\right)\frac{U'}{r}\frac{d\varrho(r)}{dr}$$

where

$$U' = -\frac{1}{r_0^5}\int J(s)\,s^4\,ds,$$

(r_0 is the usual nuclear size constant) and where ϱ is the density of all particles in the nucleus. We see that this expression for U is similar to the THOMAS one except that $V(r)$ is replaced by $\varrho(r)$ and the magnitude is determined by the new quantity called U'. The observed spin-orbit splittings have been fitted with essentially two parameters, namely U' and the surface thickness of the nucleus, i.e. the distance in which $d\varrho(r)/dr$ is appreciable. The thickness turns out to be in reasonable accord with the evidence from the analysis of electron scattering experiments. The value of U' is 5.5 MeV. If one assumes a Gaussian form for $J(s)$: $J(s)=J_0\,e^{-(s/s_0)^2}$ with $s_0=1.94\times10^{-13}$ cm, the value of J_0 must be taken as -2.0 MeV. This is very close to the value found in an analysis of the nucleon-alpha interaction with a two body vector force. If one assumes a YUKAWA well $J(s)=J_0\frac{e^{-(s/s_0)}}{(s/s_0)}$ with $s=1.37\times10^{-13}$ cm then J_0 must be -1.2 MeV. This is near with the value found in the analysis of single particle and single hole splittings in light nuclei[2]. (In this analysis, for a value of the oscillator parameter b of 1.68×10^{-13} cm, J_0 is -1.5 MeV.)

An alternative possibility is that $\boldsymbol{v}_{ij}$ may be taken to be a tensor force. In this case however, there is no net spin-orbit force in first order for the situation of a "test" particle and a closed shell. One only finds non-zero effects by going to higher order and this immediately implies considerable labour and complexity. One attempt[3] along these lines obtained a net spin-orbit force, but of the wrong sign. Another reported investigation[4] concerned itself with second order tensor force effects on the spectra of individual nuclei. We have already discussed this

[1] R. J. BLIN-STOYLE: Phil. Mag. **46**, 973 (1955).

[2] J. P. ELLIOTT and A. M. LANE: Phys. Rev. **96**, 1160 (1954).

[3] J. KEILSON: Phys. Rev. **82**, 759 (1951).

[4] A. M. FEINGOLD: Phys. Rev. **101**, 258 (1956). — E. P. WIGNER: Symposium on New Research Techniques in Physics. Rio de Janeiro 1952.

work in some detail in Chap. VII, and we only mention here the striking agreement found between the second order effects of the tensor force and the effects of a simple spin-orbit force.

As things stand at present, the question as to whether the observed spin-orbit coupling felt by a nucleon in a nucleus is a first-order effect from a two-body vector force, or a second-order effect from tensor forces, or a combination of both, remains unsolved.

We note, in conclusion, that the BRUECKNER Theory discussed in Sect. 35 can be applied to the prediction of the nuclear spin-orbit force[1]. In this theory, the nucleon-nucleon potential does not matter as much as the nucleon-nucleon scattering amplitudes, because the nuclear dynamics are described solely in terms of the latter without explicit reference to the former. As we saw in Sect. 35, it is not consistent to use the actual observed nucleon-nucleon amplitudes, at least, not in discussing the mean refractive nuclear potential, V. This is because the scattering between two nucleons inside nuclear matter may be strongly affected by the presence of nearby particles because of the PAULI principle. However, for the prediction of the spin-orbit potential, one may ignore this complication in an initial calculation. According to the BRUECKNER Theory, the two-body potentials v_{ij} are replaced by operators $t_{ij} \equiv t(\boldsymbol{r}_{ij}, \boldsymbol{p}_{ij})$ when one takes expectation values in "model wave-functions" i.e. in independent particle states. Let us suppose that $t(\boldsymbol{r}, \boldsymbol{p})$ is chosen such that $\langle \boldsymbol{k} | t(\boldsymbol{r}, \boldsymbol{p}) | \boldsymbol{k}' \rangle$ is the observed scattering amplitude for scattering between states of relative momenta, $\boldsymbol{k}$ and $\boldsymbol{k}'$. Now such an amplitude can always be written as a sum of terms involving spins and momenta in various ways. Only one of these terms is relevant for the prediction of the spin-orbit potential and this is $\frac{i}{2} \boldsymbol{S} \times (\boldsymbol{k}' \times \boldsymbol{k})\, F(k^2, k'^2, \boldsymbol{k}' \cdot \boldsymbol{k})$ where $\boldsymbol{S}$ is the vector sum of the two intrinsic nucleon spins. For a given variation of F over restricted ranges of $\boldsymbol{k}$ and $\boldsymbol{k}'$, it is usually possible to find a two body vector force that closely reproduces this variation in first order BORN approximation. Also one can probably find a tensor force to do this in second order BORN approximation.

Conversely the results of a theory of the nucleus that treats a vector force in first BORN approximation or a tensor force in second BORN approximation can be reproduced by the BRUECKNER Theory by choosing an appropriate form for F. The important advantage and power of the BRUECKNER Theory lies in its claim to still be valid in those situations when the effects of two body potentials are not at all well represented by BORN approximation. According to reported calculations, when the BRUECKNER Theory is used in conjunction with a phase-shift analysis of the observed nucleon-nucleon scattering, it predicts a very reasonable nuclear spin-orbit potential both in magnitude and form[1]. In these calculations, F was given the form

$$F(k^2, k'^2, \boldsymbol{k} \cdot \boldsymbol{k}') = a - \tfrac{1}{2}(b\, k^2 + b^*\, k'^2) + c\, \boldsymbol{k} \cdot \boldsymbol{k}'$$

with the values of the constants a, b and c determined by phase-shift analysis of observed nucleon-nucleon scattering from 40 to 100 MeV. Such a form with roughly correct values can be shown to result from a first order treatment of a short-range vector force. However, from our remarks above, it should be clear that this fact provides no immediate evidence for the existence of a two-body vector force or for the validity of a first order treatment.

[1] J. S. BELL and T. H. R. SKYRME: Phil. Mag., Ser. VIII 1, 1055 (1956).

37. Estimates of configuration mixing. The crucial task in trying to establish a theoretical basis for a shell-model is to show that the amount of configuration mixing is either small or unimportant. As we have pointed out in Sect. 33, we may consider the phenomenological absorption potential W to be a measure of this mixing. Consequently one may either try to show directly that configuration mixing is small, or else one may try to evaluate W and show that it is small. Both approaches have been attempted. In the present section, we will discuss those attempts that have been made to estimate directly the degree to which shell-model states are mixed. In the following section, we will describe attempts to do the same thing indirectly by estimating the absorption potential W.

In testing the validity of the shell-model, we are trying to assess the correctness of certain zero-order states in the presence of mixing interactions. In particular, we wish to know if these interactions can be treated as perturbations. In this connection, we note that, according to most presentations of perturbation theory, the zero-order wave-function is supposed to be "close" to the actual wave-function. It is usually not made clear whether the states must be "close" to each other in the strong or weak sense, i.e. whether they must be close in detail or only on the average. It may be so that, for perturbation theory to be valid it is necessary that the actual and zero-order wave-functions be close in the weak sense, i.e. on the average, so that they have a good overlap, but, in certain restricted regions in the configuration space of the particles, they may be completely different. The distinction between the strong and weak definitions of "closeness" is very important in testing the shell-model because the presence of strong short-range interactions between nucleons means that the actual wave-function can differ very much from the shell-model ones, but only in restricted regions of space so that the percentage impurity in the shell-model state may be small.

We will introduce below two quantities that can be said to represent the "closeness" between the actual and zero-order states. One of these, N_{00}, gives the amount of admixture to the zero-order state and so can be said to correspond to the weak meaning of "close". The other quantity $M_{00}^{(2)}$ is much more sensitive to the strong short-range deviations between the wave-functions and so can be said to correspond to the strong meaning.

α) Estimate of the total admixture to zero-order states. Let us consider a single particle moving in an orbit, say $u_0(r)$, in the presence of a "core" nucleus in its ground state ψ_0. The wave function for the entire system is then $\psi_0 u_0(r)$, which is one particular state of the complete set $\psi_c u_p(r)$. The states of an actual nucleus, X_λ say, also form a complete set and can, of course, be regarded as linear combinations of the $\psi_c u_p$:

$$X_\lambda = \sum_{cp} a_{cp}^\lambda (\psi_c u_p), \quad \text{where} \quad \sum_{cp} |a_{cp}^\lambda|^2 = 1.$$

Now if the low lying nuclear states are approximately like independent particle states, we expect that each proper state λ can be taken as a zero-order state $\psi_c u_p$. In particular, for the ground state:

$$X_0 \approx \psi_0 u_0.$$

To establish the consistency of this picture, it remains to show that configuration mixing is small, i.e. that the square integral of all the admixed states $\psi_c u_p$ is small. In terms of the a_{cp}^λ, we wish to show that:

$$N_{00} = \sum_{cp \neq 00} |a_{cp}^{\lambda=0}|^2 \ll 1.$$

The value of N_{00} can be computed, using perturbation theory to obtain the coefficients $a_{cp}^{\lambda=0}$:

$$N_{00} = \sum_{cp \neq 00} \left| \frac{(0\,0|\,H'\,|cp)}{E_{00} - E_{cp}} \right|^2$$

where H' is the perturbing interaction:

$$H' = \sum_i v(|\boldsymbol{r} - \boldsymbol{r}_i|) - V(r)$$

and where E_{cp} is the zero-order-energy of the state cp. In this expression, the first term is the actual potential seen by the particle in the orbit $u_0(r)$, and the second is the average potential which determines $u_0(r)$. On representing the state ψ_0 by a FERMI gas at zero temperature and the states u_p by plane waves, the value of N_{00} can be calculated if we are given some reasonable two-particle interaction $v(|\boldsymbol{r} - \boldsymbol{r}_i|)$. The states c that give non-zero matrix elements are those in which one particle (say i) is raised from its state $(\boldsymbol{k}_i, \chi_i)$ in the gas to a state $(\boldsymbol{k}_i', \chi_i')$ above the gas where $\boldsymbol{k}$ denotes wave-number vector and χ intrinsic spin state. We can represent the two corresponding states of the single particle by plane wave states $(\boldsymbol{k}, \chi)$ and $(\boldsymbol{k}', \chi')$. If the normalisation volume is Ω, a typical direct matrix element is:

$$(0\,0|\,H'\,|cp) = \frac{1}{\Omega^2} \int e^{i\boldsymbol{k}\cdot\boldsymbol{r} + i\boldsymbol{k}_i\cdot\boldsymbol{r}_i} \chi\chi_i v(|\boldsymbol{r} - \boldsymbol{r}_i|)\, e^{-i\boldsymbol{k}'\cdot\boldsymbol{r} - i\boldsymbol{k}_i'\cdot\boldsymbol{r}_i} \chi'\chi_i'\, d\boldsymbol{r}\, d\boldsymbol{r}_i$$

$$= \frac{1}{\Omega^2} \left[\int e^{i(\boldsymbol{k}+\boldsymbol{k}_i-\boldsymbol{k}'-\boldsymbol{k}_i')\cdot\boldsymbol{R}}\, d\boldsymbol{R} \right] \left[\int e^{i(\boldsymbol{k}-\boldsymbol{k}_i-\boldsymbol{k}'+\boldsymbol{k}_i')\cdot\frac{1}{2}\boldsymbol{\varrho}_i} \chi\chi_i v(\boldsymbol{\varrho}_i)\, \chi'\chi_i'\, d\boldsymbol{\varrho}_i \right]$$

where

$$\boldsymbol{R} = \tfrac{1}{2}(\boldsymbol{r} + \boldsymbol{r}_i), \qquad \boldsymbol{\varrho}_i = \boldsymbol{r} - \boldsymbol{r}_i$$

(The corresponding exchange matrix element can be written down similarly.) The sum over cp may be written as:

$$\sum_{cp} \rightarrow \left[\sum_{\chi'} \int \frac{\Omega\, d\boldsymbol{k}'}{(2\pi)^3} \right] \left[\sum_{\chi_i} \int \frac{\Omega\, d\boldsymbol{k}_i}{(2\pi)^3} \right] \left[\sum_{\chi_i'} \int \frac{\Omega\, d\boldsymbol{k}_i'}{(2\pi)^3} \right]$$

where the symbol $\sum_\chi$ denotes sum over the four intrinsic nucleon spin states.

$$N_{00} = (2\pi)^6 \sum_{\chi'} \sum_{\chi_i} \sum_{\chi_i'} \iiint \frac{d\boldsymbol{k}'}{(2\pi)^3} \frac{d\boldsymbol{k}_i}{(2\pi)^3} \frac{d\boldsymbol{k}_i'}{(2\pi)^3} \frac{[\chi\chi_i v(\boldsymbol{K})\chi'\chi_i']^2}{(E00 - E_{0p})^2} \cdot \delta(\boldsymbol{k} + \boldsymbol{k}_i - \boldsymbol{k}' - \boldsymbol{k}_i')$$

where $\boldsymbol{K}$ is written for $\frac{1}{2}(\boldsymbol{k} - \boldsymbol{k}_i - \boldsymbol{k}' + \boldsymbol{k}_i')$ and where $v(\boldsymbol{k})$ is defined as the three-dimensional FOURIER transform of $v(\varrho)$:

$$v(\boldsymbol{K}) = \frac{1}{(2\pi)^{\frac{3}{2}}} \int e^{i\boldsymbol{K}\cdot\boldsymbol{\varrho}}\, v(\varrho)\, d\boldsymbol{\varrho}.$$

It only remains to specify the energy denominators. If the mean potential V is assumed velocity independent, we can write

$$E_{00} - E_{cp} = \frac{\hbar^2}{2M} (k'^2 + k_i'^2 - k^2 - k_i^2).$$

If the potential is assumed velocity dependent in the manner considered in Sect. 33, $V = \alpha + \beta E$, then the separations will be increased by a factor determined by β.

The only reported evaluation[1] of N_{00} gives $N_{00}=0.14$ for a typical nucleon. There are also two related evaluations. One of these[2] considered the problem of the probability of a test particle being scattered by a set of scattering centres in an artificially fixed state that is not allowed to change. This situation is represented in the calculation by the assumed absence of momentum conservation in collisions, and the evaluation of the energy differences for the test particle only. Both of these differences will give larger values for N_{00}, mainly because they prevent the PAULI principle from acting with full power. (If momentum conservation is imposed, in each collision a target particle must be scattered right out of the gas. This eliminates a large contribution to $\int d\boldsymbol{k}_i$, and leaves only terms with large energy denominators.) Thus it is not surprising that this calculation gave a very large value for N_{00}, viz. $N_{00}\sim 2$, even after allowing a reduction for a factor of 4 for the velocity dependence of the mean potential.

The second related calculation[3] was made within the framework of the BRUECKNER Theory, and is identical with the one we describe apart from, the fact that reactance amplitudes $(\boldsymbol{k}\boldsymbol{k}_i|\,t\,|\boldsymbol{k}'\boldsymbol{k}_i')$ are used instead of the matrix elements $(\boldsymbol{k}\boldsymbol{k}_i|\boldsymbol{v}|\boldsymbol{k}'\boldsymbol{k}_i')$. The amplitudes are functions of $\boldsymbol{k}-\boldsymbol{k}_i$ and $\boldsymbol{k}'-\boldsymbol{k}_i'$ instead of just $\boldsymbol{k}-\boldsymbol{k}'=\boldsymbol{k}_i-\boldsymbol{k}_i'$ in the case of our matrix elements. The results indicate that the particles near the top of the gas are very litle excited as a result of interactions with other particles, the square integral of admixtures being only a few percent. It seems reasonable that these qualitative results would not be changed if t were replaced by $\boldsymbol{v}$ of about the same mean strength.

As a last remark, we note that the smallness of the second-order admixture to a zero-order state does not rule out absolutely the possibility of large admixtures in higher order.

β) Estimates of corrections to the energy. We now briefly consider an extension of the above calculation to include the estimation of the first and second-order corrections to the total energy of the state $\psi_0 u_0$. These have been computed[4] using just the same FERMI gas methods as those just described. One is interested in the corrections to the zero-order energy E_{00}:

$$\text{First Order:}\qquad (00|\,H'\,|00)\,,$$

$$\text{Second Order:}\qquad \sum_{cp\neq 00}\frac{|(00|\,H'\,|cp)|^2}{E_{00}-E_{cp}}\,.$$

In evaluating these terms one would normally define V as $\sum_i \boldsymbol{v}(\boldsymbol{r}-\boldsymbol{r}_i)$, thereby making $(00|\,H'\,|00)$ vanish. This definition of V then enables one to specify the energy separations $E_{00}-E_{cp}$, and evaluate the second term. In the published calculations, V has been implicitly chosen to be velocity independent, although satisfying $(00|\,H'\,|00)=0$. It was found that the second-order term was much larger than E_{00}. This does not imply necessarily that the perturbation treatment is bad. In the first place, E_{00} is fortuitously small as a result of the close cancellation of the kinetic and potential energy terms. In the second place, it has been pointed out[5] that one may actually make the sum of the first and second order terms vanish by an appropriate choice of the energy dependence of V. This

[1] S. WATANABE: Z. Physik **113**, 482 (1939).
[2] E. P. WIGNER: L. Farkas Memorial Volume. Research Council of Israel **1**, 45 (1951).
[3] K. A. BRUECKNER: Phys. Rev. **103**, 172 (1956). — L. ROSENFELD: Nuclear Forces, p. 252. Amsterdam: North Holland Publishing Co. 1948.
[4] R. HUBY: Proc. Phys. Soc. Lond. A **62**, 62, (1949).
[5] W. J. SWIATECKI: Phys. Rev. **101**, 1321 (1956).

last fact does not, of course, make the perturbation treatment necessarily valid in any real way. No basic test of the perturbation treatment is provided by the size of the energy correction up to a given order, because this can be made arbitrarily large or small by an appropriate choice of V. A more satisfactory check on the validity of the perturbation theory is to compute the ratios of the subsequent *potential* energy terms[1]; in particular, the ratio of the second-order term to the sum of the zeroth and first-order terms has been established as a function of density. This ratio is only a few percent at observed nuclear densities. Since this ratio cannot be adjusted arbitrarily in the way the total energy terms can, this result is positive evidence in favour of the perturbation theory. Of course, the most convincing check on the validity of the theory comes from the direct estimation of the admixture to the zero-order wave-function in the way described above. If such an admixture were large, shell-model wave-functions must really be very poor, although they may give a correct result for a particular observed quantity. If the admixture is small, then, always assuming that higher-order effects are small, the shell-model wave-functions are close to the actual ones on the whole.

γ) Estimate of spreading of zero-order states. We now describe a calculation that is complementary to the ones just described. Instead of estimating the degree to which other states are admixed into a given state, we now estimate the degree to which a given state is spread or dissolved amongst other states. It is clear that these two effects are expected to be connected. When the shell-model is a very good approximation both are very small; when it is bad, both are large. If other states are mixed into a zero-order state, normalisation alone $\left(\sum_\lambda |a^\lambda_{cp}|^2 = 1\right)$ implies that a corresponding part of that state must be mixed into other states. In practice we use the so-called "Second Moment", defined for any given zero-order state cp as[2]

$$M^{(2)}_{cp} = \sum_\lambda (E_{cp} - E_\lambda)^2 \, |a^\lambda_{cp}|^2$$

for representing the degree to which a given zero-order state $\psi_c u_p$ is spread amongst the proper states X_λ. Now the coefficients a^λ_{cp} are given quite generally by:

$$a^\lambda_{cp} = \frac{(cp\,|H'|\,\lambda)}{E_{cp} - E_\lambda}.$$

Thus we can use closure to obtain $M^{(2)}_{cp}$ in the form:

$$M^{(2)}_{cp} = (cp\,|H'^2|\,cp).$$

The calculated value of $M^{(2)}_{00}$ is of the order of (20 MeV)^2. If the distribution of the $|a^\lambda_{00}|^2$ as a function of E_λ is assumed not to have a long tail but rather to be fairly rectangular, then this value indicates that the zero-order state $\psi_0 u_0$ is spread over an energy range of $\sim$20 MeV by the interactions H'. This means that single particle states, which are only separated by $\sim$10 MeV, are completely mixed into each other, with the consequence that the shell-model cannot be valid.

However, it is possible to make an alternative interpretation of the large size of $M^{(2)}_{00}$. This interpretation is that the distribution of $|a^\lambda_{00}|^2$, while being confined in the main around $E_\lambda \sim E_{00}$, has long tail. Such a situation might be expected from the nature of nuclear forces, which are of short-range but very strong. Their short-range of action means that collisions are not very common (especially because of inhibition from the PAULI principle). On the other hand,

[1] W. J. SWIATECKI: Phys. Rev. **103**, 265 (1956).
[2] LANE, THOMAS and WIGNER: Phys. Rev. **98**, 1524 (1955).

the great strength implies that, when a collision does take place, it is liable to lead to violent energy exchange with a resulting admixing of states far removed in energy.

The correctness of this latter interpretation is suggested by the divergence of moments higher than the second. It is partially confirmed by a revised calculation[1] which attempts to remove the effects of the long tail from $M_{00}^{(2)}$. To do this, we work with new zero-order wave-functions in which limited correlations are allowed between the plane waves of the independent particle states. These correlations represent the strong interactions of pairs of nucleons which gave rise to the long tail in the distribution of $|a_{00}^{\lambda}|^2$. It is possible to reduce $M_{00}^{(2)}$ by a factor of about 10 by choosing an appropriate correlation function. Although the value of $M_{00}^{(2)}$ remains rather large, this important reduction gives one hope that the calculation may be reconciled with the shell-model.

38. Prediction of the absorbing potential W. In contrast to the direct prediction of configuration mixing in the last section, the prediction of the absorbing potential W is directed at definite experimental values rather than upper limits (see Sect. 34). We will begin by describing various attempts that have been made to predict W, then we will relate the forms of these predictions to the quantities representing configuration mixing from the last section.

A very simple method for estimating W is suggested by the relation between W and mean free path against collision that we mentioned in Sect. 33[2]. By analogy with classical optics, this mean free path, λ, can be evaluated as:

$$\lambda^{-1} = \varrho \langle \sigma \rangle$$

where ϱ is the volume density of nucleons in nuclear matter and $\langle \sigma \rangle$ is the nucleon-nucleon cross-section averaged over all types of colliding nucleons and relative energies of collision. This immediately gives for the absorption potential for a nucleon of energy E:

$$W(E) = \tfrac{1}{2} \hbar \, v \, \varrho \langle \sigma \rangle$$

where v is the velocity of the nucleon inside nuclear matter. One can now compute W from the observed values for ϱ and σ. This yields much too large a value for W for a nucleon with energy near the top of the Fermi gas. However this large value can be reduced by taking the Pauli principle into account. Doing this implies that, in evaluating $\langle \sigma \rangle$, one only averages over those collisions leading to unoccupied final states. This now gives very reasonable values for W, both in energy dependence and absolute magnitude[3]. In particular, it is evident that W must fall to zero as the energy of the nucleon decreases from positive energies towards the top of the Fermi gas.

We might refer to this calculation as a "classical" estimate of W since, apart from the modification due to the Pauli principle, it is just the type of estimate made in classical optics in which one equates the "macroscopic" absorption probability $2W/\hbar$ to the "microscopic" one, $v\varrho\langle\sigma\rangle$. It is important that one should not regard this estimate as providing any fundamental explanation for the smallness of W. Apart from being classical in nature, the calculation implicitly

[1] E. Vogt: Phys. Rev. **101**, 1792 (1956). — L. Verlet: unpublished. — A. M. Lane and L. Verlet: Phys. Rev. **100**, 956 (1955).

[2] M. L. Goldberger: Phys. Rev. **74**, 1269 (1948).

[3] A. M. Lane and C. F. Wandel: Phys. Rev. **98**, 693 (1955). — E. Clementel and C. Villi: Nuovo Cim. **10**, 176 (1955). — Hayakawa, Kawai and Kikuchi: Progr. Theor. Phys. **13**, 415 (1955). — Morrison, Muirhead and Murdock: Phil. Mag. **46**, 475 (1955).

assumes that W is small when it uses a FERMI gas with its implied momentum distribution of filled-up states at zero temperature. It so happens that if one repeats the calculation with an excited gas at some modest temperature[1], W is considerably increased and the PAULI principle is much less effective. The correct description of the classical estimate is that it predicts numerical values for W starting from the assumption that W is small. The fact that the calculated values are small allows us to call the calculation "self-consistent".

We now turn to a new calculation which might be described as the quantum mechanical analogue of the last one[2]. This calculation is based upon the well-known perturbation expansion for the energy E of a state that, in zero-order, has the form $\psi_0 u_0$ and energy E_{00}:

$$E = E_{00} + (00|\, H' \,|00) + \sum_{cp \neq 00} \frac{|(00|\, H' \,|cp)|^2}{E_{00} - E_{cp}}.$$

The sum in the last term is over all states allowed by the PAULI principle. If the states cp are fully anti-symmetrised, we can apply closure and express the last term conveniently as:

$$(00|\, H' \left(\frac{1}{E_{00} - H_0}\right)' H' \,|00).$$

The perturbing interaction H' leads to energy corrections to the zero-order energy E_{oo} and also introduces admixtures into the zero-order state $\psi_0 u_0$. This latter effect can be represented by an imaginary potential W in H_0 and in the energy E. The value of W can be found by taking the imaginary part of the above equation for E:

$$W = \operatorname{Im} (00|\, H' \left(\frac{1}{E_{00} - H_0 - i\,W}\right)' H' \,|00).$$

In other words:

$$W = \operatorname{Im} \sum_{cp \neq 00} \frac{|(00|\, H' \,|cp)|^2}{E_{00} - E_{cp} - i\,W}.$$

This may be evaluated in the limit

$$\sum_{cp} \to \int \varrho_{E_{cp}}\, dE_{cp}.$$

if the dependence of typical values of $|(00|\, H' \,|cp)|^2$ on E_{cp} is supposed to be small over a range of extent W about E_{00}. We then find:

$$W = \pi \int \varrho_{E_{cp}}\, dE_{cp}\, |(00|\, H' \,|cp)|^2\, \delta(E_{00} - E_{cp}).$$

If all the nucleon states are now represented by plane waves, this formula can be developed, and expressed in terms of integrals over the nucleon-nucleon interaction $v(r - r_i)$. It turns out that, if the mean potential is assumed to be velocity independent, the final expression is very similar to that used in the above "classical" calculation. The only difference is that the actual cross-sections are replaced by their BORN approximation expressions. This replacement does not prevent agreement with the experimental values of W, since there are many reasonable interaction potentials that are known to reproduce in BORN approximation the observed total cross-sections.

If the potential is allowed to be velocity dependent, the values of W as a function of particle energy are reduced by a certain factor. If the velocity dependence

[1] A. KIND and G. PATERGNANI: Nuovo Cim. **11**, 106 (1954). — K. A. BRUECKNER: Phys. Rev. **103**, 172 (1956).

[2] M. CINI and S. FUBINI: Nuovo Cim. **10**, 75 (1955). — C. BLOCH: To be published.

is of the simple form $V=\alpha+\beta E$, this factor[1] is $(1-\beta)^3$. If β has our previously quoted value of $\frac{11}{18}$, the diminution in W is by about 20. Thus the velocity dependence of Sect. 35 leads to W being under-estimated[2] by a factor of about 20. It has been pointed out that this factor can be compensated by relaxing the strict assumption of a completely degenerated FERMI gas with zero temperature[3]. As one might expect, the value of W is sensitive to this assumption and increases rapidly with any temperature in the gas. If the degeneracy of the FERMI gas is relaxed to the extent suggested by a perturbation calculation of admixtures to the plane wave states[4] (see Sect. 37), then W is found to be increased to a reasonable value.

It is interesting that the above perturbation expression for W resembles closely the "Second Moment" $M_{00}^{(2)}$ of the previous section. This latter quantity can be written:

$$M_{00}^{(2)} = (00|\, H'^2 \,|00) = \sum_{cp} |(00|\, H' \,|cp)|^2$$
$$= \int |(00|\, H' \,|cp)|^2 \varrho_{E_{cp}}\, dE_{cp},$$

and we can see that the expressions for W and $M_{00}^{(2)}$ differ only in the presence of the extra factor $\pi\delta(E_{00}-E_{cp})$ in W. We might guess that this factor tends to make W^2 much smaller than $M_{00}^{(2)}$.

This can be seen directly[1] from our previous expression for W:

$$W = \sum_{cp} |(00|\, H' \,|cp)|^2 \left(\frac{W}{(E_{00}-E_{cp})^2+W^2}\right).$$

If the quantities $(E_{00}-E_{cp})^2$ are dropped from inside the denominators, we have immediately:

$$W \ll \frac{M_{00}^{(2)}}{W} \quad \text{or:} \quad M_{00}^{(2)} \gg W^2.$$

There has been some speculation on whether $M_{00}^{(2)}$ can be regarded as an estimate of W^2. The fact that this is evidently not so when W is calculated by perturbation theory suggests that it is not correct in general. Further evidence in favour of this conclusion comes from considering an energy dependence of $\langle|a_{00}^{\lambda}|^2\rangle$ of the form:

$$\langle|a_{00}^{\lambda}|^2\rangle \propto \frac{1}{\varrho_{E_\lambda}} \left[\frac{1}{(E_{00}-E_\lambda)^2+W^2}\right]$$

where $\langle\;\rangle$ denotes averaging over levels λ in some energy interval that is much less than the single particle level spacing.

It can be shown that, if this dependence on energy is of the same form for all single particle states p, then the asymptotic form of the particle wave-function (and therefore the cross-sections for an unbound particle) are just the same as if nuclear matter is characterised by the absorption potential W. On the other

[1] BRUECKNER, EDEN and FRANCIS: Phys. Rev. **100**, 891 (1955).

[2] Such a velocity dependence also implies that the published Figure of $W(E)$ obtained from the classical calculation (A. M. LANE and C. F. WANDEL: Phys. Rev. **98**, 693 (1955), should be re-interpreted. The incident energy in the scale on the graph is equal to $E(1-\beta)-T_F-8-\alpha$, where E is the actual incident energy. With our values $\alpha=-41$ MeV, $\beta=\frac{11}{18}$, this expression becomes $\frac{7}{18}E-2$ MeV.

[3] K. A. BRUECKNER: Phys. Rev. **103**, 172(1956). — A. KIND and G. PATERGNANI: Nuovo Cim. **11**, 106 (1954).

[4] K. A. BRUECKNER: Phys. Rev. **103**, 172 (1956). — S. WATANABE: Z. Physik **113**, 482 (1939). — L. ROSENFELD: Nuclear Forces, p. 252. Amsterdam: North Holland Publishing Co. 1948.

hand, the value of $M_{00}^{(2)}$ is evidently infinite, so that $M_{00}^{(2)}$ is certainly not equal to W^2. Thus $M_{00}^{(2)}$ does not appear to be a relevant quantity for the estimation of the absorption potential W. Its main usefulness lies in estimating the nature of the energy-spreading of a state caused by the introduction of interactions.

To summarise the results of this section, we might say that the observed absorption potential W can be fitted with a rather simple calculation. However the fact that this calculation is based essentially on the perturbation approximation prevents us from claiming that it constitutes a proper theory giving us real understanding and explanation of the role of W. This is because, in the beginning of the calculation, one effectively assumes a value for W (namely $W \sim 0$) by using independent particle wave-functions. After such an assumption, the calculated value of W is found, not surprisingly, to be small, at least to second order. Whether higher order terms would preserve this result is not clear. Probably the most correct description of the calculation is that it is internally self-consistent in the sense that the assumed small value of W leads to a calculated small value.

Finally we note that W has been calculated using the BRUECKNER Theory[1]. Generally speaking, this theory is distinct from conventional perturbation theory although it has some points of similarity with it. In the published calculations of W, these similarities are very prominent and it is not clear if there is any real difference between the two approaches.

Appendix: The mathematics of the model.

Appendix 1: Surface harmonics. The physical significance[2] of the surface harmonics $Y_m^{(l)}$ is that they are eigenfunctions of the orbital angular momentum operator l_z.

We start from the spherical harmonics $v_m^{(l)}$ which are the $(2l+1)$ independent homogeneous polynomials[3] of degree l in x, y, z which satisfy the equation

$$\Delta v = 0 \tag{A1.1}$$

Here Δ is the LAPLACE operator

$$\Delta = \frac{\partial^2}{\partial x^2} + \frac{\partial^2}{\partial y^2} + \frac{\partial^2}{\partial z^2} \quad \text{in Cartesian co-ordinates}$$

or

$$\Delta = \frac{1}{r^2}\frac{\partial}{\partial r}\left(r^2\frac{\partial}{\partial r}\right) + \frac{1}{r^2}\Lambda \quad \text{in spherical polar co-ordinates}$$

where

$$\Lambda = \frac{1}{\sin\vartheta}\left\{\frac{\partial}{\partial\vartheta}\left(\sin\vartheta\frac{\partial}{\partial\vartheta}\right) + \frac{1}{\sin\vartheta}\frac{\partial^2}{\partial\varphi^2}\right\}.$$

We may write

$$v_m^{(l)} = r^l\, Y_m^{(l)}$$

where the surface harmonic $Y_m^{(l)}$ depends only on the angular co-ordinates ϑ, φ. Then substituting in (A1.1) we have

$$\Lambda Y_m^{(l)} + l(l+1)\, Y_m^{(l)} = 0. \tag{A1.2}$$

[1] BRUECKNER, EDEN and FRANCIS: Phys. Rev. **100**, 891 (1955). — K. A. BRUECKNER: Phys. Rev. **103**, 172 (1956).

[2] In group theoretical language, their significance is that for given l, the functions $Y_m^{(l)}$i spread out the $(2l+1)$ dimensional representation D_l of the rotation group R_3 in three dimensions. The connection with the angular momentum is brought about by the identity of the angular momentum operators with the infinitesimal operators of the group.

[3] H. WEYL: The Theory of Groups and Quantum Mechanics, p. 60. Dover Publications.

Since (A 1.2) is satisfied by any linear combination of the $(2l+1)$ independent functions $Y_m^{(l)}$ for a fixed l, we are free to choose a basis, defining m in a convenient way. We do this by taking a simple dependence on φ, that is by giving the $Y_m^{(l)}$ simple properties under rotation about the z-axis.

Explicitly, we define for $m = l, (l-1), (l-2) \ldots 0$

$$Y_m^{(l)} = e^{i m \varphi} (-1)^m \left\{ \frac{(2l+1)(l-m)!}{(4\pi)(l+m)!} \right\}^{\frac{1}{2}} \sin^m \vartheta \frac{d^m}{(d \cos\vartheta)^m} P_l(\cos\vartheta) \ldots \quad \text{(A 1.3)}$$

where $P_l(\cos\vartheta)$ is a LEGENDRE polynomial. For $m = -1, -2, -3 \ldots -l$, we define

$$Y_m^{(l)} = (-1)^m Y_m^{(l)*}.$$

These $Y_m^{(l)}$ are then an orthonormal set satisfying

$$\iint Y_m^{(l)*} Y_{m'}^{(l')} d\Omega = \delta(l, l') \, \delta(m, m'),$$

and their phase is that used by CONDON and SHORTLEY[1] and by RACAH[2]. The following differential properties are useful:

$$\frac{\partial}{\partial\vartheta} Y_m^{(l)} = m \cot\vartheta \, Y_m^{(l)} + \sqrt{(l+m+1)(l-m)} \, Y_{m+1}^{(l)} e^{-i\varphi} \quad \text{(A 1.4a)}$$

$$\frac{\partial}{\partial\varphi} Y_m^{(l)} = i m Y_m^{(l)}, \quad \text{(A 1.4b)}$$

together with the relation:

$$\left. \begin{aligned} m \cot\vartheta \, Y_m^{(l)} = & -\tfrac{1}{2} \sqrt{(l-m)(l+m+1)} \, Y_{m+1}^{(l)} e^{-i\varphi} - \\ & -\tfrac{1}{2} \sqrt{(l+m)(l-m+1)} \, Y_{m-1}^{(l)} e^{i\varphi}. \end{aligned} \right\} \quad \text{(A 1.4c)}$$

It is clear from (A 1.2) that the $Y_m^{(l)}$ describe the angular dependence of the eigen functions of a spherically symmetric potential, on solving the SCHRÖDINGER equation. Also, since the square of the orbital angular momentum and its z-component are simply the operators

$$\boldsymbol{l}^2 = -\hbar^2 \Lambda, \quad l_z = \left(\frac{\hbar}{i}\right) \frac{\partial}{\partial\varphi},$$

it follows from (A 1.2) and (A 1.4b) that the $Y_m^{(l)}$ are eigenfunctions of these two operators, with eigenvalues

$$\langle \boldsymbol{l}^2 \rangle = \hbar^2 l(l+1), \quad \langle l_z \rangle = \hbar m.$$

Since the surface harmonics form a complete set of polynomials of x, y, z on the unit sphere[3] it follows that we may expand

$$Y_\beta^{(b)}(\Omega) \, Y_\gamma^{(c)}(\Omega) = \sum_{a,\alpha} f(a\,b\,c, \alpha\,\beta\,\gamma) \, Y_\alpha^{(a)}(\Omega) \quad \text{(A 1.5)}$$

where, from the orthonormality of the $Y_m^{(l)}$,

$$f(a\,b\,c, \alpha\,\beta\,\gamma) = \iint Y_\alpha^{(a)*} Y_\beta^{(b)} Y_\gamma^{(c)} d\Omega.$$

This integral is of fundamental importance and its dependence on α, β, γ is described by the WIGNER, CLEBSCH-GORDON or vector-coupling coefficient

[1] E. V. CONDON and G. H. SHORTLEY: Theory of Atomic Spectra. Cambridge: Cambridge University Press 1935.
[2] G. RACAH: Phys. Rev. 62, 438 (1942).
[3] H. WEYL: The Theory of Groups and Quantum Mechanics, p. 62. Dover Publications.

discussed in Appendix 2. It may be shown that

$$f(a\,b\,c, \alpha\beta\gamma) = \left\{\frac{(2b+1)(2c+1)}{4\pi(2a+1)}\right\}^{\frac{1}{2}} (b\,c\,o\,o\,|\,b\,c\,a\,o)\,(b\,c\,\beta\,\gamma\,|\,b\,c\,a\,\alpha). \quad \text{(A1.6)}$$

Appendix 2: Coupling of angular momenta, the WIGNER coefficients. If the wave functions for two particles with angular momenta j_1, j_2 and z-components m_1, m_2 respectively be written

$$\varphi_1(j_1 m_1), \qquad \varphi_2(j_2 m_2)$$

then they may be combined to form two-particle wave functions

$$\varphi_{12}(j_1(1)\,j_2(2)\,J_{12}\,M_{12})$$

where $|j_1 - j_2| \leq J_{12} \leq (j_1 + j_2)$, $|M_{12}| \leq J_{12}$, and the bracketed number indicates that j_1 is the state of particle 1 and j_2 that of particle 2. The combined wave functions may be related to the products of the particle wave functions by the equation

$$\varphi_{12}(j_1(1)\,j_2(2)\,J_{12}\,M_{12}) = \sum_{m_1(m_2)} (j_1 j_2\, m_1\, m_2\,|\,j_1 j_2\, J_{12}\, M_{12})\, \varphi_1(j_1 m_1)\, \varphi_2(j_2 m_2) \quad \text{(A2.1)}$$

where the φ_{12} are normalised in the sense that

$$\iint \varphi_{12}^*(J_{12}\,M_{12})\,\varphi_{12}(J'_{12}\,M'_{12})\,d\tau_1\,d\tau_2 = \delta(J_{12}\,J'_{12})\,\delta(M_{12}\,M'_{12}).$$

Since the angular momentum of a wave function describes its transformation properties under rotations, the coefficients $(j_1 j_2\, m_1 m_2 |\, j_1 j_2 J_{12} M_{12})$ of (A2.1) are determined by the condition that the resulting sum has the properties corresponding to $J_{12} M_{12}$. Sometimes, an alternative notation[1] is used for the coefficients:

$$C^{j\,m}_{j_1 m_1 j_2 m_2} \equiv (j_1 j_2\, m_1\, m_2\,|\,j_1 j_2\, j\, m).$$

RACAH[2] uses the related coefficient

$$V(a\,b\,c; \alpha\,\beta\,\gamma) = (-1)^{c-\gamma}\,(2c+1)^{-\frac{1}{2}}\,(a\,b\,\alpha\,\beta\,|\,a\,b\,c - \gamma) \quad \text{(A2.2)}$$

which has simpler symmetry properties. WIGNER defined a slightly different function with even simpler symmetry relations, the $3j$ symbol,

$$\begin{pmatrix} a\,b\,c \\ \alpha\,\beta\,\gamma \end{pmatrix} = (-1)^{b+c-a}\,V(a\,b\,c; \alpha\,\beta\,\gamma) = \frac{(-1)^{b-a+\gamma}}{(2c+1)^{\frac{1}{2}}}\,C^{c-\gamma}_{a\alpha b\beta}. \quad \text{(A2.3)}$$

An even permutation if the columns of the $3j$ symbol causes no change in value while an odd permutation introduces a factor $(-1)^{a+b+c}$. There is the further relation

$$\begin{pmatrix} a\,b\,c \\ \alpha\,\beta\,\gamma \end{pmatrix} = (-1)^{a+b+c} \begin{pmatrix} a\,b\,c \\ -\alpha - \beta - \gamma \end{pmatrix}.$$

These WIGNER coefficients vanish unless $(\alpha + \beta + \gamma) = 0$ and unless the a, b, c satisfy the triangular conditions that the sum of any two is not less than the third. The general formula is complicated, but is given here for completeness,

$$V(a\,b\,c; \alpha\,\beta\,\gamma) = \left\{\frac{(a+b-c)!\,(b+c-a)!\,(c+a-b)!}{(a+b+c+1)!}\right\}^{\frac{1}{2}} v(a\,b\,c, \alpha\,\beta\,\gamma) \quad \text{(A2.4)}$$

---

[1] H. A. JAHN: Proc. Roy. Soc. Lond., Ser. A **205**, 912 (1951).
[2] G. RACAH: Phys. Rev. **62**, 438 (1942).

where

$$v(abc;\alpha\beta\gamma) = \delta(\alpha+\beta+\gamma,0)\sum_z \frac{(-1)^{z+c-\gamma}\{(a+\alpha)!\,(a-\alpha)!\,(b+\beta)!\,(b-\beta)!\,(c+\gamma)!\,(c-\gamma)!\}^{\frac{1}{2}}}{z!\,(a+b-c-z)!\,(a-\alpha-z)!\,(b+\beta-z)!\,(c-b+\alpha+z)!\,(c-\alpha-\beta+z)!}.$$

When $\alpha=\beta=\gamma=0$, as occurs in (A1.6), this simplifies considerably, giving

$$\left.\begin{aligned} &V(abc;000)=0 \quad \text{for} \quad (a+b+c) \text{ odd} \\ &V(abc;000)=(-1)^g\left\{\frac{(a+b-c)!\,(a+c-b)!\,(b+c-a)!}{(a+b+c+1)!}\right\}^{\frac{1}{2}}\frac{g!}{(g-a)!\,(g-b)!\,(g-c)!} \\ &\qquad\qquad \text{for } (a+b+c)=2g \text{ (even)}, \end{aligned}\right\} \quad \text{(A2.5)}$$

which is related to the C_{abc} tabulated by SHORTLEY and FRIED[1] by

$$C_{abc}=2\{V(a\,b\,c,000)\}^2.$$

For particular values of one of the arguments of $V(abc,\alpha\beta\gamma)$ the expression (A2.4) simplifies and the simple forms have been tabulated by various authors[2] and given in numerical form[3].

As may be imagined from their definition, the WIGNER coefficients satisfy certain orthogonality relations. For these, it is simpler to speak in terms of the coefficients in (A2.1), which describe an orthogonal transformation between the two complete sets of $(2j_1+1)\,(2j_2+1)$ states. We have

$$\sum_{m_1(m_2)}(j_1j_2m_1m_2|j_1j_2J_{12}M_{12})\,(j_1j_2m_1m_2|j_1j_2J'_{12}M'_{12})=\delta(J_{12},J'_{12})\,\delta(M_{12},M'_{12}) \quad \text{(A2.6a)}$$

$$\sum_{J_{12}(M_{12})}(j_1j_2m_1m_2|j_1j_2J_{12}M_{12})\,(j_1j_2m'_1m'_2|j_1j_2J_{12}M_{12})=\delta(m_1,m'_1)\,\delta(m_2,m'_2) \quad \text{(A2.6b)}$$

from which further relations may be deduced using the symmetry relations (A2.3).

Appendix 3: Rearrangement of angular momenta coupling, the RACAH function and 3*n*-*j* symbols. It is often necessary to change the order of coupling of the components of a combined wave function. For the two-body wave function φ_{12} of (A2.1) we have simply

$$\varphi_{12}(j_2(2)\,j_1(1)\,J_{12}M_{12})=(-1)^{j_1+j_2-J_{12}}\varphi_{12}(j_1(1)\,j_2(2)\,J_{12}M_{12}) \quad \text{(A3.1)}$$

from the symmetry properties of the WIGNER coefficients.

For more than two particles, this sort of rearrangement of coupling order introduces increasingly more complicated coefficients. Let us change notation for conciseness, using symbols $a, b, c \ldots$ to denote different j-values and suffices to denote the particle number. Consider the three-particle wave function,

$$\{\{a_1,b_2\}J_1,c_3\}JM$$

in which a state a of particle 1 is coupled to a state b of particle 2 to give resultant J_1, which in turn is coupled to a state c of particle 3 to give resultant JM.

Then, although the interchange of a_1 and b_2 is simply governed by (A3.1), the interchange of either with c_3 involves a sequence of two decouplings and two

[1] G. H. SHORTLEY and B. FRIED: Phys. Rev. **54**, 739 (1938).

[2] E. U. CONDON and G. H. SHORTLEY: Theory of Atomic Spectra. Cambridge: Cambridge University Press 1935. — M. YAMADA and M. MORITA: Progr. Theor. Phys. **8**, 431 (1952). — R. SAITO and M. MORITA: Progr. Theor. Phys. **13**, 540 (1955).

[3] A. SIMON: Oak Ridge Report, ORNL 1718.

recouplings, viz.,

$$\left.\begin{aligned}\varphi\{\{c_3, b_2\} J_1, a_1\} JM &= \sum_{M_1 m_c (m_a m_b)} (J_1 a\, M_1 m_a \,|\, J_1 a\, J\, M)\,(c\, b\, m_c m_b \,|\, c\, b\, J_1 M_1) \times \\ &\qquad \times \varphi_3(c m_c)\,\varphi_2(b m_b)\,\varphi_1(a m_a) \\ &= \sum_{M_1 m_c J_1' J' (m_a m_b M_1' M')} (J_1 a\, M_1 m_a \,|\, J_1 a\, J\, M) \times \\ &\qquad \times (c b m_c m_b \,|\, c b\, J_1 M_1)\,(a b m_a m_b \,|\, a b\, J_1' M_1') \times \\ &\qquad \times (J_1' c\, M_1' m_c \,|\, J_1' c\, J' M')\,\varphi\{\{a_1, b_2\} J_1', c_3\} J' M'\end{aligned}\right\}\text{(A3.2a)}$$

RACAH[1] has shown that this sum of products of four WIGNER coefficients taken over the variables $M_1 m_c (m_a m_b M_1')$ may be greatly reduced and he introduced the RACAH function to describe the result, viz.,

$$\left.\begin{aligned}\varphi\{\{c_3, b_2\} J_1, a_1\} JM = \delta(JJ')\,\delta(MM') \sum_{J_1'} (-1)^{a+b+c-J}\, U(c b J a; J_1 J_1') \times \\ \times \varphi\{\{a_1, b_2\} J_1', c_3\} J' M'.\end{aligned}\right\}\text{(A3.2b)}$$

Here, $U(cbJa;\, J_1 J_1')$ is the RACAH function in the notation of JAHN, related to the other definitions by

$$\left.\begin{aligned}U(cbJa; J_1 J_1') &= [(2J_1+1)(2J_1'+1)]^{\frac{1}{2}}\, W(cbJa;\, J_1 J_1') \\ &= (-1)^{a+b+c+J}\,[(2J_1+1)(2J_1'+1)]^{\frac{1}{2}} \begin{Bmatrix} c & b & J_1 \\ a & J & J_1' \end{Bmatrix}.\end{aligned}\right\}\text{(A3.3)}$$

The W-notation was originally introduced by RACAH and the curly bracket notation by WIGNER, calling it the 6-j symbol. For transormations like (A3.2) the U notation is preferable since the U functions form an orthogonal matrix with rows and columns labelled by the possible J_1 and J_1' values respectively. For exhibiting the symmetry properties of these functions, the 6-j notation is preferable, since the columns may be permuted in any way without changing the value and any two elements of the top row may be changed with those directly beneath without changing the value.

The RACAH function vanishes unless all the triads joined together in the diagrams below satisfy the triangular conditions.

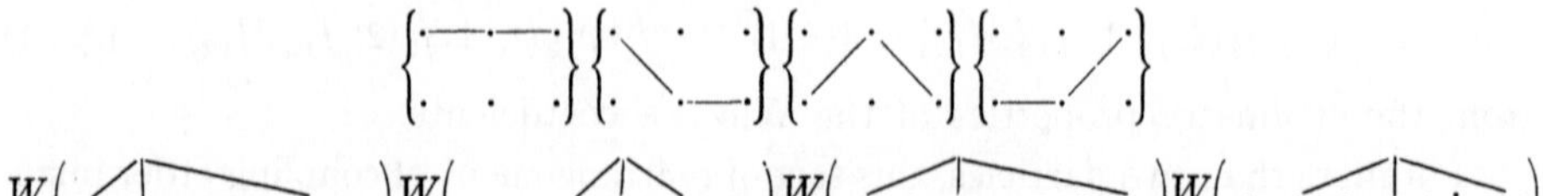

When one has states a, b, c, d of four particles coupled together, the permutation of the first with the last involves a 9-j function which is expressible either as a multiple sum over six WIGNER coefficients or a single sum over three 6-j symbols as given in (A3.6) below. Then

$$\left.\begin{aligned}&\varphi\{\{\{d_4, b_2\} J_1, c_3\} J_2, a_1\} JM \\ &= \sum_{J_1', J_2'} (-1)^{a+d+J_2+J_2'+2J}\,[(2J_1+1)(2J_2+1)(2J_1'+1)(2J_2'+1)]^{\frac{1}{2}} \begin{Bmatrix} b & a & J_1' \\ d & J & J_2' \\ J_1 & J_2 & c \end{Bmatrix} \times \\ &\qquad \times \varphi\{\{\{a_1, b_2\} J_1', c_3\} J_2', d_4\} JM.\end{aligned}\right\}\text{(A3.4)}$$

[1] G. RACAH: Phys. Rev. 62, 438 (1942).

The 9-j symbol also occurs in the transformation

$$\left.\begin{aligned}&\varphi\{\{a_1 b_2\} J_1, \{c_3 d_4\} J_2\} JM\\ &= \sum_{J_1', J_2'} [(2J_1+1)(2J_2+1)(2J_1'+1)(2J_2'+1)]^{\frac{1}{2}} \begin{Bmatrix} a & b & J_1 \\ c & d & J_2 \\ J_1' & J_2' & J \end{Bmatrix} \times \\ &\qquad\qquad \times \varphi\{\{a_1 c_3\} J_1', \{b_2 d_4\} J_2'\} JM \end{aligned}\right\} \quad (A3.5)$$

which is more frequently encountered, for example as the transformation between the two particle states in L-S and j-j coupling, see Sect. 21.

The 9-j symbol has many symmetry properties[1] and the notation is appropriate for displaying them. The value is unchanged by a reflection in either diagonal or by an even permutation of the rows or columns. The value is change by a factor $(-1)^S$ for an odd permutation of the rows and columns where S is the sum of all nine arguments in the symbol. The symbol vanishes unless the three arguments in each row and column satisfy the triangular conditions.

We give below, in a symmetric form, a list of summation properties of the 6-j symbols.

$$\left.\begin{aligned}
&\sum_x (2x+1) \begin{Bmatrix} a & b & x \\ a & b & f \end{Bmatrix} = (-1)^{2a+2b},\\
&\sum_x (-1)^{-x}(2x+1) \begin{Bmatrix} a & b & x \\ b & a & f \end{Bmatrix} = \delta(f,0)(-1)^{a+b}[(2a+1)(2b+1)]^{\frac{1}{2}},\\
&\sum_x (2x+1) \begin{Bmatrix} a & b & x \\ c & d & f \end{Bmatrix} \begin{Bmatrix} c & d & x \\ a & b & g \end{Bmatrix} = \frac{\delta(f,g)}{(2f+1)},\\
&\sum_x (-1)^{-x}(2x+1) \begin{Bmatrix} a & b & x \\ c & d & f \end{Bmatrix} \begin{Bmatrix} c & d & x \\ b & a & g \end{Bmatrix} = (-1)^{f+g} \begin{Bmatrix} a & d & f \\ b & c & g \end{Bmatrix},\\
&\sum_x (2x+1) \begin{Bmatrix} a & b & x \\ c & d & g \end{Bmatrix} \begin{Bmatrix} c & d & x \\ e & f & h \end{Bmatrix} \begin{Bmatrix} e & f & x \\ a & b & j \end{Bmatrix} = (-1)^{2a+2b} \begin{Bmatrix} e & f & j \\ d & h & e \\ g & c & b \end{Bmatrix},\\
&\sum_x (-1)^{-x}(2x+1) \begin{Bmatrix} a & b & x \\ c & d & g \end{Bmatrix} \begin{Bmatrix} c & d & x \\ e & f & h \end{Bmatrix} \begin{Bmatrix} e & f & x \\ b & a & j \end{Bmatrix}\\
&\qquad\qquad = (-1)^{a+b+c+d+e+f+g+h} \begin{Bmatrix} a & d & g \\ h & j & e \end{Bmatrix} \begin{Bmatrix} b & c & g \\ h & j & f \end{Bmatrix}.
\end{aligned}\right\} \quad (A3.6)$$

The reader will observe that for each number of 6-j symbols in the sum there are two kinds of sum, differing by a slightly different letter arrangement and a phase factor. One may generalise to the $3n$-j by considering rearrangements in a wave function of $(n+1)$ particles but one very rarely needs anything higher than the 9-j symbol. The 12-j symbol is briefly discussed by JAHN[1].

When one element of a $3n$-j symbol is zero, it reduces to some lower order symbol, viz.,

$$\begin{Bmatrix} 0 & c & c \\ a & b & b \end{Bmatrix} = \begin{Bmatrix} a & b & c \\ 0 & c & b \end{Bmatrix} = \frac{(-1)^{a+b+c}}{[(2b+1)(2c+1)]^{\frac{1}{2}}},$$

---

[1] H. A. JAHN and J. HOPE: Phys. Rev. **93**, 318 (1954).

$$\left.\begin{aligned}\begin{Bmatrix} a & b & e \\ c & d & e \\ f & f & 0 \end{Bmatrix} &= \begin{Bmatrix} 0 & e & e \\ f & d & b \\ f & c & a \end{Bmatrix} = \begin{Bmatrix} e & 0 & e \\ c & f & a \\ d & f & b \end{Bmatrix} = \begin{Bmatrix} f & f & 0 \\ d & c & e \\ b & a & e \end{Bmatrix} = \begin{Bmatrix} f & b & d \\ 0 & e & e \\ f & a & c \end{Bmatrix} \\ &= \begin{Bmatrix} a & f & c \\ e & 0 & e \\ b & f & d \end{Bmatrix} = \begin{Bmatrix} b & a & e \\ f & f & 0 \\ d & c & e \end{Bmatrix} = \begin{Bmatrix} e & d & c \\ e & b & a \\ 0 & f & f \end{Bmatrix} = \begin{Bmatrix} c & e & d \\ a & e & b \\ f & 0 & f \end{Bmatrix} \\ &= (-1)^{b+c+e+f} \left[(2e+1)(2f+1)\right]^{-\frac{1}{2}} \begin{Bmatrix} a & b & e \\ d & c & f \end{Bmatrix}. \end{aligned}\right\} \quad \text{(A3.7)}$$

The orthogonality of the 9-j symbols is expressed by

$$\sum_{g,h} (2g+1)(2h+1) \begin{Bmatrix} a & b & e \\ c & d & f \\ g & h & j \end{Bmatrix} \begin{Bmatrix} a & b & e' \\ c & d & f' \\ g & h & j \end{Bmatrix} = \frac{\delta(e,e')\,\delta(f,f')}{(2e+1)(2f+1)} \quad \text{(A3.8)}$$

and by manipulating the formulae (A3.6) or by considering alternative ways of making the same transformation, other relations may be deduced.

From the definition (A3.2a, b) of the RACAH function or 6-j symbol, one has the useful relations

$$\left.\begin{aligned}&\sum_{\delta(\varepsilon,\varphi)} (d e \delta \varepsilon | d e a \alpha)(f e \varphi \varepsilon | f e c \gamma)(d b \delta \beta | d b f \varphi) \\ &\qquad = [(2a+1)(2f+1)]^{\frac{1}{2}}\, W(a e b f; d c)\,(a b \alpha \beta | a b c \gamma)\end{aligned}\right\} \quad \text{(A3.9)}$$

and

$$\left.\begin{aligned}&\sum_{f} [(2e+1)(2f+1)]^{\frac{1}{2}} (b d \beta \delta | b d f \varphi)(a f \alpha \varphi | a f c \gamma)\, W(a b c d; e f) \\ &\quad = (a b \alpha \beta | a b e \varepsilon)(e d \varepsilon \delta | e d c \gamma).\end{aligned}\right\} \quad \text{(A3.10)}$$

Tables of values for the RACAH function have been prepared in surd form by OBI *et al.*[1]. and in decimal form by SIMON *et al.*[2], the latter using an electronic digital computer. For the 9-j symbols, restricted tables of values have been prepared[3].

Appendix 4: The algebra of tensor operators. The calculation of matrix elements of physical operators has been gratly simplified by the work of RACAH[4] and in particular by his introduction of the concept of a tensor operator.

He defines a tensor operator $\boldsymbol{T}^{(\lambda)}$ of degree λ to have $(2\lambda+1)$components $T_q^{(\lambda)}$ with $q=\lambda, (\lambda-1) \ldots -\lambda$, such that they transform under rotations like the spherical harmonics $Y_q^{(\lambda)}$, or in other words that they transform according to the representation D_λ of the rotation group R_3, or that the have the same commutation relations with the total angular momentum operator J as do the spherical harmonic operators $Y_q^{(\lambda)}$. Like the spherical harmonics, a Hermitian tensor operator satisfies the relations

$$(T_q^{(\lambda)})^* = (-1)^q\, T_{-q}^{(\lambda)}. \quad \text{(A4.1)}$$

Two tensor operators may be combined together in what is called a tensor product to form one of higher or lower degree by the use of the WIGNER coefficient,

---

[1] OBI, ISHIDZU, HORIE, YANAGAWA, TANABE and SATO: Ann. Tokyo Astro Observ. **3**, No. 3. — OBI, ISHIDZU, HORIE, YANAGAWA, TANABE and SATO: Annals of Tokyo Astro Observatory Vol. IV, No. 1.

[2] SIMON, VANDER, SLUIS and BIEDENHARN: Oak Ridge Report ORNL 1679.

[3] KENNEDY, SHARP and SEARS: Chalk River Report AECl 106 (1954). — J. M. KENNEDY and M. L. CLIFF: Chalk River Report CRT 609, 1955.

[4] G. RACAH: Phys. Rev. **62**, 438 (1942).

viz. the combination

$$T_q^{(\lambda)} = [\boldsymbol{U}^{(\lambda_1)} \times \boldsymbol{V}^{(\lambda_2)}]_q^{(\lambda)} = \sum_{q_1(q_2)} (\lambda_1 \lambda_2 q_1 q_2 | \lambda_1 \lambda_2 \lambda q)\, U_{q_1}^{(\lambda_1)} V_{q_2}^{(\lambda_2)}, \tag{A4.2}$$

where $\boldsymbol{U}$ and $\boldsymbol{V}$ are tensor operators of degree λ_1 and λ_2, will be another which we call $\boldsymbol{T}$, of degree λ where, from the WIGNER coefficient, $|\lambda_1 - \lambda_2| \leq \lambda \leq (\lambda_1 + \lambda_2)$.

For the special case of $\lambda = q = 0$, Eq. (A4.2) forms the scalar product of $\boldsymbol{U}^{(\lambda_1)}$ and $\boldsymbol{V}^{(\lambda_2)}$ which is non-zero for $\lambda_1 = \lambda_2$. Then,

$$\left.\begin{aligned} T_0^{(0)} &= \sum_{q_1(q_2)} (\lambda_1 \lambda_1 q_1 q_2 | \lambda_1 \lambda_1 0 0)\, U_{q_1}^{(\lambda_1)} V_{q_2}^{(\lambda_1)} \\ &= \sum_q \frac{(-1)^{\lambda_1 - q}}{(2\lambda_1 + 1)^{\frac{1}{2}}}\, U_q^{(\lambda_1)} V_{-q}^{(\lambda_1)} \\ &= \frac{(-1)^{\lambda_1}}{(2\lambda_1 + 1)^{\frac{1}{2}}} \sum_q U_q^{(\lambda_1)} (V_q^{(\lambda_1)})^* = \frac{(-1)^{\lambda_1}}{(2\lambda_1 + 1)^{\frac{1}{2}}} (\boldsymbol{U}^{(\lambda_1)} \cdot \boldsymbol{V}^{(\lambda_1)}) \end{aligned}\right\} \tag{A4.3a}$$

which is a multiple $\{(-1)^{\lambda_1}/(2\lambda_1 + 1)^{\frac{1}{2}}\}$ of the scalar product defined in the standard way as

$$(\boldsymbol{U}^{(\lambda)} \cdot \boldsymbol{V}^{(\lambda)}) = \sum_q U_q^{(\lambda)} (V_q^{(\lambda)})^* = \sum_q (-1)^q U_q^{(\lambda)} V_{-q}^{(\lambda)}. \tag{A4.3b}$$

Since a wave function $\varphi(jm)$ is the m component of a spherical tensor of degree j in this same sense it follows from (A4.2) that the matrix element of a tensor perator between two such states may be expressed, in RACAH'S notation, as

$$(jm | T_q^{(\lambda)} | j' m') = (j' \lambda m' q | j' \lambda j m) \frac{(j || \boldsymbol{T}^{(\lambda)} || j')}{(2j+1)^{\frac{1}{2}}} \tag{A4.4}$$

where the second factor, called the amplitude matrix or double-barred matrix, is independent of m, q and m'. A slightly different notation viz.,

$$\langle j || \boldsymbol{T}^{(\lambda)} || j' \rangle = \frac{(j || \boldsymbol{T}^{(\lambda)} || j')}{(2j+1)^{\frac{1}{2}}} \tag{A4.5}$$

has been used by ELLIOTT[1], and others. Although this latter form is more concise on occasions, RACAH'S form has the advantage that for a Hermitian operator, (A4.1),

$$(j || \boldsymbol{T}^{(\lambda)} || j') = (-1)^{j-j'} \overline{(j' || \boldsymbol{T}^{(\lambda)} || j)} \tag{A4.6}$$

where the bar denotes the conjugate complex of the matrix element.

The result (A4.4) is one of fundamental importance since it separates out the geometrical factor, which is common to all tensor operators of that degree, from the physical factor which is independent of m, q, m'. It enables one to relate the matrix elements of the various components of a tensor operator in a simple way so that the evaluation of one component, the simplest, determines the others. Furthermore, as we shall presently see, it simplifies calculations considerably to work with the amplitude matrices of the various operators in the problem, having taken out the geometrical factor at the beginning.

When one has a product tensor operator of the form (A4.2) in which the parts $\boldsymbol{U}^{(\lambda_1)}$, $\boldsymbol{V}^{(\lambda_2)}$ operate only on separate parts of the wave function it is possible to write the matrix element of the product $\boldsymbol{T}^{(\lambda)}$ in terms of the matrices of the separate parts $\boldsymbol{U}^{(\lambda_1)}$ and $\boldsymbol{V}^{(\lambda_2)}$. In fact, by repeated use of the formulae for recoupling angular momenta discussed in the preceding section, together with (A4.2) and

[1] J. P. ELLIOTT: Proc. Roy. Soc. Lond., Ser. A **218**, 345 (1953). — J. P. ELLIOTT and B. H. FLOWERS: Proc. Roy. Soc. Lond., Ser. A **229**, 536 (1955).

(A4.4), we may relate the amplitude matrices of these operators, from which any component may be derived. We find

$$\left.\begin{aligned}(\gamma j_1 j_2 j \| \boldsymbol{T}^{(\lambda)} \| \gamma' j_1' j_2' j') = [(2j+1)(2j'+1)(2\lambda+1)]^{\frac{1}{2}} \sum_{\gamma''} \begin{Bmatrix} j_1 & j_2 & j \\ j_1' & j_2' & j' \\ \lambda_1 & \lambda_2 & \lambda \end{Bmatrix} \times \\ \times (\gamma j_1 \| \boldsymbol{U}^{(\lambda_1)} \| \gamma'' j_1')(\gamma'' j_2 \| \boldsymbol{V}^{(\lambda_2)} \| \gamma' j_2')\end{aligned}\right\} \quad \text{(A4.7)}$$

where

$\boldsymbol{U}^{(\lambda_1)}$ does not operate in the space of j_2, j_2',

$\boldsymbol{V}^{(\lambda_2)}$ does not operate in the space of j_1, j_1'.

The sum over γ'' has been included to cover the case when the operators $\boldsymbol{U}$ and $\boldsymbol{V}$ both affect some other quantum number γ.

It is rarely that we use (A4.7) in its full generality; the particular cases of most use being

(i) The scalar product defined by (A4.3b) rather than (A4.3a), obtained with $k=0$, and using (A4.4)

$$\left.\begin{aligned}&(\gamma j_1 j_2 j m | (\boldsymbol{U}^{(\lambda)} \cdot \boldsymbol{V}^{(\lambda)}) | \gamma' j_1' j_2' j' m') \\ &= \delta(j,j')\,\delta(m,m')\,(-1)^{j_2+j_1'+j+2\lambda} \begin{Bmatrix} j_1 & j_2 & j \\ j_2' & j_1' & \lambda \end{Bmatrix} \sum_{\gamma''} (\gamma j_1 \| \boldsymbol{U}^{(\lambda)} \| \gamma'' j_1')(\gamma'' j_2 \| \boldsymbol{V}^{(\lambda)} \| \gamma' j_2') \\ &= \delta(j,j')\,\delta(m,m')\,(-1)^{\lambda}\, W(j j_2 j_1' \lambda; j_1 j_2') \sum_{\gamma''} (\gamma j_1 \| \boldsymbol{U}^{(\lambda)} \| \gamma'' j_1')(\gamma'' j_2 \| \boldsymbol{V}^{(\lambda)} \| \gamma' j_2')\,.\end{aligned}\right\} \quad \text{(A4.8)}$$

(ii) When $\boldsymbol{V}^{(\lambda_2)}$ is a scalar i.e. $\lambda_2 = 0$, (A4.7) yields

$$(\gamma j_1 j_2 j \| \boldsymbol{U}^{(\lambda)} \| \gamma' j_1' j_2' j') = \delta(j_2 j_2')[(2j+1)(2j'+1)]^{\frac{1}{2}}\, W(j_2 j' j_1 \lambda; j_1' j)\,(\gamma j_1 \| \boldsymbol{U}^{(\lambda)} \| \gamma' j_1'), \quad \text{(A4}$$

putting $\boldsymbol{V}^{(\lambda_2)} = 1$.

(iii) Likewise, when $\boldsymbol{U}^{(\lambda_1)}$ is a scalar, putting $\boldsymbol{U}^{(\lambda_1)} = 1$ we have

$$(\gamma j_1 j_2 j \| \boldsymbol{V}^{(\lambda)} \| \gamma' j_1' j_2' j') = \delta(j_1, j_1')\,[(2j+1)(2j'+1)]^{\frac{1}{2}}\, W(j_1 j j_2' \lambda, j_2 j')\,(\gamma j_2 \| \boldsymbol{V}^{(\lambda)} \| \gamma' j_2')\,. \quad \text{(A4.1}$$

Appendix 5: A group theoretical classification for the orbital states. In Sect. 18 we described how the set of states of a configuration $(nl)^k$ were classified according to the symmetry of the orbital and charge-spin parts separately. The set of charge-spin functions of a given symmetry are, in most cases, specified uniquely by their T and S values. For the orbital functions however when $l > 1$, there are in general a large number of states with the same symmetry and L value so that some further quantum number is necessary to separate them. RACAH[1] has developed a very powerful method for classifying these states using Group Theory and has used it to simplify the calculation of fractional parentage coefficients and the construction of the energy matrix. We give below an outline of his method.

Consider the $(2l+1)$ orbital functions φ_m^l of a single particle as the basis vectors of a $(2l+1)$ dimensional space. Now a rotation in the three dimensional physical space will induce a particular kind of transformation in this $(2l+1)$ space i.e. φ_m^l will go into $(\varphi_m^l)'$ where

$$(\varphi_m^l)' = \sum_{l,m} a_{lm}\, \varphi_{m'}^l,$$

[1] G. RACAH: Phys. Rev. **63**, 367 (1943). — G. RACAH: Group Theory and Spectroscopy. Mimeographed lecture notes. Princeton 1951.

the coefficients a_{lm} describing the rotation. One can generalise this concept to include all transformations which preserve the orthonormality of the set φ^l_m. Such transformations are referred to as the group of unitary transformations in $(2l+1)$ dimensions, denoted by U_{2l+1}. Since the rotations in three dimensions, which we shall refer to as operators of the group R_3, are a particular set of unitary transformations, it follows that R_3 is a sub-group of U_{2l+1}. Furthermore, a transformation among the φ^l_m of each single particle will induce a transformation among the complete set of states $\psi(l^k)$ of the configuration (l^k).

For a group G there are certain sets of functions which transform among themselves under all operations of the group, i.e. when a function of one set is operated on by any group element it forms some linear combination of members of that set without involving any other set. The transformation matrices in the set p, for all the group elements, consitute a representation R_p of the group and we can say that the functions of that set transform according to the representation R_p or spread out that representation. If it is impossible to reduce the set p into smaller sets which have the same invariant property that no group element operating on a member of one set produces a member of some other set, then R_p is called an irreducible representation. An irreducible representation R_p of G will in general not be irreducible as a representation of a sub-group H of G, but will break up into a number of irreducible representations of H. The concepts of an irreducible representation and an invariant subspace are obviously closely related.

From these remarks it will be seen that any complete set of functions may be classified according to the irreducible representations of certain groups of transformations. For example the functions of position on the unit sphere (i.e. the angular parts of functions of position in three dimensions) may be classified according to the $(2l+1)$ dimensional irreducible representations D_l of the group R_3. The functions which spread out this representation are, of course, the spherical harmonics $Y^{(l)}_m$ which have the desired property that the result $(Y^{(l)}_m)'$ of a rotation of $Y^{(l)}_m$ may be expanded into a sum

$$(Y^{(l)}_m)' = \sum_{m'} D^l_{m\,m'}\, Y^{(l)}_m$$

involving no other l value. In other words, labelling the wave function by its angular momentum l is equivalent to labelling it by the irreducible representation D_l of R_3 according to which it transforms. In the same way we can classify the states of the configuration $(l)^k$ by the representations of the group U_{2l+1} discussed above. Since R_3 is a sub-group of U_{2l+1}, we may classify the states *simultaneously* by the representations of those two groups according to which they transform.

Now, it so happens that the functions which spread out the irreducible representations of U_{2l+1} are just those which have definite symmetry properties as described in Sect. 18 by the label $[\lambda]$, so that this label serves also to classify the states according to U_{2l+1}. Hence our discussion of group theoretical concepts has so far given us nothing new; it has just interpreted $[\lambda]$ and L in a different way. We are not yet finished however, because RACAH observed that if we can find a group G which contains R_3 as a sub-group and is contained in U_{2l+1} as a subgroup, then we can further classify the states by irreducible representations of that group G. The search for such a sub-group may be carried out by writing down the infinitesimal operators of U_{2l+1} and looking for a sub-set of these operators which contain the operators of R_3 and satisfy the group conditions. In general it may be shown that the group R_{2l+1} of rotations in $(2l+1)$ dimensions satisfies these requirements. Hence for the d-shell one can introduce the labels $(\sigma_1\,\sigma_2)$ of the irreducible representations of R_5 to classify the states. For the f-shell in addition to R_7 there is another group contained in R_7 and containing R_3 known in the mathematical literature as G_2. As an example, we show the classification of the orbital states of the configuration d^4 in Table 27. In the d-shell this R_5 classification has definite physical significance since in the limit of very short range forces the labels $(\sigma_1\,\sigma_2)$ are good quantum numbers.

Table 27. *Group Theoretical Classification of the Orbital states of the Configuration d^4.*

U_5 $[\lambda]$	R_5 $(\sigma_1\sigma_2)$	R_3 L	U_5 $[\lambda]$	R_5 $(\sigma_1\sigma_2)$	R_3 L
[4]	(00)	S	[22]	(00)	S
	(20)	GD		(20)	GD
	(40)	$LIHGD$		(22)	$IGFDS$
[31]	(11)	FP	[211]	(11)	FP
	(20)	GD		(21)	$HGFDP$
	(31)	KIH^2GF^2DP	[1111]	(10)	D

In the j-j coupling picture this approach goes through in an analogous manner. Here, starting from the group U_{2j+1} one reduces to the sub-group Sp_{2j+1} of symplectic transformations before going to R_3. The irreducible representations of Sp_{2j+1} are labelled by two numbers $(\varrho_1\,\varrho_2)$ which are equivalent to the seniority s and reduced isotopic spin t discussed in Sect. 20. Once again this classification has physical significance since for δ-forces $(\varrho_1\,\varrho_2)$ are nearly good quantum numbers.

Having used the theory of groups to classify the states, RACAH continues to use it to show that the fractional parentage coefficients may be factorised into parts referring to adjacent groups only in the chain of sub-groups and has devised methods[1] for calculating the separate components.

Appendix 6: The calculation of the fractional parentage coefficients. In Sect. 24 we defined the concept of fractional parentage, showing how it is useful in reducing the many-body matrix to the one or two-body matrix. A number of methods have been devised for calculating the c.f.p., most of which entail a chain process, using the $(k-1|k-2)$ coefficients to determine those for the $(k|k-1)$ reduction.

The original, and simplest understood, method is very laborious for any but simple states of a few particles but is described below for completeness.

Let $\psi, \overline{\psi}, \tilde{\psi}$ denote antisymmetric states of the configurations $(l)^k$, $(l)^{k-1}$, $(l)^{k-2}$ respectively and suppose that the c.f.p. $(\overline{\psi}\{|\tilde{\psi})$ are known. To determine the coefficients $(\psi\{|\overline{\psi})$ we first expand

$$\left.\begin{aligned}\psi(1\ldots k) &= \sum_{\overline{\psi}} (\psi\{|\overline{\psi})\,\{\overline{\psi}(1\ldots k-1), \varphi(k)\}\,TSL\\ &= \sum_{\overline{\psi}}\sum_{\tilde{\psi}} (\psi\{|\overline{\psi})\,(\overline{\psi}\{|\tilde{\psi})\,\{\{\tilde{\psi}(1\ldots k-2), \varphi(k-1)\}\,\overline{\psi}, \varphi(k)\}\,TSL\end{aligned}\right\}\tag{A6.1}$$

involving the unknown coefficients. Then, rearranging the order of vector coupling in a manner similar to (A3.2b), we have

$$\left.\begin{aligned}\psi(1\ldots k) = \sum_{\overline{\psi}}\sum_{\tilde{\psi}} (\psi\{|\overline{\psi})\,(\overline{\psi}\{|\tilde{\psi})\sum_{\vartheta} U(\tilde{\psi}\varphi\psi\varphi;\overline{\psi}\vartheta)\,\times\\ \times\{\tilde{\psi}(1\ldots k-2), \vartheta(k-1,k)\}\,TSL.\end{aligned}\right\}\tag{A6.2}$$

[1] G. RACAH: Group Theory and Spectroscopy. Mimeographed lectures notes. Princeton 1951.

Here, the RACAH function with wave function symbols as arguments indicates the product of three such functions containing the T, S and L values of these functions respectively. $\vartheta(k-1, k)$ is a state formed by vector coupling $\varphi(k-1)$ and $\varphi(k)$. Now since $\overline{\psi}$ is antisymmetric in particles $1 \ldots (k-1)$ it follows that ψ will be antisymmetric in all particles $1 \ldots k$ if it is antisymmetric in particles $(k-1)$ and k. But in (A6.2) this implies that all ϑ in the sum must be antisymmetric, i.e. that the coefficients of the symmetric ϑ must be zero. Since ϑ is symmetric if $(T_\vartheta + S_\vartheta + L_\vartheta)$ is even we have the result that

$$\sum_{\overline{\psi}} (\psi\{|\overline{\psi})(\overline{\psi}\{|\tilde{\psi})\, U(\tilde{\psi}\varphi\psi\varphi; \overline{\psi}\vartheta) = 0 \quad \text{for} \quad (T_\vartheta + S_\vartheta + L_\vartheta) \ \text{even}. \qquad (A6.3)$$

This provides a set of simultaneous equations in the desired coefficients $(\psi\{|\overline{\psi})$ when the $(\overline{\psi}\{|\tilde{\psi})$ are known, and, since the $(2|1)$ coefficients are all unity, a chain calculation may be set up.

As may be imagined, this method becomes laborious and alternative methods have been derived, making use of the group theoretical ideas by which the states are classified.

The first important step is to observe that the c.f.p. may be factorised[1] into a part depending only on the charge-spin labels of the wave functions and a part depending on the orbital labels. This follows from the definition of the totally antisymmetric states by the symmetry of the orbital and charge-spin parts separately as described in Sects. 18.

One great advantage of calculating the charge-spin c.f.p. separately is that the same coefficients are applicable to the corresponding states for any orbital configuration—even for inequivalent particles, i.e. in mixed configurations.

Following up this approach and using the theory[2] of the Symmetric Group of permutations together with the theory of rearrangement of vector coupling order described in Appendix 3, it is possible to calculate the charge-spin and orbital c.f.p. separately. This approach is perfectly adequate for the charge-spin c.f.p. and for the p-shell orbital c.f.p., since in both cases the wave functions are determined by the symmetry label $[\lambda]$ and the T, S and L values.

However for $l \geq 2$, remember that in Table 27 we had to introduce other quantum numbers to classify the states, considering their properties under certain groups of transformations. RACAH has shown that, corresponding to the reduction to subgroups described in Appendix 5, there is a factorisation of the orbital c.f.p. One may then calculate these factors separately again simplifying matters, and RACAH has shown how one can use the infinitesimal operators of the various groups to derive these various factors. Further, he has shown the existence of certain reciprocal relations between sets of c.f.p. which often allow one to deduce immediately a block of coefficients from those already calculated for some smaller number of particles.

JAHN[3] and others[4] have developed a method of calculating the c.f.p. explicitly without recourse to a chain calculation, the method being particularly useful for small numbers of particles in complicated configurations. It depends essentially on the theory of the Symmetric Group, using the so called YOUNG Operators. These are permutation operators which, operating on an arbitrary function, produce a function of definite symmetry (belonging to a definite irreducible representation of the group). They are generalisations of the symmetrising and

[1] H. A. JAHN and H. VAN WIERINGEN: Proc. Roy. Soc. Lond., Ser. A **209**, 502 (1951).
[2] D. E. RUTHERFORD: Substitutional Analysis. Edinburgh: Edinburgh University Press.
[3] H. A. JAHN: Phys. Rev. **96**, 898 (1954).
[4] P. J. REDMOND: Proc. Roy. Soc. Lond., Ser. A **222**, 84 (1954). — A. HASSITT: Proc. Roy. Soc. Lond., Ser. A **229**, 110 (1955).

antisymmetrising operators

$$\mathcal{S} = \sum_P P$$

and

$$\mathcal{A} = \sum_P (-1)^P P$$

where P runs over all permutations of the particle numbers and $(-1)^P$ is the parity of P.

For mixed configurations, if we define the states genealogically as in Sect. 19, the c.f.p. may be deduced from the c.f.p. for each group of equivalent particles, together with the rearrangement theory of Appendix 3. For two kinds of particle the formula for the $(k|k-2, 2)$ c.f.p. is given by[1]

$$\left.\begin{aligned} \psi = &\left[\frac{(k-x)(k-x-1)}{k(k-1)}\right]^{\frac{1}{2}} \sum_{\tilde{\chi}\psi_{20}\vartheta_{20}} (-1)^{\chi-\psi+\psi_{20}-\tilde{\chi}}\, U(\varphi\,\tilde{\chi}\,\psi\,\vartheta_{20};\psi_{20}\chi) \times \\ &\times (\chi\{|\tilde{\chi}\vartheta_{20})\,[\psi_{20},\vartheta_{20}]\,TSL + \left[\frac{x(x-1)}{k(k-1)}\right]^{\frac{1}{2}} \sum_{\tilde{\varphi},\psi_{02},\vartheta_{02}} U(\chi\,\tilde{\varphi}\,\psi\,\vartheta_{02};\,\psi_{02}\,\varphi) \times \\ &\times (\varphi\{|\tilde{\varphi}\vartheta_{02})\,[\psi_{02},\vartheta_{02}]\,TSL + \left[\frac{2x(k-x)}{k(k-1)}\right]^{\frac{1}{2}} (-1)^{x+1} \times \\ &\times \sum_{\bar{\chi},\bar{\varphi},\psi_{11},\vartheta_{11}} [(2\chi+1)(2\varphi+1)(2\psi_{11}+1)(2\vartheta_{11}+1)]^{\frac{1}{2}} \times \\ &\times \begin{Bmatrix} \bar{\chi} & l_a & \chi \\ \bar{\varphi} & l_b & \varphi \\ \psi_{11} & \vartheta_{11} & \psi \end{Bmatrix} (\chi\{|\bar{\chi})\,(\varphi\{|\bar{\varphi})\,[\psi_{11},\vartheta_{11}]\,TSL \end{aligned}\right\} \quad \text{(A6.4)}$$

where ψ is the antisymmetrised product of a state $\chi(l_a^{k-x})$ of $(k-x)$ particles l_a, and a state $\varphi(l_b^x)$ of x particles l_b. Also, ϑ_{rs} is the antisymmetric vector coupled state of the configuration $l_a^r\, l_b^s$ and ψ_{rs} is similarly constructed from the remainder. $\bar{\chi}$ and $\bar{\varphi}$ denote states obtained on removal of one particle from χ and φ respectively, while $\tilde{\chi}$ and $\tilde{\varphi}$ denote the removal of two particles.

Appendix 7: Reduction of the many-body problem with non-central operators. In Sect. 24 we showed how the c.f.p. are used to reduce the many-body matrix to the two-body matrix for central forces (i.e. forces which are scalar in both L and S spaces). Combining these ideas with the tensor operator calculus of Appendix 4 enables us to do the same for non-central operators.

Consider the general two-body interaction $\mathcal{V} = \sum_{i<j} v_{ij}$, where

$$v_{ij} = (\mathcal{S}_{(ij)}^{(\lambda)} \cdot \mathcal{L}_{(ij)}^{(\lambda)})\, \mathcal{I}_0^{(\varrho)}(i\,j) \quad \text{(A7.1)}$$

which is of the form discussed in Sect. 23. We have slightly generalised to a non-scalar $\mathcal{I}^{(\varrho)}$ in isotopic spin space to include charge dependent interactions such as the COULOMB force. We shall of course only need the z-component $\mathcal{I}_0^{(\varrho)}$ since total charge must be conserved. Now let us evaluate the matrix element of $\mathcal{V}$ between two antisymmetric states Ψ and Ψ' of a number k of particles, not necessarily equivalent. Suffice it to define the states in terms of some fractional parentage relation

$$\Psi = \sum_{\tilde{\Psi},\vartheta} (\Psi\{|\tilde{\Psi}\vartheta)\,\{\tilde{\Psi}(1\ldots k-2), \vartheta(k-1,k)\}\,\Psi$$

[1] J. P. ELLIOTT and B. H. FLOWERS: Proc. Roy. Soc. Lond., Ser. A **229**, 536 (1955).

where $\widetilde{\Psi}$ is an antisymmetric state of the particles 1 to $(k-2)$, ϑ is an antisymmetric state of the particles $(k-1)$ and k and the notation is similar to that used in (24.4).

The result is then:

$$\left.\begin{aligned}(\Psi|\mathscr{V}|\Psi') = &\frac{k(k-1)}{2}\,\delta(J,J')\,(T'\varrho M^T o|T'\varrho T M^T)\times\\ &\times(-1)^{\lambda}[(2T'+1)(2S'+1)(2L'+1)(2S+1)(2L+1)]^{\frac{1}{2}}\times\\ &\times W(\lambda S' L J; S L')\sum_{\widetilde{\Psi},\vartheta,\vartheta'}(\Psi\{|\widetilde{\Psi}\vartheta)(\Psi'\{|\widetilde{\Psi}\vartheta')\,W(\widetilde{T}TT_2'\varrho; T_2T')\times\\ &\times W(\widetilde{S}SS_2'\lambda; S_2S')\,W(\widetilde{L}LL_2'\lambda; L_2L')\,(T_2\|\mathscr{J}^{(\varrho)}\|T_2')\times\\ &\times(S_2\|\mathscr{S}^{(\lambda)}\|S_2')\,(\gamma L_2\|\mathscr{L}^{(\lambda)}\|\gamma' L_2')\end{aligned}\right\}\quad\text{(A 7.2)}$$

in which the suffix 2 refers to the two-particle state ϑ. Notice that $\widetilde{\Psi}$ must be the same in both c.f.p. to give a non-zero contribution. When $\lambda=\varrho=0$, (A 7.2) reduces to the central force case (A 4.2). In Appendix 8 we shall discuss the two-particle matrix elements to be used in this formula for the various interactions.

For a single-particle non-central operator, such as we shall meet in (A 1.1) in calculating moments and transition strengths, one has a similar result, involving the c.f.p. for the reduction by one, viz.:

$$\Psi = \sum_{\overline{\Psi}}(\Psi\{|\overline{\Psi})\{\overline{\Psi}(1\ldots k-1),\ \varphi(k)\}\,TSL.\qquad\text{(A 7.3)}$$

For the most general symmetric operator $Q=\sum_i Q_i$ where

$$Q_i = \mathscr{J}_{\mu}^{(\varrho)}(i)\,[\mathscr{S}^{(\lambda_1)}(i)\times\mathscr{L}^{(\lambda_2)}(i)]_q^{(\lambda)}$$

n the notation of (A 4.2), we have, in a configuration $(l)^k$,

$$(\psi_{TM_TJM}|Q|\psi'_{T'M'_TJ'M'}) = (T'\varrho M^{T'}\mu|T'\varrho TM^T)(J'\lambda M' q|J'\lambda JM)\times\\ \times(\psi_{TJ}\|Q\|\psi'_{T'J'})$$

with:

$$\left.\begin{aligned}(\psi_{TJ}\|Q\|\psi'_{T'J'}) = k&\sum_{\overline{\psi}}(\psi\{|\overline{\psi})(\psi'\{|\overline{\psi})\,[(2J+1)(2J'+1)(2T+1)(2T'+1)\times\\ &\times(2S+1)(2S'+1)(2L+1)(2L'+1)(2\lambda+1)]^{\frac{1}{2}}\times\\ &\times\begin{Bmatrix}S & L & J\\ S' & L' & J'\\ \lambda_1 & \lambda_2 & \lambda\end{Bmatrix}W(\overline{T}T\tfrac{1}{2}\varrho;\tfrac{1}{2}T')\,W(\overline{S}S\tfrac{1}{2}\lambda_1;\tfrac{1}{2}S')\,W(\overline{L}Ll\lambda_2;lL')\times\\ &\times(\tfrac{1}{2}\|\mathscr{J}^{(\varrho)}\|\tfrac{1}{2})\,(\tfrac{1}{2}\|\mathscr{S}^{(\lambda_1)}\|\tfrac{1}{2})\,(l\|\mathscr{L}^{(\lambda_2)}\|l),\end{aligned}\right\}\quad\text{(A 7.4)}$$

expressing the k-particle matrix element in terms of those for the single particle.

The case of most interest is that of the one-particle spin-orbit force

$$M = \xi\sum_i(\mathbf{s}_i\cdot\boldsymbol{l}_i)\qquad\text{(A 7.5)}$$

where $\mathbf{s}$ and $\boldsymbol{l}$ are vector operators, i.e. tensor operators of degree 1.

Since the single-particle matrix elements are simply

$$(\tfrac{1}{2}\|\mathbf{s}\|\tfrac{1}{2}) = \sqrt{\tfrac{3}{2}}$$

and

$$(l\|\boldsymbol{l}\|l') = [l(l+1)(2l+1)]^{\frac{1}{2}}\,\delta(l,l')$$

we have directly, the formula

$$\left.\begin{aligned}(\Psi|M|\Psi') = -k\xi\,\delta(J,J')\times\\ \times\left[\tfrac{3}{2}l(l+1)(2l+1)(2S+1)(2L+1)(2S'+1)(2L'+1)\right]^{\frac{1}{2}}\times\\ \times W(JS'L1;L'S)\sum_{\overline{\Psi}}(\Psi\{|\overline{\Psi})(\Psi'\{|\overline{\Psi})\,W(\overline{S}S\tfrac{1}{2}1;\tfrac{1}{2}S')\,W(\overline{L}Ll1;lL')\end{aligned}\right\}\quad (\text{A}\,7.6)$$

for the matrix elements in a configuration $(nl)^k$. For states of a mixed configuration defined as in Sect. 19 there is an obvious extension[1] of (A 7.6).

Appendix 8: The two-particle interaction matrix elements. With the many-particle matrix reduced to a function of the two-particle matrix elements by means of the $(k|k-2, 2)$ c.f.p., it remains to evaluate these simple matrix elements for the various interactions. This involves a summation over the charge and spin co-ordinates and an integration over the angular and radial co-ordinates of the two particles. Since the two-particle wave function may be written as a simple product of charge, spin and orbital parts, the contributions to the matrix element from each space may be evaluated separately. For the charge and spin parts, since the operators with which we are concerned may be written in terms of the PAULI spin operators (and the corresponding operators in isotopic spin space) the matrix elements follow immediately from the properties of these PAULI operators. For the orbital part, the procedure is first to expand any function of the interparticle distance which may occur in the operator, into LEGENDRE polynomials of the angular separation of the particle vectors, with coefficients which are functions of the radial positions of the particles. The angular integration may then be carried out using the tensor operator algebra of RACAH and the radial integration by standard methods. A concise way of performing the radial integrals has been developed by TALMI[2].

The formula (A 7.2) for the many-particle matrix involves only the amplitude matrices for two particles. These can be related simply to the complete matrices by (A 4.4). One can therefore determine an amplitude matrix element from *one* element of the complete matrix of any *one* component of the tensor operator, the choice being governed purely by convenience. For a scalar, the complete and amplitude matrices are simply related by

$$(j\,m|\mathscr{I}_0^0|j\,m') = \frac{\delta(m,m')}{(2j+1)^{\frac{1}{2}}}\,(j\|\mathscr{I}^0\|j)\,.$$

Denoting the charge-spin wave function by ${}^{(2T+1)(2S+1)}\Gamma$, the two-body states are

$${}^{13}\Gamma,\ {}^{31}\Gamma \text{ which are antisymmetric in charge and spin}$$

and

$${}^{11}\Gamma,\ {}^{33}\Gamma \text{ which are symmetric in charge and spin}.$$

Consequently, in a totally antisymmetric state, the ${}^{13}\Gamma$ and ${}^{31}\Gamma$ states must have a symmetric, [2], orbital part, while the ${}^{11}\Gamma$ and ${}^{33}\Gamma$ must have an antisymmetric, [11], orbital part. Consider now the orbital wave functions for two inequivalent particles nl and $n'l'$, omitting the radial quantum numbers for

---

[1] J. P. ELLIOTT and B. H. FLOWERS: Proc. Roy. Soc. Lond., Ser. A **229**, 536 (1955).
[2] I. TALMI: Helv. phys. Acta **25**, 185 (1952).

brevity,

and

$$\left.\begin{aligned}\varphi\{(l l')\,[2]\,L M\} &= \sqrt{\tfrac{1}{2}}\,[\varphi(l_1 l_2' L M) + \varphi(l_2 l_1' L M)]\\ \varphi\{(l l')\,[11]\,L M\} &= \sqrt{\tfrac{1}{2}}\,[\varphi(l_1 l_2' L M) - \varphi(l_2 l_1' L M)],\end{aligned}\right\}\tag{A8.1}$$

where the suffix denotes particle number. The orbital matrix elements may therefore be written as

$$\left.\begin{aligned}((l l')[2]L\|\mathcal{L}_{12}\|(\tilde{l}\tilde{l}')[2]L') &= [(l_1 l_2', L\|\mathcal{L}_{12}\|\tilde{l}_1 \tilde{l}_2', L') + (l_1 l_2', L\|\mathcal{L}_{12}\|\tilde{l}_2 \tilde{l}_1', L')],\\ ((l l')[11]L\|\mathcal{L}_{12}\|(\tilde{l}\tilde{l}')[11]L') &= [(l_1 l_2', L\|\mathcal{L}_{12}\|\tilde{l}_1 \tilde{l}_2', L') - (l_1 l_2', L\|\mathcal{L}_{12}\|\tilde{l}_2 \tilde{l}_1', L')]\end{aligned}\right\}\tag{A8.2}$$

where we have considered the most general case, encountered in configurational mixing, in which the l-values on the right may both be different from those on the left. By changing the order of coupling on the right of the matrix element occurring as second term on the right-hand side of both Eqs. (A8.2) this term becomes

$$(l_1 l_2', L\|\mathcal{L}_{12}\|\tilde{l}_2 \tilde{l}_1', L') = (-1)^{\tilde{l}+\tilde{l}'-L'}\,(l_1 l_2', L\|\mathcal{L}_{12}\|\tilde{l}_1' \tilde{l}_2, L'),\tag{A8.3}$$

which is now of the same form as the first term on the right-hand side of (A8.2). It remains then to evaluate a matrix element of the form (A8.3) where the wave functions are now simply vector coupled with particle number 1 first on both sides. At this point we must consider each type of interaction separately.

α) The central force. Using the notation (23.8) for the central force

$$v_{12} = (W + M P_{12}^r - H P_{12}^\tau + B P_{12}^\sigma)\, V(r_{12})$$

the charge-spin matrix elements are simply

$$W + M(-1)^{T+S+1} + H(-1)^T + B(-1)^{S+1}.\tag{A8.4}$$

For the orbital part, expand[1]

$$V(r_{12}) = \sum_{k=0}^{\infty} P_k(\cos\omega_{12})\, V_k(r_1, r_2)\tag{A8.5}$$

where $P_k(\cos\omega_{12})$ is a LEGENDRE polynomial and ω_{12} is the angle between $\boldsymbol{r}_1$ and $\boldsymbol{r}_2$. Then the matrix element separates into angular and radial parts,

$$\left.\begin{aligned}&((n l)_1 (n' l')_2, L M \mid V(r_{12}) \mid (\tilde{n}\tilde{l})_1 (\tilde{n}'\tilde{l}')_2 L M)\\ &= \sum_{k=0}^{\infty} (l_1 l_2' L M \mid P_k(\cos\omega_{12}) \mid \tilde{l}_1 \tilde{l}_2', L M)\, F^k(n l, n' l', \tilde{n}\tilde{l}, \tilde{n}'\tilde{l}')\end{aligned}\right\}\tag{A8.6}$$

where the radial integral

$$F^k = \int_0^\infty\!\!\int_0^\infty u_{nl}(r_1)\, u_{n'l'}(r_2)\, u_{\tilde{n}\tilde{l}}(r_1)\, u_{\tilde{n}'\tilde{l}'}(r_2)\, V_k(r_1, r_2)\, r_1^2 r_2^2\, d r_1\, d r_2\tag{A8.7}$$

and $u(nl)$ is the radial part of the single particle wave function (see Chap. I) and normalised,

$$\int_0^\infty u_{nl}^2(r)\, r^2\, d r = 1.$$

From the spherical harmonic addition theorem we may write

$$P_k(\cos\omega_{12}) = \left(\frac{4\pi}{2k+1}\right) \sum_q (-1)^q\, Y_q^{(k)}(1)\, Y_{-q}^{(k)}(2)$$

---

[1] Notice that the values of k which give non-zero matrix elements are restricted by (A8.8) to a very few.

which is a scalar product of tensor operators and may be written in the notation of (A 4.3 b) as

$$P_k(\cos\omega_{12}) = \left(\frac{4\pi}{2k+1}\right)\left(\boldsymbol{Y}^{(k)}(1)\cdot\boldsymbol{Y}^{(k)}(2)\right).$$

The evaluation of the first factor on the right-hand side of (A 8.6) then follows immediately from (A 4.8), the single particle matrix elements of the form $(l\|Y^{(k)}\|\tilde{l})$ being given by (A 1.6). The general result is

$$\left.\begin{aligned}&(l_1 l_2', LM|P_k(\cos\omega_{12})|\tilde{l}_1\tilde{l}_2', LM)\\&=\frac{(-1)^{\frac{1}{2}(l+l'+\tilde{l}+\tilde{l}')+L}}{2}[(2l+1)(2l'+1)(2\tilde{l}+1)(2\tilde{l}'+1)C_{l\tilde{l}k}C_{l'\tilde{l}'k}]^{\frac{1}{2}}W(ll'\tilde{l}\tilde{l}';Lk)\end{aligned}\right\}\quad\text{(A 8.8)}$$

where the C_{abc} were defined in Appendix 2 and are tabulated[1]. This complete generality is only met in coupling terms between configurations. Specialisations of more common use are mentioned below.

When $(nl)=(n'l')=(\tilde{n}\tilde{l})=(\tilde{n}'\tilde{l}')$, as is the case in a simple configuration of equivalent particles, the process (A 8.1) of constructing states of definite symmetry is not needed, the state being symmetric or antisymmetric according as L is even or odd. We may then start immediately at (A 8.6). When $(nl)=(\tilde{n}\tilde{l})$ and $(n'l')=(\tilde{n}'\tilde{l}')$, as occurs within a mixed configuration $(nl)^p(n'l')^{p'}$, the two terms on the right hand side of (A 8.2) are referred to as direct and exchange contributions respectively, the corresponding radial integrals being, from (A 8.7),

$$F^k(nl,n'l') = \int_0^\infty\int_0^\infty u_{nl}^2(r_1)\,u_{n'l'}^2(r_2)\,V_k(r_1,r_2)\,r_1^2 r_2^2\,dr_1\,dr_2$$

and

$$G^k(nl,n'l') = \int_0^\infty\int_0^\infty u_{nl}(r_1)\,u_{n'l'}(r_1)\,u_{nl}(r_2)\,u_{n'l'}(r_2)\,V_k(r_1,r_2)\,r_1^2 r_2^2\,dr_1\,dr_2.$$

Their coefficients are given by the corresponding special cases of (A 8.8), remembering (A 8.3).

We shall describe briefly the calculation of the radial integrals in Appendix 10.

β) The spin-orbit force. Although the single-body spin orbit force $\sum_i(\boldsymbol{s}_i\cdot\boldsymbol{l}_i)$ is much used in shell model calculations, it should be remembered that it is not an interparticle force. Nevertheless it is true that such a term arises from the interaction of a particle with a closed shell through the two-body spin-orbit force introduced in Sect. 23,

$$(\boldsymbol{\sigma}_1+\boldsymbol{\sigma}_2)\cdot[(\boldsymbol{r}_1-\boldsymbol{r}_2)\times(\boldsymbol{p}_1-\boldsymbol{p}_2)]\,V_1(r_{12}).\quad\text{(A 8.9)}$$

Calculations show in fact that, even for the interaction among a number of particles outside a closed shell, the effect of (A 8.9) may be quite closely represented by the single particle form $\sum_i(\boldsymbol{s}_i\cdot\boldsymbol{l}_i)$. For accurate work however the matrix elements of (A 8.9) are needed for use in (A 7.2).

The spin part of (A 8.9) has non-zero matrix elements only in the triplet spin state and then has the value

$$(\tfrac{1}{2}\tfrac{1}{2},1\|(\boldsymbol{\sigma}_1+\boldsymbol{\sigma}_2)\|\tfrac{1}{2}\tfrac{1}{2},1)=2\sqrt{6}.$$

To evaluate the orbital part, one must, in addition to the expansion (A 8.5) of $V_1(r_{12})$, evaluate the result of the momentum operators $\boldsymbol{p}$ acting on the wave

[1] G. H. SHORTLEY and B. FRIED: Phys. Rev. 54, 739 (1938).

function using (A1.4) and after some tensor operator manipulation, carry out the resulting angular integrals using (A1.6). A general formula for this matrix element has been derived[1]. There will, of course, be slightly more complicated radial integrals than (A8.7), but they are amenable to the same method of calculation.

γ) The tensor force. Here again, the spin matrix element is non-zero only in triplet states, when

$$(\tfrac{1}{2}\,\tfrac{1}{2},1\|\mathcal{S}^{(2)}\|\tfrac{1}{2}\,\tfrac{1}{2}\,1)=\sqrt{30}\,,$$

in the notation of Sect. 23. For the orbital part, consider the $q=0$ component of $\mathcal{L}^2$,

$$\mathcal{L}_0^{(2)}(12)=\frac{V_2(r_{12})}{\sqrt{6}}\,\frac{(3z_{12}^2-r_{12}^2)}{r_{12}^2}\,.$$

It is then convenient to expand $V_2(r_{12})/r_{12}^2$ as a single function in the manner of (A8.5) and to write the remaining factor in terms of r_1, r_2, and spherical harmonics of the angles. The integrations are then carried out by the standard processes. A general expression for the two-body matrix element of the tensor force has also been derived[2].

δ) The COULOMB *force.* In the isotopic spin formalism, we may write the COULOMB force as an operator of the type Q of (A7.1) as follows. An interaction $V_1(r_{12})$ between protons only can be written as

$$\tfrac{1}{4}\big(1-\tau_z(1)\big)\big(1-\tau_z(2)\big)\,V_0(r_{12}) \tag{A8.10}$$

but this is not yet in the form (A7.1). However, by defining the tensor operators

$$T_0^{(1)}(12)=\big(\tau_z(1)+\tau_z(2)\big)\,,$$

$$T_0^{(2)}(12)=\frac{1}{\sqrt{6}}\big(2\tau_z(1)\,\tau_z(2)-\tau_x(1)\,\tau_x(2)-\tau_y(1)\,\tau_y(2)\big)\,,$$

(A8.10) becomes

$$\tfrac{1}{4}\big[1+\tfrac{1}{3}(\boldsymbol{\tau}_1\cdot\boldsymbol{\tau}_2)-T_0^{(1)}(12)+\sqrt{\tfrac{2}{3}}\,T_0^{(2)}(12)\big]\,V_0(r_{12})$$

which is of the form (A7.1) involving a combination of tensor operators in isotopic spin of degree $\varrho=0$, 1 and 2.

Appendix 9: Closed shell interactions. When one evaluates (a) the interaction energy of all particles in a closed shell, (b) the interaction energy of all particles in two different closed shells or (c) the interaction energy of an unfilled shell with a closed shell, using the interactions listed in Sect. 23, one finds particularly simple results. On general invariance grounds, the non-central forces [$\lambda\neq0$ in (A7.1)] give no contribution to (a) or (b) and the tensor force ($\lambda=2$) gives no contribution to (c) either. We shall therefore consider the central force contribution to (a), (b), and (c) and the vector-force contribution to (c).

Central Forces. We may calculate the various contributions in a straightforward way, using (A7.2) with $\lambda=\varrho=0$ together with the expressions in Appendix 8, for the two-body elements. For (a) the appropriate c.f.p. will be those for the removal of two particles from a closed shell which are simply

$$(\psi\{|\tilde{\psi})=\left\{\frac{2(2\tilde{T}+1)(2\tilde{S}+1)(2\tilde{L}+1)}{A_r(A_r-1)}\right\}^{\frac{1}{2}}$$

[1] J. HOPE and L. W. LONGDON: Phys. Rev. **102**, 1124 (1956).
[2] J. HOPE and L. W. LONGDON: Phys. Rev. **101**, 710 (1956).

where ψ is the wave function for the closed shell of A_r particles and $\tilde{\psi}$ that of a state of (A_r-2) particles with quantum numbers $\tilde{T}$, $\tilde{S}$, $\tilde{L}$, which are sufficient to specify the state. For (b) and (c) one has to construct the c.f.p. relating to the removal of one particle from one group and one from another, using the last term of (A 6.4) which simplifies considerably when closed shells are involved. The results are:

(a) Interaction energy in the $(n_r l_r)$ shell,

$$E = (2l_r+1)^2 \left\{ 8a_0 F^0(n_r l_r) - (a_0 + 3a_\sigma + 3a_\tau + 9a_{\sigma\tau}) \sum_{k=0}^{\infty} C_{l_r l_r k} F^k(n_r l_r) \right\}. \quad (A\,9.1)$$

(c) Interaction energy between a configuration $(nl)^p$ and the $(n_r l_r)$ shell,

$$E = \delta(\psi,\psi') \frac{p(2l_r+1)}{2} \left\{ 8a_0 F^0(n_r l_r, nl) - (a_0+3a_\sigma+3a_\tau+9a_{\sigma\tau}) \sum_{k=0}^{\infty} C_{l l_r k} G^k(n_r l_r, nl) \right\}. \quad (A\,9.2)$$

In (A 9.2) the $\delta(\psi,\psi')$ refers to states of the $(nl)^p$ system, so that the closed shell interaction is a single particle scalar in that system and so does not enter any calculation of the relative positions of levels of that system. The contribution (b) is a special case of (A 9.2) for $p = 4(2l+1)$.

The vector force. It has been shown by various writers[1] that the interaction between a closed $(n_r l_r)$ shell and a configuration $(nl)^p$ from the two-body vector force (A 8.9) is exactly represented by a single-body spin-orbit term within the configuration $(nl)^p$. The strength of this term depends on the wave functions $u_{n_r l_r}$ and u_{nl} and on the form of the two-body radial dependence $V_1(r_{ij})$ in a sensitive way and since these functions are not well known it is difficult to discuss the strength variation except over a small mass range. Formulae for the strength in terms of the two-particle matrix elements[2] and in terms of the radial integrals[3] have been given (see also Sect. 36).

Appendix 10: The radial integrals. The integral (A 8.7) may be reduced to a linear combination of simple integrals over the function $V(r_{12})$, without postulating the form of this function, using a method developed by TALMI[4]. Although this process is only applicable to oscillator wave functions, these are the only kind which are practicable to use. Essentially, (A 8.7) is of the form

$$\int_0^\infty \int_0^\infty r_1^\lambda r_2^\mu V_k(r_1 r_2)\, e^{-(r_1^2+r_2^2)/b^2} r_1^2 r_2^2\, dr_1\, dr_2, \quad (A\,10.1)$$

and, inverting (A 8.5) and substituting for $V_k(r_1, r_2)$ in (A 10.1) gives

$$\left(\frac{2k+1}{2}\right) \int_0^\infty \int_0^\infty \int_{-1}^{+1} r_1^\lambda r_2^\mu V(r_{12})\, P_k(\cos\omega_{12})\, e^{-(r_1^2+r_2^2)/b^2} r_1^2 r_2^2\, dr_1\, dr_2\, d(\cos\omega_{12}).$$

Now, by means of a series of transformations which are essentially those from $\boldsymbol{r}_1, \boldsymbol{r}_2$ to the relative coordinates $(\boldsymbol{r}_1+\boldsymbol{r}_2)$, $(\boldsymbol{r}_1-\boldsymbol{r}_2)$, this expression becomes

$$2^{-\frac{1}{2}(\lambda+\mu)} \left(\frac{2k+1}{2}\right) \int_0^\infty \int_0^\infty \int_{-1}^{+1} x_1^\lambda x_2^\mu V(t)\, P_k\!\left(\frac{s^2-t^2}{2x_1 x_2}\right) e^{-(s^2+t^2)/b^2} s^2 t^2\, ds\, dt\, dz \quad (A\,10.2)$$

[1] J. HUGHES and K. J. LE COUTEUR: Proc. Phys. Soc. Lond., A **63**, 1219 (1950). — C. H. BLANCHARD and R. AVERY: Phys. Rev. **81**, 35 (1951). — J. P. ELLIOTT and A. M. LANE: Phys. Rev. **96**, 1660 (1954). — R. J. BLIN STOYLE: Phil. Mag. **46**, 973 (1953).

[2] J. P. ELLIOTT: Proc. Roy. Soc. Lond., Ser. A **218**, 345 (1953).

[3] C. H. BLANCHARD and R. AVERY: Phys. Rev. **81**, 35 (1951).

[4] I. TALMI: Helv. phys. Acta **25**, 185 (1952).

where

$$x_1 = \sqrt{2}\, r_1 = \{s^2 + t^2 + 2stz\}^{\frac{1}{2}}, \qquad x_2 = \sqrt{2}\, r_2 = \{s^2 + t^2 - 2stz\}^{\frac{1}{2}}.$$

By expanding the LEGENDRE polynomial and the x_1^λ, x_2^μ terms, (A10.2) becomes a linear combination of the integrals

$$\int\limits_0^\infty V(t)\, t^{2n}\, e^{-t^2/b^2}\, dt, \tag{A10.3}$$

the coefficients being independent of $V(t)$. The integrations over s and z are simple.

For any form of the radial dependence of the interaction, (A10.3) may be readily evaluated and hence the integrals (A10.1) deduced.

Appendix 11: The matrix elements of various physical operators. In this section we collect together the operators describing the various physical quantities that we may want to calculate from our individual particle wave functions to compare with experiment. We describe the calculation of the matrix elements of those operators. As well as the static magnetic and quadrupole moments, we consider the dynamic operators concerned with transitions, for example β-decay, γ-decay and particle emission or capture. We restrict ourselves to a calculation of those factors occurring in reaction theory which depend on the internal structure of the nucleus (which our model describes).

We should remind readers of the discussion of centre-of-mass corrections in Sect. 22, in particular that so long as one is dealing with good states in the sense of Sect. 22 one gets the same result from multipole operators whether referred to the centre-of-mass or the well centre and in the usual approach the latter is, of course, simpler. For this reason we write down the operators in the well-centre form. If one has not constructed good states, then the error will be least if the operator is referred to the centre-of-mass. One can construct the centre-of-mass form simply by writing $\left(\boldsymbol{r}_i - \frac{1}{A}\sum_i \boldsymbol{r}_i\right)$ in place of $\boldsymbol{r}_i$ and likewise for $\boldsymbol{p}_i$, expanding the result into many-particle operators.

Magnetic moment. The nuclear magnetic moment μ, in units of the nuclear magneton $\frac{e\hbar}{2Mc}$, is defined as the expectation value, in the state of maximum magnetic quantum number $M = J$, of the operator

$$\mu_z = \sum_i^p l_z(i) + \mu_p \sum_i^p \sigma_z(i) + \mu_n \sum_i^n \sigma_z(i)$$

where the superscripts indicate summation over neutrons or protons, and μ_p, μ_n are the proton and neutron magnetic moments[1]. In the isotopic spin formalism this becomes

$$\mu_z = \tfrac{1}{2}\sum_i \left(1 - \tau_z(i)\right) l_z(i) + \tfrac{1}{2}\sum_i \{(\mu_n + \mu_p) + (\mu_n - \mu_p)\,\tau_z(i)\}\,\sigma_z(i), \tag{A11.1}$$

which is a combination of single particle tensor operators of degree 0 or 1 in the T, S and L spaces and may be evaluated using (A7.4) and the relation (A4.4),

$$\left.\begin{aligned} \mu &= (\psi_{JJ}|\,\mu_z\,|\psi_{JJ}) \\ &= (J1J0\,|\,J1JJ)\,\frac{(\psi_J||\boldsymbol{\mu}||\psi_J)}{\sqrt{2J+1}} = \sqrt{\frac{J}{(J+1)(2J+1)}}\,(\psi_J||\boldsymbol{\mu}||\psi_J). \end{aligned}\right\} \tag{A11.2}$$

---

[1] $\left\{\begin{aligned}\mu_n &= -1.9103\\ \mu_p &= 2.7896\end{aligned}\right\}$ in units of the nuclear magneton, $\left(\frac{e\hbar}{2Mc}\right)$.

Quadrupole moment. In a similar way, the quadrupole moment is defined as the expectation value, in the state with $M = J$, of the operator

$$e \sum_i^p \left(3 z^2(i) - r^2(i)\right) = \sqrt{\frac{16\pi}{5}}\, Q_0^{(2)}$$

where

$$Q_0^{(2)} = e \sum_i^p Y_0^{(2)}(i)\, r^2(i) = \frac{e}{2} \sum_i \left(1 - \tau_z(i)\right) Y_0^{(2)}(i)\, r^2(i), \qquad \text{(A 11.3)}$$

which again may be evaluated in a many particle state using (A 7.4).

γ-transitions. For a γ-transition we may calculate the radiation width Γ, transition probability T, or lifetime τ, related simply by

$$T = \frac{1}{\tau} = \frac{\Gamma}{\hbar}.$$

For an $M1$ transition, the relevant formula[1] for a transition of energy E from a level J to one J' is

$$\Gamma = \frac{e^2 E^3}{3 M^2 c^5 \hbar} \sum_{M'(q)} (JM|\,\mu_q\,|J'M')^2 = \frac{e^2 E^3}{3M^2 c^5 \hbar} \frac{(J\,||\boldsymbol{\mu}||\,J')^2}{(2J+1)} \sum_{M'(q)} (J' 1 M' q\,|\,J' 1\, J M)^2$$

and, carrying out the WIGNER coefficient summation using (A 2.6b).

$$\Gamma = \frac{e^2 E^3}{3 M^2 c^5 \hbar} \frac{(\psi_J ||\, \boldsymbol{\mu}\, || \psi_{J'})^2}{(2J+1)} \qquad \text{(A 11.4)}$$

where e is the electron charge, E the γ-ray energy and M the nucleon mass. The operator $\boldsymbol{\mu}$ in (A 11.4) is exactly that described above, (A 11.1). Similarly, for an $E2$ transition we have

$$\Gamma = \frac{4\pi E^5}{75 \hbar^5 c^5} \frac{(\psi_J ||\, \boldsymbol{Q}^{(2)}\, || \psi_{J'})^2}{(2J+1)} \qquad \text{(A 11.5)}$$

where $\boldsymbol{Q}^{(2)}$ is given by (A 11.3). For an $E1$ transition

$$\Gamma = \frac{16\pi}{9} \frac{E^3}{\hbar^3 c^3} \frac{(\psi_J ||\, \boldsymbol{Q}^{(1)}\, || \psi_{J'})^2}{(2J+1)} \qquad \text{(A 11.6)}$$

where

$$Q_q^{(1)} = \frac{e}{2} \sum_i \left(1 - \tau_z(i)\right) Y_q^{(1)}(i)\, r(i). \qquad \text{(A 11.7)}$$

Since the first term in (A 11.7) is simply $\sum_i \boldsymbol{r}_q(i) = A R$ and in a good state $\langle R \rangle = 0$, this term may be ignored leaving

$$Q_q^{(1)} = -\tfrac{1}{2} \sum_i \tau_z(i)\, Y_q^{(1)}(i)\, r(i)$$

which, from its vectorial behaviour in isotopic spin, vanishes between $T = 0$ states, giving the well known selection rule. For higher order transitions the procedure is very similar, using the operators already written down in Sect. 4.

β-decay. In light nuclei, most observed β-transitions are of the allowed type, for which the ft value may be written[2]:

$$ft = \frac{\mathrm{B}}{(1-x) F^2 + x G^2} \qquad \text{(A 11.8)}$$

[1] J. M. BLATT and V. F. WEISSKOPF: Theoretical Nuclear Physics, p. 595. New York: John Wiley & Sons 1952.

[2] A. BOHR and B. R. MOTTELSEN: Dan. Mat. fys. Medd. 27, No. 16 (1953).

where the square of the FERMI matrix element is given by

$$F^2 = \tfrac{1}{4}\,(\psi\,|\sum_j (\tau_x(j) \mp i\,\tau_y(j))|\,\psi')^2 = (T \mp M^{T'} + 1)\,(T \pm M^{T'})\,\delta(T,\,T')$$

and the GAMOW-TELLER term by

$$\left.\begin{aligned} G^2 &= \tfrac{1}{4}\sum_{M'} (\psi\,|\sum_j (\tau_x(j) \mp i\,\tau_y(j))\,\sigma(j)\,|\,\psi')^2 \\ &= \tfrac{1}{2}\,(T'1\,M^{T'} \mp 1\,|\,T'1\,TM^T)^2\,(\psi\,||\sum_j \boldsymbol{\tau}(j)\,\sigma(j)\,||\psi')^2/(2T+1)\,(2J+1). \end{aligned}\right\}\quad (\text{A}\,11.9)$$

The choice of sign in (A 11.9) corresponds to β^+ and β^- emission respectively, and the relevant single particle matrix elements are simply

$$(\tfrac{1}{2}\,||\boldsymbol{\tau}||\,\tfrac{1}{2}) = (\tfrac{1}{2}\,||\sigma||\,\tfrac{1}{2}) = \sqrt{6}\,.$$

The best values at present for the parameters x, B of the assumed β-interaction, are[1]

$$x = 0.560 \pm 0.012,$$
$$B = 2783 \pm 70\,.$$

Reduced nucleon widths. The nucleon width $\Gamma_{\lambda c}$ of a level λ measures the probability that the level will decay by the emission of a nucleon, through a channel c. It therefore describes directly part of the structure of the wave function for that level and is obviously a valuable quantity to use in a comparison of theory with experiment. Experimentally there are unfortunately few nuclei where these widths may be measured directly for the low levels, but the stripping reaction, once the mechanism is properly understood, can be used to deduce nucleon widths for the low levels in which we are interested.

We may first of all split the width $\Gamma_{\lambda c}$ into two factors

$$\Gamma_{\lambda c} = 2\,P_c\,\gamma_{\lambda c}^2$$

the first, P_c, being the penetrability factor, depending only on conditions outside the nucleus and on the channel energy. All dependence on the nuclear structure of the level λ is contained in the second factor $\gamma_{\lambda c}^2$ which is called the reduced width. Following LANE and THOMAS[2], the square root, $\gamma_{\lambda c}$, of the reduced width, i.e. the reduced width amplitude, is given by

$$\gamma_{\lambda c} = \left(\frac{\hbar^2}{2MR}\right)^{\frac{1}{2}} \int \psi_{\lambda JM}^*\,\varphi_{cJM}\,d\tau_c. \qquad (\text{A}\,11.10)$$

Here, $\psi_{\lambda JM}$ is the nuclear wave function for the level λ, and $r\,\varphi_{cJM}$ is the vector coupled wave function of the emitted particle and the residual nucleus with relative angular momentum l, viz:

$$r\,\varphi_{cJM} = \{\{\overline{\psi}_{\bar{J}},\,\tfrac{1}{2}\}\,j_c,\,l\}\,J\,M$$

where $\overline{\psi}$ represents the residual nucleus of particles 1 to $(k-1)$ and j_c the channel spin. The integration in (A 11.10) extends over the channel surface, that is to say over all particles but the k-th and over all co-ordinates of the k-th except the radial one which is taken at the surface value $r = R$. Since this determines the probability of emission of one particular nucleon, the k-th, we must include

[1] O. KOFOED-HANSEN: Private communication.

[2] A. M. LANE and R. G. THOMAS: Rev. Mod. Phys. **1957** (to be published).

a factor $\sqrt{k}$ into $\gamma_{\lambda c}$, i.e. a factor k in $\gamma_{\lambda c}^2$. Then, using the standard techniques, we find:

$$\left.\begin{aligned}\gamma_{\lambda c} &= \left(\frac{\hbar^2}{2MR}\right)^{\frac{1}{2}} \sqrt{k}\, R\, u_l(R)\, (\overline{T}\tfrac{1}{2} M^{\overline{T}} \pm \tfrac{1}{2} \,|\, \overline{T}\tfrac{1}{2}\, T M^T) \times \\ &\times [(2\overline{J}+1)(2j_c+1)(2S+1)(2L+1)]^{\frac{1}{2}} \times \\ &\times (\psi\{|\,\overline{\psi})\, W(S\overline{L}Jl; j_c L)\, W(\overline{L}\overline{J}S\tfrac{1}{2}; \overline{S}j_c)\end{aligned}\right\} \qquad \text{(A 11.11)}$$

where $u_l(R)$ is the radial part of the single particle wave function evaluated at $r=R$. Here we have assumed that ψ and $\overline{\psi}$ are pure L-S coupled states, but for a general state expressed as a combination of L-S states, one uses (A 11.11) to set up the matrix of γ_c in the L-S scheme and hence derives $\gamma_{\lambda c}$ as described in Sect. 25.

It is convenient to use a dimensionless reduced width

$$\vartheta_{\lambda c}^2 = \left(\frac{MR^2}{\hbar^2}\right) \gamma_{\lambda c}^2$$

and also to use the ratio $\Lambda_{\lambda c}^2 = \vartheta_{\lambda c}^2 / \vartheta_0^2$ of the reduced width to the single particle width $\vartheta_0^2 = \frac{R^3}{2} u_l^2(R)$. In this way $\Lambda_{\lambda c}^2$ depends on nothing but the structure of the level λ. Finally, one sums over all channels to get, in the general case,

$$\Lambda_\lambda^2 = k(\overline{T}\tfrac{1}{2} M^{\overline{T}} \pm \tfrac{1}{2} \,|\, \overline{T}\tfrac{1}{2}\, T M^T)^2 (2\overline{J}+1) \sum_{j_c} (2j_c+1) \times$$
$$\times \Big[\sum_{\overline{S},\overline{L},S,L} (2S+1)^{\frac{1}{2}} (2L+1)^{\frac{1}{2}} (\Psi|\psi)(\psi\{|\,\overline{\psi})(\overline{\Psi}|\overline{\psi})\, W(S\overline{L}Jl; j_c L)\, W(\overline{L}\overline{J}S\tfrac{1}{2}, \overline{S}j_c)\Big]^2$$

where the general state Ψ is expressed as a linear combination of the complete L-S coupled set ψ by the expansion

$$\Psi = \sum_\psi (\Psi|\psi)\, \psi$$

and similarly for the residual nucleus $\overline{\Psi}$.

In principle one may also calculate the widths for break-up into components having k_1 and $k_2 = k - k_1$ nucleons respectively. For a discussion of such processes, together with a more complete description of the simple process than that given here, see the review article by LANE and THOMAS[1].

Acknowledgement.

Our task in writing the present review has been considerably lightened by the existence of a number of earlier reviews and papers devoted to various aspects of the shell-model. The names of the many authors to whom we are indebted on this score are to be found in the references accompanying this review. We would like to record here a few papers from which we have derived especial benefit. These are:

Elementary Theory of Nuclear Shell Structure, by M. G. MAYER and J. H. D. JENSEN, New York: John Wiley & Sons 1955;

Shell Theory of the Nucleus, by E. FEENBERG, Princeton: Princeton University Press 1955;

Theories of Nuclear Moments, by R. J. BLIN-STOYLE, Rev. Mod. Phys. **28**, 75 (1956).

The Energy Levels and Structure of Light Nuclei, by D. R. INGLIS, Rev. Mod. Phys. **25**, 390 (1953).

[1] A. M. LANE and R. G. THOMAS: Rev. Mod. Phys. **1957** (to be published).

Models of Nuclear Structure.

By

S. A. MOSZKOWSKI.

With 44 Figures.

Introduction and brief survey of the nuclear models.

Nuclear physics is a young science dating back only to the discovery of radioactivity in 1896. The development of nuclear physics can be divided into two stages. During the first stage (1896—1932) many fundamental facts about atomic nuclei were discovered. Thus Lord RUTHERFORD's experiments showed that the nucleus occupies only a very small fraction of the atomic volume. Also, isotopes were discovered for the first time, the field of mass spectroscopy was developed, and a few nuclear reactions were induced in the laboratory.

The two most important theoretical developments of this period were the formulations of the theory of relativity and of quantum theory. Many problems in atomic structure were solved by the application of quantum mechanics, and in nuclear physics, the phenomenon of α-decay was explained, at least qualitatively. Attempts to understand details of nuclear structure were not successful, however, since it was widely thought that nuclei are made up of protons and electrons, the only fundamental particles known at that time.

The discovery of the neutron in 1932 ushered in the period of modern nuclear physics. This discovery was quickly followed by the introduction of the general neutron-proton model which forms the underlying basis of all subsequent theories of nuclear structure.

During the last twenty-five years the progress in nuclear physics has been very rapid. The development of new experimental techniques, especially the development of high-energy accelerators, has resulted in very extensive studies of nuclear phenomena involving stable as well as artifically radioactive nuclei. A large amount of information regarding nuclear states, level spectra, and reactions[1-4] has been accumulated. Much information regarding the interactions between individual nucleons has been gained both from experimental studies, e.g. the deuteron and nucleon-nucleon scattering, and from theoretical investigations based on meson theory[5]. An outstanding feature of the nuclear forces is

[1] For a summary of information on nuclear masses, spins and moments, see the articles by A. H. WAPSTRA (masses), LAUKIEN and KELLY (spins and magnetic moments) and C. H. TOWNES (quadrupole moments), all in Vol. XXXVIII of this Encyclopedia.

[2] The available experimental information on nuclear level spectra and reactions is summarized by W. E. BURCHAM (light nuclei) and B. B. KINSEY (heavy nuclei) in Vol. XL.

[3] The theory of nuclear reactions is discussed by BREIT *et al.* in Vol. XLI.

[4] Radioactive decay processes and their application to nuclear structure are discussed by E. GREULING and L. W. NORDHEIM (theory of beta-decay) and by O. KOFOED-HANSEN (experimental information on beta-decay) in Vol. XLI, and further by ALBURGER (nuclear isomers), PERLMAN and RASMUSSEN (alpha-decay), DEVONS and GOLDFARB (angular correlation) and BLIN-STOYLE and GRACE (oriented nuclei) in Vol. XLII.

[5] For a complete discussion of the two-nucleon problem see the article by L. HULTHÉN and M. SUGAWARA in this volume.

their great strength and short range in contrast with the long-ranged electrostatic forces in atoms.

Consider now the problems to be faced in developing the theory of nuclear structure. First of all, even if the exact character of the nuclear forces were known, it would still not be possible to solve the resulting many body problem exactly. In general only two-body problems are soluble. The major task then is to make approximations which reduce the exact but insoluble problem to a soluble one, while introducing as little error as possible. In physical terms we construct a model system which approximates the actual system as closely as possible, but is mathematically tractable.

The development of nuclear models has followed along two different lines. Side by side with the so-called "*strong interaction models*" which treat the nucleus as an assemblage of closely coupled particles, are the various "*independent particle models*", in which the nucleons are assumed to move rather independently in the average nuclear field. Recent evidence suggests that the assumptions of the independent particle models are the more nearly correct, at least for low energy phenomena, though the coupling of nucleons by virtue of their mutual interactions can by no means be neglected.

In this article, the main features of several nuclear models are described[1]. These models, most of which were formulated during the early 1930's, have yielded considerable insight into nuclear structure, proven valuable for correlating nuclear data and stimulated further work. In the present article attention is given mainly to the study of the principles underlying the nuclear models. Applications are given primarily to illustrate these principles. In particular we consider especially nuclear ground states and states of low excitation. Information, both experimental and theoretical, is much more complete for these than it is for highly excited states.

The liquid drop model, perhaps the simplest of all nuclear models, is discussed in Chap. A. The analogy between a nucleus and a liquid drop can be used to understand such features as nuclear binding, at least in a qualitative way.

Chap. B deals with the Fermi gas model. In this model the nucleons are regarded as moving approximately independently in the nucleus in spite of the strong interactions between free nucleons. The nucleus is regarded as effectively infinite so that the wavefunctions of individual nucleons are plane waves. With this model it is possible to correlate empirical data on nuclear binding energies in a more precise way than with the liquid drop model.

In Chap. C we discuss the optical model which treats the nucleus as a refractive medium. This model has recently proven very successful for the interpretation of nucleon-nuclei scattering data.

A related low-energy nuclear model, the compound nucleus model, is more appropriately treated in connection with nuclear reactions, and is discussed only briefly in this chapter.

Chap. D briefly surveys the alpha-particle model, whose validity seems to be restricted to a few special nuclei.

The nuclear shell model, based in large part on the electron shell model of atoms, has been very fully treated in the preceding article of this volume. For this reason the discussion of the shell model in Chap. E is confined to a few essentials and is given mainly to illustrate the relation of this model to the other nuclear models.

[1] For complete treatments of theoretical nuclear physics, including some of the nuclear models, see Ref. [*1*] to [*3*] in the bibliography at the end of this article. For shorter surveys of nuclear models see Ref. [*4*] to [*6*].

Table 1. *Assumptions of nuclear models.*

Model	Type	Assumptions	See Chapter
Independent particle models (IPM)		Nucleons move nearly independently in a common nuclear potential.	B, C, E, F
Strong interaction models (SIM)		Nucleons are strongly coupled to each other because of their strong and short range interactions.	A, C, D
Liquid drop model	SIM	Nucleus is regarded as a liquid drop with nucleons playing the role of molecules.	A
Fermi gas model (uniform model)	IPM	Nucleons move approximately independently in the nucleus and their individual wave functions are taken to be plane waves.	B
Potential well model	IPM	Nucleus is regarded as a simple real potential well.	C
Compound nucleus model	SIM	Whenever an incident nucleon enters the nucleus, it is always absorbed ,and a compound nucleus is formed. The mode of disintegration of the compound nucleus is independent of the specific way in which it has been formed.	C
Optical model (Cloudy crystal ball model) (Complex potential well model)	IPM	A modification of the potential well model in which the potential is made complex to account for elastic scattering as well as nuclear reactions. The latter effectively remove nucleons from the beam of incident particles.	C
Alpha particle model	SIM	Alpha particles can be regarded as stable subunits inside the nucleus.	D
Shell model	IPM	Nucleons move nearly independently in a common static spherical potential which follows the nuclear density distribution.	E
Single particle shell model		Same as shell model *and:* Specific properties of odd-A nuclei are due to the last unpaired nucleon.	
Many particle shell model		Same as shell model *and:* Coupling between loosely bound nucleons due to mutual interactions is taken into account.	
j-j coupling model		Same as many particle shell model *and:* Each nucleon is characterized by a definite value of angular momentum j.	
Unified model	IPM	Nucleons move nearly independently in a common, slowly changing, non-spherical potential. Both excitations of individual nucleons *and* collective motions involving the nucleus as a whole are considered.	F
Collective model		Same as unified model *except:* Only collective motions involving the nucleus as a whole are considered.	
Rotational model (Strong Coupling model)		Same as unified model *and:* Nuclear shape remains invariant. Only rotations and particle excitations are assumed to occur.	
Spheroidal core model		Same as unified model *and:* Deformation of the nucleus into a spheroidal shape results from the polarization of the core, the bulk of nucleons in filled shells, by a few nucleons in unfilled shells.	

Finally, in Chap. F, the unified nuclear model, encompassing both excitations of individual nucleons and collective motions involving the nucleus as a whole, is discussed. This recently developed model is dealt with in considerably more detail than the others, so that many of its essential points may be covered.

Assumptions underlying the various nuclear models are briefly listed in Table 1. The models may be differentiated according to whether they presuppose strongly correlated or nearly independent motions of nucleons and also with respect to emphasis on nuclear spectra or on nuclear reactions. This classification, shown in Fig. 1, makes it possible to get some idea of the relationship between the different models. Besides the seven main models mentioned above, we list also seven models which are special forms of one or another of the main models.

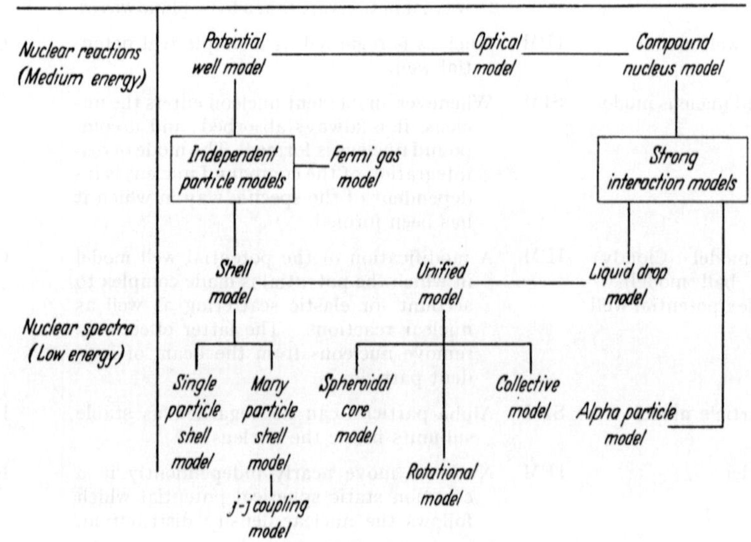

Fig. 1. Models of the nucleus.

The different kinds of relations between the nuclear models are interesting. To list a few examples discussed in this article: (i) the shell model is a refinement of the Fermi gas model, (ii) the Fermi gas model and liquid drop model are based on diametrically opposite assumptions, but they can *both* account for certain nuclear properties, such as binding energies, (iii) the optical model is essentially a hybrid between two other models, the potential well and compound nucleus models, (iv) the shell model and unified model appear to be quite different at first sight, but they are probably essentially equivalent.

The very fact that we have many different *nuclear models*, each of which is applicable only to certain nuclear phenomena, rather than a single consistent *nuclear theory* shows that much theoretical work remains to be done in this field.

A. The liquid drop model.

1. Introduction. The liquid drop model is perhaps the simplest of all nuclear models. The analogy between nuclear matter and a liquid drop is immediately suggested by the saturation property, the fact that both density and average binding energy per particle are approximately the same for all but the lightest nuclei. The basic assumption which underlies the liquid drop model is that the mean free path of nucleons in nuclear matter must be small compared to the nuclear diameter in the same way that the mean free path of molecules in a liquid drop is small compared to the diameter of the drop. This assumption seemed plausible enough when it was first proposed, in view of the known large strength of the nuclear forces. It implies that the nucleus must be regarded as a system of closely coupled particles, [11], [12] quite different from any picture which presupposes independent motions, i.e., in which the constituent particles move in fairly well-defined orbits. The liquid drop model seemed very plausible in view of the success of the compound nucleus theory and perhaps, even more dramatically, the discovery and interpretation of nuclear fission [13]. However, it proved incapable of explaining any of the finer details of nuclear structure. Furthermore, the assumption of a short mean free path seems now to be contradicted by the success of the various independent particle models as is discussed in the following chapters. Nevertheless because of its simplicity the liquid drop model provides considerable insight into many aspects of nuclear structure. The independent particle description is more realistic, but also more complicated to apply. In this chapter we briefly consider nuclear binding energies, nuclear radii, and collective oscillations. For all these phenomena a hydrodynamical description is instructive although not accurate.

2. Nuclear binding energies. The development of mass-spectroscopy around 1920 made extensive and accurate measurements of nuclear masses possible long before there was an adequate nuclear theory to account for the results. The nuclear masses are directly related to the masses of the constituent particles and the nuclear binding energies by the well known equation:

$$M_{\text{Nucl}} = Z M_p + N M_n - c^{-2} E_{\text{Bind}}. \tag{2.1}$$

Nuclear binding energies, in contrast to atomic and molecular binding energies, are sufficiently large to significantly affect the masses. During the last 20 years, a huge amount of additional information regarding nuclear binding energies[1] has been obtained by studies of various nuclear reactions and decay schemes.

The most striking overall feature of nuclear binding energies is their approximate constancy for different nuclei, contrasting sharply with the trend of atomic binding energies. A more detailed study shows that the binding energy per nucleon reaches a maximum for mass numbers around 60, and decreases slightly for both heavier and lighter elements. However, for all except the lightest nuclei the binding energies are quite close to 8 Mev per particle.

The empirical binding energies, except for the lightest nuclei, can be quite accurately reproduced by a semi-empirical formula, first given by VON WEIZSÄCKER [14]. Of the terms which appear in this formula, three are just of the form suggested by the liquid drop model. Thus the volume energy is the average energy due to saturated bonds between the nucleons; strictly speaking it is the value which would result for a nucleus of the prescribed density but infinite extent. The surface energy represents the decrease in the binding energy due to the presence of unsaturated bonds formed by nucleons at the surface. The Coulomb energy is just the electrostatic energy of the nucleus which is regarded as a uniformly charged droplet. The liquid drop model was a valuable guide in constructing the semi-empirical mass formula. However, it requires more detailed models to relate the magnitudes of the various terms to the basic interactions between nucleons. A fourth important component of nuclear energies which has no classical interpretation is the symmetry energy, the tendency of nuclei to have

[1] For a summary of these results see A. H. WAPSTRA, The Masses of Nuclei, Vol. XXXVIII of this Encyclopedia.

approximately equal numbers of neutrons and protons in opposition to the Coulomb energy. Finally, a pairing term must be included to reproduce the special stability of even-even nuclei and the almost complete absence of stable odd-odd nuclei.

The complete semi-empirical formula for nuclear binding energies reads:

$$\left.\begin{aligned} -E_{\text{Bind}} &= -C_{\text{vol}} A + C_{\text{surf}} A^{\frac{2}{3}} + C_{\text{coul}} Z^2 A^{-\frac{1}{3}} + \\ &\quad + C_{\text{sym}} (A-2Z)^2 A^{-1} + C_{\text{pair}} A^{-\varepsilon}\delta \\ \text{where}\quad \delta &= +1 \quad \text{for odd-odd nuclei} \\ &= 0 \quad \text{for odd-}A\text{ nuclei} \\ &= -1 \quad \text{for even-even nuclei.} \end{aligned}\right\} \tag{2.2}$$

A recent analysis of empirical data[1] gives the following values for the various coefficients:

$$C_{\text{vol}} = 15.75 \text{ Mev}, \tag{2.3a}$$

$$C_{\text{surf}} = 17.8 \text{ Mev}, \tag{2.3b}$$

$$C_{\text{coul}} = 0.710 \text{ Mev}, \tag{2.3c}$$

$$C_{\text{sym}} = 23{,}7 \text{ Mev}, \tag{2.3d}$$

$$C_{\text{pair}} = 34 \text{ Mev}, \tag{2.3e}$$

$$\varepsilon = \tfrac{3}{4}. \tag{2.3f}$$

The semi-empirical mass formula is accurate to within a few Mev for most cases, and it is very useful for reproducing all important trends, except for the lightest nuclei. However certain details are not adequately accounted for; e.g. the special stability of certain "magic-number" nuclei and fluctuations of the pairing energies.

3. Nuclear charge distribution[2]. Interest in the sizes of nuclei dates back to the early days of nuclear physics and, indeed, the results of the first measurements of nuclear radii by scattering of α-particles were surprisingly close to the currently accepted values. However, accurate measurements of nuclear radii were made only very recently, by studies of elastic scattering of high energy electrons by nuclei[3], and also of X-rays emitted by μ-mesic atoms[4]. The analysis of electron scattering has, in fact, made it possible to plot the variation of the charge distribution within the nucleus quite accurately. We confine ourselves to a brief discussion of the results from electron scattering[3] and from analysis of nuclear binding energies.

α) Results from electron scattering. The charge distribution seems to be approximately uniform in the interior of the nucleus as is expected in view of saturation and it falls to zero in a fairly thin surface region. The thickness t of the surface layer, defined as the distance over which the density falls from 90 to 10% of its central value, is about 2.4×10^{-13} cm for a wide range of nuclei. The "half-fall-off radius" $R_{\frac{1}{2}}$, defined as the distance from the center to the point where the density has fallen by half, has the value[5] $1.07\,A^{\frac{1}{3}}$, where the

[1] Cf. Ref. [5], p. 287.

[2] For a more detailed discussion of the distribution of charge and matter in nuclei, see the article by D. L. Hill in this volume, K. W. Ford and D. L. Hill, Ann. Rev. Nucl. Sci. **5**, 25 (1955), and J. M. C. Scott, Progr. Nucl. Phys. **5**, 157 (1956).

[3] For a summary of this work see R. Hofstadter, Rev. Mod. Phys. **28**, 214 (1956).

[4] V. L. Fitch and J. Rainwater: Phys. Rev. **92**, 789 (1953). — L. N. Cooper and E. M. Henley: Phys. Rev. **92**, 801 (1953).

[5] A length for which no units are given is understood to be in units of 10^{-13} cm.

constant does not change appreciably from nucleus to nucleus. The charge density can be quite well fitted to a Fermi distribution

$$\varrho = \varrho_0 \left[\exp\left(\frac{r - R_{\frac{1}{2}}}{a}\right) + 1\right]^{-1} \tag{3.1}$$

where $a \approx 0.23t \approx 0.55$. However, other distributions uniform in the interior and with the same radius $R_{\frac{1}{2}}$ and surface thickness fit the experimental data about equally well.

Also of interest is the "Coulomb radius" R_c, which is the radius of a uniform distribution of the same total charge Ze and Coulomb energy E_{coul} as the actual distribution.

This Coulomb energy is given by the well-known formula:

$$E_{\text{coul}} = \frac{3}{5} \frac{Z^2 e^2}{R_c}. \tag{3.2}$$

For the Fermi type distribution (3.1) it can be verified on basis of standard electrostatics that:

$$R_c = R_{\frac{1}{2}} + \frac{7}{6} \frac{\pi^2 a^2}{R_{\frac{1}{2}}} \tag{3.3}$$

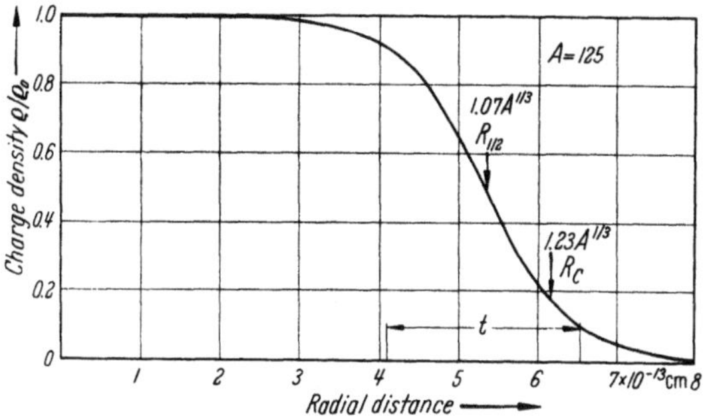

Fig. 2. Nuclear charge distribution.

to second order in a.

It can also be shown that the mean square radius is to a very good approximation given by[1]:

$$\langle r^2 \rangle \approx \tfrac{3}{5} R_c^2. \tag{3.4}$$

The Coulomb radius deduced from Eq. (3.3) is $1.23\,A^{\frac{1}{3}}$ for $A = 125$. The coefficient of $A^{\frac{1}{3}}$ actually varies somewhat with mass number. Its value is 1.3 for $A = 50$ and 1.18 for $A = 200$. Fig. 2 shows the charge distribution (3.1) for a nucleus with $A = 125$. Because of the appreciable surface thickness, the Coulomb radius is always considerably larger than the half-fall off radius, e.g. by almost 10^{-13} cm for $A = 125$. There is no single well defined "nuclear radius" as there is for a uniform distribution.

β) Results from nuclear binding energies. Coulomb radii may also be deduced from systematics of nuclear masses. The magnitude of the Coulomb energy term (2.3c) implies that the coefficient 1.22 gives the best overall fit of Coulomb radii to an $A^{\frac{1}{3}}$ law, in agreement with the results from electron scattering. The energy difference between mirror nuclei has been for a long time used to obtain information regarding Coulomb radii. Interpretation of the empirical data using (2.2) implies that $R_c \approx 1.4\,A^{\frac{1}{3}}$. In fact this value of the nuclear radius agrees with early determinations from nuclear reactions and alpha decay, and was not seriously questioned until about 1952. More recent studies[2], stimulated by the results of the other electromagnetic measurements of the nuclear radius, have shown that quantum mechanical corrections to Eq. (2.2) may be quite important for light nuclei. Thus an approximate formula for the energy difference between mirror nuclei of atomic number Z and $Z+1$ is:

$$\Delta E = (1.2Z - 0.6Z^{\frac{1}{3}})\, e^2 R_c^{-1} + (M_p - M_n)\, c^2 \text{ in Mev} \tag{3.5}$$

[1] Hahn, Ravenhall and Hofstadter: Phys. Rev. **101**, 1131 (1956).

[2] B. G. Jancovici: Phys. Rev. **95**, 389 (1954). — B. C. Carlson and I. Talmi: Phys. Rev. **96**, 436 (1954). — D. C. Peaslee: Phys. Rev. **95**, 717 (1954).

although the coefficient of the "exchange" term (the term in $Z^{\frac{1}{3}}$) is rather uncertain. Interpretation of the empirical energy differences using (3.5) leads to $R_c \approx 1.3\,A^{\frac{1}{3}}$, consistent with the results from high energy electron scattering.

4. Collective oscillations. According to the liquid drop model any energy acquired by a nucleon is quickly shared. Thus nuclear excitations involve collective displacements of many nucleons. Not much error is made in treating nuclear matter as incompressible and having a sharp surface.

In the liquid drop model, the nuclear wavefunction is described entirely in terms of the position of the nuclear surface. The motions of individual nucleons are not considered explicitly (in contrast to the treatment in the independent particle models).

α) Energy of static deformed nucleus. Suppose that the nucleus would have a surface of radius R_0 if it were spherical. Then, for small deviations from sphericity, the equation for the surface may be written [*33*] as:

$$R(\vartheta,\varphi) = R_0\left[1 + \sum_\lambda \sum_\mu \alpha_{\lambda\mu}\, Y^*_{\lambda\mu}(\vartheta,\varphi)\right] \tag{4.1}$$

where the $\alpha_{\lambda\mu}$ are deformation parameters whose values determine the nuclear shape[1] and $Y_{\lambda\mu}$ are spherical harmonics. According to (4.1), if $\alpha_{00}=0$, then the nuclear volume is independent of the deformation parameters, at least to first order, as required by the assumption of incompressibility.

In Eq. (4.1) the index μ ranges from $-\lambda$ to $+\lambda$, giving altogether $2\lambda+1$ modes of deformation of order λ. In particular any mode with $\mu=0$ represents an axially symmetric shape, i.e. symmetry with respect to arbitrary rotation about the z axis. For surface oscillations, the order λ is 2 or larger. For $\lambda=2$ (quadrupole) we have 5 independent modes which represent essentially ellipsoidal shapes, provided the deformation parameters are relatively small.

According to the liquid drop model, the energy of a nucleus is simply the sum of volume energy, surface energy and Coulomb energy. The volume energy is independent of the nuclear shape (at least on the assumption of incompressibility). The surface energy is smallest for spherical shape and increases with deviation from spherical symmetry. On the other hand, the coulomb repulsion between the protons tends to favor asymmetric shapes [*13*]. Altogether, for small deviations from sphericity the nuclear energy, according to the liquid drop model, is of the form:

$$E(\boldsymbol{\alpha}) = E(0) + \tfrac{1}{2}\sum_\lambda C_\lambda \sum_\mu |\alpha_{\lambda\mu}|^2. \tag{4.2}$$

In this equation, the term $\boldsymbol{\alpha}$ denotes the set of deformation variables, $E(0)$ is the energy for spherically symmetric shape, and C_λ are the deformability coefficients which measure the resistance of the nucleus against deformation. It has been shown [*13*] on the basis of simple geometric and electrostatic theory that:

$$C_\lambda = \frac{\lambda-1}{4\pi}\left[(\lambda+2)\,E_s(0) - \frac{10}{(2\lambda+1)}\,E_c(0)\right] \tag{4.3}$$

where $E_s(0)$ and $E_c(0)$ denote, respectively, the surface energy and Coulomb energy in the limit of spherical shape. Thus according to to the liquid drop model, nuclei should have spherical equilibrium shapes and be stable against deformation (of order $\lambda \geqq 2$), provided $E_c(0) < 2E_s(0)$. This condition (i.e. $Z^2/A \lesssim 48$) is satisfied by all known nuclei.

---

[1] Since R is real, we must have $\alpha_{\lambda\mu} = (-1)^\mu \alpha^*_{\lambda-\mu}$. The deformation parameters defined here are the complex conjugates of the coresponding quantities defined in Ref. [*33*].

In a more detailed treatment, even within the framework of the liquid drop model, one must often take into account the energy terms proportional to the third and higher powers of the deformation parameters. This is absolutely necessary in the analysis of the nuclear fission process[1].

β) Monopole and dipole oscillations. Let us briefly mention monopole ($\lambda=0$) and dipole ($\lambda=1$) oscillations. These can not occur at all in a strictly incompressible fluid. The single $\lambda=0$ mode is a radial compressive oscillation[2]. The 6.06 Mev state of O^{16} may be the lowest excited state for such an oscillation[3]. The three $\lambda=1$ modes represent oscillations of the center of mass. It is possible to have the center of masses of the protons and neutrons separately perform such oscillations always with opposite phase so that the center of mass of the whole nucleus remains at rest[4]. There is strong evidence for the existence of such dipole oscillations in many nuclei (cf. discussion of giant photonuclear resonances, Sect. 27) but the energy of the corresponding lowest excited state is rather large, between 10 and 20 Mev.

γ) Surface oscillations. Let us now consider the effect of variation in the deformation parameters. Although the parameters are quantum variables, it appears plausible, at least for some applications, to treat them as classically time dependent[5]. If the nuclear surface changes slowly in a prescribed way, there will occur a collective transport of nuclear matter in the interior. If the flow is assumed to be irrotational, i.e. free of vortices, then the velocity field at every point inside the surface is given by:

$$\boldsymbol{v} = R_0^2 \sum_{\lambda} \frac{1}{\lambda} \sum_{\mu} \dot{\alpha}_{\lambda\mu} \operatorname{grad}\left[\left(\frac{r}{R}\right)^{\lambda} Y_{\lambda\mu}\right] \tag{4.4}$$

to first order in the $\dot{\alpha}_{\lambda\mu}$. Note that the velocity is largest at the surface.

The total kinetic energy of the mass transport throughout the nucleus is then of the form:

$$T = \tfrac{1}{2} \sum_{\lambda} B_{\lambda} \sum_{\mu} |\dot{\alpha}_{\lambda\mu}|^2 \tag{4.5}$$

for slow changes in the deformation parameters. The quantities B_{λ}, the mass parameters, measure the inertial effect of the nucleus with respect to changes in deformation. On the above assumption of irrotational flow and using (4.4) we find that:

$$B_{\lambda} = \frac{3}{4\pi} \frac{M A R_0^2}{\lambda} \tag{4.6}$$

where M denotes the mass of a nucleon and A is the mass number of the nucleus.

The Hamiltonian of the oscillations is given by:

$$H = T + E(\boldsymbol{\alpha}) \tag{4.7}$$

where the collective kinetic energy T is the excess of the actual energy over the value $E(\boldsymbol{\alpha})$ which would result if the nucleus were static; the term $E(\alpha)$ plays

---

[1] The fission phenomenon is discussed by J. A. WHEELER in Vol. XLI of this Encyclopedia.

[2] Because of the relative incompressibility of nuclear matter this mode is expected to have a large frequency. See Ref. [13] and W. J. SWIATECKI, Phys. Rev. **83**, 178 (1951).

[3] Cf. Sect. 25.

[4] M. GOLDHABER and E. TELLER: Phys. Rev. **74**, 1046 (1948). — J. H. D. JENSEN and P. JENSEN: Z. Naturforsch. **5**a, 343 (1950). — H. STEINWEDEL and J. H. D. JENSEN: Z. Naturforsch. **5**a, 413 (1950).

[5] S. FLÜGGE: Ann. d. Physik (5) **39**, 373 (1941). — M. FIERZ: Helv. phys. Acta **16**, 365 (1943).

the role of a potential energy for the collective motion. For small deformations (4.7) becomes

$$H = E(0) + \sum_{\lambda} \sum_{\mu} H_{\lambda\mu} \tag{4.8}$$

where

$$H_{\lambda\mu} = \tfrac{1}{2} B_\lambda |\dot{\alpha}_{\lambda\mu}|^2 + \tfrac{1}{2} C_\lambda |\alpha_{\lambda\mu}|^2. \tag{4.9}$$

The oscillations of the system are the same as that of a particle in a many-dimensional harmonic oscillator potential. The classical oscillation frequencies are given by:

$$\omega_\lambda = \left(\frac{C_\lambda}{B_\lambda}\right)^{\frac{1}{2}}. \tag{4.10}$$

The equation of motion may be quantized in the usual way by introducing momenta $\pi_{\lambda\mu}$ conjugate to the $\alpha_{\lambda\mu}$ so that instead of (4.9) we obtain:

$$H_{\lambda\mu} = \frac{|\pi_{\lambda\mu}|^2}{2B_\lambda} + \frac{1}{2} C_\lambda |\alpha_{\lambda\mu}|^2. \tag{4.11}$$

The energy levels of the complete Hamiltonian (4.8) are then given by

$$E = E(0) + \sum_{\lambda} \sum_{\mu} (n_{\lambda\mu} + \tfrac{1}{2}) \hbar \omega_\lambda. \tag{4.12}$$

Each of the $n_{\lambda\mu}$ may be regarded as the number of oscillator quanta in the mode $\lambda\mu$.

δ) Lowest two states. The wavefunction for the ground state of the nucleus (all $n_{\lambda\mu} = 0$) has the form

$$\psi_0 = \text{const} \times \exp\left[-\sum_{\lambda} \frac{B_\lambda \omega_\lambda}{2\hbar} \sum_{\mu} (\alpha_{\lambda\mu})^2\right]. \tag{4.13}$$

This state is non-degenerate and must therefore have angular momentum 0. The state for which one of the $n_{\lambda\mu} = 1$ but all other modes are unexcited has the following wavefunction:

$$\psi_{n_{\lambda\mu}=1} = \text{const} \times \alpha_{\lambda\mu} \psi_0, \tag{4.14}$$

and its excitation energy with respect to the ground state is simply $h\omega_\lambda$. This state is $(2\lambda+1)$ fold degenerate. It is expected therefore that its angular momentum is λ. Indeed it has been shown[1] [*33*] that the oscillator quanta can be regarded as Bose-Einstein particles each of which has angular momentum λ. It is also of interest to determine the parity of the above states[2]. A state of even parity has its wavefunction unaltered by the transformation:

$$\boldsymbol{r} \to -\boldsymbol{r} \tag{4.15}$$

---

[1] See footnote 5, p. 419.

[2] There is every reason to believe that parity is conserved in nuclear and electromagnetic interactions [T. D. Lee and C. N. Yang, Phys. Rev. **104**, 254 (1956)], i.e. that these interactions are invariant with respect to space inversion. Recent experiments [Wu, Ambler, Heyward, Hoppes and Hudson, Phys. Rev. **105**, 1413 (1957); Garwin, Lederman and Weinrich, Phys. Rev. **105**, 1415 (1957); J. I. Friedman and V. L. Telegdi, Phys. Rev. **105**, 1681 (1957)], have indicated that parity is not conserved in reactions involving neutrinos, e.g. beta decay. However the relevant interactions are extremely weak in comparison with the others mentioned above, and their effect on the nuclear structure can be safely neglected. Consequently we make practically no error by assuming the every nuclear state is characterized by a definite (even or odd) parity.

while for a state of odd parity the wavefunction is multiplied by -1. Now it may be seen from (4.1) (and elementary properties of spherical harmonics) that the transformation (4.15), has the same effect on the nuclear shape as the transformation

$$\alpha_{\lambda\mu} \rightarrow (-1)^{\lambda} \alpha_{\lambda\mu}. \tag{4.16}$$

We then see from (4.13) that the ground state has even parity and that the first excited state has even or odd parity according to whether λ is even or odd.

The excitation energies corresponding to the various modes may be readily calculated using (4.3), (4.6) and (4.10). Neglecting the Coulomb energy, it is found that:

$$\hbar\,\omega_{\lambda} \approx 13\,\lambda^{\frac{3}{2}} A^{-\frac{1}{2}}\,\mathrm{Mev}. \tag{4.17}$$

Inclusion of the Coulomb energy will lower these values slightly. Since $\lambda \geqq 2$, the ground state and first excited state should both have even parity and angular momentum 0 and 2, respectively. This is indeed found to be the case for almost all even-even nuclei. Since *all* the nucleons contribute to the collective oscillation, one would expect that the strength of the electric quadrupole transition between these two states should be much larger than the strength resulting from transition of a single nucleon. This is also found to be the case experimentally (cf. Sects. 31 and 34 for further discussion). However, here the agreement between empirical data and the liquid drop model ends. For example, the observed excitation energies of the lowest 2^+ state in even-even nuclei depend quite strongly on the nucleon number and are, in general, only a few hundred kev while the liquid drop model implies a value of 3 Mev for $A = 100$ and no strong dependence on nucleon number.

As we will see in Chap. F there is convincing evidence for the existence of collective oscillations in nuclei; but while these oscillations resemble those of a liquid drop in some respects, there are also many profound differences.

B. The Fermi gas model.

5. Introduction. The Fermi gas model is the simplest of the class of independent particle models, which presuppose that the nucleons move approximately independently within a nucleus, i.e. that the effective mean free path of nucleons in nuclear matter is at least comparable to the nuclear diameter. This crucial assumption is the very opposite of the basic assumption underlying the strong-interaction models (i.e. the liquid drop model).

Also according to the independent particle models, the lowest modes of excitation involve a change in the wavefunctions of only one or a few particles, while in the strong interaction models, the easiest excitable degrees of freedom involve a large number of particles.

Work on the independent particle models of nuclei dates back to the early 1930's [*1*]. The early studies were greatly stimulated by the success of the Hartree self consistent field method in accounting for many features of atomic structure. It was, however, expected from the outset that this description would be less accurate for nuclei than for atoms, because of the large and shortranged interactions between nucleons and also because of the absence of a common center of force. Indeed, the apparent success of the liquid drop and compound nucleus models [*11*] to [*13*] appeared to contradict the validity of the independent particle approach entirely. During the last ten years, it has, however, become clear that the idea of independent particle motion is applicable after all, at least for low energy nuclear properties. This was most strikingly indicated by the dramatic successes of the shell model (Chap. E) and optical model (Chap. D).

In this chapter, we will discuss the Fermi gas model. The additional assumption of the Fermi gas model is that the wavefunctions of individual nucleons are taken to be plane waves[1]. This enormously simplifies many calculations and makes it possible to gain much insight into the mechanics of nuclear saturation, the nuclear potential and also other properties which do not vary strongly from nucleus to nucleus. However, more detailed aspects of nuclear structure, e.g. nuclear shell effects, cannot be treated in the framework of the Fermi gas model.

In Sects. 6 to 9 of this chapter we apply the Fermi gas model to the analysis of nuclear binding energies and the nuclear potential. It will be tentatively assumed that correlations due to the nucleon interactions may be neglected.

In Sects. 10 to 13, the Fermi gas model is used to study a few specific nuclear properties.

In the last three sections of this chapter, we briefly mention some refinements of the model. In particular, in Sect. 15 it is suggested that the effect of correlations (due to internucleon forces) on the nuclear binding energy is relatively small, and that the independent particle picture owes its approximate validity largely to the action of the Pauli exclusion principle.

I. Nuclear saturation and the nuclear potential.

6. Nuclear saturation. — General remarks. The existence of stable nuclei makes it obvious that the nuclear interactions are predominantly attractive. Yet if the interactions between *all* nucleons were attractive there would be nothing to prevent the nucleus from collapsing to a state of much higher density and binding energy than observed. As the density of the nucleus increases, so does the average kinetic energy of the nucleons inside, since it is proportional to $\varrho^{\frac{2}{3}}$. However, because of the short range of the nuclear interactions, the average potential energy would vary linearly with the number of bonds formed between close-lying nucleons, which gives an average contribution per nucleon proportional to $-\varrho$. Clearly this system would shrink until all nucleons can form bonds with each other; i.e. the nuclear radius would be independent of A and the binding energy proportional to A^2.

A phenomenon of this type actually seems to happen for the very lightest nuclei. The total binding energies of H^2, He^3 and He^4 are 2.2, 7.6, and 28.1 Mev, respectively. For these nuclei, each particle evidently forms attractive bonds with all the others. In fact, the binding energy increases even faster than the number of bonds (respectively 1, 3 and 6), suggesting that in He^4 the nucleons spend a larger fraction of time within the range of their mutual interactions than they do in H^2. Beyond He^4, however, the binding energy per particle no longer increases significantly. Thus the total binding energy of He^5 is actually less than that of He^4. Apparently each nucleon can form only a limited number of attractive bonds with other nucleons. This could come about in several ways. It might happen that the nuclear interactions actually become repulsive at very small distances, e.g. as for van der Waals forces. Another possibility is that the nuclear interactions contain exchange terms, i.e. they depend on the spin and spatial symmetry of the nucleon pair so that they are attractive between some pairs and repulsive between others, somewhat analogous to the case of chemical valence bonds. These as well as other possibilities are discussed in Sect. 14.

[1] This model is very similar to Wigner's "Uniform model", Ref. [*18*].

Let us briefly consider the corresponding situation in atoms. For these, equilibrium is established essentially because of the balance between the kinetic energy of the electrons (proportional to r^{-2} per particle, where r is the atomic radius), and the long range attraction of each electron by the charged nucleus (proportional to Z/r per particle). If the nucleus-electron potential were proportional to $1/r^3$, rather than $1/r$, the system would collapse, in a manner similar to the situation of a hypothetical nucleus with purely attractive short range interactions.

Note, however, that electrons in atoms do not saturate. Thus we see from the above argument that the atomic radius actually shrinks[1] (approximately proportional to $Z^{-\frac{1}{3}}$) as we go to heavy elements, while the total energy is roughly proportional to $Z^{\frac{7}{3}}$. Also, the density of electrons in an atom falls off sharply away from the nucleus and is not even approximately uniform.

7. Interaction energy of a nucleon pair[2]. To study the Fermi gas model in more detail, it is necessary to consider the interactions between nucleons. We assume at the outset that many-body forces are not present. It is also supposed, in this section at least, that the nuclear interactions are sufficiently non singular that their effects on the wavefunction are small. Thus all energies will be calculated by use of first order perturbation theory. Refinements to this approach will be discussed in Sects. 15 and 16, and also in Sects. 18 and 19 of the next chapter.

The total nuclear energy is given by:

$$E = \langle\psi|\sum_i T_i + \sum_{i<j}\sum V_{ij}|\psi\rangle \tag{7.1}$$

where T_i and V_{ij} denote kinetic energies and internucleon potentials respectively. The potential V_i seen by nucleon i is given by:

$$V_i = \sum_j V_{ij}. \tag{7.2}$$

On the assumption of independent particle motion ψ is essentially a product type wavefunction, except, of course for the antisymmetrization required by the exclusion principle. It is most convenient to let each of the individual particle wavefunctions be a plane wave:

$$\psi_i = (\Omega)^{-\frac{1}{2}} e^{i\boldsymbol{k}_i\cdot\boldsymbol{r}_i}\{\uparrow \text{ or } \downarrow\}\{P \text{ or } N\} \tag{7.3}$$

where the quantity in the first parenthesis is a spin function (spin up or down), while the charge function in the second parenthesis specifies whether the particle is a proton or a neutron. The nuclear volume is denoted by Ω. Strictly speaking, the plane wave assumption is valid only for an hypothetical nucleus of infinite extent. However, it is all right for finite nuclei too, as long as we do not consider the effect of the nuclear surface.

For any plane wave, the propagation constant k_i and associated momentum is a constant of the motion. The kinetic energy of a nucleon with momentum $\hbar k_i$ is given by[3]:

$$T_i = \frac{\hbar^2}{2M} k_i^2 \tag{7.4}$$

---

[1] The statistical model of the atom is discussed by P. GOMBÁS in Vol. XXXVI of this Encyclopedia.

[2] Cf. Ref. [2], [3].

[3] We are treating the nucleons non-relativistically. Relativistic corrections are expected to be quite small at low energies and are difficult to treat consistently. See, however, H. P. DUERR: Phys. Rev. **103**, 469 (1956).

and the total energy of the nucleus may be written in the form:

$$E = \sum_i T_i + \sum_{i<j}\sum \Delta E_{ij} \tag{7.5}$$

where ΔE_{ij} is the interaction energy of a pair of nucleons in states i and j:

$$\Delta E_{ij} = \langle \psi | V_{ij} | \psi \rangle . \tag{7.6}$$

While the expression (7.6) involves the coordinates of all nucleons, integration over all coordinates other than i and j results in the equation:

$$\Delta E_{ij} = \langle \psi_{ij} | V_{ij} | \psi_{ij} \rangle . \tag{7.7}$$

where ψ_{ij} is the wavefunction of two particles in states i and j.

For the following calculation of interaction energies we will first consider a system of identical particles e.g. neutrons only, rather than the actual nuclei composed of both neutrons and protons. This simplifies the calculations without loss of any essential physical features. The results can then easily be modified so as to apply to real nuclei.

α) Two-particle wavefunctions. The two-particle wavefunctions are essentially products of the one-particle functions:

$$\psi_{ij} = \Omega^{-1} e^{i(\boldsymbol{k}_i\cdot\boldsymbol{r}_i + \boldsymbol{k}_j\cdot\boldsymbol{r}_j)} \{\uparrow\uparrow, \uparrow\downarrow, \downarrow\uparrow \text{ or } \downarrow\downarrow\}\{NN\}. \tag{7.8}$$

Now the spatial part of the wavefunction is conveniently expressed in terms of center of mass and relative coordinates and momenta:

$$\boldsymbol{r} = \boldsymbol{r}_i - \boldsymbol{r}_j, \tag{7.9}$$

$$\boldsymbol{k} = \tfrac{1}{2}(\boldsymbol{k}_i - \boldsymbol{k}_j). \tag{7.10}$$

The two particle wavefunction can thus be rewritten in the form:

$$\psi_{ij} = \psi_{\text{C.M.}} \Omega^{-\frac{1}{2}} e^{i\boldsymbol{k}\cdot\boldsymbol{r}} \{\text{spins}\} \tag{7.11}$$

where the first factor is just the wavefunction of the center of mass. The charge function $\{NN\}$ has been suppressed.

The above wavefunction must still be modified in one respect—it is to be antisymmetric with respect to exchange of the two nucleons so as to obey the Pauli exclusion principle. This requires that we take certain linear combinations of the above two-particle functions. Suppose first that the two nucleons have their spins pointing in the same direction, e.g. up. Then the desired wavefunction is given by

$$\psi_{ij}^{=} = \psi_{\text{C.M.}} (2\Omega)^{-\frac{1}{2}} \{e^{i\boldsymbol{k}\cdot\boldsymbol{r}}\uparrow\uparrow - e^{-i\boldsymbol{k}\cdot\boldsymbol{r}}\uparrow\uparrow\}. \tag{7.12}$$

For this case, the spin function is symmetric, and the spatial wavefunction is antisymmetric in the two particles[1]. On the other hand, the two particles may have their spins pointing in opposite directions. The antisymmetrical wavefunction is then of the form

$$\psi_{ij}^{\mp} = \psi_{\text{C.M.}} (2\Omega)^{-\frac{1}{2}} \{e^{i\boldsymbol{k}\cdot\boldsymbol{r}}\uparrow\downarrow - e^{-i\boldsymbol{k}\cdot\boldsymbol{r}}\downarrow\uparrow\}. \tag{7.13}$$

This last wavefunction is not symmetric or antisymmetric in the spatial coordinates alone. As a matter of fact it contains space-symmetric and space-antisymmetric components with equal weight.

---

[1] In the event that the two nucleons are in the same spatial state ($k = 0$) this two-particle state cannot be formed at all. However only a very small fraction of nucleon pairs are of this kind so that little error is committed by not treating this special case separately.

β) Two-particle interactions. Let us now consider the possible kinds of interactions between two nucleons of the same kind. We assume for simplicity that the interactions are central and independent of relative velocity. It is well known [2] that under these conditions invariance with respect to rotations limits the interactions to two possible kinds. These are (i) an ordinary interaction $V_0(r)$, whose strength depends only on the distance between the particles, and (ii) an exchange interaction of the form $P_x V_e(r)$, where P_x denotes the space exchange operator:

$$P_x \psi(R, \boldsymbol{r}) = \psi(R, -\boldsymbol{r}). \tag{7.14}$$

Thus, if the exchange interaction is attractive in space-symmetric states, it will be repulsive, but of equal magnitude, in space-antisymmetric states. Since exchanging the particles (which changes the sign of the wavefunction) involves exchanging both spatial and spin wavefunctions, we see that the exchange interactions can also be written in the form $-P_\sigma V_e(r)$ where P_σ denotes the spin exchange operator with eigenvalue $= +1$ for spin-symmetric states (total spin $= 1$) and -1 for spin-antisymmetric states (total spin $= 0$). In terms of the well-known Pauli matrices, we have

$$P_\sigma = \tfrac{1}{2}\left[1 + \boldsymbol{\sigma}_i \cdot \boldsymbol{\sigma}_j\right]. \tag{7.15}$$

The diagonal matrix elements of the above two interactions with respect to both kinds of two-particle wavefunctions are easily calculated, and we find:

$$\langle \psi_{ij}^{=} | V_0 | \psi_{ij}^{=} \rangle = \Omega^{-1} \int (1 - \cos 2\boldsymbol{k} \cdot \boldsymbol{r}) V_0(r)\, d^3 r, \tag{7.16a}$$

$$\langle \psi_{ij}^{=} | P_x V_e | \psi_{ij}^{=} \rangle = \Omega^{-1} \int [-1 + \cos 2\boldsymbol{k} \cdot \boldsymbol{r}) V_e(r)\, d^3 r, \tag{7.16b}$$

$$\langle \psi_{ij}^{\neq} | V_0 | \psi_{ij}^{\neq} \rangle = \Omega^{-1} \int V_0(r)\, d^3 r, \tag{7.16c}$$

$$\langle \psi_{ij}^{\neq} | P_x V_e | \psi_{ij}^{\neq} \rangle = \Omega^{-1} \int \cos 2\boldsymbol{k} \cdot \boldsymbol{r}\, V_e(r)\, d^3 r. \tag{7.16d}$$

Consideration of the two different kinds of nucleons in actual nuclei alters the above results somewhat. First of all the two body wavefunctions now acquire a slightly more complicated form. Then also, an exchange of the two nucleons now involves interchanging not only their spatial and spin wavefunctions, but also their charge wavefunctions. The space exchange and spin exchange operations may have quite different effects on the wavefunctions, in contrast to the case of identical nucleons, and they can be regarded as independent operations. Consequently, there are now four possible kinds of central velocity-independent interactions [2]: the forms $V_0(r)$ (Wigner interaction) and $P_x V_e(r)$ (Majorana interaction) considered above and two other forms, (Heisenberg and Bartlett interactions) which involve exchange of spins in addition. We will restrict ourselves to a consideration of the first two interactions mainly because these are the easiest to treat. (The calculations can also be carried through for spin dependent interactions but the present results will not be altered in any essential way). The equations (7.16) remain unchanged except for one point: the wavefunction $\psi_{ij}^{=}$ now signifies that the two nucleons have the same spin direction and the same charge, while $\psi_{ij}^{\neq}$ denotes a wavefunction for which the nucleons differ with respect to at least one of these quantities.

Suppose that the nucleus contains an equal number of neutrons and protons and that both groups have an equal number of nucleons with spin up and spin down. Then of all the nucleon pairs $\frac{1}{4}$ will have two nucleons of the same spin direction and charge (i.e. will be of the form $\psi_{ij}^{=}$) while $\frac{3}{4}$ of the pairs will be characterized by $\psi_{ij}^{\neq}$. For two nucleons in specified spatial states the average value of the matrix elements will be [from Eqs. (7.16)]:

$$\langle\psi_{ij}|\,V_0\,|\,\psi_{ij}\rangle = \Omega^{-1}\int\left(1-\tfrac{1}{4}\cos 2\boldsymbol{k}\cdot\boldsymbol{r}\right)V_0(r)\,d^3r\,, \tag{7.17a}$$

$$\langle\psi_{ij}|\,P_x V_e\,|\psi_{ij}\rangle = \Omega^{-1}\int\left(-\tfrac{1}{4}+\cos 2\boldsymbol{k}\cdot\boldsymbol{r}\right)V_e(r)\,d^3r\,, \tag{7.17b}$$

and the average interaction energy of a nucleon pair is:

$$\Delta E_{ij} = \Omega^{-1}\left[\int\left(V_0-\tfrac{1}{4}V_e\right)d^3r + \int\left(V_e-\tfrac{1}{4}V_0\right)\cos 2\boldsymbol{k}\cdot\boldsymbol{r}\,d^3r\right]. \tag{7.18}$$

It is interesting that the value of the interaction energy (7.18) depends on the relative momentum of the two particles. This results mainly by virtue of the exchange interaction. However, there is some momentum dependence even for ordinary interactions, because of the antisymmetrization of the wavefunctions. From the considerations preceding Eqs. (7.17) we see that only $\frac{3}{8}$ of all nucleon pairs have space symmetric wavefunctions while $\frac{5}{8}$ of the pairs are space antisymmetric. If the two particle wavefunctions had not been antisymmetrized, just half of them would be space-symmetric and the interaction energy would take the simple form:

$$\Delta E_{ij} = \Omega^{-1}\left[\int V_0\,d^3r + \int V_e\cos 2\boldsymbol{k}\cdot\boldsymbol{r}\,d^3r\right]. \tag{7.19}$$

8. Nuclear binding energy. We next wish to calculate the total energy of the nucleus in the ground state. This is given, in first order perturbation theory, by Eq. (7.5). To make the summation over all particles we use the fact that the density of states in k-space is given by:

$$\varrho_k = (2\pi)^{-3}\,(4\Omega)\,. \tag{8.1}$$

The factor 4 allows for the two possible orientations of spin and for the two possible kinds of nucleons. In the ground state all single particle states with momentum less than a certain value $\hbar k_F$, the Fermi momentum, are occupied by nucleons, but all higher states are empty. Since the occupied volume in k-space is $\frac{4\pi}{3}k_F^3$, we see from Eq. (8.1) that the nucleon density is given by:

$$\varrho \equiv \frac{A}{\Omega} = \frac{2}{3\pi^2}\,k_F^3\,. \tag{8.2}$$

Now the nucleon density is related to the nuclear radius R by[1]:

$$\varrho = \frac{3A}{4\pi R^3} = \frac{3}{4\pi r_0^3} \tag{8.3}$$

where the nuclear radius is assumed to be of the form

$$R = r_0 A^{\frac{1}{3}}\,. \tag{8.4}$$

Combining Eqs. (8.2) and (8.3) we find that

$$k_F = \left(\frac{9\pi}{8}\right)^{\frac{1}{3}}\frac{1}{r_0} \approx \frac{1.52}{r_0}\,. \tag{8.5}$$

The maximum kinetic energy of any nucleon (i.e. the Fermi kinetic energy) in the nucleus is given by:

$$T_F = \frac{\hbar^2}{2M}\,k_F^2 \tag{8.6}$$

---

[1] Strictly speaking, this is true only for a uniform density distribution. See, however, the remarks following (9.12).

while the average kinetic energy per particle is

$$T_{AV} = \frac{3}{5}\, T_F. \tag{8.7}$$

α) Conventional interactions. The total potential energy of the nucleus is obtained by summing the interaction energies (7.18) over all nucleon pairs. First consider the term involving $\cos 2\boldsymbol{k}\cdot\boldsymbol{r}$. At very high densities, ($k_F \times$ range of forces $\gg 1$) this term would not give any large contribution, since for most nucleon pairs the cosine factor will change signs many times in the integration and cancel out in the integral. On the other hand, each nucleon pair will contribute the same amount to the first term. There are $\frac{1}{2}A(A-1) \approx \frac{1}{2}A^2$ such nucleon pairs. Taking into account Eq. (8.2), the binding energy per particle is given by:

$$-\frac{E_{\text{Bind}}}{A} = T_{\text{Av}} + \frac{A}{2}\langle \Delta E_{ij}\rangle_{\text{Av}} = \frac{3}{10}\,\frac{\hbar^2}{M}\,k_F^2 + \frac{1}{3\pi^2}\,k_F^3 \int \left(V_0 - \frac{1}{4}V_e\right) d^3r \tag{8.8}$$

at high densities. The factor $\frac{1}{2}$ is required in order not to count the interactions twice.

For the moment let us suppose that neither V_0 nor V_e change sign as function of distance. We know from the stability of the deuteron that the nuclear interaction is attractive in S-states, i.e., states with relative orbital angular momentum 0. If the interactions are restricted to the ordinary and space-exchange forms, then the interactions must be equally attractive, of strength $V_0 + V_e$, in *all* space-symmetric states (i.e. for all even values of relative orbital angular momentum). Now if the potential $V_0 - \frac{1}{4}V_e$ is attractive at all distances, then according to (8.8) the binding energy increases indefinitely with increasing density and no saturation can occur [*1*] to [*3*]. This conclusion holds also if we drop the assumption that first order perturbation theory is valid. It is well known that the ground state energy of any system is lower than (or possibly equal to) the value calculated by first order theory. Thus the actual binding energy of the above system will be even larger than implied by Eq. (8.8), so that there is certainly no saturation. On the other hand, if the exchange interaction is more than four times as strong as the ordinary interaction and of the same sign, then saturation occurs. In this case, any attractive contribution to the binding energy can come only from the second term of Eq. (7.18).

From analyses of nucleon-nucleon scattering cross sections[1] it appears that the two-body interactions are weak in P-states (a space-antisymmetric state with orbital angular momentum 1). Combined with the previous remarks regarding the S-state interaction, this would require that $V_0 = V_e$ and that the two body interaction is of the form

$$V = \frac{1}{2}(1 + P_x)\, V_s \tag{8.9}$$

i.e., of magnitude V_s in space-symmetric states and 0 in space-antisymmetric states. This particular mixture of ordinary and exchange interactions is commonly designated as the "Serber Interaction" and it comes fairly close to fitting empirical nucleon-nuclear scattering cross sections. Assuming that the interaction is of the form (8.9), it follows from Eq. (7.18) that the average interaction energy of a nucleon pair is

$$\Delta E_{ij} = \frac{3}{8}\Omega^{-1}\int V_s\, d^3r + \frac{3}{8}\Omega^{-1}\int V_s \cos 2\boldsymbol{k}\cdot\boldsymbol{r}\, d^3r = \frac{3}{4}\Omega^{-1}\int V_s \cos^2 \boldsymbol{k}\cdot\boldsymbol{r}\, d^3r. \tag{8.10}$$

It is readily seen from Eqs. (8.8) and (8.10) that the Serber Force cannot lead to saturation, at least not if V_s is attractive at all distances.

---

[1] Cf. Footnote 5, p. 411.

We may modify the above treatment in several directions so as to get saturation. These possibilities are briefly treated in Sect. 14.

β) Saturating pseudo-potential. To see how one can obtain saturation in the framework of the present discussion, let us make a power series expansion of the cosine term in Eq. (8.10)

$$\Delta E_{ij} = \tfrac{3}{4}\Omega^{-1}\int V_s[1-(\boldsymbol{k}\cdot\boldsymbol{r})^2\ldots]\,d^3r. \tag{8.11}$$

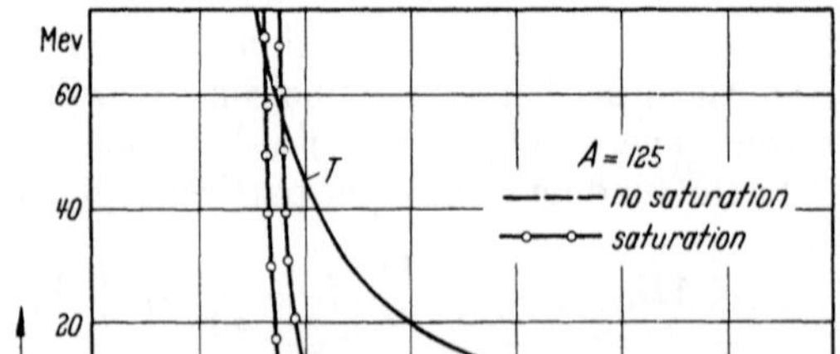

Fig. 3. Average energy (kinetic, potential and total) per nucleon vs. nuclear radius. The kinetic energy T per particle is assumed to be 24 Mev for a nuclear radius $1.1\,A^{\frac{1}{3}}\times 10^{-13}$ cm [cf. Eq. (9.13)] and to vary as R^{-2}. The curves for the case "saturation" have been drawn using (8.16) with values of the coefficients given by (9.20) and (9.21). To obtain the case "no saturation", we have merely omitted the last term in (8.16).

The value of $(\boldsymbol{k}\cdot\boldsymbol{r})^2$ averaged over all angles is $\frac{1}{3}k^2r^2$. Thus we may rewrite Eq. (8.11) in the form:

$$\Delta E_{ij} = \tfrac{3}{4}\Omega^{-1}\left[\int V_s\,d^3r - \tfrac{1}{3}k^2\int V_s r^2\,d^3r\ldots\right]. \tag{8.12}$$

Suppose we now drop all terms of order k^4, k^6 etc. (cf. Sect. 14). Strictly speaking, this corresponds to the assumption that the only non-vanishing even-order moments of the interaction are $\int V\,d^3r$ and $\int V r^2\,d^3r$. This, in turn, implies an interaction of the form:

$$\left(\int V_s\,d^3r\right)\delta(\boldsymbol{r}) - \tfrac{1}{3}\left(\int V_s r^2\,d^3r\right)\delta(\boldsymbol{r})\,\nabla^2_{\mathrm{Rel}} \tag{8.13}$$

where $-i\hbar\nabla_{\mathrm{Rel}}$ is the operator which represents the relative momentum of the two particles. This interaction does not correspond to physical reality. It is, however, useful as a simple "pseudo potential"[1,2] which gives rise to saturation even in a first order perturbation theory. While the use of this interaction does nothing toward explaining the origin of saturation, it proves useful for correlating

[1] The idea of pseudo potentials appears to have been first discussed by E. P. WIGNER, Phys. Rev. **43**, 253 (1933) and used by E. FERMI, Ricerca Sci. **7**, 13 (1936).

[2] The pseudopotential (8.13) is very similar to the one used in Ref. [23].

various properties related to the saturation phenomenon. Eq. (8.12) may be written with the help of (7.10) as:

$$\Delta E_{ij} = \Omega^{-1}[-U_0 + (\boldsymbol{k}_i - \boldsymbol{k}_j)^2 U_2] \tag{8.14}$$

where

$$U_0 = -\tfrac{3}{4}\int V_s\, d^3r\,, \tag{8.15a}$$

$$U_2 = -\tfrac{1}{16}\int V_s\, r^2\, d^3r \tag{8.15b}$$

are both positive. The binding energy per particle is now given by:

$$-\frac{E_{\text{Bind}}}{A} = T_{\text{Av}} + \frac{A}{2}\langle \Delta E_{ij}\rangle_{\text{Av}} = \frac{3}{10}\frac{\hbar^2}{M}k_F^2 + \frac{A}{\Omega}\left[-\frac{1}{2}U_0 + \frac{3}{5}k_F^2 U_2\right]. \tag{8.16}$$

Thus, the "pseudo potential" (8.13) does lead to saturation. For sufficiently large densities, the dominant term in the expression for the energy is proportional to $\varrho^{\frac{5}{3}}$ and positive, so that the nucleus cannot collapse. On the other hand, we will always obtain a bound state ($E_{\text{Bind}} > 0$) unless the constant U_0 is very small. These features are illustrated qualitatively in Fig. 3.

9. Depth of the nuclear potential. Values of both the nuclear potential and the nuclear energy depend on the detailed form of the interaction, and in the previous treatment also on the validity of the Born approximation. On the other hand, it appears that certain relations between the potential and the energy are much less dependent on the detailed assumptions. In so far as this is the case we may deduce certain properties of the potential directly from empirical saturation data, without explicit reference to the two body forces. First of all, the potential seen by a nucleon of given momentum may be calculated from (7.2) and (8.14):

$$V(k_i) = \sum_j \Delta E_{ij} = \Omega^{-1} A\left[-U_0 + \left(k_i^2 + \tfrac{3}{5}k_F^2\right) U_2\right]. \tag{9.1}$$

Thus the potential can be approximately written in the form:

$$V(k) = V(0) + \beta\, T(k). \tag{9.2}$$

where β is a positive number, $T(k)$ is the appropriate kinetic energy, and $V(0)$ must be negative to give any net binding. In principle, the potential also contains terms of order k^4, k^6, but more detailed calculations based on a realistic interaction (Sect. 14) indicate that their effect is probably quite small, at least for nucleons below the top of the Fermi sea.

On the other hand, the term of order k^2 plays a very important role, being essential for the existence of saturation. According to Eq. (9.2), the nuclear potential seen by a nucleon is velocity-dependent, becoming less attractive with increasing nucleon kinetic energy[1] [*15*], [*16*]. For sufficiently energetic nucleons, the potential actually turns repulsive so that the nucleus cannot collapse.

α) Empirical argument for velocity dependence of the nuclear potential[2]. The energy of an individual nucleon is conveniently defined by:

$$E_i = T_i + V_i. \tag{9.3}$$

For bound nucleons E_i is negative and the energy required to just remove the nucleon from the nucleus is $-E_i$. Assuming a potential of the form (9.2), this energy may be expressed as follows:

$$E_i = \frac{k_i^2}{2M^*} + V(0) \tag{9.4}$$

---

[1] This appears to have been first pointed out by J. H. van Vleck, Phys. Rev. **48**, 367 (1935).

[2] Cf. also Ref. [*15*], [*24*], and V. F. Weisskopf, Nucl. Phys. **3**, 423 (1957).

where

$$M^* = \frac{M}{1+\beta}. \tag{9.5}$$

The particular velocity dependence expressed by Eq. (9.2) is thus equivalent to a reduction in the mass of each nucleon. However, if the potential also contains terms of order k^4, the motion of nucleons will be affected in a more complicated way.

The energy of an individual nucleon on top of the Fermi sea is given by:

$$E_F = T_F + V_F. \tag{9.6}$$

On the other hand, the average energy per nucleon is:

$$E_{\text{Av}} = T_{\text{Av}} + \tfrac{1}{2} V_{\text{Av}} \tag{9.7}$$

where V_{Av} denotes the average strength of the potential felt by the nucleons. The factor $\frac{1}{2}$ is required so as to avoid counting the basic two body interactions twice. Minimization of the total nuclear energy [see Eq. (8.16)] as a function of the density leads to the result:

$$E_F = E_{\text{Av}}. \tag{9.8}$$

This equation holds irrespective of the form of V as function of k provided first order perturbation theory is valid.

The physical interpretation of Eq. (9.8) is quite simple: starting with a nucleus of mass number A at equilibrium, we remove a nucleon from the top of the Fermi sea. This requires an input of $-E_F$. The residual nucleus now has slightly too low a density. However since it is very near to equilibrium to start with, the energy gained in contraction to equilibrium is very small (of order T_F/A) and can be safely neglected. The energy difference between the initial and final state of this process is thus essentially just $-E_F$, but we also know that it is equal to $-E_{\text{Av}}$.

That the potential must indeed be velocity dependent may be seen by taking an appropriate linear combination of (9.6) and (9.7):

$$E_F - 2E_{\text{Av}} = T_F - 2T_{\text{Av}} + V_F - V_{\text{Av}}. \tag{9.9}$$

Substituting the results from (8.7) and (9.8) we find

$$\tfrac{1}{5} T_F - E_F = V_F - V_{\text{Av}}. \tag{9.10}$$

Since the left hand side of this equation must be positive, we see that a particle on top of the Fermi sea is subject to a weaker than average potential, i.e. the potential must be velocity dependent. The values of the two parameters V_0 and β can be determined from empirical radii and average binding energies.

β) The depth parameters of the nuclear potential. The Fermi kinetic energy can be expressed directly in terms of the density. Using Eqs. (8.2) and (8.6) we find:

$$T_F = \frac{\hbar^2}{2M}\left(\frac{3\pi^2}{2}\right)^{\frac{2}{3}} \varrho^{\frac{2}{3}}. \tag{9.11}$$

If the nucleus were a sphere of uniform density and radius $r_0 A^{\frac{1}{3}}$, the kinetic energy would be given, using (8.3), by:

$$T_F = 48.7\, r_0^{-2}\ \text{Mev} \tag{9.12}$$

where r_0 is in units of 10^{-13} cm. Now for the actual non-uniform nuclear density distribution the term "nuclear radius" may be defined in several different ways, each being relevant for a particular application, e.g. Coulomb radius, half-fall-off radius etc. For the present purpose the radius should be defined in such a way that (8.3) gives the correct particle density well inside the nucleus. Assuming a density distribution of the form (3.1), this radius is essentially equal to the half-fall-off radius. We will adopt a value $r_0 = 1.1$, slightly larger than the radius of the charge distribution, since there is some reason (cf. Sect. 21) to believe that the neutron distribution extends slightly beyond the proton distribution. Thus from Eq. (9.11) we have[1]

$$T_F = 40 \text{ Mev} \tag{9.13}$$

and from (8.5) it follows that:

$$k_F = 1.4 \times 10^{13} \text{ cm}^{-1}. \tag{9.14}$$

For the nuclear binding energy we will use the value 8 Mev per particle:

$$E_F = -8 \text{ Mev}. \tag{9.15}$$

Assuming $V(k)$ is of the form (9.2), we can rewrite Eq. (9.6) in the form:

$$E_F = (1+\beta)\, T_F + V(0), \tag{9.16}$$

and Eq. (9.7) becomes, with the help of (8.7) and (9.8):

$$E_F = \left(\tfrac{3}{5} + \tfrac{3}{10}\beta\right) T_F + \tfrac{1}{2} V(0). \tag{9.17}$$

We now substitute the empirical values of T_F and E_F into these equations and solve for β and $V(0)$. The resulting magnitude of the nuclear potential is given by:

$$V(k) = (-88 + T) \text{ Mev}. \tag{9.18}$$

It is interesting to note that:

$$\beta = 1 \tag{9.19}$$

which is equivalent to an effective nucleon mass M^* just half of the free nucleon mass.

The values of the parameters in Eq. (9.1) are:

$$\Omega^{-1} U_0 A = 112 \text{ Mev}, \tag{9.20}$$

$$\Omega^{-1} U_2 k_F^2 A = 40 \text{ Mev}. \tag{9.21}$$

We may also express V in terms of E, the total energy of the nucleon, more specifically, its kinetic energy outside the nucleus. With the help of Eq. (9.3) we find:

$$V = \left(-44 + \tfrac{1}{2} E\right) \text{ Mev}. \tag{9.22}$$

In the above treatment we have not attempted to separate the binding energy into contributions from volume, surface, symmetry and Coulomb effects. Thus the estimate (9.22) refers to the *average* of the potential wells seen by a neutron and a proton and for the latter it includes the average Coulomb potential. Analyses of the elastic scattering cross sections for low and medium energy protons by nuclei by means of the optical model (Chap. D, esp. Sect. 17) indicate that the potential well felt by protons is given approximately by $(-55 + \frac{1}{2}E)$ Mev[2] for E up to 30 Mev. This, however, does not include the Coulomb potential, which amounts to 10 to 15 Mev for medium heavy nuclei. The *total* potential seen by a proton at zero energy is thus about 40 to 45 Mev deep inside the nucleus.

[1] Until a few years ago, the accepted value of T_F was around 25 Mev corresponding to a nuclear radius $1.4\, A^{\frac{1}{3}}$.

[2] Melkanoff, Moszkowski, Nodvik and Saxon: Phys. Rev. **101**, 507 (1956).

It appears, both from analysis of neutron-nuclei scattering cross sections[1], and theoretical considerations (Sect. 11) that the nuclear potential felt by neutrons is 5 or 10 Mev less deep than for protons (without the Coulomb potential), i.e. 45 to 50 Mev deep at zero energy. We see that the empirical and theoretical estimates for the average nuclear potential agree very well at low energies.

II. Further applications.

In the next four sections, we will apply the Fermi gas model to the analysis of some other important nuclear properties. This model proves quite helpful for understanding the physical principles underlying each property, but it cannot account for the magnitude of some of the relevant quantities.

10. Nuclear surface energy. Although for the study of many nuclear properties it is satisfactory to treat the nucleus as an infinite medium, the existence of a nuclear surface gives rise to characteristic effects [*16*], [*23*]. Thus the nuclear density falls from its constant interior value to zero within a rather thin surface layer. Closely related to the variable nucleon density at the surface is the nuclear surface energy which arises from the lack of maximum binding for the nucleons on the surface.

To study the origin of this nuclear surface energy let us suppose that the nucleons move within a box[2], i.e. a potential well with infinitely high walls. For simplicity let the box be a cube of side L. The wavefunctions are then given by:

$$\psi = (2/L)^{\frac{3}{2}} \sin k_x x \cdot \sin k_y y \cdot \sin k_z z \tag{10.1}$$

where $k_i = (n_i \pi/L)$, $i = x, y, z$ and the n_i are positive integers. As might be expected, the density of available states in k-space turns out to be essentially the same as for an infinite medium [Eq. (8.1)]. There is however, one important change: states with vanishing momentum along any of the three axes cannot occur for the box potential. Such states can occur for an infinite medium; the corresponding wavefunctions are independent of the coordinate involved, however, these cannot satisfy the required boundary conditions of the finite medium.

For the box potential the density of states in k space is effectively given by:

$$\left.\begin{aligned} \varrho_k &= \frac{4\Omega}{(2\pi)^3} \quad \text{provided all} \quad |k_i| \geq \frac{\pi}{2L}, \\ &= 0 \qquad\qquad \text{if any} \quad |k_i| \leq \frac{\pi}{2L}. \end{aligned}\right\} \tag{10.2}$$

Averaging (10.2) over all angles, but keeping the magnitude of k fixed, we obtain:

$$\varrho_k = \frac{4\Omega}{(2\pi)^3} [1 - f(k)] \tag{10.3}$$

where Ω is the volume of the box and $f(k)$ is the fractional solid angle (in k space) subtended by the forbidden region. For the present case we have, using Eq. (10.2)

$$f(k) = \frac{3\pi}{2Lk} \tag{10.4}$$

neglecting a small correction of order k^{-2}. If the cube is replaced by a rectangular parallelipiped, the above conclusions remain essentially unchanged, except that

---

[1] J. R. Beyster, M. Walt and E. W. Salmi: Phys. Rev. **104**, 1319 (1956). — F. E. Bjorklund, S. Fernbach and N. Sherman: Phys. Rev. **101**, 1832 (1956).

[2] See Ref. [*16*] and W. J. Swiatecki: Proc. Phys. Soc. Lond. A **64**, 226 (1951).

the expression for $f(k)$ is generalized to read

$$f(k) = \frac{\pi S}{4\Omega k} \tag{10.5}$$

where S is the surface area of the parallelipiped.

It is expected that Eqs. (10.3) and (10.5) hold regardless of the shape of the enclosure in which the particles are assumed to move. Thus for a spherical box of radius R, we would have:

$$f(k) = \frac{3\pi}{4Rk}. \tag{10.6}$$

Integrating Eq. (10.3) over momenta we obtain the following relation:

$$\frac{A}{\Omega} = \frac{2}{3\pi^2} k_F^3 \left[1 - \frac{3}{2} f(k_F)\right]. \tag{10.7}$$

This result may be interpreted as follows: While the nucleon density is approximately constant in the interior of the box, it must vanish at the surface because of the boundary conditions. Thus there is a small region near the surface where the density falls from its constant central value to zero. To compensate for this, the density in the central region must be slightly larger than A/Ω, or equivalently, the particles are effectively distributed over a volume which is slightly smaller than Ω.

If we define this effective volume by:

$$\Omega' = \Omega \left[1 - \tfrac{3}{2} f(k_F)\right] \tag{10.8}$$

then we have

$$\varrho = \frac{A}{\Omega'} = \frac{2}{3\pi^2} k_F^3 \tag{10.9}$$

the same relation as for an infinite medium. Eq. (10.8) can be justified by direct calculation of the particle density in a box as function of position.

Now from Eq. (10.5) we see that the suppression of states is relatively more important for low momenta. Therefore, the distribution of the nucleons in the box will be slightly weighted toward higher momenta compared to the situation in an infinite medium. Thus the average kinetic energy is now given by:

$$T_{\text{Av}} = \tfrac{3}{5} T_F \left[1 + \tfrac{1}{4} f(k_F)\right]. \tag{10.10}$$

The first term in this equation is the value for an infinite medium, while the second term is the kinetic contribution to the surface energy given by:

$$T_{\text{surf}} = \frac{3\pi}{80} \frac{T_F}{k_F} \frac{S}{\Omega} A \tag{10.11}$$

when summed over all particles. There is also a contribution to the surface energy of potential origin. In the framework of the independent particle model, this arises from the velocity dependence of the nuclear potential. Since states of high momenta interact less than the average, the favoring of high momentum states due to the surface has the effect of decreasing the average interaction. Using Eq. (9.2) it can be shown that:

$$V_{\text{surf}} = \beta\, T_{\text{surf}} \tag{10.12}$$

thus the total surface energy is given by:

$$E_{\text{surf}} = (1 + \beta) \frac{9\pi}{80} \frac{T_F}{k_F R} A \tag{10.13}$$

assuming the enclosure is a sphere of radius R. Substituting in the empirical values $T_F = 40$ Mev, $\beta = 1$ derived previously and with the aid of Eqs. (8.4) and (8.5), we find:

$$E_{\text{surf}} = 18.5\, A^{\frac{2}{3}}\, \text{Mev}. \tag{10.14}$$

While the agreement between this estimate and the empirical value $19.1\, A^{\frac{2}{3}}$ Mev is interesting, it should not be taken too seriously. This caution is suggested by the fact that the simple model of particles in a box considerably underestimates the surface thickness, a quantity which seems to be closely related to the surface energy[1]. Thus from Eq. (10.8) and with the aid of Eq. (10.6) we see that the effective radius of the sphere is given by:

$$R' = R - \frac{3\pi}{8 k_F}. \tag{10.15}$$

R' is approximately equal to the half-fall-off radius of the density distribution. Assuming a linear fall-off of the density near the surface, the 90 to 10% fall-off distance is given by:

$$t \approx 1.8 \cdot \frac{3\pi}{8 k_F}. \tag{10.16}$$

Substituting the result of Eq. (9.14) for k_F we find that t equals 1.3×10^{-13} cm, only about half of the empirical surface thickness. The empirical fall-off distance can be reproduced by postulating a potential well with sloping sides, but it is difficult to see why the surface energy should then not also be changed.

11. The nuclear symmetry energy. An important aspect of nuclear structure is the symmetry effect, the tendency of nuclei to have approximately equal numbers of neutrons and protons. Indeed for the lightest nuclei, the most stable isobars of given A have $N = Z$ as nearly as possible. For heavier nuclei, the Coulomb effect shifts the balance somewhat toward a neutron excess. The symmetry energy is the energy which could be gained by changing a given nucleus into a nucleon of the same A but with $N = Z$, neglecting coulomb effects and the neutron-proton mass difference. Like the surface energy, the symmetry energy contains both kinetic and potential terms [*16*].

Let us define the "asymmetry parameter"

$$\varepsilon = \frac{N - Z}{A}. \tag{11.1}$$

Consider a nucleus with a small neutron excess ($\varepsilon \ll 1$). Its total kinetic energy will be slightly larger than for a nucleus of the same A but with $N = Z$, as may be seen in the following way: Suppose we have a symmetric nucleus with maximum kinetic energy T_F. To change it to a nucleus of asymmetry ε, we must convert $\frac{1}{2} A \varepsilon$ protons into neutrons. This requires an input of kinetic energy since the neutrons must be put into levels above T_F, all lower levels being already occupied. Suppose the average spacing of neutron levels(and also of proton levels) near the top of the Fermi sea is given by d_F. Then each particle must be raised on the average by the amount $\frac{1}{2} A \varepsilon\, d_F$, so that the total energy input is

$$T_{\text{sym}} = \tfrac{1}{4} A^2\, d_F\, \varepsilon^2. \tag{11.2}$$

---

[1] For a treatment of the surface energy directly in terms of the variable nucleon density, i.e. a surface energy density proportional to $(\nabla \varrho)^2$, see Ref. [*14*], [*23*] and R. A. Berg and L. Wilets, Phys. Rev. **101**, 201 (1956).

Now the statistical density of single particle levels is given by:

$$\frac{dN}{dT} = \frac{3}{4}\,\frac{A}{T_F^{\frac{3}{2}}}\,T^{\frac{1}{2}}. \tag{11.3}$$

The value of the coefficient in (11.3) is chosen so that the total number of neutrons with kinetic energy below T_F is just $A/2$. It follows from Eq. (11.3) that the spacing of neutron levels near the top of the Fermi sea[1] is:

$$d_F = \left(\frac{dN}{dT}\right)_{T_F}^{-1} = \frac{4}{3}\,\frac{T_F}{A}. \tag{11.4}$$

Consequently the kinetic part of the symmetry energy is given by:

$$T_{\text{sym}} = \tfrac{1}{3}\,T_F A\,\varepsilon^2 \tag{11.5}$$

to order ε^2.

The contribution of potential energy to the symmetry effect may be separated into several parts[2]. To begin with, we have an effect which also contributes to the surface energy. Associated with the increased kinetic energies is a decrease in the effective binding—due to the velocity dependence of the potential[3] [*16*]. Also, in an asymmetric nucleus, more than half of all possible bonds are formed between like particles. Now the average interaction between like nucleons is less than it is between neutron and protons for two reasons: First, the Pauli principle inhibits the interaction of like-particles by forbidding some of the two-body states (e.g. spatially symmetric states in which the two nucleons have their spins lined up parallel) in which an interaction would otherwise occur; on the other hand the interaction between unlike particles is not suppressed. Second, a small effect in the same direction is expected to result from the known spin dependence of the nuclear interactions.

Omitting the very last contribution, which is hard to estimate quantitatively, the total symmetry energy is calculated to be:

$$E_{\text{sym}} = \tfrac{1}{3}\,A\,T_F\left[1 + \beta + \tfrac{2}{5}\right]\varepsilon^2. \tag{11.6}$$

The last two terms are the contributions of potential energy. The numerical estimate for the symmetry energy, using the values of T_F and β from Sect. 9, is $32\,\varepsilon^2$ Mev, in fairly good agreement with the empirical value (2.3*d*), $24\,\varepsilon^2$ Mev.

A possible way to remove the discrepancy between the two values might be the assumption of a slightly larger effective volume Ω_N for the neutrons as compared to the protons (Ω_P). (The existence of a small effect of this kind is, in fact, suggested by the calculations of Sect. 21.) Basically, the symmetry energy depends on the difference between neutron and proton *densities* rather than their absolute numbers. In terms of the fractional difference of volumes defined by:

$$\Delta = \Omega^{-1}(\Omega_N - \Omega_P) \tag{11.7}$$

it is readily shown that the dominant first two terms in Eq. (11.6) remain unchanged except for the replacement of ε by $\varepsilon - \frac{1}{2}\Delta$. Thus for a nucleus with $\varepsilon = 0.2$, even a 5% difference in the volumes will reduce the calculated symmetry energy by about 25%.

[1] The spacing of *single-particle* levels (excitations of either neutrons *or* protons) is, of course, half as large $= 2\,T_F/3\,A$.

[2] See Ref. [*24*] and also, H. Bethe, Physica, **22**, 941 (1956).

[3] M. H. Johnson and E. Teller: Phys. Rev. **98**, 783 (1955).

The same effects which contribute to the potential part of the symmetry energy also lead to a difference in the nuclear potentials seen by a neutron and a proton of the same energy. Thus in a neutron rich nucleus, the neutrons on top of the Fermi sea are subject to a weaker potential than the protons, partially because of their larger kinetic energy and partially because there are fewer particles of the opposite kind (protons) to interact with. To relate this difference to the symmetry energy we may argue as follows: If in a certain nucleus, we change one proton at the top of the Fermi sea into a neutron, the potential energy changes simply by the amount $V_N - V_P$. Alternatively we may say that since the potential energy is proportional to ε^2, it changes by an amount

$$\delta V_{\text{sym}} = 2 V_{\text{sym}}\, \delta\varepsilon/\varepsilon \tag{11.8}$$

The asymmetry will increase by the fraction $2/A$, (if it is small to begin with). Thus we find that the difference of potentials is given by:

$$V_N - V_P = \frac{4 V_{\text{sym}}}{\varepsilon A}. \tag{11.9}$$

We will use an empirical value of $11 A\varepsilon^2$ for V_{sym}. Combined with the $13 A\varepsilon^2$ estimated for T_{sym}, this gives the empirical value (2.3d) of E_{sym}. Thus for a medium weight nucleus ($A\approx 50$, $\varepsilon\approx 0.1$) the proton potential should be about 5 Mev deeper than the neutron potential. For heavy nuclei ($A\approx 200$, $\varepsilon\approx 0.2$) the difference should be 10 Mev. Well depths deduced from nucleon-nuclei scattering (Sect. 9) do indeed show a difference consistent with the above estimates.

12. The pairing effect. The level spacing in the Fermi gas model is essentially statistical except for a four-fold degeneracy associated with spins and charge. For heavier nuclei, this degeneracy is reduced to a two-fold one (spin up or down) by virtue of the Coulomb effect. The existence of any degeneracy whatever gives rise to a small pairing energy. In terms of the average single particle level spacing d_F in absence of any interaction, the average spacing between the doubly degenerate levels must be $2d_F$. On the other hand, the average energy does not change. Thus this effect alone does not alter the energy of a pair of nucleons in a degenerate state. However, for a nucleon added to an even-even configuration, the lowest unoccupied state is $\frac{1}{2} d_F$ higher in energy than it would be without any degeneracy. Thus from kinetic energy considerations alone we expect to find an odd-even difference given by

$$E_{\text{pair}} = \frac{1}{2} d_F = \frac{2}{3}\frac{T_F}{A}. \tag{12.1}$$

For $A\approx 100$ this implies a value of about $\frac{1}{4}$ Mev, whereas the empirical values are ~ 1 to 2 Mev.

It is possible to raise the theoretical estimate to $\frac{1}{2}$ Mev by taking into account the velocity dependence of the potential but this is still too small by at least a factor of 2. This discrepancy evidently reflects the effect of correlations. The nucleon in paired orbits will overlap more than the average and thus will interact especially strongly. Thus it is not surprising that the empirical pairing energy is considerably larger than the value predicted on the basis of a pure statistical model[1].

13. Density of nuclear levels. It is well known that the density of levels in nuclei increases rapidly with increasing excitation energy. This is mainly due to

[1] Cf. Ref. [2], p. 276.

the fact that as the excitation energy increases, so does the number of ways of dividing it up among the nucleons.

According to the Fermi gas model the level density of a many particle system at any given excitation energy E is proportional to the number of ways in which this energy can be partitioned up among excitations of individual particles. Now it is known from the theory of numbers that the numbers of ways of partitioning a large integer n is given approximately by[1]:

$$\varrho(n) = \frac{1}{\sqrt{48}\,n} \exp\left(\pi \sqrt{\frac{2n}{3}}\right). \tag{13.1}$$

The number of single particle excitations available is:

$$n = \frac{3A}{2T_F} E \tag{13.2}$$

since (Sect. 11) the average level spacing for single particle excitations of either neutrons or protons is given by $2T_F/3A$ for $N=Z=A/2$. Keeping only the exponential term in Eq. (13.1) we may write:

$$\varrho(E) \approx \exp\left(\sqrt{4CE}\right) \tag{13.3}$$

where

$$C = \frac{\pi^2 A}{4T_F}. \tag{13.4}$$

For a nucleus with $A=120$, Eq. (13.4) gives $C=7.4\ \mathrm{Mev}^{-1}$. Empirically[2], the level densities for nuclei of about this mass number are well fitted by Eq. (13.3) with $C=8\ \mathrm{Mev}^{-1}$. There is, however, no justification for neglecting the velocity dependence of the potential which had not been included in the above argument (or more basically, the exchange interactions which underly this velocity dependence). The exchange effects[3] inevitably increase the spacing of individual particle levels and consequently, decrease the average level density.

It has, however, been pointed out that the statistical model cannot be accurate for predictions of average level spacings, at least at relatively small excitation energies. Because of the approximate spherical nuclear symmetry, the single particle states are not plane waves (of degeneracy 2) but eigenstates of angular momentum (of degeneracy $2j+1 \geq 2$). This degeneracy effectively *decreases*[4] the average single particle level spacing in nuclei away from closed shells, at least for low energy excitations.

The above argument works the other way for closed shell or "magic number" nuclei (cf. Chap. E). For those, the relevant average single particle level spacing is in fact increased even further compared to the statistical value so that especially *low* level densities result. Indeed, the exceptionally small neutron cross sections found for some "magic number" nuclei[5] (which imply, in turn, anomalously small level densities), constitute important evidence for the existence of shell structure in nuclei.

---

[1] G. H. HARDY and S. RAMANUJAN: Proc. Lond. Math. Soc. **17**, 75 (1918). — C. VAN LIER and G. E. UHLENBECK: Physica, Haag **4**, 531 (1937).

[2] Cf. Ref. [*2*], p. 372.

[3] J. BARDEEN: Phys. Rev. **51**, 399 (1937).

[4] C. BLOCH: Phys. Rev. **94**, 1089 (1954), N. ROSENZWEIG, Phys. Rev. in press.

[5] Ref. [*29*], p. 35.

III. Refinements of the model.

In all of the previous arguments, two assumptions have been made without prior justification. First of all, it has been supposed that the nuclear many-body problem can be treated by first order perturbation theory, at least for the calculation of binding energies and related quantities. The other major assumption concerned the dependence of the interaction energy of a pair of particles on their relative momentum. In the above treatment, terms of order k^4, k^6 etc. have been dropped. In the next three sections we will discuss the validity and limitations of these two assumptions in the light of empirical properties of the nucleon-nucleon interactions.

14. Realistic two-body interactions. *α) Requirements for saturation.* It was pointed out previously that ordinary attractive forces between all pairs of nucleons would lead to collapse of the nucleus. However, other, more complicated, forms of interactions can lead to saturation. (i) First of all, of the interactions were predominantly (at least $\frac{4}{5}$) of the exchange (Majorana) form, then saturation would result very naturally. The empirically determined interactions between nucleons do indeed contain appreciable components of exchange terms (perhaps about $\frac{1}{2}$), but not enough by themselves to give saturation. (ii) Another possibility is that the interactions turn strongly repulsive at very short distances like the interactions between atoms and between molecules. This by itself would lead to saturation with an average spacing between particles essentially equal to the distance for which the interactions become repulsive. (More correctly, at a slightly larger distance, because of the zero point kinetic energies.) Empirically, the nucleon-nucleon interactions are believed to turn repulsive at short distances of order 0.5×10^{-13} cm. Thus the empirical short range repulsion may not be the main factor in leading to saturation at the observed nuclear density, though they would become very effective at about 10 times the normal density. (iii) It has been proposed that there might be in nuclei direct many-body forces acting between more than two nucleons at a time[1]. If there are such forces, say, three-body forces which act in such a way as to keep the particles apart, then we might possibly obtain saturation even though all two-body forces are attractive. This interesting possibility cannot be ruled out at the present time; however, there is no experimental evidence that such many-body forces are of great importance. (iv) Next there is the possibility that the interaction is not central. In particular tensor forces do not contribute at all to the nuclear binding in first order, though they strongly influence properties of the nucleon-nucleon system. Thus, since the nucleon-nucleon interactions are known to contain an appreciable fraction of tensor forces the task of obtaining saturation is eased somewhat in the sense that only *part* of the total interaction is now capable of causing collapse of the nucleus. It must be noted, however, that in *higher order*, the tensor forces lead to polarization of the nucleus[2], e.g. the positive quadrupole moment of the deuteron. (v) Finally, the interactions may not be static but dependent on the relative velocity of the nucleons. (The well known spin-orbit interactions, which exist between pairs of nucleons, are of this type.) A particularly simple kind of velocity dependent interaction which leads to saturation is one which is effective *only* for S waves, [*15*] i.e. states of relative orbital angular momentum zero. At high densities, the maximum angular momentum formed by nucleon pairs close enough to interact would be approximately

$$L_{\max} = k_F a \tag{14.1}$$

[1] S. D. Drell and K. Huang: Phys. Rev. **91**, 1527 (1953).

[2] J. M. Volkoff: Phys. Rev. **62**, 126, 134 (1942).

where a is the range of the nuclear forces. Since for each value of L there are $2L+1$ different spatial states, the total number of possible states with $L \leq L_{\max}$ is proportional to $(k_F a)^2$, but only one of these states is an S state. Assuming that each of these states occur with equal a-priori probability, the fraction of S states formed is only of order $(k_F a)^{-2}$ so that the interaction energy per particle is no longer proportional to k_F^3 as for purely attractive forces, but to k_F itself. The kinetic energy of the nucleons, proportional to k_F^2 per particle, is thus enough to keep the nucleus from collapsing even if the S wave interaction is purely-attractive.

β) Empirical evidence. Recent analyses of high energy nucleon-nucleon scattering experiments suggest that the two body interaction has some similarity to a central S-wave interaction plus a tensor interaction[1]. However this conclusion is by no means established and is not readily understood in the present framework of meson theory[2]. Experimentally, (as deduced from phase shift analyses) the interactions seem to be strongly attractive in S states. The interactions tend to cancel in P states when averaged over all spin states (as expected for a Serber interaction), though they are appreciable in individual spin states. The important feature which points toward a velocity dependence of the nuclear forces, and its possible important role in leading to saturation, is the interaction in D states[3]. From the analyses of scattering cross sections it appears that the central part of the D-wave interaction is small[4]. (Actually, the effective D-wave interactions appear to be somewhat *repulsive*, but it can be shown that tensor forces may be responsible for this.) Any static central potential, regardless of the exchange mixture, can only depend on the spins, charges and spatial symmetry of the wavefunction. In particular, since the central forces are known to be attractive for S waves they would also have to be attractive for D-waves, if they were static, probably in contradiction to experiment. Little is known regarding the interactions in waves of large L beyond the fact that they are very small for F waves. However these waves are not expected to be important at normal density. Using Eq. (14.1) with $k_F = 1.4 \times 10^{13}\,\text{cm}^{-1}$ and $a = 2 \times 10^{-13}$ cm we find $L_{\max} \approx 3$, the largest value of L for which interactions may be appreciable at normal density.

γ) Velocity-dependence of nuclear potential. Closely related to the problem of saturation is the dependence of the pair interaction energy ΔE_{ij} between nucleons on their relative momentum. To get saturation with purely attractive interactions it is necessary that ΔE_{ij} vanishes or becomes positive repulsive for sufficiently high momenta. Thus, for the saturating S-wave interaction discussed above, we see that ΔE_{ij} is proportional to k_F^{-2} in the limit of large momenta (on the other hand, for the non saturating Serber interaction, ΔE_{ij} falls only to half its low-momentum value). One would therefore expect that the average nuclear potential seen by a nucleon should vanish at high energy. Empirically the nuclear potential is about 20 Mev deep for nucleons of 100 Mev kinetic energy

---

[1] Y. Yamaguchi: Phys. Rev. **95**, 1628 (1954).

[2] In fact, better fits to the two-body data up to 150 Mev have been obtained [P. S. Signell and R. E. Marshak, Phys. Rev. **106**, 832 (1957)] by using a static potential deduced from meson theory [S. Gartenhaus, Phys. Rev. **100**, 900 (1955)] plus a spin-orbit coupling term. The central part of this potential is strongly singular at short distances; e.g. for even-parity states it has essentially a repulsive core of radius $\approx 0.5 \times 10^{-13}$ cm. It is likely, however, that the effect of the actual highly singular potential can be simulated quite well by a less singular bur velocity dependent potential, at least at fairly low energies.

[3] C. De Dominicis and P. C. Martin: Phys. Rev. **105**, 1419 (1957).

[4] H. Feshbach and E. Lomon: Phys. Rev. **102**, 891 (1956).

and about 10 Mev at nucleon energies above 200 Mev, but it is not clear whether or not it actually goes to zero[1]. In any case, the observed nuclear potential is much weaker at high energies than at low energies, e.g. about 50 Mev deep at about 0 Mev energy.

Finally, we remark on the accuracy of the so-called effective-mass approximation, the neglect of all terms of order k^4, k^6 in the expression for ΔE_{ij}.

The effective mass M^* may be defined by:

$$\frac{M}{M^*} = 1 + \frac{dV}{dT} = 1 + \frac{M}{\hbar^2} \frac{1}{k} \frac{dV}{dk} \tag{14.2}$$

as was discussed above.

If the effective mass (quadratic) approximation were strictly valid, M^* would be independent of k. On the basis of calculations which have been made using the two body forces discussed above, it appears that this approximation is quite good at low energy. Thus, M^* may be about $\frac{1}{2}M$ for the bound nucleons of average kinetic energy but slightly larger, perhaps $0.6M$, for nucleons at the top of the Fermi sea [*16*]. On the other hand, at high energies the potential is relatively independent of k, and M^* is nearly equal to M.

15. Higher order corrections to binding energy. In the previous discussion it has been assumed that the nuclear interactions affect only the ground state energy but not the wavefunction. However the interactions also change the nuclear wavefunction by inducing correlations between motions of nucleons. Thus, the probability of finding two nucleons within the range of their mutually attractive interactions will be larger than it would be in the absence of interaction, i.e. according to any independent particle model. The binding energy is always increased relative to the first order value by an amount known as the correlation energy. Presumably, if the correlation energy, calculated with second order perturbation theory, is small compared to the first order interaction energy, then the independent particle description should be a good approximation. In this section we will make a qualitative estimate of the ratio of these energies. It will be assumed for mathematical simplicity that the nuclear interactions are Serber Forces of Gaussian radial dependence [*17*].

$$V(r) = -V_0 \exp(-r^2/a^2). \tag{15.1}$$

As was already discussed, this interaction does not lead to saturation. Also it does not have the strong short range singularities of the empirical potential (cf. next section). Thus it is expected that the estimates of the correlation energy made in this section have only qualitative significance.

In first order the interaction energy is given by:

$$E^{(1)} = \sum_{i<j} \langle ij | V | ij \rangle. \tag{15.2}$$

The sum ij goes over all occupied nucleon states (k_i, $k_j \leq k_F$), care being taken to count each pair only once.

α) *Second order energy.* The second order contribution to the energy is given by the well-known equation

$$E^{(2)} = -\sum_{i'j'} \sum_{i<j} \frac{|\langle i'j' | V | ij \rangle|^2}{E_{i'j'} - E_{ij}}. \tag{15.3}$$

---

[1] W. B. RIESENFELD and K. M. WATSON: Phys. Rev. **102**, 1157 (1956). — A. KIND and L. JESS: Nuovo Cim. **4**, 595 (1956). — W. E. FRAHN: Nuovo Cim. **5**, 393 (1957).

Each term of the double sum represents a "virtual excitation" of a nucleon pair from states i, j, to states i', j'. The sum $i' j'$ goes over pairs of unoccupied nucleon states only, $(k_{i'}, k_{j'} > k_F)$, since the exclusion principle forbids more than one nucleon from occupying any given state. The matrix elements of the interaction are of the form

$$\langle i' j' | V | i j \rangle = \int V \, e^{i(\boldsymbol{k}-\boldsymbol{k}')\cdot \boldsymbol{r}} d^3 r \, \delta_{\boldsymbol{K}\boldsymbol{K}'} = - V_0 \pi^{\frac{3}{2}} a^3 \, e^{-\frac{(k'-k)^2}{4}} \delta_{\boldsymbol{K}\boldsymbol{K}'} \tag{15.4}$$

for a potential of Gaussian shape, where K, K' denote center of mass momenta in the two states involved and k, k' are the relative momenta [Eq. (7.10)] in the two states. Evidently, the center of mass momenta must be the same in the two states in order that the matrix element be finite. The relative momenta may however differ in magnitude and direction, with the vector difference representing a momentum transfer. The matrix element is small for momentum transfers larger than about $2/a$. The energy difference in Eq. (15.3) is readily shown to be given by:

$$E_{i' j'} - E_{i j} = \frac{\hbar^2}{M} (k'^2 - k^2) \tag{15.5}$$

where M is the nucleon mass.

β) Low density case. First consider the low density case, i.e. average spacing of nucleons large compared to the range of nuclear forces. In the ground state, all the nucleons will have small momenta:

$$k_F a \ll 1. \tag{15.6}$$

Explicit calculations [*17*] show that the ratio of second order to first order energies is given by:

$$\frac{E^{(2)}}{E^{(1)}} = \frac{1}{2^{\frac{3}{2}}} \frac{M a^2}{\hbar^2} V_0. \tag{15.7}$$

We may define the strength s of the potential by:

$$V_0 = s \cdot 2.68 \frac{\hbar^2}{M a^2}. \tag{15.8}$$

A Gaussian potential of strength $s = 1$ is just strong enough so that the ground state is at zero energy. Thus Eq. (15.8) becomes:

$$\frac{E^{(2)}}{E^{(1)}} \approx 0.95 \, s. \tag{15.9}$$

Since the internucleon potentials are strong enough to give a bound state for the two body system (the deuteron) we see that the Born approximation is very poor at low density. Indeed the entire perturbation theory expansion does not converge at all. The interpretation of this is very simple: a system of nucleons placed in a large enclosure will cluster into nuclei rather than moving independently.

γ) High density case. The situation is quite different at high densities. Using EULER's formulas we readily obtain the following result:

$$\frac{E^{(2)}}{E^{(1)}} = \frac{0.70 \, s}{(k_F a)^3}. \tag{15.10}$$

The reduction of the correlation energy below the low density value is due to two factors. First, the particles will have larger momenta than at low density and as result the energy denominators will be larger on the average. This is the same effect which tends to make the Born approximation for scattering problems more accurate with increasing energy. Probably even more important than the above is the effect of the Pauli exclusion principle: Many momentum transfers are forbidden because they would leave one or both nucleons in an occupied state below the top of the Fermi sea. If both nucleons are well below the top of the Fermi sea then the only allowed terms involve large momentum transfers and these are small as is seen in Eq. (15.4).

To make a numerical estimate for the correlation energy, we use the average values $s = 1.22$, $a = 1.75 \times 10^{-13}$ cm deduced from analyses of the two body problem and $k_F = 1.4 \times 10^{13}$ cm^{-1} from Eq. (9.14).

Thus we have

$$k_F a \approx 2.5. \tag{15.11}$$

Substitution of these numbers into Eq. (15.10) gives:

$$\frac{E^{(2)}}{E^{(1)}} \approx 0.05. \tag{15.12}$$

More detailed calculations starting from the same assumptions as above, but in which we also keep terms of higher order in $(k_F a)^{-1}$, rather than using (15.10), give essentially the same value for the above ratio. Thus according to the present estimate, first order perturbation theory gives a relatively accurate value for the binding energy[1]. With somewhat more singular interactions, e.g. a Yukawa potential, or with potential exchange mixtures designed to give saturation at the observed density the ratio turns out to be somewhat larger, perhaps 10 or 15%. These results speak much more in favor of an independent particle model than was the case when it was first studied (at that time, the above ratio was thought to amount to about 30%). The main reason for this change is the reduction in the nuclear radius constant from 1.4 to 1.1, i.e. the currently accepted value of the nuclear density is twice as large as the value assumed before about 1953. One might argue that a better measure of the correlations is given by the ratio P.E.$^{(2)}$/P.E.$^{(1)}$ of potential energy terms[2]. Now the first order energy $E^{(1)}$ is precisely P.E.$^{(1)}$, while for the second order term it is readily shown that

$$\frac{\text{P.E.}^{(2)}}{\text{P.E.}^{(1)}} = 2\,\frac{E^{(2)}}{E^{(1)}}. \tag{15.13}$$

The kinetic energy is actually increased by the correlations, but the increase in potential binding energy is twice as large. Even the ratio (15.13) is expected to be quite small[2].

δ) Third order energy. Finally we should briefly mention the third order terms in the expansion for the energy. One class of third-order terms describes correlations involving three particles at a time. These have been estimated to amount to about 10% of the second order terms [*15*]. Another class of terms represents the effect of the velocity dependent potential on the correlations between two particles. In terms of Eq. (15.5), the energies of single particle levels are spread out by the velocity dependence of the nuclear potential, so that the energy denominators are increased. If the potential were a quadratic function of momentum at all energies, then the correlation energy would be reduced to a fraction $M^*/M = \frac{1}{2}$ of the previously calculated value. However, the potential is much less velocity dependent for nucleons in states $i'j'$, which are mostly between 0 and 100 Mev

---

[1] This conclusion has also been reached by Bethe, Ref. [*15*], and by W. J. Swiatecki, Phys. Rev. **103**, 265 (1956).

[2] W. J. Swiatecki: Phys. Rev. **103**, 265 (1956).

above the top of the Fermi sea, than for bound nucleons. On the basis of detailed calculations it appears that the velocity dependence of the potential reduces the correlation energy by at most 20%[1].

16. Effects of correlations. Before discussing correlations in nucleon motions due to the internucleon interactions it must be pointed out that there are important correlations due to the exclusion principle, even in the absence of interactions. For example, if a nucleon occupies a given state, then no other nucleon of the same spin and charge can occupy the same state or be at the same position. Thus, nucleons of the same spin and charge are effectively kept apart even without any explicit interactions between them. These correlations are, however, already included in the antisymmetrical product wavefunction assumed in the Fermi gas model. Only if the wavefunction were a simple product of one particle functions and not antisymmetrized, would the particles move *truly* independently.

α) *Effect on ground state wavefunction.* Consider now the additional correlations due to the interactions between nucleons. We have already seen that these increase the nuclear potential energy only by a small fraction. On the other hand, some other nuclear properties are quite strongly affected by the interactions. To consider an extreme case let us mention the effect of correlations on the nuclear ground state wave function itself.

To first order in the interactions, the wavefunction is given by:

$$\psi \approx \psi_0 - \sum_{i'j'} \sum_{i<j} \frac{\langle i'j'|V|ij\rangle}{E_{i'j'} - E_{ij}} \psi_{ij\to i'j'} \tag{16.1}$$

where ψ_0 is the unperturbed wavefunction and $\psi_{ij\to i'j'}$ is a wavefunction in which a pair of nucleons have been excited from states ij to $i'j'$. Eq. (16.1) can be written in the form:

$$\psi = \psi_0 + \sum_v \alpha_v \psi_v \tag{16.2}$$

where α_v is the fractional admixture of state $\psi_v = \psi_{ij\to i'j'}$ in the wavefunction. If the effect of interaction were *really* small, we should have

$$\sum_v |\alpha_v|^2 \ll 1. \tag{16.3}$$

Using Eq. (15.5) we may readily express the sum in terms of the contribution $E_v^{(2)}$ of each term to the bindig energy. Thus

$$\sum_v |\alpha_v|^2 = \sum_v \frac{E_v^{(2)}}{E_v - E_0} \approx \frac{E^{(2)}}{\Delta E_{\mathrm{Av}}} \tag{16.4}$$

where ΔE_{Av} is an average energy denominator.

A reasonable estimate for this sum is[2]:

$$\sum_v |\alpha_v|^2 \approx 0.1\, A. \tag{16.5}$$

The factor A occurs here since $E^{(2)}$ is proportional to A while ΔE_{Av} is independent of mass number. Thus ψ_0 is actually only a very *small* part of the actual nuclear wavefunction. At first sight, this result implies that the independent particle model is a very poor approximation, especially for heavy nuclei where the Fermi gas model should be the most nearly applicable.

---

[1] C. De Dominicis and P. C. Martin: Phys. Rev. **105**, 1417 (1957).
[2] S. Watanabe: Z. Physik **113**, 482 (1939).

β) Effect on physical observables. For most physically observable quantities the situation is somewhat more favorable for the independent particle model, though not as much as might be thought from the discussion of nuclear binding energies in the last section. Thus it can be shown [*15*] that matrix elements of single particle operators, i.e. operators of the form

$$\mathfrak{M} = \sum_i \mathfrak{M}(i) \tag{16.6}$$

where the sum extends over all nucleons, are altered only by a fraction of order unity, independent of A. The most important nuclear properties, besides the energy, involve one particle operators (e.g. moments, transition probabilities, momentum distributions).

Consider now the probability of finding a nucleon with momentum p. This is given by:

$$N(p) = \frac{N_0(p) + \sum_v |\alpha_v|^2 N_v(p)}{1 + \sum_v |\alpha_v|^2} \tag{16.7}$$

where $N_0(p)$ and $N_v(p)$ are the desired probabilities for states ψ_0 and ψ_v, respectively. We have:

$$\left.\begin{aligned} N_0(p) &= 1 \quad \text{for} \quad p < p_F, \\ &= 0 \quad \text{for} \quad p > p_F \end{aligned}\right\} \tag{16.8}$$

appropriate to a degenerate Fermi gas at zero temperature. Thus in the absence of correlations, the wavefunction contains no components with momenta larger than p_F. The effect of interactions is to smooth out the momentum distribution so that it resembles that of a Fermi gas at finite temperature. In particular the wavefunction now contains appreciable components of higher momenta. The existence of high momentum components in the nuclear wavefunctions has been verified by several kinds of experiments[1]. One of these is the pickup process in which a proton of large momentum picks up a neutron from inside the nucleus so as to form a deuteron. This process can occur only if the neutron has a large momentum while inside the nucleus. Closely related to the existence of high momentum components are the spatial correlations in the wavefunction. Thus there is increased probability of finding two nucleons at close range. Incidentally, the large amounts of high-momentum components in the nuclear wavefunction inferred from experiments are consistent with the assumption that interactions between nucleons in nuclear matter are approximately the same as in free space. According to proposed "meson potential" theories[2] in which the interactions in nuclei occur not via direct two body interactions, but via an overall meson field, the direct two body correlations between nucleons would be small, contrary to the above evidence.

γ) Singular interactions. In all the above treatment it has been assumed that the nuclear interactions are not especially singular, i.e. do not contain repulsive cores etc. For repulsive core potentials, the perturbation expansion fails completely—all matrix elements become infinite. While the nuclear interactions are indeed believed to turn repulsive at sufficiently short distances, it may be permissible to neglect this effect at least for properties involving wavelengths larger than the radius of the repulsive core $\approx 0.5 \times 10^{-13}$ cm, i.e. energies less than

[1] The results of these experiments have been summarized by BRUECKNER, EDEN and FRANCIS, Phys. Rev. **98**, 1445 (1955).

[2] L. I. SCHIFF: Phys. Rev. **84**, 1, 10 (1951); **86**, 856 (1952). — M. H. JOHNSON and E. TELLER: Phys. Rev. **98**, 783 (1955).

100 Mev. Thus low energy nucleons can only feel variations of the potential occurring over 10^{-13} cm. However, the existence of a repulsive core implies extremely strong correlation at very short distances and greatly increases the amplitude of high momentum components in the wavefunction.

The many body problem with singular interactions can be studied by a method proposed by BRUECKNER[1], which amounts to replacing the actual two-body potentials by "pseudopotentials"[2]. In this treatment, correlations involving two particles at a time may be arbitrarily large. It is only required that correlations involving three or more particles be relatively small. On the other hand, the success of a perturbation treatment of the nuclear problem requires that *all* correlations due to interactions be relatively small.

In the Brueckner treatment the nuclear wavefunction ψ is written in the form[3]:

$$\Psi = F\,\psi_0 \tag{16.9}$$

where ψ_0 is an independent particle wavefunction, the so-called "model wavefunction" of the same energy of ψ, and F is a unitary transformation operator. If we only consider two-body correlations, then F can be expressed in terms of the so-called "reaction matrix" for two-body collisions, which is, in turn, given in terms of the nucleon-nucleon phase shifts for nucleon-nucleon scattering. There is a close connection between this treatment and the derivation of the optical potential by multiple scattering theory (Sect. 18 to 19). The actual wavefunction may quite different from the model wave-function. However, to calculate physically observable quantities we may often use the model wavefunction. Thus consider the evaluation of matrix elements of an operator $\mathfrak{M}$ which commutes with F, between two states ψ_a and ψ_b. We then have

$$\int \psi_b^* \,\mathfrak{M}\, \psi_a \,d r = \int (\psi_0)_b^* \,\mathfrak{M}\, (\psi_0)_a \,d r. \tag{16.10}$$

It may be of interest to mention the case of He^3. This substance, like a hypothetical infinite nucleus, is a nearly degenerate Fermi gas, actually a liquid at low temperatures[4]. The interactions between He atoms have qualitatively the same behavior, strongly repulsive at short range, attractive at longer range, as do the nuclear interactions. However, unlike the case in nuclei, the average spacing between He atoms is only very slightly larger than the effective radius of the repulsive core. Thus it is expected that correlations are much more important in He than in nuclei.

In summary, it appears that the independent particle description may be a good approximation to average nuclear binding energies and as we will see later, also other low energy nuclear properties. However, correlations are extremely important for different aspects of nuclear structure, especially high energy phenomena.

C. The optical model.

17. Introduction[5]. *α) Early models of nuclear reactions.* Nuclear reactions began to be studied extensively in the 1930's and the interaction of nucleons with nuclei has been a subject of great interest since then. The first model to be

---

[1] K. A. BRUECKNER, C. A. LEVINSON and H. M. MAHMOUD: Phys. Rev. **95**, 217 (1954) and succeeding papers, in particular: K. A. BRUECKNER and W. WADA, Phys. Rev. **103**, 1008 (1956). These papers are summarized in Ref. [*15*]. See also R. J. EDEN, Theory of the Nucleus as a Many-Body System, to appear in Nuclear Reactions, edited by ENDT and DEMEUR, North Holland Publishing Company Amsterdam.

[2] K. HUANG and C. N. YANG: Phys. Rev. **105**, 767 (1957). Also, cf. Footnote 1, p. 428.

[3] R. J. EDEN and N. C. FRANCIS: Phys. Rev. **99**, 1366 (1955).

[4] Cf. K. MENDELSSOHN: Liquid Helium, Vol. XV of this Encyclopedia.

[5] Cf. Ref. [*20*], [*22*].

considered seriously for the understanding of these processes envisaged the nucleus as acting like a simple potential well for the incident particle[1]. On the basis of this model, elastic scattering should predominate and cross sections show maxima whenever standing waves can develop in the nucleus. The energy spectrum of the cross sections would thus be characterized by a few widely spaced resonances.

The discovery of resonances for low-energy neutrons contradicted this potential well picture completely. The resonances were closely spaced in energy and characterized by large cross sections for capture and other reactions. As was pointed out by Bohr [*11*] these features suggested the validity of a strong-coupling description (quite opposite to the potential well model) in which nucleons incident on the nucleus are immediately assimilated into it so that a compound nucleus is formed. The disintegration mode of the compound nucleus depends only on its overall properties, such as its energy and angular momentum, but is independent of the specific way in which it was formed. Cross sections in the vicinity of individual resonances were soon well explored and successfully described by the Breit-Wigner dispersion formula[2].

Although the cross sections show rapid variations as functions of energy because of the detailed resonance structure, averages of the cross sections over an energy range including a fairly large number of resonances show a much more regular behavior. The average cross sections can be calculated on the basis of a special form of the compound nucleus model, the so-called "continuum" model[3], in which the nucleus is regarded as an opaque sphere; this implies that there are no outgoing waves anywhere on the nuclear surface. On the basis of these assumptions, it was shown that the total cross sections should be monotonic functions of energy and nuclear radius.

β) Success of the optical model. Although qualitative features of the cross sections were reproduced by the above theory, the evidence soon pointed to a partial transparency of the nucleus. This was first noted[4] at high energies (90 Mev) where cross sections vary systematically about the geometric value which would be expected according to the continuum model. These data were satisfactorily interpreted on the basis of an optical model. In this model the nucleus is considered as a homogeneous medium of complex refractive index, or alternatively, as a complex potential well. This picture differs from the earlier potential well model in that the potential contains an imaginary term[5], whose strength is related to the cross section for all non-elastic processes, i.e. absorption, reactions, inelastic scattering. All non-elastic processes are lumped together as "absorption" in so far as they effectively remove particles from the incident beam. The optical model was soon successfully applied to low and medium energy reactions and found to reproduce certain cross section data (total cross sections and elastic scattering cross sections vs. angle) quite well[6]. Most strikingly, it was found that for scattering of 0 to 3 Mev neutrons, the cross sections[7] (averaged over individual resonances, of course) show a behavior quite similar to that predicted

---

[1] H. Bethe: Phys. Rev. **47**, 747 (1935). — Fermi *et al.*: Proc. Roy. Soc. Lond. **149**, 522 (1935).

[2] G. Breit and E. Wigner: Phys. Rev. **49**, 519 (1936).

[3] H. Feshbach and V. F. Weisskopf: Phys. Rev. **76**, 1550 (1949).

[4] Fernbach, Serber and Taylor: Phys. Rev. **75**, 1352 (1949).

[5] H. Bethe: Phys. Rev. **57**, 1125 (1940).

[6] K. W. Ford and D. Bohm: Phys. Rev. **79**, 745 (1950). — R. E. LeLevier and D. S. Saxon: Phys. Rev. **87**, 40 (1952).

[7] H. H. Barshall: Phys. Rev. **86**, 431 (1952).

by the simple potential well model [19]. Thus, maxima in the averaged cross-sections occur at about those energies where the potential well model would predict them to occur.

Analyses of scattering cross sections lead to the conclusion that the absorption potential, W_0, the imaginary part of the optical potential, is very small for incident nucleons of low energy, e.g. only about 1 or 2 Mev (compared to a refraction potential V_0 of about 50 Mev) for nucleons of energies below 5 Mev. This implies that low energy nucleons have an effective mean free path of about 20×10^{-13} cm in nuclear matter, much larger than the nuclear diameter, so that they have a good chance of traversing the nucleus without losing energy. This feature, which is very much in line with ideas of independent particle motion, has been recently accounted for along the lines indicated in Sect. 19[1]. Evidently most of the collisions by which an incident nucleon might lose energy in nuclear matter would have either this nucleon, or the one with which it has collied in a state below the top of the Fermi sea. All these states are, however, already occupied by other nucleons, so that such processes are forbidden because of the Pauli principle[2-4]. While the nuclear opacity may thus be small when averaged over an energy range including *many* resonances, it is large in the vicinity of each *individual* resonance.

At higher incident energies, a substantial fraction of collisions leave both nucleons above the top of the Fermi sea. The Pauli principle is here of less importance and the probability of allowed collisions is essentially the same as between free nucleons. Since free nucleons interact quite strongly, it is expected that the absorption potential is now much larger than at low energies. For example, for nucleons of 30 Mev incident energy, we have $W_0 = 15$ Mev corresponding to a mean free path of only 3×10^{-13} cm. Under these conditions, the optical model gives results not very different from the predictions of the continuum model. At even higher energies the Pauli principle becomes quite ineffective, but since the individual nucleon-nucleon cross sections decrease with energy, so does the absorption potential; for 100 Mev nucleons the mean free path is about 6×10^{-13} cm.

γ) The nuclear reaction scheme[5]. While the optical model is extremely useful for the analysis of scattering cross sections, it does not tells us anything about the detailed reactions which occur when a nucleon is "caught" by the nucleus besides the total cross section for all such processes. According to the compound nucleus model any nucleon which is caught immediately forms a compound nucleus in which its energy has been shared with the other nucleons. One important consequence of this assumption is that any particles emitted in the decay of this compound nucleus usually have much smaller energy than the incident particle, their energy spectrum is, in fact, expected to approximate a Maxwellian exponential distribution. It would also follow that the angular distribution of

---

[1] A. M. LANE and C. F. WANDEL: Phys. Rev. **98**, 1524 (1955). — E. CLEMENTEL and C. VILLI: Nuovo Cim. **2**, 176 (1955). — MORRISON, MUIRHEAD and MURDOCH: Phil. Mag. **46**, 795 (1955).

[2] V. F. WEISSKOPF: Science, Lancaster, Pa. **113**, 101 (1951).

[3] If the exclusion principle is not taken into account, a much larger value of W_0 is obtained.

[4] The detailed interpretation of W_0 in terms of the nuclear many-body problem is still in a less satisfactory state and a number of discrepancies have yet to be resolved. See: LANE, THOMAS and WIGNER, Phys. Rev. **98**, 693 (1955); M. CINI and S. FUBINI, Nuovo Cim. **2**, 443 (1955); A. KIND, Nuovo Cim. **2**, 443 (1955); BRUECKNER, EDEN and FRANCIS, Phys. Rev. **100**, 891 (1955).

[5] V. F. WEISSKOPF: Physica, **22**, 952 (1956).

the emitted particles is isotropic. In many cases, these predictions are verified experimentally[1], that is to say, the nucleons which are absorbed do indeed appear to form a compound nucleus in which all memory regarding the mode of formation has been lost. However, in a number of reactions it is found that the independence hypothesis regarding the mode of disintegration fails[2]. Thus in some reactions, the particles are emitted preferentially in the forward direction. Also, inelastic scatterings involving only small energy losses sometimes occur much more frequently than would be expected on the compound nucleus model[3]. These features suggest that nucleons may be caught by nuclei forming a "compound system" (C.S.) but not necessarily a compound nucleus (C.N.). For example the incident nucleon may interact with only a few nucleons in the nucleus, rather than being entirely assimilated. In particular direct interactions at the nuclear surface may be significant competitors to the compound nucleus formation[4]. Other processes which can produce absorption without compound nucleus formation are (i) direct interactions of the incident nucleon with a particular nucleon in the target nucleus. (ii) Multiple collisions of the incident nucleon and (iii) excitations of collective oscillations (Chap. F). Nevertheless, compound nucleus formation is, in general, expected to predominate in many cases. The different stages in a nuclear reaction are illustrated in Fig. 4. Note that part of the total (observed) elastic scattering occurs through formation of a compound nucleus. The optical model only gives information regarding the "shape-elastic" scattering which occurs prior to any absorption. The effect of "compound-elastic" scattering is negligible at higher energies but it may be quite significant at low energies (up to a few Mev).

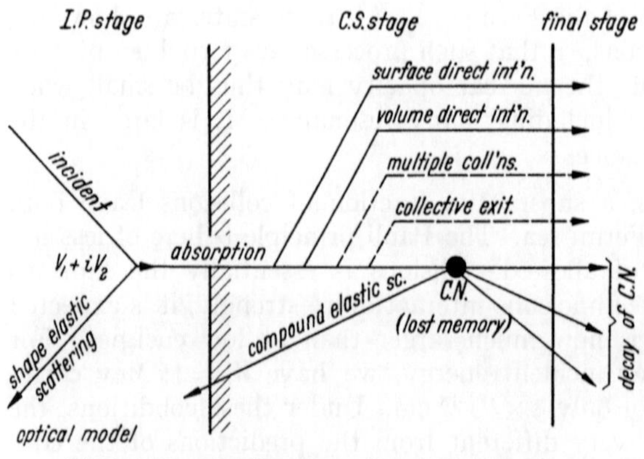

Fig. 4. Nuclear reaction scheme [from V. F. Weisskopf: Physica, Haag 22, 952 (1956)].

The optical model is somewhat of a hybrid between the potential well and compound nucleus models. In the limit of zero absorption ($W=0$) it reduces exactly to the potential well model. On the other hand, if W is permitted to become arbitrarily large, we have essentially the condition of the strong interaction model, of course without any resonance fine structure.

In this Chapter we discuss various topics relevant to the optical model. The nect two sections deal with the derivation of the optical potential from the two-body scattering properties. This is followed in Sect. 20 by a brief summary of empirical information of the radius of the optical potential. It appears that this radius is considerably larger than the radius of the density distribution. This feature is discussed and qualitatively explained in Sects. 21 to 23.

---

[1] E. R. Graves and L. Rosen: Phys. Rev. **89**, 343 (1953). — E. R. Graves and R. W. Davis: Phys. Rev. **97**, 1205 (1955).

[2] P. C. Gugelot: Physica **22**, 1019 (1956).

[3] B. Cohen: Phys. Rev. (to be published).

[4] L. R. D. Elton and L. C. Gomes: Phys. Rev. **105**, 1027 (1957).

I. Multiple scattering and the complex potential.

In Sect. 18 we derive an expression for the refractive index of an infinite medium. We regard the refraction of the incident wave by the particles in the medium as a multiple scattering process. This method is valid even if the Born approximation fails. We do, however, make the assumption that the scattering of the wave by *each* particle is not affected by the configuration of the remaining particles (except in an average way).

The treatment is applied in Sect. 19 to a derivation of the potential seen by a nucleon in an infinite nucleus. The above assumptions correspond to the condition that while two-body correlations may be quite large, correlations involving three or more particles are neglected.

The fact that two-body correlations are considered gives rise to some important differences between the potential derived here and the results of first order perturbation theory. Thus, the strength of the potential is no longer proportional to the strength of the fundamental two body interactions, and also, the potential is no longer real but complex, at least for nucleons of positive energy. This treatment is closely related to the Brueckner theory of nuclear structure, though the latter theory contains many refinements not considered here.

18. The refractive index[1]. The refraction of a wave in a medium can be regarded as a multiple scattering process, each scattering involving the interaction of the wave with one of the particles in the medium. The refractive index of the medium may be expressed in terms of properties of the fundamental scattering processes.

We picture an incident wave as traveling trough the medium of randomly located scatterers. Suppose that in free space the wave is a plane wave of propagation constant k. Before taking up the multiple scattering problem, consider the effect of one isolated scatterer on the incident wave. In this section we neglect recoil effects. If the scatterer, assumed to give rise to a spherically symmetric field $U(r)$, is located at the origin, the propagation of the wave is given by an equation of the form:

$$(\nabla^2 + k^2)\,\psi_k(0; r) = U(r)\,\psi_k(0; r) \tag{18.1}$$

where $\psi_k(0; r)$ is the amplitude of the wave. The subscript denotes the propagation constant far away from the scatterer, while 0 describes the position of the scatterer.

The asymptotic solution of Eq. (18.1) is well known to be given by[2]:

$$\psi_k(0; r) \xrightarrow[r\to\infty]{} \mathrm{e}^{i\boldsymbol{k}\cdot\boldsymbol{r}} + f(\boldsymbol{k}, \boldsymbol{k}')\,\mathrm{e}^{ikr}/r. \tag{18.2}$$

The quantity $f(\boldsymbol{k}, \boldsymbol{k}')$ is the "scattering amplitude" in the direction of the vector $\boldsymbol{k}'$. For elastic scattering, k and k' must have the same magnitude. This scattering amplitude is given by:

$$f(\boldsymbol{k}, \boldsymbol{k}') = -\frac{1}{4\pi}\int \mathrm{e}^{-i\boldsymbol{k}'\cdot\boldsymbol{r}}\,U(r)\,\psi_{\boldsymbol{k}}(0; r)\,d^3r. \tag{18.3}$$

We now take the Fourier transform of this equation with respect to k' and substitute into (18.1) to find:

$$(\nabla^2 + k^2)\,\psi_{\boldsymbol{k}}(0; r) = -\frac{1}{2\pi^2}\int f(\boldsymbol{k}, \boldsymbol{k}')\,\mathrm{e}^{i\boldsymbol{k}'\cdot\boldsymbol{r}}\,d^3k'. \tag{18.4}$$

[1] Cf. Ref. [*21*].

[2] We normalize the wave to one photon (or particle) per unit volume.

Next, suppose that the scatterer is centered at a point $\boldsymbol{r}_i$, so that its field is of the form $U(|\boldsymbol{r}-\boldsymbol{r}_i|)$. Then by use of a simple coordinate transformation we may show that:

$$(\nabla^2+k^2)\,\psi_{\boldsymbol{k}}(\boldsymbol{r}_i;\boldsymbol{r}) = -\frac{1}{2\pi^2}\int f(\boldsymbol{k},\boldsymbol{k}')\,e^{i(\boldsymbol{k}-\boldsymbol{k}')\cdot\boldsymbol{r}_i}\,e^{i\boldsymbol{k}'\cdot\boldsymbol{r}}\,d^3k'. \tag{18.5}$$

The righthand side of this equation can be regarded as the product of an integral operator and the unperturbed incident plane wave.

Consider now the propagation of the wave under the influence of a large number A of scatterers, located at points $\boldsymbol{r}_1, \boldsymbol{r}_2, \ldots, \boldsymbol{r}_A$. The amplitude of the wave at each point is now denoted by $\psi_{\boldsymbol{k}}(\boldsymbol{r}_1,\ldots,\boldsymbol{r}_A;\boldsymbol{r})$. We must sum the righthand side of Eq. (18.5) over the position of all scatterers. For any scatterer i, the effective incident wave denoted by $\psi_{\boldsymbol{k}'}^{(i)}(\boldsymbol{r}_i,\ldots,\boldsymbol{r}_A;\boldsymbol{r})$, is not the unperturbed wave, but includes the contribution from all *other* scatterers.

Thus we obtain the set of coupled integral-differential equations

$$\left.\begin{aligned}&(\nabla^2+k^2)\,\psi_{\boldsymbol{k}}(\boldsymbol{r}_1,\ldots,\boldsymbol{r}_A;\boldsymbol{r})\\ &\quad= -\frac{1}{2\pi^2}\int\sum_i f(\boldsymbol{k},\boldsymbol{k}')\,e^{i(\boldsymbol{k}-\boldsymbol{k}')\cdot\boldsymbol{r}_i}\,\psi_{\boldsymbol{k}'}^{(i)}(\boldsymbol{r}_1,\ldots,\boldsymbol{r}_A;\boldsymbol{r})\,d^3k'\end{aligned}\right\} \tag{18.6}$$

which is the formally exact representation of the multiple scattering problem.

We now make two important approximations. First, since there are so many scattering particles, the effect of any *one* particle on the wave is small. Consequently we commit little error in replacing the wavefunction on the right hand side of Eq. (18.6) by the complete wavefunction (including the effect of *all* scatterers) with incident propagation constant k'. Second, the scatterers are not located at fixed positions, as in a crystal but are essentially randomly distributed. We are thus not interested in the wavefunction for a specified configuration of scatterers, but in the wavefunction averaged over all possible configurations of scatterers. The only condition on the distribution is that the density averaged over a region including several scatterers should be constant. This configurational average of the wavefunction is denoted here by $\langle\psi_{\boldsymbol{k}}(\boldsymbol{r})\rangle$. We may now replace the sum over scatterers by a volume integral in which the dependence of the wavefunction on $\boldsymbol{r}_i$ is neglected[1]. Thus

$$\sum_i f(\boldsymbol{k},\boldsymbol{k}')\,e^{i(\boldsymbol{k}-\boldsymbol{k}')\cdot\boldsymbol{r}_i} = N\int f(\boldsymbol{k},\boldsymbol{k}')\,e^{i(\boldsymbol{k}-\boldsymbol{k}')\cdot\boldsymbol{r}_i}\,d^3r_i = (2\pi)^3 N f(\boldsymbol{k},\boldsymbol{k}')\,\delta(\boldsymbol{k}-\boldsymbol{k}') \tag{18.7}$$

where N is the number of scatterers per unit volume and $\delta(\boldsymbol{k}-\boldsymbol{k}')$ is the Dirac delta function. Substituting this result into Eq. (18.6) we obtain:

$$(\nabla^2+k^2)\langle\psi_{\boldsymbol{k}}(\boldsymbol{r})\rangle = -4\pi N f(\boldsymbol{k},\boldsymbol{k})\langle\psi_{\boldsymbol{k}}(\boldsymbol{r})\rangle, \tag{18.8}$$

or equivalently:

$$[\nabla^2+k^2+4\pi N f(0)]\langle\psi_{\boldsymbol{k}}(\boldsymbol{r})\rangle = 0 \tag{18.9}$$

where $f(0)\equiv f(\boldsymbol{k},\boldsymbol{k})$, the forward scattering amplitude for the incident wave.

The average effect of the medium is thus to change the propagation constant of the wave from k to a modified value k' given by

$$k'^2 = k^2 + 4\pi N f(0). \tag{18.10}$$

---

[1] This assumption is essentially equivalent to saying that the scattering of the wave by *one* particle is not influenced by the configuration of the *other* particles, except in so far as they modify the average propagation of the wave. Equivalently the range of the field due to each scatterer is assumed to be small compared to the average distance between scatterings.

(Strictly speaking, the scattering amplitude should be evaluated for this modified propagation constant k' in the medium.) The refractive index of the medium is defined by:

$$n = \frac{k'}{k} = \left[1 + \frac{4\pi N f(0)}{k^2}\right]^{\frac{1}{2}}. \tag{18.11}$$

If $f(0)$ were proportional to k^2, the refractive index would be independent of k, and there would be no dispersion. To a good approximation this holds for low frequency electromagnetic radiation. However, dispersion is an important effect for many cases of interest, in particular, for the nuclear potential.

19. The complex nuclear potential. The considerations of the preceding section can be readily applied to study the propagation of nucleons in nuclei, well away from the surface region. Thus we would like to know how the presence of nuclear matter alters the propagation characteristics of a nucleon. This nucleon is just the wave discussed in the last section and the nucleons in the nucleus act as "scatterers"[1]. The effect of the nuclear medium is, of course, to give rise to an effective potential of the form:

$$V = T_{\text{out}} - T_{\text{in}} = \frac{\hbar^2}{2M}(k^2 - k'^2) \tag{19.1}$$

where T_{out} and T_{in} denote kinetic energies of the nucleon outside and inside the nucleus, respectively and k, k' are the associated propagation constants. Using (18.10) we have:

$$V = -\frac{\hbar^2}{2M} 4\pi N f(0). \tag{19.2}$$

This equation would be correct if the recoil of the scatterers could be neglected (i.e., if they had infinite mass) as was done in the last section. The fact that incident and scattering nucleons have the same mass, requires us to replace, in Eq. (19.2), the nuclear mass M by the reduced mass $M/2$ of the two particle system involved in each scattering process. Thus we obtain:

$$V = -\frac{\hbar^2}{M} 4\pi N f(0). \tag{19.3}$$

In general $f(0)$ and therefore V is complex. We will express the potential in the form:

$$V = -(V_0 + i W_0). \tag{19.4}$$

The depth of the potential (i.e. the "refraction potential") is given by:

$$V_0 = \frac{\hbar^2}{M} 4\pi N \,\text{Re}\, f(0). \tag{19.5}$$

If (i) all effects of antisymmetrization could be neglected, and (ii) no exchange forces are considered, and (iii) we make the Born approximation (neglect the distortion of the incident wave due to the interaction), then Eq. (18.3) becomes:

$$f(0) = -\frac{1}{4\pi}\int U(r)\, d^3r = -\frac{M}{4\pi\hbar^2}\int V(r)\, d^3r \tag{19.6}$$

where $V(r)$ is the assumed ordinary two-body potential, and

$$V_0 = -N \int V(r)\, d^3r. \tag{19.7}$$

[1] For a more rigorous treatment of this problem see N. C. Francis and K. M. Watson, Phys. Rev. **92**, 291 (1953).

This result had already been obtained directly in Sect. 7 by use of first order perturbation theory. The virtue of the present method is that it can be used even if the Born approximation fails, i.e. if two-body correlations are large. Under these conditions the potential will no longer be proportional to the strength of the basic interactions, and it will usually be velocity dependent even apart from any effect of exchange forces or antisymmetrization.

The absorption potential W_0 is given by:

$$W_0 = \frac{\hbar^2}{M} 4\pi N \operatorname{Im} f(0). \tag{19.8}$$

Now, using the well known "cross section theorem"[1],

$$\sigma = \frac{4\pi}{k} \operatorname{Im} f(0) \tag{19.9}$$

where σ denotes the total cross section for the two-body scattering process and k is the relative momentum, we find that:

$$W_0 = \frac{\hbar^2}{M} N k \sigma. \tag{19.10}$$

Note that the absorption potential vanishes in the Born approximation. Like the cross sections, it is in lowest order, proportional to the square of the two-body interactions. Strictly speaking Eq. (19.10) is valid only if all relative momenta are the same. A more realistic expression for W_0 is obtained from the above with the aid of (7.10):

$$W_0(i) = \frac{\hbar^2}{2\Omega M} \sum_i (\boldsymbol{k}_i - \boldsymbol{k}_j)\,\sigma_{ij} = \frac{\hbar}{2\Omega} \sum_j (v_i - v_j)\,\sigma_{ij} \tag{19.11}$$

where Ω denotes the nuclear volume, and the sum extends over all nucleons in the nucleus.

Eq. (19.11) can also be derived directly in the following way[2]: If the potential contains an imaginary term, the total number of particles is not conserved. In particular for a potential of the form (19.4) we have:

$$\frac{\partial}{\partial t}(\psi^* \psi) = -\frac{1}{T}(\psi^* \psi) \tag{19.12}$$

where T is the mean life of a particle in the well, given by:

$$\frac{1}{T} = \frac{2W_0}{\hbar}. \tag{19.13}$$

Now if any two-body scattering is regarded as an absorption, we have from simple classical conditions:

$$\frac{1}{T} = \frac{1}{\Omega} \sum_j |v_i - v_j|\,\sigma_{ij}. \tag{19.14}$$

In fact, (19.14) is essentially a definition of the cross section. Combining the last two equations we obtain Eq. (19.11).

---

[1] H. S. W. MASSEY: Theory of Atomic Collisions, Vol. XXXVI of this Encyclopedia, p. 278.

[2] R. SERBER: Phys. Rev. **72** 1114 (1947). — M. L. GOLDBERGER: Phys. Rev. **74**, 1269 (1948).

For calculating the absorption potential we must, of course, consider only these two-body scattering processes which do not violate the Pauli principle, i.e. in which both nucleons have final states above the top of the Fermi sea.

II. Relation between the nuclear density and the nuclear potential.

20. Radius of the nuclear potential. Much information on the optical potential parameters has been obtained recently, especially from detailed analyses of proton-nuclei elastic scattering cross sections. It is found that good fits to the data can be obtained with a potential of the same form[1] as the electric charge distribution:

$$V = \frac{(V_0 + i\,W_0)}{(\exp[(r - R_v)/a] + 1} \tag{20.1}$$

where V_0 and W_0 denote the strengths of the refraction potential and absorption potential in the interior of the nucleus. The values of the size parameters which give the best fits to the data are:

$$R_v = 1.33\,A^{\frac{1}{3}}, \qquad a_v = 0.5 \tag{20.2}$$

in units of 10^{-13} cm. Thus the radius of the potential is seen to be considerably larger than the radius of the charge distribution (Sect. 4), though the surface thicknesses are about the same[2]. For a nucleus of medium weight, say $A = 125$, the difference between the radii is about 1.3×10^{-13} cm. This value is still subject to some uncertainty, however[3].

In the next three sections, we will analyze the difference between charge radius and potential radius and attempt to account for it, at least qualitatively, on the basis of some simple calculations. It will be seen that ordinary attractive interactions between nucleons would not give this difference in radii besides, of course, not giving saturation. Some of the very features which give saturation also lead to a potential extending beyond the density. In fact, crude estimates below give a radius difference not far from the empirical value.

21. Analysis with ordinary potential. *α) Calculation of the density from the potential.* For a study of the relation between nuclear density and potential let us first try to deduce the density distribution, given a reasonable potential. In this section we will not take into account the velocity dependence of the potential.

For qualitative estimates, it seems adequate to calculate the density using the Thomas-Fermi statistical method which has also proved very useful in atomic problems[4]. The density is assumed to be given by the same expression as in an

---

[1] Cf. Footnote 2, p. 431.

[2] Due to the mathematical complications involved, calculations involving a diffuse surface potential have been made only since 1954. [R. D. Woods and D. S. Saxon, Phys. Rev. **95**, 577 (1954) — Elastic Proton Scattering Cross-sections; J. M. C. Scott, Phil. Mag. **45**, 941 (1954) — Total Neutron Scattering Cross sections.] Previous calculations, in which the nuclear potential was assumed to have a sharp surface, were not, in general, able to give good fits to elastic scattering cross sections as function of angle [D. M. Chase and F. Rohrlich, Phys. Rev. **94**, 81 (1954)] though they could roughly account for the trends of total low energy neutron cross sections as function of energy and mass number [*19*].

[3] Elastic scattering cross sections can also be fitted quite well with different values of R_v, provided V_0 is rescaled so that $V_0 R^2$ is roughly constant. Melkanoff, Nodvik, Saxon and Woods: Phys. Rev. **106**, 793 (1957). It will require detailed analyses of reaction cross sections which depend essentially only on R_v to fix the radius more precisely. Still, the best fits to the data seem to be obtained with the values given above and we shall use them here.

[4] See the article by P. Gombás in Vol. XXXVI of this Encyclopedia.

infinite medium:

$$\varrho = \frac{2}{3\pi^2 \hbar^3} (2M T_F)^{\frac{3}{2}}. \tag{21.1}$$

However since the potential is a function of position, so is the maximum kinetic energy:

$$T_F(r) = E_F - V(r). \tag{21.2}$$

Thus the statistical density is given in terms of the potential at each point by[1]:

$$\varrho = \text{const} \times [E_F - V(r)]^{\frac{3}{2}}. \tag{21.3}$$

Let us take $V(r)$ to be of the form (20.1) except, of course, that the absorption potential W_0 vanishes for the bound states considered here. Using $E_F = -8$ Mev we find

$$\varrho = \text{const} \times \left[\frac{V_0}{\exp[(r - R_v)/a_v] + 1} - 8\right]^{\frac{3}{2}}. \tag{21.4}$$

We now calculate the "half-fall-off radius" R_ϱ of the matter distribution and compare it with the corresponding quantity R_v of the potential. Using Eq. (21.4) we have

$$\left[\frac{V_0}{\exp[(R_\varrho - R_v)/a_v] + 1} - 8\right]^{\frac{3}{2}} = \frac{1}{2} [V_0 - 8]^{\frac{3}{2}}. \tag{21.5}$$

Using the values $a_v = 0.5$ and $V_0 = 48$ Mev [the well depth for a nucleon on top of the Fermi sea, according to (9.22)], the difference in radii is given by:

$$R_v - R_\varrho = 0.40. \tag{21.6}$$

β) Distribution of neutrons and protons. Next we compare the "matter radius" R_ϱ with the "charge radius" which is assumed to coincide with the radius of the proton distribution and is denoted by R_p. The quantities R_ϱ and R_p are equal unless the protons and neutrons have different density distributions. A slight difference in the density distributions is, in fact, expected to occur because of the Coulomb interaction between protons. We will see that there are two separate (and opposite) effects.

First of all, the Coulomb forces tend to push protons away from the center of the nucleus toward the surface. If the nuclear potential (excluding the Coulomb interaction) were a box potential:

$$\left.\begin{aligned} V &= -V_0, \quad r < a, \\ V &= \infty, \quad\;\; r > a \end{aligned}\right\} \tag{21.7}$$

this would in fact be the only effect of the Coulomb forces. The density of protons (but not of neutrons) would then be larger near the surface than near the center, and thus the charge radius would exceed the matter radius.

There is, however, another effect, which results directly from the gradual fall-off of the potential at the surface and tends to cancel the above effect[2]. At the surface, all particles are subject to a force of magnitude dV/dr and directed inward. Now, in a neutron-rich nucleus the protons have smaller average kinetic

[1] Detailed calculations of the nuclear density, by solution of the Schrödinger equation and summation of the densities of individual particles, have been made [Ross, Mark and Lawson, Phys. Rev. **102**, 1613 (1956), A. E. S. Green, Phys. Rev. **99**, 1410 (1955); A. E. S. Green, K. Lee and R. J. Berkley, Phys. Rev. **104**, 1625 (1956)] and the results do not differ substantially from the results of the statistical method.

[2] M. H. Johnson and E. Teller: Phys. Rev. **93**, 357 (1954). — K. Wildermuth: Z. Naturforsch. **9**a, 1047 (1954). — P. Mittelstaedt: Z. Naturforsch. **10**a, 379 (1955).

energies than the neutrons. Thus a proton moving outward will, on the average, be more readily turned back by this force than a neutron and will not penetrate as far into the outer part of the nuclear surface. Thus we expect to find a layer of neutrons on the nuclear surface. Clearly this effect tends to make the matter radius larger than the charge radius.

In Fig. 5 we have sketched the calculated potential and density distribution of both neutrons and protons for a nucleus of $A=125$. Note that, as expected from the above considerations, the proton distribution has a slight hump in the inner part of the surface, and the neutron distribution extends considerably beyond the proton distribution. The half fall-off radii for neutrons and protons differ only slightly however[1]. Thus we have

$$R_n - R_\varrho \approx 0.2 \qquad (21.8)$$

and

$$R_\varrho - R_p \approx 0.1\,. \qquad (21.9)$$

Adding (21.7) and (21.9) we obtain,

$$R_v - R_p \approx 0.5\,, \qquad (21.10)$$

an appreciable difference in the right direction, but less than half of the empirical value.

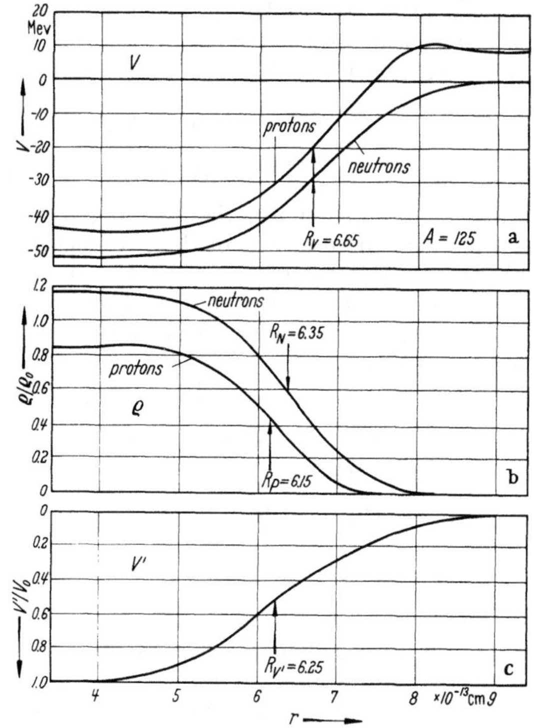

Fig. 5 a—c. Potentials and densities of neutrons and protons assuming ordinary two-body interactions. The assumed potentials for neutrons and protons are plotted in part (a). The neutron and proton densities calculated on basis of the statistical theory are shown in part (b). The potential calculated from the above densities assuming short range attractive forces is sketched in part (c). Also shown are the radii at which the various quantities are half as large as at the center of the nucleus.

γ) Calculation of the potential from the density. It is also of interest to start with an assumed density, e.g. the density obtained just previously, and calculate the resulting potential. If the assumed interactions are realistic, the calculations should be self-consistent, i.e. the potential should come out the same as the potential assumed originally for the calculation of the density. Let us take the interactions between nucleons as purely attractive and of finite range. Then it is evident that the potential will extend out somewhat further than the density. It may be readily shown, however, that the radius of the potential and the density are essentially the same. The potential simply has a larger surface thickness than the density[2,3]. This is illustrated qualitatively in Fig. 5c.

The above calculations, besides not giving a sufficiently large difference between R_v and R_p, are not self-consistent. The potential V' which results from the nuclear density distribution, has a smaller radius than the original potential V assumed for the calculation of this density[3]. This is evidently connected with the fact that ordinary short-range attractive interactions do not lead to saturation.

[1] The same conclusion has been reached by L. Wilets, Phys. Rev. **101**, 1805 (1956).
[2] S. D. Drell: Phys. Rev. **100**, 97 (1955).
[3] J. D. Talman: Phys. Rev. **102**, 455 (1956).

22. The velocity-dependent nuclear potential. In view of the apparent failure of ordinary potentials to account for the empirical difference between matter radius and potential radius let us introduce a velocity-dependence into the potential[1] [*23*]. We have seen in Sect. 9 that the existence of nuclear saturation does in fact require the potential to be velocity-dependent. To derive such a potential we will make the essentially same assumptions as in Sects. 7 to 9, except for one point: The wavefunctions of individual nucleons are no longer taken as plane waves. This makes it possible to study regions of variable density, i.e. the nuclear surface. To begin with, the two particle wavefunctions, analogous to those given in (7.12) and (7.13) are now of the form:

$$\psi_{ij}^{=} = 2^{-\frac{1}{2}} [\psi_i(\boldsymbol{r})\, \psi_j(\boldsymbol{r}')\uparrow\uparrow - \psi_j(\boldsymbol{r})\,\psi_i(\boldsymbol{r}')\uparrow\uparrow], \tag{22.1a}$$

$$\psi_{ij}^{\mp} = 2^{-\frac{1}{2}} [\psi_i(\boldsymbol{r})\, \psi_j(\boldsymbol{r}')\uparrow\downarrow - \psi_j(\boldsymbol{r})\,\psi_i(\boldsymbol{r}')\downarrow\uparrow] \tag{22.1b}$$

where ψ_i (or ψ_j) denotes the spatial part of the single particle wavefunction and $\boldsymbol{r}, \boldsymbol{r}'$ are the positions of the two nucleons. Again we assume a Serber type interaction (8.9). For a moment let us consider the earlier expression (8.10) for the interaction energy of a nucleon pair assuming the single particle wavefunctions are plane waves. We can write this energy as a sum of direct (D) and exchange (X) terms:

$$\Delta E_{ij} = \Delta E_{ij}^{(D)} + \Delta E_{ij}^{(X)}. \tag{22.2}$$

Eq. (22.2) holds also for the present case, at least in first order perturbation theory. Thus we find:

$$\Delta E_{ij}^{(D)} = \tfrac{3}{8} \iint \psi_i^*(\boldsymbol{r})\, \psi_j^*(\boldsymbol{r}')\, V_s(\boldsymbol{r}-\boldsymbol{r}')\, \psi_j(\boldsymbol{r}')\, \psi_i(\boldsymbol{r})\, d^3r'\, d^3r \tag{22.3a}$$

and

$$\Delta E_{ij}^{(X)} = \tfrac{3}{8} \iint \psi_i^*(\boldsymbol{r})\, \psi_j^*(\boldsymbol{r}')\, V_s(\boldsymbol{r}-\boldsymbol{r}')\, \psi_j(\boldsymbol{r})\, \psi_i(\boldsymbol{r}')\, d^3r'\, d^3r. \tag{22.3b}$$

The derivation of Eqs. (22.3) from the two-particle wavefunctions (22.1) is practically identical to the derivation of Eq. (8.10) from the plane wave two-particle functions (7.12), (7.13).

As before, the total interaction energy of a particle in state i, i.e. the "test particle", with all other particles is given by $\sum_j \Delta E_{ij}$. On the other hand this same interaction energy can also be written in the form $\int \varrho_i V_i\, d^3r$ where ϱ_i is the matter density of the test particle. Thus a reasonable definition of the potential is implicitly given by the equation:

$$\int \psi_i^*(\boldsymbol{r})\, V(\boldsymbol{r}, \nabla)\, \psi_i(\boldsymbol{r})\, d^3r = \sum_j \Delta E_{ij}. \tag{22.4}$$

As is indicated in Eq. (22.4) the potential involves not only coordinates but also gradients, i.e. it is velocity-dependent. Eq. (22.4) can be justified more rigorously. Given the form of the two-body interaction, we obtain (by a variational calculation) that set of single particle wavefunctions which minimizes the total nuclear energy, the latter calculated using a single Slater determinant of product type wavefunctions. It is found that these single particle wave-functions obey a Schrödinger equation in which the potential is given by the same expression as in (22.4). This potential is closely related to the Hartree-Fock potential[2,3] which, acting on a wavefunction, gives $\int V(\boldsymbol{r}, \boldsymbol{r}')\, \psi(\boldsymbol{r}')\, d^3r'$.

---

[1] K. A. Brueckner: Phys. Rev. **103**, 1121 (1956).

[2] F. Hund: Quantenmechanik der Atome, Vol. XXXVI of this Encyclopedia, pp. 107 to 108.

[3] For a discussion of the relation between these two potentials see P. A. M. Dirac, Proc. Cambridge Phil. Soc. **26**, 376 (1930).

From Eq. (22.3a) we can immediately write down the direct part of the potential (which does not involve gradients)

$$V^{(D)}(\boldsymbol{r}) = \tfrac{3}{8} \int \sum_j \psi_j^*(\boldsymbol{r}')\, V_s(\boldsymbol{r}-\boldsymbol{r}')\, \psi_j(\boldsymbol{r}')\, d^3r' . \tag{22.5}$$

Now, since the total matter density at any point is defined by

$$\varrho = \sum_j |\psi_j|^2 , \tag{22.6}$$

Eq. (22.6) becomes

$$V^D(\boldsymbol{r}) = \tfrac{3}{8} \int \varrho(\boldsymbol{r}')\, V_s(\boldsymbol{r}-\boldsymbol{r}')\, d^3r' . \tag{22.7}$$

We desire to express the potential directly in terms of the density and its derivatives at $\boldsymbol{r}$. This can be accomplished by making a Taylor series expansion:

$$\varrho(\boldsymbol{r}') = \varrho(\boldsymbol{r}) + \sum_\mu (\boldsymbol{r}_\mu - \boldsymbol{r}'_\mu) \frac{\partial \varrho}{\partial \boldsymbol{r}'_\mu}(\boldsymbol{r}) + \frac{1}{2} \sum_\mu \sum_\nu (\boldsymbol{r}_\mu - \boldsymbol{r}'_\mu)(\boldsymbol{r}_\nu - \boldsymbol{r}'_\nu) \frac{\partial^2 \varrho}{\partial \boldsymbol{r}'_\mu \partial \boldsymbol{r}'_\nu}(\boldsymbol{r}) . \tag{22.8}$$

Substitution of this expansion into Eq. (22.7) gives a series involving various moments of V_s. Since the two-body potential is spherically symmetric, all the odd moments, and many of the even moments, vanish. Thus for the second moments we have:

$$\int (\boldsymbol{r}'_\mu - \boldsymbol{r}_\mu)(\boldsymbol{r}'_\nu - \boldsymbol{r}_\nu)\, V_s(\boldsymbol{r}-\boldsymbol{r}')\, d^3r' = \tfrac{1}{3} \delta_{\mu\nu} \int V_s(\boldsymbol{r}-\boldsymbol{r}')(\boldsymbol{r}-\boldsymbol{r}')^2\, d^3r' . \tag{22.9}$$

As was done in Sect. 8, we do not consider moments higher than the second. The result is:

$$V^{(D)}(\boldsymbol{r}) = -\tfrac{1}{2} U_0 \varrho - U_2 \nabla^2 \varrho \tag{22.10}$$

where U_0 and U_2 are defined by (8.15) and both ϱ and $\nabla^2\varrho$ are evaluated at $\boldsymbol{r}$. Of the two terms on the right hand side of (22.10) the first term is essentially a volume interaction, while the second term represents the lowest order effect of the finite range of nuclear forces.

The expression for the exchange potential $V^{(X)}$ is not as immediately evident from Eq. (22.3b) as was the case for the direct potential. However, we can bring this equation into the required form (22.4) by making Taylor series expansions of both $\psi_j^*(\boldsymbol{r}')$ and $\psi_i(\boldsymbol{r}')$ about their values at $\boldsymbol{r}$. Again we obtain a series involving moments of V_s. Retaining the same moments as in the derivation of $V^{(D)}$, we find:

$$\left.\begin{aligned} \int \psi_i^*(\boldsymbol{r})\, V^{(x)}(\boldsymbol{r}, \nabla)\, \psi_i(\boldsymbol{r})\, d^3r = & -\tfrac{1}{2} U_0 \int \psi_i^* \sum_j (\psi_j^* \psi_j)\, \psi_i\, d^3r - \\ & - U_2 \Big\{ \int \psi_i^* \sum_j [(\nabla^2 \psi_j^*)\, \psi_j]\, \psi_i\, d^3r + \\ & + 2 \int \psi_i^* \sum_j [(\nabla \psi_j^*)\, \psi_j]\, \nabla \psi_i\, d^3r + \\ & + \int \psi_i^* \sum_j (\psi_j^* \psi_j)\, \nabla^2 \psi_i\, d^3r \Big\} \end{aligned}\right\} \tag{22.11}$$

where all quantities are evaluated at $\boldsymbol{r}$. Let us now suppose that all the ψ_j are real. Then we have

$$\sum (\nabla \psi_j^*)\, \psi_i = \tfrac{1}{2} \nabla \varrho \tag{22.12}$$

and

$$\sum_i (\nabla^2 \psi_j^*)\, \psi_j = \tfrac{1}{2} \nabla^2 \varrho - \tau \tag{22.13}$$

where τ is defined as:

$$\tau = \sum_j |\nabla \psi_j|^2. \tag{22.14}$$

From (22.11) we can write down an expression for the exchange part of the potential. With the aid of Eqs. (22.6), (22.12) to (22.14) we find:

$$V^{(X)}(\boldsymbol{r}, \nabla) = -\tfrac{1}{2} U_0 \varrho - U_2 [\tfrac{1}{2} \nabla^2 \varrho - \tau + \nabla\varrho \cdot \nabla + \varrho \nabla^2]. \tag{22.15}$$

This velocity dependent potential may be expressed in slightly different form to exhibit its hermiticity:

$$V^{(X)}(\boldsymbol{r}, \nabla) = -\tfrac{1}{2} U_0 \varrho - U_2 [\tfrac{1}{2}(\varrho \nabla^2 + \text{h.c.}) - \tau] \tag{22.16}$$

where h.c. is the hermitean conjugate of the operator ∇^2. Addition of (22.10) and (22.16) yields the total potential:

$$V(\boldsymbol{r}, \nabla) = -U_0 \varrho - U_2 [\tfrac{1}{2}(\varrho \nabla^2 + \text{h.c.}) - \tau + \nabla^2 \varrho]. \tag{22.17}$$

Of the terms on the right hand side, the first represents the volume interaction, the next two are velocity dependent terms signifying a dependence of V of the kinetic energy of the "test particle". This is followed by a term representing the dependence of V on the average kinetic energy of the nucleons in the nucleus, and finally we have a term which depends specifically on the spatial variation of the nuclear density.

23. Analysis with velocity-dependent potential. We begin the analysis by defining the density ratio:

$$y = \frac{\varrho(r)}{\varrho_0} \tag{23.1}$$

where ϱ_0 is the matter density well in the interior of the nucleus. Thus we find, using (22.14):

$$\tau = \varrho k_{\text{Av}}^2 \approx \tfrac{3}{5} \varrho_0 k_F^2 y^{\frac{5}{3}} \tag{23.2}$$

where k_F is understood to be the Fermi momentum for normal density. Eq. (22.17) takes the form:

$$V(\boldsymbol{r}, \nabla) = -U_0 \varrho_0 y - U_2 \varrho_0 [\tfrac{1}{2}(y \nabla^2 + \text{h.c.}) - \tfrac{3}{5} k_F^2 y^{\frac{5}{3}} + \nabla^2 y]. \tag{23.3}$$

The values of U_0 and U_2 are given by (9.20) and (9.21) so that[1]

$$V(\boldsymbol{r}, \nabla) = -112 y - 40 k_F^{-2} [\tfrac{1}{2}(y \nabla^2 + \text{h.c.})] + 24 y^{\frac{5}{3}} - 40 k_F^{-2} \nabla^2 y \quad \text{in Mev}. \tag{23.4}$$

We may define the strength of the potential at each point by:

$$V(\boldsymbol{r}) = \frac{\psi^*(\boldsymbol{r})\, V(\boldsymbol{r}, \nabla)\, \psi(\boldsymbol{r})}{\psi^*(\boldsymbol{r})\, \psi(\boldsymbol{r})} \tag{23.5}$$

and similarly

$$T(\boldsymbol{r}) = \frac{\psi^*(\boldsymbol{r})\, T(\nabla)\, \psi(\boldsymbol{r})}{\psi^*(\boldsymbol{r})\, \psi(\boldsymbol{r})} = -\frac{40}{k_F^2} \frac{\psi^*(\boldsymbol{r})\, \nabla^2 \psi(\boldsymbol{r})}{\psi^*(\boldsymbol{r})\, \psi(\boldsymbol{r})} \quad \text{in Mev} \tag{23.6}$$

for the kinetic energy at each point. The last three equations may be combined to give:

$$V(\boldsymbol{r}) = -112 y + T(\boldsymbol{r})\, y + 24 y^{\frac{5}{3}} - 40 k_F^{-2} \nabla^2 y \quad \text{in Mev}. \tag{23.7}$$

[1] The velocity-dependent potential used by ROSS, LAWSON and MARK, Phys. Rev. **104**, 401 (1956) for the calculation of single particle energy levels, is of the same form as (23.4) except for the absence of the last two terms.

Now consider the potential seen by a nucleon of given energy near the nuclear surface. Since the wavefunction of this nucleon satisfies the Schrödinger equation:

$$[T(\nabla) + V(\mathbf{r}, \nabla)]\, \psi(\mathbf{r}) = E\, \psi(\mathbf{r}) \tag{23.8}$$

we see from (23.5) and (23.6) that

$$T(r) + V(r) = E. \tag{23.9}$$

Eliminating $T(r)$ between (23.7) and (23.9) we obtain an expression for the potential in terms of the density alone:

$$V(r) = \frac{-112y + Ey + 24y^{\frac{5}{3}} - 40\, k_F^{-2}\, \nabla^2 y}{1+y} \quad \text{in Mev} \tag{23.10}$$

which reduces to the value (9.22) well in the interior of the nucleus ($y = 1$).

To exhibit the important features of the nuclear potential, Fig. 6 shows plots of $y = \varrho/\varrho_0$ and V/V_0 calculated for the case $E = 0$, as function of distance from the center of the nucleus. In this calculation, the curvature of the nuclear surface has been neglected, i.e., we have replaced ∇^2 by $d^2\varrho/dr^2$. It is seen that the potential extends almost 10^{-13} cm beyond the density, essentially the amount required by the empirical evidence. The strength of the potential in the interior, where the density has the normal value, is calculated to be 44 Mev. As we move outward through the nuclear surface, the potential decreases, but slower than the density. Thus, about 10^{-13} cm beyond R_ϱ the density has fallen to 12% of ϱ_0, but the potential is still about 40% as strong as at the center. Very far out, the potential is indeed proportional to the density, but with a ratio V/ϱ more than four times as large as normal density. The above result, that the potential radius significantly exceeds the matter radius, is due to two related effects[1].

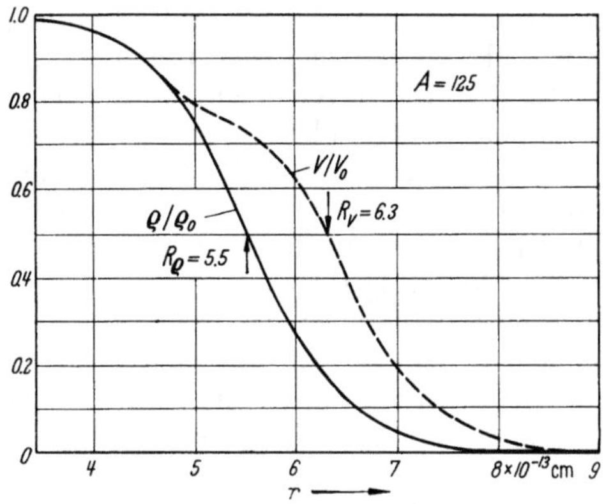

Fig. 6. Calculated density and potential distribution near the nuclear surface, assuming saturating two-body interactions

First, even if we disregard any spatial variation of the density, the relation between potential and density is non linear[2]—the potential falls to zero much slower than the density. This non-linearity is, in fact, required by the saturation phenomena. The potential does not increase indefinitely with increasing density.

The variation of density near the nuclear surface, in particular the term proportional to ∇^2, also contributes to the difference between matter radius and potential radius. Qualitatively, we may say that the relation between V and ϱ is everywhere the same as in an infinite medium, provided by ϱ we now mean not the local density, but the density averaged over a small region ($\approx 10^{-13}$ cm) around the point in question. This is also true for ordinary interactions as may be seen from (22.10). The main effect of this "finite range effect" is to further increase the potential in the outer part of the nuclear matter distribution since the "averaged density" is here about twice as large at each point as the local density.

[1] Cf. Footnote 1, p. 456.
[2] R. A. Berg and L. Wilets: Phys. Rev. **101**, 201 (1956).

On the other hand, in the inner part of the nuclear surface region, the finite range effect has the opposite sign and in fact it just about cancels the non-linearity effect. Consequently ϱ is nearly proportional to V in this region.

Finally, since the potential discussed in this section leads to the empirical saturation properties, it is also expected to be "self-consistent[1]". The total density of particles calculated using this potential should be essentially the same as the originally assumed density and this appears to be the case [*23*].

D. The alpha particle model.

24. Introduction. The phenomenon of alpha-decay has been known since the early days of nuclear physics. The fact that no other heavy particles are emitted in radioactive decay, indicates that alpha particles have special stability. Indeed, it was thought at one time that perhaps α-particles exist as stable subunits of nuclei, but the development of the neutron-proton model of nuclei made this view point untenable.

The fact that α-particles have exceptional stability is easily explained. The two protons and two neutrons which make up an α-particle can go into the same spatial state with the four different spin-charge combinations $p\uparrow p\downarrow n\uparrow n\downarrow$, without violating the exclusion principle. The wavefunction of the nucleons will thus overlap completely, giving rise to a large binding energy (the empirical value is 29 Mev). An especially important property of the α-particle is its stability with respect to any form of excitation. Thus it requires about 20 Mev to remove a nucleon from an α-particle (as compared to about 8 Mev for most nuclei). Also, no excited states of the α-particle have been found within 28 Mev of the ground state[2]. Altogether, it seems justified to treat the free α-particle as a structureless unit at least for some applications and provided the energies involved are less than 20 Mev.

A special stability exists also for heavier even-even nuclei with $N=Z$, the so-called "α-nuclei". This feature is connected with the stability of the α-particle. In the ground state of an "α-nucleus", each nucleon has three partners in the same spatial state with which it can interact especially strongly, i.e., there is maximum spatial symmetry.

The basic assumption of the α-particle model is that α-particles can be regarded as stable subunits of nuclei. At the very least this requires that there is a pronounced clustering of nucleons into such subunits, i.e. that there is a large probability of finding a group of nucleons close together and well separated from other such groups. This kind of correlation is, in fact, expected by virtue of the interaction between nucleons and results of recent shell model calculations for α-nuclei give strong support for this view[3]. In its strict form, however, the α-particle model requires in addition that the nucleons are localized in individual α-particles. In other words, exchanges of nucleons between different alpha particles should occur comparatively infrequently, i.e. the frequency of nucleon exchange should be small compared to the frequency of vibrations[4].

---

[1] A different kind of self-consistency calculation has been made by M. ROTENBERG, Phys. Rev. **100**, 439 (1955), who used two body interactions which give saturation but only by virtue of containing a large fraction of exchange interactions (Sect. 8). It was found that such interactions, which are also in conflict with empirical information regarding the two-nucleon problem (cf. Sect. 14), do not yield the required radius-potential difference. See also Footnote 2, p. 455.

[2] R. M. EISBERG: Phys. Rev. **102**, 1104 (1956).

[3] J. K. PERRING and T. H. R. SKYRME: Proc. Phys. Soc. Lond. **69**, 600 (1956).

[4] J. A. WHEELER: Phys. Rev. **52**, 1083 (1937).

Crude estimates[1] indicate that the vibrational and exchange frequencies are comparable, but arguments[2] can be given that the α-particle model should still be valid. Also, in spite of the questionable foundation of the α-particle model, properties of several states in "α-nuclei" can be understood as resulting from vibration and rotation of an α-particle structure (cf. Sect. 25). If the structure is nearly rigid, i.e. if the distance between α-particles is nearly fixed, then the motion approximately separates into vibrations and much slower rotations. The frequencies of these motions are small compared to the frequency of the nucleonic motion within an α-particle.

The basic interactions between α-particles are believed to be similar to the van der Waals forces in molecules—except for the Coulomb interactions at large distances. Thus at medium distances we have an attractive α-α-interaction because of the attractions between the individual nucleons. Possibly this interaction is a resonance effect, i.e. it arises from exchange of nucleons between the α-particles, in the same way that chemical bonds arise from the exchange of electrons between atoms. At short distances the interaction turns repulsive. Part of this repulsion evidently is due to the Coulomb interaction but perhaps most of it is due to the exclusion principle. Nucleons of the same charge and spin (on the different α-particles) cannot be in at the same position, and effectively repel one another even without any explicit interactions between them. At large distances, the nuclear interaction vanishes and only the repulsive Coulomb interaction remains. We see then that there is an equilibrium distance between α-particles and presumably the zero-point amplitude of vibration is small compared to this distance.

The easily excited degrees of freedom in "α-nuclei" probably involve collective excitations of α-particle-like clusters, rather than excitations of individual nucleons, hence these collective motions are most easily described in terms of the α-particle model. However, it seems likely that the shell model (with configuration mixing) can also account for these same features and even more accurately though it does not provide as much insight into the dynamics involved, and is more complicated to apply.

Finally, it is known that the properties of odd-A nuclei cannot be well accounted for with the α-particle model[3]. Excitations of individual nucleons, which are outside the framework of the α-particle model, are of great importance for these nuclei.

25. Successes of the alpha particle model. *α) Binding energies.* An argument which has sometimes been advanced in support of the α-particle model is that about 90% of the binding energy in light nuclei results from the internal binding within the constituent α-particles. The interactions between the α-particles constitute only 10% of the binding energies. Empirically in α-nuclei, this residual binding energy is closely proportional to the number of classical "bonds" between α-particles, with about 2.4 Mev per bond[4]. (Be^8, being barely unstable for decay into two α-particles, is an exception to this rule.) These features are consistent with an α-particle model. There are good reasons to believe that the nuclear interactions between α-particles are additive. Furthermore both Coulomb energy

---

[1] B. GRÖNBLOM and R. MARSHAK: Phys. Rev. **55**, 229 (1939). — H. MARGENAU: Phys. Rev. **59**, 37 (1941).

[2] A. HERZENBERG: Nuovo Cim. **1**, 906, 1008 (1955). — Nucl. Phys. **3**, 1 (1957).

[3] L. R. HAFSTAD and E. TELLER: Phys. Rev. **54**, 681 (1938). — C. KITTEL: Phys. Rev. **52**, 109 (1942).

[4] W. WEFELMEIER: Naturwiss. **25**, 525 (1937). — Z. Physik **107**, 332 (1937).

and the zero point vibrational energy, which must be added to the nuclear interaction energy, are also proportional to the number of bonds[1]. Thus, one would indeed expect that the residual binding energy should exhibit the observed proportionately. The interpretation of binding energies in terms of an α-particle model, is however, not required. A more plausible explanation is, in fact, that the average binding energy of nucleons happens to be only slightly smaller in the α-particle than in heavier nuclei.

According to a strict α-particle model, any extra nucleons which do not form α-particles should give a rather small binding energy (as for H^2 and He^3). Thus the binding energy as functions of atomic number should increase slowly from $A=4n$ to $4n+3$ and then very strikingly as a new α-particle is completed at $A=4n+4$. A trend of this kind, is indeed observed between He^4 and Be^8. However, for heavier nuclei, the binding energies increase more nearly in a uniform manner, although nuclei with $A=4n$ still have special stability.

β) Level schemes. We consider here the nuclei Be^8, C^{12}, O^{16}, the only nuclei for which the α-particle model has had a fair amount of success[2]. In calculating theoretical levels for comparison with the empirical data, we assume that the α-particles have a nearly fixed distance from each other as if they were connected by springs. The potential is assumed to vary quadratically with deviations of distance from equilibrium. (This is, of course, valid only if the deviations from equilibrium are small. For large separation, the potential vanishes except for the Coulomb interaction.) The system will then have a number of normal modes of frequency ω_i, and the vibrational energy corresponding to each mode will be an integral multiple of $\hbar\omega_i$. In addition the system may undergo rotation, essentially as a solid body. The total energy is the sum of vibrational and rotational energies, provided we can neglect the coupling between these motions. Strictly speaking, this assumption is valid only if the vibrational frequencies are much larger than the rotational frequencies. The energies of this system may be found essentially classically. However, the number of allowed states is greatly reduced by the requirement that the wavefunction be symmetric with respect to interchange of any two α-particles. In so far as the underlying assumptions are valid, this problem is very similar to the study of rotation and vibration of molecules.

Fig. 7 shows the empirical level schemes[3] for Be^8, C^{12}, O^{16} and the best fits obtained with the α-particle model. Most of the observed levels can be fitted quite well, escpecially in O^{16}. According to the shell model, this nucleus consists of filled s and p shells so that it forms a spherically symmetric distribution, which is not very different from the tetrahedral symmetry predicted by the α-particle model, especially when correlations between nucleons are taken into account. There are, of course, several arbitrary parameters in the theory, which makes the fits less significant. The α-particle model is seen to acount for all levels shown, and to predict a few additional levels whose existence is doubtful, though not definitely ruled out.

The prediction of 0^+ vibrational levels is a definite success of the α-particle model. These states do not occur in the simple version of the shell model, in which it is assumed that all states belong to the same configuration $s^4 p^{A-4}$.

[1] L. HAFSTAD and E. TELLER: Phys. Rev. **54**, 681 (1938).

[2] Some interesting regularities have also been noted in the level spectra of heavier alpha-nuclei (Ne^{20}, Mg^{24}, Si^{28}) by H. MORINAGA, Phys. Rev. **101**, 254 (1956).

[3] For Be^8, see the following discussion. — For C^{12}, see A. E. GLASSGOLD and A. GALONSKY, Phys. Rev. **103**, 701 (1956). — For O^{16}, see D. M. DENNISON, Phys. Rev. **96**, 378 (1954); S. L. KAMENY, Phys. Rev. **103**, 358 (1956).

The electric monopole matrix elements[1] between these 0^+ states and the ground state (in O^{16} and C^{12}) are known to be an order of magnitude larger than the values expected on the basis of a single particle model. This can be most simply understood as resulting from simple dilational collective motion involving all the nucleons[2]. The α-particle model is a simple model which generates such motion. No doubt more complicated collective models can probably accomplish the same thing and besides fit empirical lifetimes better than the α-particle model does.

γ) Be^8 and α-α-scattering. The Be^8 level scheme is of special interest since it yields information on the effective α-α interaction. Consider, first the energy

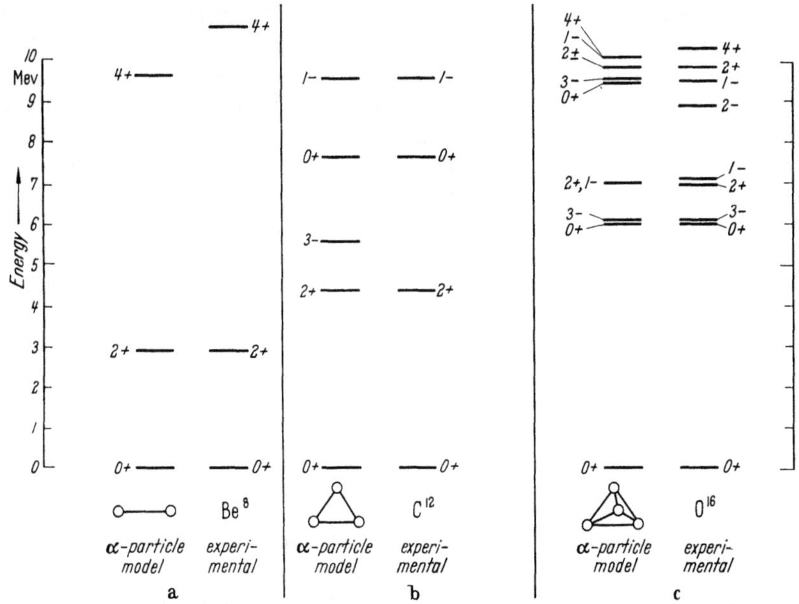

Fig. 7 a—c. Level schemes of Be^8, C^{12}, and O^{16}, and comparison with alpha particle model.

ration E_{4+}/E_{2+} whose empirical value is 3.7. According to a rigid rotator model this ratio should be just $\frac{10}{3}$[3]. However, the centrifugal forces are expected to increase the moment of inertia for states with large angular momentum which results in a lowering of the theoretical value of the energy ratio. Thus, the empirical energy ratio is not satisfactorily explained on the basis of the α-particle model.

Another feature of interest is the absence of any excited 0^+ state below 10 Mev[4]. This, together with the known energies of the 2^+ and 4^+ states strongly implies

[1] The magnitudes of the electric monopole matrix elements $\int \psi_b^* \sum_i (e_i r_i^2) \psi_a d^3r$ may be deduced from cross sections for inelastic scattering of high energy electrons [R. A. FERRELL and V. M. VISSCHER, Phys. Rev. **104**, 475 (1956)], or from life times for pair emission (no direct gamma transitions between two states of angular momentum zero can occur). See E. L. CHURCH and J. WENESER: Phys. Rev. **103**, 1035 (1956).

[2] This problem has also been investigated from the shell model point of view with considerable success. L. I. SCHIFF: Phys. Rev. **98**, 1281 (1955). — P. J. REDMOND: Phys. Rev. **101**, 751 (1956). — J. P. ELLIOTT: Phys. Rev. **101**, 1212 (1956). — B. J. SHERMAN and D. G. RAVENHALL: Phys. Rev. **103**, 949 (1956).

[3] The shell model with the additional assumption of a pure s^4p^4 configuration (i.e. no configuration interaction, cf. Sect. 30) also predicts an energy ratio of $\frac{10}{3}$ if spin orbit coupling is neglected and a smaller value if it is taken into account.

[4] A number of other low lying states have been reported in the literature. Cf. F. AJZENBERG and T. LAURITSEN: Rev. Mod. Phys. **27**, 77 (1955). However, the results of recent experimental investigations [e.g. HOLLAND, INGLIS, MALM and MOORING, Phys. Rev. **99**, 92 (1955)] strongly indicate that there are no levels in Be^8 between 3 and 10 Mev.

that the interaction between 2 α-particles becomes repulsive at small distances. If the interaction could instead be represented by a square well, we would expect the 0^+ state slightly above the known 2^+ state. The presence of a short range repulsion raises the excited 0^+ states more than the lowest 2^+ and 4^+ states, since in the latter the particles are kept apart anyway by their centrifugal interaction and so do not come close enough to feel the repulsion. Of course, the above argument must be refined to include the fact that all the states of Be^8 are virtual, i.e. unstable with respect to decay into two alphas; however, this should not change the essential conclusions.

Important information on the effective potential between α-particles is provided by the interpretation of α-α scattering data[1]. In recent years much detailed and extensive information on α-α scattering cross-sections has become available[2]. An analysis of these data suggests strongly that the interaction actually must turn repulsive at short distances. Furthermore, the scattering data seem to require an equilibrium distance of about 3.5×10^{-13} cm [3]. This is considerably smaller than the value 4.5×10^{-13} cm deduced from the position of the 2^+ state on the assumption that it represents a simple rotational excitation. One way to remove this discrepancy is to have the well depth or range of the potential increase with increasing angular momentum; it this becomes possible to approximately fit both the scattering cross-sections and the excitation energies of 2^+ and 4^+ states with the smaller of the two above equilibrium distances. The scattering data available at present do not, however, fix the width or depth of the attractive well, except within wide limits.

E. The shell model[4].

26. Introduction. The possibility that atomic nuclei might exhibit shell structure was already considered in the early 1930's [*1*]. Underlying this model is the fundamental assumption that the nucleons move approximately independently within the nucleus, in spite of the strong interactions known to exist between free nucleons[5]. In the shell model the nuclear potential is taken as static and spherically symmetric and it generates a complete orthonormal set of single particle functions $\psi^{\nu_i}(r_i)$ where the r_i indicate position, spin and charge variables of nucleons and ν_i denotes the quantum numbers of the state. Any function obeying the same boundary conditions as these eigenfunctions can be uniquely expressed as a linear combination of them. Although in principle almost any potential may be used to generate the basic single particle functions, it is, of course, desirable to have this generating potential resemble as closely as possible the average potential seen by each nucleon. Now suppose certain prescribed single particle states of a nucleus are occupied by nucleons. Then the nuclear

[1] H. MARGENAU: Phys. Rev. **59**, 37 (1941). — R. R. HAEFNER: Rev. Mod. Phys. **23**, 228 (1951).

[2] N. P. HEYDENBURG and G. M. TEMMER: Phys. Rev. **104**, 123 (1956). — RUSSELL, PHILLIPS and REICH: Phys. Rev. **104**, 135 (1956). — NILSON, KERMAN, BRIGGS and JENTSCHKE Phys. Rev. **104**, 1673 (1956). — The empirical level assignment shown in Fig. 7 is, in fact, based on the analysis of α-α scattering.

[3] C. H. HUMPHREY: Ph. D. Dissertation, University of California, Los Angeles 1956.

[4] For a detailed account of the shell model, see the preceding article in this volume.

[5] Thus each of the nucleons moves in the potential due to the remaining nucleons. However, the nucleons may instead be treated as moving in the *same* potential. (This procedure formally introduces "self-interaction" terms; these, however, vanish if the antisymmetry of the complete many-particle wavefunction is properly taken in to account.)

wavefunction is given by the following Slater determinant:

$$\Psi^{\nu}(r) = \begin{vmatrix} \psi^{\nu_1}(\boldsymbol{r}_1) & \dots & \psi^{\nu_1}(\boldsymbol{r}_A) \\ \vdots & & \vdots \\ \psi^{\nu_A}(\boldsymbol{r}_1) & \dots & \psi^{\nu_A}(\boldsymbol{r}_A) \end{vmatrix}. \tag{26.1}$$

The validity of this wavefunction implies that the nucleons move essentially independently, the only correlations between nucleons being due to the antisymmetry requirement.

α) Motion of nucleons in the nuclear potential. The fact that the shell-model potential is a function of position represents an improvement over the Fermi gas model. The potential roughly follows the nuclear density distribution, being constant inside the nucleus, and vanishing outside. The single particle eigenfunctions are now no longer plane waves but more complicated functions, which must, in general, be determined by numerical integration. In any case, however, the wavefunctions are automatically small outside the nuclear surface. Thus we reproduce at least some of the correlations due to the interactions in the sense that it is improbable for two nucleons to be more than one nuclear diameter apart.

The spherical symmetry of the shell model potential also has important consequences: Each nucleon has a definite orbital angular momentum l and the single particle wave function factors into radial and angular parts.

Thus we have:

$$\psi_{nlm}^{(\text{space})}(\boldsymbol{r}) = R_{nl}(r)\, Y_{lm}(\vartheta, \varphi). \tag{26.2}$$

Here m is the component of orbital angular momentum along the z-axis. Only the radial functions depend on the detailed shape of the potential; n is the number of radial nodes plus 1. Both the radial functions and the single particle energy levels are independent of m. Thus, since m can have any integral value between $-l$ and l, we have a $2l+1$ fold degeneracy characteristic of the spherical symmetry. Of course, each state also has a two fold spin-degeneracy and, if the Coulomb field is neglected, a twofold charge degeneracy in addition. An important consequence of these degeneracies is that the average spacing between distinct particle levels (denoted here as orbitals) is considerably larger than the statistical spacing.

In addition to the degeneracies discussed above, there may be "accidental" degeneracies between different levels for special forms of the potential. A well-known example of such a potential is the isotropic harmonic oscillator, which is, in fact, a useful first approximation to the nuclear potential. For different potential shapes, one may still obtain "near-degeneracies" with certain groups of orbitals closely spaced and large gaps between different shells, i.e. groupings of close-lying levels. Since the Pauli principle only permits one nucleon to go into any given state, the energetically lowest configuration of A particles will have the lowest A states occupied. Closed shell configurations which possess special stability occur when there is a large gap between the highest occupied and lowest unoccupied states.

The density distribution of any single particle state with $l \neq 0$ state is not isotropic. However, a filled orbital always has a spherically symmetric density distribution. Since the potential is determined by the particles themselves, it is thus spherically symmetric only for filled-shell or filled orbital configurations[1]

[1] Because of the relative ease with which these nucleons may be excited to different orbitals (in the same shell) this spherical symmetry of a filled orbital is sometimes not very stable. Thus if there are many nucleons outside filled shells, the nucleus may acquire a non-spherical shape (see Chap. F).

Still, if the nucleus contains only a few nucleons outside of, or missing from, a closed shell configuration, the nuclear potential will still be approximately central and the basic shell-model assumption remains well satisfied. This difficulty of exciting the closed shell configurations tends to stabilize the spherical shape of the potential.

Early attempts to interpret properties of nuclear states in terms of a shell model met with only limited success. True, reasonable nuclear potentials gave closed shells at N or $Z=2$, 8 and 20 and the observed special stability of nuclei with these nucleon numbers was considered evidence for shell structure. However, detailed properties of nuclear states were not well accounted for. At the time it was widely thought that perhaps the basic assumption, of nucleons moving approximately independently in the nucleus, was not justified. These doubts were strengthened after 1935 by the discovery of the nuclear resonances and later by the success of the strong interaction description [*11*], [*12*] in accounting for qualitative features of nuclear reactions. Finally, the phenomenon of nuclear fission, so easily understood on the basis of the liquid drop model, seemed to be quite outside the scope of the shell model. It is thus not surprising that not much work on the nuclear shell model was done between 1939 and 1948.

β) The Mayer-Jensen shell model. The present form of the nuclear shell model came into being in 1948. By that time it had become quite clear that nuclei with N or $Z=50$, 82, 126, have a special stability[1]. During the following year it was shown by MAYER [*28*] and independently by HAXEL, JENSEN and SUESS [*26*] that the existence of closed shells at these "magic numbers" could be accounted for by a nuclear potential of reasonable shape, provided, however, that in addition there is a strong spin-orbit interaction of opposite sign to the corresponding electronic interaction.

With this one additional ad-hoc assumption it became at once possible to account also for many other features which had not been explained by the earlier shell model. The tremendous success which the new shell model achieved in correlating and predicting a large number of nuclear data is surely one of the outstanding developments in nuclear physics[2]. Thus it became possible to understand, at least qualitatively, not only the special stability of the nuclei possessing "magic" numbers of neutrons or protons, but also ground state spins and parities of most odd-A nuclei, and many features of nuclear decay schemes. In almost all cases, the empirical ground state spins and parities turn out to agree with the shell model assignments for the last unpaired nucleon. This implies the validity of a single particle shell model, an especially simple form of the shell model, in which the paired nucleons couple to form an inert core of spin 0. A coupling of this kind is indeed expected in view of the short range attractive nature of the nuclear interactions[3].

It may be of interest to point out the similarities and differences between the nuclear shell model and the electron shell model of the atom. Table 2 gives such a comparison of the essential features of these two models.

In its simplest "single particle" form, the shell model accounts only qualitatively for properties of the nuclear wavefunction. A more detailed picture is required for the quantitative interpretation of level schemes, moments and transition rates.

---

[1] M. G. MAYER: Phys. Rev. **74**, 235 (1948). — Some evidence along these lines had been previously noted by W. ELSASSER, J. Phys. Radium **5**, 625 (1934) and K. GUGGENHEIMER, J. Phys. Radium **5**, 253 (1934).

[2] It is, however, not yet quite clear *why* this model works as well as it does.

[3] M. G. MAYER: Phys. Rev. **78**, 22 (1950).

Table 2. *Comparison of atomic and nuclear shell models.*

Atoms	Nuclei
1. Electrons move nearly independently in a common potential due to the nucleus and the other electrons—a screened Coulomb potential.	1. Nucleons move approximately independently in a common potential due to the other nucleons—a well with a flat bottom and sloping sides.
2. The nucleus provides a common center of force which helps to stabilize the spherical symmetry of the system. Thus the potential is essentially isotropic and due to the density distribution averaged over all possible orientations.	2. There is no common center of force. The potential has spherical symmetry only by virtue of the spherically symmetric density of the filled shells. For nuclei with many nucleons in unfilled shells, the potential is, in general, not spherically symmetric (unified model).
3. There are large variations in electron binding energies.	3. There are only small variations in nucleon binding energies, except for the lightest nuclei.
4. The electronic spin-orbit coupling is weak and is dominated by the interactions between electrons. (L-S coupling) except for some heavy elements.	4. The nuclear spin-orbit coupling is appreciable and it frequently dominates over the interactions between nucleons (j-j coupling).
5. Closed shell configurations which have exceptional stability occur for $Z=2$, 10, 18, 36, 54, 86.	5. Closed shell configurations of slightly greater than average stability occur for N or $Z=2$, 8, 20, 50, 82, 126 (the magic numbers).
6. Interactions (long ranged, repulsive)—favor states of minimum spatial symmetry.	6. Interactions (short ranged, attractive)—favor states of maximum spatial symmetry
7. Ground states of atoms tend to have maximum possible spin: $S=$ half the number of particles outside of (or missing from) closed shells (HUND's rule).	7. Ground states of even-even nuclei have angular momentum $I=0$. Each pair of identical nucleons coupling to $I=0$ has special stability (the pairing effect). Ground states of most odd-A nuclei have $I=j$ of the odd unpaired nucleon (the single particle model).

In actual nuclei, the internucleon interactions always give rise to additional correlations. There are basically two methods, the many-particle shell model[1], and the unified model (Chap. F), for generalizing the determinental form of the wavefunction so as to include such correlations.

In the many-particle shell model we represent the actual wavefunction as a linear combination of different Slater determinant eigenfunctions of the same Hamiltonian:

$$\Psi(\boldsymbol{r}) = \sum c_\nu \Psi^\nu(\boldsymbol{r}). \tag{26.3}$$

The different determinants form a complete orthonormal set of functions and in principle it is possible to represent any many particle wave function as an expansion of the form (26.3) (provided only the actual wavefunction is subject to the same boundary conditions as is each of the basic functions).

Now for many nuclear properties, e.g. level spectra, nuclear moments, it is the *low-lying* excitations, i.e. admixtures of low-lying excited states to Ψ, which are of primary importance. On the other hand, some other properties, such as the probability of finding a nucleon with very large momentum, are much more sensitive to high energy excitations, which in turn depend especially on the short range correlations due to the interactions.

[1] Another frequently used name for this model e.g. in [*24*], is "Individual Particle Model".

For nuclei near closed shells, there are relatively few possible low-energy excitations, i.e. excitations of the nucleons within the last unfilled shell, thus Ψ contains relatively few determinants and the shell model is comparatively easy to apply. Much work on the many-particle shell model has been recently done [*24*]. In this way, it has been possible to account for properties of many low-lying levels in nuclei with only a few nucleons outside of (or missing from) closed shells. On the other hand, far away from closed shells, there are effectively more loosely bound nucleons and consequently more low-energy modes of excitation. The shell model approach in which all these excitations are treated explicitly, is still valid in principle; however the calculations become too complicated since the wave-function is now a superposition of a large number of Slater determinants.

Since the shell model is treated in great detail in the preceding article, we will limit ourselves to a discussion of a few points which illustrate the relation of this model to the other nuclear models.

Some aspects of the single particle shell model dealing with the level ordering of individual particle states are mentioned in Sect. 27/28. Some features of the many-particle shell model are briefly treated in the last two sections of this chapter.

I. The single particle shell model.

In Sect. 27 we consider the isotropic harmonic oscillator potential. This potential is useful for studying some aspects of nuclear shell structure. The level ordering of single particle states in actual nuclei, using a realistic nuclear potential including a spin-orbit coupling term, is considered in Sect. 28.

27. The isotropic harmonic oscillator potential. The nuclear potential is fairly well approximated by a three dimensional isotropic harmonic oscillator potential, except outside the nucleus. Thus the bound state single particle wavefunctions generated by an oscillator potential do not differ significantly from the corresponding functions calculated with a more realistic potential.

On the other hand, the Schrödinger equation for the harmonic oscillator potential is solved very easily. Thus simple expressions for wave-functions and energies are readily obtained analytically without the imposition of any special boundary conditions at the nuclear surface (as is necessary for example, with a square well potential).

The harmonic oscillator potential may be written as

$$V = -V_0 + \tfrac{1}{2} M \omega_0^2 r^2. \tag{27.1}$$

The value of V_0 is not of interest here and will be set equal to zero since we are interested only in wavefunctions and excitation energies, not in absolute energies.

α) The magic numbers. The single particle energies are given by:

$$\varepsilon = \hbar \omega_0 (N + \tfrac{3}{2}) \tag{27.2}$$

where N, the principal quantum number, is given by $2n-2+l$, is the orbital angular momentum, and n is the number of nodes in the radial wavefunction plus 1.

Single particle levels in the oscillator potential are widely separated in energy and highly degenerate, in sharp contrast to the statistical distribution of levels in the Fermi gas model. Thus up to $(N+1)\,(N+2)$ neutrons and the same number of protons can go into the states of principal quantum number N. For light nuclei, the oscillator potential gives approximately the correct level ordering and the observed magnic numbers 2, 8 and 20. For heavier nuclei, this potential implies closed shells for nucleon numbers 40, 70, 112, while the observed magic numbers are 50, 82, and 126.

β) Distribution of nucleons. Let us now briefly mention the spatial variation of the nuclear density. The density is known to be approximately constant throughout the nucleus and falls off rather sharply at the surface. On the other hand, the oscillator potential yields a density which is peaked at the center and falls to zero approximately linearily with distance from the center.

Although details of the empirical nuclear density can thus not be reproduced with an harmonic oscillator potential, it is of interest to note what value of the constant ω_0 yields the correct nuclear radius. To begin with, an important property of the harmonic oscillator potential is that the expectation values of kinetic and potential energies (the latter measured from the bottom of the well) in any state are equal (and thus equal to half of the single particle energy). The sum of the single particle energies of all nucleons is thus given by:

$$\sum \varepsilon_i = M\,\omega_0^2\,A\langle r^2\rangle \tag{27.3}$$

where $\langle r^2\rangle$ is the mean square radius. From the considerations of Sect. 3, we have

$$\langle r^2\rangle \approx \tfrac{3}{5}R_c^2 \tag{27.4}$$

where R_c is the Coulomb radius, which averages about $1.2\,A^{\frac{1}{3}}$ for heavy nuclei. Thus we find:

$$\langle r^2\rangle \approx 0.86\,A^{\frac{2}{3}}\times 10^{-26}\ \mathrm{cm}^2. \tag{27.5}$$

An approximate value for the sum $\sum \varepsilon_i$ in the nuclear ground state can be readily obtained. Suppose all states with N up to and including a certain value N_0 are occupied by protons and neutrons ($N=Z=\frac{1}{2}A$). Then we have from the previous considerations:

$$A = \sum_{N=0}^{N_0} 2\,(N+1)\,(N+2) \tag{27.6}$$

and

$$(\hbar\,\omega_0)^{-1}\sum \varepsilon_i = \sum_{N=0}^{N_0} 2\,(N+1)\,(N+2)\,(N+\tfrac{3}{2}). \tag{27.7}$$

It is easy to show by straightforward summation that:

$$A \approx \tfrac{2}{3}(N_0+2)^3 + 0\times(N_0+2)^2\ldots \tag{27.8}$$

and similarly

$$(\hbar\,\omega_0)^{-1}\sum \varepsilon_i \approx \tfrac{1}{2}(N_0+2)^4 - \tfrac{1}{3}(N_0+2)^3\ldots. \tag{27.9}$$

Keeping only the terms of highest order in (N_0+2) and eliminating N_0 from Eqs. (27.7) and (27.8) we find:

$$\sum \varepsilon_i = \tfrac{1}{2}\left(\tfrac{3}{2}\right)^{\frac{4}{3}}\hbar\,\omega_0\,A^{\frac{4}{3}} \approx 0.86\,\hbar\,\omega_0\,A^{\frac{4}{3}}. \tag{27.10}$$

Equating Eqs. (27.3) and (27.10) we find, with the help of Eq. (27.5),

$$\hbar\,\omega_0 \approx 41\,A^{-\frac{1}{3}}\ \mathrm{Mev} \tag{27.11}$$

for the spacing of harmonic oscillator levels.

γ) Spacing of oscillator levels. If the oscillator approximation were exactly valid, all excitation energies would be integral multiples of $h\omega_0$. In particular, all electric dipole radiation would have frequency ω_0, just as expected classically[1]. Thus the cross section for electric dipole ($E1$) gamma ray absorption should have a sharp peak and the energy $h\omega_0$. Indeed, giant photonuclear resonances, which probably involve $E1$ radiation, have been observed in many nuclei[2]. The relatively small observed widths of the resonance peak (3 to 6 Mev) in each nucleus imply that the simple oscillator model is a fairly good approximation. The finite width of the peak is evidently due in part to the fact that the actual nuclear potential deviates somewhat from the oscillator potential and in part to interactions between nucleons in unfilled shells.

The resonance peaks are known to occur at energies 22, 17, and 14 Mev for $A = 16$, 100, 200 respectively. These values are almost twice as large as the predictions (27.11) and vary as $A^{-0.2}$ rather than as $A^{-0.33}$. The agreement between empirical and theoretical values can be greatly improved by taking into account the velocity dependence of the effective nuclear potential. As was seen in Sect. 9, the data on saturation suggests that the effective nucleon mass inside nuclei is appreciably less than the free nucleon mass. One can easily modify the oscillator potential to take this into account. We merely replace M by M^* everywhere in the Hamiltonian and the spacing constant by $\omega^* = (M/M^*)\,\omega$; all energy spacings being proportional to $h\omega^*$, will thus be increased by the factor M/M^*[3], but the wavefunctions, which depend only on the parameter $(M^*\omega^*/\hbar)^{\frac{1}{2}} r$, will remain unchanged. The empirical data imply that $M^*/M \approx 0.6$, consistent with the remarks at the end of Sect. 14. For a somewhat more detailed treatment we must keep in mind that the coefficient "1.2" of the Coulomb radius is known to be somewhat larger for light nuclei (cf. Sect. 3). In this way we can at least qualitatively account for the empirical dependence of the resonance energy on A.

28. Realistic nuclear potential. We consider here briefly the single particle energy levels in the actual nuclear potential. This "diffuse surface" potential is essentially flat in the nuclear interior and falls off rather sharply near the edge. It is intermediate between the simple harmonic oscillator and square well potentials [*25*], [*29*].

Considerable insight into the nuclear level order may be gained by treating the deviation of the nuclear potential from the oscillator potential as a perturbation. In going from a oscillator to a square well, we must raise the potential near the center and lower it just inside the nuclear edge. This has the effect of lowering the energies of those states which have large wavefunctions at the edge, i.e. states of large l, at least relative to states of small l. The potentials also differ outside the nucleus but since the bound state wavefunctions are small there, the effect of this difference is small. The above argument also applies in going from an harmonic oscillator to a diffuse well, though the change is less marked.

Inspection of calculated energies for a square well or diffuse well shows that levels which would be degenerate for an harmonic oscillator are split, but for both potentials the energies can be approximately fitted to the form

$$E_l = c_0 - c_1\, l(l+1) \tag{28.1}$$

[1] Matrix elements of Cartesian coordinates x_i and of their associated momenta vanish, except when taken between states of principal quantum numbers differing by unity.

[2] A summary of information regarding nuclear photodisintegration has been given by D. H. WILKINSON, Physica, Haag **22**, 1039 (1956).

[3] S. RAND: Phys. Rev. **99**, 1620 (1955).

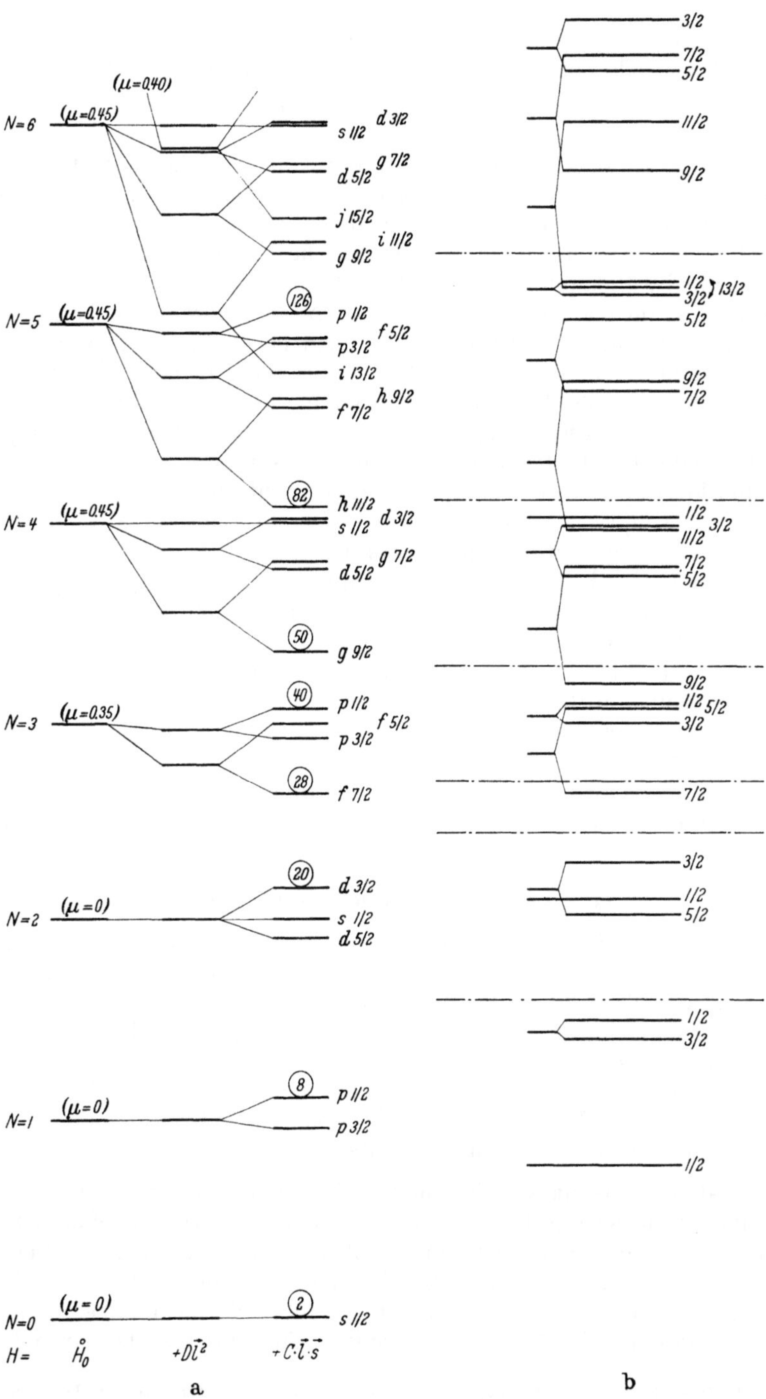

Fig. 8 a and b. Level order of neutron orbits in an isotropic simulated diffuse well potential (with spin-orbit coupling) compared with empirical level order. The calculated level order [Eq. (28.2)] (a) is taken from Ref [42]. The empirical level scheme (b) is reproduced from P. F. A. KLINKENBERG, Rev. Mod. Phys. 24, 63 (1952). The quantity μ equals $-20\,D$ as defined in (28.2).

where c_0 and c_1 depend on the shape of the potential and the particular shell under consideration. In addition the effect of the nuclear spin orbit coupling must be considered. A simple nuclear potential which approximates the "realistic" single particle potential, at least for level ordering, is [42]

$$V = \tfrac{1}{2} m \omega_0^2 r^2 + \hbar \omega_0 [D \boldsymbol{l} \cdot \boldsymbol{l} + c \boldsymbol{l} \cdot \mathbf{s}] \tag{28.2}$$

where D and C are negative constants whose values are then adjusted to fit the empirical data and s is the spin vector of the nucleon. The calculated and empirical single particle level order for neutrons is shown in Fig. 8. Note that states of large l are affected the most by the deviations from the oscillator potential and by the spin orbit coupling. In particular, of all the subshells in a given shell N, the one with $n=1$, $l=N$, $j=N+\frac{1}{2}$ is lowered by both perturbations, and for $N \geq 4$, it is pushed into the next lower shell. Other subshells are also affected but tend to stay together. Thus there is a definite grouping of levels into major shells, though not as extreme as for the oscillator potential. In this way the empirical magic numbers can be accounted for. For protons the extra Coulomb potential tends to favor states of larger l, for heavier nuclei, the proton level order differs slightly from the neutron order.

While the simple potential (27.2) is satisfactory for some applications, it does not take into account certain important features. Thus the splitting between $j=l+\frac{1}{2}$ and $j=l-\frac{1}{2}$ states is probably not constant $= h\omega_0 C (l+\frac{1}{2})$ as given by Eq. (27.2), but is larger for states with large amplitude near the nuclear surface (i.e. states of small n). This feature can be qualitatively accounted for if the spin-orbit interaction has a radial dependence proportional to dV/dr as is indeed expected on the basis of plausible theories[1].

Regarding the form of a reasonable diffuse surface central potential we might first consider form (20.1) (without the imaginary part since we are dealing with bound states). Detailed calculations of level ordering in such a diffuse well with spin-orbit coupling have been recently made and they reproduce the empirical results quite well[2].

II. The many-particle shell model.

In Sect. 29, we discuss the connection between the general independent particle models (and in particular, the many-particle shell model) to the many-body problem. We will see that the validity of any independent particle description demands that the potential seen by each particle does not fluctuate too much about its average value.

In Sect. 30, we make a few general remarks on the particularly simple form of the many-particle shell model: the j-j coupling model, with respect to energy levels, wavefunctions and nuclear moments.

29. Relation to the many-body problem. Let us first consider a system of non-interacting fermions subject to a common potential V. The wavefunction of the system is, of course, a Slater determinant of the appropriate single particle wavefunctions, ψ_i. We may equivalently describe the motion of the particles by the behavior of the various occupation numbers N_i (the number of particles occupying state ψ_i, $N_i=0$ or 1) with time[3]. Thus for the above problem, we

[1] R. J. Blin-Stoyle: Phil. Mag. **46**, 973 (1955). — J. S. Bell and T. H. R. Skyrme: Phil. Mag. **1**, 1055 (1956). — L. S. Kisslinger: Phys. Rev. **104**, 1077 (1956).

[2] Ross, Lawson and Mark: Phys. Rev. **102**, 1613 (1956); **104**, 401 (1956). It seems, however, that the required surface thickness of the potential is somewhat larger than the empirical value.

[3] This approach was suggested to the author by Professor W. G. McMillan. The considerations of this section are meant to illustrate the general principles underlying the independent particle models, rather than for detailed calculations.

have

$$H_{\text{I.P.}} = \sum_i N_i (T_i + V_i) \tag{29.1}$$

and each of the occupation numbers is a constant of the motion.

Now consider the more complicated nuclear problem. Assuming only two-body forces we can write the Hamiltonian in the form:

$$H = \sum_i N_i T_i + \sum_{i<k} \sum N_i N_k V_{ik} \tag{29.2}$$

where the N_i's are again occupation numbers of single particle levels. For this case, the instantaneous potential governing the motion of particle i is given essentially by:

$$V_i' = \sum_k N_k V_{ik}. \tag{29.3}$$

Since this potential depends on the motion of the other particles (i.e. the values of N_k), it is subject to fluctuations. Now the basic condition for the approximate realization of an independent particle description is that the fluctuations in the potential be small compared to its average value. The validity of this assumption depends, of course, on the dynamics of the system. Thus, if the forces between the particles give rise to clustering, or if the potential depends sensitively on the state of a few particles, then there are strong correlations and the independent particle description fails.

On the other hand, the independent particle description works well for nuclei, in spite of the strong internucleon forces, only because the Pauli principle reduces the fluctuations in the occupation numbers. Thus most of the nucleons, especially the tightly bound ones, are constrained to stay in their orbits—most of the orbits to which collisions might lead being already occupied by other nucleons [cf. Eq. (16.8)].

If the occupation numbers fluctuate rather slightly, then the instantaneous potential seen by particle i will always be close to an average value:

$$V_i = \sum_k \overline{N}_k V_{ik} \tag{29.4}$$

where the N_k represent average occupation numbers, e.g. the average over all nuclear orientations.

However, some fluctuations of the occupation numbers denoted by:

$$\delta N_i = N_i - \overline{N}_i \tag{29.5}$$

will occur[1]. It is convenient to use the identity:

$$N_i N_k = \overline{N}_i N_k + N_i \overline{N}_k - \overline{N}_i \overline{N}_k + \delta N_i \, \delta N_k. \tag{29.6}$$

for the purpose of rewriting the total Hamiltonian in the form:

$$H = \sum_i N_i T_i + \tfrac{1}{2} \sum_i \sum_k N_i N_k V_{ik} - \tfrac{1}{2} \sum_i N_i^2 V_{ii} \tag{29.7}$$

and

$$H = \sum_i N_i T_i + \sum_i \sum_k N_i \overline{N}_k V_{ik} - \tfrac{1}{2} \sum_i \sum_k \overline{N}_i \overline{N}_k V_{ik} + \tfrac{1}{2} \sum_i \sum_k \delta N_i \, \delta N_k V_{ik} - \tfrac{1}{2} \sum_i N_i^2 V_{ii} \tag{29.8}$$

where $\sum_i \sum_k$ denotes independent summation of indices i and k over all states.

---

[1] We may regard N_i as the occupation number (0 or 1) of state i for a *particular* Slater determinant Ψ^ν in the wavefunction Ψ (26.3), while $\overline{N}_i$ refers to the *average* occupation number of the state i for the wavefunction Ψ. Thus there are no fluctuations if the wavefunction consists of a *single* Slater determinant.

The first two terms of (29.8) are identical with the independent particle Hamiltonian $H_{\text{I.P.}}$ (29.1). The remainder represents the various corrections arising from the two-body nature of the interactions. Thus the third term is constant, and it corrects for counting all interactions twice in $H_{\text{I.P.}}$. The energy of each state is shifted by the same large amount, but wavefunctions and excitation energies are not affected.

The "residual interactions", i.e. those interactions not included in the average potential, are described by the fourth term of Eq. (29.8). These represent fluctuations of occupation numbers from about their average values $\bar{N}_i$. Now the potentials which are used to construct the single particle wavefunctions (e.g. a harmonic oscillator) always differ somewhat from the "optimum" potential which comes closest to reproducing the motion of individual particles. Thus the single particle wavefunctions calculated with these potentials differ somewhat from the "optimal" single particle functions. The residual terms thus contain both corrections to the single particle motions to make them "optimal" and also direct correlations between the motions of different particles due to their mutual interactions[1]. These two corrections are difficult to disentangle.

To return to (29.8), the last term represents an apparent self energy. However, this vanishes for the case of Fermions because of the antisymmetry of the wavefunction.

In summary, we see that the many body Hamiltonian be rewritten as an independent particle Hamiltonian plus correction terms:

$$H = \sum_i (T_i + V_i) + \text{const} + \sum_{i<k}\sum V_{\text{Resid}}. \tag{29.9}$$

30. Remarks on *j-j* coupling[2]. *α) Nuclear coupling schemes.* Most properties of low-lying nuclear states are determined by the coupling of the loosely bound nucleons. Under the influence of the central average potential (including the strong spin-orbit coupling), each nucleon moves in a definite orbital characterized by l, j and a radial quantum number n. The nucleons in the last unfilled shell can combine their individual angular momenta in various ways, so that in general there are a number of possible values of I, the total angular momentum. In the absence of "residual interactions", these states would all be degenerate. However, there must be some residual interactions, since the nuclear potential is due to the nucleons themselves and therefore fluctuates somewhat about its average value. Thus states of different I will, in general, have different energies. In the j-j coupling model, it is assumed that the interactions do not mix different orbitals, i.e. that each nucleon still has a definite angular momentum j. Of course, for detailed studies of nuclear states it is necessary to take into account deviations from j-j coupling[3]. Nevertheless, the j-j coupling scheme often is a reasonable first approximation and it gives, for example, correct ground state angular momenta for most odd-A nuclei.

To illustrate some essential features of the nuclear fine structure, consider the various possible states which may be formed by identical nucleons in a

[1] Ref. [*2*], p. 292.

[2] See [*29*], appendix A-II.

[3] One possibility is that the l of each nucleon, but not its j, may remain a good quantum number. This situation is called "intermediate coupling". A limiting case of this, called L-S coupling, occurs if the total orbital angular momentum L and its spin S are good quantum numbers. It may also happen that neither the l nor j of individual nucleons remain good quantum numbers. In this case we are dealing with "configuration interaction". Many properties of some of the lightest nuclei ($4 \leq A \leq 16$) are well described by the intermediate coupling approximation, [*24*], [*27*] but configuration interaction is of importance for most other nuclei.

$j=\frac{5}{2}$ subshell. A single nucleon must give $j=\frac{5}{2}$, two nucleons can couple to $I=0$, 2 or 4, and three nucleons may combine their angular momenta to a resultant of $\frac{3}{2}$, $\frac{5}{2}$, or $\frac{9}{2}$. Since it requires $2j+1=6$ nucleons to fill this subshell, a configuration of $6-n$ particles is equivalent to n holes. In general, excitation spectra (but not absolute energies) are the same for $(j)^n$ and $(j)^{2j+1-n}$ configurations.

β) Normal coupling. For short range attractive interactions, the ground state always has angular momentum $I=0$ (even n) or $I=j$ (odd n). (In the limit of zero range, the level scheme for identical particles is independent of the exchange character of the interaction.) The wavefunctions of these states (denoted here as states of "normal coupling") can be constructed in terms of the wavefunction for the $(j^2)_{I=0}$ configuration

$$\Psi(j^2)_{I=0} = \text{const} \times \sum_m (-1)^m \varphi_{j,m}(1)\, \varphi_{j-m}(2). \tag{30.1}$$

Each time a pair of nucleons is added, the wavefunction is multiplied by $\Psi(j^2)_{J=0}$ and the product is antisymmetrized with respect to all the particles. Thus for the $(\frac{5}{2})^3_{\frac{3}{2}}$ state we have:

$$\Psi(\tfrac{5}{2})^3_{I=\frac{3}{2}, M=\frac{3}{2}} = \text{const} \times [\varphi_{\frac{3}{2}}\, \varphi_{\frac{3}{2}}\, \varphi_{-\frac{3}{2}} - \varphi_{\frac{5}{2}}\, \varphi_{\frac{1}{2}}\, \varphi_{-\frac{1}{2}}) \tag{30.2}$$

where each "product" denotes a Slater determinant and the subscript j has been suppressed in each φ.

γ) Magnetic moments. For the normal state of an odd number of identical nucleons, the magnetic moment has the same value as for a single particle. It is this feature which has made it possible to use a single particle model for the qualitative interpretation of empirical magnetic moments. Of course the actual nuclear wavefunctions often differ considerably from the idealized forms above (e.g. due to configuration mixing) so that the magnetic moments often deviate somewhat from the single particle values.

δ) Quadrupole moments. The situation is quite different for nuclear quadrupole moments defined by:

$$Q = e\,(3z^2 - r^2)\,\varrho(r)\,d^3r\big|_{M=I}. \tag{30.3}$$

The quadrupole moment of a single proton in a state of angular momentum j is given by

$$Q_1 = -\frac{2j}{2j+3}\, e\,\langle r^2\rangle \tag{30.4}$$

where $\langle r^2\rangle$ denotes the expectation value of r^2 for the state. For the normal state $(j^n)_{I=j}$ of protons we have:

$$Q_n = Q_1\left(\frac{2j+1-2n}{2j-1}\right). \tag{30.5}$$

According to (30.5) the quadrupole moment is a linear function of the number of particles in the shell, and cannot exceed in absolute magnitude the value for a single particle. In particular, for a configuration containing $2j$ nucleons (i.e. filled subshell minus one particle), the quadrupole moment has the same value as for a single particle, except for sign. Even for other kinds of configurations in which $I=j$, it is still difficult to obtain quadrupole moments larger than the single particle values. Only for states with $I=j-1$ can large quadrupole moments occur, i.e. for $(\frac{9}{2})^3_{\frac{7}{2}}$, we have[1]

$$\frac{Q_3}{Q_1} = \frac{121}{90}. \tag{30.6}$$

[1] Cf. Ref. [*34*], p. 54.

The closed shells have a spherically symmetric density distribution and do not contribute to the quadrupole moment. Empirical quadrupole moments of nuclei consisting of closed shells plus (or minus) one proton are uniformly negative (or positive) in agreement with predictions of the single particle model. However, even for these nuclei, the quadrupole moments usually exceed the single particle values. More strikingly, the quadrupole moments increase as one moves away from closed shells. In particular, quadrupole moments of some rare earths reach 30 times the single particle values. These trends clearly require strong configuration interaction. A simpler way of describing these features is to regard the nucleus as being permanently non-spherical (see Chap. F).

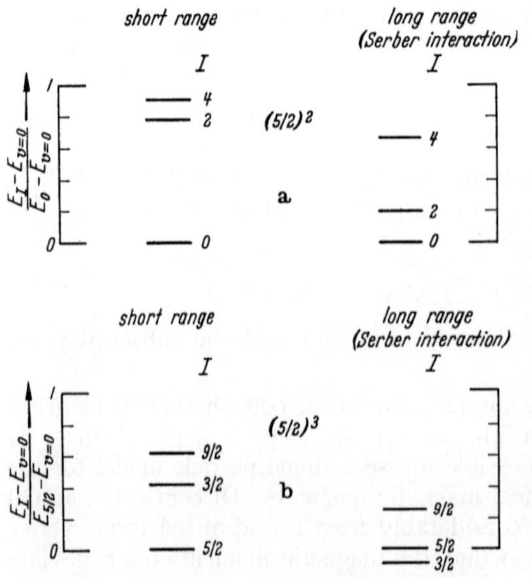

Fig. 9 a and b. Excitation spectra for configurations of interacting $j = \frac{5}{2}$ nucleons. The energy of a state of angular momentum I is denoted by E_I. The term $E_{v=0}$ refers to the energy in absence of interactions. All energies are calculated on the assumption of j-j coupling.

ε) *Energy spectra.* The theoretical energy spectra of $(\frac{5}{2})^n$ configuration of identical particles are shown in Fig. 9 for two limiting interactions, the δ-function interaction discussed above, and also Serber forces of very long range. In the latter limit, the levels are ordered according to I. Thus there is no longer a special favoring of the normal state. In fact, for three identical nucleons, the lowest state always has $I = \frac{3}{2}$. The situation in actual nuclei is intermediate between these extremes but probably somewhat closer to the short range limit.

F. The unified model.

31. Introduction[1]. α) *Nuclear quadrupole moments and transitions.* It has been known since the early days of the shell model that a number of odd-A nuclei in the rare earth region far away from magic numbers have large spectroscopic quadrupole moments[2] as much as 30 times the values (30.4) expected according to a single particle model. Evidently the coupling between the nucleons issuch that their individual quadrupole moments tend to add coherently. It was recognized (see Sect. 30) that this situation cannot occur if the angular momentum of each individual nucleon remains a good quantum number, e.g. a j-j coupling scheme. Thus configuration interaction is very much in evidence here[3].

Although in principle a shell model description of the large quadrupole moments should be possible, a much simpler description, first suggested by Rainwater [43] is to regard these nuclei as having permanently non-spherical

[1] See Ref. [32] and [35].

[2] Townes, Foley and Low: Phys. Rev. **76**, 1415 (1949).

[3] Many nuclei which are not far removed from closed shell configurations also have quadrupole moments several times large as the single particle values. Cf. the articles by R. J. Blin-Stoyle, Rev. Mod. Phys. **28**, 75 (1956) and by Way, Kundu, McGinnis and van Lieshout, Ann. Rev. Nucl. Sci. **6**, 129 (1956). Some progress has been made toward accounting for these large quadrupole moments on the basis of configuration interaction. See H. Horie and A. Arima: Phys. Rev. **99**, 778 (1955).

essentially spheroidal shapes. The resulting quadrupole moments may then be quite large even for relatively small deviations from sphericity, due to the large number of protons in the nucleus. It was postulated by RAINWATER that nuclear deformations result from the polarizing action of one or several loosely bound nucleons on the rest of the nucleus. While this "spheroidal core" model has since been modified in several important respects (Sects. 37, 38) it is still in essence accepted as the basic mechanism underlying the large nuclear quadrupole moments for nuclei far removed from closed shell configurations.

The large spectroscopic quadrupole moments of the rare earths are evidently of collective origin; however perhaps the most impressive evidence for the existence of collective effects in nuclei is provided by the low energy excitation spectra of even-even nuclei.

Many transitions between low-lying nuclear states throughout the periodic table proceed by electric quadrupole ($E\,2$) radiations. In the last few years, a large number of these transitions have been investigated by the Coulomb excitation process. It is very striking that most $E\,2$ transitions, unlike known transitions of other multipolarities, have strengths greatly exceeding those which would be expected if they involved the excitation of a single proton. The enhancement factors are largest, more than 100, for nuclei far away from closed shells[1]. Transitions in nuclei near magic numbers are not as strongly enhanced but even *their* strengths usually exceed the single proton values by at least an order of magnitude. It is evident that these $E\,2$ transitions are of a cooperative nature.

β) Energy spectra. Consider now the energy spectra. In all known even-even nuclei the ground state has zero spin and even parity and in almost all of them the first excited state is 2^+. The excitation energy of this first excited state varies in a rather regular fashion with nucleon number reaching maxima at closed shells and minima in between[2]. These features are shown in Fig. 10 and in more detail for the nuclei above *Pb* in Fig. 11. Especially small values of the 2^+ excitation energy occur at the rare earths and the actinides[3]. The higher excited states of many nuclei have also been investigated but the systematics are not as complete as for the first excited states[4].

The excitation spectra of even-even nuclei can be roughly divided into three classes, (i) closed shell region, (ii) vibrational, and (iii) rotational. The position of the energy levels in class (i) spectra can be quite well accounted for by the direct coupling of the nucleons in unfilled shells. The observed enhancement of the $E\,2$ transitions evidently results from polarization of the core by the outer particles[5]. This polarization does not alter the level schemes significantly, but it effectively increases the charge of each outer nucleon.

As we move further away from magic numbers, the shell model approach becomes too complicated from a calculational standpoint. However, the excitation spectra acquire a fairly simple form. For nuclei moderately far removed from closed shell configurations, the spectra are most simply described as collective vibrations about a spherical equilibrium shape [*44*]. One important feature

[1] G. M. TEMMER and N. P. HEYDENBURG: Phys. Rev. **99**, 1609 (1955).

[2] G. SCHARFF-GOLDHABER: Phys. Rev. **90**, 587 (1953).

[3] I. PERLMAN and F. ASARO: Ann. Rev. Nucl. Sci. **4**, 157 (1954).

[4] The empirical information regarding the lowest odd-parity state in even-even nuclei has been summarized by H. MORINAGA, Phys. Rev. **103**, 503 (1956).

[5] K. W. FORD and C. LEVINSON: Phys. Rev. **100**, 1 (1955). — W. W. TRUE: Phys. Rev. **101**, 1342 (1956). — F. C. BARKER: Phil. Mag., Ser. VIII **1**, 329 (1956). — F. AMADO and R. J. BLIN-STOYLE: To be published.

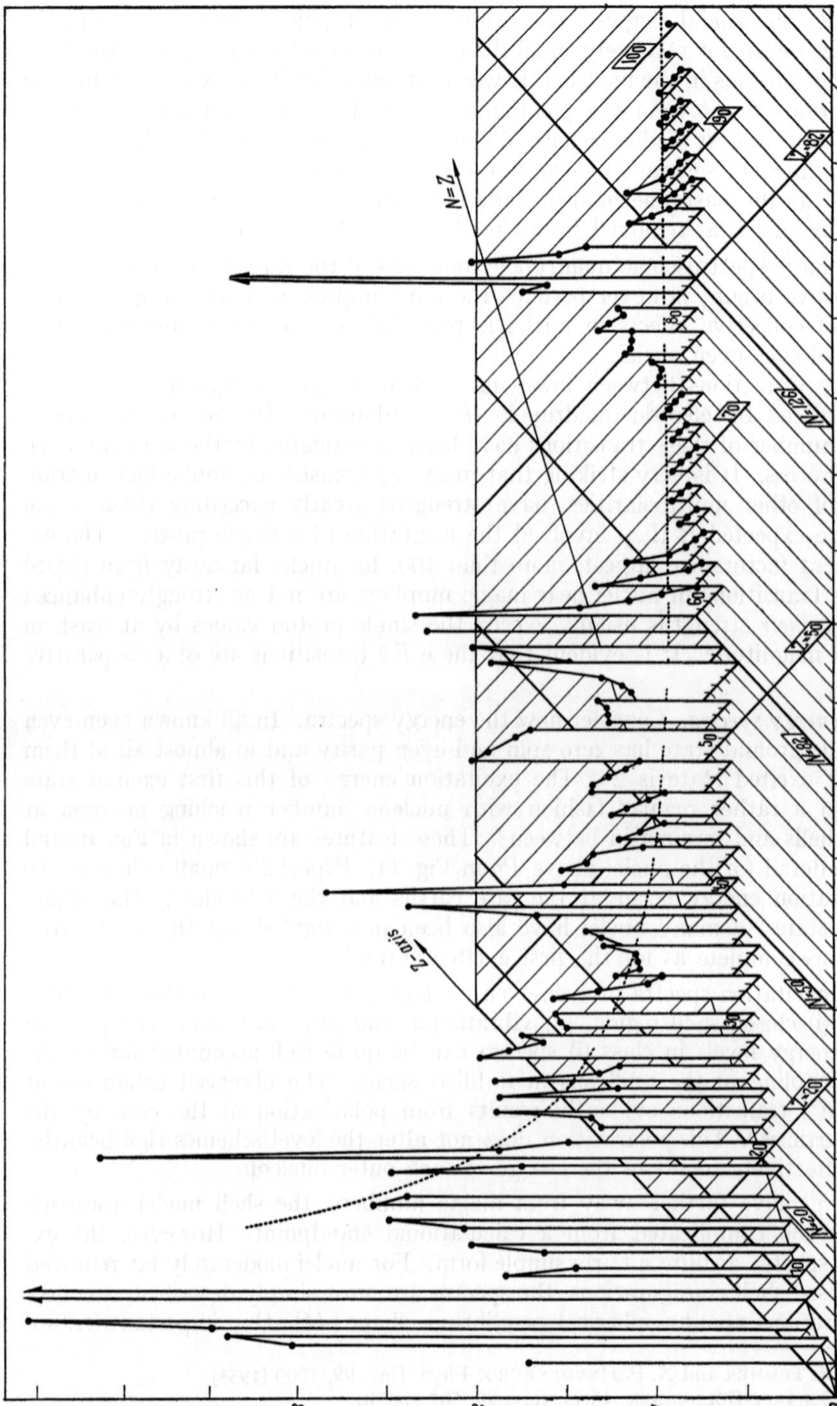

Fig. 10. Excitation energies of first excited 2^+ states in even-even nuclei. From Ref. [32]. The rotational spectra (Sect. 33) occur in the regions farthest from closed shells, where the excitation energies are lowest. In other regions, the excitations have the character of collective quadrupole excitations (Sect. 41). The separation between these two regions is approximately given by the criterion (43.27) and is indicated in this figure by the dashed line. For references to experimental energies see the above article.

of these vibrational spectra is that the second excited state (usually found to be 2^+) occurs at approximately twice the energy of the first excited state (see Sect. 41).

For even-even nuclei very far from closed shells ($155 < A < 185$, $A > 225$), the spectra show very striking regularities with respect to energy spacings.

Thus the second excited state (always found to be 4'), occurs ar very nearly $\frac{10}{3}$ of the energy of the first excited state. Most of these spectra can be very accurately represented as rotational excitations (see Sect. 33). It happens that nuclei in these regions have permanently non-spherical shape but axial, (essentially spheroidal) symmetry.

γ) Assumptions of the unified model. The phenomena mentioned above as well as many others can be accounted for (and indeed many were predicted) by the "Unified Nuclear Model", developed by A. Bohr [*33*] and applied by Bohr and Mottelson [*34*], which involves the assumption that the nucleons move nearly *independently* in a common *slowly changing* potential. As in the shell model we consider explicitly the degrees of freedom associated with one or a few of the loosely bound nucleons. On the other hand, we also borrow from the liquid drop model in explicitly considering collective excitations as variations of the shape and orientation of the nucleus as a whole[1].

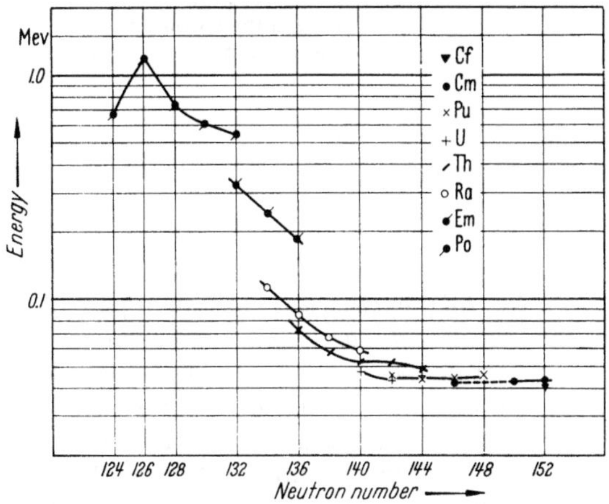

Fig. 11. Excitation energies of first excited 2^+ states in even-even nuclei with $84 \leq Z \leq 98$. From I. Perlman and J. O. Rasmussen: Alpha-Decay, Vol. XLII of this Encyclopedia. Most of these data were obtained from alpha spectra. Cf. Table 1 of the above reference.

While the *unified model* is essentially a hybrid of shell model and liquid drop model, it is much closer to the first one as regards the physical conditions, i.e. that the nucleons move approximately independently rather than being strongly-coupled as supposed in the liquid drop model. In principle it is possible to describe collective excitations by means of the many-particle shell model, though this is no doubt a very complicated calculational problem, and the physical character of the nuclear motion would probably be obscured in such a treatment. Another connection between shell model and unified model is suggested by the fact that while collective oscillations in nuclei involve *all* the nucleons to some extent, the most loosely bound ones (just the ones which are considered in the shell model treatment) have proportionately the most effect [*40*].

If the nuclear potential changes sufficiently slowly, we have an approximate separation of the nuclear motion into intrinsic and collective motions. The first of these represents the motions of the nucleons in a fixed potential while the second is associated with variations in the shape and orientation of the nuclear field. This separation is, in many respects, analogous to the separation between electronic and nuclear motion in molecules. In many cases the intrinsic motion is strongly coupled to the collective field, i.e. it follows the changing field nearly adiabatically. However, if the intrinsic structure possesses degenerate or low lying energy levels, it may become partially decoupled from the collective motion. An approximate separation into intrinsic and collective

[1] The interplay between individual particle and collective motion was also extensively studied by D. L. Hill and J. A. Wheeler, Ref. [*37*], with special reference to the nuclear fission process, and by K. W. Ford, Phys. Rev. **90**, 29 (1953).

motions occurs not only in nuclei and molecules but also in metals. In Table 3 we compare characteristics of these motions for the above three kinds of systems.

Table 3. *Intrinsic and collective motions in metals, molecules and nuclei.*

Property	Metals[1]	Molecules[2]	Nuclei
Intrinsic motion	The electrons move nearly independently in the field of the lattice and the electrons.	The electrons move nearly independently in the field of nuclei and the remaining electrons.	The nucleons move nearly independently in the collective field due to the nucleons themselves.
Collective motion	Collective oscillation of all electrons due to long range Coulomb interaction.	Motion of nuclei.	Response of nucleons to variable nuclear deformation. Excitation of all nucleons but mainly the loosely bound ones.
Coupling of intrinsic motion to collective field.	Weak	Strong	Moderate to strong.
$\omega_{\text{coll}}/\omega_{\text{particles}}$	> 1	$\ll 1$	$\lesssim 1$

δ) The rotational model. The main part of this chapter is devoted to the study of the "*rotational model*" a special form of the unified model in which we regard the nuclear shape as fixed. Thus the only kind of collective motions considered are rotations; i.e. we neglect vibrations, however excitations of the intrinsic structure are also treated. The rotational or "strong-coupling" model has proven very successful in the interpretation of level spectra and other properties of deformed nuclei. In deformed even-even nuclei, nearly all the low lying excitations are collective (i.e. rotational), in character. In deformed odd-A nuclei, low-lying intrinsic levels involving essentially only the last unpaired nucleon also often occur [*41*]. Each of these levels form the ground states of a rotational band.

ε) Contents of chapter. In the next few sections, we discuss some of the essential features of nuclear rotational states such as wavefunctions (Sect. 32), energy levels (Sect. 33), electromagnetic moments (Sect. 34), transition probabilities (Sect. 35) and moments of inertia (Sect. 36).

Some aspects of the intrinsic motions in non-spherical but axially symmetric potentials are discussed in Sects. 37 to 40. In Sect 37 we discuss the origin of nuclear deformations according to the spheroidal core model and in Sect. 38 a more detailed treatment, using an anisotropic harmonic oscillator binding potential, is given. The classification of nucleon states in deformed nuclei (Sect. 39) has many important similarities to the corresponding problem for spherical nuclei—although specific assignments for any given nucleus are usually different.

The next two sections deal with collective vibrations in even-even nuclei. Vibrations of both spherical and spheriodal nuclei are treated (Sect. 41, 42). To understand these vibrations we use the "*collective model*", another special form of the unified model in which we consider both collective rotations and vibrations, but do not treat excitations of individual particles.

[1] See D. PINES: Solid State Physics, Vol. 1, p. 368. 1955.

[2] G. HERZBERG: Spectra of Diatomic Molecules. Princeton, New Jersey: Van Nostrand Publishing Co. 1950. Rotational spectra in nuclei are analogous in many respects to rotational spectra of homonuclear diatomic molecules.

While the consideration of only collective degrees of freedom makes this collective model resemble the liquid drop model, there are also many profound differences. Thus the values of the parameters which characterize the collective motion is nuclei are quite different from the hydrodynamical values.

In the last three sections we consider the nature of the collective motions and their relation of the intrinsic nuclear structure in more detail. The inertial parameters for collective oscillations are discussed in Sect. 43, and in the following section we study the perturbations of the intrinsic motion due to the nuclear rotation. Finally Sect. 45 contains some general remarks concerning the relation between the unified model and the shell model.

I. Nuclear rotational states.

In the next five sections we will study rotational states in atomic nuclei. It will be *assumed* that the nuclei are non-spherical but essentially spheroidal, and in addition that the nuclear motion separates into intrinsic and rotational motions. Vibrational motions are here not treated explicitly but are included in the intrinsic motion. In these sections we will discuss different aspects of rotational states—the rotational part of the wavefunction, energies, electromagnetic moments, transition probabilities, and the moments of inertia. In succeeding sections, we will study the theoretical basis of the above assumptions.

32. Nuclear wave functions[1]. *α) The symmetric top.* The collective rotation of an axially symmetric nucleus has some similarities to the rotation of a symmetric to which we will study prior to the consideration of nuclei. We denote the set of mutually perpendicular axes in the rotating body—fixed reference frame as 1, 2, 3 axes (or x', y', z' axes), the last of which is taken to point along the symmetry axis. The orientation of the body fixed reference frame with respect to the laboratory fixed reference frame (x, y, z axes) is specified by three Euler angles. Because of the axial symmetry, two of the principal moments of inertia are equal. The Hamiltonian of the symmetric top is given by:

$$H_{\text{Sym. Top}} = \sum_{\nu=1}^{3} \frac{\hbar^2}{2\mathfrak{J}_\nu} I_\nu^3 = \frac{\hbar^2}{2\mathfrak{J}} (I^2 - I_3^2) + \frac{\hbar^2}{2\mathfrak{J}_3} I_3^2 \tag{32.1}$$

where $\mathfrak{J}_3$ and $\mathfrak{J}$ denote respectively the moments of inertia for rotations about the symmetry axis and about an axis perpendicular to the symmetry axis, and I_ν denote components of angular momentum along the body fixed axes.

The eigenfunctions of the symmetric top are the $\mathfrak{D}$ functions discussed in Appendix I. In terms of angular momentum operators we have

$$I_z \mathfrak{D}^I_{MK} = M\,\mathfrak{D}^I_{MK}, \tag{32.2}$$

$$I_3 \mathfrak{D}^I_{MK} = K\,\mathfrak{D}^I_{MK}, \tag{32.3}$$

$$I^2 \mathfrak{D}^I_{MK} = I(I+1)\,\mathfrak{D}^I_{MK}. \tag{32.4}$$

Thus the solutions $\mathfrak{D}^I_{MK}$ correspond to a total angular momentum I, angular momentum M along the z axis and angular momentum K along the 3 axis (both $|K|$ and $|M|$ must $\leq I$).

β) Rotations of nuclei. The rotation of a nucleus differs from that of a symmetric top in two important respects. First, since the nuclear moment of inertia

[1] See Ref. [*33*].

is associated with the collective reponse of the nucleons to variations of the nuclear field, and since this field in invariant with respect to arbitrary rotations about the symmetry axis, we must have $\mathfrak{J}_3 = 0$[1]. If the nuclear shape were spherical, then no collective rotations could occur at all, and *all* moments of inertia would vanish.

Secondly, we may have to treat explicitly the degrees of freedom associated with excitations of the very loosely bound nucleons, e.g. the unpaired nucleon in odd-A nuclei. The intrinsic wavefunction, i.e. the wavefunction assuming a static binding potential, is not an eigenfunction of the angular momentum since this potential is not spherically symmetrical. However because of the axial symmetry of the potential, the 3-component of the intrinsic angular momentum is a constant of the motion denoted by Ω. We designate the intrinsic wavefunction by χ_Ω^τ where τ refers to all the quantum numbers besides Ω which characterize the intrinsic state. This wavefunction is, of course, a function of the particle coordinates in the *body-system*. We now divide up the total angular momentum into intrinsic and rotational parts, $\boldsymbol{j}$ and $\boldsymbol{I}-\boldsymbol{j}$, respectively. The nuclear Hamiltonian can be expressed as follows:

$$H = H_{\text{Intr}}(\boldsymbol{r}') + T_{\text{Rot}} \tag{32.5}$$

where

$$T_{\text{Rot}} = \sum_\nu \frac{\hbar^2}{2\mathfrak{J}_\nu}(I_\nu - j_\nu)^2 = \frac{\hbar^2}{2\mathfrak{J}}\left[(\boldsymbol{I}-\boldsymbol{j})^2 - (I_3 - j_3)^2\right] \tag{32.6}$$

is the total kinetic energy associated with the rotational motion.

We may further express T_{Rot} as a sum of two sets of terms:

$$T_{\text{Rot}} = T_{\text{Rot}}^0 + T_{\text{coupl}} \tag{32.7}$$

where

$$T_{\text{Rot}}^0 = \frac{\hbar^2}{2\mathfrak{J}}\left[(\boldsymbol{I}^2 + \boldsymbol{j}^2) - (I_3 - j_3)^2\right] \tag{32.8}$$

and

$$T_{\text{coupl}} = \frac{\hbar^2}{2\mathfrak{J}}\left[-2\boldsymbol{I}\cdot\boldsymbol{j}\right]. \tag{32.9}$$

The first set of terms are ordinary rotational terms, in the sense that they are diagonal with respect to both χ_Ω^τ and $\mathfrak{D}_{MK}^I$; the other term represents the coupling between intrinsic and rotational motions. In most of the following considerations except Sect. 44, we will neglect the coupling terms, so that the nuclear Hamiltonian is written as:

$$H = H_{\text{Intr}}(\boldsymbol{r}') + \frac{\hbar^2}{2\mathfrak{J}}\left[(I^2 + j^2) - (I_3 - j_3)^2\right]. \tag{32.10}$$

A normalized eigenfunction of this Hamiltonian is [cf (I.5), p. 544]:

$$\Psi = \left(\frac{2I+1}{8\pi^2}\right)^{\frac{1}{2}} \chi_\Omega^\tau(r')\,\mathfrak{D}_{MK}^I(\vartheta_i). \tag{32.11}$$

γ) *Symmetry properties of the nuclear wavefunction.* The above wavefunction must be refined somewhat to take into account the assumed symmetry properties of the nuclear shape. First of all, the existence of axial symmetry implies that the nuclear wavefunction must be invariant with respect to arbitrary rotation of the body-fixed reference frame about the symmetry axis. Such a

[1] On the other hand, since a symmetric top rotates like a rigid body, its moment of inertia about the symmetry axis does not vanish (except in the limiting case that all its matter is concentrated on the symmetry axis).

rotation, say through an angle φ has the effect of transforming the Euler angles α, β, γ into $\alpha, \beta, \gamma+\varphi$ (see Appendix I) and

$$\mathfrak{D}^I_{MK} \to e^{iK\varphi}\,\mathfrak{D}^I_{MK} \tag{32.12}$$

and

$$\chi^\tau_\Omega \to e^{-i\Omega\varphi}\,\chi^\tau_\Omega. \tag{32.13}$$

The exponents in the last two equations have opposite signs, because the $\mathfrak{D}$ function is the wavefunction of the body system with respect to the laboratory system, while the intrinsic function refers to the wavefunction with respect to the body system.

The above invariance requirement clearly requires that

$$K=\Omega. \tag{32.14}$$

Another important symmetry property of a spheroid is its invariance with respect to rotations by 180° about any axis going through the center. Now a rotation by 180° about the x' axis transforms the Euler angles α, β, γ into $\pi+\alpha$, $\pi-\beta, \gamma$. This transformations changes

$$\mathfrak{D}^I_{MK} \to e^{\pi i(I+K)}\,\mathfrak{D}^I_{M-K}. \tag{32.15}$$

The effect of this rotation on the intrinsic function is more complicated. Suppose we decompose the intrinsic wavefunction into eigenfunctions of j.

$$\chi^\tau_\Omega = \sum_j c_j\,\chi^{\tau,j}_\Omega. \tag{32.16}$$

Then the rotation changes

$$\chi^{\tau,j}_\Omega \to e^{-\pi i(j+\Omega)}\,\chi^{\tau,j}_{-\Omega}. \tag{32.17}$$

The reason for the different signs in (32.15) and (32.17) is the same as before. We consider now separately the cases $K=\Omega\neq 0$ and $K=\Omega=0$. For the first case, invariance with respect to the above rotations requires that the wavefunction be a linear combination of two products. In particular the proper normalized wavefunction is given by:

$$\Psi=\left(\frac{2I+1}{16\pi^2}\right)^{\frac{1}{2}}\left[\chi^\tau_K(\boldsymbol{r}')\,\mathfrak{D}^I_{MK}(\vartheta_i)+(-1)^{I-j}\chi^\tau_{-K}(\boldsymbol{r}')\,\mathfrak{D}^I_{M-K}(\vartheta_i)\right] \tag{32.18}$$

where the term $(-1)^j$ is understood to act separately on each j-component of the intrinsic wavefunction [Eq. (32.16)].

If $K=\Omega=0$, then the product function (32.11), now of the form[1]

$$\Psi=(2\pi)^{-\frac{1}{2}}\chi^\tau_0(r')\,Y_{IM}(\vartheta_i), \tag{32.19}$$

satisfies the required symmetry condition, but only certain values of I can occur. In particular, if the intrinsic wavefunction only contains even values of j, then the total I of the nucleus must be even (the wavefunction vanishes if I is odd). This indeed occurs for the low-lying states in even-even nuclei.

The above symmetrization of the wavefunction guarantees that the parity π of the nuclear state, defined by:

$$\Psi(-r)=\pi\,\Psi(r)=\pm\Psi(r). \tag{32.20}$$

[1] The factor $(2\pi)^{-\frac{1}{2}}$ in this equation is required since the angular integration goes not only over all solid angles (α and β) but also over the Euler angle γ.

i.e. $\pi = \pm 1$, is the same as the parity of the intrinsic wavefunction. An inversion of the coordinates of all particles is equivalent to (i) an inversion of all the coordinates which appear in the intrinsic wavefunction and (ii) an inversion of the nuclear surface through the origin. The first operation leaves the wavefunction unchanged or changes its sign according to whether the parity of the intrinsic wavefunction is even or odd. Incidentally, the intrinsic wavefunction has a unique parity (even or odd) because of the assumed inversion symmetry of the potential. The second operation is essentially equivalent to a rotation of the surface by 180° about the x' axis, which leaves the wavefunction invariant.

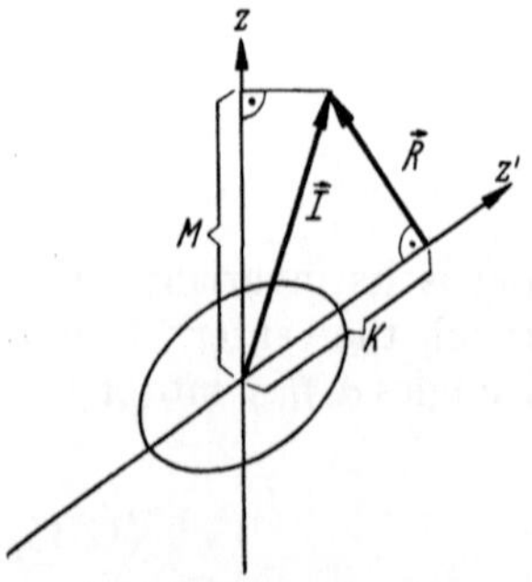

Fig. 12. Coupling scheme for strongly deformed nuclei. From Ref. [32].

The coupling scheme for axially symmetric deformed nuclei is shown in Fig. 12.

We have

$$\boldsymbol{I} = K\,\boldsymbol{e}_3 + \boldsymbol{R} \tag{32.21}$$

where $\boldsymbol{e}_3$ is a unit vector in the direction of the symmetry axis, and $\boldsymbol{R}$ is the rotational angular momentum. From Eq. (32.14) we see that the angular momentum parallel to the symmetry axis is all of intrinsic origin, so that $\boldsymbol{R}$ is perpendicular to the symmetry axis. This provides a simple geometrical interpretation of the result $I \geq |K|$.

δ) Some refinements. Let us now consider the consequences of dropping the assumptions of axial symmetry or of inversion symmetry.

First, if the system does not have axial symmetry[1], but has inversion symmetry the intrinsic wavefunction is no longer characterized by a unique Ω. Suppose, however, that the deviation from axial symmetry is small so that we still have an approximate symmetry axis, and Ω is approximately a good quantum number. (32.15) no longer holds, but if the nucleus has ellipsoidal symmetry, the nuclear shape will be invariant with respect to rotation of the body-fixed reference frame by 180° about the approximate symmetry axis. From (32.12) and (32.13) we see that

$$K - \Omega = \text{even}. \tag{32.22}$$

The above case is of interest in connection with the so-called "γ-vibrations" of spheroidal nuclei (Sect. 42).

Finally consider the case that the nucleus has axial symmetry but not inversion symmetry, i.e. it is pearshaped. This is of interest in connection with low-lying odd-parity states in heavy even-even nuclei (Sect. 42). Because of the axial symmetry, Eq. (32.14) still holds, but in view of the absence of inversion symmetry, each intrinsic state no longer has unique parity. We will suppose, however, that the deviation from inversion symmetry is small so that the intrinsic states still are approximately characterized by unique parities. The nuclear shape may be specified by a parameter p which denotes the deviation from reflection symmetry and where the shape "$-p$" is simply the mirror image of "p". Since

$$H(-p) = H(p) \tag{32.23}$$

the wavefunction must be either even or odd with respect to the transformation $p \to -p$.

The nuclear wavefunction is no longer invariant with respect to rotation by 180° about the x' axis. However such a rotation must have the same effect

[1] The rotation of nuclei not possessing axial symmetry has been studied by C. Marty, Nuclear Phys. **1**, 85 (1956); **3**, 193 (1957). The spectra of such nuclei would not be expected to follow the $I(I+1)$ law.

on the wavefunction as a sign-change of p. In particular if $K=\Omega=0$, the wavefunction is of the form

$$\Psi=(2\pi)^{-\frac{1}{2}}\chi_0^\tau(r')\,Y_{IM}(\vartheta)\,\Phi(p) \tag{32.24}$$

where $\Phi(p)$ is a so-called "vibrational function". Assuming j = even, the above rotation multiplies the wavefunction by $(-1)^I$. As in the spheroidal case, even value of I are allowed and are now associated with even vibrational functions. We may however, also have I odd, provided the associated vibrational function is odd. The parity of the states is $(-1)^I$, provided the state with $I=0$ has even parity. This may be seen from the argument following (32.20).

33. Energy spectra. The rotational model described in the last section has proven extremely successful in accounting for many features of the spectra found in nuclei with $155\leq A\leq 185$ and $A\geq 225$[1] [*35*]. Consider the properties of a rotational band, a sequence of states which have the same intrinsic wavefunction and differ only in their rotational wavefunctions. If the Hamiltonian were of the form (32.10), and the wavefunction of the form (32.18) or (32.19) the energy spectrum would be given by:

$$E_K(I)=E_K^0+\frac{\hbar^2}{2\mathfrak{J}}I(I+1) \tag{33.1}$$

where E_K^0 is a constant term independent of I.

The above treatment is not quite complete in that we still have to include the coupling term (32.9). It turns out, however (see appendix II), that this does not alter the form fo the energy spectrum except when $K=\frac{1}{2}$, a case we will consider below, or possibly if there are low-lying intrinsic states (see Sect. 44). Rotational spectra are usually easily identified on the basis of their characteristic spin sequence and energy ratios. According to Eq. (33.1) the rotational band has a ground state spin $I_0=K$ and an energy spectrum, relative to the ground state, given by:

$$E_I=E_{I_0}(I)-E_{I_0}(I_0)=\frac{\hbar^2}{2\mathfrak{J}}\left[I(I+1)-I_0(I_0+1)\right]. \tag{33.2}$$

α) Even-even nuclei. For the lowest rotational band in even-even nuclei we have $I_0=0$, and the spin sequence is 0, 2, 4 ..., odd values of I being forbidden (as long as the nuclear shape has inversion symmetry). The theoretical energy ratios are given by

$$\frac{E_4}{E_2}=\frac{10}{3}, \tag{33.3a}$$

$$\frac{E_6}{E_2}=7 \tag{33.3b}$$

and

$$\frac{E_8}{E_2}=12. \tag{33.3c}$$

Empirical energy ratios are shown in Fig. 13 and are seen to agree remarkably well with the predictions of the simple rotational model (though the empirical ratios start to decrease as we approach closed shells). Whatever small deviations

[1] Nuclei with $A\approx 25$ also appear to have rotational spectra [Litherland, Paul, Bartholomew and Gove: Phys. Rev. **102**, 208 (1956); Gove *et al.*: Physica, Haag **22**, 1141 (1956); R. K. Sheline: Nucl. Phys. **2**, 382 (1956/57); G. Rakavy: Dan. Mat.-Fys. Med. (in press)] but perturbations of the intrinsic structure (Sect. 44) are usually more pronounced here than in heavy nuclei. Consequently the spectra in these light nuclei sometimes deviate appreciably from the simple $I(I+1)$ law discussed in this section.

there are well away from closed shells can be attributed to "rotation-vibration interaction", an effect which also occurs in molecular spectra, and is discussed in Sect. 44. Fig. 14 shows the spectrum of Hf^{180} for which the 5 members of the lowest rotational band (even values of I up to 8) are known and the energy levels have been very accurately measured[1].

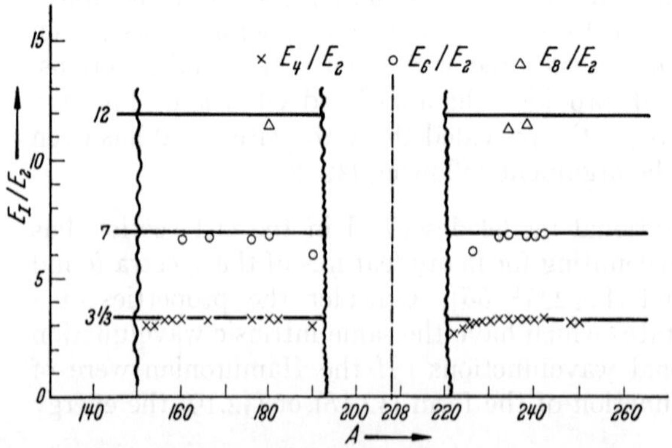

Fig. 13. Energy ratios of rotational excitations in even-even nuclei. From Ref. [32].

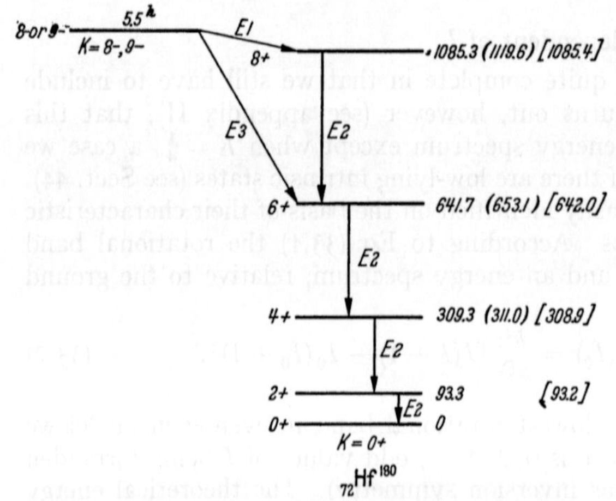

Fig. 14. Energy spectrum of Hf^{180}. Revised figure from A. Bohr and originally from Ref. [31]. The numbers in parenthesis are those obtained from (33.1) with the moment of inertia fitted to the first excited state. The observed minor deviations from these energies show a systematic trend and are of the sign and magnitude expected from the rotation-vibration interaction (44.9). The energies in square brackets are calculated rotational energies with the inclusion of the rotation-vibration effect.

β) Odd-A nuclei. Odd-A nuclei also exhibit striking rotational spectra although they are somewhat more sensitive to details of the intrinsic motion than even-even nuclei. According to (33.2), the sequence of spins is $I_0, I_0+1, I_0+2, \ldots$ The level scheme of Ta^{181}, one of the earliest to be extensively studied, is shown in Fig. 15. Except for the special case $K=\frac{1}{2}$ the ratio of excitation energies is given by:

$$\frac{E_{I_0+2}}{E_{I_0+1}} = 2 + \frac{1}{I_0+1} \qquad (33.4)$$

so that an accurate measurement of this ratio does, in fact, determine the ground state spin. As in even-even nuclei, the energy ratios are usually slightly smaller than the simple theoretical values, presumably due to rotation-vibration interaction. However specific perturbations associated with low-lying intrinsic states can affect the level scheme in a more complicated way (cf. Sect. 44).

γ) Level scheme for $K=\frac{1}{2}$. For the special case of $K=\frac{1}{2}$, there is usually a substantial deviation from the simple rotational spectrum. For this case the term (32.9) is of importance, i.e. the intrinsic motion becomes partially decoupled from the rotational motion. Using (II.7) and (II.9) it is easily shown that the energy spectrum now has the form:

$$E_K(I) = E_K^0 + \frac{\hbar^2}{2\mathfrak{J}}\left[I(I+1) + a(-1)^{I+\frac{1}{2}} I + \frac{1}{2}\right] \qquad (33.5)$$

---

[1] J. W. Mihelich, G. Scharff-Goldhaber and M. McKeown: Phys. Rev. **94**, 794 (A), (1954). — G. Scharff-Goldhaber, M. McKeown and J. W. Mihelich: Bull. Amer. Phys. Soc. II **1**, 206 (1956).

where a is the so-called decoupling parameter [34]. The value of a depends on the intrinsic motion of the odd nucleon, and is given by

$$a = \sum_j (-1)^{j-\frac{1}{2}} |c_j|^2 (j + \tfrac{1}{2}) \tag{33.6}$$

for an intrinsic wavefunction of the form (32.16).

Fig. 16 shows the theoretical $K = \frac{1}{2}$ level scheme as function of a. For all known rotational spectra in heavy nuclei with $K = \frac{1}{2}$, a is between -1 and $+1$, (e.g. -0.77 for Tm^{169}, 0.19 for W^{183}) so that the order of levels is not affected by the decoupling. For light nuclei, on the other hand, large values of a are known to occur; thus the level ordering may be partially inverted,

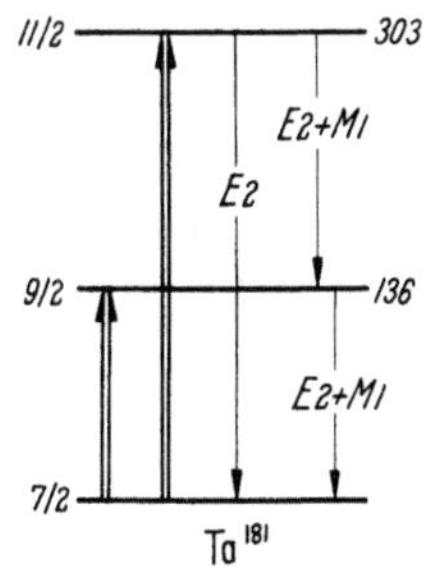

Fig. 15. Rotational spectrum of Ta^{181}. From Ref. [32].

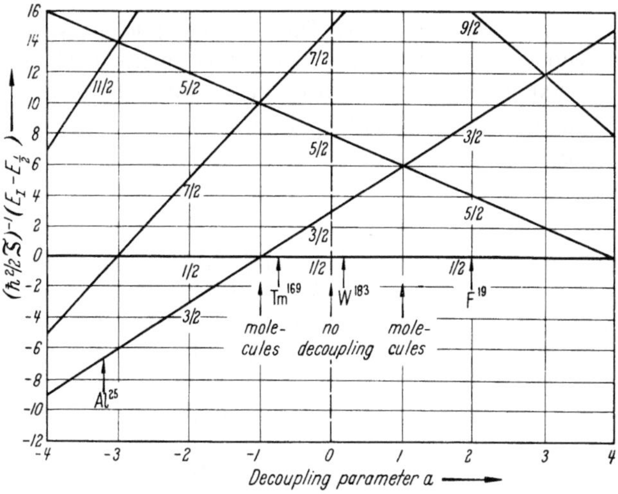

Fig. 16. Rotational spectrum of $K = \frac{1}{2}$ band as function of decoupling parameter.

e.g. one of the $K = \frac{1}{2}$ bands in Al^{25} has $a = -3.2$ and in F^{19} it has $a \approx 2$. Decoupling effects of this kind also occur in molecules, e.g. for $^2\Sigma$ states one finds two rotational bands of opposite parity with $a = +1$ and -1 [1].

34. Nuclear moments. The separation of the nuclear wavefunction into intrinsic and rotational parts implies that the angular momentum is shared between intrinsic and rotational motions in a very simple way [2].

Thus for the nuclear ground state ($I = K = \Omega$) we find, (except for the case $K = \frac{1}{2}$):

$$\langle j_z \rangle = \frac{(\boldsymbol{j} \cdot \boldsymbol{I}) \langle I_z \rangle}{\boldsymbol{I} \cdot \boldsymbol{I}} = \frac{(\boldsymbol{j} \cdot \boldsymbol{e}_3)(\boldsymbol{e}_3 \cdot \boldsymbol{I}) I_z}{\boldsymbol{I} \cdot \boldsymbol{I}} = \frac{\Omega K M}{I(I+1)} = \frac{I}{I+1} M. \tag{34.1}$$

and:

$$\langle R_z \rangle = \langle I_z \rangle - \langle j_z \rangle. \tag{34.2}$$

α) Magnetic moments. The magnetic moments are given by:

$$\mu = \mu_z|_{M=I} = \frac{(\boldsymbol{\mu} \cdot \boldsymbol{I}) I}{I(I+1)}. \tag{34.3}$$

Now, keeping in mind Eq. (32.21), let us write:

$$\boldsymbol{\mu} = g_K K \boldsymbol{e}_3 + g_R \boldsymbol{R} \tag{34.4}$$

[1] G. Herzberg: Spectra of Diatomic Molecules, p. 222. Princeton, New Jersey: Van Nostrand Publishing Co. 1950.

[2] A. Bohr: Phys. Rev. **81**, 134 (1951).

where g_K and g_R denote gyromagnetic ratios of intrinsic and rational motions, respectively. The values of both g_K and g_R, but especially the former, depend upon details of the intrinsic structure. Combining (34.1), (34.2) and (34.4), and substituting the result into (34.3) it is found that:

$$\mu = (I+1)^{-1}\left[(g_K - g_R)\,K^2 + g_R I(I+1)\right]. \tag{34.5}$$

For the ground state we obtain:

$$\mu = \frac{I}{I+1}\,(g_K I + g_R). \tag{34.6}$$

β) Quadrupole moments. The deformed nucleus is characterized by an intrinsic quadrupole moment Q_0, the quadrupole moment which would result if the symmetry axis were always pointed in the direction of the angular momentum.

Now the nuclear quadrupole moment Q in the laboratory system is related to Q_0 by:

$$Q = \frac{3K^2 - I(I+1)}{(I+1)(2I+3)}\,Q_0. \tag{34.7}$$

provided the nuclear wavefunction is of the form (32.18) or (32.19). This equation is derived in Appendix II. The coefficient of Q_0 is commonly known as the "projection factor". Note that if $K = I \to \infty$, then $Q \to Q_0$. However, because of the uncertainty principle, there are bound to be fluctuations in the direction of the nuclear symmetry axis. These always tend to reduce the actual Q below Q_0 and will even change the sign if $I \gg K$. For the ground state Eq. (34.7) reduces to:

$$Q = \frac{I(2I-1)}{(I+1)(2I+3)}\,Q_0. \tag{34.8}$$

The nuclear quadrupole moment Q thus vanishes for $I = 0$ and $\frac{1}{2}$ even though the intrinsic quadrupole moment may be different from zero.

γ) Nuclear deformation. The intrinsic quadrupole moment may be calculated very simply by regarding the nucleus as a uniformly charged spheroid with the three principal axes of lengths $R_0\,\mathrm{e}^{-\frac{\alpha}{2}}$, $R_0\,\mathrm{e}^{-\frac{\alpha}{2}}$, $R_0\,\mathrm{e}^{\alpha}$, respectively. Using the definition (30.3) we find

$$Q_0 = \tfrac{6}{5}\,Z\,e\,R_0^2\,\alpha \tag{34.9}$$

to first order in α.

We may introduce another deformation variable a_0, appropriate to a spheroidal nucleus by expressing the fractional elongation along any direction in the form:

$$\delta R = R(\vartheta,\varphi) - R_0 = R_0\,a_0\,Y_{20}(\vartheta,\varphi) \tag{34.10}$$

in the body fixed reference frame, where R_0 is the mean nuclear radius. $\left(\text{Volume} = \frac{4\pi}{3}R_0^3\right)$. The deviation from sphericity is given by the variable β defined by:

$$\beta = |a_0|. \tag{34.11}$$

A more convenient definition of β, equivalent to (34.10) in first order, is

$$\beta = \left(\frac{16\pi}{45}\right)^{\frac{1}{2}} \frac{\Delta R}{R_0} \approx 1.06\,\frac{\Delta R}{R_0} \tag{34.12}$$

where ΔR is the difference between major and minor axes.

Spectroscopic measurements show that essentially all strongly deformed odd-A nuclei (with $I > \frac{1}{2}$) have positive quadrupole moments, i.e. the nuclei are essentially football-shaped. For such nuclei, the intrinsic quadrupole moments are given by [*36*]:

$$Q_0 = \frac{3}{(5\pi)^{\frac{1}{2}}} Z e R_0 \beta [1 + 0.16\beta \ldots]. \tag{34.13}$$

Another deformation variable, δ, essentially equal to $\Delta R/R_0$, has been introduced in Ref. [*42*] and is used in some of the figures below. Some relations between the three deformation parameters α, β, and δ are:

$$\alpha \approx 0.63\beta [1 - 0.16\beta + 0(\beta^2)], \tag{34.14}$$

$$\delta \approx 0.95\beta [1 - 0.48\beta + 0(\beta^2)]. \tag{34.15}$$

35. Transition probabilities. The approximate separation of the nuclear motion into intrinsic and rotational motion which occurs in deformed nuclei, has important consequences for transition rates of nuclear processes. In this section we will essentially restrict ourselves to a consideration of electromagnetic transitions, although beta-transitions will also be briefly mentioned.

Gammy-ray emission is a simple kind of electromagnetic process with a transition probability given by[1, 2] [*34*]:

$$T(\lambda) = \frac{8\pi(\lambda+1)}{\lambda[(2\lambda+1)!!]^2} \frac{1}{\hbar} \left(\frac{\Delta E}{\hbar c}\right)^{2\lambda+1} B(\lambda) \tag{35.1}$$

where λ denotes the multipole order, E is the energy difference between initial and final states, and $B(\lambda)$ is the so-called "reduced transition probability". In this expression, only $B(\lambda)$ depends upon the details of the nuclear structure.

During the last few years, the process of "Coulomb excitation" of the nucleus by the electromagnetic field of moving charged particles has been extensively studied, both experimentally, and theoretically[3] [*32*]. The cross section for this process depends on various parameters, e.g. the charge of the target nucleus and of the projectile, as well as the velocity of projectile before and after the excitation. However the cross section is dependent on the detail nuclear structure only by virtue of being proportional to $B(\lambda)$, essentially the same quantity as occurs in the expression for the gamma-decay transition probability.

Conservation of angular momentum in a nuclear transition implies the following rigorous selection rule for electromagnetic transitions:

$$|I_i - I_f| \leq \lambda \leq I_i + I_f. \tag{35.2}$$

Conservation of parity gives

$$\left.\begin{aligned} \pi_i \pi_f &= (-1)^{\lambda} && \text{for electric transitions,} \\ \pi_i \pi_f &= (-1)^{\lambda+1} && \text{for magnetic transitions} \end{aligned}\right\} \tag{35.3}$$

where $\pi = +1$, or -1 denotes even or odd parity, respectively.

α) Transitions within the same rotational band. Let us first make some qualitative remarks concerning electromagnetic transitions between states of the same rotational family. All known electromagnetic transitions within a rotational

[1] For a review of electromagnetic transitions in nuclei, see the article by D. ALBURGER in Vol. XLII of this Encyclopedia; also M. GOLDHABER and J. WENESER: Ann. Rev. Nucl. Sci. **5**, 1 (1955).

[2] $(2\lambda+1)!! = 1 \times 3 \times 5 \ldots \times (2\lambda+1)$.

[3] N. P. HEYDENBURG and G. M. TEMMER: Ann. Rev. Nucl. Sci. **6**, 77 (1956).

band are of multipolarities $M1$ and $E2$ or a mixture of these two. Because of the angular momentum selection rules such transitions can occur only between states with $|\Delta I| \leq 2$. If $|\Delta I| = 2$, the radiation is pure $E2$. For $|\Delta I| = 1$, both $M1$ and $E2$ can contribute. On the basis of a single particle model[1], one would expect the radiation of lower multipole order, i.e. $M1$, to dominate. However the strong $E2$ transitions within a rotational band must involve the collective effects of many nucleons, combining their individual matrix elements coherently. This same effect is also responsible for the sizable nuclear deformations necessary for the existence of rotational spectra. Although $M1$ transitions can also involve a large number of nucleons, their individual magnetic moments do not add coherently and no significant enhancement of the $M1$ transition strengths over the single particle values occurs. It is then not surprising that $M1$ and $E2$ components often occur with comparable intensities.

For transitions within a rotational band the initial and final state have the same intrinsic wavefunction. Matrix elements for electromagnetic transitions will then depend only on the parameters which describe the collective behavior of the system, i.e. Q_0, g_K, g_R and statistical factors, but not on details of the intrinsic motion itself. The symmetrization of the wavefunctions does not alter this conclusion except for bands with $K = \frac{1}{2}$, where the partial decoupling between intrinsic and rotational motions has important effects.

Turning now to more quantitative considerations, the reduced transition probability for either electric or magnetic 2^λ pole radiation can be written as follows:

$$B(\lambda; i \to f) = \sum_{M_f, \mu} |\langle f| \mathfrak{M}(\lambda, \mu) |i\rangle|^2 \tag{35.4}$$

in terms of the matrix elements of the multipole operator $\mathfrak{M}(\lambda, \mu)$ between an initial state i and final state f with magnetic quantum numbers M_i, M_f. The reduced transition probability is independent of the value of M_i.

Since the statistical weight of a state with angular moment I is $2I+1$, it follows quite generally that

$$\frac{B(f \to i)}{B(i \to f)} = \frac{2I_i + 1}{2I_f + 1}. \tag{35.5}$$

For many pairs of nuclear states both the "upward transition"—e.g. Coulomb excitation and the "downward transition"—e.g. gamma-emission, have been studied. The reduced transition probabilities for these two processes are related by Eq. (35.5).

α1) Electric quadrupole transitions. The electric multipole operator is given by:

$$\mathfrak{M}_e(\lambda, \mu) = \sum_{p=1}^{A} e_p r_p^\lambda Y_{\lambda\mu}^*(\vartheta, \varphi). \tag{35.6}$$

In particular the electric quadrupole operator is simply related to the components of the quadrupole tensor whose $\mu = 0$ component, the conventional quadrupole moment, is defined by Eq. (30.3). Thus we have:

$$\mathfrak{M}_e(2, \mu) = \left(\frac{5}{16\pi}\right)^{\frac{1}{2}} e\, Q_\mu \tag{35.7}$$

and (35.4) becomes

$$B(E2; i \to f) = \frac{5}{16\pi} \sum_{M_f, \mu} |\langle f| Q_\mu |i\rangle|^2. \tag{35.8}$$

---

[1] Ref. [2], Chap. XII; S. A. Moszkowski: Multipole Radiation, Chap. XIII of Beta and Gamma-ray Spectroscopy, edit. by K. Siegbahn. Amsterdam, Netherlands: North Holland Publishing Company.

Eq. (35.8) holds quite generally independent of any specific assumptions regarding the nuclear wavefunction. Now consider a transition between two members of the same rotational band with quantum number K. For this case, the intrinsic wavefunction is the same in both states, only the rotational function is different. We then find (see Appendix II)

$$B(E2; i \to f) = \frac{5}{16\pi} e^2 Q_0^2 |\langle I_i 2 K 0 | I_i 2 I_f K \rangle|^2 \tag{35.9}$$

where the quantity in brackets is a Clebsch-Gordan coefficient[1].

Electromagnetic methods provide more accurate determinations of nuclear quadrupole moments than do spectroscopic methods since the interpretation of the latter requires knowledge of the gradient of the electric field at the nucleus, a quantity often subject to considerable uncertainty.

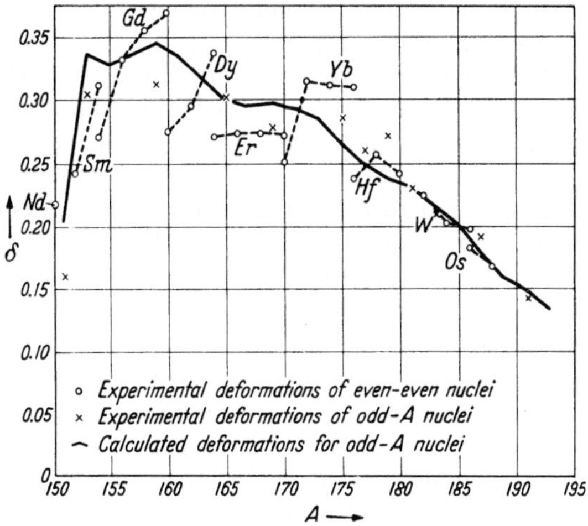

Fig. 17. Nuclear deformations for $150 \lesssim A \lesssim 190$. From Ref. [41]. This figure shows nuclear deformations deduced from empirical intrinsic quadrupole moments (cf. Sect. 34, 35) and calculated nuclear deformations (cf. Sect. 38, 39).

Nuclear quadrupole moments are known to reach maximum values farthest away from closed shells. The trend of deformations of nuclei with $150 < A < 190$ as function of nucleon number is shown in Fig. 17. Typical large values reached by many nuclei in the rare earth region are: $\beta = 0.3$, $Q_0 = 8 \times 10^{-24}\,\text{cm}^2$. There is no noticeable difference between values of Q_0 for even-even and neighboring odd-A nuclei[2].

α2) Magnetic dipole transitions. Next we consider magnetic dipole transitions. The component of the magnetic dipole operator is given by:

$$\mathfrak{M}_m(1, \mu) = \frac{e\hbar}{2Mc} \left(\frac{3}{4\pi}\right)^{\frac{1}{2}} \mu_\mu \tag{35.10}$$

where

$$\mu = \sum_{P=1}^{A} \mu_P \tag{35.11}$$

and Eq. (35.4) takes the form

$$B(M1; i \to f) = \frac{3}{4\pi} \left(\frac{e\hbar}{2Mc}\right)^2 \sum_{M_f, \mu} |\langle f | \mu_\mu | i \rangle|^2. \tag{35.12}$$

If the nucleus has axial symmetry we can write (cf. Sect. 34)

$$\mu_\mu = (g_K - g_R)(I_3)_\mu + g_R I_\mu. \tag{35.13}$$

where $(I_3)_\mu$ denotes the component of that part of the angular momentum vector which is parallel to the symmetry axes.

---

[1] E. U. Condon and G. H. Shortley: Theory of Atomic Spectra, Chap. 3. Cambridge: Cambridge University Press 1951.

[2] Of course, a measurement of $E2$ transition rates only determines the magnitude of Q_0 and not its sign. It has been *assumed* that all intrinsic quadrupole moments in this region of mass numbers are positive. This is a plausible assumption in view of (i) evidence regarding spectroscopic quadrupole moments (Sect. 34) and (ii) theoretical considerations (Sect. 39).

For an $M1$ transition between two members of the same rotational band we employ the same method used for $E2$ transitions and find:

$$\left.\begin{aligned} B(M1; i\to f) &= \left(\frac{3}{4\pi}\right)\left(\frac{e\hbar}{2Mc}\right)^2 (g_K-g_R)^2 K^2 |\langle I_i 1 K 0 | I_i 1 I_f K\rangle|^2. \\ &\text{except for } K=\tfrac{1}{2}. \end{aligned}\right\} \tag{35.14}$$

The second term in (35.13) does not contribute since matrix elements of the angular momentum operator $\boldsymbol{I}$ between different (non-degenerate) states vanish.

Measurements of $M1$ transition rates within a rotational band together with the ground state magnetic moment [Eq. (34.6)] gives important information concerning the gyromagnetic ratios[1] [*32*]. The quantities g_K and g_R are not determined uniquely, but limited to one of two pairs of numbers. Thus in Hf^{179}, the interpretation of measured transition rates and ground state magnetic moments implies that either (a) $g_K=-0.2$, $g_R=0.2$, or (b) $g_K=0$, $g_R=-0.4$. It requires, in addition, the measurement of the relative phase between $M1$ and $E2$ radiation to tell which of these choices is correct. From a theoretical point of view one would expect g_R not to differ very much from the value Z/A which would result if neutrons and protons contribute proportionately the same to the collective motion [*36*]. Thus for the above case, this argument would favor choice (a).

β) Transitions between different rotational bands. Consider now transitions between different rotational families. Since the intrinsic states are now different, one expects such transitions to involve only a few particles. Thus the transition rates generally do not exceed the single particle values. Possible exceptions are transitions between "vibrational states", another form of collective excitation, discussed in Sect. 41/42. In fact, such transitions usually proceed considerably slower than expected from a single particle model. Although a large change occurs only in the wavefunction of one or a few nucleons, the nuclear shape may be slightly different for initial and final states. Thus the wavefunctions of all the remaining nucleons have to change a little, and the lack of perfect overlap can reduce the transition rate considerably, often by 1 or 2 orders of magnitude below the single particle value.

β1) K-selection rules. If the nuclear wavefunction can be expressed as a simple product of intrinsic and rotational functions, i.e. if the simple rotational coupling scheme is applicable, then the quantity K, the component of angular momentum parallel to the nuclear symmetry axis, is a constant of the motion.

The transition probability between states in different rotational bands depends on the value of the integral

$$\int \mathfrak{D}^{I_f}_{M_f K_f}\, \mathfrak{D}^{\lambda}_{\mu\nu}\, \mathfrak{D}^{I_i}_{M_i K_i}\, d\Omega. \tag{35.15}$$

For this integral not to vanish, it is necessary to satisfy not only the angular momentum selection rule (35.2) but also the condition:

$$|K_i - K_f| \leq \lambda. \tag{35.16}$$

Eq. (35.16) is a "K-selection rule", which would hold rigorously if the above assumption were exactly satisfied [*31*].

Empirically, transitions for which the K selection rules are violated do occur, but they are usually much weaker than expected on basis of a single particle model. Thus it appears that the wavefunctions for one or both of the states between which the transition occurs are to a good approximation essentially of the simple product form.

[1] N. P. Heydenburg and G. M. Temmer: Phys. Rev. **104**, 981 (1956).

An extreme example of K forbiddenness occurs in Hf^{180} where the $E1$ decay of the isomeric state involves $|\Delta K| = 8$ or 9 with a transition probability of only about 10^{-15} of the single particle value for the same multipole order[1]. The $E3$ transition is also retarded for the same reason, but by a somewhat smaller factor, about 10^9. Other less extreme aspects of K-forbiddenness are illustrated in the decay scheme of W^{182} (Fig. 18). The assignment of quantum numbers in the figure (K, I, π) provides an interpretation of the observation that the $|\Delta I| = 0$ and 1 transitions between the $K = 2^-$ and $K = 0^+$ rotational bands appear to

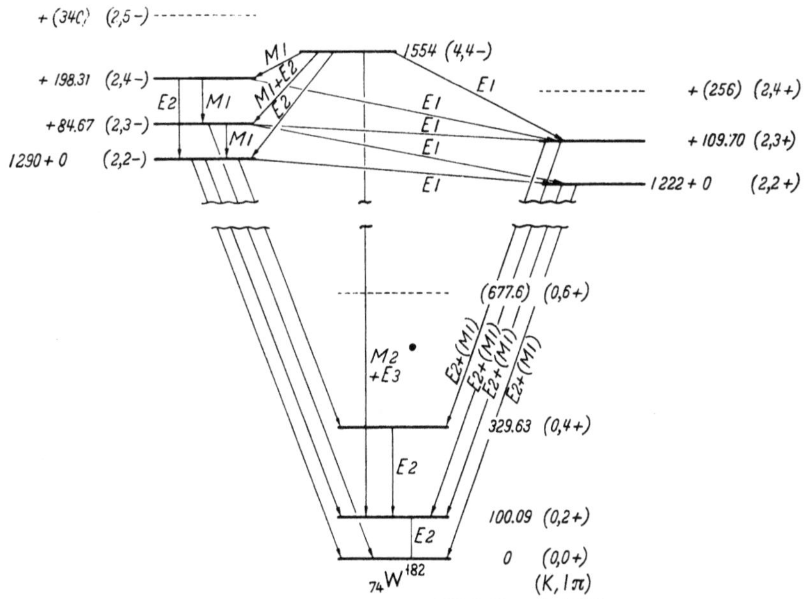

Fig. 18. Energy spectrum of W^{182}. From Ref. [31].

be mainly of $M2$ and $E3$ type rather than $E1$ which is K forbidden. Such a K forbiddenness may also account for the fact that the $|\Delta I| = 1$ transitions between the $K = 2^+$ and 0^+ are rather pure $E2$, rather than involving an $M1 - E2$ mixture. Also indicated in Fig. 18 are the predicted positions of some other, as yet unobserved, rotational levels[2].

β2) Branching ratios. Another consequence of the simple rotational coupling scheme is the validity of certain "intensity rules" governing branching ratios[3] [31]. Consider the transitions leading from a state characterized by K_i to any one of the states in the rotational band K_f. (The initial state may or may not belong to the same rotational band as the final states.) For transitions with $\lambda \leq K_i + K_f$ or if K_i or $K_f = 0$, the electromagnetic transitions satisfy the relation:

$$\frac{B(\lambda; I_i \to I_f')}{B(\lambda; I_i \to I_f)} = \frac{|\langle I_i \lambda K_i K_f - K_i | I_i \lambda I_f' K_f \rangle|^2}{|\langle I_i \lambda K_i K_f - K_i | I_i \lambda I_f K_f \rangle|^2}. \tag{35.17}$$

This may be seen to be true for $E2$ radiation Eq. (35.9) and a similar derivation shows it is true for all multipole orders. As an example, for $E2$ Coulomb

[1] G. Scharff-Goldhaber, M. Mc Keown and J. W. Mihelich: Bull. Amer. Phys. Soc., Ser. II **1**, 206 (1956).

[2] These as well as a number of other levels have been recently found by C. J. Gallagher jr., and J. O. Rasmussen.

[3] Evidence for similar intensity rules in alpha-decay has been discussed by Bohr, Froman and Mottelson, Dan. Mat.-Fys. Medd **29**, 10 (1955).

excitation starting from the Ta^{181} ground state ($I_0 = \frac{7}{2}$) we have [35]

$$\frac{B(E\,2;\,\frac{7}{2}\to\frac{11}{2})}{B(E\,2;\,\frac{7}{2}\to\frac{9}{2})} = \begin{cases} 0.26 & \text{from (35.17)} \\ 0.24 & \text{from experiment.} \end{cases} \tag{35.18}$$

In W^{182}, tbe 1222 kev state ($K=2$, $I=2$) decays to the $I=0$, 2, and 4 states of the ground state rotational band by $E\,2$ radiation. The relative values of $B(E\,2)$ for decay to these three states are calculated to be: $1:\frac{10}{7}:\frac{1}{14}$, while the empirical values are given by $1:1.6:<0.2$.

Intensity ratios also hold for beta-decays[1]; here the reduced transition probability B is replaced by the quantity $(ft)^{-1}$. As an example, in the decay of Tm^{170} (shown in Fig. 19) we have [31]:

$$\frac{''B''(1\to 0)}{''B''(1\to 2)} = \frac{ft(1\to 2)}{ft(1\to 0)} = \begin{cases} 2 & \text{from (35.17),} \\ 1.9 & \text{from experiment.} \end{cases} \tag{35.19}$$

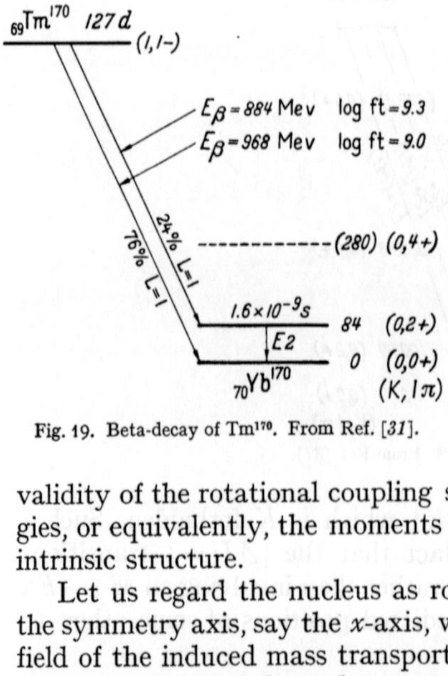

Fig. 19. Beta-decay of Tm^{170}. From Ref. [31].

Most known branching ratios agree well with the theoretical values though differences of 10% are quite common. Such differences imply a partial breakdown of the simple rotational coupling scheme. These effects are, however, small for most nuclei exhibiting rotational spectra (except for $K=\frac{1}{2}$), as is also seen by the close fit of the empirical energies to an $I(I+1)$ schemes, and by the approximate validity of K selection rules.

36. Moments of inertia. While some striking properties of rotational spectra (e.g. energy ratios) depend only on the validity of the rotational coupling scheme, the absolute values of rotational energies, or equivalently, the moments of inertia, depend upon further details of the intrinsic structure.

Let us regard the nucleus as rotating slowly about an axis perpendicular to the symmetry axis, say the x-axis, with prescribed frequency ω. Then the velocity field of the induced mass transport everywhere will be proportional to ω in first order. Its detailed dependence on position depends, of course, on the dynamics of the system. The angular momentum carried by this collective motion is given by:

$$L_{\text{Rot}} = \mathfrak{J}\,\omega. \tag{36.1}$$

where $\mathfrak{J}$ is the moment of inertia. The collective kinetic energy of rotation is:

$$T_{\text{Rot}} = \tfrac{1}{2}\mathfrak{J}\,\omega^2. \tag{36.2}$$

Thus, in general, we have

$$T_{\text{Rot}} = \frac{\hbar^2}{2\mathfrak{J}}\,L_{\text{Rot}}^2, \tag{36.3}$$

a relation not involving the rotation frequency explicitly and depending on the detailed motion only through $\mathfrak{J}$.

α) Simple models of collective rotation. Before discussing the empirical data let us consider some simple models of rotation, viz. rigid and irrotational, each

[1] Cf. the articles on beta decay in Vol. XLI of this Encyclopedia.

of which has relevance for the nuclear case. Some characteristics of the flow for each model are illustrated in Fig. 20. For rigid rotation, the particles follow the rotating orientation exactly. The moment of inertia is given by:

$$\mathfrak{J}_{\text{Rig}} = \sum_i M_i (y_i^2 + z_i^2) \approx \tfrac{2}{5} M A R_0^2 (1 + \tfrac{1}{2}\alpha \ldots) \approx \tfrac{2}{5} M A R_0^2 (1 + 0.31\beta \ldots) \tag{36.4}$$

for a spheroid, in terms of α and β defined in Sect. 34.

On the other hand, if the flow is irrotational, the collective motions take place mainly at the surface and the moment of inertia is much less than the value for rigid rotation:

$$\mathfrak{J}_{\text{Irrot}} \approx \mathfrak{J}_{\text{Rig}} \frac{9}{4} \alpha^2 = \mathfrak{J}_{\text{Rig}} \frac{45}{16\pi} \beta^2 \tag{36.5}$$

for a spheroid, to lowest order in the deformation[1].

β) Systematics of empirical data. Empirical moments of inertia as deduced from rotational energy spacings increase as we move away from closed shells. The absolute magnitudes are intermediate between the values for irrotational flow and rigid rotation [*36*]. Thus for Yb^{170}, we have:

$$\mathfrak{J}_{\text{Emp}} \approx 5.5\, \mathfrak{J}_{\text{Irrot}} \approx 0.45\, \mathfrak{J}_{\text{Rig}}. \tag{36.6}$$

The ratios of the empirical moments to the rigid rotation moments for nuclei in the rare earth region $150 < A < 190$ are plotted against N in Fig. 21. According to (33.2) the excitation energy of the first excited state in even-even nuclei is given by

$$E_2 = 3\hbar^2/\mathfrak{J} \tag{36.7}$$

while for odd-A nuclei it is:

$$E_{I_0+1} = (I_0 + 1)\, \hbar^2/\mathfrak{J} \tag{36.8}$$

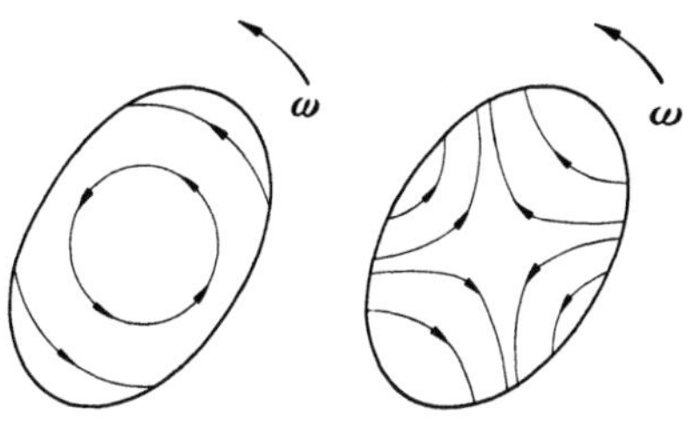

a b

Fig. 20a and b. Velocity fields for collective rotations. From Ref. [*35*]. Part (a) illustrates the rotation of a rigid body, while in part (b) the velocity field for the wave-like rotation of an irrotational fluid is shown.

except when $K = \frac{1}{2}$. It is seen that the moments of inertia, unlike quadrupole moments, are significantly larger for odd-A nuclei than for neighboring even-even nuclei. Fig. 22 shows ratios $\mathfrak{J}_{\text{Emp}}/\mathfrak{J}_{\text{Rig}}$ for even-even nuclei with $150 < A < 190$, plotted against deformation, the latter deduced in most cases from measured cross-sections for Coulomb excitation. The empirical moments increase with deformation much faster than expected for rigid rotation, but slower than for the irrotational case. Thus the ratio $\mathfrak{J}_{\text{Emp}}/\mathfrak{J}_{\text{Irrot}}$ decreases from 7 for the nuclei of smallest deformations, but still exhibiting rotational spectra, i.e. $\beta \approx 0.2$, to 3 for the largest deformations, $\beta \approx 0.45$.

Deformations of the actinide elements have not been measured as extensively as those of the rare earths but are believed to reach $\beta \approx 0.3$. Above $A = 220$ the values of $\hbar^2/2\mathfrak{J}$ rapidly decrease to the limit 7 kev for even-even nuclei (cf. Fig. 11) corresponding to $(\mathfrak{J}/\mathfrak{J}_{\text{Rig}} \approx 0.5)$, and are somewhat smaller (5 to 6 kev) for neighboring odd-A nuclei.

γ) Qualitative theoretical considerations. In early studies of collective motion in nuclei a model of irrotational flow was often used. It had been pointed out [*34*] that this assumption could well be expected to have only *qualitative* significance for nuclei. However the liquid drop model on which this idea is based, does,

[1] More detailed studies of the irrotational flow model, including the effect of higher mu tipoles in the nuclear shape, have been made by T. GUSTAFSON, Dan. Mat.-Fys. Medd. **30**, 5 (1955).

after all, account qualitatively for some collective aspects of nuclear structure (Sect. 4). In particular, it was noted that the nucleus resembles a liquid drop (as opposed to a solid body), at least in so far as the individual nucleons can freely traverse the entire volume of the nucleus. The inadequancy of the simple irrotational flow model in accounting for empirical moments of inertia[1] stimulated further studies of this problem. This has led to an increased understanding of the dynamics of the collective motion in nuclei (cf. Sect. 43).

An essential feature of a hydrodynamical description is that the mean free path of the particles involved be small compared to the dimensions of the system itself. The success of the shell model and optical model shows, on the other hand,

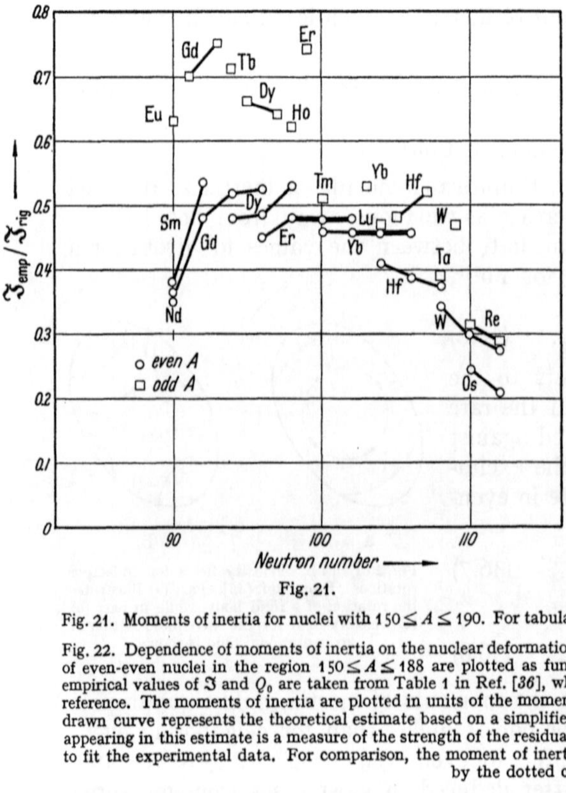

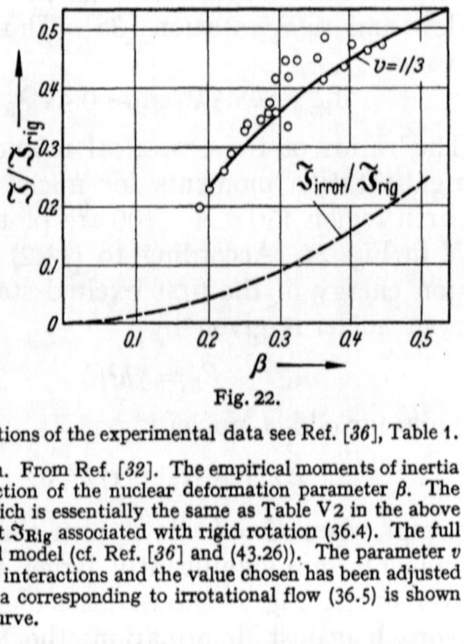

Fig. 21. Fig. 22.

Fig. 21. Moments of inertia for nuclei with $150 \leq A \leq 190$. For tabulations of the experimental data see Ref. [*36*], Table 1.

Fig. 22. Dependence of moments of inertia on the nuclear deformation. From Ref. [*32*]. The empirical moments of inertia of even-even nuclei in the region $150 \leq A \leq 188$ are plotted as function of the nuclear deformation parameter β. The empirical values of $\mathfrak{J}$ and Q_0 are taken from Table 1 in Ref. [*36*], which is essentially the same as Table V 2 in the above reference. The moments of inertia are plotted in units of the moment $\mathfrak{J}_{\mathrm{Rig}}$ associated with rigid rotation (36.4). The full drawn curve represents the theoretical estimate based on a simplified model (cf. Ref. [*36*] and (43.26)). The parameter v appearing in this estimate is a measure of the strength of the residual interactions and the value chosen has been adjusted to fit the experimental data. For comparison, the moment of inertia corresponding to irrotational flow (36.5) is shown by the dotted curve.

that the mean free path for nucleons in nuclear matter is at least comparable to the nuclear diameter; thus it is not surprising that a hydrodynamical model cannot account for details of the collective motion.

Consider now the collective motion of a system of particles moving independently in a slowly rotating potential well of arbitrary shape except that it does not have axial symmetry about the rotation axis [*36*]. (If there is such axial symmetry, e.g. if the potential is isotropic, then the field seen by the particles does not change with time and no collective motion can occur.) We assume that the number of particles is so large that a statistical approximation has validity. In this limit we may describe the state of the system classically by the density of nucleons in phase space $\varrho(\boldsymbol{r}, \boldsymbol{\varrho})$. Consider first the system in the absence of rotation. At any point, the velocity distribution is isotropic, thus the net velocity must vanish everywhere. Now let the system rotate slowly with frequency ω. In the rotating frame of reference, the equations of motion for each nucleon become

$$\ddot{\boldsymbol{r}} = -M^{-1}\nabla V - 2(\boldsymbol{\omega}\times\dot{\boldsymbol{r}}) - \boldsymbol{\omega}\times(\boldsymbol{\omega}\times\boldsymbol{r}). \tag{36.9}$$

[1] K. Ford: Phys. Rev. **95**, 1250 (1954).

The terms on the right hand side represent effects due to potential, Coriolis force, and centrifugal force respectively. The Coriolis forces disturb the motion of individual particles, but since they are proportional to $\boldsymbol{r}$, they cannot alter the isotropy of the velocity distribution, at least to first order in ω. The action of centrifugal force results in a current directed away from the axis of symmetry, but its magnitude is only of order ω^2. Thus in the rotating coordinate system the velocity distribution at any point remains isotropic, and the velocity field of the collective rotation in this reference frame vanishes up to terms linear in ω[1]. In the fixed frame of reference this motion is simple rigid rotation. Thus, although the dynamics of the above system differs from the rotation of a solid body (i.e., the individual nucleons are not fixed to certain positions in the rotating reference frame) the resulting collective flow and thus the moment of inertia is the same for the two systems[2].

Deviations of the empirical moments of inertia from the solid body value result from a combination of two effects [*36*]. First the residual interactions which give rise to correlations tend to "slow down" the collective motion and reduce the moment of inertia. If the interactions were strong enough to result in an effective nucleon mean free path very small compared to the nuclear size, we would in fact have just the conditions of the liquid drop model and the collective motion would be irrotational[3] [*40*].

Second, since the statistical limit is only approximated one expects some fluctuations in the moments associated with details of the intrinsic level scheme. Only for the special case of a pure oscillator binding potential (see Sect. 44) would we find $\mathfrak{J} = \mathfrak{J}_{\text{Rig}}$ in the absence of residual interactions[4].

The analysis of the empirical data yields interesting information regarding the above effects. This is further discussed in Sect. 43.

II. Intrinsic structure of deformed nuclei.

In the next four sections, we discuss certain aspects of the intrinsic motion of nucleons in deformed nuclei. The two main problems considered, together with some of the relevant empirical data, are (i) the origin of nuclear deformation and (ii) the effect of asphericity on the motion of individual nucleons. As in the previous sections it will be assumed that the adiabatic approximation holds, i.e. that the nucleus rotates so slowly that the intrinsic structure is not

[1] Eq. (36.9) is the same as the equation of motion of a charged particle in a magnetic field with ω corresponding to $eH/2Mc$. Consequently the absence of an induced flow in the rotating coordinate system corresponds to the Bohr-van Leeuwen theorem: A classical electron gas has no magnetic susceptibility. (J. H. VAN VLECK: The Theory of Electric and Magnetic Susceptibilities, Chap. IV. Oxford: University press 1932.)

[2] A somewhat similar problem has been considered by A. M. SESSLER and H. M. FOLEY, Phys. Rev. **96**, 366 (1954). These authors find that a Thomas-Fermi treatment of an atom with a net angular momentum leads to a net collective motion corresponding to rigid rotation.

[3] This would be only true of the interactions are short ranged—e.g. the interactions between molecules in a liquid drop. For long range interactions between particles the situation may be quite different. Thus a system of particles interacting pairwise via HOOKE's law forces (i.e. by springs) will rotate essentially as a solid body, since each particle is nearly restricted to a fixed equilibrium position in the body-fixed reference frame.

[4] It should be noted that a system of interacting particles moving in a slowly translating (rather than rotating) potential well:

$$V(\boldsymbol{r}, t) = f(\boldsymbol{r} - \boldsymbol{v}\,t) \qquad (36.10)$$

undergoes a collective translation, whose properties are independent of and details of the intrinsic structure. The velocity of the collective motion is $\boldsymbol{v}$ everywhere, thus there is no difference between solid-body motion and irrotational motion for this case.

appreciably perturbed by the Coriolis and centrifugal forces. Most all of the discussion will be restricted to spheroidal shapes, since the empirical evidence for the rare earths suggests that this is a good approximation.

In all except the last section, the nucleons will be treated as moving independently. Effects of correlation due to the residual interactions will be briefly considered in Sect. 40.

37. The spheroidal core model. The existence of large quadrupole moments and rotational spectra in certain nuclei forces one to the conclusion that these nuclei are appreciably aspherical. However, there is still the poblem of *why* these nuclei acquire a non-spherical, and, in particular, an axially symmetric shape. Since the nuclear binding is due to forces between the nucleons themselves, it is clear that the nuclear shape must depend in an important way on the nucleon configuration, i.e. on the intrinsic structure. Also, while the occurrence of rotational spectra in deformed nuclei depends only on the validity of the adiabatic condition and on the existence of axial symmetry, more specific aspects of these spectra, among them ground state spins, moments of inertia and decoupling parameters, depend upon finer details of the intrinsic motion.

The above two problems are of course closely related to each other. In this section we will consider them from a qualitative point of view, a somewhat more detailed treatment will be given in the next section.

α) General considerations. The pioneer model for interpreting the occurrence of large quadrupole moments, was given by Rainwater [*43*]. This spheroidal core model was later developed into the rotational model by Bohr and Mottelson [*33*], [*34*]. First of all we regard the nucleons as moving independently in a deformable but static potential well. For sake of simplicity, let us assume a square well and restrict ourselves to spheroidal deformation, i.e. the boundary is given by the spheroid:

$$e^{\alpha}(x^2+y^2)+e^{-2\alpha}z^2=R_0^2 \tag{37.1}$$

with $V=-V_0$ and 0 inside and outside the boundary, respectively. According to (37.1) the volume enclosed by the boundary is independent of deformation, consistent with the relative incompressibility of nuclear matter.

To first order in α, the boundary is given by:

$$R(\vartheta)=R_0\left[1+\alpha P_2(\vartheta)\right]. \tag{37.2}$$

Since the potential has axial symmetry, the component Ω_i of angular momentum along the symmetry axis for each nucleon i is a constant of the motion. However, because of the lack of spherical symmetry, the j of each nucleon is no longer a good quantum number. On the other hand, if the deformation is sufficiently small, its only effect (i.e. to first order in α) will be to break up the $2j+1$ fold degeneracy of states which would be present in a central field (with spin orbit coupling). For larger deformations, the nucleon wavefunctions can no longer be characterized by unique values of j. However, throughout the present section we will assume that j remains a good quantum number. This simplifying assumptions helps one to gain some insight into the mechanism of nuclear deformations even though it is not valid for strongly deformed nuclei. A more refined treatment is discussed in the following two sections.

The nuclear deformation results from a competition between two effects: (i) the "outer" nucleons in unfilled shells exert a centrifugal pressure on the nuclear surface. In the absence of any restoring force, the wall would acquire essentially the same strongly nonspherical shape as the density distribution of the outer nucleons. (ii) The core, i.e. all the nucleons in filled shells, have a spherical density distribution and provide a restoring force which opposes the deformation. In the spheroidal core model we treat both of these effects in a simplified way. The energy of the core is written in the form:

$$E_{\mathrm{core}}(\alpha) = E_{\mathrm{core}}(0) + \tfrac{1}{2} C \alpha^2 \tag{37.3}$$

to second order in α. The energy of the outer particles may be found in a simple way by treating the deviation from sphericity by first perturbation theory. This, of course, is valid only if the deformation is sufficiently small, i.e. if the effect of deformation on the energy is small compared to splitting between states of different j.

β) Single nucleon in unfilled shell. For a nucleon characterized by the quantum numbers j and Ω we have[1]

$$\varepsilon_{j\Omega}(\alpha) = \varepsilon_j(0) - V_0 \int\limits_{R_0}^{R(\vartheta)} |\chi^j_\Omega|^2 \, d r \tag{37.4}$$

where $\varepsilon_j(0)$ is the energy of each nucleon in the limit of spherical field ($\alpha = 0$). We now factor the single particle wavefunction as follows:

$$\chi^j_\Omega = R_{nl}(r)\, \Theta_{lj\Omega}(\vartheta, \varphi, \text{spins}). \tag{37.5}$$

The first term is the radial wavefunction. The second term is the wellknown angle and spin (j-j coupling) wavefunction—for a particle of spin $s = \frac{1}{2}$. If we neglect the variation of R with position and use Eq. (37.2), we can rewrite Eq. (37.4) as

$$\Delta \varepsilon_{j\Omega} = E_{j\Omega}(\alpha) - \varepsilon_j(0) = -\tfrac{1}{3} V_0 \alpha R^2_{nl}(R_0)\, R_0^3 \langle \Theta_{lj\Omega} | P_2(\cos\vartheta) | \Theta_{lj\Omega} \rangle . \tag{37.6}$$

The term inside the integral is easily evaluated and is given by:

$$\frac{j(j+1) - 3\Omega^2}{4j(j+1)} . \tag{37.7}$$

Thus we may rewrite (37.5) as follows:

$$\Delta \varepsilon_{j\Omega} = - c_\Omega \alpha = - c_j \alpha \left[j(j+1) - 3\Omega^2 \right]. \tag{37.8}$$

The interpretation of (37.8) is simple: If the nucleus is deformed, say into an oblate shape, ($\alpha < 0$), a nucleon will find it energetically favorable to go into an orbit whose density distribution is also oblate, especially an orbit with $|\Omega| = j$. States which differ *only* with respect to the sign of Ω, i.e. the sense of rotation about the symmetry axis, are degenerate. The total energy of core and outer particle is now given by

$$E(\alpha) = E(0) - c_\Omega \alpha + \tfrac{1}{2} C \alpha^2 . \tag{37.9}$$

The nuclear equilibrium deformation may be obtained by minimizing the energy. Thus we find:

$$\alpha_{\mathrm{eq}} = \frac{c_\Omega}{C} \tag{37.10}$$

and

$$E_{\min} = E(0) - \frac{1}{2} \frac{c_\Omega^2}{C} . \tag{37.11}$$

Thus according to the spheroidal core model, a single nucleon will always (except if $j = \frac{1}{2}$, which implies $c_\Omega = 0$) deform the nuclear shape by virtue of its non-spherical density distribution. To obtain an absolute minimum of the energy,

[1] E. Feenberg and K. C. Hammack: Phys. Rev. **81**, 285 (1951).

it is in addition necessary to choose as large as possible a value of $|c_\Omega|$. For a single particle with $j > \frac{3}{2}$, this is achieved by having $\Omega = \pm j$. The resultant intrinsic quadrupole moment of the nucleus is negative like that of the single particle but it will be much larger than the single particle quadrupole moment if the restoring term C is small.

γ) Several nucleons in unfilled shell. The above considerations are readily generalized to configurations containing several outer particles. The energy is plausibly assumed to be a sum of the single particle energies so that we merely replace c_Ω by $\Sigma\, c_\Omega$ summed over particles in the shell. Note that this sum vanishes if all states in the shell are occupied, as is expected since a filled shell always has a spherically symmetric density distribution and thus can exert no deforming influence.

For two particles outside filled shells it is energetically most favorable to put them into the paired $\Omega = +j$ and $\Omega = -j$ orbits. Thus for two particles (and indeed for any even number of particles) we have $\Omega = 0$ in the ground state. If a third particle is added, it can not go into an orbit with $|\Omega| = j$ since these are now filled, but it may have $|\Omega| = (j-1)$. In general, the configuration giving minimum energy always has all the orbits $|\Omega| = j$, $(j-1)$, $(j-2)$ ... down to a certain $|\Omega|$, or all the orbits $|\Omega| = \frac{1}{2}, \frac{3}{2}, \ldots$ up to a certain $|\Omega|$ occupied. The first kind of configuration always gives negative quadrupole moment, and according to the spheroidal core model it is favored if thes hell is less than half filled. On the other hand, for a more than half filled shell the second kind of configuration, corresponding to positive quadrupole moments, gives the lowest energy. According to this model the deformations reach maximum values in the middle of the shell, and there is a symmetry between configurations of positive and negative quadrupole moments.

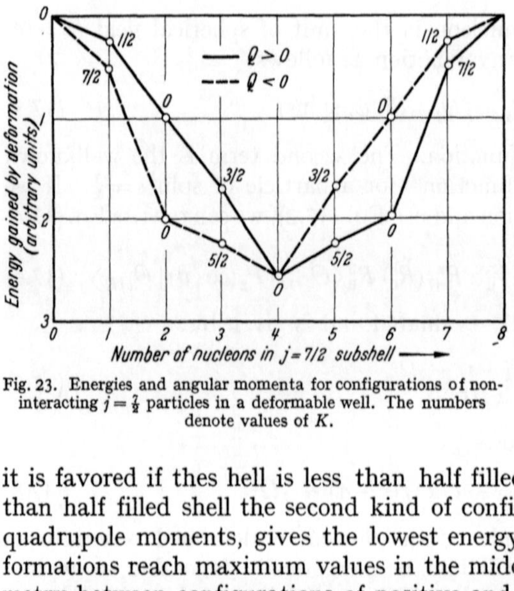

Fig. 23. Energies and angular momenta for configurations of non-interacting $j = \frac{7}{2}$ particles in a deformable well. The numbers denote values of K.

Fig. 23 shows the calculated energies and angular momenta $K (= \Omega)$ of the lowest states of positive and negative quadrupole moment for a simple case —identical particles of $j = \frac{7}{2}$. From the arguments of the previous sections the ground state spin I_0 is equal to K.

While the spherical core model provided considerable insight into the mechanism of nuclear deformation it could not account quantitatively for the empirical trends of nuclear quadrupole moments[1]. [The restoring term C was identified with the nuclear surface tension, and the constant c_j could be calculated explicitly from (37.6).] It was found that calculated quadrupole moments always are much larger than the experimental values, except for the rare earths, where the experimental quadrupole moments are exceptionally large.

To make this model more realistic we should take into account the fact that, because of the large deviations from sphericity, the j of each particle is not even approximately a good quantum number (cf. Sect. 38, 39). It is also necessary

[1] R. VAN WAGENINGEN and J. DE BOER: Physica **18**, 369 (1952). — R. VAN WAGENINGEN: Physica **19**, 1004 (1953).

to take into account terms of higher order terms in α which were neglected in the present consideration. Thus the trend of ground state angular momenta in nuclei exhibiting rotational spectra is completely different from the simple sequence shown in Fig. 23. On the other hand for some odd-A nuclei not too far from closed shells, the spheroidal core model does in fact predict the correct ground state spins. These nuclei are probably not strongly deformed so that it may be at least qualitatively reasonable to ascribe to each nucleon a definite j.

Thus consider some of the nuclei with odd nucleon number between 20 and 28, i.e. filling up the $f\,\frac{7}{2}$ shell[1]. The detailed interpretation of the level schemes in these nuclei by means of the shell model may be fairly complicated. However, at least the ground state I_0 in some of these nuclei can be accounted for with the simple spheroidal core model. Ground states with $I_0 = j - 1$ would usually be expected to occur when the even-particle configuration is relative far away from closed shells. Thus both ${}_{22}\mathrm{Ti}^{47}_{25}$ and ${}_{25}\mathrm{Mn}^{55}_{30}$ have ground state spins $I_0 = j - 1 = \frac{5}{2}$. On the other hand, if the even particles form a closed shell configuration, then pairing effects will dominate (cf. Sect. 40) and we obtain $I_0 = j$. This is exemplified by the nuclei ${}_{20}\mathrm{Ca}^{43}_{23}$ and ${}_{23}\mathrm{V}^{51}_{28}$, for both of which $I_0 = j = \frac{7}{2}$.

38. Anisotropic harmonic oscillator potential. To study some important aspects of nucleon motion in deformed nuclei it is very useful to approximate the potential by an anisotropic harmonic oscillator potential. This type of potential has the great advantage of being soluble exactly by separation of variables. On the other hand, it comes fairly close to representing the actual nuclear potential, and in any case, the difference can be approximately corrected for, as will be seen in Sect. 39. As before we restrict ourselves to spheroidal deformations and specify that the nuclear volume be independent of deformation. Accordingly we choose an anisotropic oscillator potential of the form[2]:

$$V(\boldsymbol{r}) = \tfrac{1}{2} m\, \omega_0^2 \left[\mathrm{e}^{\alpha}(x^2 + y^2) + \mathrm{e}^{-2\alpha} z^2\right]. \tag{38.1}$$

It is seen that any surface of given constant V is a spheroid with major and minor axis proportional to e^{α} and $\mathrm{e}^{-\alpha/2}$ respectively, and with an enclosed volume independent of α. The waveequation for a particle moving in this potential can be solved exactly by separation of variables in either Cartesian or cylindrical coordinates.

α) Single particle levels. In Cartesian coordinates, the problem is equivalent to that of three independent oscillators:

$$\chi_{n_1, n_2, n_3} = u_{n_1}\left(\mathrm{e}^{\frac{\alpha}{4}} x\right) u_{n_2}\left(\mathrm{e}^{\frac{\alpha}{4}} y\right) u_{n_3}\left(\mathrm{e}^{-\frac{\alpha}{2}} z\right) \tag{38.2}$$

where each of the functions u_n is the solution of the one dimensional problem, n denoting the number of oscillator quanta along the appropriate axis. The single particle energies are given by:

$$\varepsilon_{n_1, n_2, n_3} = \hbar\,\omega_0 \left[(n_1 + n_2 + 1)\,\mathrm{e}^{\frac{\alpha}{2}} + (n_3 + \tfrac{1}{2})\,\mathrm{e}^{-\alpha}\right]. \tag{38.3}$$

For an anisotropic oscillator potential the l or j of each nucleon is not even approximately a good number. However, as long as the potential is axially

---

[1] For a summary of the empirical level schemes in these nuclei see R. H. Nussbaum, Rev. Mod. Phys. **28**, 423 (1956).

[2] D. Pfirsch: Z. Physik **132**, 409 (1952). — S. Gallone and C. Salvetti: Nuovo Cim. **10**, 145 (1953).

symmetric, there are special degeneracies associated with the fact that the energy depends only upon the quantity

$$n_1 + n_2 = n_\perp = N - n_3 \tag{38.4}$$

the number of oscillator quanta perpendicular to the symmetry axis, rather than upon n_1 and n_2 separately. (N is the principal quantum number.)

To take better advantage of the axial symmetry it is convenient to separate the wave equation in cylindrical coordinates (ϱ, z, φ). The resulting single particle wavefunctions which are, of course, linear combinations of the Cartesian functions (38.2), can be written as:

$$\left.\begin{aligned}\chi_{N,n_3,\Lambda} &= \mathfrak{R}_{n,\Lambda}\left(e^{\frac{\alpha}{4}}\varrho\right)\times\\ &\times u_{n_3}\left(e^{-\frac{\alpha}{2}}z\right)e^{i\Lambda\varphi}\end{aligned}\right\} \tag{38.5}$$

i.e., they are eigenfunctions of the operator l_3, the component of orbital angular momentum parallel to the symmetry axis, with eigenvalues Λ. The $\mathfrak{R}$ functions are solutions of the 2 dimensional harmonic oscillator problem. The single particle energies now have the form:

$$\left.\begin{aligned}\varepsilon_{N,n_3,\Lambda} = \hbar\,\omega_0\,[(N - n_3 + 1)\,e^{\alpha/2} +\\ + (n_3 + \tfrac{1}{2})\,e^{-\alpha}].\end{aligned}\right\} \tag{38.6}$$

The energy does not depend on Λ, which can take on the values $\pm(N - n_3)$, $\pm(N - n_3 - 2)$, ± 1 or 0, nor, of course, on the direction of spin ($\Sigma = \pm\frac{1}{2}$).

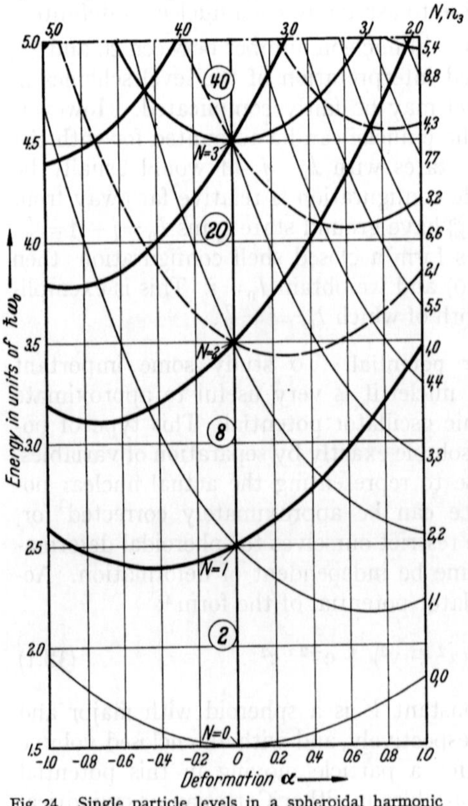

Fig. 24. Single particle levels in a spheroidal harmonic oscillator potential as function of deformation. The energies are calculated from Eq. (38.6), and each level is characterized by the pair of quantum numbers N and n_3. Also indicated are the magic numbers associated with an isotropic harmonic oscillator potential.

The separation of the nucleon motion into components parallel and perpendicular to the 3-axis implies the existence of special selection rules for certain processes[1]. Thus the electric dipole matrix elements of the form $\langle\chi_f|r_\nu|\chi_i\rangle$. vanish unless either

$$\Delta n_3 = \pm 1\,, \quad \Delta\Lambda = \Delta n_\perp = 0\,, \qquad \Delta N = \pm 1 \tag{38.7a}$$

or

$$\Delta n_3 = 0\,, \qquad \Delta\Lambda = \pm 1, \Delta n_\perp = \pm 1\,, \quad \Delta N = \pm 1\,. \tag{38.7b}$$

Since α is usually rather small compared to unity for cases of interest, it is satisfactory to expand the energy powers of α and keep only the lowest terms. We then obtain the following approximation:

$$\varepsilon_{N,n_3,\Lambda} = \hbar\,\omega_0\left[\left(1 + \frac{\alpha^2}{4}\right)\left(N + \frac{3}{2}\right) - \frac{\alpha}{2}\,(3\,n_3 - N)\right]. \tag{38.8}$$

Single particle energy levels for $N = 0$ to 3 plotted against deformation are shown in Fig. 24.

[1] G. Alaga: Phys. Rev. **100**, 432 (1956).

β) The deformation potential. Consider now the behavior of intrinsic energy with deformation. For each value of the deformation α the "ground state" intrinsic wavefunction (an antisymmetrized product of single particle functions) is obtained by filling up nucleon orbits in order of increasing energy consistent with the Pauli principle. The "ground state intrinsic energy" $E_L(\alpha)$ is essentially the sum of the single particle energies (apart from corrections due to residual interactions, cf. Sect. 40, and a correction required since we are counting the basic two-body interactions twice, cf. the remarks at the end of Sect. 39). As will be seen in Sects. 41 and 42, E_L is the potential energy of the collective motion which we call the "deformation potential". Fig. 25 shows the calculated energies E_L vs. deformation for the cases of 20, 24, 30 and 36 identical nucleons. In the limit of zero deformation, the first three ($N = 0$, 1, 2) shells are completely filled and the $N = 3$ shell is in various stages of being filled. Because of the extensive crossing of single particle levels, especially for large deformations, each of the curves shown in Fig. 25 is scalloped rather than smooth[1] [37]. Each segment of the curve corresponds to a fixed set of quantum numbers N, n_3 for the individual particles. Its energy is given in terms of the deformation by:

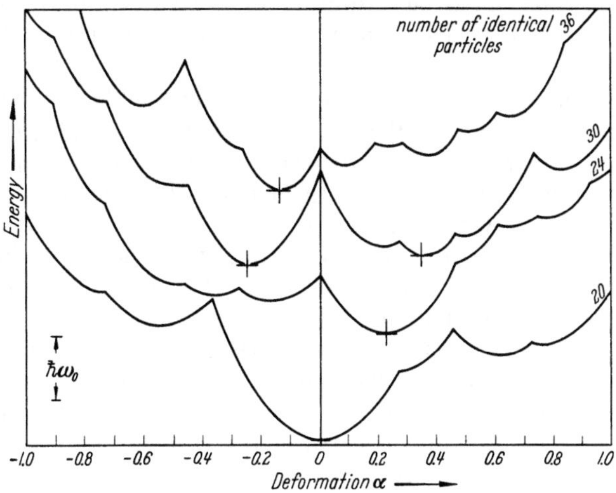

Fig. 25. Deformation potentials for various stages of shell filling-spheroidal harmonic oscillator binding potential.

$$E(\alpha) = E(0)\left[1 - \tfrac{1}{2}\alpha\,\alpha_{\mathrm{eq}} + \tfrac{1}{4}\alpha^2\right] \tag{38.9}$$

where α_{eq}, the equilibrium deformation resulting in minimum energy is:

$$\alpha_{\mathrm{eq}} = \frac{\sum_i (3n_3 - N)_i}{\sum_i (N + \frac{3}{2})_i}. \tag{38.10}$$

The sum in the numerator with respect to all the nucleons in a filled shell vanishes. Thus, as expected, only the nucleons in unfilled shells can give rise to a deformation. Discontinuities in the shape of the curve represent transitions of individual nucleons from one orbit to another as these become energetically favored. Note that there are usually several deformations for which the energy has a minimum. For some cases, i.e. 30 particles, the energy difference between these minima is quite small, and not much perturbation would be required to alter the ordering. This always occurs for the oscillator potential if the last shell is about half full as it is for 30 particles. Further calculations show that little or no energy can usually be gained by going to ellipsoidal shapes with all three principal axes of different lengths. As a shell begins to fill, positive quadrupole moments are favored, while in the second half of the filling, the shape giving lowest energy

[1] S. GALLONE and C. SALVETTI: Nuovo Cim. **10**, 145 (1953).

has negative quadrupole moment. Superficially, at least, this trend of deformations with nucleon number is just the opposite of the trend expected from the simple spheroidal core model (Sect. 37). Actually the situation is more complicated. In the spheroidal core model we have assumed that the j of each particle is a good quantum number, i.e. we fill the subshells (given l and j) *one* at a time. In the oscillator model, there is no subshell structure at all (because l and j are no longer good quantum numbers); however the distinction between different major shells (given N) remains. Thus the number of deforming particles can be much larger in the oscillator model than in the spheroidal core model. On the other hand, according to the oscillator model the restoring force exerted by the closed shells is found to be much larger than the classical surface tension term (proportional to A rather than to $A^{\frac{2}{3}}$) so that the deformations calculated using the two models turn out to be roughly of the same order of magnitude.

According to the oscillator model, the spherical shape is expected to be stable with respect to deformation only when the nucleon configuration consists entirely of filled shells. In fact, for the cases (e.g. 20 particles), the energy increases very strongly as we move away from spherical shape. No "cross-overs" occur for small deformations and other energy minima are at least about 1 $\hbar\omega_0$ unit above the ground state value. For other configurations, the energy (averaged over the fluctuations) increases rather slowly with increasing deformation due to crossovers between orbits. Thus the "average curvature" $(d^2 E_L/d\alpha^2)$ is small compared to the curvature of an individual segment. Strictly speaking, any crossovers would be forbidden for a pure harmonic oscillator potential since there is no coupling between the crossing levels. However, even weak perturbations, e.g. residual interactions, deviations from axial symmetry etc., make it possible for the system to remain in the lowest intrinsic state, provided the deformation changes sufficiently slowly.

γ) Relation between density and potential. The harmonic oscillator potential is useful for providing insight into the relation between the non-spherical nuclear potential and the anisotropy of the density distribution. We define the anisotropy of the density as follows [*40*]:

$$\langle\alpha^\varrho\rangle = \frac{\left\langle\sum_i (3z^2 - r^2)_i\right\rangle}{\left\langle\sum_i r_i^2\right\rangle} \tag{38.11}$$

where the brackets denote expectation values with respect to the intrinsic wavefunction. In the statistical limit of very many particles, the total density at each point is a function of V alone. It is readily shown that in this limit we always have

$$\langle\alpha^\varrho\rangle = \alpha \tag{38.12}$$

to first order in α, provided only that the surfaces of constant V are spheroids and independent of the specific form of the potential.

Of course, the existence of a shell structure is contrary to the statistical approximation. However, for the case of an oscillator potential, there is a relation similar to Eq. (38.12). Consider the quantity $dE_L/d\alpha$. We have

$$\frac{dE_L}{d\alpha} = \frac{d}{d\alpha}\langle\chi_L|\,H(\alpha)\,|\chi_L\rangle = \langle\chi_L|\,\frac{dH(\alpha)}{d\alpha}\,|\chi_L\rangle + \left[\left\langle\frac{d\chi_L}{d\alpha}\right|H\left|\chi_L\right\rangle + \text{c.c.}\right] \tag{38.13}$$

where c.c. denotes complex conjugate.

The sum of the last two terms is equal to $E_L \frac{d}{d\alpha} \langle \chi_L | \chi_L \rangle$ and vanishes. Using (38.1) we have:

$$\frac{dH}{d\alpha} = \frac{d}{d\alpha} \sum_i V_i(\alpha) = m\,\omega_0^2 \sum_i \left[e^{\alpha}(x^2 + y^2) - 2e^{-2\alpha} z^2 \right]_i \tag{38.14}$$

which in the limit of small α reduces to

$$-m\,\omega_0^2 \sum_i (3z^2 - r^2)_i . \tag{38.15}$$

Using (38.11) and (38.13) we then find:

$$\frac{dE_L}{d\alpha} = m\,\omega_0^2 \left\langle \sum_i r_i^2 \right\rangle \left[\alpha - \langle \alpha^{\varrho} \rangle\right]. \tag{38.16}$$

According to (38.13), this equation is also valid even if the nuclear Hamiltonian contains additional terms, e.g. the terms introduced in the next section, or residual pairing interactions, provided only that these terms do not involve α explicitly. Thus we see that the anisotropies of density and potential are equal [(38.12) holds] at equilibrium, a reasonable result. Of course, the detailed spatial distribution of the density may deviate considerably from the statistical value.

Now suppose that the deformation changes with time, but so slowly that the nucleons can follow the change adiabatically. As we move along any of the segments shown in Fig. 25, the quantum numbers characterizing the states of the individual nucleons remain constant. As may be seen from (38.2) or (38.5), the wavefunction of each nucleon undergoes a simple linear transformation, a partial adjustment to the changing deformation, which we define here as „distortion". Note that the exponents appearing in Eq. (38.5) for the wavefunctions are only half as large as the corresponding exponents in (38.6). As a result the anisotropy of the density distribution changes only half as fast as the anisotropy of the potential, provided the nucleons stay in orbits of definite N and n_3— (no cross-overs). If the potential, assumed to be spheroidal at equilibrium becomes spherical, the anisotropy of the density is still half as large as it was at equilibrium. However, the density distribution of the nucleons in filled shells is now isotropic so that the residual anisotropy *must* be due to the relatively few nucleons in the last unfilled shell. Thus the bulk of the nucleons in filled shells contribute altogether only about half of the intrinsic nuclear quadrupole moment by virtue of the collective distortion of their wavefunctions, and the relatively few nucleons in the last unfilled shell account for the other half.

According to Fig. 25, if the intrinsic structure is to remain in its *lowest* state over fairly wide changes in the deformation, it is necessary that cross-overs, i.e. transitions of individual loosely bound nucleons from one orbit to another, occur. Thus at each cross-over point, the quantity $dE_L/d\alpha$ and thus $\alpha - \langle \alpha^{\varrho} \rangle$ are discontinuous. We describe this kind of adjustment—large changes in the wavefunctions of only a few loosely bound nucleons—as "mixing" since it somewhat resembles the configurational mixing in the shell model. If residual interactions are included, the mixing process is more complicated than a simple crossover (cf. Sect. 40). However, in any case the combined effect of distortion and mixing is to keep $\langle \alpha^{\varrho} \rangle$ nearly equal to α for all deformations. In this way, the slope $dE_L/d\alpha$ never becomes very large and the intrinsic energy does not depend very strongly on deformation (except, of course, for local fluctuations). Strictly speaking, the distinction between "distortion" and "mixing" is valid only for an oscillator potential; however it is expected to hold, at least in an approximate sense, regardless, of the detailed form of the nuclear potential.

δ) Ellipsoidal deformations. The above considerations can be readily extended to take into account ellipsoidal deformations. The potential (38.1) may then be generalized to read:

$$V(r) = \frac{1}{2} m \omega_0^2 r^2 \left[1 + \frac{5}{4\pi} \sum_\mu |\alpha_\mu|^2 - 2 \sum_\mu (\alpha_\mu Y_{2\mu})\right] \tag{38.17}$$

for small deformations, and the equipotential surfaces are now given by:

$$r(\vartheta, \varphi) = \text{const}\left[1 + \sum_\mu \alpha_\mu^* Y_{2\mu} - \frac{1}{4\pi} \sum_\mu |\alpha_\mu|^2\right]. \tag{38.18}$$

The sum over μ extends from -2 to 2 corresponding to the 5 parameters which characterize an ellipsoid of given volume. The multipoles of the density now are of the form

$$\langle \alpha_\mu^\varrho \rangle = \frac{4\pi}{5} \frac{\left\langle \sum_i (r^2 Y_{2\mu})_i \right\rangle}{\left\langle \sum_i r_i^2 \right\rangle}, \tag{38.19}$$

a generalization of (38.11), which guarantees that:

$$\langle \alpha_\mu^\varrho \rangle = \alpha_\mu \tag{38.20}$$

in the statistical limit.

The previous results then hold without essential modification. Thus the self consistency condition (38.20) holds quite generally at equilibrium, and the remarks concerning the roles of distortion and mixing are still applicable.

39. Realistic nuclear potential. The potential seen by an individual nucleon is essentially flat inside the nucleus and falls to zero rather sharply at the surface. Thus the exact calculation of single particle wavefunctions and energy levels in a deformed nucleus requires the solution of the Schroedinger equation in a diffuse-edged non-spherical well, an exceedingly difficult problem even if we restrict ourselves to spheroidal shapes.

α) Spheroidal box potential. To obtain further insight regarding the motion of nucleons in the nonspherical nuclear potential let us first consider a spheroidal box potential.

$$\left.\begin{array}{l} V = 0 \\ V = \infty \end{array}\right\} \text{if} \quad e^{\alpha}(x^2 + y^2) + e^{-2\alpha} z^2 \left\{\begin{array}{l} \le R_0^2, \\ > R_0^2. \end{array}\right. \tag{39.1}$$

The level scheme in a diffuse potential is expected to be roughly intermediate between those of the box and harmonic oscillator potentials. For small nucleon numbers, say $N < 40$, all three schemes are known to be rather similar, at least in the spherical limit.

We assume, as in the previous section, that (i) the depth of the potential is independent of deformation, (ii) the volume enclosed by the box is independent of deformation, and (iii) the total intrinsic energy E_L is the sum of single particle energies. The Schroedinger equation for the spheroidal box potential has been solved both by separation in spheroidal coordinates[1], and by a technique which treats the deviation from sphericity as a perturbation[2]. As is seen in Fig. 26, the trend of the calculated equilibrium deformations as function of nucleon number is quite different from that for the harmonic oscillator. The calculations

[1] R. D. SPENCE: Phys. Rev. **83**, 460 (1951). — D. L. HILL and J. A. WHEELER: To be published.

[2] S. A. MOSZKOWSKI: Phys. Rev. **99**, 803 (1955).

show that the symmetry between states of positive and negative Q_0 which was present for the harmonic oscillator potential is now upset and the states giving lowest energy now always have positive quadrupole moments except near the beginning and end of a shell where they are very small. Indeed, nearly all measured quadrupole moments of odd-A are positive, except for nuclei just above magic numbers[1].

Calculations [*42*] also indicate that the maximum deformation occurs somewhat before the last shell is half-filled. The experimental data support this conclusion. For example, in the rare earth region, the maximum value of Q_0 is attained for ${}_{64}Gd_{96}^{160}$. It is interesting to note that this maximum occurs slightly before the major shell ($Z=51$ to 82, $N=83$ to 126) is half filled. There is also evidence of similar behavior in the actinides and in the deformed nuclei with $A\approx 25$.

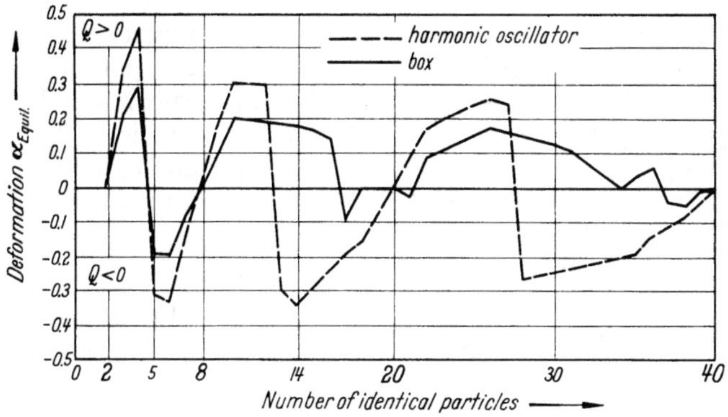

Fig. 26. Equilibrium deformations for configurations of non-interacting particles in deformable wells. The equilibrium deformation α_{eq} corresponds to the absolute minimum of the deformation potential for the appropriate number of particles. We have considered only the two minima corresponding to the *smallest* values of positive and of negative α, i.e. we assume that there is no crossover between $\alpha=0$ and $\alpha=\alpha_{eq}$. For the binding potentials considered in this figure, there are only a few exceptions to this but for a more realistic nuclear potential neither of the first minima on either side of $\alpha=0$ gives the lowest energy.

For nucleon numbers above 40, not shown in Fig. 26, the level scheme of the box approaches the statistical form, and little or no shell structure remains. For very large N, the deformation potential fluctuates little about the statistical limit[2]:

$$E(\alpha)=E(0)+\tfrac{1}{2}C\alpha^2+0(\alpha^3)\ldots \tag{39.2}$$

and calculated equilibrium deformations become small although they remain predominantly prolate.

We should remark here that the equilibrium deformations of actual nuclei are expected to be roughly equal to the number of particles (or holes, whichever is less) in the last unfilled shell divided by the total number of nucleons. The maximum number of particles in the unfilled shell is roughly proportional to $A^{\frac{2}{3}}$, so that equilibrium deformations are expected to be approximately proportional as $A^{-\frac{1}{3}}$, in addition to their variation as function of shell filling. Such a trend is known empirically. The deformations in the rare earths reach a maximum $\beta\approx 0.45$ while in the actinides they probably do not exceed $\beta\approx 0.35$.

[1] S. A. Moszkowski and C. H. Townes: Phys. Rev. **93**, 306 (1954).

[2] According to both the statistical (Fermi gas) and the liquid drop models, the energy depends upon deformation only by virtue of the surface area and the Coulomb energy (see Sect. 4 and 10). Thus both models predict the same dependence of total energy upon deformation, even though the basic assumptions underlying the models are quite different—indeed contradictory (cf. Table 1). It is essentially because of this circumstance that the liquid drop model proved so successful in accounting for important features of the nuclear fission process [*13*].

β) The Nilsson potential. To account for detailed features of heavy strongly deformed nuclei, it is evidently necessary to consider the nucleon motion in a diffuse well potential with strong spin orbit coupling. Such a potential can be simulated quite well by starting with an anisotropic oscillator and adding terms of the form $D\boldsymbol{l}\cdot\boldsymbol{l}$ favoring large l values (due to the flatness of the well) and $C\boldsymbol{l}\cdot\boldsymbol{s}$, a simple form for the spin orbit coupling. The constants C and D are adjusted so as to reproduce the observed level order in spherical nuclei (Sect. 28). Extensive calculations with this simulated potential have been made by NILSSON [42].

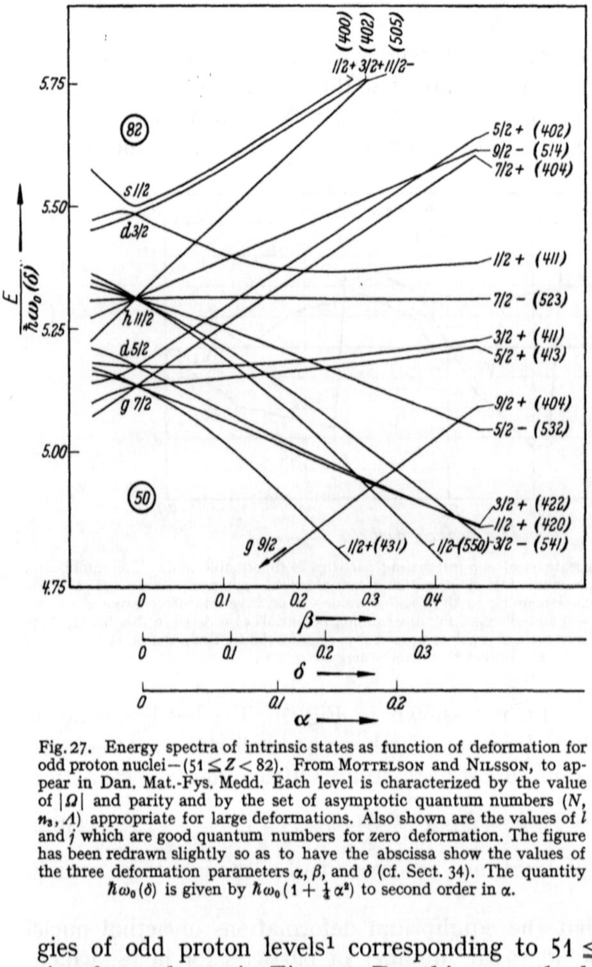

Fig. 27. Energy spectra of intrinsic states as function of deformation for odd proton nuclei—($51 \leq Z < 82$). From MOTTELSON and NILSSON, to appear in Dan. Mat.-Fys. Medd. Each level is characterized by the value of $|\Omega|$ and parity and by the set of asymptotic quantum numbers (N, n_3, Λ) appropriate for large deformations. Also shown are the values of l and j which are good quantum numbers for zero deformation. The figure has been redrawn slightly so as to have the abscissa show the values of the three deformation parameters α, β, and δ (cf. Sect. 34). The quantity $\hbar\omega_0(\delta)$ is given by $\hbar\omega_0(1+\frac{1}{4}\alpha^2)$ to second order in α.

Energy levels and wavefunctions of single particles in this potential can be readily calculated by matrix mechanics. To a very good approximation the Hamiltonian can be written in the form:

$$\left.\begin{aligned} \frac{H}{\hbar\omega_0} &= \left(1+\frac{\alpha^2}{4}\right)\times \\ &\times\left(N+\frac{3}{2}\right)- \\ &-\alpha\frac{m\omega_0}{\hbar}r^2 P_2(\vartheta)+ \\ &+D\boldsymbol{l}\cdot\boldsymbol{l}+C\boldsymbol{l}\cdot\boldsymbol{s} \end{aligned}\right\} \qquad (39.3)$$

where it is further understood that only the coupling between states of the *same* principal quantum number N is taken into account. In the absence of $\boldsymbol{l}\cdot\boldsymbol{l}$ and $\boldsymbol{l}\cdot\boldsymbol{s}$ terms, the resulting single particle energies would be identical with those given by (38.8). The calculated energies of odd proton levels[1] corresponding to $51 \leq Z < 82$ and prolate deformations[2] are shown in Fig. 27. For this case, the best fit to the level ordering in spherical nuclei is obtained by setting $C = -0.10$ and $D = -0.0275$.

In the limit of very small deformation, the single particle states are simultaneous eigenfunctions of N, l, j and Ω as in the spheroidal core model (Sect. 37). As the deformation increases, neither l nor j remain good quantum numbers, and the level scheme becomes complicated, but there soon appears an approximate separation of the nucleon motion into components parallel and perpendicular

[1] Somewhat similar calculations have been made by K. GOTTFRIED, Phys. Rev. **103**, 1017 (1956) and by M. RICH, University of California Radiation Laboratory Report UCRL 3587.

[2] Calculations show that in the first half of a major shell, the single particle states of positive Q_0 contain, on the average, a larger fraction of high angular momentum components than states of negative Q_0. Thus it is not surprising that the addition of the negative $\boldsymbol{l}\cdot\boldsymbol{l}$ term to the oscillator potential leads to the observed preponderance of positive quadrupole moments.

to the symmetry axis, as for a pure spheroidal oscillator potential. In the limit of large deformation, the states are characterized by the quantum numbers N, n_3, Λ, and Ω associated with the pure oscillator potential. Although the deformation causes much interweaving between states of the same major shell, there is usually rather little interaction between different major shells. Thus levels which would correspond to $Z > 82$ in the spherical limit do not seem to make their appearance in nuclei with $50 < Z < 82$ at least for the lowest levels.

To calculate the intrinsic excitation spectrum for nuclei one must also know the equilibrium deformation. This may be obtained from measured $E2$ transition rates or spectroscopic quadrupole moments (cf. Fig. 17).

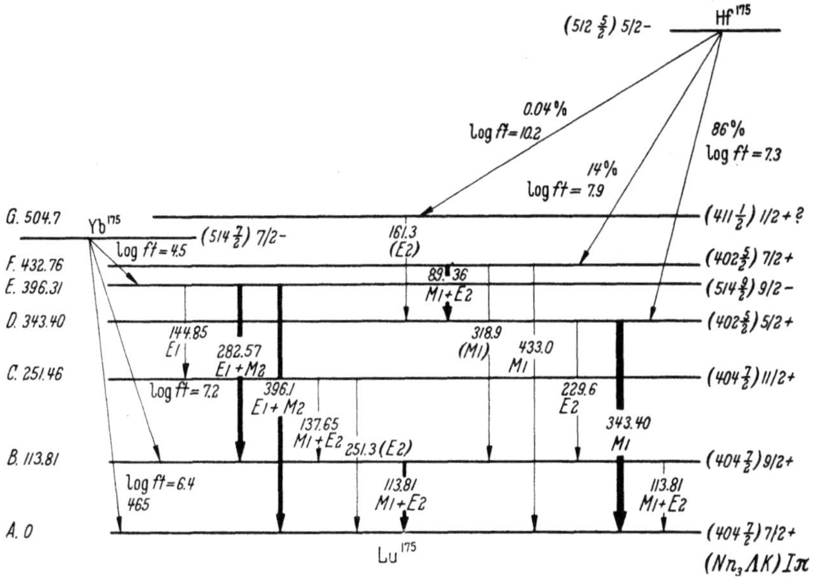

Fig. 28. Level spectrum of Lu^{175}. From HATCH, BOEHM, MARMIER and DUMOND: Phys. Rev. **104**, 745 (1956). The figure has been redrawn slightly so as to exhibit the asymptotic quantum numbers $N\, n_3\, \Lambda$ appropriate to each state.

γ) The level spectrum of Lu^{175}. Since nucleons fill orbits pairwise, the ground state of an odd A nucleus should be essentially just the state of the last unpaired nucleon[1]. To illustrate this consider the nucleus ${}_{71}Lu^{175}_{104}$ [2] which is known to have a ground state deformation of $\beta \sim 0.35$. At this deformation the 71st proton should go into the $^7/_2{}^+$ orbit characterized by the "asymptotic quantum numbers" ($N=4$, $n_3=0$, $\Lambda=4$) or 404 $^7/_2{}^+$ in abbreviated notation. Indeed the ground state spin is measured as $^7/_2$, an assignment supported by the energy ratio $E_2/E_1 = 2.21$ [Eq. (33.4)] of the first two rotational states. For strongly deformed nuclei the single particle wavefunctions are much closer to the eigenfunctions of the anisotropic pure harmonic oscillator potential than to the eigenfunctions of the spherical potential. Thus the single particle level ordering in deformed nuclei differs markedly from the ordering in spherical nuclei.

Consider now the excitation spectrum of Lu^{175} shown in Fig. 28. Besides the rotational states, we note the existence of low-lying intrinsic states with assignments $^5/_2{}^+$, $^9/_2{}^+$ and $^1/_2{}^+$. These states are precisely the ones expected theoretically the first two by exciting the last odd proton, the last by transferring another proton from $^1/_2{}^+$ to $^7/_2{}^+$ orbits. The ground state assignments of Yb^{175} and Hf^{175}

[1] Ref. [*41*]; B. R. MOTTELSON and S. G. NILSSON: Dan. Mat.-Fys. Medd. (in press).

[2] HATCH, BOEHM, MARMIER and DU MOND: Phys. Rev. **104**, 745 (1956).

are required by experimental evidence and are consistent with theoretical predictions. (The level ordering in odd-N nuclei with $82 < N < 126$ has also been given by Mottelson and Nilsson [*41*]).

The Lu^{175} level scheme illustrates the operation of selection rules regarding the "asymptotic quantum numbers"[1]. First of all, the allowed beta transition from the Yb ground state to the 396 Mev state in Lu has an anomalously low $\log ft$ value of 4.5 as compared to 5 or 5.5 for most allowed beta transitions. This is evidently related to the large overlap between the initial and final wavefunctions implied by their identical asymptotic quantum numbers. The decay of Yb to the Lu ground state rotational band is first forbidden with typical $\log ft$ values. The decay to the $^5/_2{}^+$ band is also first forbidden and thus should be somewhat weaker than the decay to the ground state because of the smaller available energy. On the other hand the asymptotic quantum number must change by two units. If the intrinsic motion were completely separated into components parallel and perpendicular to the symmetry axis, then this beta-transition should, in fact, be 3rd forbidden instead of 1st forbidden so that the selection rules strictly applied give a hindrance factor of about 10^{10} for the decay rate. In fact neither this transition nor the K capture decay of Hf to the Lu ground state band is observed. (The latter transitions would be favored over the decay to the $^5/_2{}^+$ band on the basis of energy considerations alone.)

In a number of other nuclei, the hindered beta transitions are observed and their decay rates imply that the reduction factors are usually of order 10^2 suggesting an approximate rather than complete separation of the intrinsic motion into components parallel to and perpendicular to the symmetry axis. Of course, the separation is not expected to be complete since the actual nuclei potential differs somewhat from the pure oscillator potential (i.e. the last two terms in Eq. (39.3)].

Another striking feature shown in the level scheme concerns the gamma decay of the 396 kev state of Lu^{175}. Although this state can decay to the ground state by $E1$ radiation, the observed decay proceeds by a mixture of $E1$ and $M2$ modes. This implies a very substantial reduction of the $E1$ transition rate below the single particle strength. Indeed measured lifetimes of some wother states which decay by pure $E1$ radiation imply that hindrance factors for $E1$ transitions between low-lying states are frequently as large as 10^6 and never smaller than 10^3 [2]. Asymptotic selection rules for $E1$ transitions have been given in (38.7). One or both of these rules is *always* violated for transitions between close-lying states and the reduction in transition probability is usually even greater than for beta transitions[3].

δ) Remarks on other nuclei. Turning now to a consideration of intrinsic level schemes of other odd-Z odd-A nuclei, shown in Fig. 29, we se that there is in general good agreement with the Nilsson scheme, at least regarding the level ordering thought not necessarily with respect to absolute energies. The scheme has also proven quite successful for classifying intrinsic states in the actinide region[4]. However the interpretations are usually less clear cut than for the lighter elements because of the larger number of available orbits.

[1] G. Alaga: Phys. Rev. **100**, 432 (1955).
[2] O. Strominger and J. O. Rasmussen: Nucl. Phys. **3**, 197 (1957).
[3] It is interesting to note that low-lying $E1$ transitions do not occur in spherical nuclei either, since the required pairs of single particle states (different parity and $\Delta j = 0$ or ± 1) are always well separated in energy.
[4] B. R. Mottelson and S. G. Nilsson: Dan. Mat.-Fys. Medd. (in press).

Decoupling factors [Eq. (33.6)] for states with $K=\frac{1}{2}$ may be calculated in terms of the intrinsic structure by Eq. (32.16). The value of a is usually between -1 and 1, implying no inversion of the rotational level ordering (Fig. 16) except when $n_3 = N$. The sign of a is usually given by $(-1)^{n_3}$. Thus for the ground state of Tm^{169} characterized by the quantum numbers $411\ {}^1/_2{}^+$, we have $a = -0.77$, while for the $330\ {}^1/_2{}^+$ band of Al^{25} the decoupling factor is -3.2. Calculated values of decoupling factors are in good agreement with the values deduced from rotational energies.

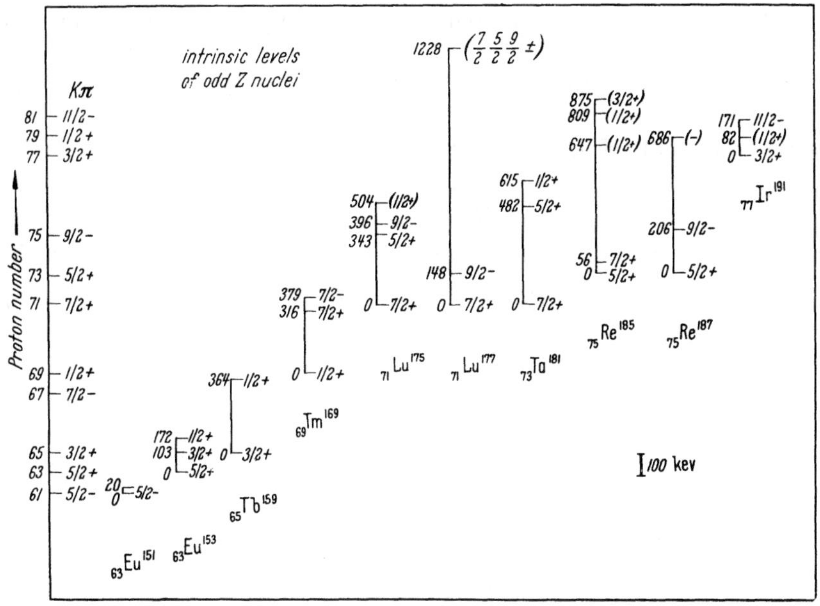

Fig. 29. Intrinsic levels of odd proton nuclei—($63 \leq Z \leq 77$). From HATCH, BOEHM, MARMIER and DUMOND: Phys. Rev. **104**, 745 (1956). The levels are compared with the calculated proton level scheme (Fig. 27) assuming a deformation parameter $\delta = 0.28$ ($\beta \approx 0.35$).

ε) Two-body forces and the deformation potential. In the previous calculations, the total intrinsic energy has been taken simply as the sum of the single particle energies. Such a procedure might be critized on the grounds that in this way we are counting the basic two-body interactions twice [*42*]. This problem was already discussed in Sect. 29 and we showed there that the above criticism is justified only as far as *absolute* energies are concerned. Now in the unified model we are interested not in absolute energies but in wavefunctions and *energy differences,* i.e. excitation spectra and the *variation* of the total intrinsic energy with deformation. It was shown in Sect. 29, that wavefunctions and excitation energies are determined essentially by a Hamiltonian of the form $\sum_i (T_i + V_i)$ apart from residual interactions.

To make it plausible that the variation of energy with deformation is also given by this Hamiltonian we might argue as follows: For a given shape of the nucleus, the potential seen by nucleon i can be written as[1]

$$V_i(\alpha, \boldsymbol{r}_i) = \sum_k \int \varrho_k(\alpha, \boldsymbol{r}_k)\, V(\boldsymbol{r}_i - \boldsymbol{r}_k)\, d^3 r_k \qquad (39.4)$$

[1] We ignore any exchange effects (e.g. due to either anti-symmetrization or exchange forces), in this qualitative treatment and also assume the validity of first order perturbation theory.

where $\varrho_k(\alpha, \boldsymbol{r}_k)$ is the matter density associated with a nucleon in state k (and at $\boldsymbol{r}_k$) and deformation α. The total intrinsic energy is given by:

$$E(\alpha) = \sum_i \int \varrho_i(\alpha, \boldsymbol{r}_i)\, T_i(\boldsymbol{r}_i)\, d^3 r_i + \sum_{i<k}\sum \varrho_i(\alpha, \boldsymbol{r}_i)\, \varrho_k(\alpha, \boldsymbol{r}_k)\, V(\boldsymbol{r}_i - \boldsymbol{r}_k)\, d^3 r_i \cdot d^3 r_k \tag{39.5}$$

or, using (39.4)[1]:

$$E(\alpha) = \sum_i \int \varrho_i(\alpha, \boldsymbol{r}_i)\, T_i(\boldsymbol{r}_i)\, d^3 r_i + \tfrac{1}{2} \sum_i \int \varrho_i(\alpha, \boldsymbol{r}_i)\, V_i(\alpha, \boldsymbol{r}_i)\, d^3 r_i . \tag{39.6}$$

Next let the deformation vary be a small amount. Then both the matter distribution of particles *and* the potential changes. Using the previous equation we find:

$$\delta E = \sum_i \int \delta\varrho_i\, T_i\, d^3 r_i + \tfrac{1}{2} \sum_i \int \delta\varrho_i\, V_i\, d^3 r_i + \tfrac{1}{2} \sum_i \int \varrho_i\, \delta V_i\, d^3 r_i . \tag{39.7}$$

Taking into account (39.4) and its corollary:

$$\delta V_i = \sum_k \delta\varrho_k\, V(\boldsymbol{r}_i - \boldsymbol{r}_k)\, d^3 r_k \tag{39.8}$$

we see immediately that the last two terms in (39.7) are equal, so that we have:

$$\delta E = \sum_i \int \delta\varrho_i\, (T_i + V_i)\, d^3 r_i . \tag{39.9}$$

Now the single particle energy may be written as:

$$\varepsilon_i(\alpha) = \int \varrho_i(\alpha, \boldsymbol{r}_i)\, [T_i(\boldsymbol{r}_i) + V_i(\alpha, \boldsymbol{r}_i)]\, d^3 r_i . \tag{39.10}$$

If we *do not permit* V_i to vary when the density distribution of the other particles changes, i.e. keep the well depth fixed, we find

$$\delta\varepsilon_i = \int \delta\varrho_i\, (T_i + V_i)\, d^3 r_i \tag{39.11}$$

so that

$$\delta E = \sum_i \delta\varepsilon_i . \tag{39.12}$$

The interpretation of this above result seems to be the following: the change in the total energy with deformation is the same as the change in the sum of single particle energies provided we keep the *well depth* of the potential independent of deformation. Both of these assumptions were made in the previous treatment. If instead we wish to write the energy in the form (39.6), we must take into account the fact that the well depth of the actual nuclear potential depends slightly upon the deformation. This effect is very small for *any one* nucleon, but it is large enough to appreciably affect the change of *total* energy with deformation.

40. Effects of residual interactions. While the nucleons move nearly independently in the deformed nucleus, the effects of residual interactions cannot be neglected entirely. A small part of the nuclear potential energy cannot be conveniently represented as the interaction of individual nucleons with an overall potential and must be treated separately. Thus the interaction is especially strong between nucleons whose quantum numbers differ only with respect to sign of Ω, and it gives rise to a special stability of the $K=0$ ground state configuration in even-even nuclei—the pairing effect. These correlations are expected to modify the form of the deformation potential.

While Ω remains a good quantum number even in the presence of residual interactions (as long as the nuclear shape stays axially symmetric), different

[1] Apart from self energy terms which will be ignored.

states of the same Ω are usually coupled by the residual interactions. Now it is well known that two interacting levels can never be degenerate, provided the coupling to other levels can be neglected. The energy difference between two levels is at least equal to twice the matrix element of the coupling term between them. Thus the residual interaction almost always prevent states of the same Ω from crossing. (The only exception to this occurs if the coupling between the levels happens to vanish at the deformation where they are degenerate.) In the absence of residual interaction we might have expected to find low lying excited states in which a pair of nucleons has been promoted to a higher orbit, [41] e.g. an excited $K = {}^7/_2{}^+$ state in Lu^{175} which is formed from the ground state by transferring both nucleons from the ${}^1/_2{}^+$ orbit into the unoccupied ${}^5/_2{}^+$ orbit. No such states have been observed and their absence can be attributed to residual interactions which keep different states of the same K and parity well separated in energy.

α) *General effect on the deformation potential.* The effect of residual interactions near a crossing point of two orbits [37] is illustrated in Fig. 30. To understand the variation of the intrinsic wavefunction with deformation, it is helpful to study the behavior of the nodes. In the idealized case shown in this figure it is supposed that each state has only a single node. As the deformation of the nucleus goes through the cross-over point (c), the configuration of lowest energy changes from "1", characterized by one radial node to "2" with one angular node. Also given are the coefficients c_1 and c_2, the approximate amplitudes of the limiting configurations in the state.

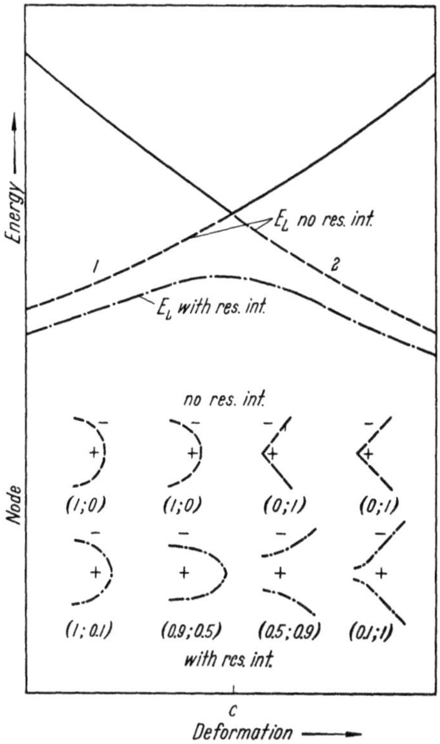

Fig. 30. Effect of residual interactions in mixing two otherwise orthogonal states.

In the absence of residual interactions, the energy levels corresponding to configurations 1 and 2 (which are assumed to differ with respect to the wavefunction of one nucleon) cross at (c). As the deformation changes but not through (c), the wavefunction and thus the node is slowly distorted, but the system remains in the same configuration. However, as the deformation goes through (c), there is a sudden change of configurations to enable the system to remain in the lowest energy state. Now if there are residual interactions the two levels do not cross and the configuration changes gradually. The stronger the interaction, the less sudden will be the transition between the different configurations.

The trend of ground state intrinsic energy versus deformation is illustrated in Fig. 31. We see that the residual interactions smooth out the at the crossing points. It is expected (cf. discussion below) that the effect of residual interactions is largest for spherical shape because of the especially large overlap of the wavefunctions in a spherical field and since the residual interactions are attractive. The energy is thus lowered the most for small deformations [36], [40]. Consequently, the equilibrium deformation of the first energy minimum away from spherical shape is decreased[1]. If the residual interactions were very strong (and short ranged), the nuclear shell structure would be destroyed and the fluctuations in the deformation potential would disappear. The resultant dependence of energy on deformation would be just that of the liquid drop model; in particular, stable equilibrium would occur only for a spherical shape.

β) The spheroidal core model. To obtain more detailed information concerning the effect of residual interactions, we use a special version of the spheroidal core model discussed in Sect. 37[2]. We suppose that the nucleus contains only one half-filled subshell with $j = \frac{3}{2}$ and consider only nucleons of one kind, i.e. two nucleons in this subshell. This "two-nucleon model" is, of course, not realistic since it implies that at most two nucleons can act to deform the nucleus, while in real nuclei the deformation is due to a sizable number of loosely bound nucleons.

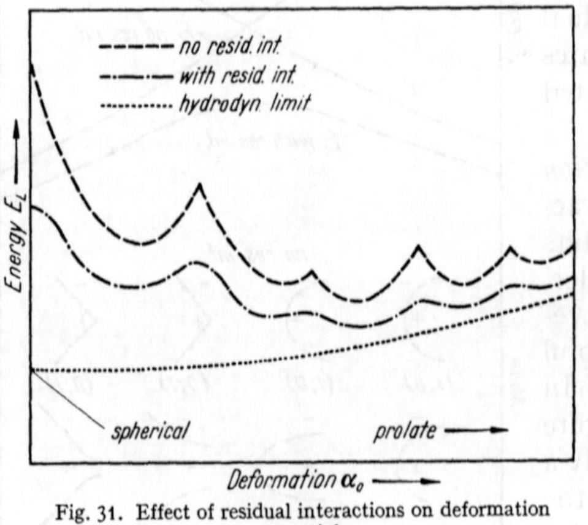

Fig. 31. Effect of residual interactions on deformation potential.

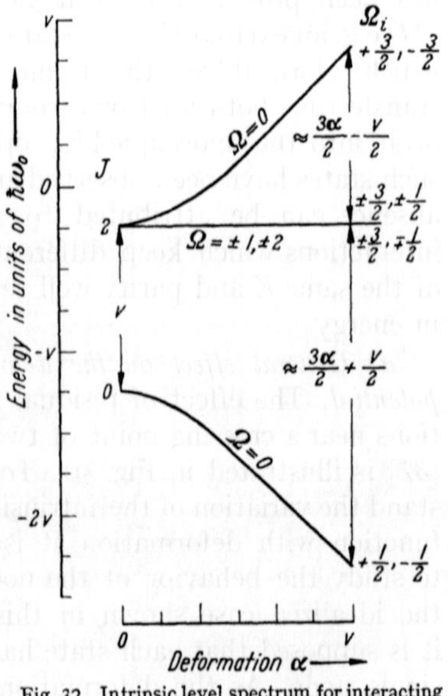

Fig. 32. Intrinsic level spectrum for interacting $j = \frac{3}{2}$ nucleons in a spheroidal potential as function of deformation.

However, we can still obtain large deformations in this model by letting the core of filled shells be easily deformable. The two nucleons in the $j = \frac{3}{2}$ subshell are assumed to interact via a short range (δ function) attractive potential. As in Sect. 37 we assume that the j of each particle remains a good quantum number. The strength of the residual interaction is measured by v, the splitting between the $J = 0$ and $J = 2$ states of the (j^2) configuration for $\alpha = 0$ (in $\hbar\omega_0$ units). It is further postulated that j-j coupling holds, that j and the radial wavefunction of each nucleon remains a good quantum number in spite of the deformation and residual interactions.

The total perturbation due to deformation and residual interactions may thus be written in the form:

$$H' = -\text{const}\, x\, \alpha \sum_i [r^2 P_2(\vartheta)]_i - \text{const}\, x\, v\, \delta(\boldsymbol{r}_1 - \boldsymbol{r}_2). \tag{40.1}$$

We assume that the energy splitting between $|\Omega| = \frac{3}{2}$ and $\frac{1}{2}$ single particle orbits is $\frac{3}{2}\hbar\omega_0\alpha$, the same as the splitting between a pair of states of the same N and $\Delta n_3 = 1$ in an oscillator potential. In principle we may deduce the values of the two coefficients in (40.1) from these conditions but this is not necessary. We are

[1] However this is not necessarily the case for the other energy minima, e.g. the case illustrated in Fig. 31.

[2] This problem was studied in Ref. [36] using a slightly different form of the spheroidal core model in which the nucleons are assumed to move in a p-shell.

interested only in the matrix elements of the perturbations and these may be expressed directly in terms of α and v:

$$\langle \chi_{J=0,\Omega=0} | H' | \chi_{J=0,\Omega=0} \rangle = -\frac{5v}{4}, \tag{40.2a}$$

$$\langle \chi_{J=2,\Omega=0} | H' | \chi_{J=0,\Omega=0} \rangle = \frac{3\alpha}{2}, \quad \text{in units of } \hbar\omega_0, \tag{40.2b}$$

$$\langle \chi_{J=2,\Omega} | H' | \chi_{J=2,\Omega} \rangle = -\frac{v}{4}. \tag{40.2c}$$

The resulting level spectrum for prolate deformations ($\alpha \geqq 0$) is shown in Fig. 32.

The residual interactions have a marked effect on the (j^2) configuration, in particular they further lower the energy of the $\Omega=0$ ground state relative to the other states. Incidentally the $\Omega=0$ states are eigenstates of angular momentum only for $\alpha=0$. In general, they can be expressed as mixtures of $J=0$ and $J=2$ states. In the limit of large deformation both angular momenta appear with equal probability for each state. The effect of the pairing interaction is especially strong for the $J=0$ state (five times as strong as for $J=2$) since a pair of nucleons coupling to a resultant $J=0$ have maximum overlap of their wavefunctions[1]. Thus the energy of the $\Omega=0$ ground state is lowered relatively more near spherical shape than for large deformation.

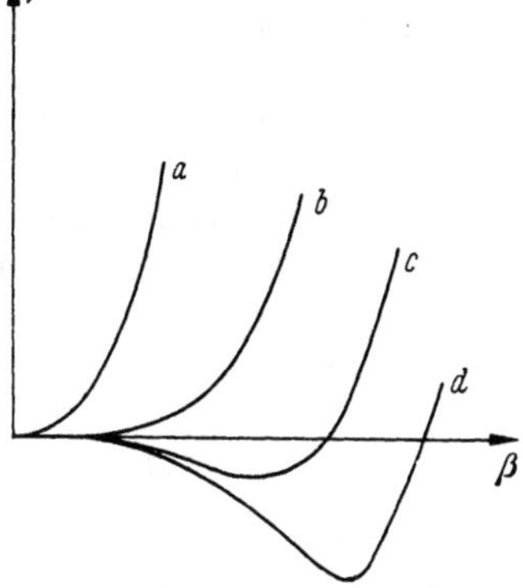

Fig. 33. Deformation potentials for even-even nuclei—Transition from spherical to spheroidal equilibrium shapes. From Ref. [32].

Consider now the effect of the residual interactions on the deformation potential. We must add a restoring term proportional to α^2 so as to represent the resistance of the core against deformation. As we begin to fill up a major shell in an actual nucleus, the outer nucleons will tend to deform the nucleus more and more until we are about 40% through the shell (cf. previous section). This effect can be simulated in the framework of the spheroidal core model, by letting the restoring force of the core vary but keeping the configuration of outer nucleons (and v) fixed. The deformation potential is calculated to be:

$$E_i(\alpha) = \hbar\omega_0 \left\{ -\frac{3v}{4} - \left[\left(\frac{3\alpha}{2}\right)^2 + \left(\frac{v}{2}\right)^2 \right]^{\frac{1}{2}} + \frac{1}{2\alpha_0} \left(\frac{3\alpha}{2}\right)^2 \right\}. \tag{40.3}$$

The first term is the calculated effect of deformation and residual interactions on the $\Omega=0$ ground state [cf. (40.2)]. The coefficient of the second term, representing the resistance of the nucleus against deformation, is chosen so that in the absence of residual interactions, the equilibrium deformation is α_0. According to the spheroidal core model, the residual interaction lowers the equilibrium deformation. Thus differentiating (40.3) with respect to α and setting the result equal to zero we find that:

$$\alpha_{\text{eq}} = (\alpha^2 - \tfrac{1}{9} v^2)^{\frac{1}{2}} \quad \text{provided} \quad \tfrac{1}{3} v < \alpha_0. \tag{40.4}$$

If $v/3 > \alpha_0$, the shape giving minimum energy is spherical.

γ) Filling of a major shell. The process of filling a major shell can be simulated by an appropriate variation of α_0, keeping v constant. The transition between the stability of spherical and nonspherical shapes is further illustrated in Fig. 33.

[1] M. G. MAYER: Phys. Rev. **78**, 22 (1950).

Note that the spherical shape is always an equilibrium shape since the deformation does not alter the energy of the $J=0$ ground state in first order, but it is unstable if $v/3 < \alpha_0$. We will see in the next section that stable spherical shapes are associated with vibrational spectra, e.g. (a) and (b) in the figure. The relation between permanent deformations, e.g. (c) and (d) and rotational spectra has been discussed above. To obtain a non-spherical equilibrium shape for an even-even nucleus we must evidently have a sufficiently large number of nucleons in unfilled shells cooperating to overcome the effect of the pairing interactions[1]. In particular, we must be approximately in the middle of shell-filling for both protons and nucleons [*32*]. This is realized for the nuclei around $A \approx 25$ where neutrons and protons are both filling up the 8 to 20 shell. On the other hand for mass numbers above 40, the Coulomb effect brings the shell fillings for neutrons and protons "out of phase". Thus nuclei with $N \approx 65$, half way between $N=50$ and 82, have $Z \approx 50$, a closed shell. In addition, the level spectrum has a gap at nucleon number 28 due to the spin orbit coupling. This feature also tends to oppose deformation. As a result of these effects, no well developed rotational spectra exist for $40 \lesssim A \lesssim 150$. For mass numbers above 150 and up to the heaviest known nuclei, the neutrons and protons are again approximately "in phase" though now one shell apart, as far as the filling of levels is concerned. Thus for heavy nuclei well away from the double magic number Pb^{208}, the protons and neutrons cooperate to give large deformations.

III. Collective vibrational excitations.

In the next two sections we treat vibrational excitations involving oscillations of the nuclear shape about its equilibrium form. Both spherical and spheroidal nuclei are discussed.

We must point out at the outset that the characterization of excitation level spectra as vibrational is not expected to be as accurate as was the case for rotational spectra. The vibrational energies involved (several hundred kev to 1 Mev) are no longer very small compared to typical intrinsic excitation energies (1 or 2 Mev), thus interaction between the intrinsic motion and the vibration may now be rather significant. Crude as it is, the description of certain collective excitations as vibrational is nevertheless very useful.

41. Vibrations of spherical nuclei. There is a large class of nuclei in which there are sufficiently many nucleons in unifilled shells to give much configuration mixing, but not enough to cause permanent deformation. Such nuclei occur mainly in the regions $60 \lesssim A \lesssim 150$ and $190 \lesssim A \lesssim 220$.

α) Systematics of empirical data. Even-even nuclei. Excitation spectra of even-even nuclei in this class show regularities not as striking as those of rotational spectra, but nevertheless very significant [*44*]. Thus the energy of the first excited state (2^+ except in a very small number of cases) is largest for magic number nuclei, and smallest far away from closed shells (see Fig. 10). The character of the second excited state, where known, is mostly 2^+, though 4^+ also occurs rather frequently and for a few cases it is 0^+, 1^- or 3^-.

Fig. 34 shows the ratio of energies E_2/E_1 in even-even nuclei as function of neutron number. For the nuclei with $N \lesssim 88$ these ratios never deviate very

[1] While the energy gained by forming a $(j^2)_{J=0}$ pair is large, the total pairing energy of a configuration is proportional to the number of pairs, i.e. to the number of particles n in the shell. On the other hand, the deformation and thus the coupling energy of each particle to the nuclear symmetry axis is proportional to n, so that the total energy gained by deforming the nucleus is proportional to n^2.

much from the value 2.2[1] [44], except for the cases where the spin of the second excited state is odd or when N or Z is a magic number. These features are also illustrated by a plot of E_2/E_1 as function of E_1 (Fig. 35). There seems to be a slight decrease of the energy ratio with increasing E_1. All points with $E_1 > 1200$ kev

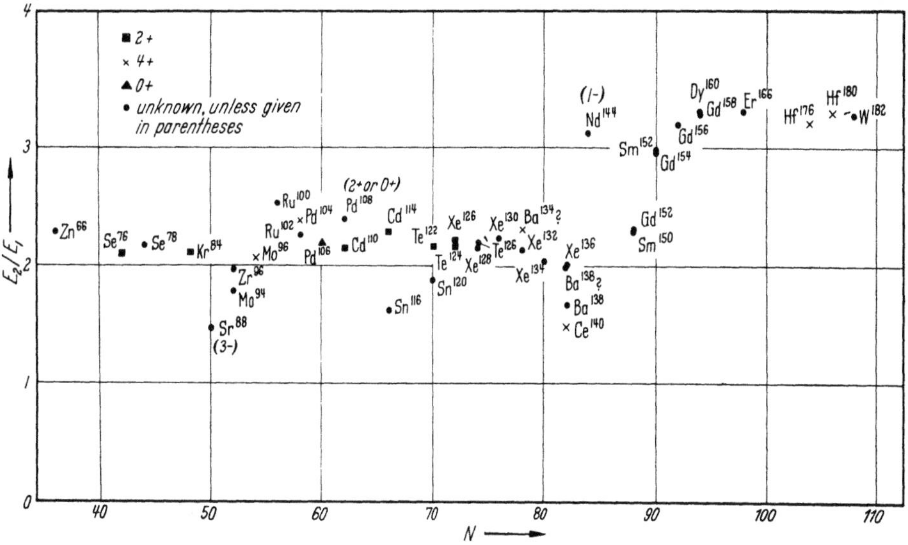

Fig. 34. Energy ratios of first two excited states E_2/E_1 of even-even nuclei as function of neutron number N ($36 \leq N \leq 108$). From Ref. [44].

refer to magic number nuclides. Between $N=88$ and 90 there is a rapid transition, e.g. for Gd^{152} ($N=88$) we have $E_1 = 344$ kev, $E_2/E_1 = 3.0$. A similar though slightly less rapid transition, also occurs between $Z=86$ and 88[2].

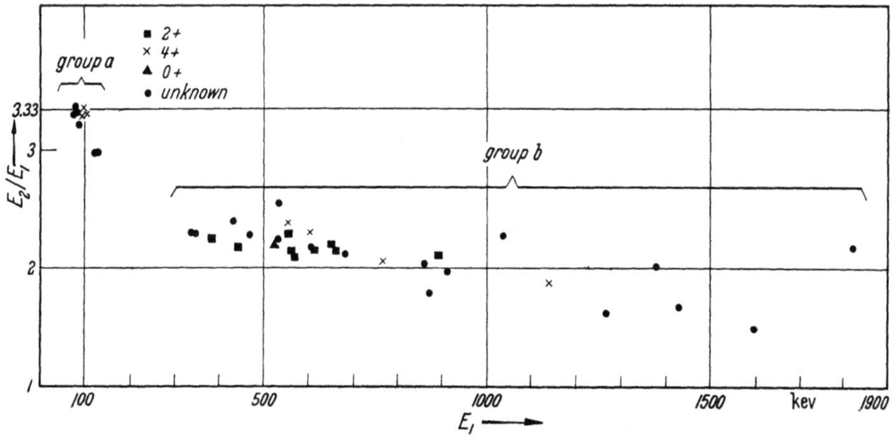

Fig. 35. Energy ratios E_2/E_1 of even-even nuclei as function of E_1 ($36 \leq N \leq 108$). From Ref. [44]. Points for nuclei with odd spins for the second excited states are omitted from this figure. Group (a) consists of nuclei with rotational states ($90 \leq N \leq 108$).

Empirical information of transition probabilities also shows interesting regularities. Thus the $E2$ transition probability between ground and first excited state is usually at least one order of magnitude larger than expected on the basis of a single particle model. According to a single particle model we have

[1] M. Nagasaki and T. Tamura: Progr. Theor. Phys. **12**, 248 (1954).
[2] G. Scharff-Goldhaber: Phys. Rev. **103**, 837 (1956).

from (35.4) and (35.6) [1]:

$$B_{\text{S.P.}}(E\,\lambda) = (4\pi)^{-1}\, e^2\, R_0^{2\lambda}\, S\, \mathfrak{R} \tag{41.1}$$

where S is a statistical factor whose value depends on the angular momenta involved, ($S=1$ if the transition leads to a state of $I=0$), and $\mathfrak{R}$ is a dimensionless factor of order $\frac{1}{4}$ arising from integration over radial wavefunctions. For cases where the second excited state is 2^+, this state (denoted here by $2'$) decays primarily to the 1st excited state rather than the ground state, although the opposite would be expected on energy considerations alone. Furthermore, the decay $2' \to 2$ takes place mainly by $E2$ emission with only a small $M1$ admixture.

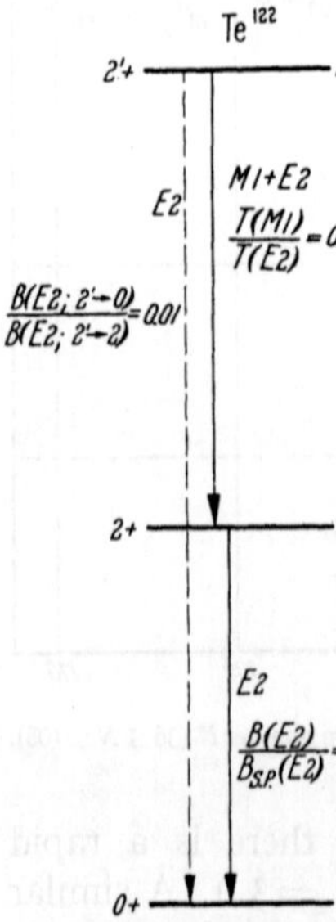

Fig. 36. Level spectrum of Te^{122}. The data on which this figure is based are taken from Table V 6 of Ref. [32].

The operation of the above rules is well illustrated by a typical example, the level scheme of Te^{122} [2] shown in Fig. 36. The form of level spectra such as that of Te^{122} suggests that we are here dealing with collective ellipsoidal vibrations about a spherical equilibrium shape. We have already seen that the equilibrium shape is expected to be spherical if the deforming influence of the outer nucleons is insufficient to overcome the effect of the pairing forces, and this may well be the case for the above nuclei.

β) Deformation parameters for ellipsoidal shapes. Before discussing the vibrational motion, let us briefly consider the description of ellipsoidal shapes in terms of the deformation parameters [33]. Quite generally, the surface of an ellipsoid $R(\vartheta, \varphi)$ deviating slightly from a sphere of radius R_0 of the same volume can be characterized by the five deformation parameters α_μ where [3]

$$\delta R(\vartheta, \varphi) = R(\vartheta, \varphi) - R_0 = \sum_{\mu=-2}^{2} \alpha_\mu^*\, Y_{2\mu}. \tag{41.2}$$

Since δR is a real function of angle, we have

$$\alpha_{-\mu} = (-1)^\mu\, (\alpha_\mu)^*. \tag{41.3}$$

Thus there are five independent deformation parameters—two specifying the shape and the other three decribing the orientation. Of interest here are the two shape parameters. These are conveniently defined in terms of the fractional elongation along the three principal axes:

$$\delta R_\nu = \left(\frac{5}{4\pi}\right)^{\frac{1}{2}} \beta\, R_0 \cos\left(\gamma - \frac{2\pi\nu}{3}\right) \tag{41.4}$$

where $\nu = 1, 2, 3$. The quantity β measures the deviation from sphericity and the value of the parameter γ determines the specific shape of the nucleus. Thus, $\gamma=0$, describes a prolate (football shaped) spheroid whose symmetry axis coincides with the z axis, while $\gamma=\pi$ corresponds to an oblate (disk shaped) spheroid,

[1] Footnote 1, p. 490.

[2] Cf. Ref. [32], Table V6.

[3] The deformation parameters used here are the complex conjugates of the parameters defined in [33].

again with a symmetry axis along the z direction. Other values of γ which are integral multiples of $\pi/3$ also describe spheroidal shapes but with the symmetry axis coinciding with the x or y axis. If γ is not a multiple of $\pi/3$, the corresponding shape is an ellipsoid with three unequal princiapl axes coinciding with the coordinate axes. In any case the sum of the elongations along the three principal axes vanishes, so that the volume is not changed by the deformation, at least in first order. In the body fixed reference frame (coordinate axes coinciding with principal axes), the deformation parameters, here denoted by a_μ instead of α_μ, are given by:

$$a_0 = \beta \cos\gamma, \tag{41.5a}$$

$$a_1 = a_{-1} = 0, \tag{41.5b}$$

$$a_2 = a_{-2} = \sqrt{\tfrac{1}{2}}\,\beta \sin\gamma. \tag{41.5c}$$

Since the deformation parameters in the two systems are related by:

$$\sum_\mu a_\mu^* Y_{2\mu}(\vartheta,\varphi) = \sum_\nu (a_\nu)^* Y_{2\nu}(\vartheta,\varphi). \tag{41.6}$$

it is seen, using (I.1), that:

$$\alpha_\mu = \sum_\nu a_\nu D^2_{\mu\nu}(\vartheta_i). \tag{41.7}$$

Eqs. (41.5) and (41.7) define the coordinate transformation from the parameters β, γ, ϑ_i to the five α_μ.

γ) The collective Hamiltonian. In our study of vibrational excitations we will again assume that the intrinsic structure follows the changing nuclear field adiabatically. We will, however, not consider the intrinsic wavefunction explicitly, and regard *all* excitations as collective. The assumptions constitute what we call the "collective model"[1]. We further restrict ourselves to quadrupole (ellipsoidal) vibrations. According to this model, the collective nuclear Hamiltonian can be written in the from:

$$H_{\mathrm{Coll}} = \tfrac{1}{2} \sum_\mu B_2(\alpha)\,|\dot\alpha_\mu|^2 + V(\alpha) \tag{41.8}$$

where α_μ are the deformation variables, collectively denoted by α, which characterize the shape and orientation of the nucleus. The first term on the right hand side of (41.8) is the total kinetic energy associated with the collective motion assuming the adiabatic condition is satisfied (Sect. 43), B_2 is referred to as the mass parameter for collective quadrupole oscillations, and V is the deformation potential or the collective potential energy (Sect. 40).

We first consider those features which do not depend upon the explicit form of $V(\alpha)$ and B_2. In even-even nuclei, the intrinsic state carries no angular momentum ($I_0 = 0$), thus any angular momentum and also the magnetic moment must be due to the collective motion. Consequently the magnetic moment is parallel to the angular momentum:

$$\mu = g_R \boldsymbol{I} \tag{41.9}$$

where g_R is the gyromagnetic ratio for the collective motion. Now since the operator I connects a state essentially only to itself, its matrix elements between different, (non-degenerate) states must vanish and if (41.9) holds, no $M1$ radiation can occur. Physically speaking, according to (41.9) the magnetic moment is classically stationary. However, each state of collective excitation will have

[1] This name has also been frequently used as an alternative designation of the unified model, e.g. in Ref. [37].

a static magnetic moment (unless I or $g_R=0$). Turning to a consideration of $E2$ transitions, we have, from (34.13) and (35.8)

$$B(E2; i\to f) = \left(\frac{3}{4\pi} Z e R_0^2\right)^2 \sum_\mu |\langle f|\alpha_\mu|i\rangle|^2. \tag{41.10}$$

Now we may sum over all possible final states and apply the rule of matrix nultiplication to find:

$$\sum_f B(E2; i\to f) = \left(\frac{3}{4\pi} Z e R_0^2\right)^2 \langle i|\beta^2|i\rangle. \tag{41.11}$$

Calculations for various forms of the collective Hamiltonian (41.8)[1] show that if i represents the ground state of an even-even nucleus then the sum is nearly or completely "used up" by the transition to the first excited state. In so far as this is the case, direct $E2$ transitions between the ground state and any but the first excited state will be strongly inibited (and may even have smaller than single particle strength), in agreement with the empirical data mentioned above.

δ) Harmonic surface oscillations. Next we study some features which depend on the specific form of the collective Hamiltonian. We have seen in the last section that even-even nuclei not very far from magic numbers are expected to have spherical equilibrium shapes. As a first approximation we may take the collective Hamiltonian for such nuclei of the same form as in the liquid drop model (with $\lambda=2$) [*33*], i.e.

$$H_{\text{Coll}} = \frac{1}{2B_2}\sum_\mu |\pi_\mu|^2 + \frac{1}{2} C_2 \sum_\mu |\alpha_\mu|^2 \tag{41.12}$$

where as before, π_μ is the momentum conjugate to α_μ. Eq. (41.12) involves two assumptions: (i) a harmonic deformation potential and (ii) a mass parameter independent of deformation. The treatment is just the same as that in Sect. 4, except that B_2 and C_2, the effective surface tension, are now regarded as arbitrary parameters to be deduced from the interpretation of empirical data. The resulting energy spectrum is harmonic with a uniform spacing:

$$\hbar\omega_2 = \hbar\left(\frac{C_2}{B_2}\right)^{\frac{1}{2}} \tag{41.13}$$

between non-degenerate levels. A state of energy

$$E^{(n)} = \left(n+\tfrac{5}{2}\right)\hbar\omega_2 \tag{41.14}$$

may be characterized as containing n "phonons" each of which carries an angular momentum of 2 and even parity [*33*]. Thus all the excited states are expected to have even parity. The energy spectrum assuming a simple collective Hamiltonian of the form (41.12) is shown in part a of Fig. 37 up to the $n=2$ state. The second excited "2 phonon" state is multiply degenerate according to this simple model and $E2$ transition between the $2'^+$ (2 phonon) state and the ground state are forbidden (since they involve only one phonon). For the group of nuclei considered in this section, the ratios E_2/E_1 of excitation energies are usually rather close to 2, which implies that the harmonic approximation has at least some validity.

Calculations have also been made on the assumption that the deformation potential is anharmonic[1], but a function of β alone, and not of γ (no favoring

[1] L. WILETS and M. JEAN: Phys. Rev. **102**, 788 (1956).

of prolate or oblate shapes). The mass parameter B_2 is still regarded as a constant. According to these calculations the degeneracy of the second excited state is partially lifted but the $2'^{+}$ and 4^{+} state remain at the same energy, as shown in part b of Fig. 37. The $E2$ transition between the $2'$ excited state and ground state is however expected to remain forbidden. For the form of $V(\alpha)$ suggested by the considerations of Sect. 40, the energy ratio E_2/E_1 is expected to be somewhat larger than 2, in agreement with experiment. The average value 2.2 of the empirical energy ratios E_2/E_1 implies that there is indeed some anharmonicity, but with the spherical shape at least on the borderline of being stable. The empirical trend of E_2/E_1 as function of E_1 is also reasonable. As we move further away from closed shells, the anharmonicity in V is expected to increase so that E_2/E_1 increases. However, soon there appears a special lowering of prolate spheroidal shapes. Any favoring of prolate (or oblate) shapes will split the degeneracy between 2^{+} and 4^{+} state, but always so as to lower the 4^{+} state, et least if the adiabatic condition holds, and if B_2 is independent of deformation. The frequent appearance of 2^{+} second excited states (part c of Fig. 37) may be due to non-adiabatic perturbation of the vibrational motion (vibration-particle coupling). This effect might be significant since the vibrational energies are not much smaller than excitation energies of individual nucleons. It is also possible that the frequently observed 0-2-2 (rather than 0-2-4) sequence of spins might result from a dependence of B_2 upon the nuclear shape (cf. Sect. 43).

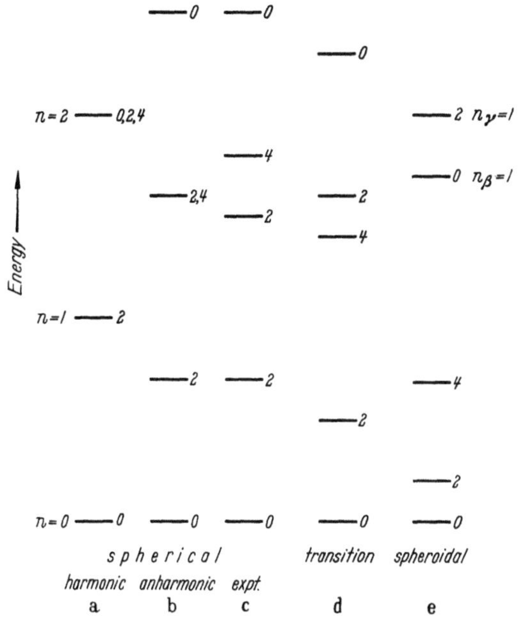

Fig. 37 a—e. Collective excitation spectra of even-even nuclei—Transition from spherical to spheroidal equilibrium shapes. For parts (a), (b), (c) see Sect. 41, while parts (d) and (e) are discussed in Sect. 42.

Assuming that the collective excitations are harmonic vibrations, we can readily deduce the values of B_2 and C_2 from the energy difference and the $E2$ transition probability between ground state and first excited state[2] [32]. A general property of any harmonic oscillator Hamiltonian is that the expectation value of the potential energy is just half of the total energy (both measured from the bottom of the well). Applying these considerations to the ground state of (41.12) we find:

$$C_2 \langle i | \beta^2 | i \rangle = \tfrac{5}{2} \hbar \omega_2 \tag{41.15}$$

which, in conjunction with (41.13) gives B_2 and C_2 in terms of $h\omega_0$ and $\langle i | \beta^2 | i \rangle$.

It is useful to express the empirical mass parameter in terms of the "hydrodynamic" value (4.6) which would result if the collective flow were irrotational. Fig. 38 shows a plot of $B_2/(B_2)_{\text{Irrot}}$ (denoted by B_2/B_2^* in the figure) as function

---

[1] Cf. Footnote 1, p. 520.
[2] G. M. TEMMER and N. P. HEYDENBURG: Phys. Rev. **104**, 967 (1956).

of neutron number. We see that this ratio is largest near closed shells and decreases away from magic numbers. It is interesting to note that the same ratio for rotational spectra (written here as $\mathfrak{J}/\mathfrak{J}_{\text{Irrot}}$) also decreases as we move toward the most strongly deformed nuclei. In fact, the smallest values, about 8, for vibrational spectra join nicely onto the largest values, 7, for rotational spectra, at least for mass numbers below 180.

Finally, analysis of vibrational spectra also permits determination of the effective surface tension C_2. Values of C_2 deduced from Eq. (41.15) are plotted in Fig. 39. As is indicated, the surface tension is especially large near closed shells, an order of magnitude larger than the liquid drop value (Eq. 4.3) which is shown in the figure. Away from closed shells, the surface tension decreases as is expected from the simple model of Sect. 40 and is often much less than the hydrodynamical value. Very near to closed shells, the collective description of the excitation becomes inaccurate, as the vibrational energies become comparable to, or even larger than, the energies for particle excitation. Thus the large values of B_2 and C_2 found near magic number nuclei have only qualitative significance.

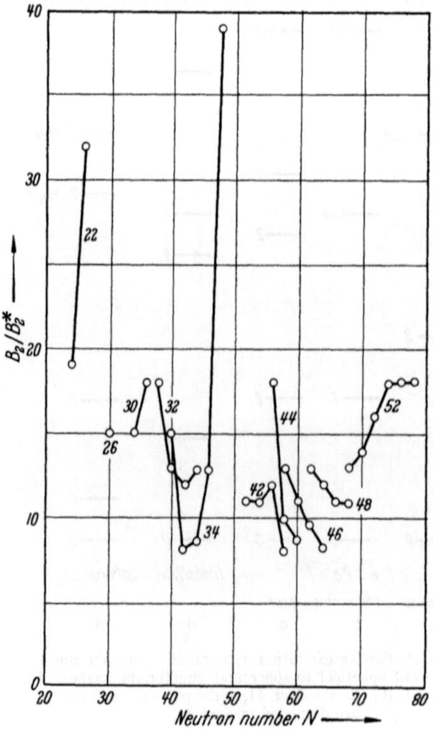

Fig. 38. Vibrational mass parameters for even-even nuclei ($22 \leq Z \leq 52$). From G. M. TEMMER and N. P. HEYDENBURG: Phys. Rev. 104, 967 (1956).

ε) *Spectra of odd-A nuclei*[1]. Many odd-*A* nuclei in the regions moderately far from closed shell configurations have rather complicated level spectra. The low-energy spectrum of such nuclei often involves not only collective excitations as in neighboring even-even nuclei, but also excitations of the unpaired odd-nucleon, e.g. the case of nuclear isomers[2].

The degrees of freedom of the unpaired nucleon are coupled to the collective oscillations, since the later involve variations in the nuclear field. If this coupling could be neglected, then associated with each intrinsic state there would be a collective vibrational excitation spectrum. On the basis of the spheroidal core model, it has been shown [34] that the strength of this coupling depends essentially on the value of a "coupling parameter" q given by

$$q = \left(\frac{5}{16\pi}\right)^{\frac{1}{2}} \frac{k}{(\hbar\,\omega_2\,C_2)^{\frac{1}{2}}} \tag{41.16}$$

where k has the order of magnitude of the average potential energy per nucleon. As we move away from closed shells, both the effective surface tension and the vibrational frequencies decrease, consequently the coupling parameter is expected to increase. If the coupling strength is estimated using the empirical vibrational

[1] This material is taken from Ref. [32].
[2] Cf. ALBURGER: Nuclear Isomers, Vol. XLII of this Encyclopedia and M. GOLDHABER and R. D. HILL: Rev. Mod. Phys. 24, 179 (1952).

parameter for neighboring even-even nuclei, it is found that $1 < q < 3$ for most nuclei indicating an intermediate coupling situation. The resulting level system is a rather complicated superposition of particle and collective excitations[1]. In a few regions ($A \sim 20$, 75, 105, 150 and 192) one obtains $q < 4$. Even-even nuclei in these regions seem to barely have stable spherical shapes as may be seen by a comparison of the first excited state energies with the critical value drawn in Fig. 10. Addition of an unpaired nucleon to such an even-even nucleus, may be sufficient to polarize the nuclear shape. It is thus not surprising that the coupling schemes of some of the odd-A nuclei in these regions (e.g. Se^{77}, Ag^{107})[2] are approximately rotational in character.

ζ) Higher order vibrations. Finally it should be mentioned that vibrations of higher order, e.g. octupole or 2^4 pole, may also occur, although their characteristic frequencies are expected to exceed those of quadrupole vibration [cf. Eq. (4.10)]. The one phonon state for octupole vibration in even-even nuclei is expected to have $I = \lambda = 3$ $\pi = (-1)^\lambda = \text{odd}$ (Sect. 4). Indeed a rather low lying 3-state (1.12 Mev) has been observed in the level spectrum of Gd^{152}[3] and may possibly represent an octupole excitation. Empirical evidence for higher order vibrational excitations in spherical nuclei is still very scarce, however.

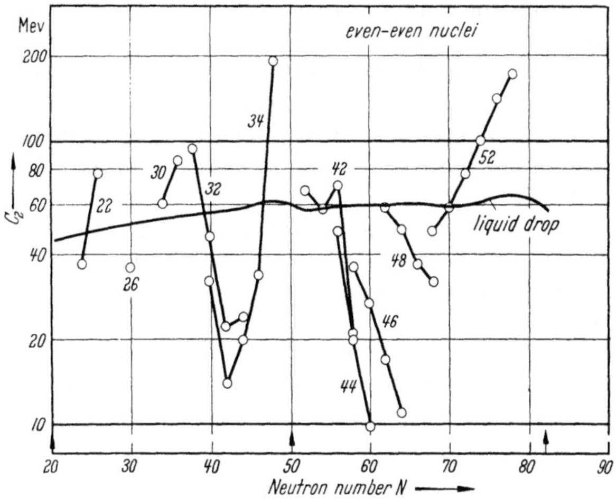

Fig. 39. Surface tension parameters for even-even nuclei ($22 \leq Z \leq 52$). From G. M. TEMMER and N. P. HEYDENBURG: Phys. Rev. 104, 967 (1956).

42. Vibrations of spheroidal nuclei. In a strongly deformed nucleus, the lowest states of collective excitation involve rotations without change in the nuclear shape. One would expect, however, that vibrations about about the equilibrium shape can occur, resulting in collective excitations of higher energy.

α) Quadrupole vibrations. It is likely that quadrupole vibrations, in which the nuclear shape remains ellipsoidal, are the easiest to excite. From the considerations of Sect. 41 we see that an axially symmetric (say prolate) nucleus may perform two kinds of quadrupole vibrations [*35*]. We may have oscillations of the eccentricity about its equilibrium value, but with preservation of axial symmetry: vibrations of β about β_0, (γ fixed at 0), or equivalently of a_0 about β_0 with all other a's remaining at 0. These vibrations evidently carry no angular momentum about the symmetry axis. The other possible kind of vibrations involves oscillation of the nuclear shape about axial symmetry; β is fixed at β_0

[1] Cf. footnote 5, p. 477, Ref. [*34*] and D. C. CHOUDHURY: Dan. Mat.-Fys. Medd. **28**, No. 4 (1954); L. L. FOLDY and F. J. MILFORD: Phys. Rev. **80**, 751 (1950); K. FORD: Phys. Rev. **90**, 29 (1953); P. GOLDHAMMER: Phys. Rev. **101**, 1375 (1956); F. J. MILFORD: Phys. Rev. **93**, 1297 (1954); B. J. RAZ: Ph. D. dissertation, University of Rochester, 1955; A. REIFMAN: Z. Naturforsch. **8**a, 505 (1953); M. TROCHERIS: J. Phys. Radium **14**, 635 (1954).

[2] Se^{77}: G. M. TEMMER and N. P. HEYDENBURG: Phys. Rev. **104**, 967 (1956). — Ag^{107}: Ref. [*31*].

[3] H. KENDALL and L. GRODZINS: Bull. Amer. Phys. Soc., Ser. II **1**, 164 (1956).

but γ oscillates about 0, i.e. $a_2 = a_{-2}$ oscillate about 0 while the other a's remain fixed at least in first order. These "γ-vibrations" carry two units of angular momentum parallel to the symmetry axis.

Superimposed upon each vibrational state one expects to find a rotational band. Thus in even-even nuclei the rotational band associated with the lowest mode of β vibration ($n_\beta = 1$, $n_\gamma = 0$) is expected to have the same form as the ground band: $K = 0$, $I = 0, 2, 4$; while the band corresponding to the first excited state of γ vibration ($n_\beta = 0$, $n_\gamma = 1$) should have $K = 2$, $I = 2, 3, 4$. Incidentally since any ellipsoidal shape is invariant under reflection through the origin and the intrinsic structure in an even-even nucleus has even parity, we see from the considerations following (32.20) that all quadrupole vibrational as well as rotational excitations in even-even nuclei have even parity.

Consider now the transition between spherical and spheroidal equilibrium shapes.

For the first case, the empirical spectra in even-even nuclei have the form illustrated in part c of Fig. 37. As the prolate equilibrium shape becomes more and more favored energetically the ratio E_2/E_1 increases toward $\frac{10}{3}$ and the 4^+ state drops below the second 2^+ state, as is shown in part d of Fig. 37. Such a trend is believed to occur for nuclei of mass numbers around 190 and 220. This 4^+ state finally becomes the second excited rotational state (part e). On the other hand, the energy of the second excited 2^+ state increases with deformation and it goes over into the lowest state associated with excitation of γ vibrations (cf. Sect. 42). The lowest excited 0^+ state which occurs at comparatively high energy in vibrational spectra seems to drop somewhat and to become the lowest excited state associated with β-vibrations.

For all strongly deformed even-even nuclei the lowest even-parity nonvibrational states are always 0^+ or 2^+ as far as is known and in a number of cases the associated rotational bands have been investigated. These vibrational excitations usually occur at around 1 Mev. One of the few nuclei for which both kinds of bands have been identified is Pu^{238} [1] and its level scheme is shown in Fig. 40. Note that for this case both excited bands have nearly the same moments of inertia as the ground state band[2].

β) Octupole vibrations. A number of deformed even-even nuclei with $A \approx 230$, e.g. Pu^{238}, have a rather low lying odd-parity state of spin 1 [3]. In a few cases, odd parity states of higher spin have also been identified. A good illustration of this is provided by the spectrum of Ra^{226} [4] shown in Fig. 41. The 1-state which occurs in this spectrum can decay to both 0^+ and 2^+ states in the ground state band by means of $E1$ radiation.

To investigate the value of K of the 1-state (which might be 0 or 1) we calculate the branching ratio $B(1 \to 0)/B(1 \to 2)$ using Eq. (35.17), and compare with experimental values. The theoretical branching ratio is 0.50 for $K = 0$ and 2.00 for $K = 1$. Empirical branching ratios are always very close[5] to 0.50 supporting the assignment $K = 0$ for the states and presumably also for the other odd parity states.

A detailed quantitative interpretation of the odd-parity states, and of the reasons why they occur only on the low A side of the two rotational regions has

[1] I. PERLMAN and J. O. RASMUSSEN: Alpha Decay, Vol. XLII, of this Encyclopedia.

[2] It is probably also significant that in this spectrum, as in W^{182} (Fig. 18), a near-degeneracy exists between states of the same I in the excited bands. This implies that the relevant vibrational frequencies are nearly identical.

[3] STEPHENS, ASARO and PERLMAN: Phys. Rev. **96**, 1568 (1954).

[4] STEPHENS, ASARO and PERLMAN: Phys. Rev. (to be published).

[5] STEPHENS, ASARO and PERLMAN: Phys. Rev. **100**, 1543 (1955).

not been given so far. However it seems likely that these states represent pear shaped (octupole) vibrations about the prolate spheroidal equilibrium shape[1] because the states of odd I have odd parity. Since $K=0$, these vibrations carry no angular momentum parallel to the nuclear symmetry axis. In Sect. 32 we showed that if a pear shape is actually stable, there results a single rotational spectra containing both even and odd states. For this case there are two possible pear shaped equilibrium configurations, mirror images of each other, and one changes into the other comparatively infrequently (i.e. the tunneling frequencies are small compared to the other frequencies of the system). This limit is frequently realized in molecules, i.e. the inversion spectrum of NH_3, but the coupling scheme in nuclei is apparently always less extreme, though sometimes approaching it.

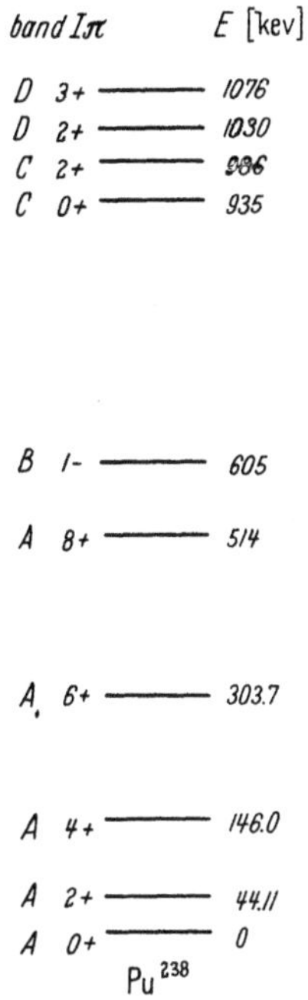

Fig. 40.

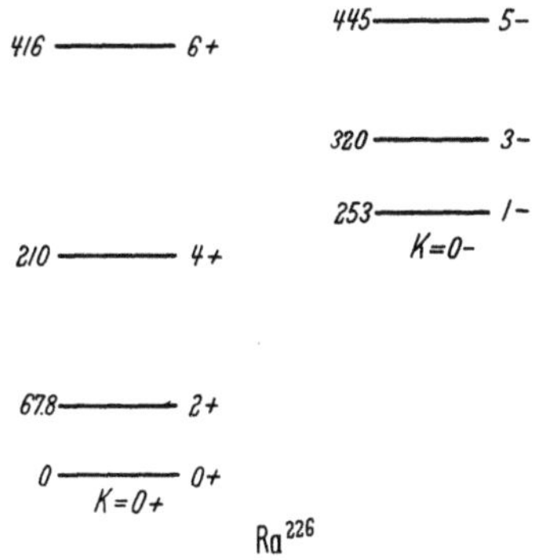

Fig. 41.

Fig. 40. Level spectrum of Pu^{238}. Based on Fig. 29 of I. PERLMAN and J. O. RASMUSSEN, Alpha-Decay, Vol. XLII of this Encyclopedia.

Fig. 41. Level spectrum of Ra^{226}. From Ref. [32].

Finally, while the energy spacings within the odd parity band follow the rotational law very nicely, the empirical moments of inertia are considerably larger than for the ground state band. For the case of Ra^{226} we have $\hbar^2/2\mathfrak{J}=$ 11.3 kev for the even parity band but only 6.7 kev for the odd parity band, about the same value as for nearby odd-A nuclei.

IV. Relation between intrinsic and collective motions.

In the previous sections we have treated both collective and intrinsic motion, but relatively little has been said about their relation to each other. In the next three sections we discuss some aspects of this problem. In Sect. 43 we discuss the dependence of the inertial parameters, and in particular the nuclear moment of inertia, upon the intrinsic structure. In Sect. 44 we consider treatments of perturbations of the intrinsic structure due to the rotations, the rotational coupling between particle states due to Coriolis forces, and rotation-vibration interaction which is due to centrifugal forces. In the last section we conclude with some remarks concerning the relation between the unified and shell model.

[1] R. F. CHRISTY: Unpublished data 1954.

43. Inertial parameters for collective nuclear oscillations. The character of the collective motion in nuclei depends quite significantly upon the detailed nuclear motion. As in the previous treatment, we assume the approximate validity of the adiabatic condition. However, the fact that the nuclear binding potential is not static but variable in time, even if slowly, prevents the intrinsic structure from being *exactly* in any stationary state, i.e. there are bound to be *some* non-adiabatic effects. Even if the effect of non-adiabaticity on the motion of any individual nucleon can be neglected, the combined effect on the nucleus as a whole is very appreciable.

If the nucleus would be in the *lowest* intrinsic state u_L of energy E_L in a *static* field, the time variation of the field will "admix" to the wavefunction states of higher energy. It is precisely these admixtures which give rise to collective motion[1]. In particular, the collective kinetic energy is just the extra energy of the system, above E_L which is entirely due to non-adiabaticity.

First consider a static field characterized by the single deformation parameters α[2], and the set of stationary states k, wavefunctions $\chi_k(\alpha, \boldsymbol{r})$, and energy levels $E_k(\alpha)$. In this field the state of lowest energy would be simply the lowest stationary state L.

α) Calculation of the inertial parameters by time-dependent perturbation theory. To derive expressions for the inertial parameters characterizing the collective motion, it was proposed by INGLIS that the deformation, while variable, can be treated classically, i.e., that it can be regarded as changing with time in a prescribed way[3] [*38*]. (Of course the motion of the nucleons in the deformed potential is treated quantum mechanically.) Thus suppose the field changes slowly with time. Using time dependent perturbation theory we find that the wavefunction Φ_L which corresponds to χ_L in the static limit is given by:

$$\Phi_L = \chi_L + i\hbar \sum_k (E_k - E_L)^{-1} \left(\int \chi_k^* \frac{\partial \chi_L}{\partial t} d^3 r \right) \chi_k \tag{43.1}$$

---

[1] On the other hand, in molecules almost all of the collective kinetic energy is due to the kinetic energy of the heavy nuclei and only a very small fraction results from non-adiabatic perturbations of the electronic motions.

[2] The treatment can be readily generalized to take into account several deformation variables.

[3] Attempts to calculate the inertial parameters, in particular the moment of inertia, by a strictly quantum mechanical treatment may be divided into two groups. — One approach is to find a canonical transformation which changes the many-particle nuclear Hamiltonian into the form of the rotational model. If we define the collective coordinates explicitly in terms of the particle coordinates, e.g. by (45.1) the required transformation introduces a collective motion of irrotational character. This kind of collective motion is designated as "distortion" in the present article (cf. Sect. 38, and discussion below). Various aspects of this transformation have been discussed in Ref. [*36*] and by: A. BOHR: Rotational States in Atomic Nuclei. Copenhagen: Ejnar Munksgaard 1954; G. SÜSSMANN: Z. Physik **139**, 543 (1954); H. A. TOLHOEK: Physica, Haag **21**, 1 (1955); S. TOMONAGA: Progr. Theor. Phys. **13**, 467, 482 (1955); F. COESTER: Phys. Rev. **99**, 170 (1955) and Nuovo Cim. **4**, 1307 (1956); MARUMORI, YUKAWA and TANAKA: Progr. Theor. Phys. **13**, 442 (1955); T. MARUMORI and E. YAMADA: Progr. Theor. Phys. **13**, 557 (1955); T. MARUMORI: Progr. Theor. Phys. **14**, 608 (1955); LIPKIN, DE SHALIT and TALMI: Nuovo Cim. (10) **2**, 773 (1955) and Phys. Rev. **103**, 1773 (1956); T. MIYAZIMA and T. TAMURA: Progr. Theor. Phys. **15**, 255 (1956); T. TAMURA: Nuovo Cim. (10) **4**, 713 (1956); MARUMORI, SUEKANE and YAMAMOTO: Progr. Theor. Phys. **16**, 320 (1956); R. S. NATAF: Nucl. Phys. **2**, 497 (1957); F. VILLARS: Nucl. Phys. **3**, 240 (1957); S. HAYAKAWA and T. MARUMORI: Progr. Theor. Phys. **17**, 43 (1957). — The other approach is to apply a variational principle to a wavefunction of the form (45.7). This method has been discussed by: J. YOCCOZ: J. Phys. Radium **7**, 517 (1956); T. H. R. SKYRME: Proc. Roy. Soc. Lond. **239**, 399 (1957); T. H. R. SKYRME: Proc. Phys. Soc. Lond. **70**, 433 (1957); R. E. PEIERLS and J. YOCCOZ: Proc. Phys. Soc. Lond. **70**, 381 (1957); J. YOCCOZ: Proc. Phys. Soc. Lond. **70**, 388 (1957); J. J. GRIFFIN and J. A. WHEELER: Phys. Rev. (in press).

to first order in $\partial\chi/\partial t$, apart from a time dependent exponential phase factor of modulus unity. We may express this equation in terms of the intrinsic structure alone to find:

$$\Phi_L = \chi_L + i\hbar\dot{\alpha}\sum_k (E_k - E_L)^{-1}\langle k|\frac{\partial}{\partial\alpha}|L\rangle\chi_k \tag{43.2}$$

to first order in $\dot{\alpha}$, where

$$\langle k|\frac{\partial}{\partial\alpha}|L\rangle = \int \chi_k^*(\alpha,\boldsymbol{r})\frac{\partial\chi_L}{\partial\alpha}(\alpha,\boldsymbol{r})\,d^3r. \tag{43.3}$$

Associated with the changing field is a collective flow, a mass transport of nuclear matter whose velocity fields is proportional to $\dot{\alpha}$. The extra energy of the system above E_L, the total kinetic energy of the collective flow throughout the nucleus is given by:

$$T_{\text{Coll}} = \langle\Phi_L|\,H - E_L\,|\Phi_L\rangle = \tfrac{1}{2}B\dot{\alpha}^2 \tag{43.4}$$

where

$$B = 2\hbar^2\sum_k (E_k - E_L)^{-1}\left|\langle k|\frac{\partial}{\partial\alpha}|L\rangle\right|^2 \tag{43.5}$$

is the inertial parameter corresponding to the deformation variable α. If the deformation variable is one of the α_μ, the quantity B is designated as the mass parameter.

As an important example, if the nuclear potential rotates slowly with frequency ω, the collective rotational energy is given by (36.2) where the nuclear moment of inertia is of the form[1]

$$\mathfrak{J} = 2\hbar^2\sum_k (E_k - E_L)^{-1}|\langle k|\,J_\perp\,|L\rangle|^2 \tag{43.6}$$

and

$$J_\perp = i\hbar\frac{\partial}{\partial\vartheta_\perp} \tag{43.7}$$

the component of angular momentum about the axis of rotation[2]. For the axially symmetric nuclei considered here, the moment of inertia is understood to refer to rotations perpendicular to the nuclear symmetry axis. (For rotations about the symmetry axis, $\mathfrak{J}_3$ vanishes, since rotation of the field has no effect on the intrinsic motion, i.e. the operator j_3 is diagonal with respect to the intrinsic wavefunctions and does not have matrix elements between different states.) The collective angular momentum is, of course, given by (36.1).

The relation between $\mathfrak{J}$ and B_2 for an axially symmetric nucleus is simple to obtain [*33*].

Since

$$\tfrac{1}{2}\mathfrak{J}\dot{\vartheta}^2 = \tfrac{1}{2}B_2\sum|\dot{\alpha}_\mu|^2, \tag{43.8}$$

we have

$$\mathfrak{J} = B_2\sum_\mu\left|\frac{\partial\alpha_\mu}{\partial\vartheta}\right|^2. \tag{43.9}$$

In the body system, all deformation parameters except a_0 vanish, since the shape is axially symmetric. In the laboratory system, the shape is characterized by

---

[1] D. R. INGLIS: Phys. Rev. **96**, 1059 (1954).

[2] Note that $\partial/\partial\vartheta$ refers here to the rotation of the body system with respect to the particles. In the usual definition of J, the factor $\partial/\partial\vartheta$ refers to the rotation of the particles with respect to the body system and its coefficient is then $-i\hbar$.

the parameters α_μ. Suppose that the symmetry axis, at first coincident with the z axis, is then rotated by an arbitrary angle. The change in the α_μ is then given by:

$$\delta\alpha_\mu = \alpha_\mu - \beta\,\delta_{\mu 0} \tag{43.10}$$

from which it can be shown that

$$\sum_\mu |\delta\alpha_\mu|^2 = -2\beta\,\delta\alpha_0. \tag{43.11}$$

Now using (41.7) we have

$$\alpha_0 = a_0\,\mathfrak{D}^2_{00}(\vartheta_i) = \beta\,P_2(\vartheta) \tag{43.12}$$

where ϑ is the angle between z and z' axes. For a small change of ϑ away from 0, (43.12) gives:

$$\delta\alpha_0 = \beta\,[P_2(\vartheta) - 1]_{\vartheta\to 0} \to -\tfrac{3}{2}\beta\,(\delta\vartheta)^2. \tag{43.13}$$

Substitution of (43.11) and (43.13) into (43.9) gives the result:

$$\mathfrak{J} = 3\,B_2\,\beta^2. \tag{43.14}$$

Note that while the φ_L contains admixed terms of order α, the density $(\Phi_L)^2$ and all multipole moments are unaffected by the time dependence except for terms of order α^2. In other words, while the time dependence of the potential gives rise to a collective velocity field at each point proportional to α in first order, the density distribution adjusts almost adiabatically to the changing deformation.

β) Dynamics of collective motion. Suppose that the intrinsic Hamiltonian is an anisotropic harmonic oscillator potential plus arbitrary correction terms independent of deformation (e.g. the Hamiltonian used by Nilsson plus residual interactions). The expression for the mass parameter can then be rewritten in a form which provides further insight into the dynamics of collective motion [*40*]. Before taking up this problem let us consider how the intrinsic wavefunction responds to changes in the deformation (assumed to take place extremely slowly). Associated with a change of the Hamiltonian form $H(\alpha_\mu, \boldsymbol{r})$ to $H(\alpha_\mu + \delta\alpha_\mu, \boldsymbol{r})$ is a change in the intrinsic wavefunction from $\chi_L(\alpha_\mu, \boldsymbol{r})$ to $\chi_L(\alpha_\mu + \delta\alpha_\mu, \boldsymbol{r})$. Now on the one hand we can express the "perturbed" wavefunction by a Taylor expansion about α_μ and use matrix mechanics, i.e.

$$\chi_L(\alpha_\mu + \delta\alpha_\mu, \boldsymbol{r}) = \chi_L(\alpha_\mu, \boldsymbol{r}) + \delta\alpha_\mu \sum_{k\neq L} \langle k|\frac{\partial}{\partial\alpha_\mu}|L\rangle\,\chi_k(\alpha_\mu, \boldsymbol{r}) \tag{43.15}$$

to first order in $\delta\alpha_\mu$. Alternatively, we may find the "perturbed" wavefunction by first order perturbation theory

$$\chi_L(\alpha_\mu + \delta\alpha_\mu, \boldsymbol{r}) = \chi_L(\alpha_\mu, \boldsymbol{r}) - \delta\alpha_\mu \sum_{k\neq L} (E_k - E_L)^{-1} \langle k|\frac{\partial H}{\partial\alpha_\mu}|L\rangle\,\chi_k(\alpha_\mu, \boldsymbol{r}). \tag{43.16}$$

Equating the last two formulas term by term we have:

$$\langle k|\frac{\partial H}{\partial\alpha_\mu}|L\rangle = -(E_k - E_L)\,\langle k|\frac{\partial}{\partial\alpha_\mu}|L\rangle \tag{43.17}$$

provided $k \neq L$.

For an intrinsic Hamiltonian of the form assumed above, (38.17) enables us to express the above result as follows:

$$\langle k|\sum_i r^2\,Y_{2\mu}|L\rangle = (m\,\omega_0^2)^{-1}\,(E_k - E_L)\,\langle k|\frac{\partial}{\partial\alpha_\mu}|L\rangle. \tag{43.18}$$

Now consider the effect of the change in α_μ on the multipole parameters $\langle\alpha_\mu^\varrho\rangle$ defined by (38.19). We obtain:

$$\left.\begin{aligned}\delta\langle\alpha_\mu^\varrho\rangle &= \left(\frac{4\pi}{5}\sum_i r_i^2\right)\left[\langle\delta\chi_L^*|\sum_i r^2 Y_{2\mu}|\chi_L\rangle + \text{c.c.}\right]\\ &= \delta\alpha_\mu\left[\frac{4\pi}{5}\sum_i r_i^2\right]\left[\langle k|\sum_i r^2 Y_{2\mu}|L\rangle\langle k|\frac{\partial}{\partial\alpha_\mu}|L\rangle + \text{c.c.}\right].\end{aligned}\right\}\qquad(43.19)$$

The change in $\sum_i r^2$ can be shown to be of order $|\delta\alpha_\mu|^2$ and has been neglected. With the help of (43.18) we may rewrite (43.19) in the form

$$\frac{\partial\langle\alpha_\mu^\varrho\rangle}{\partial\alpha_\mu} = \sum_k w_k \qquad (43.20)$$

where

$$w_k = \left[\frac{8\pi}{5} m w_0^2 \sum_i r_i^2\right](E_k - E_L)\left|\langle k|\frac{\partial}{\partial\alpha_\mu}|L\rangle\right|^2. \qquad (43.21)$$

The quantity w_k is the contribution of the excited intrinsic state k to the quantity $\delta\langle\alpha_\mu^\varrho\rangle/\delta\alpha_\mu$. Expressing the mass parameter in terms of the w_k we have

$$B_2 = \left[\frac{4\pi}{5} m\hbar^2\omega_0^2\sum_i r_i^2\right]\sum_k (E_k - E_L)^{-2} w_k. \qquad (43.22)$$

Now the mass parameter for irrotational flow is given by (4.6) for a uniform distribution and by

$$(B_2)_{\text{Irrot}} = \frac{5}{8\pi} m \sum_i r_i^2$$

for an arbitrary density distribution, so that we find:

$$\frac{B_2}{(B_2)_{\text{Irrot}}} = 2(\hbar\omega_0)^2\sum_k (E_k - E_L)^{-2} w_k. \qquad (43.23)$$

γ) "Distortion" and "mixing". We have seen in Sect. 38 that $\langle\alpha_\mu^\varrho\rangle$ and α_μ tend to stay approximately equal. Thus the sum of w_k over all excited states must be nearly equal to unity. Also it was shown that (according to the oscillator model) simple distortion of the nucleus, i.e. a linear transformation of the wavefunction of each nucleon, contributes about half of this effect. The remainder is due to the more complicated "mixing" process which involves large changes in the wavefunction of a few loosely bound nucleons. Now the excited states involved in distortion occur at the large energy, $2\hbar\omega_0$ above the ground state, at least for small deformations, i.e. they are mainly in the next major shell of the same parity as the ground state. As a simple example, consider a single particle in the ground state of the harmonic oscillator potential, a $1s$ state in the spherical limit. Any small quadrupole distortion of this state is equivalent to an admixture of $1d$ states.

On the other hand, the excited states involved in mixing are at comparatively low energies as long as the independent particle model is a good approximation. They usually occur in the same major shell as the ground state. As a first approximation we suppose that all these states occur at the same energy $\hbar\omega_0\zeta$ above the ground state and assume that $\zeta \ll 1$. In actual nuclei the states will of course occur at various energies, so that the above is to be regarded as an average excitation energy. From the above considerations it then follows that:

$$\frac{B_2}{(B_2)_{\text{Irrot}}} = \frac{\mathfrak{J}}{\mathfrak{J}_{\text{Irrot}}} = \zeta^{-2} \gg 1. \tag{43.24}$$

so that the inertial parameters are determined almost entirely by the mixing process. (The contribution of distortion to the above ratio of $\frac{1}{4}$, and it has been neglected.)

Residual interactions have the effect of keeping the intrinsic states apart, thus increasing the average excitation energy and decreasing the inertial parameters and the collective kinetic energy [36], [40]. In the limit of strong residual interactions, i.e. $\zeta \approx 1$, one expects that the distinction between distortion and mixing disappears and that the collective flow becomes irrotational. In fact in this hypothetical situation (mean free path small compared to nuclear diameter) the nucleus would behave very much like a liquid drop. On the other hand, in actual nuclei, where the nucleons move nearly independently, the collective flow is irrotational only under special conditions which are not realized in deformed nuclei. One such case occurs if all nucleons are in the lowest single particle state of an arbitrary potential[1]. We also obtain irrotational flow if the potential is of the anisotropic oscillator type and the particle configuration consists entirely of closed shells [38]. For this case the system performs simple distortional vibrations about a spherical equilibrium shape. However, both the particle excitation energies and collective vibration energies now turn out to be very large, equal to $2\hbar\omega_0$ according to the oscillator model; thus the adiabatic condition fails. Oscillations which may resemble this kind of vibration seem to occur in actual nuclei, and they result in enhanced $E2$ transition probabilities even for light nuclei near magic numbers [32]. As a consequence of "mixing", the mass parameter for collective oscillations is greatly increased, and also the effective surface tension, i.e. resistance against deformation is reduced (cf. Sect. 38). The combination of these two effects results in collective vibration energies smaller than average nucleonic excitation energies [40]. If all excited intrinsic states involved in mixing were at the same energy, the adiabatic condition would always be satisfied for the lowest states of collective vibration and even more strongly so for rotational states. However, it frequently happens, especially in odd-A nuclei, that one or more of these intrinsic states occurs at much lower energy, so that the adiabatic condition sometimes fails even for the lowest states (cf. next section).

Let us apply the above considerations to a system of particles moving independently in an anisotropic oscillator potential. States coupled to the ground state by the matrix elements of $\partial/\partial\vartheta$ must have n_3 differing from the ground state value by unity. Further it can be shown that the states involved in mixing have the same principal quantum number N as the ground state. These are all degenerate and their excitation energy is given by $\hbar\omega_0 \cdot \frac{3\alpha}{2}$ for small deformation. From (36.5) and (43.24) we find that the resulting moment of inertia has the same value as for solid body rotation of the whole nucleus although it is almost entirely due to the relatively few nucleons in unfilled shells. It has been shown that this result holds quite generally for particles moving in a slowly rotating anisotropic harmonic oscillator potential provided the deformation is such as to give minimum energy[1] [36].

[1] G. WICK: Phys. Rev. 73, 51 (1948).

δ) Effect of residual interactions in even-even nuclei. Next let us study the effect of residual interactions on the inertial parameters. For the study of even-even nuclei we use the simple version of the spheroidal core model discussed in Sect. 40 [*36*]. We will only treat rotations. The excited intrinsic state has $\Omega = \pm 1$ and its excitation energy relative to the $\Omega = 0$ ground state is given by:

$$\zeta = [(\tfrac{3}{2}\alpha)^2 + (\tfrac{1}{2}v)^2]^{\frac{1}{2}} + \tfrac{1}{2}v. \tag{43.25}$$

The pairing interaction, whose strength is denoted by v, increases this average energy denominator, as may be seen using Eqs. (40.1) or equivalently, from Fig. 32. Using (36.5) and (43.24) we have:

$$\frac{\mathfrak{J}}{\mathfrak{J}_{\text{Rig}}} = \left\{\left[1 + \left(\frac{v}{3\alpha}\right)^2\right]^{\frac{1}{2}} + \frac{v}{3\alpha}\right\}^{-2}. \tag{43.26}$$

If residual interactions were absent ($v = 0$), the moment of inertia would reduce to the rigid rotation value, as expected. If v is set equal to unity, we obtain the moment of inertia characteristic of irrotational flow. This corresponds to having the residual interactions sufficiently strong to destroy the nuclear shell structure. The effective nucleon mean free path would then be so small that the nucleus would actually behave like a liquid drop.

The empirical moments of inertia are intermediate between these two limiting cases. Empirical moments of the even-even nuclei with $150 < A < 190$ can be fitted quite well [*36*] with (43.26) provided we set v equal to $\frac{1}{3}$. This corresponds to a strength of the pairing forces of the right order of magnitude to account for the empirical even-odd staggering of nuclear masses[2].

Let us briefly turn to a qualitative consideration of collective vibrations about spherical shape. We expect that with increasing $\langle\beta^2\rangle$, the nuclear shape will be substantially deformed a larger fraction of time, which implies in turn that the average excitation energy of the admixed intrinsic states increases (cf. Fig. 32). Thus according to Eq. (43.24) the vibrational mass parameters should vary inversely with $\langle\beta^2\rangle$ and such a trend is indeed implied by the analysis of the empirical data (Sect. 42).

From the considerations of Sect. 40, one expects to find a rotational spectrum if the ratio α/v is larger than some critical value. If this ratio is too small, the zero-point oscillations will "wash out" any slight favoring of aspherical shape and we obtain instead a vibrational spectrum. According to (43.26), rotational spectra should also have the ratio $\mathfrak{J}/\mathfrak{J}_{\text{Rig}}$ larger than a certain critical value. From the analysis of spectra in the rare earth region, it appears that this critical ratio is approximately 0.23 or equivalently

$$(E_2)_{\text{crit}} \approx \frac{13\hbar^2}{\mathfrak{J}_{\text{Rig}}}. \tag{43.27}$$

---

[1] D. Inglis: Phys. Rev. **103**, 1786 (1956). — J. G. Valatin: Proc. Roy. Soc. Lond. **238**, 132 (1956).

[2] Qualitatively, at least, the increased energy differences between intrinsic states can be attributed to a velocity dependence of the nuclear potential or, equivalently, a reduced effective nucleon mass in nuclear matter [W. J. Swiatecki: Phys. Rev. **101**, 1321 (1956)]. We have seen in Sect. 9 that such an effect is in fact required by the existence of nuclear saturation. It is interesting that empirical moments of inertia for the most strongly deformed even-even nuclei are about half of the rigid values. The same result would be obtained if each nucleon had actually only half of the free mass but was moving independently in the rotating nuclear potential [R. J. Blin Stoyle: Nuclear Phys. **2**, 169 (1956)]. It appears that the concept of a reduced effective nucleon mass is a crude way of representing at least some of the effects of correlations on the average level spacing, though it does not lead to any positional correlations in the wavefunction.

Spectra for which $E_2 < (E_2)_{\text{crit}}$, are expected to be rotational, while vibrational spectra are associated with $E_2 > (E_2)_{\text{crit}}$.

ε) Odd-A nuclei. Moments of inertia in odd-*A* nuclei are almost always larger than in neighboring even-even nuclei, yet consistently smaller than the rigid rotation values. These features can be qualitatively understood as follows [*36*]: The last odd nucleon is especially easy to excite without breaking any pairs. This unpaired nucleon can contribute significantly to the moment of inertia, usually about 10 to 20%. This extra contribution is, of course, absent for even-even nuclei. The largest contribution to the moment of inertia comes from the loosely bound paired nucleons in the last unfilled shell. The tightly bound nucleons in filled shells are expected to contribute altogether only about 10 to 20%, no more than the single unpaired nucleon.

44. Rotational perturbations of the intrinsic structure. The simple $I(I+1)$ law comes close to fitting most rotational spectra very well, but strictly speaking, it represents a limiting situation: namely that the rotational motion is so slow that the intrinsic structure is not perturbed appreciably. As the rotational frequency increases, both distortion of the nuclear shape due to centrifugal forces (rotation-vibration interaction) and non-adiabatic perturbations of the particle structure due to Coriolis forces (rotation-particle coupling) can become important.

α) Rotation-vibration interaction. The nuclear rotation-vibration interaction [*33*] to [*35*] is analogous to an effect of the same name occurring in moleculer spectra and is associated in both cases with an increase of the moment of inertia with angular momentum.

To study this effect more quantitatively let us specialize to spheroidal shapes for the moment. Suppose that in the absence of any rotation the nucleus would have an equilibrium deformation. Then for small deviations from equilibrium the deformation potential is of the form:

$$V(\beta) = E_0 + \tfrac{1}{2} C_\beta (\beta - \beta_0)^2 \tag{44.1}$$

where E_0 is the intrinsic energy at equilibrium, and C is the restoring term. The nucleus can vibrate about this equilibrium shape. Assuming that the mass parameter β is independent of deformation, the vibrational frequency is given by:

$$\omega_\beta = \left(\frac{C_\beta}{B_2}\right)^{\frac{1}{2}}. \tag{44.2}$$

For strongly deformed nuclei, it is much easier to excite rotations than vibrations. For a spheroidal nucleus, the rotational moment of inertia is given by (43.14).

To find the equilibrium shape of a rotating nucleus, we must add the energy of rotation to Eq. (44.1). The result is:

$$E(\beta) = E_0 + \frac{B\,\omega_\beta^2}{2} (\beta - \beta_0)^2 + \frac{\hbar^2}{6B\,\beta^2} I(I+1) \tag{44.3}$$

which has a minimum at:

$$\beta = \beta_0 \left[1 + 12 \left(\frac{\hbar^2}{2\mathfrak{J}_0}\right)^2 \frac{I(I+1)}{(\hbar\,\omega_\beta)^2}\right] \tag{44.4}$$

to first order in $I(I+1)$ where the quantity $\mathfrak{J}_0$ refers to the "unperturbed" moment of inertia, the value for $\beta = \beta_0$. Using (43.14) we have, to the same order of approximation

$$\mathfrak{J} = \mathfrak{J}_0 \left[1 + 24 \left(\frac{\hbar^2}{2\mathfrak{J}}\right)^2 \frac{I(I+1)}{(\hbar\,\omega_\beta)^2}\right]. \tag{44.5}$$

The rotational energy spectrum now deviates slightly from the $I(I+1)$ law so that we have:

$$E_I = \left(\frac{\hbar^2}{2\mathfrak{J}_0}\right) I(I+1) + E_\beta^{(2)} I^2 (I+1)^2. \tag{44.6}$$

where

$$E_\beta^{(2)} = -\frac{12}{(\hbar\omega_\beta)^2}\left(\frac{\hbar^2}{2\mathfrak{J}_0}\right)^3. \tag{44.7}$$

The rotational motion is also perturbed by γ vibrations, i.e. ellipsoidal vibrations about the equilibrium shape with fixed β. Detailed calculations [*33*] show that this correction is somewhat smaller than (44.7) giving a contribution:

$$E_\gamma^{(2)} = -\frac{4}{(\hbar\,\omega_\gamma)^2}\left(\frac{\hbar^2}{2\mathfrak{J}_0}\right)^3 \tag{44.8}$$

to $E^{(2)}$.

The perturbation of the rotational spectrum is small if the rotational energies are small compared to vibrational energies. Substituting (44.7) and (44.8) into (44.6) we obtain [*35*]:

$$\frac{\Delta E}{E_{\mathrm{Rot}}} = -\frac{1}{I(I+1)}\left[\frac{12}{(\hbar\,\omega_\beta)^2} + \frac{4}{(\hbar\,\omega_\gamma)^2}\right] E_{\mathrm{Rot}}^2. \tag{44.9}$$

We see that for states of large I the relative correction is small even when these energies are comparable, though, of course, the *absolute* deviation from the $I(I+1)$ law is larger than for states of small I. Note that the corrections to the energies due to rotation-vibration interaction are always negative.

Finally consider the spectrum of Pu^{238} (Fig. 40). We may determine the values of the two constants in Eq. (44.6) from the observed excitation energies of the 2^+ and 4^+ states as the ground state band. This gives $\hbar^2/2\mathfrak{J}_0 = 7.37$ kev and $E^{(2)} = 0.0034$ kev. Using (44.6) the 6^+ state is then predicted to occur at 303.4 kev in very good agreement with the empirical value of 303.7 kev. From (44.9) and assuming $\omega_\beta = \omega_\gamma$, we find $\hbar\omega = 1.35$ Mev which fits quite well with the empirical average vibrational energy of 1 Mev.

β) Rotation-particle coupling. The rotation of the nuclear symmetry axis gives rise to a partial decoupling of the intrinsic motion from the instantaneous collective field. (If the rotation were sufficiently rapid, the nucleons would feel not the instantaneous field but the field averaged over time.) As a result of this partial decoupling, the quantity K is no longer a good quantum number. Thus there will be at least a partial breakdown of the K selection rules and intensity rules, and deviations from the simple $I(I+1)$ law will occur.

To study this effect we go back to the original Hamiltonian (32.5). In most of the previous considerations, the coupling term (32.9) between I and j has been neglected. In this approximation, the nuclear spectrum separated into rotational bands, each of which is characterized by a unique intrinsic wave-function.

Let us now take the coupling term into account [*39*]. For sake of simplicity we consider the simplest possible case for which the rotational coupling can have an important effect—the case that there is a single low lying intrinsic configuration coupled to the ground state by the Coriolis term (32.9). The component of angular momentum parallel to the symmetry axis must differ by unity for any coupling to occur, and we will suppose here that the ground state has the lower value K, i.e. the excited state has a value $K+1$. We neglect here any additional decoupling which will generally occur if $K = \frac{1}{2}$. Suppose that in the absence of coupling the rotational states would occur at energies $E_K(I)$ and

$E_{K+1}(I)$ respectively. The matrix element of the rotational coupling between these states is given by (II.8)

$$\langle \psi^I_{K+1} | -\frac{\hbar^2}{\mathfrak{J}} \boldsymbol{I} \cdot \boldsymbol{j} | \psi^I_K \rangle = -[(I-K)(I+K+1)]^{\frac{1}{2}} A_K \tag{44.10}$$

where

$$A_K = \frac{\hbar^2}{2\mathfrak{J}} \langle K+1 | j_1 + i j_2 | K \rangle = \frac{\hbar^2}{\mathfrak{J}} \langle K+1 | j_\perp | K \rangle . \tag{44.11}$$

(Only states of the same I can be coupled to each other.)

β 1) Effect on the energy spectrum. The energy levels of the system are given by:

$$\left.\begin{aligned} E(I) &= \tfrac{1}{2}[E_{K+1}(I) + E_K(I)] \pm \\ &\pm \tfrac{1}{2}\{[E_{K+1}(I) - E_K(I)]^2 + 4A_K^2 (I-K)(I+K+1)\}^{\frac{1}{2}} . \end{aligned}\right\} \tag{44.12}$$

To consider a limiting case; if the two states are degenerate in the absence of coupling, we have:

$$E(I) = E_K(I) \pm A_K [(I-K)(I+K+1)]^{\frac{1}{2}} . \tag{44.13}$$

In this case, or if the two states are close lying, the perturbation can have a large effect on the rotational spectrum, possibly enough to make it unrecognizable as such. The well known decoupling for $K = \frac{1}{2}$ can be regarded as this kind of effect [corresponding to $K = -\frac{1}{2}$ in (44.13)], but the symmetry requirements on the wavefunction introduce some additional complications here.

When the energy of the two configurations is very different, i.e. the energy difference is large compared to the magnitude of the coupling term (44.11), we may find the energy of coupled system by a perturbation expansion or for the above case, by an expansion of (44.12) in powers of A_K^2.

In first order, the energy of the state which becomes $\psi_K(I)$ in the absence of coupling is:

$$E(I) = E_K(I) - \frac{I(I+1) - K(K+1)}{E_{K+1}(I) - E_K(I)} A_K^2 . \tag{44.14}$$

The most important effect of the coupling on the energy spectra is to reduce the coefficient of the term proportional to $I(I+1)$. Suppose that in the absence of coupling the energy levels of the ground and excited state bands are:

$$E_K(I) = E_K^0 + \frac{\hbar^2}{2\mathfrak{J}_K^0} I(I+1) , \tag{44.15a}$$

$$E_{K+1}(I) = E_{K+1}^0 + \frac{\hbar^2}{2\mathfrak{J}_{K+1}^0} I(I+1) . \tag{44.15b}$$

Then in lowest order, the coupling does not alter the form of Eqs. (44.15) but merely increases the effective moment of inertia of the ground state band by the amount:

$$\Delta\mathfrak{J} = \mathfrak{J}_K^0 \frac{2A_K^2}{E_{K+1}^0 - E_K^0} = 2\hbar^2 \frac{|\langle K+1 | j_\perp | K \rangle|^2}{E_{K+1}^0 - E_K^0} . \tag{44.16}$$

The moment of inertia of the excited state band is decreased by the same amount. This is precisely the same result as would be obtained by using time dependent perturbation theory[1] (43.6). In other words, the rotational coupling between the two levels can be taken into account either explicitly as was done in the

[1] G. Lüders: Z. Naturforsch. **11**a, 617 (1956).

above treatment or "collectively" in the sense that we include *only* its contribution to the nuclear moment of inertia, i.e. as in the approach of the last section.

In lowest order the two approaches give the same results. Of course even in the present treatment we are still considering the effect of all *other* levels "collectively". These contribute a moment of inertia $\mathfrak{J}_K^0$ to the ground state band. Although strictly speaking *any* rotational coupling is non-adiabatic, we would treat explicitly (as a specific non-adiabatic perturbation) only the coupling between the configuration K and $K+1$ in the present discussion. In the simple version of the rotational model we do not consider these couplings at all (provided $K \neq \frac{1}{2}$) except in so far as they contribute to the moment of inertia, i.e. we regard the intrinsic structure as following the rotation adiabatically.

In addition to "renormalizing" the moments of inertia, the rotational coupling also gives rise to deviations from the $I(I+1)$ law. In lowest order the correction term can be shown to be of the form:

$$\frac{\Delta\mathfrak{J}}{2\mathfrak{J}^4}\,\frac{1}{E_{K+1}^0 - E_K^0}\,(\mathfrak{J}_K - \mathfrak{J}_{K+1} - \Delta\mathfrak{J})\, I^2(I+1)^2. \tag{44.17}$$

In this expression $\mathfrak{J}_K$ and $\mathfrak{J}_{K+1}$ denote the effective moments of inertia, calculated by summing (43.6) over all excited states. It is assumed that $\mathfrak{J}_K$ and $\mathfrak{J}_{K+1}$ are not very different and their average value is denoted by $\mathfrak{J}$. Note that

$$\mathfrak{J}_K - \mathfrak{J}_{K+1} - \Delta\mathfrak{J} = \mathfrak{J}_K^0 - \mathfrak{J}_{K+1}^0 + \Delta\mathfrak{J} \tag{44.18}$$

thus if the effective moments of inertia are equal, the coefficient of the $I^2(I+1)^2$ term is negative, the same sign as for rotation-vibration interaction. The moment of inertia increases slightly with angular momentum. On the other hand, if the *unrenormalized* moments of inertia are equal, the coefficient of the $I^2(I+1)^2$ term is positive.

β 2) Effect on transition probabilities. The explicit treatment of rotational coupling between low lying states makes it possible to account for small deviations of transition probabilities from the simple rules of Sect. 35. The number K is no longer a constant of the motion. If the perturbations are small, the wave-functions of states in the lowest rotational band are now given by[1] [*39*]

$$\psi^I = (1 - a_{K+1}^2)^{\frac{1}{2}}\psi_K^I + a_{K+1}\psi_{K+1}^I \tag{44.19}$$

where

$$a_{K+1} = \left[\frac{I(I+1) - K(K+1)}{E_{K+1} - E_K}\,\frac{\Delta\mathfrak{J}}{2\mathfrak{J}^2}\right]^{\frac{1}{2}} \tag{44.20}$$

to lowest order in the perturbation.

To illustrate the effect of this perturbation, consider the $E2$ transition probabilities between the nuclear ground state $K = I_0$ and one of the excited states I in the lowest band, assuming that the perturbing band has the higher value of $K(=I_0+1)$. The ground state is not affected in this case, but each excited state now is "diluted" by an admixture of the intrinsic state $K+1$. Since the collective quadrupole operator only couples states of the same K, its matrix elements will be reduced to a fraction

$$\frac{B(E2)}{B^0(E2)} = 1 - a_{K+1}^2 \tag{44.21}$$

of their original values. On the other hand, transitions between bands differing in K one unit will be greatly enhanced by the same coupling since in a sense the transition can now partially take place without change of K.

[1] See also G. Lüders: Z. Naturforsch. **12**a, 353 (1957).

β 3) The level spectrum of W^{183}. The effect of coupling between rotational bands is well illustrated by the accurately determined level scheme of W^{183} shown in part d of Fig. 42. There are two close-lying rotational bands with $K=\frac{1}{2}$ and $K=\frac{3}{2}$ and both have even parity. The rotational energies in both bands can be quite nicely fitted by treating the bands as independent (part a). For the $K=\frac{1}{2}$ ground state band, we must, of course include the decoupling effect. The parameters which give the best fit on this basis are:

$$\frac{\hbar^2}{2\mathfrak{J}_{\frac{1}{2}}}=13.0\text{ kev},\qquad a=0.19,\qquad \frac{\hbar^2}{2\mathfrak{J}_{\frac{3}{2}}}=16.7\text{ kev}. \tag{44.22}$$

According to the simple rotational model, collective $E2$ transition should occur only within a band. Thus, Coulomb excitation of the ground state should lead

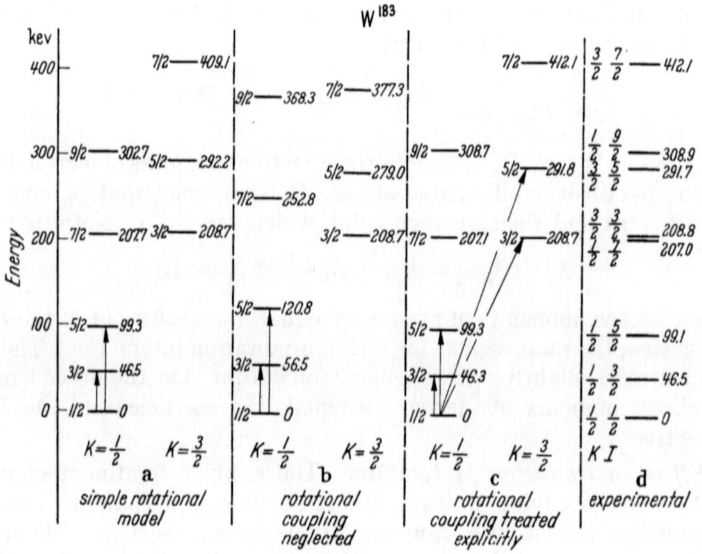

Fig. 42 a—d. Level spectrum of W^{183} illustrating rotation-particle coupling. The data on which this figure is based are taken from Ref. [39].

only to states of the $K=\frac{1}{2}$ band. However, in W^{183} the Coulomb excitation is known to excite the $K=\frac{3}{2}$ band with appreciable probability. Thus we have for the branching ratio for transition to the two $I=\frac{3}{2}$ states

$$\frac{B(E2;\frac{1}{2}\frac{1}{2}\rightarrow\frac{3}{2}\frac{3}{2})}{B(E2;\frac{1}{2}\frac{1}{2}\rightarrow\frac{1}{2}\frac{3}{2})}=0.06 \tag{44.23}$$

and the corresponding branching ratio for transition to the $I=\frac{5}{2}$ states is 0.14.

A more careful examination of the energy spectra also reveals some unusual features. Thus while the first four states of the $K=\frac{1}{2}$ band fit very nicely to the simple rotational law corrected only for $K=\frac{1}{2}$ decoupling, the $K=\frac{9}{2}$ state is found to be 6 kev above the predicted values.

The comparatively strong $E2$ transitions between the two rotational bands as well as the slight deviation from the simple rotational law can be quantitatively accounted for as resulting from Coriolis coupling between these bands . If the rotational coupling between the two bands is arbitrarily *not* considered, at all, the moment of inertia of the two bands is appreciably changed, as is shown in (b) If we now consider the rotational coupling between these two states explicitly we obtain the spectrum shown in (c), which is similar to (a), but in better agreement with experimental data. The analysis of the empirical data requires only one additional parameter, the coupling strength A_K. As we have seen, K is no longer a good quantum number. The impurity of each state in the ground state band (and also of the $K=\frac{3}{2}$ band) is given by $a_{\frac{3}{2}}^2$ defined in (44.19). The calculated values of this admixture are 0, 0.06, 0.12, 0.23, and 0.26 for $I=\frac{1}{2}, \frac{3}{2}, \frac{5}{2}, \frac{7}{2}, \frac{9}{2}$, respectively. The effect of coupling evidently becomes more pronounced for states of large angular momentum. On the other hand, the ground state is not perturbed at all.

The "unrenormalized" moments of inertia are calculated to be:

$$\frac{\hbar^2}{2\mathfrak{J}_{\frac{1}{2}}^0} = 15.9 \text{ kev}, \qquad \frac{\hbar^2}{2\mathfrak{J}_{\frac{3}{2}}^0} = 14.0 \text{ kev}. \tag{44.24}$$

The rotational coupling increases the effective moment of inertia of the ground state band by about 20% and decreases it by the same amount in the other band. We note that $\mathfrak{J}_{\frac{1}{2}}^0$ is much closer to $\mathfrak{J}$ in the neighboring even-even nucleus W^{182} ($\hbar^2/2\mathfrak{J}=16.7$ kev), than is $\mathfrak{J}_{\frac{1}{2}}$. The remaining discrepancy might easily be a result of other higher order effects or the coupling to another $K=\frac{3}{2}$ band at higher energy. Note also that the unrenormalized moments (44.24) differ by a smaller amount than the effective moments (44.22). In view of (44.17) this implies that the $I^2(I+1)^2$ correction terms are positive, in agreement with experiment.

β 4) Self-consistency of the rotational model. It has been previously pointed out that not only deviations from the rotational law, but the entire rotational structure itself is due to non-adiabatic excitations. Thus, in principle, we should consider the coupling not only between close lying levels as in W^{183}, but between *all* levels, i.e. the combined effect of all non-adiabatic effects, at least to convince ourselves that these do not perturb the rotational structure appreciably, i.e. that the rotational model is self-consistent.

First let us look at transition probabilities within a rotational band. With the help of (44.20) we can show that the Coriolis coupling changes the transition probabilities between different bands by a fraction of the order of magnitude

$$\frac{\delta B(E2)}{B^0(E2)} = \sum_k \frac{\Delta\mathfrak{J}^k}{\mathfrak{J}} \frac{E_{\text{Rot}}}{E_k - E_L} \approx \frac{E_{\text{Rot}}}{(E_k - E_L)_{\text{Av}}} \tag{44.25}$$

where $\delta B(E2)$ refers to the change in the transition probability between two different bands and $B^0(E2)$ is a typical reduced transition probability between states within the same band. The term $(E_K-E_L)_{\text{Av}}$ denotes an appropriate average of the intrinsic excitation energies, E_{Rot} is a rotational energy and $\Delta\mathfrak{J}^k$ the contribution of state K to the moment of inertia.

Consider now the effect of the perturbation on energy spectra. From the result Eq. (44.17) it seems plausible that fractional deviations from the $I(I+1)$ law will be of the following order of magnitude:

$$\frac{\delta E_{\text{Rot}}}{E_{\text{Rot}}} \approx \sum_k \left(\frac{\Delta\mathfrak{J}^k}{\mathfrak{J}}\right)^2 \frac{E_{\text{Rot}}}{E_k - E_L} = \frac{\Delta\mathfrak{J}^{\text{Av}}}{\mathfrak{J}} \frac{E_{\text{Rot}}}{(E_k - E_L)_{\text{Av}}}. \tag{44.26}$$

where $\Delta\mathfrak{J}^{\text{Av}}$ is the average contribution of each state to $\mathfrak{J}$. From the considerations of Sect. 43 it follows that in heavy nuclei $(E_K-E_L)_{\text{Av}}$ is several Mev, while the rotational energies are only a few hundred kev. Consequently, both of the above deviations, especially the second, are expected to be small, at least for heavy nuclei. In particular, since many states are expected to contribute to $\mathfrak{J}$; i.e. $\Delta\mathfrak{J}_{\text{Av}} \ll \mathfrak{J}$ whenever there are many nucleons outside filled shells, the energy spectra should deviate less from the prediction of a simple rotational model than do transition probabilities. This is in agreement with experimental evidence.

Finally, it may sometimes be difficult to tell to what extent deviations from an $I(I+1)$ law are due to rotation particle coupling or to rotation vibration interaction. However, the latter effect is believed to predominate in even-even nuclei; excited intrinsic states usually occur here at higher energy than vibrational states.

45. Relation of the unified model to the shell model. *α) Successes of the two models.* While most known rotational spectra occur in heavy nuclei, their existence is by no means restricted to this group. Thus at least some of the excited states in a number of medium weight odd-A nuclei, e.g. Se^{77}, Ag^{107}, can also be described as rotational, at least to a fair degree of accuracy[1]. Even more striking is the existence of well defined rotational bands recently found in odd-A nuclei of mass numbers around 25[2,3].

It may at first seem somewhat surprising that light nuclei should be capable of having collective excitation spectra in view of the rather small number of nucleons in the unfilled major shell ($8 \leq N$ or $Z \leq 20$). However these nucleons constitute an appreciable fraction of the total, in fact, a larger fraction than in heavy nuclei. Further, in these light nuclei, N and Z are approximately equal so that neutrons and protons can act coherently in deforming the nucleus.

In view of these considerations, it is expected that the nuclei of mass number around 25 are appreciably non-spherical and indeed their deformations may well be even larger than those of the heavy nuclei. However since the nuclear rotation is due almost exclusively to the loosely bound nucleons, of which there are very few in these nuclei, the frequency of this rotation may be comparable to the characteristic frequencies of individual nucleons. Thus, even if a spectrum is basically rotational, non-adiabatic effects (Sect. 44) may perturb its level structure to a much larger extent than in most heavy nuclei.

Let us briefly mention the regions of nuclei where the shell model has had success. First the ground state spins and parities of almost all odd-A nuclei (except those in the region of heavy strongly deformed nuclei) are consistent with predictions of the single particle shell model. This is, indeed, one of the greatest accomplishments of the shell model. On the other hand, to understand further details of nuclear states, e.g. energy spectra, moments, transition probabilities in the framework of the shell model, it is usually necessary to take into account the direct coupling of the loosely bound nucleons through their mutual interactions [*24*]. This approach, the many-particle shell model, appears to be mathematically feasible only for nuclei with but a few nucleons outside of (or missing from) closed shell configurations and for the group of light nuclei with $4 \leq A \leq 16$, but it has proven very successful in accounting for many detailed properties of just such nuclei [*24*]. The regions of nucleon numbers for which the shell model or rotational model has proven successful in accounting for details

---

[1] Cf. footnote 2, p. 521.

[2] Litherland, Paul, Bartholomew and Gove: Phys. Rev. **102**, 208 (1956).

[3] In addition, there is some evidence that Li^7 and Be^8 are also well described by the rotational model. It is interesting to note that mutual spin-independent interactions of $1p$ nucleons would by themselves (L-S coupling) always lead to a set of rotational bands. However, because of the exclusion principle, no L-values larger than 4 can occur. Anything which splits the single particle states (in the average central potential), e.g. spin-orbit coupling, tends to destroy the rotational structure, but only in higher order. Thus the $1p$ shell nuclei with mass numbers above 10 (for which the empirical spin-orbit coupling is appreciably stronger than for the lighter nuclei, cf. Ref. 24), do not exhibit any evident rotational structure.

of level spectra are illustrated in Fig. 43. For many other even-even nuclei some of the low lying states can be approximately interpreted as vibrational excitations (Sect. 41).

β) The level spectrum of F^{19}. There is one nucleus, F^{19}, for which the properties of ground state and low lying states have been successfully interpreted on the basis of both the shell model and the unified model, in particular the rotational

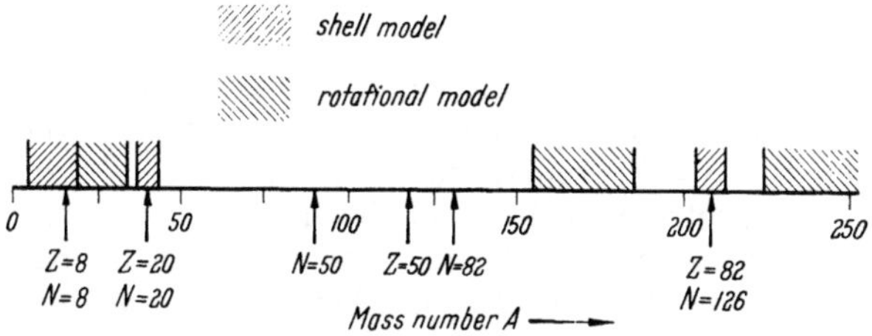

Fig. 43. Nuclei for which detailed shell-model or rotational-model analyses of level schemes have been made.

model. We have here three particles outside the O^{16} core of closed shells. The analysis in the level spectrum is illustrated in Fig. 44. The shell model approach was carried through by ELLIOTT and FLOWERS[1] and independently by REDLICH[2].

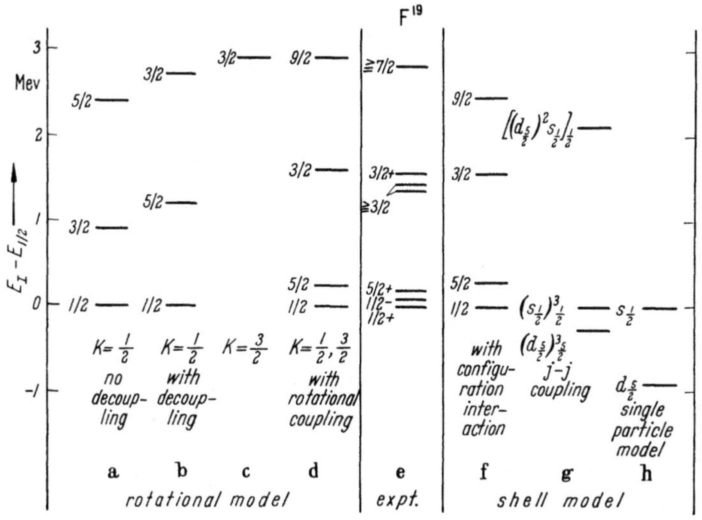

Fig. 44 a—h. Alternative interpretations of F^{19} level scheme on basis of shell model and rotational model. Parts (a) to (e) are taken from Fig. 1 of E. B. PAUL, Phil. Mag. **15**, 311 (1957). Only levels below 3 Mev are shown in this figure and calculated energy levels are shown only for states of even parity. Part (f) is taken from Fig. 6 of J. P. ELLIOTT and B. H. FLOWERS, Proc. Roy. Soc. Lond. **229**, 536 (1955), with V_c set equal to 40 Mev, and part (h) is taken from Fig. 1 of this reference. Part (g) is based on Table V of M. G. REDLICH, Phys. Rev. **99**, 1427 (1955).

The single particle shell model does not give the correct ground state spins and predicts that there should be only one excited state below 5 Mev (as in O^{17}) as is shown in part h. As may be seen in part g the situation is only slightly improved by taking into account the interactions between the three nucleons in the unfilled shell if we insist that the angular momentum j of each particle remains a good quantum number. However if configuration interaction is taken into account, the character of the spectrum changes completely and good agreement with experiment may be attained (part f) by suitable choice of the interaction parameters.

[1] J. P. ELLIOTT and B. H. FLOWERS: Proc. Roy. Soc. Lond., Ser. A **229**, 536 (1955).
[2] M. REDLICH: Phys. Rev. **99**, 1427 (1955).

Recently, Paul[1] has given an alternative interpretation of the F^{19} excitation spectrum on basis of the rotational model. The nucleons are assumed to move essentially independently in a rotating spheroidal well. The relevant single particle energies and wavefunctions have been calculated by Nilsson [*42*], and it is found that the ground state has K, the component of angular momentum along the symmetry axis, equal to $\frac{1}{2}$. The actual level scheme does not even approximately follow the simple $I(I+1)$ law (part a), however. First of all we have the well-known decoupling effect which is always present in $K=\frac{1}{2}$ bands. For the present case it is in fact large enough to alter the level ordering which improves the agreement with experiment (part b). Secondly, there is rotational coupling between the $K=\frac{1}{2}$ ground state band and $K=\frac{3}{2}$ excited band with further changes the spectrum to the form shown in part c so as to bring it into good agreement with the empirical one. The nonadiabatic perturbations, i.e. deviations from a simple $I(I+1)$ law, are larger here than for heavier nuclei mainly because we have fewer particles contributing to the deformation. However, these perturbations can be calculated theoretically so that the spectrum can be accounted for on basis of the rotational model, although the rotational structure is now practically unrecognizable.

The evident success of both shell and rotational models in accounting for the energy spectrum of F^{19} (and also transition probabilities) leads one to speculate that these models may be basically much more similar than might have been thought considering the very apparent differences between them.

γ) Significance of the large inertial parameters. The large values of the inertial parameters for low energy collective oscillations as compared to the values characteristic of irrotational flow (cf. Sect. 43) imply that the loosely bound nucleons play the predominant role in the collective motion. In connection with this point, it is of interest to consider the behavior of the multipole parameters α_μ^ϱ defined by:

$$\alpha_\mu^\varrho = \frac{4\pi}{5} \frac{\sum_i (r^2 Y_{2\mu})_i}{\sum_i r_i^2}. \tag{45.1}$$

We have already seen (Sect. 38) that the expectation values of each multipole parameter should be approximately equal to the corresponding deformation parameter α_μ of the binding potential, provided the potential is of the oscillator form. However, the dynamical behavior (i.e. the time derivatives) of α_μ^ϱ and of α_μ may be quite different.

Thus, consider the nuclear Hamiltonian

$$H = \sum_i \frac{P_i^2}{2M} + \sum_{i<k}\sum V_{ik} \tag{45.2}$$

where V_{ik} are the two body potentials.

Suppose we neglect all exchange effects and velocity dependence and treat $\sum_i r_i^2$ as a constant. Then the multipole parameter commutes with the potential, so that, independent of the detailed form of interactions, we can readily evaluate the following double commutator:

$$[\alpha_\mu^\varrho, [H, \alpha_\mu^\varrho]] = \frac{4\pi\hbar^2}{M\left\langle\sum_i r_i^2\right\rangle} \tag{45.3}$$

to lowest (zero) order in α_μ^ϱ.

[1] E. B. Paul: Phil. Mag. **15**, 311 (1957).

On the other hand, according to the simple collective model (no intrinsic excitations) the Hamiltonian is given by:

$$H_{\mathrm{coll}} = \sum_{\mu} \frac{\pi_\mu^2}{2 B_2} + V(\alpha_\mu) \tag{45.4}$$

with B_2 set equal to a constant and $V(\alpha_\mu)$ denoting the deformation potential. Thus we find

$$[\alpha_\mu, [H_{\mathrm{coll}}, \alpha_\mu]] = \frac{5}{2} \frac{\hbar^2}{B_2} \tag{45.5}$$

independent of the form of $V(\alpha_\mu)$. If α_μ and α_μ^ϱ were strictly equal, we would expect the two commutators (45.3) and (45.5) also to be equal. Actually we find

$$\frac{[\alpha_\mu^\varrho, (H, \alpha_\mu^\varrho)]}{[\alpha_\mu, (H, \alpha_\mu)]} = \frac{B_2}{(B_2)_{\mathrm{Irrot}}}. \tag{45.6}$$

The fact that this ratio is larger than unity seems to be closely connected with fluctuations of α_μ^ϱ about α_μ. These fluctuations are in turn an essential consequence of the approximately independent motions of nucleons in the nucleus. Thus the collective motion in nuclei is of a rather complicated character not describable by any potential flow model. It is true that nuclear matter prefers a certain average density; however for the density to remain *rigorously* constant it would be necessary that as a nucleon moves through the nucleus, the nucleons in front of it move out of the way. This in turn would imply that the interactions between particles are very strong, e.g. as in a liquid drop. Indeed in this limiting case we would find $B_2 = (B_2)_{\mathrm{Irrot}}$.

Besides collective oscillations, the unified model also encompasses low-energy excitations[1] of individual loosely bound nucleons. Thus according to the unified model, the low energy properties of nuclei, whether involving collective or individual particle motions, turn out to be mainly due to the loosely bound nucleons. In the shell model, we *assume* that this is the case, i.e. we attempt to represent each low-lying nuclear state as a linear configuration of a relatively small number of Slater determinants.

δ) Degrees of freedom in the unified model. An important problem in the unified model is the question of the number of degrees of freedom. It is well known that an A-particle system has $3A-3$ internal degrees of freedom. However in nuclei it is only the degrees of freedom associated with the loosely bound nucleons which are of significance at least for most low energy phenomena. In principle we could account for all degrees of freedom either by treating the particle excitations explicitly (i.e. an extended shell model refined to take into account *all* configuration interaction). We could also do this by treating all excitations as variations of the nuclear shape and orientation—the response of the nucleons to the time dependence of the nuclear binding potential, provided we include an arbitrary large number of deformation parameters. Actually neither of the above approaches can be applied because of the enormous mathematical difficulties involved.

[1] It may be of interest to mention the possible kinds of excitation in two other many-body systems. A degenerate electron gas (e.g. in a metal) is subject to both single particle excitations and collective excitations. [D. PINES: Solid State Physics **1**, 368 (1955).] The latter involve a large number of particles, i.e. they are essentially sound waves, in contrast to the nuclear case. In liquid helium at low temperature there are also two kinds of excitation both of them collective. [R. P. FEYNMAN and M. COHEN: Phys. Rev. **102**, 1189 (1956).] One of them, the "phonon" excitations, are again sound waves involving many particles. The second, "roton" excitations, involve motions of only a few (perhaps 3 to 6) particles at a time.

On the other hand, in the various versions of the unified model we consider some modes of excitation of each kind. There remain, however, the problem of being sure that each important degree of freedom is counted *once*, either as particle or as collective.

ε) Expression of the rotational wavefunction in shell model language. Let us consider the nuclear wavefunction according to the rotational model. The intrinsic wavefunction determines the nature of the single particle excitations while the rotational function refers to the collective rotational motion. In so far as the nucleons move approximately independently in the well, the intrinsic function is essentially a single Slater determinant. [Of course, it may be necessary to take some residual interactions (Sect. 40) into account. Furthermore in some cases, coupling between intrinsic and rotational motion must be considered cf. Sect. 44].

How can the simple rotational function (32.11) be expressed in the language of the shell model? A plausible approach is to regard the intrinsic function as referring to the wavefunction of *all* the particles in a fixed potential while the rotational function is the probability amplitude of finding the intrinsic structure in a field of the specified orientation. The nuclear wavefunction is then obtained by integration of (32.11) over all orientation angles[1,2]. Thus we have:

$$\psi_M^I(\boldsymbol{r}) = \text{const} \int \chi_K^\tau(\boldsymbol{r}') \, \mathfrak{D}_{MK}^I(\vartheta) \, d\vartheta. \tag{45.7}$$

In general, the intrinsic function is not an eigenfunction of angular momentum, however if we write it in the form (32.16) and use (I.10) to transform the intrinsic function into the laboratory reference frame we find:

$$\psi_M^I(\boldsymbol{r}) = \text{const} \sum_{jK'} c_j \, \chi_{K'}^{\tau j}(\boldsymbol{r}) \int \mathfrak{D}_{K'K}^{j*}(\vartheta) \, \mathfrak{D}_{MK}^I(\vartheta) \, d\vartheta. \tag{45.8}$$

With the help of (I.6) we obtain

$$\psi_M^I(\boldsymbol{r}) = \sum_{j,K'} c_j \, \chi_{K'}^{\tau j} \, \delta_{Ij} \, \delta_{K'M} = c_I \, \chi_M^{\tau I}. \tag{45.9}$$

In other words, integration of the wavefunction over all orientation angles selects out of the intrinsic function the component with the appropriate angular momentum[3].

This integration also introduces correlations into the wavefunctions, even though the nucleons move essentially independently for any *fixed* nuclear orientation. Preliminary studies indicate that the wavefunctions of the rotational model reproduce the shell model wavefunctions at least fairly well[4]. In Appendix III we illustrate this procedure by the very simple example of particles bound to each other by a harmonic oscillator force. For this case, the wavefunctions can

---

[1] In the corresponding case of molecular spectra, the wavefunctions also have this form. Here, the r again refer to intrinsic, i.e. electronic coordinates, while the collective variables R are extra parameters which specify the positions of the nuclei. For this case however, the latter represent additional degrees of freedom over and above these of the electronic motion so that we should not integrate the molecular wavefunction over nuclear coordinates.

[2] This approach has also been used to study vibrational excitations of filled shells, e.g. in O^{16} [J. GRIFFIN: Ph. D. Dissertation, Princeton University, 1955; R. FERRELL and VISSCHER: Phys. Rev. **102**, 450 (1956)]. These occur at energies of 10 to 20 Mev while excitations involving unfilled shells usually occur at about 1 Mev.

[3] A similar approach has been used in the last six references given in footnote 3, p. 526.

[4] The wavefunctions are, in fact, equivalent for the case of L-S coupling in the $1p$ shell. More generally, the energy levels and wavefunctions of the rotational model can be reproduced if the central potential is of the harmonic oscillator type, and if the mutual interactions between the nucleons are essentially of the form: $V_{ik} = r_i^2 r_k^2 P_2(\cos\vartheta_{ik})$ (J. P. ELLIOTT, to be published).

be found either directly by treating the particles as moving independently under the influence of a common but variable center of force. On the other hand, explicit calculations show that if we start from the shell model and take configuration interaction into account, the mutual interactions do, in fact, give rise to collective effects of the kind implied by the unified model, i.e. enhancement of $E2$ transitions relative to the single particle strength[1].

In conclusion, it is interesting to see that according to all indications, the shell model and unified model are not fundamentally different, but are, in fact, merely alternative ways of describing the low energy properties of atomic nuclei.

Acknowledgements.

I wish to thank R. J. FINKELSTEIN, B. STECH, A. WINTHER, D. C. CHOUDHURY and L. KOMAI for many helpful comments and criticisms on this article.

I am also very grateful to many colleagues, especially A. BOHR and B. R. MOTTELSON for making the results of their work available before publication.

Appendix I.

Properties of the $\mathfrak{D}$ functions[2].

We list here some properties of the $\mathfrak{D}$-functions which are relevant to the material in Chap. F of this article.

First of all, the $\mathfrak{D}$ functions can be defined as the transformation functions for spherical harmonics under finite rotations. Thus, suppose a fixed point P is described by the coordinates r, ϑ, φ in a right handed reference frame R and by r', ϑ', φ' in another reference frame R' which is obtained from R by a rotation through Euler angles[3] α, β, γ. More specifically, we perform (i) a counterclockwise (right-handed) rotation $\alpha\,(0 \leq \alpha \leq 2\pi)$ about the z axis, followed by (ii) a rotation $\beta\,(0 \leq \beta \leq \pi)$ about the *new* y axis and finally (iii) a rotation $\gamma\,(0 \leq \gamma \leq 2\pi)$ about the new z axis. Then the spherical harmonics Y_{lm} transform as follows:

$$Y_{lm}(\vartheta, \varphi) = \sum_{m'=-l}^{l} Y_{lm'}(\vartheta', \varphi')\, \mathfrak{D}^{l}_{mm'}(\alpha\beta\gamma). \tag{I.1}$$

The $\mathfrak{D}$ functions are also the irreducible representations of the $2l+1$ dimensional rotation group. Both m and m' can independently take any of the $2l+1$ values $-l, -l+1, \ldots l$. The $\mathfrak{D}$ functions have the form

$$\mathfrak{D}^{l}_{mm'}(\alpha\beta\gamma) = \mathrm{e}^{im\alpha}\, \Theta_{lmm'}(\beta)\, \mathrm{e}^{im'\gamma} \tag{I.2}$$

where the second term on the right hand side is a rather complicated function of β containing the so-called Jacobi-polynomials. Spherical harmonics and Legendre polynomials are special forms of $\mathfrak{D}$ functions

$$\mathfrak{D}^{l}_{m0}(\alpha\beta\gamma) = \left(\frac{4\pi}{2l+1}\right)^{\frac{1}{2}} Y_{lm}(\beta, \alpha) \tag{I.3}$$

and

$$\mathfrak{D}^{l}_{00}(\alpha\beta\gamma) = P_l(\cos\beta). \tag{I.4}$$

---

[1] This is due to the fact that the mutual interactions tend to maximize the spatial symmetry of the low-lying nuclear states, i.e. to put as many particles as possible into the same spatial state or closely overlapping orbits.

[2] We follow the notation of A. R. EDMONDS: Angular Momentum in Quantum Mechanics, C. E. R. N. 55–26, 1955 except that our $\mathfrak{D}$ functions are the complex conjugates of his.

[3] In Ref. [*33*] these angles are denoted by φ, ϑ, ψ, respectively.

The $\mathfrak{D}$ functions have the following normalization properties:

$$\int \mathfrak{D}^{J*}_{MN}(\vartheta)\, \mathfrak{D}^{j}_{mn}(\vartheta)\, d\Omega = \delta_{Mm}\, \delta_{Nn}\, \delta_{Jj} \frac{8\pi^2}{2J+1} \tag{I.5}$$

where

$$d\Omega = \sin\beta\, d\beta\, d\alpha\, d\gamma. \tag{I.6}$$

ϑ symbolizes the set of Euler angles and the δ_{Mm} etc. are Kronecker deltas.

Other relations are:

$$\sum_{m'} \mathfrak{D}^{j}_{mm'}(\vartheta)\, \mathfrak{D}^{j*}_{m''m'}(\vartheta) = \sum_{m'} \mathfrak{D}^{j}_{m'm}(\vartheta)\, \mathfrak{D}^{j*}_{m'm''}(\vartheta) = \delta_{mm''}. \tag{I.7}$$

Multiplication of two $\mathfrak{D}$ functions yields the following

$$\left.\begin{aligned} \mathfrak{D}^{J}_{MN}(\vartheta)\, \mathfrak{D}^{j}_{mn}(\vartheta) = \sum_{K=|J-j|}^{J+j} \mathfrak{D}^{K}_{M+m\,N+n}(\vartheta)\, \langle Jj\,M\,m \mid Jj\,K\,M+m\rangle \times \\ \times \langle Jj\,N\,n \mid Jj\,K\,N+n\rangle \end{aligned}\right\} \tag{I.8}$$

where the quantities in products are Clebsch-Gordan coefficients[1]. Thus matrix elements involving three $\mathfrak{D}$ functions are given by:

$$\left.\begin{aligned} \int \mathfrak{D}^{K*}_{M+m\,N+n}\, \mathfrak{D}^{J}_{MN}\, \mathfrak{D}^{j}_{mn}\, d\Omega = \frac{8\pi^2}{2K+1} \langle JjM\,m \mid JjKM+m\rangle \times \\ \times \langle Jj\,N\,n \mid JjK\,N+n\rangle. \end{aligned}\right\} \tag{I.9}$$

Though strictly speaking, the above relations apply only for integral values of l and j, all except (I.3) and (I.4) can also be used for half integral values of angular momentum. Thus the $\mathfrak{D}$ functions can be defined by:

$$\chi^{j}_{m}(\mathbf{r}) = \sum_{m'} \chi^{j}_{m'}(\mathbf{r}')\, \mathfrak{D}^{j}_{mm'}(\vartheta) \tag{I.10}$$

where $\chi^{j}_{m}(\mathbf{r})$ is the wavefunction of integral *or* half integral angular momentum j and component along the z axis, and $\chi^{j}_{m'}(\mathbf{r}')$ is the same kind of wavefunction except that it refers to the body reference frame and has a component m' along the z' axis.

The transformation from the body frame to the laboratory frame is given by:

$$\chi^{j}_{m}(\mathbf{r}') = \sum_{m'} \chi^{j}_{m'}(\mathbf{r})\, D^{j*}_{m'm}(\vartheta) \tag{I.11}$$

as may be verified with the help of (I.7).

Appendix II.

Mathematical relations applicable to nuclear rotational states.

We list here some mathematical relations, equations and values of matrix elements which are applicable to rotational states in nuclei.

α) Commutation rules for angular momenta. The components of total angular momentum I and intrinsic angular momentum j satisfy the following commutation rules [*33*]:

$$I_x I_y - I_y I_x = i\, I_z \tag{II.1}$$

[1] We use the notation of CONDON and SHORTLEY.

where the subscripts x, y, z refer to the components in the laboratory fixed reference frame and

$$I_1 I_2 - I_2 I_1 = -i I_3, \tag{II.2}$$

$$j_1 j_2 - j_2 j_1 = i j_3 \tag{II.3}$$

where the subscripts 1, 2, 3 denote the components in the body fixed reference frame. Cyclic permutations of the above relations also hold. Furthermore, all components of j commute with those of I.

Eqs. (II.1) and (II.3) have the usual form for angular momentum commutation relations. The occurrence of the opposite sign in Eq. (II.2) can be understood if we consider the fact that an arbitrary rotation of the body fixed reference frame has exactly the same effect on the $\mathfrak{D}$ function (upon which the I's operate) as a rotation of the laboratory fixed reference frame about the same axis and by the same angle but in the opposite direction.

β) Matrix elements of angular momenta. Using Eqs. (32.3) and (II.2) one can show that

$$(I_1 \pm i I_2)\, \mathfrak{D}^I_{MK} = [(I \pm K)(I \mp K + 1)]^{\frac{1}{2}}\, \mathfrak{D}^I_{MK\mp 1} \tag{II.4}$$

and verify (32.4). The intrinsic wavefunctions satisfy the relation

$$j_3 \chi^j_K = K \chi^j_K \tag{II.5}$$

and

$$(j_1 \pm i j_2)\, \chi^j_K = [(j \mp K)(j \pm K + 1)]^{\frac{1}{2}}\, \chi^j_{K\pm 1}. \tag{II.6}$$

Using the above relations we can calculate the matrix elements of the quantity $\boldsymbol{I}\cdot\boldsymbol{j}$:

$$\langle \chi^j_K \mathfrak{D}^I_{MK} |\, \boldsymbol{I}\cdot\boldsymbol{j}\, | \chi^j_K \mathfrak{D}^I_{MK} \rangle = K^2 \tag{II.7}$$

$$\langle \chi^j_{K_2} \mathfrak{D}^I_{MK_2} |\, \boldsymbol{I}\cdot\boldsymbol{j}\, | \chi^j_{K_1} \mathfrak{D}^I_{MK_1} \rangle = [(I-K)(I+K+1)(j-K)(j+K+1)]^{\frac{1}{2}} \tag{II.8}$$

if $|K_2 - K_1| = 1$ (K is the larger of K_1, K_2).

All other matrix elements of $\boldsymbol{I}\cdot\boldsymbol{j}$ vanish. For the case $K = \frac{1}{2}$, (II.8) gives

$$\langle \chi^j_{\frac{1}{2}} \mathfrak{D}^I_{M\frac{1}{2}} |\, \boldsymbol{I}\cdot j\, | \chi^j_{-\frac{1}{2}} \mathfrak{D}^I_{M-\frac{1}{2}} \rangle = (I + \tfrac{1}{2})(j + \tfrac{1}{2}). \tag{II.9}$$

γ) Expectation values of operators. Let us now calculate the expectation values of operators according to the rotational model. Suppose the operator in question is a tensor of rank λ. This tensor has $2\lambda + 1$ components, with $\mu = -\lambda \dots \lambda$. However only the $\mu = 0$ component, e.g. only the z component of the angular momentum vector ($\lambda = 1$), has finite expectation value (since the nuclear state is characterized by a unique M). Now, using (I.10) we can express the components of the tensor in the laboratory reference frame in terms of the components in the body-fixed reference frame:

$$T^\lambda_0(r) = \sum_\mu T^\lambda_\mu(\boldsymbol{r}')\, \mathfrak{D}^\lambda_{0\mu}(\vartheta). \tag{II.10}$$

Assuming the wavefunction is of the form (32.11)[1] (and setting $K = \Omega$) we find for the expectation value:

$$\langle \psi |\, T^\lambda_0(\boldsymbol{r})\, | \psi \rangle = \langle \chi^\tau_K(\boldsymbol{r}') |\, T^\lambda_0\, | \chi^\tau_K(\boldsymbol{r}') \rangle \frac{2I+1}{8\pi^2} \int (\mathfrak{D}^I_{MK})^* \mathfrak{D}^\lambda_{00} \mathfrak{D}^I_{MK}\, d\Omega. \tag{II.11}$$

[1] Strictly speaking the wavefunction is of the form (32.18) except when $K = 0$. However so-called "cross-terms", which couple the two components of the wavefunction, appear only for the case $2K \leq \lambda$, which we will not consider further.

The first term on the right hand side is a matrix element involving only "intrinsic" functions—i.e. in the body frame of reference, and we shall denote it by $\langle T_0^\lambda(\boldsymbol{r}')\rangle$. The second term involving integration over angles may be evaluated using Eq. (I.9). The result is:

$$\langle T_0^\lambda(\boldsymbol{r})\rangle = \langle T_0^\lambda(\boldsymbol{r}')\rangle \langle I\,\lambda\,M\,0\,|\,I\,\lambda\,I\,M\rangle \langle I\,\lambda\,K\,0\,|\,I\,\lambda\,I\,K\rangle. \tag{II.12}$$

Some values of the Clebsch-Gordon coefficients appearing in Eq. (II.12) are[1]

$$\langle I\,1\,M\,0\,|\,I\,1\,I\,M\rangle = \frac{M}{\sqrt{I(I+1)}}, \tag{II.13}$$

$$\langle I\,2\,M\,0\,|\,I\,2\,I\,M\rangle = \frac{3M^2 - I(I+1)}{\sqrt{I(I+1)(2I-1)(2I+3)}}. \tag{II.14}$$

To illustrate the application of Eq. (II.12) suppose first that T is the angular momentum operator ($\lambda = 1$). Then we find with the help of Eq. (II.13)

$$\langle j_z\rangle = \langle j_3\rangle \frac{MK}{I(I+1)} \quad \left(\text{except if } K = \tfrac{1}{2}\right) \tag{II.15}$$

essentially the same result as (34.1). If T is the quadrupole moment tensor, Eq. (II.12) becomes with the help of Eq. (II.14).

$$\langle Q\rangle = Q_0 \frac{[3M^2 - I(I+1)]\,[3K^2 - I(I+1)]}{I(I+1)(2I-1)(2I+3)}. \tag{II.16}$$

Setting $M = I$, this result reduces to (34.7).

δ) Transition probabilities for electric quadrupole radiation. Consider now the calculation of reduced electric quadrupole transition probabilities between members of the same rotational band [Eq. (35.7)]. The matrix element of the quadrupole operator is given by:

$$\left.\begin{aligned} \langle f|\,Q_\mu(\boldsymbol{r})\,|i\rangle &= \left\langle \left(\frac{2I_f+1}{8\pi^2}\right)^{\frac{1}{2}} \chi_K^\tau\, \mathfrak{D}_{M_f K}^{I_f} \right| Q_\mu(\boldsymbol{r}) \left| \left(\frac{2I_i+1}{8\pi^2}\right)^{\frac{1}{2}} \chi_K^\tau\, \mathfrak{D}_{M_i K}^{I_i} \right\rangle \\ &\qquad \text{except if } K \leq 1. \end{aligned}\right\} \tag{II.17}$$

Using Eq. (II.10) we find

$$\langle f\,|\,Q_\mu(\boldsymbol{r})\,|i\rangle = \langle \chi_K^\tau|\,Q_0(\boldsymbol{r}')\,|\chi_K^\tau\rangle \frac{(2I_i+1)^{\frac{1}{2}}(2I_f+1)^{\frac{1}{2}}}{8\pi^2} \int \mathfrak{D}_{M_f K}^{I_f *}\, \mathfrak{D}_{\mu 0}^{2}\, \mathfrak{D}_{M_i K}^{I_i}\, d\Omega. \tag{II.18}$$

The first term on the right hand side is the intrinsic quadrupole moment. The second term may be evaluated using Eq. (I.9). The result is

$$\langle f|\,Q_\mu(\boldsymbol{r})\,|i\rangle = Q_0 \left(\frac{2I_i+1}{2I_f+1}\right)^{\frac{1}{2}} \langle I_i\,2\,M_i\,\mu\,|\,I_i\,2\,I_f\,M_f\rangle \langle I_i\,2\,K\,0\,|\,I_i\,2\,I_f\,K\rangle. \tag{II.19}$$

Using Eq. (35.8) and the relation

$$\sum_{M_1, M_2, M} |\langle I_1\,I_2\,M_1\,M_2\,|\,I_1\,I_2\,I\,M\rangle|^2 = 2I+1, \tag{II.20}$$

we readily verify Eq. (35.9). The same method may be used to calculate $M1$ transition probabilities between members of the same rotational band [Eq. (35.11)].

[1] Page 76 and 77 of CONDON and SHORTLEY.

Appendix III.

The unified model applied to a simple problem.

In this appendix we consider a very simple problem, the motion of two particles under the influence of a harmonic oscillator potential between them. We treat the problem exactly, and also from the standpoint of the unified model, in which we assume that the particles move independently in a variable potential. It is hoped that this example provides some insight into the method of the unified model, though, of course, the application of this model to actual nuclei is much more complicated.

Consider two mass points, free to move along the x axis, except for a harmonic oscillator potential between them:

$$V = \tfrac{1}{2}\mu\,\omega^2\,x^2 \tag{III.1}$$

where μ is the reduced mass of the system, x is the distance between the particles, and ω is the classical oscillation frequency of the system. The solution of this problem is trivial. Assuming the relative motion to be in the ground state, we find for the total energy:

$$E = \frac{P^2}{2M} + \frac{\hbar\,\omega}{2} \tag{III.2}$$

representing the energy of the center of mass and relative motion, respectively. The total mass and momentum of the system are denoted by M and P. The corresponding wavefunction, neglecting any exchange effects or spin, is given by:

$$\psi(X, x) = \mathrm{e}^{i\,P X/\hbar}\,\mathrm{e}^{-(\mu/2\hbar)\,\omega^2\,x^2} \tag{III.3}$$

where X is the position of the center of mass, and apart from a normalization constant. The motion of the two particles is evidently strongly correlated because of the binding potential between them.

Now consider the same physical problem from the standpoint of the unified model. At any *particular* time, the force on each particle i, due to the other is exactly the same as if it were instead bound to a center of force located at $\varphi = X$ and subject to the following potential:

$$V_i = \tfrac{1}{2}\,m_i\,\omega^2\,(x_i - \varphi)^2 \tag{III.4}$$

where m_i and x_i are the mass and position of particle i (1 or 2). If this fictitious center of force were fixed at φ for *all* time, then the particles would move independently. The resulting "intrinsic wavefunction" of the ground state is, apart from normalization, of the form:

$$\chi = \mathrm{e}^{-\frac{\omega}{2\hbar}\,[m_1(x_1-\varphi)^2 + m_2(x_2-\varphi)^2]} \tag{III.5}$$

which can also be written as:

$$\chi = \mathrm{e}^{\frac{\omega}{2\hbar}\,(M\,Y^2 + \mu\,X^2)} \tag{III.6}$$

where

$$Y = X - \varphi\,. \tag{III.7}$$

The "intrinsic energy" of the system equals $\hbar\omega$. Note that according to (III.6) the center of mass of the two particles no longer coincides with φ, only its expectations value does. The intrinsic wavefunctions, unlike the correct wavefunction is not an eigenfunction of total momentum.

Now in the actual system under consideration the center of mass is not fixed, but moves along with the particles. Let us then correct the intrinsic wavefunction to take into account the motion of the fictitious center of force. If this center of force moves with velocity $\dot{\varphi}$, there is induced a collective momentum

$$P = M\dot{\varphi} \tag{III.8}$$

and a collective kinetic energy

$$T_{\text{coll}} = \frac{1}{2} M \dot{\varphi}^2 = \frac{P^2}{2M}. \tag{III.9}$$

The collective wavefunction is given by:

$$\varphi_{\text{coll}} = e^{iP\varphi/\hbar}. \tag{III.10}$$

For this case the collective motion is, essentially equivalent to the center of mass motion of the actual system. The total energy of the system is now given by

$$E = \frac{P^2}{2M} + \hbar\omega \tag{III.11}$$

which is larger than the exact result by an amount $\hbar\omega/2$. [This difference results from the "spurious" oscillations of the center of mass inherent in the intrinsic wavefunction (III.6). Such an oscillation does not occur in the actual physical system described by wavefunction (III.3).]

The unified model wavefunction is then of the form

$$\psi(\varphi, Y, X) = e^{iP\varphi/\hbar} e^{-(\omega/2\hbar)(MY^2+\mu X^2)}. \tag{III.12}$$

This wavefunction describes the intrinsic motion of the two particles about a fixed fictitious center of force. Note that it depends on three variables though here we have only two degrees of freedom. In this simple example, the intrinsic and collective motions are uncoupled from each other.

The above wavefunction may be expressed in terms of particle coordinates alone in either of two ways. We may define the collective variable φ explicitly in terms of particle coordinates, i.e.

$$\varphi = X, \quad Y = 0. \tag{III.13}$$

Then the above wavefunction reduces precisely to the correct wavefunction (III.3). However, this method seems to work only whenever the collective flow is irrotational, e.g. in the present simple case of collective translation[1].

The other approach, more plausible on physical grounds, is to integrate the wavefunction over all collective coordinates, i.e. over φ. This integration picks out of the intrinsic function the Fourier component of total momentum P. We then obtain:

$$\psi(X, x) = \text{const}\, e^{-(P^2/2M\hbar\omega)} e^{iPX/\hbar} e^{-(\mu/2\hbar)\omega X^2}. \tag{III.14}$$

The resulting "unified model" wavefunction agrees with the exact wavefunction except that it is not normalized to unity. By performing the above integrations we have introduced correlations into the wavefunctions, and in fact, for the case considered, these correlations are precisely the same as those resulting from the actual interaction between the particles.

For short range interactions, rather than harmonic oscillator interactions between particles, the unified model approach reproduces some but not all of the correlations in the system. Thus it is necessary to add "residual interactions" into the intrinsic Hamiltonian (Sect. 40). Also, the collective motion may perturb the intrinsic structure somewhat, as is discussed in Sect. 44.

---

[1] Cf. footnote 4, p. 497.

General references.

Detailed treatments of nuclear models containing extensive bibliographies.

[1] Bethe, H. A., and R. F. Bacher: Nuclear Physics; A. Stationary states of Nuclei. Rev. Mod. Phys. **8**, 82—229, Chap. II, V, VI (1936).
[2] Blatt, J., and V. F. Weisskopf: Theoretical Nuclear Physics, Chap. III, VI, VII. New York: Wiley Publishing Co. 1952.
[3] Rosenfeld, L.: Nuclear Forces, Chap. IX—XIV. New York: Interscience Publishers 1948.

Shorter surveys of nuclear models.

[4] Eden, R. J.: Nuclear Models to appear in Progress of Nuclear Physics. Volume 6, 1957.
[5] Green, A. E. S.: Nuclear Physics, Chap. 8—11. New York: McGraw Hill Book Co. 1955.
[6] Sachs, R. G.: Nuclear Theory, Chap. 8, 9. Cambridge: Addison Wesley Publishing Co. 1953.

Recent conferences relevant to nuclear models.

[7] Proceedings of the 1954 Glasgow conference on Nuclear and Meson Physics, edit. by E. H. Bellamy and R. G. Moorhouse. London: Pergamon Press 1955.
[8] Conference on Statistical Aspects of the Nucleus, Jan. 1955, BNL 331 (C-21) Brookhaven National Laboratory, Upton, New York.
[9] Proceedings of the Internal Conference on the Peaceful Uses of Atomic Energy, held in Geneva, August 1955 — Vol. 2: Physics; Research Reactors. New York: United Nations 1956.
[10] Proceedings of the International Conference on Nuclear Reactions, Amsterdam, July 1956; Published in Physica, Haag **22** (Nov. 1956).

Liquid drop model.

[11] Bohr, N.: Neutron Capture and Nuclear Constitution. Nature, Lond. **137**, 344 (1936).
[12] Bohr, N., and F. Kalckar: On the Transmutation of Atomic Nuclei by Impact of Material Particles. I. General Theoretical Principles. Dan. Mat.-Fys. Medd. **14**, No. 10 (1937).
[13] Bohr, N., and J. A. Wheeler: The Mechanism of Nuclear Fission. Phys, Rev. **56**, 426 (1936).
[14] Weizsäcker, C. F.: Zur Theorie der Kernmassen. Z. Physik **96**, 431 (1935).

Fermi gas model.

[15] Bethe, H. A.: The Nuclear Many Body Problem. Phys. Rev. **103**, 1353 (1956).
[16] Brueckner, K.: Two-body Forces and Nuclear Saturation. III. Details of the Structure of the Nucleus. Phys. Rev. **97**, 1353 (1955).
[17] Euler, H.: Über die Art der Wechselwirkung in schweren Atomkernen. Z. Physik **105**, 553 (1937).
[18] Wigner, E. P.: On the Structure of Nuclei beyond Oxygen. Phys. Rev. **51**, 947 (1937).

Optical model.

[19] Feshbach, H., C. E. Porter and V. F. Weisskopf: Model for Nuclear Reactions with Neutrons. Phys. Rev. **96**, 448 (1954).
[20] Friedman, F. L., and V. F. Weisskopf: The Compound Nucleus, from: Niels Bohr and the Development of Physics, pp. 134—162. London: Pergamon Press 1955.
[21] Lax, M.: Multiple Scattering of Waves. Rev. Mod. Phys. **23**, 287 (1951).
[22] Peaslee, D. C.: Nuclear Reactions of Intermediate Energy Heavy Particles. Ann. Rev. Nucl. Sci. **5**, 99 (1955).
[23] Skyrme, T. H. R.: The Nuclear Surface. Phil. Mag. **1**, 1003 (1956).

Shell model.

[24] Elliott, J. P., and A. M. Lane: The Nuclear Shell Model. Preceding article of this volume.
[25] Feenberg, E.: Shell Theory of the Nucleus. Princeton, New Jersey: Princeton University Press 1955.
[26] Haxel, O., J. H. D. Jensen and H. E. Suess: Modellmäßige Deutung der ausgezeichneten Nukleonen-Zahlen im Kernbau. Z. Physik **128**, 295 (1950).

[27] Inglis, D. R.: The Energy Levels and the Structure of Light Nuclei. Rev. Mod. Phys. **25**, 390 (1953).
[28] Mayer, M. G.: Nuclear Configurations in the Spin Orbit Coupling Model. I. Empirical Evidence. Phys. Rev. **78**, 16 (1950).
[29] Mayer, M. G., and J. H. D. Jensen: Elementary Theory of Nuclear Shell Structure. New York: Wiley Publishing Co. 1955.
[30] Pryce, M. H. L.: Nuclear Shell Structure. Rep. Progr. Phys. **17**, No. 1 (1954).

The unified model.

[31] Alaga, Alder, Bohr and Mottelson: Intensity Rules for Beta and Gamma Transitions to Nuclear Rotational States. Dan. Mat.-Fys. Medd. **29**, No. 9 (1955).
[32] Alder, Bohr, Huus, Mottelson and Winther: Study of Nuclear Structure by Electromagnetic Excitations with Accelerated Ions, Chap. V. Rev. Mod. Phys. **28**, 432 (1956).
[33] Bohr, A.: The Coupling of Nuclear Surface Oscillations to the Motion of Individual Nucleons. Dan. Mat.-Fys. Medd. **26**, 14 (1952).
[34] Bohr, A., and B. R. Mottelson: Collective and Individual Particle Aspects of Nuclear Structure. Dan. Mat.-Fys. Medd. **27**, No. 16 (1953).
[35] Bohr, A., and B. R. Mottelson: Collective Nuclear Motion and the Unified Model, Chap. XVII of Beta and Gamma Ray Spectroscopy, edit. by K. Siegbahn. Amsterdam: North Holland Publishing Co. 1955.
[36] Bohr, A., and B. R. Mottelson: Moments of Inertia of Rotating Nuclei. Dan. Mat.-Fys. Medd. **30**, No. 1 (1955).
[37] Hill, D. L., and J. A. Wheeler: Nuclear Constitution and the Interpretation of Fission Phenomena. Phys. Rev. **89**, 1102 (1953).
[38] Inglis, D.: Dynamics of Nuclear Deformation. Phys. Rev. **97**, 701 (1955).
[39] Kerman, A. K.: Rotational Perturbations in Nuclei—Application to Wolfram 183. Dan. Mat.-Fys. Medd. **30**, No. 15 (1956).
[40] Moszkowski, S. A.: Inertial Parameters for Collective Nuclear Oscillations. Phys. Rev. **103**, 1328 (1956).
[41] Mottelson, B. R., and S. G. Nilsson: Classification of Nucleonic States in Deformed Nuclei. Phys. Rev. **99**, 1615 (1955).
[42] Nilsson, S. G.: Binding States of Individual Nucleons in Strongly Deformed Nuclei. Dan. Mat.-Fys. Medd. **29**, No. 16 (1955).
[43] Rainwater, J.: Nuclear Energy Level Argument for a Spheroidal Nuclear Model. Phys. Rev. **79**, 432 (1950).
[44] Scharff-Goldhaber, G., and J. Weneser: System of Even-even Nuclei. Phys. Rev. **98**, 212 (1955).

Sachverzeichnis.

(Deutsch-Englisch.)

Bei gleicher Schreibweise in beiden Sprachen sind die Stichwörter nur einmal aufgeführt.

Abschneiden des Potentials, *cut-off of potential* 29, 30.
Absorptionsanteil des Potentials (Imaginärteil), *absorption potential* 369, 381, 385, 446.
Absorptionsprozeß, *absorption process* 446.
Abstoßungskraft, *repulsive force* 438.
Abstoßungspotential s. hard core, *repulsive potential see hard core.*
adiabatisches Abbrechen, *adiabatic switching off* 136.
äquivalentes Zweikörperproblem für drei Nucleonen, *equivalent two-body problem for three nucleons* 172.
α-Streuung, Bestimmung der Kerngröße, *α-particle scattering, determination of nuclear size* 216, 217.
α-Teilchen-Kern, *α-nucleus* 460, 461.
α-Teilchen-Modell, *α-particle model* 366, 460 bis 464.
α-Zerfall, Bestimmung der Kerngröße, *α-decay determinations of nuclear size* 218.
α-α-Wechselwirkung aus Streuversuchen, *α-α-interaction from scattering* 464.
Anregungsenergien der ersten angeregten Zustände, *excitation energies of first excited states* 260.
Anregungsspektren der Kerne, *excitation spectra of nuclei* 477—479.
Asymmetrie-Koeffizient, Vorwärtsrichtung, *asymmetry coefficient, forward* 6, 113.
Asymmetrie-Parameter, *asymmetry parameter* 434.
Austauschströme und magnetische Momente von H^3 und He^3, *exchange currents and magnetic moments of* H^3 *and* He^3 177.
Austausch-Stromkorrektur, *exchange current correction* 98, 114, 115, 116, 130.

Bahndrehimpulse, Kopplung, *angular momenta, coupling* 390, 391.
Bahn-Zustände, gruppentheoretische Klassifikation, *orbital states, group theoretical classification* 396.
BARTLETTsche Wechselwirkung, BARTLETT *interaction* 425.
β-Übergänge, *β-transitions* 296.
β-Zerfall (s. auch *ft*-Werte), *β-decay (see also ft values)* 270f.
—, Matrixelemente, *β-decay, matrix elements* 408.
Beugungstheorie, *diffraction theory* 105f.
Bindungseffekt, chemischer, *chemical binding effect* 43, 46.
Bindungsenergie des Deuterons, *binding energy of the deuteron* 2, 32, 91.
— eines Kernes im FERMI-Modell, *binding energy of a nucleus in the* FERMI *model* 426—429.
— —, Korrekturen höherer Ordnung, *binding energy of a nucleus, higher order corrections* 440—442.
Bindungsenergien, *binding energies* 258, 295, 370, 371, 373.
— aus dem α-Teilchen-Modell, *binding energies from the α-particle model* 461—464.
— von H^3 und He^3, *binding energies of* H^3 *and* He^3 153.
— der Kerne, halbempirische Formel, *binding energies of nuclei, semiempirical formula* 415, 416.
Brechungsanteil des Potentials (Realteil), *refractive (real) potential* 369, 372, 447, 449—451.
Brechungsindex, *refractive index* 449—451.
BREIT-WIGNERsche Dispersionsformel, BREIT WIGNER *dispersion formula* 446.
BRUECKNERsche Methode, BRUECKNER'S *method* 374, 380, 445.

CASIMIR-Operator, CASIMIR *operator* 291.
c.f.p. (coefficients of fractional parentage) 332, 398, 399.
chemische Valenzbindung, *chemical valence bond* 422.
CLEBSCH-GORDAN-Koeffizienten, CLEBSCH-GORDAN-*coefficients* 389, 390.
Compound-Kern, *compound nucleus* 446, 448.
Compound-System, *compound system* 448.
Corioliskraft bei Kernrotation, *coriolis force in nuclear rotation* 497, 537.
COULOMB-Anregung, COULOMB *excitation* 224, 495, 498.
COULOMB-Effekte bei Proton-Deuteron-Streuung, COULOMB *effects in proton-deuteron scattering* 175.
COULOMB-Energie, COULOMB *energy* 370.
COULOMB-Kräfte, COULOMB *forces* 289.
COULOMB-Radius, COULOMB *radius* 417, 430.
COULOMB-Streuphase, COULOMB *phase shift* 58, 59.
COULOMB-Wall, Bestimmung der Kerngröße, COULOMB *barrier determinations of nuclear size* 217, 218.

Dämpfungsparameter, *damping parameter* 22, 27.
Dämpfungstheorie, *damping theory* 139.
Deformationsparameter, *deformation parameters* 219, 518.
Deformationspotential, *deformation potential* 503, 522, 513.
— und Zweikörperkräfte, *deformation potential and two-body forces* 511, 512.
Deformierbarkeit der Kerne, *deformability of nuclei* 237.
Deuteron, angeregte Zustände, *deuteron, excited states* 33, 35.
—, Bindungsenergie, *deuteron, binding energy* 1, 32, 91.
—, effektive Reichweite, *deuteron, effective range* 86, 91.
—, elektrisches Quadrupolmoment, *deuteron, electric quadrupol moment* 67, 91.
—, Grundzustand, *deuteron, ground state* 32, 36, 66.
—, magnetisches Moment, *deuteron, magnetic moment* 32, 69, 74.
—, — —, Austauschstromkorrektur, *deuteron, magnetic moment, exchange current correction* 71, 72.
—, — —, nicht additive Korrektur, *deuteron, magnetic moment, non-additivity correction* 71, 72, 73.
—, — —, relativistische Korrektur, *deuteron, magnetic moment, relativistic correction* 71, 72, 73.
—, Wellenfunktion, *deuteron, wave function* 33, 66, 74, 76, 91.
Deuteronradius, *deuteron radius* 35, 40, 75.
Deuteronspin, *deuteron spin* 2, 32.
Dichte, Berechnung aus Potential, *density, calculation from potential* 453, 454.
— der Kerne, Theorie, *density of nuclei, theory* 231, 233.
— der Kernmaterie, *density, nuclear* 178, 416, 431.
— — unter der Oberfläche, *density, nuclear, below the surface* 433, 459.
—, statistische, angeregter Niveaus, *statistical density of excited levels* 437.
—, —, der Einteilchen-Niveaus, *statistical density of single particle levels* 435.
— im Zentrum des Kerns, *density, central, of the nucleus* 193.
Dichteanisotropie, aus dem Potential berechnet, *density anisotropy, calculated from potential* 504.
Dichten gemäß individueller Teilchenzustände, *densities from individual particle states* 235.
Dipolmoment, elektrisches, *electric dipole moment* 68.
—, magnetisches, *magnetic dipole moment* 252, 277, 294, 311, 407.
Dipolschwingung eines Kernes, *dipole oscillation of a nucleus* 419.
Drehimpuls einer kollektiven Schwingung, *angular momentum of collective oscillation* 420.
Dreikörperkräfte, *three-body forces* 156, 352.
$3n$-j-Symbole, *$3n$-j-symbols* 391.
DYSON-Transformation, DYSON *transformation* 19.
D-Zustand, Anteil im Deuteron, *D-state probability in the deuteron* 70, 73, 74, 91.

Ebene Wellen im Kern, *plane waves within the nucleus* 422.
effektive Reichweite, *effective range* 38, 51, 54, 59—61, 83, 85, 91.
Einteilchen-Modell, *individual (independent) particle model* 236, 245, 257, 317.
—, Anwendung auf leichte Kerne, *individual particle model, application to light nuclei* 335.
elektrischer Multipolübergang, *electric multipol transition* 100, 102, 103, 104, 110.
— Quadrupolübergang, *electric quadrupole transition* 490, 546.
Elektronenstreuung, *electron scattering* 416.
—, Bestimmung der Kerngröße, *electron scattering, determinations of nuclear size* 192 bis 195.
— an Kernen, *electron scattering by nuclei* 186. 228f.
— —, BORNsche Näherung, *electron scattering by nuclei*, BORN *approximation* 187.
— —, hohe Energien, *electron scattering by nuclei, high energies* 191.
— —, niedrige Energien, *electron scattering by nuclei, low energies* 188.
— —, unelastische, *inelastic electron scattering by nuclei* 197.
— —, Wirkungsquerschnitte, *electron scattering by nuclei, cross sections* 195.
— an einer nicht-kugelförmigen Ladung, *electron scattering from aspherical charge* 228.
elektromagnetische Momente, *electromagnetic moments* 252.
— Übergänge für H^3 und He^3, *electromagnetic transitions for* H^3 *and* He^3 176.
ellipsoidale Schwingungen, *ellipsoidal vibrations* 518, 533.
Energiematrix, *energy matrix* 332.
Energietermverschiebung gebundener Zustände, *energy shift of bound states* 199.
EULERsche Winkel, EULER *angles* 483.

Feinstruktur, optische, *optical fine structure* 202.
FERMI-Gas-Modell der Kerne, FERMI *gas model of nuclei* 421.
FERMI-Impuls, FERMI *momentum* 426.
FERMI-Näherung, FERMI *approximation* 45.
Flüssigkeitsspiegel, Reflexion von Neutronen, *liquid mirror reflection of neutrons* 3, 49.
Fractional parentage coefficients 332, 398, 399.
Freiheitsgrade im kombinierten Kernmodell, *degrees of freedom in unified model* 541.
ft-Werte, einfach verbotene Übergänge in ungeraden Kernen, *ft-values, first forbidden transitions in odd nuclei* 274.

ft-Werte, erlaubte Übergänge in ungeraden Kernen, *ft-values, allowed transitions in odd nuclei* 273.
—, übererlaubte Übergänge in ungeraden Kernen, *ft-values, superallowed transitions in odd nuclei* 272.
—, zweifach verbotene Übergänge, *ft-values, second forbidden transitions* 276.

γ-Schwingungen sphäroidaler Kerne, *γ-vibrations of spheroidal nuclei* 544—546.
γ-Übergänge, Matrixelemente, *γ-transitions, matrix elements* 408.
γ-Zerfall, *γ-decay* 223.
GAUSSsche Potentialform, GAUSS *shape* 356.
g-g-Kerne, *even-even nuclei* 293.
Größe der Kerne, Bestimmung mit α-Teilchen, *size of nuclei, determination by α-particles* 216—218.
— —, — aus Elektronenstreuung, *size of nuclei, determination from electron scattering* 192—195.
— —, — durch gebundene Elektronen, *size of nuclei, determination by bound electrons* 199.
— —, — aus Neutronenstreuung, *size of nuclei, determination by neutron scattering* 213, 214.
— —, — aus Protonenstreuung, *size of nuclei, determination by proton scattering* 215.
— —, frühe Untersuchungen, *size of nuclei, early studies* 178.
— sehr leichter Kerne, *size of very light nuclei* 199.
Grundzustand des H^3 und He^3, empirische Daten, *ground states of H^3 and He^3, empirical data* 149.

Häufigkeit, gesamte, *total abundance* 259.
—, relative, *relative abundance* 260.
HAMILTON-Funktion der kollektiven Schwingung, *Hamiltonian of collective vibration* 519.
Hard core (Abstoßungspotential), *(repulsive potential)* 30, 31, 43, 53, 56, 63, 76, 88, 90, 92, 96, 117, 118, 122, 124, 127, 130, 156.
HARTREE-FOCK-Methode des self consistent field, HARTREE-FOCK *self consistent field method* 376.
HARTREE-FOCKsches Potential, HARTREE FOCK *potential* 456.
H^3, Bindungsenergie, H^3 *binding energy* 153.
—, Einfluß der Tensorkraft auf die Bindung, H^3, *role of tensor force in binding* 155.
—, elektromagnetische Übergänge, H^3, *electromagnetic transitions* 176.
—, Grundzustand, empirische Daten, H^3, *ground state, empirical data* 149.
—, magnetisches Moment, H^3, *magnetic moment* 150, 177.
He^3, Bindungsenergie, He^3, *binding energy* 153.
—, Einfluß der Tensorkraft auf die Bindung, He^3, *role of tensor force in binding* 155.
He^3, elektromagnetische Übergänge, He^3, *electromagnetic transitions* 176.
—, Grundzustand, empirische Daten, He^3, *ground state, empirical data* 149.
—, magnetisches Moment, He^3, *magnetic moment* 150, 177.
HEISENBERGsche Wechselwirkung, HEISENBERG *interaction* 425.
HULTHÉN-Potential, HULTHÉN *potential* 31.
HULTHÉNsche Wellenfunktion, HULTHÉN *wave function* 33, 34, 35, 140.
HUNDscher Operator, HUND *operator* 329.

Imaginärteil des Potentials, *imaginary part of potential* 446, 451, 452.
Impulse approximation 165.
Impulsraumdarstellung, *momentum space representation* 139.
Impulsübertragung in dichten Kernen, *momentum transfer in dense nuclei* 441, 442.
Interferenz-Minimum, *interference minimum* 7, 8, 59.
isomere Übergänge, *isomeric transitions* 296.
— Zustände, *isomeric states* 282.
isotoper Spin, *isotopic spin* 289, 319.
— —, reduzierter, *reduced isotopic spin* 289, 326.
Isotopieverschiebung, optische, *optical isotope shift* 202, 224.
Isotopie-Verschiebung bei Röntgenstrahlen, *isotope shift of x-rays* 201.
Isotropiefaktor, *isotropy factor* 14.
Isotropie-Koeffizient, *isotropy coefficient* 6, 114, 129.

j-*j*-Kopplung, *j-j coupling* 3, 25, 474.
— und *L-S*-Kopplung, Transformation ineinander, *j-j coupling and L-S coupling, transformation between them* 327.

K-Auswahlregeln, *K-selection rules* 492.
Kerndichte s. Dichte, *nuclear density see density.*
Kerngestalt s. Gestalt des Kerns, *nuclear shape see shape of nucleus.*
Kerngröße s. Größe des Kerns, *nuclear size see size of nucleus.*
Kernkräfte, Ladungssymmetrie bei hohen Energien, *nuclear forces, charge symmetry at high energies* 170.
Kernladungsverteilung, *nuclear charge distribution* 195, 196, 416, 431.
Kernmodell, α-Teilchen-Modell, *model of nucleus, α-particle model* 366, 460—464.
—, Einteilchen-Modell, *model of nucleus, independent (individual) particles* 236, 245, 257, 317, 335, 412, 421.
—, FERMI-Gas-Modell, *model of nucleus*, FERMI *gas model* 421.
—, Flüssigkeitstropfen, *model of nucleus, liquid drop* 362, 414.
—, kollektives, *collective model of nucleus* 350, 361, 480.
—, kombiniertes, *unified model of nucleus* 349, 362, 476—543.

Kernmodell, mittelstarke Kopplung, *model of nucleus, intermediate coupling* 335, 338, 354.
—, optisches, *optical model of nucleus* 372, 445—459.
— von RAINWATER, *model of nucleus by* RAINWATER 477, 498.
—, Rotationsmodell, *model of nucleus, rotational model* 480, 537.
—, Schalenmodell, *model of nucleus, shell model* 230, 236, 242, 368, 464—476.
—, sphäroidaler Rumpf, *model of nucleus, spheroidal core* 477, 498—501, 514.
—, starke Kopplung, *model of nucleus, strong coupling* 446.
Kernmodelle, kurze Übersicht, *models of nuclei, brief survey* 412—414.
— mit starker Wechselwirkung, *models of nuclei with strong interaction* 412.
Kernpotential s. Potential, *nuclear potential see potential.*
Kernprozeß-Streuphasen, *nuclear phase shift* 58, 59.
Kernradius s. Radius, *nuclear radius see radius.*
Kernspaltung, *fission* 231.
Kern-Wechselwirkung, *nuclear interaction* 330.
kinetische Energie einer Konfiguration, *kinetic energy of a configuration* 334.
Körnigkeit der Kernladung, *granularity of nuclear charge* 198.
körperfestes System, *body system* 482, 519, 527.
kollektive kinetische Energie, *collective kinetic energy* 419.
— Rotation, *collective rotation* 494.
— Schwingungen, klassische, *classical collective oscillations* 418, 419.
— —, quantisiert, *quantized collective oscillations* 420.
kollektives Modell, *collective model* 350, 361, 480.
kombiniertes Kernmodell, *unified model of nucleus* 349, 362, 476—543.
— — und Schalenmodell, Beziehungen, *unified model and shell model, relations* 538 bis 543.
komplexes Potential, *complex potential* 446, 449f.
Kompressibilität der Kerne, *nuclear compressibility* 203, 232.
Konfigurations-Mischung, *configuration mixing* 310, 357, 358, 381.
Kontinuums-Modell s. optisches Modell, *continuum model see optical model.*
Kopplung von Bahndrehimpulsen, *coupling of angular momenta* 390.
—, normale, *normal coupling* 475.
—, schwache, *weak coupling* 314.
—, starke, *strong coupling* 314.
Kopplungskonstanten (ps-ps und ps-pv), *coupling constants (ps-ps and ps-pv)* 15, 27.
Kopplungsschemata in Kernen, *coupling schemes in nuclei* 474, 484.
Korrelationseffekte, *correlation effects* 422, 443—445.
Kraftradius des Kernes, s. auch Radius, Größe usw., *force radius of nucleus, see also radius, size etc.* 212f., 236.
Kugelflächenfunktionen, *surface harmonics* 388.
Kugelgestalt, Abweichungen davon, *departure from sphericity* 314.
Kurzreichweitengrenze, *short range limit* 291.

Ladungskoordinaten, *charge coordinates* 16.
Ladungsraum, *charge space* 16.
Ladungsspin s. Isotopiespin, *charge spin see isotopic spin* 16.
Ladungssymmetrie der Kernkräfte bei hohen Energien, *charge symmetry of nuclear forces at high energies* 170.
Ladungsverteilung, *charge distribution* 195, 196, 416, 431.
LAMB-Verschiebung, LAMB *shift* 201.
Lebensdauer, reduzierte, *reduced lifetime* 284.
leichteste Kerne (bis Helium), *lightest nuclei (up to helium)* 422.
LIPPMANN-SCHWINGERsche Gleichung, LIPPMANN-SCHWINGER *equation* 135.
L-S-Kopplung, *L-S-coupling* 117, 119.
— und *j-j*-Kopplung, Transformation ineinander, *L-S-coupling and j-j-coupling, transformation between them* 327.

magische Zahlen, *magic numbers* 249, 466, 468.
— —, Anomalien, *magic numbers, anomalies* 258.
magnetische Momente der Kerne, *magnetic moments of nuclei* 475, 487.
— — stabiler Kerne ungerader Massenzahl im Grundzustand (tabuliert), *magnetic moments of stable nuclei of odd mass number in the ground state (table)* 263ff.
— Multipolübergänge, *magnetic multipole transitions* 100, 102, 103, 104, 110.
magnetischer Dipolübergang, *magnetic dipole transition* 491.
magnetisches Moment des Deuterons, *magnetic moment of the deuteron* 2, 32, 69.
— — von H^3 und He^3, *magnetic moments of* H^3 *and* He^3 150, 177.
— —, Matrixelemente, *magnetic moment, matrix elements* 407.
MAJORANAsche Wechselwirkung, MAJORANA *interaction* 425, 438.
Masse, effektive, *effective mass* 234.
—, —, gebundene Nucleonen, *effective mass of bound nucleons* 440.
Massen der Kerne, halbempirische Formel, *masses of nuclei, semiempirical formula* 231, 233.
Massenformel, halbempirisch, *semiempirical mass formula* 370.
Mehrkörperproblem mit nichtzentralen Operatoren, *many-body problem with noncentral operators* 400.

Mehrteilchenwellenfunktionen, *multiparticle wave functions* 288.
Meson-Atome, Bindungsenergien, *mesonic atoms, binding energies* 206f.
—, Dublettaufspaltung, *mesonic atoms, doublet splitting* 206f.
Mesonpaartermdämpfung, *pair damping* 22, 95.
Mesonpaarterme, *meson pair terms* 20, 21.
Mesontheorie mit geladenen Mesonen, *charged meson theory* 17, 18, 24.
—, neutrale, *neutral meson theory* 17, 18.
—, symmetrische, *symmetrical meson theory* 17, 18, 24, 90, 91.
Mesonterme, Quadrupoleffekte, *mesonic levels, quadrupole effects* 227, 228.
Mischung, *mixing* 505, 529—530.
Mischungsparameter, *mixing parameter* 79, 80, 81, 82, 85.
mittelstarke Kopplung, Abweichungen, *intermediate coupling, departures* 355.
— —, Kernmodell, *intermediate coupling model* 335, 338, 354.
mittlere freie Weglänge, *mean free path* 369, 421.
Modell-Wellenfunktion von BRUECKNER, *model wavefunction of* BRUECKNER 445.
Modelle s. Kernmodelle, *nuclear models see models*.
Moment, zweites, *second moment* 384, 387.
Monopolschwingung (Kompressionsschwingung) eines Kernes, *monopole (compressive) oscillation of a nucleus* 419.
Multipolmomente, *multipole moments* 67ff., 113.
Multipolübergänge, Auswahlregeln und Matrixelemente, *multipole transitions, selection rules and matrix elements* 100, 102, 103, 104, 110.
Myonen als Kern-Test-Teilchen, *muons as nuclear probe particles* 204.

neutrale Theorie, *neutral theory* 17, 18.
Neutronenstreuung, Bestimmung der Kerngröße, *neutron scattering, determination of nuclear size* 213, 214.
Neutron-Deuteron-Streuung, empirische Daten, *neutron-deuteron scattering, empirical data* 158.
— bei niederen Energien, *neutron-deuteron scattering at low energies* 170.
—, Störungstheorie, *neutron-deuteron scattering, perturbation theory* 159.
Neutroneneinfang durch Deuteronen, *neutron capture by deuterons* 176.
Neutronen-Einfangquerschnitt, *neutron capture cross section* 262.
Neutronen-Emission, verzögerte, *delayed neutron emission* 259.
Neutron-Proton-Einfang, thermischer, *thermal neutron-proton capture* 5, 115.
Neutron-Proton-Streuung, doppelte, *double neutron-proton scattering* 11, 121, 124.
— bei hohen Energien, *neutron-proton scattering at high energies* 9, 116.
Neutron-Proton-Streuung bei niedrigen Energien, *neutron-proton scattering at low energies* 3, 4, 83.
nichtzentrale Kräfte, s. auch Tensorkraft, *noncentral forces, see also tensor force* 23, 24.
— Operatoren, *non-central operators* 400.
NILSSON-Potential, NILSSON *potential* 508.
NORDHEIMsche Zahl, NORDHEIM *number* 298.
Nukleonenbreite, reduzierte, *reduced nucleon width* 409.
Nukleonstreuung, *nucleon scattering* 246, 371.
Nullreichweite, *zero range* 306.

Oberfläche des Kerns, deformierte, *deformed surface of nucleus* 418.
Oberflächen-Dichteabfall, *surface slope of density* 193, 233.
Oberflächenenergie eines Kernes, *surface energy of nucleus* 233, 370, 415, 432—434.
Oberflächenschwingungen der Kerne, *surface oscillations of nuclei* 419, 520.
Oktopol-Anregung, *octupole excitation* 523 bis 525.
Oktopol-Moment, magnetisches, *magnetic octupole moment* 255, 281.
Opazität der Kerne, *nuclear opacity* 447.
optisches Kernmodell, *optical model of nucleus* 372, 445—459.
Ortho-Wasserstoff, *ortho-hydrogen* 48, 49.
Oszillatorpotential, anisotropes, *anisotropic oscillator potential* 501—506.
—, harmonisches, *infinite harmonic oscillator well* 246.
— für Nukleonen, *oscillator potential for nucleons* 468—470.

Paarungsenergie, *pairing energy* 268, 436.
Para-Wasserstoff, *para-hydrogen* 48, 49.
Parität, *parity* 25.
— kollektiver Schwingungen, *parity of collective oscillations* 420.
—, nicht-normale, *non-normal parity* 354.
—, normale, *normal parity* 337.
Paritätserhaltung, *parity conservation* 489.
PAULI-Term, PAULI *term* 71, 97, 98, 130.
Phasenwinkel, *phase shift* 4, 5, 36, 37, 78, 79, 80, 81, 82.
Phasenwinkelanalyse, *phase shift analysis* 13, 57, 77.
Photoeffekt am Deuteron bei hohen Energien, *photodisintegration of the deuteron at high energies* 13, 128.
— — bei niedrigen Energien, *photodisintegration of the deuteron at low energies* 5, 108—114.
Pick-up-Reaktionen, *pick-up reactions* 218.
Pionen als Test-Teilchen, *pions as probe particles* 218, 219.
Polarisation, *polarization* 10, 12, 31, 120—122.
— des Kerns durch einen Mesonterm, *polarization of nucleus by a mesonic level* 210.
Potential, abgeschnittenes, *cut-off potential* 29, 30.
—, Berechnung aus Dichte, *potential, calculation from density* 455.

Potential der Kerne, *nuclear potential* 30, 422, 446.
— —, Geschwindigkeitsabhängigkeit, *nuclear potential, velocity dependence* 429, 439, 442, 456—459.
— —, Parameterwerte, *nuclear potential, parameter values* 431.
— — für ein Proton, *nuclear potential for a proton* 431.
— —, Topftiefe, *nuclear potential, well depth* 429.
—, realistisches, des Kerns, *realistic potential of nucleus* 470, 506—512.
—, vierter Ordnung, *potential of fourth order* 22, 27, 96.
—, zweiter Ordnung, *potential of second order* 21, 27, 94.
Potentiale, singuläre, *singular potentials* 31, 56.
potentialformabhängiger Parameter, *shape dependent parameter* 40, 52, 53, 60, 63, 85, 87.
potentialformunabhängige Näherung, *shape independent approximation* 40, 51.
Potentialtopf, unendlicher, *infinite square well* 248.
Potentialtopftiefe, *well depth parameter* 41, 52.
Projektionsfaktor, *projection factor* 220, 488.
Proton-Deuteron-Streuung, COULOMB-Effekte, *proton-deuteron scattering,* COULOMB *effects* 175.
—, empirische Daten, *proton-deuteron scattering, empirical data* 158.
Proton-Proton-Streuung, doppelte, *double proton-proton scattering* 11, 121, 122, 124.
— bei hohen Energien, *proton-proton scattering at high energies* 7, 116, 118.
— bei niederen Energien, *proton-proton scattering at low energies* 97.
—, Winkelverteilung bei hohen Energien, *proton-proton scattering, angular distribution at high energies* 124.
Protonenstreuung, Bestimmung der Kerngröße, *proton scattering, determination of nuclear size* 215.
Pseudopotential, *pseudopotential* 45, 48.
— von BRUECKNER, *pseudopotential of* BRUECKNER 445.
—, sättigendes, *saturating pseudopotential* 428, 429.
pseudoskalare Mesontheorie, *pseudoscalar meson theory* 14.

Quadrupolmoment, *quadrupole moment* 219, 222, 475, 488.
—, eigentliches, *intrinsic quadrupole moment* 488.
—, elektrisches, *electric quadrupole moment* 254, 295, 298, 313, 317, 380.
—, elektrisches, des Deuterons, *electric quadrupole moment of the deuteron* 2, 67, 68.
—, Matrixelemente, *quadrupole moment, matrix elements* 408.
Quadrupolmomente stabiler Kerne ungerader Massenzahl im Grundzustand (tabuliert), *quadrupole moments of stable nuclei of odd mass numbers in the ground state (table)* 263 ff.
Quadrupolschwingung, *quadrupole vibration* 523, 524.
Quadrupolverzerrung, *quadrupole distortion* 270, 281.

RACAH-Funktion, RACAH *function* 391.
Radialabhängigkeit der Kerndichte, *radial dependence of nuclear density* 231.
Radialmomente, *radial moments* 189.
Radius, äquivalenter, *equivalent radius* 181.
— des Kerns, *radius of nucleus* 416, 417.
— —, aus Massendefekten bestimmt, *radius of nucleus, determined from mass defects* 212.
— —, durch Neutronenstreuung bestimmt, *radius of nucleus, determined by neutron scattering* 213, 214.
— —, s. auch Größe des Kerns, *radius of nucleus, see also size of nucleus.*
— der Kernmaterie, s. auch Größe usw., *matter radius of nucleus, see also radius, size etc.* 186f., 454, 455.
— des Kernpotentials, *radius of nuclear potential* 453.
— der Ladungsträger, *charge radius* 454, 455.
RAINWATERsches Kernmodell, RAINWATER'S *nuclear model* 477, 498.
RAMSAUER-Effekt, nuklearer, *nuclear* RAMSAUER *effect* 186.
Reaktionsschema, *reaction scheme* 447, 448.
Reichweite, effektive, *effective range* 37, 38, 40, 51, 54, 59—61, 83, 85, 91.
—, —, gemischte, *effective range, mixed* 39, 51.
—, eigentliche, *intrinsic range* 41, 52.
relativistische Stromkorrektur, *relativistic current correction* 98, 114, 115, 116, 127, 128, 130.
Resonanzniveaus für Neutroneneinfang, *resonance levels for neutron capture* 446.
restliche Wechselwirkungen, *residual interactions* 512—516, 530, 531.
Retardierungsfaktor, *retardation factor* 14, 129.
Riesenresonanzen, *giant resonances* 470.
Röntgen-Isotopieverschiebung, *x-ray isotope shift* 201.
Röntgenstrahlenfeinstruktur, *x-ray fine structure* 199.
— von μ-Meson-Atomen, *x-rays from μ-mesonic atoms* 204, 227.
ROSENFELDsche Wechselwirkung, ROSENFELD *interaction* 337.
Rotation-Teilchen-Kopplung, *rotation-particle coupling* 533—535.
rotationsfreie Strömung, *irrotational flow* 495, 531.
Rotationsmodell, *rotational model* 480
—, Widerspruchsfreiheit, *rotational model, self-consistency* 537.
Rotations-Schwingungs-Wechselwirkung, *rotation-vibration interaction* 486, 532.
Rotationsspektren, *rotational spectra* 485.

Rotationszustände der Kerne, *rotational states of nuclei* 222, 229, 481—516.
— —, Matrixelemente, *rotational states of nuclei, matrix elements* 481—516.
Rumpf, sphäroidaler, plus einzelnes Nukleon, *spherical core plus single nucleon* 499.
—, —, — mehrere Nukleonen, *spheroidal core plus several nucleons* 500.
RUSSELL-SAUNDERS-Kopplung, RUSSELL-SAUNDERS *coupling* 319.
— und *j-j*-Kopplung, Transformation ineinander, RUSSELL-SAUNDERS-*coupling and j-j-coupling, transformation between* 327.

Sättigung der Kernbindung, *saturation of nuclear binding* 422.
Sättigungserfordernisse, *saturation requirements* 438.
Schalen, abgeschlossene, *closed shells* 220, 221.
— des Kerns, *shell model of nucleus* 230, 236, 464—476.
—, Geschichte, *shell model, history* 242.
— mit einem Teilchen, *shell model with a single particle* 468—472.
— mit mehreren Teilchen, *shell model with many particles* 472.
—, theoretische Grundlage, *shell model, theoretical basis* 368.
—, Vergleich von Atomen und Kernen, *shell models, comparison of atoms and nuclei* 467.
SCHMIDT-Diagramme (SCHMIDT-Linien), SCHMIDT *diagrams (*SCHMIDT *lines)* 277, 295, 316, 407.
SCHMIDTsches Moment, SCHMIDT *moment* 311.
schwache Kopplung, *weak coupling* 314.
Schwerpunktsbewegung, Korrektion, *centre of mass motion, correction* 329.
Schwingungsanregung, kollektive, *collective vibrational excitation* 516—525.
Schwingungsfunktion, *vibrational function* 485.
Seniority number 289, 326.
SERBER-Kraft, SERBER *force* 117, 337.
SERBERsche Wechselwirkung, SERBER *interaction* 428, 439.
Singulettzustände (gerade und ungerade), *singlet states (even and odd)* 27, 30, 31.
SLATER-Determinante, SLATER *determinant* 467, 475.
SLATER-Integrale, SLATER *integrals* 291.
Spektren, Berechnung, *spectra, calculation* 334.
— von Zuständen, *spectra of states* 291.
sphäroidales Kastenpotential, *spheroidal box potential* 506, 507.
— Rumpfmodell, *spheroidal core model* 477, 498—501, 514.
Spiegelkerne, *mirror nuclei* 211, 417.
Spin, isotoper, *isotopic spin* 289, 319.
Spin-Bahn-Kopplung, *spin orbit coupling* 23, 250, 325, 404, 471, 472.
—, Beitrag von zwei Teilchen, *two-body spin-orbit forces* 332, 379, 404.
Spin-Bahn-Potential, *spin orbit potential* 369, 378.
Spineigenzustände für drei Nukleonen, *spin-eigenstates for three nucleons* 145.
Spins stabiler Kerne ungerader Massenzahl im Grundzustand, *spins of stable nuclei of odd mass number in the ground state* 263ff.
— unstabiler Kerne ungerader Massenzahl im Grundzustand, *spins of unstable nuclei of odd mass number in the ground state* 270.
Spurious states 329.
starke Kopplung, *strong coupling* 314.
— —, Modell von N. BOHR („Sandsackmodell"), *strong coupling model of* N. BOHR 446.
statische Näherung für Nukleonen, *static approximation for nucleons* 18, 20, 23.
Stöße, unelastische, *inelastic collisions* 168.
Strahlungskorrekturen, *radiative corrections* 198.
— der μ-Mesonniveaus, *radiative correction of μ-mesonic levels* 204.
Streulänge, *scattering length* 3, 37, 38, 39, 47, 51, 60, 61, 85, 91.
—, kohärente, *coherent scattering length* 3, 47, 48, 49, 51.
Streumatrix, *scattering matrix* 78, 79, 82, 135.
Streuung von α-Teilchen, *scattering of α-particles* 464.
—, compound-elastische, *compound-elastic scattering* 448.
—, elastische, Potentialstreuung, *shape elastic scattering* 448.
— von Nukleonen, *scattering of nucleons* 246, 371.
Supermultiplett von WIGNER, *supermultiplet of* WIGNER 322.
Symmetrie der Kernwellenfunktion, *symmetry of nuclear wave functions* 482—485.
Symmetrieenergie eines Kernes, *symmetry energy, nuclear* 233, 370, 371, 434—436.
Symmetrieoperatoren für drei Nukleonen, *symmetry operators for three nucleons* 145.
symmetrischer Kreisel, *symmetric top* 481.
symplektische Gruppe, *symplectic group* 326.

Tensorkraft, *tensor force* 23, 24, 26, 27, 30, 31, 64ff., 77, 83, 87, 96, 117, 332, 337, 355, 360, 405, 438.
— bei der Bindung von H^3 und He^3, *tensor force in the binding of* H^3 *and* He^3 155.
Tensorkraftmodell, singuläres, *singular tensor force model* 117, 119, 122, 123, 124, 127.
Tensorkräfte-Effekt zweiter Ordnung, *tensor force effect of second order* 379.
Tensoroperator, *tensor operator* 23, 394.
Termdichte, *level density* 262.
Termschema für Zr^{90}, *level diagram of* Zr^{90} 309.
Termschemata nach dem α-Teilchen-Modell, *level schemes from α-particle model* 462 bis 464.
THOMAS-Kopplung, THOMAS *coupling* 378.

Trägheitsmoment, *moment of inertia* 494 bis 497, 527.
Trägheitsparameter, *inertial parameters* 526—532.
Triplett-Zustände (gerade und ungerade), *triplet states (even and odd)* 27, 30, 31.
Tröpfchenmodell, *liquid drop model* 362, 414.

Übergänge, begünstigt, *favoured transitions* 276.
—, isomere, *isomere transitions* 296.
— zwischen Rotationszuständen, *transitions between rotational states* 490—493.
—, unbegünstigt, *unfavoured transitions* 276.
Übergangsmatrix, *transition matrix* 138.
Übergangswahrscheinlichkeit, reduzierte, *reduced transition probability* 489.
Übergangswahrscheinlichkeiten, *transition probabilities* 252, 255, 317.
Unschärfe der Kernbegrenzung, *diffuseness of nuclear boundary* 236.
Unterschied der Neutronen- und Protonenverteilung, *difference of neutron and proton distribution* 435, 454, 455.
u-u-Kerne, *odd-odd nuclei* 293, 298.
—, elektromagnetische Momente, *odd-odd nuclei, electromagnetic moments* 298.

Vakuumpolarisations-Effekte bei μ-Meson-Termen, *vacuum polarization effects in μ-mesonic levels* 209.
Van der Waals-Kraft zwischen α-Teilchen, Van der Waals *force between α-particles* 461.
Variationsprinzipien von Hulthén und Schwinger, *variational principles of* Hulthén *and* Schwinger 131ff.
Vektor-Kopplungs-Koeffizienten, *vector-coupling coefficients* 389.
Verzerrung, *distortion* 505, 529, 530.
Verzweigungsverhältnis bei Rotationsübergängen, *branching ratios in rotational transitions* 493.
Vielfachstöße, *multiple collisions* 448, 450.
Vielfachstreuung, *multiple scattering* 445, 450.
virtueller ^1S-Zustand, *virtual ^1S state* 56.
Volumenergie, *volume energy* 370, 373.
Vorzeichen der Tensorkraft, *sign of tensor force* 27.

Wechselwirkung abgeschlossener Schalen, *closed shell interaction* 405.
—, ladungsunabhängige, *charge independent interaction* 17, 88, 91, 117, 124, 125, 126, 127.
—, nichtstatische, *non-static interaction* 23.
—, statische, *static interaction* 17, 23.
Wechselwirkungsdarstellung, *interaction representation* 135.
Wechselwirkungskräfte, endliche Reichweite, *interaction forces, finite range* 307.
„Weinflaschen"-Potential, *"wine-bottle" potential* 250.
Wellenfunktionen der Kreiselbewegung des Kernes, *wave functions of nuclear top motion* 481—485, 543.
Wignersche Koeffizienten, Wigner *coefficients* 389, 390.
— Wechselwirkung, Wigner *interaction* 425.
Winkelabhängigkeit der Kernausdehnung, *angular shape of nuclei* 229, 237.
— — in Nähe abgeschlossener Schalen, *angular shape of nuclei near closed shells* 237.
Wirkungsquerschnitt, photoelektrischer, *photo-electric cross section* 6.
—, photomagnetischer, *photomagnetic cross section* 6.
— zur Streuenergie Null, *cross section at energy zero* 43, 46, 51.
Wirkungsquerschnitte, gemittelte, *averaged cross sections* 426.

Youngsches Bild, Young *tableau* 321.
Yukawa-Potential, Yukawa *potential* 31, 356.

Zentralkraft bei der Zweiteilchen-Wechselwirkung, *central force in the two-particle interaction* 403.
Zentralkräfte, *central forces* 23, 24, 27, 30, 31, 32, 96, 290, 332, 403, 405.
Zentralpotential, effektives, *effective central potential* 74, 87.
Zustände von Kernen in der Nähe von Pb^{208}, *states of nuclei in the region of* Pb^{208} 302.
—, Klassifikation, *classification of states* 290.
— mehrerer Teilchen in derselben Schale, *states of several particles in the same shell* 287.
— — — in verschiedenen Schalen, *states of several particles in different shells* 296.
— normaler Parität, *states of normal parity* 337.
— ungleicher Teilchen, *states of inequivalent particles* 323.
Zweikörper-Vektorwechselwirkung, *two-body vector interaction* 346.
Zwei-Teilchen-Wechselwirkung, *two-particle interaction* 405, 423, 425, 438.
Zwei-Teilchen-Wechselwirkung, Coulomb-Kraft, *two-particle interaction,* Coulomb *force* 405.
—, Radialintegral, *two-particle interaction, radial integral* 403.
—, Spin-Bahn-Kopplung, *two-particle interaction, spin-orbit force* 404.
—, Zentralkraft, *two-particle interaction, central force* 403.
Zweiteilchen-Wechselwirkungs-Matrixelement, *two-particle interaction matrix element* 402.
Zwei-Teilchen-Wellenfunktion, *two-particle wavefunction* 424.
—, relativistische, *relativistic two-body wavefunction* 140.

Subject Index.

(English-German.)

Where English and German spelling of a word is identical the German version is omitted.

Absorption potential, *Absorptionsanteil des Potentials (Imaginärteil)* 369, 381, 385, 446.
— process, *Absorptionsprozeß* 446.
Abundance, relative, *relative Häufigkeit* 260.
— total, *gesamte Häufigkeit* 259.
Adiabatic switching off, *adiabatisches Abbrechen* 136.
α-α interaction from scattering, *α-α-Wechselwirkung aus Streuversuchen* 464.
α decay determinations of nuclear size, *α-Zerfall: Bestimmung der Kerngröße* 218.
— nucleus, *α-Teilchen-Kern* 460, 461.
— particle model, *α-Teilchen-Modell* 366, 460—464.
— — scattering: determination of nuclear size, *α-Streuung: Bestimmung der Kerngröße* 216, 217.
Angular momenta, coupling, *Bahndrehimpulse, Kopplung* 390, 391.
— momentum of collective oscillation, *Drehimpuls einer kollektiven Schwingung* 420.
— shape of nuclei, *Winkelabhängigkeit der Kernausdehnung* 229, 237.
— — of nuclei near closed shells, *Winkelabhängigkeit der Kernausdehnung in Nähe abgeschlossener Schalen* 237.
Asymmetry coefficient, forward, *Asymmetrie-Koeffizient, Vorwärtsrichtung* 6, 113.
— parameter, *Asymmetrie-Parameter* 434.

BARTLETT interaction, BARTLETT*sche Wechselwirkung* 425.
β-decay (see also *ft* values), *β-Zerfall (siehe auch ft-Werte)* 270 seq.
—, matrix elements, *β-Zerfall, Matrixelemente* 408.
β-transitions, *β-Übergänge* 296.
Binding effect, chemical, *chemischer Bindungseffekt* 43, 46.
— energies, *Bindungsenergien* 258, 295, 370, 371, 373.
— — from the α-particle model, *Bindungsenergien aus dem α-Teilchen-Modell* 461 to 464.
— — of nuclei, semiempirical formula, *Bindungsenergien der Kerne, halbempirische Formel* 415, 416.
— energy of a nucleus, higher order corrections, *Bindungsenergie eines Kernes, Korrekturen höherer Ordnung* 440—442.
— — of a nucleus in the FERMI model, *Bindungsenergie eines Kernes im* FERMI-*Modell* 426—429.
Binding energy of the deuteron, *Bindungsenergie des Deuterons* 2, 32.
— — of H^3 and He^3, *Bindungsenergien von* H^3 *und* He^3 153.
Body system, *körperfestes System* 482, 519, 527.
Branching ratios in rotational transitions, *Verzweigungsverhältnis bei Rotationsübergängen* 493.
BREIT-WIGNER dispersion formula, BREIT-WIGNER*sche Dispersionsformel* 446.
BRUECKNER's method, BRUECKNER*sche Methode* 374, 380, 445.

CASIMIR operator, CASIMIR-*Operator* 291.
Central forces, *Zentralkräfte* 23, 24, 27, 30, 31, 32, 96, 290, 332, 403, 405.
— force in the two-particle interaction, *Zentralkraft bei der Zweiteilchen-Wechselwirkung* 403.
— potential, effective, *effektives Zentralpotential* 74, 87.
Centre of mass motion, correction, *Schwerpunktsbewegung, Korrektion* 329.
c.f.p. (coefficients of fractional parentage) 332, 398, 399.
Charge coordinates, *Ladungskoordinaten* 16.
— distribution, *Ladungsverteilung* 195, 196, 416, 431.
— independent interaction, *ladungsunabhängige Wechselwirkung* 17, 88, 91, 117, 124, 125, 126, 127.
— radius, *Radius der Ladungsträger* 454, 455.
— space, *Ladungsraum* 16.
— spin see isotopic spin, *Ladungsspin, s. Isotopiespin* 16.
— symmetry of nuclear forces at high energies, *Ladungssymmetrie der Kernkräfte bei hohen Energien* 170.
Charged meson theory, *Mesontheorie mit geladenen Mesonen* 17, 18, 24.
Chemical valence bond, *chemische Valenzbindung* 422.
CLEBSCH GORDAN-coefficients, CLEBSCH-GORDAN-*Koeffizienten* 389, 390.
Closed shell, *abgeschlossene Schale* 220, 221.
— — interaction, *Wechselwirkung abgeschlossener Schalen* 405.
Collective kinetic energy, *kollektive kinetische Energie* 419.
— model, *kollektives Modell* 350, 361, 480.
— oscillations, classical, *klassische kollektive Schwingungen* 418, 419.

Collective oscillations, quantized, *quantisierte kollektive Schwingungen* 420.
— rotation, *kollektive Rotation* 494.
Complex potential, *komplexes Potential* 446, 449 seq.
Compressibility, nuclear, *Kompressibilität der Kerne* 203, 232.
Compound-elastic scattering, *compound-elastische Streuung* 448.
— nucleus, *Compound-Kern* 446, 448.
— system, *Compound-System* 448.
Configuration mixing, *Konfigurations-Mischung* 310, 357, 358, 381.
Continuum model see optical model, *Kontinuums-Modell s. optisches Modell.*
Coriolis force in nuclear rotation, *Coriolis-kraft bei Kernrotation* 497, 537.
Correlation effects, *Korrelationseffekte* 422, 443—445.
Coulomb barrier determinations of nuclear size, Coulomb-*Wall, Bestimmung der Kerngröße* 217, 218.
— effects in proton-deuteron scattering, Coulomb-*Effekte bei Proton-Deuteron-Streuung* 175.
— energy, Coulomb-*Energie* 370.
— forces, Coulomb-*Kräfte* 289.
— excitation, Coulomb-*Anregung* 224, 495, 498.
— phase shift, Coulomb-*Streuphase* 58, 59.
— radius, Coulomb-*Radius* 471, 430.
Coupling constants (ps-ps and ps-pv), *Kopplungskonstanten* (ps-ps *und* ps-pv) 15, 27.
— of angular momenta, *Kopplung von Bahndrehimpulsen* 390.
—, normal, *normale Kopplung* 475.
—, strong, *starke Kopplung* 314.
—, weak, *schwache Kopplung* 314.
— schemes in nuclei, *Kopplungsschemata in Kernen* 474, 484.
Cross sections, averaged, *gemittelte Wirkungsquerschnitte* 426.
— — at energy zero, *Wirkungsquerschnitt zur Streuenergie Null* 43, 51, 46.
Cut-off of potential, *Abschneiden des Potentials* 29, 30.

Damping parameter, *Dämpfungsparameter* 22, 27.
— theory, *Dämpfungstheorie* 139.
Deformability of nuclei, *Deformierbarkeit der Kerne* 237.
Deformation parameters, *Deformationsparameter* 219, 518.
— potential, *Deformationspotential* 503, 522, 513.
— — and two-body forces, *Deformationspotential und Zweikörperkräfte* 511, 512.
Degrees of freedom in unified model, *Freiheitsgrade im kombinierten Kernmodell* 541.
Density anisotropy, calculated from potential, *Dichteanisotropie, aus dem Potential berechnet* 504.
Density, calculation from potential, *Dichte, Berechnung aus Potential* 453, 454.
— from individual particle states, *Dichte gemäß individueller Teilchenzustände* 235.
—, central, of the nucleus, *Dichte im Zentrum des Kerns* 193.
—, nuclear, *Dichte der Kernmaterie* 178, 416, 431.
—, —, below the surface, *Dichte der Kernmaterie unter der Oberfläche* 433, 459.
— of nuclei, theory, *Dichte der Kerne, Theorie* 231, 233.
—, statistical, of excited levels, *statistische Dichte angeregter Niveaus* 437.
—, —, of single particle levels, *statistische Dichte der Einteilchen-Niveaus* 435.
Deuteron, binding energy, *Deuteron, Bindungsenergie* 1, 32, 91.
—, effective range, *Deuteron, effektive Reichweite* 86, 91.
—, electric quadrupol moment, *Deuteron, elektrisches Quadrupolmoment* 67, 91.
—, excited states, *Deuteron, angeregte Zustände* 33, 35.
—, ground state, *Deuteron, Grundzustand* 32, 36, 66.
—, magnetic moment, *Deuteron, magnetisches Moment* 32, 69, 74.
—, — —, exchange current correction, *Deuteron, magnetisches Moment, Austauschstromkorrektur* 71, 72.
—, — —, non-additivity correction, *Deuteron, magnetisches Moment, nicht additive Korrektur* 71, 72, 73.
—, — —, relativistic correction, *Deuteron, magnetisches Moment, relativistische Korrektur* 71, 72, 73.
— radius, *Deuteronradius* 35, 40, 75.
— spin, *Deuteronspin* 2, 32.
—, wave function, *Deuteron, Wellenfunktion* 33, 66, 74, 76, 91.
Difference of neutron and proton distribution, *Unterschied der Neutronen- und Protonenverteilung* 435, 454, 455.
Diffraction theory, *Beugungstheorie* 165 seq.
Diffuseness of nuclear boundary, *Unschärfe der Kernbegrenzung* 236.
Dipole moment, electric, *elektrisches Dipolmoment* 68.
— —, magnetic, *magnetisches Dipolmoment* 252, 277, 294, 311, 407.
Dipole oscillation of a nucleus, *Dipolschwingung eines Kernes* 419.
Distortion, *Verzerrung* 505, 529, 530.
D-state probability in the deuteron, *D-Zustand, Anteil im Deuteron* 70, 73, 74, 91.
Dyson transformation, Dyson-*Transformation* 19.

Effective range, *effektive Reichweite* 38, 51, 54, 59—61, 83, 85, 91.
Electric multipol transition, *elektrischer Multipolübergang* 100, 102, 103, 104, 110.
— quadrupole transition, *elektrischer Quadrupolübergang* 490, 546.

Electromagnetic moments, *elektromagnetische Momente* 252.
— transitions for H^3 and He^3, *elektromagnetische Übergänge für* H^3 *und* He^3 176.
Electron scattering, *Elektronenstreuung* 416.
— — by nuclei, *Elektronenstreuung an Kernen* 186, 228 seq.
— — —, BORN approximation, *Elektronenstreuung an Kernen*, BORN*sche Näherung* 187.
— — —, cross sections, *Elektronenstreuung an Kernen, Wirkungsquerschnitte* 195.
— — —, inelastic, *unelastische Elektronenstreuung an Kernen* 197.
— — —, high energies, *Elektronenstreuung an Kernen, hohe Energien* 191.
— — —, low energies, *Elektronenstreuung an Kernen, niedrige Energien* 188.
— —, determinations of nuclear size, *Elektronenstreuung, Bestimmung der Kerngröße* 192—195.
— — from aspherical charge, *Elektronenstreuung an einer nichtkugelförmigen Ladung* 228.
Ellipsoidal vibrations, *ellipsoidale Schwingungen* 518, 533.
Energy matrix, *Energiematrix* 332.
— shift of bound states, *Energietermverschiebung gebundener Zustände* 199.
Equivalent two-body problem for three nucleons, *äquivalentes Zweikörperproblem für drei Nukleonen* 172.
EULER angles, EULER*sche Winkel* 483.
Even-even nuclei, *g-g-Kerne* 293.
Exchange current correction, *Austausch-Stromkorrektur* 98, 114, 115, 116, 130.
— currents and magnetic moments of H^3 and He^3, *Austauschströme und magnetische Momente von* H^3 *und* He^3 177.
Excitation energies of first excited states, *Anregungsenergien der ersten angeregten Zustände* 260.
— spectra of nuclei, *Anregungsspektren der Kerne* 477—479.

FERMI approximation, FERMI-*Näherung* 45.
— gas modell of nuclei, FERMI-*Gas-Modell der Kerne* 421.
— momentum, FERMI-*Impuls* 426.
Fine structure, optical, *optische Feinstruktur* 202.
Fission, *Kernspaltung* 231.
Force radius of nucleus, see also radius, size etc., *Kraftradius des Kernes, s. auch Radius, Größe usw.* 212 seq., 236.
Fractional parentage coefficients 332, 398, 399.
ft-values, allowed transitions in odd-nuclei, *ft-Werte, erlaubte Übergänge in ungeraden Kernen* 273.
—, superallowed transitions in odd nuclei, *ft-Werte, übererlaubte Übergänge in ungeraden Kernen* 272.
—, first forbidden transitions in odd nuclei, *ft-Werte, einfach verbotene Übergänge in ungeraden Kernen* 274.
—, second forbidden transitions, *ft-Werte, zweifach verbotene Übergänge* 276.

γ-decay, *γ-Zerfall* 223.
γ-transitions, matrix elements, *γ-Übergänge, Matrixelemente* 408.
γ-vibrations of spheroidal nuclei, *γ-Schwingungen sphäroidaler Kerne* 544—546.
GAUSS shape, GAUSS*sche Potentialform* 356.
Giant resonances, *Riesenresonanzen* 470.
Granularity of nuclear charge, *Körnigkeit der Kernladung* 198.
Ground states of H^3 and He^3, empirical data, *Grundzustand des* H^3 *und* He^3, *empirische Daten* 149.

Hamiltonian of collective vibration, HAMILTON-*Funktion der kollektiven Schwingung* 519.
Hard core cut-off, see cut-off.
— — (repulsive potential), *(Abstoßungspotential)* 30, 31, 43, 53, 56, 63, 76, 88, 90, 92, 96, 117, 118, 122, 124, 127, 130, 156.
HARTREE-FOCK potential, HARTREE-FOCK*sches Potential* 456.
— self consistent field method, HARTREE-FOCK-*Methode des self consistent field* 376.
H^3, binding energy, H^3, *Bindungsenergie* 153.
—, electromagnetic transitions, H^3, *elektromagnetische Übergänge* 176.
—, ground state, empirical data, H^3, *Grundzustand, empirische Daten* 149.
—, magnetic moment, H^3, *magnetisches Moment* 150, 177.
—, role of tensor force in binding, H^3, *Einfluß der Tensorkraft auf die Bindung* 155.
He^3, binding energy, He^3, *Bindungsenergie* 153.
—, electromagnetic transitions, He^3, *elektromagnetische Übergänge* 176.
—, ground state, empirical data, He^3, *Grundzustand, empirische Daten* 149.
—, magnetic moment, He^3, *magnetisches Moment* 150, 177.
—, role of tensor force in binding, He^3, *Einfluß der Tensorkraft auf die Bindung* 155.
HEISENBERG interaction, HEISENBERG*sche Wechselwirkung* 425.
HULTHÉN potential, HULTHÉN-*Potential* 31.
— wave function, HULTHÉN*sche Wellenfunktion* 33, 34, 35, 140.
HUND-operator, HUND*scher Operator* 329.

Imaginary part of potential, *Imaginärteil des Potentials* 446, 451, 452.
Impulse approximation 165.
Independent (individual) particle model, *Einteilchen-Modell* 236, 245, 257, 317.
Individual (independent) particle model, *Einteilchen-Modell* 236, 245, 257, 317.
— particle model, application to light nuclei, *Einteilchen-Modell, Anwendung auf leichte Kerne* 335.
Inelastic collisions, *unelastische Stöße* 168.
Inertial parameters, *Trägheitsparameter* 526—532.
Interaction forces, finite range, *Wechselwirkungskräfte, endliche Reichweite* 307.
—, non-static, *nichtstatische Wechselwirkung* 23.

Interaction representation, *Wechselwirkungsdarstellung* 135.
Interference minimum, *Interferenz-Minimum* 7, 8, 59.
Intermediate coupling, departures, *mittelstarke Kopplung, Abweichungen* 355.
— — model, *mittelstarke Kopplung, Kernmodell* 335, 338, 354.
Irrotational flow, *rotationsfreie Strömung* 495, 531.
Isomeric states, *isomere Zustände* 282.
— transitions, *isomere Übergänge* 296.
Isotope shift of x-rays, *Isotopie-Verschiebung bei Röntgenstrahlen* 201.
— —, optical, *optische Isotopieverschiebung* 202, 224.
— spin, *isotoper Spin* 209, 319.
— —, reduced, *reduzierter isotoper Spin* 289, 326.
Isotropy coefficient, *Isotropie-Koeffizient* 6, 114, 129.
— factor, *Isotropiefaktor* 14.

j-j coupling, *j-j-Kopplung* 3, 25, 474.
— and *L-S* coupling transformation between them, *j-j-Kopplung und L-S-Kopplung, Transformation ineinander* 327.

Kinetic energy of a configuration, *kinetische Energie einer Konfiguration* 334.
K-selection rules, *K-Auswahlregeln* 492.

LAMB shift, LAMB-*Verschiebung* 201.
Level density, *Termdichte* 262.
— diagram of Zr^{90}, *Termschema für* Zr^{90} 309.
— schemes from α-particle model, *Termschemata nach dem α-Teilchen-Modell* 462—464.
Lifetime, reduced, *reduzierte Lebensdauer* 284.
Lightest nuclei (up to helium), *leichteste Kerne (bis Helium)* 422.
LIPPMANN-SCHWINGER equation, LIPPMANN-SCHWINGER*sche-Gleichung* 135.
Liquid drop model, *Tröpfchenmodell* 362, 414.
— mirror reflection of neutrons, *Flüssigkeitsspiegel, Reflexion von Neutronen* 3, 49.
L-S-coupling, *L-S-Kopplung* 117, 119.
— and *j-j*-coupling, transformation between them, *L-S-Kopplung und j-j-Kopplung, Transformation ineinander* 327.

Magic numbers, *magische Zahlen* 249, 466, 468.
— —, anomalies, *magische Zahlen, Anomalien* 258.
Magnetic dipole transition, *magnetischer Dipolübergang* 491.
— moment, matrix elements, *magnetisches Moment, Matrixelemente* 407.
— — of H^3 and He^3, *magnetisches Moment von* H^3 *und* He^3 150, 177.
— — of nuclei, *magnetische Momente der Kerne* 475, 487.
— — of stable nuclei of odd mass number in the ground state (table), *magnetische Momente stabiler Kerne ungerader Massenzahl im Grundzustand (tabuliert)* 263 seq.
Magnetic moment of the deuteron, *magnetisches Moment des Deuterons* 2, 32, 69.
— multipole transitions, *magnetische Multipolübergänge* 100, 102, 103, 104, 110.
MAJORANA interaction, MAJORANA*sche Wechselwirkung* 425, 438.
Many-body problem with non-central operators, *Mehrkörperproblem mit nichtzentralen Operatoren* 400.
Mass, effective, *effektive Masse* 234.
—, —, of bound nucleons, *effektive Masse, gebundene Nukleonen* 440.
— formula, semiempirical, *halbempirische Massenformel* 370.
Masses of nuclei, semiempirical formula, *Massen der Kerne, halbempirische Formel* 231, 233.
Matter radius of nucleus, see also radius, size etc., *Radius der Kernmaterie s. auch Größe usw.* 186 seq., 454, 455.
Mean free path, *mittlere freie Weglänge* 369, 421.
Mesonic atoms, binding energies, *Meson-Atome, Bindungsenergien* 206 seq.
— —, doublet splitting, *Meson-Atome, Dublettaufspaltung* 206 seq.
— levels, quadrupole effects, *Mesonterme, Quadrupoleffekte* 227, 228.
Meson pair terms, *Mesonpaarterme* 20, 21.
— theory, neutral, *neutrale Mesontheorie* 17, 18.
— —, symmetrical, *symmetrische Mesontheorie* 17, 18, 24, 90, 91.
Mirror nuclei, *Spiegelkerne* 211, 417.
Mixing, *Mischung* 505, 529, 530.
— parameter, *Mischungsparameter* 7, 9, 80, 81, 82, 85.
— —, α-particle model, *Kernmodell, α-Teilchen-Modell* 366, 460—464.
— — by RAINWATER, *Kernmodell von* RAINWATER 477, 498.
— —, collective, *kollektives Kernmodell* 350, 361, 480.
— —, FERMI gas model, *Kernmodell,* FERMI-*Gas-Modell* 421.
— —, independent (individual) particles, *Kernmodell, Einteilchen-Modell* 236, 245, 257, 317, 335, 412, 421.
— —, intermediate coupling, *Kernmodell, mittelstarke Kopplung* 335, 338, 354.
— —, liquid drop, *Kernmodell, Flüssigkeitstropfen* 362, 414.
— —, optical, *optisches Kernmodell* 372, 445—450.
— —, rotational model, *Kernmodell, Rotationsmodell* 480, 537.
— —, shell model, *Kernmodell, Schalenmodell* 230, 236, 242, 368, 464—476.
— —, spheroidal core, *Kernmodell, sphäroidaler Rumpf* 477, 498—501, 514.
— —, strong coupling, *Kernmodell, starke Kopplung* 446.
— —, unified, *kombiniertes Kernmodell* 349, 362, 476—543.

Models of nuclei, brief survey, *Kernmodelle, kurze Übersicht* 412—414.
— — with strong interaction, *Kernmodelle mit starker Wechselwirkung* 412.
Model wavefunction of BRUECKNER, *Modell-Wellenfunktion von* BRUECKNER 445.
Moment of inertia, *Trägheitsmoment* 494 to 497, 527.
—, second, *zweites Moment* 384, 387.
Momentum space representation, *Impulsraumdarstellung* 139.
— transfer in dense nuclei, *Impulsübertragung in dichten Kernen* 441, 442.
Monopole (compressive) oscillation of a nucleus, *Monopolschwingung (Kompressionsschwingung) eines Kernes* 419.
Multiparticle wave functions, *Mehrteilchenwellenfunktionen* 288.
Multiple collisions, *Vielfachstöße* 448, 450.
— scattering, *Vielfachstreuung* 445, 450.
Multipole moments, *Multipolmomente* 67 seq., 113.
— transitions: selection rules and matrix elements, *Multipolübergänge: Auswahlregeln und Matrixelemente* 100, 102, 103, 104, 110.
Muons as nuclear probe particles, *Myonen als Kern-Test-Teilchen* 204.

Neutral theory, *neutrale Theorie* 17, 18.
Neutron capture by deuterons, *Neutroneneinfang durch Deuteronen* 176.
— — cross section, *Neutronen-Einfangsquerschnitt* 262.
Neutron-deuteron scattering at low energies, *Neutron-Deuteron-Streuung bei niedrigen Energien* 170.
— —, empirical data, *Neutron-Deuteron-Streuung, empirische Daten* 158.
— —, perturbation theory, *Neutron-Deuteron-Streuung, Störungstheorie* 159.
Neutron emission, delayed, *verzögerte Neutronen-Emission* 259.
Neutron-proton capture, thermal, *thermischer Neutron-Proton-Einfang* 5, 115.
— scattering, double, *doppelte Neutron-Proton-Streuung* 11, 121, 124.
— — at high energies, *Neutron-Proton-Streuung bei hohen Energien* 9, 116.
— — at low energies, *Neutron-Proton-Streuung bei niedrigen Energien* 3, 4, 83.
Neutron scattering: determination of nuclear size, *Neutronenstreuung: Bestimmung der Kerngröße* 213, 214.
NILSSON potential, NILSSON-*Potential* 508.
Non-central forces, see also tensor force, *nichtzentrale Kräfte, s. auch Tensorkraft* 23, 24.
— operators, *nichtzentrale Operatoren* 400.
NORDHEIM number, NORDHEIM*sche Zahl* 298.
Nuclear charge distribution, *Kernladungsverteilung* 195, 196, 416, 431.
— density see density, *Kerndichte s. Dichte.*
Nuclear forces, charge symmetry at high energies, *Kernkräfte, Ladungssymmetrie bei hohen Energien* 170.
— interaction, *Kern-Wechselwirkung* 330.
— models see models, *Modelle s. Kernmodelle.*
— phase shift, *Kernprozeß-Streuphasen* 58, 59.
— potential see potential, *Kernpotential s. Potential.*
— radius see radius, *Kernradius s. Radius.*
— shape see shape of nucleus, *Kerngestalt s. Gestalt des Kerns.*
— size see size of nucleus, *Kerngröße s. Größe des Kerns.*
Nucleon scattering, *Nukleonstreuung* 246, 371.
— width, reduced, *reduzierte Nukleonenbreite* 409.

Octupole excitation, *Oktopol-Anregung* 523—525.
— moment, magnetic, *magnetisches Oktopol-Moment* 255, 281.
Odd-odd nuclei, *u-u-Kerne* 293, 298.
— —, electromagnetic moments, *u-u-Kerne, elektromagnetisches Moment* 298.
Opacity, nuclear, *Opazität der Kerne* 447.
Optical model of nucleus, *optisches Kernmodell* 372, 445—459.
Orbital states, group theoretical classification *Bahn-Zustände, gruppentheoretische Klassifikation* 396.
Ortho-hydrogen, *Ortho-Wasserstoff* 48, 49.
Oscillator potential, anisotropic, *anisotropes Oszillatorpotential* 501—506.
— — for nucleons, *Oszillatorpotential für Nukleonen* 468—470.
— well, infinite harmonic, *harmonisches Oszillatorpotential* 246.

Pair damping, *Mesonpaartermdämpfung* 22, 95.
Pairing energy, *Paarungsenergie* 268, 436.
Para-hydrogen, *Para-Wasserstoff* 48, 49.
Parity, *Parität* 25.
— conservation, *Paritätserhaltung* 489.
—, non-normal, *nicht-normale Parität* 354.
—, normal, *normale Parität* 337.
— of collective oscillations, *Parität kollektiver Schwingungen* 420.
PAULI term, PAULI-*Term* 71, 97, 98, 130.
Phase shift, *Phasenwinkel* 4, 5, 36, 37, 78, 79, 80, 81, 82.
— — analysis, *Phasenwinkelanalyse* 36, 57, 77.
Photodisintegration of the deuteron at high energies, *Photoeffekt am Deuteron bei hohen Energien* 13, 128.
— — at low energies, *Photoeffekt am Deuteron bei niedrigen Energien* 5, 108—114.
Photo-electric cross section, *photoelektrischer Wirkungsquerschnitt* 6.
Photo-magnetic cross section, *photomagnetischer Wirkungsquerschnitt* 6.
Pick-up reactions, *Pick-up Reaktionen* 218.
Pions as probe particles, *Pionen als Test-Teilchen* 218, 219.

Plane waves within the nucleus, *ebene Wellen im Kern* 422.
Polarization, *Polarisation* 10, 12, 31, 120 to 122.
— of nucleus by a mesonic level, *Polarisation des Kerns durch einen Mesonenterm* 210.
Potential, calculation from density, *Potential, Berechnung aus Dichte* 455.
— cut-off, *abgeschnittenes Potential* 29, 30.
—, nuclear, *Potential der Kerne* 30, 422, 446.
—, —, parameter values, *Potential der Kerne, Parameterwerte* 431.
—, —, for a proton, *Potential der Kerne für ein Proton* 431.
—, —, velocity dependence, *Potential der Kerne, Geschwindigkeitsabhängigkeit* 429, 439, 442, 456—459.
—, —, well depth, *Potential der Kerne, Topftiefe* 429.
— of fourth order, *Potential vierter Ordnung* 22, 27, 96.
— of second order, *Potential zweiter Ordnung* 21, 27, 94.
—, realistic, of nucleus, *realistisches Potential des Kerns* 470, 506—512.
Potentials, singular, *singuläre Potentiale* 31, 56.
Projection factor, *Projektionsfaktor* 220, 488.
Proton-deuteron scattering, COULOMB effects, *Proton-Deuteron-Streuung, COULOMB-Effekte* 175.
— —, empirical data, *Proton-Deuteron-Streuung, empirische Daten* 158.
Proton-proton scattering, angular distribution at high energies, *Proton-Proton-Streuung, Winkelverteilung bei hohen Energien* 124.
— —, double, *doppelte Proton-Proton-Streuung* 11, 121, 122, 124.
— — at high energies, *Proton-Proton-Streuung bei hohen Energien* 7, 116, 118.
— — at low energies, *Proton-Proton-Streuung bei niederen Energien* 97.
Proton scattering, determination of nuclear size, *Protonenstreuung, Bestimmung der Kerngröße* 215.
Pseudopotential, *Pseudopotential* 45, 48.
—, of BRUECKNER, *Pseudopotential von BRUECKNER* 445.
—, saturating, *sättigendes Pseudopotential* 428, 429.
Pseudoscalar meson theory, *pseudoskalare Mesontheorie* 14.

Quadrupole distortion, *Quadrupolverzerrung* 270, 281.
— moment, *Quadrupolmoment* 219, 222, 475, 488.
— —, electric, *elektrisches Quadrupolmoment* 254, 295, 298, 313, 317, 380.
— —, —, of the deuteron, *elektrisches Quadrupolmoment des Deuterons* 2, 67, 68.
— —, intrinsic, *eigentliches Quadrupolmoment* 488.
— —, matrix elements, *Quadrupolmoment, Matrixelemente* 408.
— moments of stable nuclei of odd mass numbers in the ground state (table), *Quadrupolmomente stabiler Kerne ungerader Massenzahl im Grundzustand (tabuliert)* 263seq.
Quadrupole vibration, *Quadrupolschwingung* 523, 524.

RACAH function, RACAH-*Funktion* 391.
Radial dependence of nuclear density, *Radialabhängigkeit der Kerndichte* 231.
— moments, *Radialmomente* 189.
Radiative corrections, *Strahlungskorrekturen* 198.
— — of μ-mesonic levels, *Strahlungskorrekturen der μ-Mesonniveaus.*
Radius, equivalent, *äquivalenter Radius* 181.
— of nuclear potential, *Radius des Kernpotentials* 453.
— of nucleus, *Radius des Kerns* 416, 417.
— —, determined by neutron scattering, *Radius des Kerns, durch Neutronenstreuung bestimmt* 213, 214.
— —, — from mass defects, *Radius des Kerns, aus Massendefekten bestimmt* 212.
— —, see also size of nucleus, *Radius des Kerns, s. auch Größe des Kerns.*
RAINWATER's nuclear model, RAINWATER-*sches Kernmodell* 477, 498.
RAMSAUER effect, nuclear, *nuklearer* RAMSAUER-*Effekt* 186.
Range, effective, *effektive Reichweite* 37, 38, 40, 51, 54, 59—61, 83, 85, 91.
—, —, mixed, *gemischte, effektive Reichweite* 39, 51.
—, intrinsic, *eigentliche Reichweite* 41, 52.
Reaction scheme, *Reaktionsschema* 447, 448.
Refractive index, *Brechungsindex* 449—451.
— (real) potential, *Brechungsanteil des Potentials (Realteil)* 369, 372, 447, 449—451.
Relativistic current correction, *relativistische Stromkorrektur* 98, 114, 115, 116, 127, 128, 130.
Repulsive force, *Abstoßungskraft* 438.
— potential see hard core, *Abstoßungspotential s. hard core.*
Residual interactions, *restliche Wechselwirkungen* 512—516, 530, 531.
Resonance levels for neutron capture, *Resonanzniveaus für Neutroneneinfang* 446.
Retardation factor, *Retardierungsfaktor* 14, 129.
Rotational model, *Rotationsmodell* 480.
— —, self-consistency, *Rotationsmodell, Widerspruchsfreiheit* 537.
— spectra, *Rotationsspektren* 485.
— states of nuclei, *Rotationszustände der Kerne* 222, 229, 481—516.
— — of nuclei, matrix elements, *Rotationszustände der Kerne, Matrixelemente* 481 to 516.
Rotation-particle coupling, *Rotation-Teilchen-Kopplung* 533—535.
Rotation-vibration interaction, *Rotations-Schwingungs-Wechselwirkung* 486, 532.
ROSENFELD interaction, ROSENFELD*sche Wechselwirkung* 337.

RUSSELL-SAUNDERS coupling, RUSSELL-SAUNDERS-*Kopplung* 319.
— — and *j-j*-coupling, transformation between, RUSSELL-SAUNDERS-*Kopplung und j-j-Kopplung, Transformation ineinander* 327.

Saturation of nuclear binding, *Sättigung der Kernbindung* 422.
— requirements, *Sättigungserfordernisse* 438.
Scattering length, *Streulänge* 3, 37, 38, 39, 47, 51, 60, 61, 85, 91.
— —, coherent, *kohärente Streulänge* 3, 47, 48, 49, 51.
— matrix, *Streumatrix* 78, 79, 82, 135.
— of α-particles, *Streuung von α-Teilchen* 464.
— of nucleons, *Streuung von Nukleonen* 246, 371.
SCHMIDT diagrams (SCHMIDT lines) SCHMIDT *Diagramme (*SCHMIDT *Linien)* 277, 295, 316, 407.
— moment, SCHMIDT*sches Moment* 311.
Seniority number 289, 326.
SERBER force, SERBER-*Kraft* 117, 337.
— interaction, SERBER*sche Wechselwirkung* 428, 439.
Shape dependent parameter, *Potentialformabhängiger Parameter* 40, 52, 53, 60, 63, 85, 87.
— elastic scattering, *elastische Streuung, Potentialstreuung* 448.
— independent approximation, *Potentialformunabhängige Näherung* 40, 51.
Shell model, history, *Schalenmodell, Geschichte* 242.
— — of nucleus, *Schalenmodell des Kerns* 230, 236, 464—476.
— —, theoretical basis, *Schalenmodell, theoretische Grundlage* 368.
— — with a single particle, *Schalenmodell mit einem Teilchen* 468—472.
— — with many particles, *Schalenmodell mit mehreren Teilchen* 472.
— models, comparison of atoms and nuclei, *Schalenmodelle, Vergleich von Atomen und Kernen* 467.
Short range limit, *Kurzreichweitengrenze* 291.
Single particle model, see individual particle model.
Singlet states (even and odd), *Singulettzustände (gerade und ungerade)* 27, 30, 31.
Sign of tensor force, *Vorzeichen der Tensorkraft* 27.
Size of nuclei, determination by α-particles, *Größe der Kerne, Bestimmung mit α-Teilchen* 216—218.
— —, — by bound electrons, *Größe der Kerne, Bestimmung durch gebundene Elektronen* 199.
— —, — by neutron scattering, *Größe der Kerne, Bestimmung aus Neutronenstrauung* 213, 214.
Size of nuclei, determination by proton scattering, *Größe der Kerne, Bestimmung aus Protonenstreuung* 215.
— —, — from electron scattering, *Größe der Kerne, Bestimmung aus Elektronenstreuung* 192—195.
— —, early studies, *Größe der Kerne, frühe Untersuchungen* 178.
— of very light nuclei, *Größe sehr leichter Kerne* 199.
SLATER determinant, SLATER-*Determimante* 467, 475.
— integrals, SLATER *Integrale* 291.
Spectra, calculation, *Spektren, Berechnung* 334.
— of states, *Spektren von Zuständen* 291.
Sphericity, departure from, *Kugelgestalt, Abweichungen* 314.
Spheroidal box potential, *sphäroidales Kastenpotential* 506, 507.
— core model, *sphäroidales Rumpfmodell* 477, 498—501, 514.
— — plus several nucleons, *sphäroidaler Rumpf plus mehrere Nukleonen* 500.
Spherical core plus single nucleon, *sphäroidaler Rumpf plus einzelnes Nukleon* 499.
Spin eigenstates for three nucleons, *Spineigenzustände für drei Nukleonen* 145.
—, isotopic, *isotoper Spin* 289, 319.
— orbit coupling, *Spin-Bahn-Kopplung* 23, 250, 325, 404, 471, 472.
— — potential, *Spin-Bahn-Potential* 369, 378.
Spins of stable nuclei of odd mass number in the ground state, *Spins stabiler Kerne ungerader Massenzahl im Grundzustand* 263 seq.
— of unstable nuclei of odd mass number in the ground state, *Spins unstabiler Kerne ungerader Massenzahl im Grundzustand* 270.
Spurious states 329.
Square well, infinite, *unendlicher Potentialtopf* 248.
States, classification, *Klassifikation von Zuständen* 290.
— of normal parity, *Zustände normaler Parität* 337.
— of nuclei in the region of Pb^{208}, *Zustände von Kernen in der Nähe von* Pb^{208} 302.
— of inequivalent particles, *Zustände ungleicher Teilchen* 323.
— of several particles in different shells, *Zustände mehrerer Teilchen in verschiedenen Schalen* 296.
— — — in the same shell, *Zustände mehrerer Teilchen in derselben Schale* 287.
Static approximation for nucleons, *statische Näherung für Nukleonen* 18, 20, 23.
— interaction, *statische Wechselwirkung* 7, 18, 23.
Straight cut-off see cut-off.
Strong coupling, *starke Kopplung* 314.
— — model of N. BOHR, *starke Kopplung, Modell von* N. BOHR *(„Sandsackmodell")* 446.

Supermultiplet of WIGNER, *Supermultiplett von* WIGNER 322.
Surface energy of nucleus, *Oberflächenenergie eines Kernes* 233, 370, 415, 432—434.
Surface harmonics, *Kugelflächenfunktionen* 388.
— of nucleus, deformed, *Oberfläche des Kerns, deformierte* 418.
— oscillations of nuclei, *Oberflächenschwingungen der Kerne* 419, 520.
— slope of density, *Oberflächen-Dichteabfall* 193, 233.
Symmetric top, *symmetrischer Kreisel* 481.
Symmetry energy, nuclear, *Symmetrieenergie eines Kernes* 233, 370, 371, 434—436.
— of nuclear wave functions, *Symmetrie der Kernwellenfunktion* 482—485.
— operators for three nucleons, *Symmetrieoperatoren für drei Nukleonen* 145.
Symplectic group, *symplektische Gruppe* 326.

Tensor force, *Tensorkraft* 23, 24, 26, 27, 30, 31, 64ff., 77, 83, 87, 96, 117, 332, 337, 355, 360, 405, 438.
— — in the binding of H^3 and He^3, *Tensorkraft bei der Bindung von* H^3 *und* He^3 155.
— — model, singular, *singuläres Tensorkraftmodell* 117, 119, 122, 123, 124, 127.
— force effect of second order, *Tensorkräfte-Effekt zweiter Ordnung* 379.
— operator, *Tensoroperator* 23, 394.
THOMAS coupling, THOMAS-*Kopplung* 378.
Three-body forces, *Dreikörperkräfte* 156, 352.
3*n-j*-symbols, 3*n-j-Symbole* 391.
Transition matrix, *Übergangsmatrix* 138.
— probabilities, *Übergangswahrscheinlichkeiten* 252, 255, 317.
— probability, reduced, *reduzierte Übergangswahrscheinlichkeit* 489.
Transitions between rotational states, *Übergänge zwischen Rotationszuständen* 490 to 493.
—, favoured, *begünstigte Übergänge* 276.
—, isomere, *isomere Übergänge* 296.
—, unfavoured, *unbegünstigte Übergänge* 276.
Triplet states (even and odd), *Triplett-Zustände (gerade und ungerade)* 27, 30, 31.
Two-body spin-orbit forces, *Spin-Bahn-Kopplung, Beitrag von zwei Teilchen* 332, 379, 404.
— vector interaction, *Zweikörper-Vektorwechselwirkung* 346.
— wavefunction, relativistic, *relativistische Zwei-Teilchen-Wellenfunktion* 140.
Two-particle interaction, *Zwei-Teilchen-Wechselwirkung* 405, 423, 425, 438.
— —, central force, *Zwei-Teilchen-Wechselwirkung, Zentralkraft* 403.
— —, COULOMB force, *Zwei-Teilchen-Wechselwirkung,* COULOMB-*Kraft* 405.
Two-particle interaction, matrix element, *Zwei-Teilchen-Wechselwirkung, Matrixelement* 402.
— —, radial integral, *Zwei-Teilchen-Wechselwirkung, Radialintegral* 403.
— —, spin-orbit force, *Zwei-Teilchen-Wechselwirkung, Spin-Bahn-Kopplung* 404.
— wavefunction, *Zwei-Teilchen-Wellenfunktion* 424.

Unified model of nucleus, *kombiniertes Kernmodell* 349, 362, 476—543.
— — and shell model, relations, *kombiniertes Kernmodell und Schalenmodell, Beziehungen* 538—543.

VAN DER WAALS force between α-particles, VAN DER WAALS-*Kraft zwischen α-Teilchen* 461.
Variational principles of HULTHÉN and SCHWINGER, *Variationsprinzipien von* HULTHÉN *und* SCHWINGER 131 seq.
Vacuum polarization effects in μ-mesonic levels, *Vakuumpolarisations-Effekte bei μ-Meson-Termen* 209.
Vektor-coupling coefficients, *Vektor-Kopplungs-Koeffizienten* 389.
Vibrational excitation, collective, *kollektive Schwingungsanregung* 516—525.
— function, *Schwingungsfunktion* 485.
Virtual ^1S state, *virtueller* ^1S*-Zustand* 56.
Volume energy, *Volumenenergie* 370, 373.

Wave functions of nuclear top motion, *Wellenfunktionen der Kreiselbewegung des Kernes* 481—485, 543.
Weak coupling, *schwache Kopplung* 314.
Well depth parameter, *Potentialtopftiefe* 41, 52.
WIGNER coefficients, WIGNER*sche Koeffizienten* 389, 390.
— interaction, WIGNER*sche Wechselwirkung* 425.
"Wine-bottle" potential, „*Weinflaschen*"-*Potential* 250.

X-ray fine structure, *Röntgenstrahlenfeinstruktur* 199.
X-rays from μ-mesonic atoms, *Röntgenstrahlen von μ-Meson-Atomen* 204, 227.
— isotope shift, *Röntgen-Isotopieverschiebung* 201.

YOUNG tableau, YOUNG*sches Bild* 321.
YUKAWA potential, YUKAWA-*Potential* 31, 356.

Zero cut-off see cut-off.
— energy cross section see cross section.
— range, *Nullreichweite* 306.

MIX
Papier aus verantwortungsvollen Quellen
Paper from responsible sources
FSC® C105338

If you have any concerns about our products,
you can contact us on
ProductSafety@springernature.com

In case Publisher is established outside the EU,
the EU authorized representative is:
Springer Nature Customer Service Center GmbH
Europaplatz 3, 69115 Heidelberg, Germany

Printed by Libri Plureos GmbH
in Hamburg, Germany